A SOURCE BOOK IN ANIMAL BIOLOGY

SOURCE BOOKS IN THE HISTORY OF THE SCIENCES

EDWARD H. MADDEN, GENERAL EDITOR

Ordo primus.

Degit aut iuxta Nilum, ab Oceano ascendens: & cum apud Indos Bouis nomine dicatur, apud Nilū nominatur Rhinoceros, Scriptor Græcus recentior innominatus. Oppianus hoc animal Oryge non multo maius esse scribit. Strabo
magniudinē Tauri ei tribuit. Plinius Elephanto longitudine propè parem facit: alij uel parem, uel pauló longiorem, hu
miliorē tamen, & breuioribus cruribus. ¶ Pictura quā exhibeo, Alberti Dureri est, qua clariss. ille pictor Rhinocerotē
Emanueli Lusitaniæ regi anno Salutis 1515. è. Cambaia Indiæ regione Vlysbonā aduectum, perpulchrè expressit.
GERM. Ein groß frömbd thier vß India/mit einem kleinen hörnle/vff dem halß oben: vnd einem anderen starcken
horn vff der nasen/mit welchem er fräfenlich wider den Helfanten streyt/reyßt jm den bauch auff/vnd überwindet
jn. Mag ein Naßhorn genent werden.

ELEPHAS Græcum uocabulum, Latinis etiam receptum est, dicitur & Elephantus à Cicerone

Illustrations from the *Historia animalium* (Zürich, 1551) of Konrad GESNER (Swiss
physician and encyclopedic zoologist, 1516–1565).   Photographed from the collected
Icones animalium in the Rare Book Room of the Library of Congress.   The note ac-

Quadrup. ferorum

sub finem Decembris, nunquã se maiorem cœli caloris uim sensisse testatũ. Cornua huius quadrupedis Bernæ in curia uisuntur. nos etiam Augustæ in ædibus amplissi mi uiri Ge. Fuggeri uidimus.   GALLICE Rangier, Ranglier.

GERMANICE Rein/Reen/Reyner/Rainger/Reinßthier. Wirt mit grossen scharen gefunden in Norwegen vnd anstossenden lande: auch in Massouia/welche ist ein gegne im künigrych Poland/aber zů vnserer zeyt vast wenig. Ich halt daß sein nam ein Teütschen vrsprung habe von dem rennen har: dañ diß thier ist also schnäll/daß es eins tags vff xxx. Teütscher meylen louffen mag/ɩc.   POLONICE Renscheron.

RHINOCEROS, uel Taurus Aethiopicus Pausaniæ: nam alij sunt Aethiopici illi tauri, quorũ Plinius & Aelianus me minerunt. Rhinocerotis nomen Græcum in linguis alijs plerisϛ seruatur. Latinè dici posset Naricornis. ¶ Rhinocerotis eius picturã uidimus, cuius cadauer è naufragio eiectum est in Tyrrhenum litus. Capite est suillo, tergore munitus scutulato. cornu gemino, altero pusillo in fronte, altero in nare robustissimo, quo audacissimè pugnat ac uincit Elephantum, Scaliger. Indorũ lingua hoc animal Sandabenam uocat. ¶ Rhinoceros magnitudine par est Equo fluuiatili.

Degit

companying the plate of the Rhinoceros explains that the picture is from the hand of Albrecht Dürer.

A SOURCE BOOK IN ANIMAL BIOLOGY

Edited by **THOMAS S. HALL**

Professor of Zoology,
Washington University

Harvard University Press
Cambridge, Massachusetts

GENERAL EDITOR'S PREFACE

The Source Books in this series are collections of classical papers that have shaped the structures of the various sciences. Some of these classics are not readily available and many of them have never been translated into English, thus being lost to the general reader and frequently to the scientist himself. The point of this series is to make these texts readily accessible and to provide good translations of the ones that either have not been translated at all or have been translated only poorly.

The series was planned originally to include volumes in all the major sciences from the Renaissance through the nineteenth century. It has been extended to include ancient and medieval Western science and the development of the sciences in the first half of the present century. Many of these books have been published already and several more are in various stages of preparation.

The Carnegie Corporation originally financed the series by a grant to the American Philosophical Association. The History of Science Society and the American Association for the Advancement of Science have approved the project and are represented on the Editorial Advisory Board. This Board at present consists of the following members.

Marshall Clagett, History of Science, University of Wisconsin
I. Bernard Cohen, History of Science, Harvard University
Ernest Mayr, Zoology, Harvard University
Ernest A. Moody, Philosophy, University of California at Los Angeles
Ernest Nagel, Philosophy, Columbia University
Harlow Shapley, Astronomy, Harvard University
Harry Woolf, History of Science, Johns Hopkins University

The series was begun and sustained by the devoted labors of Gregory D. Walcott and Everett W. Hall, the first two General Editors. I am indebted to them, to the members of the Advisory Board, and Joseph D. Elder, Science Editor of Harvard University Press, for their indispensable aid in guiding the course of the Source Books.

Edward H. Madden, General Editor

Department of Philosophy
State University of New York at Buffalo
Buffalo, New York

PREFACE

The two main aims of this book are to increase the general availability of classical contributions to animal biology and to present the development of thought in this field in the words of those who produced it.

The first of these aims is realized by assembling in one volume works previously scattered and, in some cases, rather rare. It is hoped, moreover, that the usefulness of the materials may be enhanced by placing them before the reader in English, several for the first time.

The second object, that of tracing the principal patterns of development in the field, is made possible through the selection of appropriate materials and the inclusion of brief critical comments indicating the historical position of each work and its author. Exhaustive textual commentary has not been attempted. With careful reading most of the papers will be found to be self-contained, though bibliographical notes have been included to guide the reader to further aids in interpretation.

As to criteria for selection, works were weighed chiefly on the basis of their effect upon subsequent developments in their several areas. Other criteria, such as priority, definitiveness, and literary excellence, were also considered. The problem of selection from so vast a literature is a thorny one, and each historically minded biologist may find some particular friend missing from the collection. Space was limited, however, and undoubtedly no two individuals would ever perfectly agree on who or what should be left out. After the painful process of exclusion is completed, we are still left with a wonderfully rich treasure of the products of the scientific imagination.

Most of the items are separate papers, single chapters, or sizable excerpts, and most of these have been reprinted in their entirety. Otherwise, as much was included as was, in the editor's opinion, fully adequate to an understanding of the subject. Most of the works turn out to be arguments in defense of a thesis, and the attempt was made to preserve the integrity of such arguments. In doubtful cases, we have tried to err on the side of inclusion rather than exclusion of materials within each separate paper, even though this cuts down the total number of titles which may be brought into the book. The dates covered were set by a general policy covering the entire series of source books, of which this is one.

Grateful acknowledgment is made to Professor Gregory D. Walcott, General Editor of the series of Source Books in the History of the Sciences; to McGraw-Hill Book Company, Inc., for guaranteeing the cost of publication;

and to the American Philosophical Association for advancing funds to finance the preparation of the manuscript.

Many associates and friends gave helpful advice concerning the selection of materials, especially Professors Viktor Hamburger, John S. Nicholas, John F. Fulton, G. Evelyn Hutchinson, Alexander Petrunkevitch, Jane M. Oppenheimer, and others. Their advice was not always followed, however, and responsibility for shortcomings rests entirely with the editor. The notes of Professor E. M. Smallwood, temporary editor of this volume, proved useful in many instances.

Preparation of the manuscript involved the cooperation of both general and departmental libraries at Yale, Purdue, Harvard, and Washington Universities and the University of Chicago; also of the Libraries of Congress, the Surgeon General's Office, the Department of Agriculture; and of the John Crerar Library, the Boston Athenaeum, and the Boston Medical Library. Acknowledgment of permissions to reprint is made with each item. Most of these were generously granted without request for payment.

Special indebtedness is acknowledged to Martha and Viktor Hamburger and Liselotte Mezger for their careful translations from the German, and to Margaret Hackford for her invaluable editorial assistance.

THOMAS S. HALL

St. Louis, Mo.
August, 1950

CONTENTS

I. THE ORGANIZATION OF ANIMAL LIFE

Papers relating to comparative and systematic zoology and the organization of the individual

REVIVAL OF BIOLOGICAL INQUIRY

da VINCI, Leonardo (Florentine painter, sculptor, poet, inventor, engineer, and anatomist, 1452–1519). From *The note books of Leonardo da Vinci*; arr. and tr. by E. MacCurdy, New York, 1939; by permission of Reynal & Hitchcock, Inc., publisher.

The evaluation of Leonardo as zoologist is complicated by the fragmentary and unsystematic nature of his records. The worth of his observations is enhanced when one considers how early he appears in the intellectual renaissance and that zoology was for him but one of an almost unbelievable number of activities of several of which he was the chief exponent of an illustrious century. The following fragments are prophetic of the physiological approach to anatomy which later characterized Italian biology (see Vesalius).

Nature of Muscles. The tendons of the muscles are of greater or less length as a man's fleshy excrescence is greater or less. And in leanness the fleshy excrescence always recedes towards the point at which it starts from the fleshy part. And as it puts on fat it extends towards the beginning of its tendon. . . . The end of each muscle becomes transformed into tendon, which binds the joint of the bone, to which this muscle is attached. . . .

The number of the tendons which successively one above the other cover each other and all together cover and bind the joint of the bones to which they are joined, is as great as the number of the muscles which meet in the same joint. . . . These muscles have a voluntary and an involuntary movement seeing that they are those which open and shut the lung. When they open they suspend their function which is to contract, for the ribs which at first were drawn up and compressed by the contracting of these muscles then remain at liberty and resume their natural distance as the breast expands. And since there is no vacuum in nature the lung which touches the ribs from within must necessarily follow their expansion; and the lung therefore opening like a pair of bellows draws in the air in order to fill the space so formed. . . .

In all the parts where man has to work with greater effort nature has made the muscles and tendons of greater thickness and breadth. . . . With the muscles it happens almost universally that they do not move the limb where they are fixed but move that where the tendon that starts from the muscle is joined, except that which raises and moves the side in order to help respiration.

All these muscles serve to raise the ribs and as they raise the ribs they dilate the chest, and as the chest becomes dilated the lung is expanded, and the expansion of the lung is the indrawing of the air which enters by the mouth into this lung as it enlarges. . . .

3

If any muscle whatsoever be drawn out lengthwise a slight force will break its fleshy tissue; and if the nerves of sensation be drawn out lengthwise slight power tears them from the muscles where their ramification weaves them together and spreads and consumes itself; and one sees the same process enacted with the sinewy covering of the veins and arteries which are mingled with these muscles. What is therefore the cause of so great a force of arms and legs which is seen in the actions of any animal whatsoever? One cannot say other than that it is the skin which clothes them; and that when the nerves of sensation thicken the muscles these muscles contract and draw after them the tendons in which their extremities become converted; and in this process of thickening they fill out the skin and make it drawn and hard; and it cannot be lengthened out unless the muscles become thinner; and not becoming thinner they are a cause of resistance and of making strong the before mentioned skin, in which the swollen muscles perform the function of a wedge. · · ·

Changes of the arteries, hepatic veins, and abdominal organs in the old. The artery and the vein which in the old extend between the spleen and the liver, acquire so great a thickness of skin that it contracts the passage of the blood that comes from the mesaraic veins, through which this blood passes over to the liver and the heart and the two greater veins, and as a consequence through the whole body; and apart from the thickening of the skin these veins grow in length and twist themselves after the manner of a snake, and the liver loses the humour of the blood which was carried there by this vein; and consequently this liver becomes dried up and grows like frozen bran both in colour and substance, so that when it is subjected even to the slightest friction this substance falls away in tiny flakes like sawdust and leaves the veins and arteries.

And the veins of the gall and of the navel which entered into this liver by the gate of the liver all remain deprived of the substance of this liver, after the manner of maize or Indian millet when their grains have been separated.

The colon and the other intestines in the old become much constricted, and I have found there stones in the veins which pass beneath the fork of the breast, which were as large as chestnuts, of the colour and shape of truffles or of dross or clinkers of iron, which stones were extremely hard, as are these clinkers, and had formed bags which were hanging to the said veins after the manner of goitres.

And this old man, a few hours before his death, told me that he had lived a hundred years, and that he did not feel any bodily ailment other than weakness, and thus while sitting upon a bed in the hospital of Santa Maria Nuova at Florence, without any movement or sign of anything amiss, he passed away from this life.

And I made an autopsy in order to ascertain the cause of so peaceful a death, and found that it proceeded from weakness through failure of blood and of the artery that feeds the heart and the other lower members, which I found to be very parched and shrunk and withered; and the result of this autopsy I wrote down very carefully and with great ease, for the body was devoid of either fat or moisture, and these form the chief hindrance to the knowledge of its parts.

The other autopsy was on a child of two years, and here I found everything the contrary to what it was in the case of the old man.

The old who enjoy good health die through lack of sustenance. And this is brought about by the passage to the mesaraic veins becoming continually restricted by the thickening of the skin of these veins; and the process continues until it affects the capillary veins, which are the first to close up altogether; and from this it comes to pass that the old dread the cold more than young, and that those who are very old have their skin the colour of wood or of dried chestnut, because this skin is almost completely deprived of sustenance.

And this network of veins acts in man as in oranges, in which the peel becomes thicker and the pulp diminishes the more they become old. And if you say that as the blood becomes thicker it ceases to flow through the veins, this is not true, for the blood in the veins does not thicken because it continually dies and is renewed.   . . .

How the body of the animal continually dies and is renewed. The body of anything whatsoever that receives nourishment continually dies and is continually renewed. For the nourishment cannot enter except in those places where the preceding nourishment is exhausted, and if it is exhausted it no longer has life. Unless therefore you supply nourishment equivalent to that which has departed, the life fails in its vigour; and if you deprive it of this nourishment, the life is completely destroyed. But if you supply it with just so much as is destroyed day by day, then it renews its life just as much as it is consumed; like the light of this candle formed by the nourishment given to it by the fat of this candle, which light is also continually renewed by swiftest succour from beneath, in proportion as the upper part is consumed and dies, and in dying becomes changed from radiant light to murky smoke. And this death extends for so long as the smoke continues; and the period of duration of the smoke is the same as that of what feeds it, and in an instant the whole light dies and is entirely regenerated by the movement of that which nourishes it; and its life receives from it also its ebb and flow, as the flicker of its point serves to show us. The same process also comes to pass in the bodies of the animals by means of the beating of the heart, whereby there is produced a wave of blood in all the veins, and these are continually either enlarging or contracting, because the expansion occurs when they receive the excessive quantity of blood, and the contraction is due to the departure of the excess of blood they have received; and this the beating of the pulse teaches us, when we touch the aforesaid veins with the fingers in any part whatsoever of the living body.

But to return to our purpose, I say that the flesh of the animals is made anew by the blood which is continually produced by that which nourishes them, and that this flesh is destroyed and returns by the mesaraic arteries and passes into the intestines, where it putrifies in a foul and fetid death, as they show us in their deposits and steam like the smoke and fire which were given as a comparison.   . . .

On the Eye.   . . . How the images of any object whatsoever which pass to the eye through some aperture imprint themselves on its pupil upside down and the understanding sees them upright:

The pupil of the eye which receives through a very small round hole the images of bodies situated beyond this hole always receives them upside down and the visual faculty always sees them upright as they are.   And this proceeds from the fact that the said images pass through the centre of the crystalline sphere situated in the middle of the eye; and in this centre they unite in a point and then spread themselves out upon the opposite surface of this sphere without deviating from their course; and the images direct themselves upon this surface according to the object that has caused them, and from thence they are taken by the impression and transmitted to the common sense where they are judged.   . . .

The pupil of the eye changes to as many different sizes as there are differences in the degrees of brightness and obscurity of the objects which present themselves before it:

In this case nature has provided for the visual faculty when it has been irritated by excessive light by contracting the pupil of the eye, and by enlarging this pupil after the manner of the mouth of a purse when it has had to endure varying degrees of darkness.   And here nature works as one who having too much light in his habitation blocks up the window half-way or more or less according to the necessity, and who when the night comes throws open the whole of this window in order to see better within this habitation.   Nature is here establishing a continual equilibrium, perpetually adjusting and equalising by making the pupil dilate or contract in proportion to the aforesaid obscurity or brightness which continually presents itself before it.   You will see the process in the case of the nocturnal animals such as cats, screech-owls, long-eared owls and suchlike which have the pupil small at midday and very large at night.   And it is the same with all land animals and those of the air and of the water but more, beyond all comparison, with the nocturnal animals.

And if you wish to make the experiment with a man look intently at the pupil of his eye while you hold a lighted candle at a little distance away and make him look at this light as you bring it nearer to him little by little, and you will then see that the nearer the light approaches to it the more the pupil will contract.   . . .

How the five senses are the ministers of the soul.   The soul apparently resides in the seat of the judgment, and the judgment apparently resides in the place where all the senses meet, which is called the common sense; and it is not all of it in the whole body as many have believed, but it is all in this part; for if it were all in the whole, and all in every part, it would not have been necessary for the instruments of the senses to come together in concourse to one particular spot; rather would it have sufficed for the eye to register its function of perception on its surface, and not to transmit the images of the things seen to the sense by way of the optic nerves; because the soul—for the reason already given—would comprehend them upon the surface of the eye.

Similarly, with the sense of hearing, it would be sufficient merely for the voice to resound in the arched recesses of the rock-like bone which is within the ear, without there being another passage from this bone to the common sense, whereby the said mouth might address itself to the common judgment.

The sense of smell also is seen to be forced of necessity to have recourse to this same judgment.

The touch passes through the perforated tendons and is transmitted to this sense; these tendons proceed to spread out with infinite ramifications into the skin which encloses the body's members and the bowels. The perforating tendons carry impulse and sensation to the subject limbs; these tendons passing between the muscles and the sinews dictate to these their movement, and these obey, and in the act of obeying they contract, for the reason that the swelling reduces their length and draws with it the nerves, which are interwoven amid the particles of the limbs, and being spread throughout the extremities of the fingers, they transmit to the sense the impression of what they touch.

The nerves with their muscles serve the tendons even as soldiers serve their leaders, and the tendons serve the common sense as the leaders their captain, and this common sense serves the soul as the captain serves his lord.

So therefore the articulation of the bones obeys the nerve, and the nerve the muscle, and the muscle the tendon, and the tendon the common sense, and the common sense is the seat of the soul, and the memory is its monitor, and its faculty of receiving impressions serves as its standard of reference.

How the sense waits on the soul, and not the soul on the sense, and how where the sense that should minister to the soul is lacking, the soul in such a life lacks conception of the function of this sense, as is seen in the case of a mute or one born blind. . . .

How the nerves sometimes work of themselves, without the command of other agents or of the soul:

This appears clearly for you will see how paralytics or those who are shivering or benumbed by cold move their trembling limbs such as the head or the hands without permission of the soul; which soul with all its powers cannot prevent these limbs from trembling. The same happens in the case of epilepsy or with severed limbs such as the tails of lizards.

Function of liver, bile, and intestines: The liver is the distributor and dispenser of vital nourishment to man.

The bile is the familiar or servant of the liver which sweeps away and cleans up all the dirt and superfluities left after the food has been distributed to the members by the liver.

The intestines.   As to these you will understand their windings well if you inflate them.   And remember that after you have made them from four aspects thus arranged you then make them from four other aspects expanded in such a way that from their spaces and openings you can understand the whole, that is, the variations of their thicknesses.   . . .

Contrast between the perfection of the body and the coarseness of the mind in certain men.   Methinks that coarse men of bad habits and little power of reason do not deserve so fine an instrument or so great a variety of mechanism as those endowed with ideas and with great reasoning power, but merely a sack wherein their food is received, and from whence it passes away.

For in truth one can only reckon them as a passage for food; since it does not seem to me that they have anything in common with the human race except speech and shape, and in all else they are far below the level of the beasts.

RENAISSANCE OF ANATOMY

VESALIUS, Andreas (Belgian-born, later Italian, anatomist, 1514–1564).   From *De vivorum sectione nonnulla* in (*Historiae**) *de humani fabrica corporis, etc.*, Basel, 1543; tr. by S. W. Lambert of chap. 19 in Book 7 as *Concerning some dissection of the living* in his *A reading from Vesalius*, in *Proceedings of the Charaka Club*, vol. 8, p. 3, 1935; reprinted in *Bulletin of the New York Academy of Science*; copyright 1935 by Columbia University Press; by permission of the editor.

By "fabric" Vesalius meant not only that our body is a *product* of fabrication but also that it is the *scene* of fabrication.   To think of Vesalius too exclusively as anatomist would be to do him an injustice.   Correlation of structure with function is, next to his dramatic departure from traditionalism, possibly the chief virtue of his approach.   Constantly he draws distinctions between organs whose action can be correctly grasped only by vivisection and those whose functions are apparent by examination after death.   Dr. Lambert's translations amply illustrate the nature of the method.†

ON DISSECTION OF THE LIVING

What May Be Learned by Dissection of the Dead and What of the Living

Quite as the dissection of the dead teaches well the number, position and shape of each part, and most accurately the nature and composition of its material substance, thus also the dissection of a living animal clearly demonstrates at once the function itself, at another time it shows very clearly the rea-

---

* The word *Historiae* is included in the title to the first and many subsequent editions but omitted from some.

† The preface and final chapter of the *De fabrica* have been translated by B. Farrington in *Proceedings of the Royal Society of Medicine*, vol. 25, p. 1357, 1932, and *Transactions of the Royal Society of South Africa*, vol. 20, p. 1, 1931, respectively.

sons for the existence of the parts. Therefore, even though students deservedly first come to be skilled in the study of dead animals, afterward when about to investigate the action and use of the parts of the body they must become acquainted with the living animal.

On the other hand since very many small parts of the body are endowed with different uses and functions, it is fitting that no one doubt that dissections of the living present also many contradictions. . . .

The Use and Function of the Muscles Which May Be Seen

In a proper dissection thou shalt see the function of the muscles; notice during their own action they contract and become thick where they are most fleshy, and again they lengthen and become thin according as they in combination draw up a limb, either letting themselves go back and having been drawn out permit the limb to be pulled in an opposite direction by another muscle, or at other times indeed they do not put in action their own combination. . . .

Examination of the Uses of the Veins and Arteries

Also when inquiring into the use of the veins the work is scarcely one for the dissection of the living, since we shall become sufficiently acquainted in the case of the dead with the fact that these veins carry the blood through the whole body and that any part is not nourished in which a prominent vein has been severed in wounds.

Likewise concerning the arteries we scarcely require a dissection of the living although it will be allowable for anyone to lay bare the artery running into the groin and to obstruct it with a band, and to observe that the part of the artery cut off by the band pulsates no longer.

And thus it is observed by the easy experiment of opening an artery at any time in living animals that blood is contained in the arteries naturally.

In order that on the other hand we may be more certain that the force of pulsation does not belong to the artery or that the material contained in the arteries is not the producer of the pulsation, for in truth this force depends for its strength upon the heart. Besides, because we see that an artery bound by a cord no longer beats under the cord, it will be permitted to undertake an extensive dissection of the artery of the groin or of the thigh, and to take a small tube made of a reed of such a thickness as is the capacity of the artery and to insert it by cutting in such a way that the upper part of the tube reaches higher into the cavity of the artery than the upper part of the dissection, and in the same manner also that the lower portion of the tube is introduced downward farther than the lower part of the dissection, and thus the ligature of the artery which constricts its caliber above the canulla is passed by a circuit.

To be sure when this is done the blood and likewise the vital spirit run through the artery even as far as the foot; in fact the whole portion of the artery replaced by the cannula beats no longer. Moreover when the ligature has been cut, that part of the artery which is beyond the canulla shows no less pulsation than the portion above.

We shall see next how much force is actually carried to the brain from the heart by the arteries. Now in this demonstration thou shalt wonder greatly at a vivisection of Galen in which he advises that all things be cut off which are common to the brain and heart, always excepting the arteries which seek the head through the transverse processes of the cervical vertebrae and carry also besides a substantial portion of the vital spirit into the primary sinuses of the dura mater and also in like manner into the brain. So much so that it is not surprising that the brain performs its functions under these conditions for a long time, which Galen observed could easily be done, for the animal breathes for a long time during this dissection, and sometimes moves about. If indeed it runs, and therefore requires much breath, it falls not long afterwards although the brain will still afterwards receive the essence of the animal spirit from those arteries which I have closely observed seek the skull through the transverse processes of the cervical vertebrae. . . .

We see that the peritoneum is a wrapper for all the organs enclosed in it; that the omentum and likewise the mesentery serve in the best manner for the conduction and distribution of the blood vessels; that the stomach prepares the food and drink, and passes these through the stomach onward.

Examination of the Function of Those Parts Which Are Contained in the Peritoneum

And however nothing may prevent our taking living dogs which have consumed food at less or greater intervals previously and examine them alive, and thus investigating the functions of the intestines. But on the contrary we are able to behold the functioning of the liver as also of the spleen or of the kidneys or of the bladder during the dissection of the living, scarcely better than in that of the dead; unless someone shall wish to excise the spleen in the living dog which I have once done, and have preserved the dog (alive) for some days. . . .

In truth these operations, just as the dislocations and fractures of bones, which we sometimes do on brute beasts, serve more for training the hands and for determining correct treatment rather than for investigating the functions of organs. . . .

Examination of the Fetus

Quite pleasing is it in the management of the fetus to see how when the fetus touches the surrounding air it tries to breathe. And this dissection is performed opportunely in a dog or pig when the sow will soon be ready to drop her young.

To be sure if thou dividest the abdomen of such an animal down to the cavity of the peritoneum, and then thou also openest the uterus at the site of a single fetus, and when the secundines have been separated from the uterus thou shalt place the fetus on a table, thou shalt see through its coverings and its transparent membrane how the fetus attempts in vain to breathe, and dies just as if suffocated. If in fact thou shalt perforate the covering of the fetus, and shalt free its head from the coverings, thou shalt soon see that the fetus as it were comes to life again and breathes finely.

And when thou shalt have investigated this in one fetus thou shalt turn to another which thou shalt not free from the uterus but thou shalt invert the uterus opened by the same management as that of the fetuses just described, and shalt turn the edges of the dissection already made backward until the lower part or site of the coverings of another and nearest fetus shall appear, and thou shalt free this site from the uterus even up to that place where the uterus is fused in the exterior covering of the fetus, and where its abundant flesh will possess a substance similar to the spleen, which interweaves vessels stretching out from the uterus into the external coverings of the fetus.

For that network of vessels must be preserved intact during this manipulation, and the remaining external covering must be removed from the uterus in order that thou shalt see through the transparent covering of the fetus that the arteries distributed by the covering and running to the umbilicus pulsate in the rhythm of the arteries running to the uterus, and that the naked fetus attempts and struggles for respiration and thereupon when the coverings are punctured and broken thou shalt see that then the fetus breathes and that pulsations of the arteries of the fetal membranes and of the umbilicus stop. Up to this moment the arteries of the uterus are beating in unison with the rest of the arteries outside of itself. . . .

And we consider many things concerning the functions of the heart and lungs; of course the motion of the lungs, and whether their rough artery takes to itself a portion of those things which are inhaled; next, the dilation and contraction of the heart, and whether the pulsation of the heart and arteries act in the same rhythm; also in what manner the venous artery is dilated and contracted, and whereby an animal continues to live after its heart has been excised. For making these investigations we particularly require an animal endowed with a wide breastbone and possessing membranes dividing the thorax separated in such a way that when the sternum has been divided lengthwise we may be able to carry the dissection between these membranes even to the heart itself without causing a perforation of the cavities of the thorax in which the lungs are contained.

Examination of the Function of the Heart and of the Parts Ministering to the Heart Itself

. . . Indeed, when no such animal except man or a tailless ape is to be had, these dissections must be carried out in dogs and pigs and those animals of which an abundance is furnished us, as thou hast seen all this mentioned above.

Therefore and in like manner the lung follows the movement of the thorax. It is evident from this way, when a cut is made in an intercostal space penetrating the thoracic wall, that a portion of the lung on the injured side collapses, and distends with the thorax no longer. While until then the other lung following up to this time the movement of the chest, also soon collapses if thou shalt make a cut penetrating even into the cavity of the thorax on the other side. And then the animal, even if it moves the chest for some time, will die nevertheless just as if it were suffocated.

In fact it ought to be observed in this demonstration that thou shalt make the cut as close as possible to the upper edge of any rib lest thou mayest direct the incision by chance along the lower edge and perforate the vessels stretched along there. For the blood will flow forth thence at once, which rendered frothy in consequence of the air inhaled and exhaled through the wound, will appear to thee falsely as the lung, for this froth will appear to be the lung to those who dissect carelessly, and they will think that the lung is distended on such occasions by its own inherent force.

In order that thou mayest see the natural relationship of the lung to the thorax, thou shalt cut the cartilages of two or three median ribs from the opposite side, and when the incisions have been carried through the interspaces of these ribs thou shalt bend the separate ribs outward and break them where thou shalt decide to be the proper place through which thou seekest to observe the lung of the uninjured side, for since the membranes dividing the thorax in dogs are sufficiently translucent, it is very easy to inspect through them the portion of the lung which is following up to this time the movements of the thorax, and when these membranes have been perforated slightly, to consider in what manner that part of the lung collapses.

Before thou perforatest these membranes it will have been useful to grasp with the hands the branches of the venous artery in that part of the lung which now collapses, and in some other portion to free the substance of the lung from those vessels in order that thou mayest learn whether these vessels and the heart are moved in like manner.

For the movement of the heart is evident to thee even here especially if thou shalt divide the covering of the heart and shalt uncover the heart from it on that side where thou are carrying on this operation.

It will be fitting to try this on the left side as of course the right part of the lung may be lifted up to the point and at one time thou mayest conveniently grasp the main trunk of the venous artery easily in the hand. Also thou mayest in an operation of this kind carefully handle the base of the heart, and may cut off swiftly and at one time and with one ligature the vessels from their origins, and then thou mayest cut to excise the heart under the ligature, and when the bands have been loosened with which the animal has been tied, thou mayest allow it to run about.

Indeed we have at times seen dogs, but especially cats, treated in this manner, run around.

Besides thou wilt see the movement of the heart and also of the arteries more accurately if soon thou shalt bind the dog to a plank, and shalt carry the incision on both sides with a very sharp knife from the clavicle through the cartilages of the ribs where they are continuous with the bones, and thou shalt make a third incision even into the cavity of the peritoneum transversely from the end of one of the above incisions to the end of the other; and when the sternum has been lifted with its cartilages pressed to it, and when the transverse incision has been freed from them, thou shalt turn the sternum upward towards the head of the animal and soon when the covering of the heart has been opened thou shalt grasp the heart with one hand and with the other thou shalt take hold of the great artery extending into the back.

BEGINNINGS OF COMPARATIVE ANATOMY

BELON du Mans (Bellonius), Pierre (French zoologist, 1517–1564). From *L'histoire de la nature des oyseaux avec leurs descriptions & naifs portraicts retirez du naturel*, Paris, 1555; tr. of a fragment from chap. 12, by T. S. Hall and S. Trocmé for this volume.

The comparative method was first extensively employed by Aristotle, who in the *De partibus* distinguished between the encyclopedic and comparative approaches, selecting the latter for his treatise. Only with Belon, however, was the method developed sufficiently to produce the concept of homology and so to play its historic role in the growth of biology: namely (after its revival by Buffon, Goethe, Cuvier, *et al.*), to evoke the evolution concept as an explanatory hypothesis. Belon published, also, his travels* and two important monographs on fishes;† was friend of Ronsard and protégé of Henry II; was assassinated by ruffians in the Bois de Boulogne.

Birds of different kinds have their members differently fashioned. Just as the external aspect displays limbs relatively larger or smaller, so the bones forming the internal foundation correspond to what one sees from the outside. Birds of prey have stronger bones than swamp- or land-birds. Never has a bird fallen into our hands but we dissected it if we could. Thus we have in-

---

* *Observations de plusieurs singularitez, etc., trouvez en Grèce, etc.*, Paris, 1555.

† *L'histoire naturelle des éstranges poissons marins, etc.*, Paris, 1551, and *La nature et diversités des poissons, etc.*, Paris, 1555.

LJVRE I. DE LA NATVRE

Portraiɕt de l'amas des os humains , mis en comparaiſon
de l'anatomie de ceux des oyſeaux, faiſant que les
lettres d'icelle ſe raporteront à ceſte cy, pour
faire apparoiſtre combien l'affinité eſt
grande des vns aux autres.

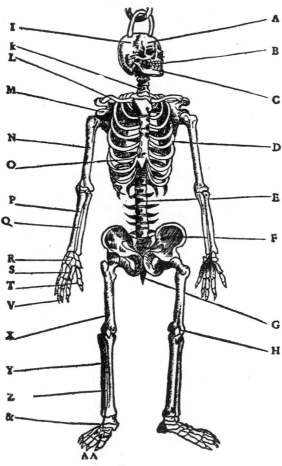

Fig. 1

DES OYSEAVX, PAR P. BELON.

La comparaiſon du ſuſdit portraiƈt des os humains monſtre combien ceſtuy cy qui eſt d'vn oyſeau, en eſt prochain.

Portraiƈt des os de l'oyſeau.

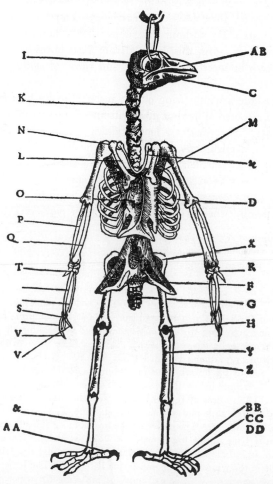

A B *Les Oyſeaux n'ont dents ne leures , mais ont*
   le bec tranchant fort ou foible, plus ou moins ſe-
   lon l'affaire qu'ils ont eu à mettre en pieces ce
   dont ils viuent .

M *Deux pallerons longs & eſtroiƈts, vn en chaſ-*
   cun coſté.

ꝛ   *L'os qu'on nommé la Lunette ou Fourchette*
   n'eſt trouué en aucun autre animal , hors mis en
   l'oyſeau.

D *Six coſtes , attachees au coffre de l'eſtomach par*
   denãt, & aux ſix vertebres du dos par derriere.

F *Les deux os des hanches ſont longs , car il n'y a*
   aucunes vertebres au deſſoubs des coſtes ;

G *Six oſſelets au cropion.*

H *La rouelle du genoil .*

I *Les ſutures du teſt n'apparoiſſent gueres ſinon*
   qu'il ſoit bouilly.

k *Douze vertebres au col, & ſix au dos.*

FIG. 2

(*For explanation of Figs.* 1 *and* 2 *see p.* 16.)

spected the internal parts of 200 species of birds. It must not be considered strange, therefore, that we now describe the bones of birds and portray them so exactly.

Whoever will observe two-footed animals and compare them with four-footed ones will find none which, while resting or sleeping, does not lie on its side except only the birds which remain always on their feet. Admittedly they support themselves on their breast; nevertheless, there are some which can sleep on one foot, standing without any support, or which kneel, as in the case of those with long legs. But this whole matter rests entirely upon the distinctions I have drawn between birds of prey and those of swamp, earth, woodland, and bush.

Whoever will take a whole wing or leg and thigh of a bird and compare it with that of a four-footed animal or man will find the bones as if in correspondence to each other. For, just as a man walking on his 'claws,' on tiptoe that is, would have the heel directly in line above the bones of the foot, so four-footed animals walking on their claws and having the heel in line with fingers or toes appear comparable in leg structure to the birds.

But in order to present this so that all can understand it and so as not to lose time in explaining the parts, let us name each particular bone and com-

---

FIG. 1.   *Translation of text accompanying figure:* "Portrayal of the totality of the bones of man placed in anatomical comparison with those of birds, making the letters of the former correspond to those of the latter, so as to make apparent how great is the affinity between them."

FIG. 2.   *Translation of complete text accompanying figure:* "Comparison with the aforesaid portrayal of human bones shows how close to it is the following, which is of the bird. *AB:* The birds have neither teeth nor lips but have a prominent beak, strong or weak more or less according to the trouble they have in breaking up their food. *M:* Two long and narrow pallerons, one to each rib. *κ:* The bone called the lunette or forked bone is found only among the birds. *D:* Six ribs attached to the casing of the stomach in front and to six dorsal vertebrae behind. *F:* The two hip bones are long since there are no vertebrae below the ribs. *G:* The six ossicles of the rump. *H:* The knee sprocket. *I:* The sutures of the skull are almost invisible unless it be boiled. *K:* Twelve vertebrae of the neck and six of the back. *L:* The bones of the two 'clefs.' *N:* Arm or shoulder bone. *O:* The coffer of the stomach. *P:* Small bones of the lower arm. *Q:* The large bone of the lower arm. *R:* The wrist bone known as the carpus. *S:* The joints and articulations called chondyles. *T:* The wing tip called the appendix which corresponds in the wing to the thumb in the hand. *T:* The bone coming after the wrist called the metacarpus. *V:* The extremity of the wing tip which is like the finger in us. *V:* More bones at the end of the wing, two of which are in the form of rape-seeds, one larger, one smaller, which correspond in the bird to the hollow of the hand in us, called the *thenar* in Greek and in Latin *palma*. *X:* The great bone of the lower leg. *Z:* The small bone of the lower leg. *&:* The bone acting as leg in birds which corresponds to our heel. *AA:* Just as we have four toes in the feet so birds have four digits of which the posterior corresponds to the big toe in us. *BB:* Four joints on the outside fingers. *CC:* Three joints on this finger. *DD:* Two joints on this finger as on the posterior one."

pare it with those of other animals and of man. The general description of
the bones of the human body is necessary so as to learn to distinguish the regions needing medication when a patient presents himself for treatment. But
we need not say much about all this here: for, since it has already been described and figured by so many, we do not pretend to add another exposition,
except, as far as it can be generally shown, of what is needed to exhibit the diverse play of nature in her works almost as though the form of one animal was
a product of that of another. Wherefore, we hope it will be understood that
we are making here a comparison of human bones with those of birds only,
promising to do likewise for other animals each in its proper place in our com-
mentaries upon Dioscorides, in the tongue selected for that work.

Whoever kills any bird whatever and carefully scrubs the skull (for it is
with the head that we desire to begin our dissection) will see no seams or sutures visible there; this is not to say, however, birds do not have them. For
if you will take the head of a boiled bird and dismember it you will be able to
distinguish the six bones corresponding to ours with their frontal, sagittal, and
occipital sutures, as well as the petrous portion of the temporal; and you will
recognize the frontal or coronal, the petrous portion of the temporal, the
parietal bones on the top of the head, the one forming the back called the occipital which is connected at the base of the brain, and, above the palate, the
basilary bone.

For mastication, they have the beak, since they have no teeth except in certain river birds with a dentellated beak. Whereas most land birds have two
ossicles within the root of the tongue, these birds have them at the sides, by
virtue of which they extend and retract the tongue.

Next after the head come the vertebrae or *rouelles* of the neck which in
French could well be called *pesons* but which the Latins called *Vertebrae*, the
Greeks *Spondyli*. Birds do not follow the pattern of other animals as regards the vertebrae of the neck. For, where other animals have but seven,
the birds have twelve. Very different in shape from those in the neck are the
next six in the spine of the back, to which are attached on each side six ribs: for
the birds have only twelve entire ribs and one little one on each side under the
wing, but all are meshed crosswise with other ossicles along the spine.

One finds in birds two large bones which we call 'plats,' or sacra, through
which there is an aperture on each side and a socket into which fits the thigh-
bone which we call the haunch. But the chest is of a different sort than in
other animals. For to those animals which must develop great strength in the
wings nature gave large and strong muscles reinforced by a large bone in the
chest in which the lungs are found. On the sides of this are the clavicles,
joined posteriorly to the scapula to hold the wings firmly in place. They have
in addition another bone which is called the *lunette* (eye-glass) or *fourchette*
(small fork) since one often places it over the nose as with eye glasses, or one

calls it the *bruchet* since it originates in front of the stomach and is joined to the ends of the two clavicles in the shoulder region and, on the other side, to the corselet, that is the chest bone. It is shaped like a fork.

Behind the large bones called the *sacra* comes the rump composed of six small bones which can be separated.

In the wings one finds almost the same bones as in the arms of men or in the forelegs of the 4-footed animals. For the large arm-bone called in Latin *Os adiuvatori,* which we call *avant-bras* which originates at the blades of the fork and the 'clefs' (see Fig. 1, *L*), is recognized as corresponding to that in other animals and man having the same protuberances, concavities, and rotundities. To this the other two bones of the arm are attached. There is no every day name for the latter. The ancients named the larger the *ulna,* the smaller the *radius:* we will call all three indifferently the bones of the arm especially since we have already called the big one the *avant-bras.*

BEGINNINGS OF MODERN ORNITHOLOGY

WILLUGHBY, Francis (British naturalist, 1635–1672) and RAY, John (*q.v.*). From *Ornithologiae libri tres, etc. Totum opus recognovit, digessit, supplevit Joannes Raius,* London, 1676; tr. by J. Ray as *The ornithology of Francis Willughby of Middleton in the county of Warwick wherein all the birds, etc.,* London, 1678.

"Willughby's" *Ornithologiae libri tres* are viewed by most historians as laying the foundations of the modern science of ornithology. Omitting dubious birds reported in travels or described in fables, the book is mainly based upon, and supplies a workable key to, species actually collected and observed by Ray and Willughby during extended European travels. Ray credits Willughby with senior authorship either because of actual leadership by Willughby in the work's preparation or because of an understanding that in their projected collaborations Willughby would take precedence in the zoological, Ray in the botanical, sections. Willughby, Ray's student, benefactor, and collaborator, died at thirty-seven leaving the work for Ray to complete. Representative and significant fragments are reprinted here.

OF THE DIVISION OF BIRDS

Birds in general may be divided into *Terrestrial* and *Aquatic,* or *Land* and *Water-fowl.*

Terrestrial are such as seldom frequent waters, but for the most part seek their food on dry land.

Aquatic are such as are much conversant in or about waters, and for the most part seek their food in watery places; of which we will treat [in] *Book III.*

Terrestrial Birds are either such as have *crooked Beak* and Talons, called by the *Grecians* ταμψώνυχες, or such as have more *streight Bills* and Claws.

Those that have *crooked* Bills and Claws, called ταμψώνυχες, are either

Rapacious and carnivorous, such as we call *Birds of prey,* or more *gentle* and frugivorous, as *Parrots.*

Rapacious and carnivorous are either *Diurnal,* such as prey by day-light, or *Nocturnal,* such as prey by night.

Rapacious diurnal Birds are usually divided according to their magnitude into the *greater* and *lesser* kind.

The *greater* kind are either the *more generous,* which have their Beaks hooked almost from the root, and are called *Eagles,* or the *sluggish* and *less generous,* having their Beaks streight for a good space from the root, and hooked only toward the point, called *Vultures.*

The *lesser* kind, called in Latine *Accipitres,* may be again subdivided into the *more generous,* which are usually reclaimed and trained up for fowling, properly called *Hawks;* and the *more cowardly* or *less generous,* such as are neglected by *Falconers,* as being of no use for fowling; and therefore permitted to live at large, which may be called *wild Hawks.*

Hawks properly so called are divided by Falconers into *long-winged* and *short-winged.*

Long-winged Hawks are such the tips of whose wings when closed reach almost to the end of the train: *Short-winged* are such the tips of whose wings when shut or withdrawn fall much short of the end of the train.

Birds that have *more streight* bills and claws are either the *greater* or the *lesser,* which we call *small birds.* Under the title of *greater* we comprehend all that do exceed or equal the common *Thrush* or *Mavis* in bigness. Yet to some kinds of bigger Birds (as for example *Woodpeckers*) by reason of the agreement of the characteristic notes we are forced to add one or two Birds lesser than *Thrushes.* The *greater* are either such as have *large, strong, streight, and long Bills,* or *lesser and shorter ones.* The first are either such as feed promiscuously upon Flesh, Insects and fruit (or grain) or at least Insects and fruit; or such as feed upon Insects only. Those in respect of colour may be divided into two kinds, viz. 1. *The Crow-kind,* whose body is for the most part of one colour and black: 2. *The Pie-kind,* whose body is covered with party-coloured feathers. Of these, [that feed only on Insects] there is but one kind, v.g. *Woodpeckers.* Such as have *lesser* and *shorter bills* may be distinguished by the colour of their flesh, into such as have *white flesh,* and such as have *black flesh.* Those that have *white flesh* are the Poultry kind, *Hens, Peacocks, Turkeys,* &c. Those that have *black flesh* are either the *greater,* that lay but two Eggs at a time, as *Pigeons;* or the *lesser,* which lay more than two Eggs at once, as *The Thrush kind.* The *lesser* sort of Birds with *streighter bills,* such as we usually call *small birds,* may be divided according to their Bills, into such as have *slender bills,* and such as have *thick and short bills.* Of both kinds there be many subalternate species; of which when we come to treat of *small Birds.*

Of the Fowling-piece, and Stalking-horse

The best Fowling-pieces are the long-barrelled [of five and a half or six foot] of an indifferent bore [somewhat under Harquebuse] for they hold the best charges, and carry the furthest level; and such as have Fire-locks.

The charge must be round hail-shot, of bigness according to the Game you shoot at.

As near as you can shoot with the wind, and sideways of, or behind the Fowl: And if possible under the shelter of some hedge, bank, or tree, &c. sometimes (if need be) creeping on your hands and knees. Chuse rather to shoot at a rank or file than a single fowl; and then send your Dog for what you have strucken. You must have your Dog in such true obedience as not to stir from your heels till you bid him go.

Where you have no shelter use a Stalking-horse, which is any old Jade trained up for that purpose; which being stript naked, and having nothing but a string about the nether Chap, of two or three yards long, will gently, and as you have occasion to urge him, walk on the banks of Brooks and Rivers, or Meadows and Moors, or up and down in the water, which way you please, flodding, and eating on the grass and weeds that grow therein; and so hardy as not to take any affright at the report of your Piece. You shall shelter your self and your Piece behind his fore-shoulder, bending your body down low by his side, and keeping his body still full between you and the Fowl. Then having chosen your mark, take your level from before the fore-part of the Horse, shooting as it were between the horses neck and the water, which is more safe than taking the level under the horses belly, and much less to be perceived; the shoulder of the horse covering the body of the man, and his legs also the mans legs. Whiles you are stalking you may leave your Dog with your Bags, &c. where he may lie close, and never stir till you have shot, and then upon the least call (but not before) come to you, and fetch forth what you have killed.

For want of a live-horse you may make an artificial stalking-horse of Canvas, either stuft, or hollow, and stretcht upon splints of wood or strong Wires, with his head bending down, as if he grazed, of due shape, stature, and bigness, painted of the colour of a horse [the darker the less apt to be discovered.] Let it be fixt in the middle to a staff with a pick of Iron, to stick it in the ground while you shoot.

Instead of a horse you may make and use the shape of an Oxe, Stag, or any other horned beast, painted of the usual colour of beasts in that Country, and having the natural horn or head.

N. These Engines are to be employed in those places where the birds are used to see, and be acquainted with the beasts they represent.

N. 2. These Engines are fitter for Water than Land, the water hiding their imperfections.

When you have so much beaten the fowl with the Stalking horse that they begin to find your deceit, and will not sit: Then you may otherwhiles use your Oxe-engine, till the Horse be forgotten, and so by change of your Engines make your sport last. The shape of a Stag may be useful in such places where Stags commonly feed, and are familiar with the Fowl, but they are subject to quicker discovery.

Some stalk with dead Engines, as an artificial Tree, Shrub, or Bush, or a dead Hedge. But these are not so useful for the stalk as the stand: It being unnatural for dead things to move, and the Fowl will not only apprehend, but eschew it. Therefore if you use them, you must either not move them at all, or so slowly as that their motion shall not be perceived.

SEVERAL WAYS OF TAKING THEM BY NIGHT
I. OF TAKING BIRDS WITH THE LOW-BELL

This is of use chiefly in Champain Countries, and that from the end of *October* till the end of *March* following.

About eight of the clock at night the Air being mild, and the Moon not shining, take your Low-bell of such size as a man may well carry it in one hand, having a deep, hollow, and sad sound; and with it a Net of small Mash, at least twenty yards deep, and so broad as to cover five or six ordinary Lands, or more, according as you have company to carry it: and go into a Stubble-field [a Wheat stubble is the best.] He that carries the Bell must go foremost, and toll it as he goeth along as solemnly as may be, letting it but now and then knock on both sides. Then shall follow the Net born up at each corner, and on each side. Another must carry a pan of live coals, but not blazing. At these, having pitcht your Nets where you think any Game is, you must light bundles of Hay, Straw, or Stubble, or else Links and Torches, and with noises and poles beat up all the Birds under the Net, that they may rise, and entangle themselves in it, and you take them at pleasure. Which done extinguish your Lights and proceeding to another place, do as before.

N. The sound of the *Low-bell* astonies the Birds, and makes them lie close; and the blaze of light dazling their eyes affrights them, and causes them to rise and make to it.

N. 2. In this past time all must be done with great silence, no noise being heard but the *Low-bell* only, till the Nets be placed, and the Lights blazing, and then you may use your pleasure: Which once extinguished, a general silence must be again made.

II. OF TAKING BIRDS WITH THE TRAMMEL

The Trammel is much like the Lowbelling Net, only it may be somewhat longer, but not much broader. This Net, when you come to a fit place, spread on the ground, and let the hinder end thereof, being plummed with lead lie loose on the ground, but the foremost end at the two corners be born up by the strength of men, a full yard or more from the ground, and so trail the Net along the ground. On each side the Net some must carry great blazing Lights of fire, and by the Lights others must march with long Poles, to beat up the Birds as you go, and as they rise so take them. In this sort you may go over a whole field, or any other champain ground.

III. HOW THEY TAKE BIRDS IN ITALY BY NIGHT WITH LIGHT AND A NET CALLED LANCIOTOIA

This sport is most used in the Champain of *Rome*. The Net is of the Mash of an ordinary Lark-net. It is fastened to two green sticks of pliant wood, twice so big as ones greatest finger, and two or three [Roman] yards long. These sticks must be fastened to the end of a square baston of two yards and half long in two holes, a little distant the one from the other, and covered with the same Net. [This Baston serves for a handle to carry and mannage the Net with, and may be as well round as square, and then the whole instrument will somewhat resemble a Racket, such as they play at Tennis with.] These two sticks serve to extend the Net at top to about four yards breadth. This Net the Fowler carries on his shoulder, holding the handle of it in one hand, and a Lanthorn called *Frugnuolo*, with a Lamp burning in it in the other; and when by the light he discovers any Bird within his reach, he claps his Net upon it, and covers it. Besides the Lanthorn the Fowler carries a Bell either at his Girdle, or his Knee, (like our Low-bell) the better to secure the birds to himself. This exercise cannot be used at all times, but only in Autumn or Winter, not beginning before one hour of the night. Whether the weather be cloudy or clear it is all one, so the Moon shine not.

The *Frugnuolo* is a sort of Lanthorn made of Latten (commonly, but falsely, called Tin) all close but the fore-side. Its Base about a *Roman* Palm and half long, and at the aperture about a Palm broad, or a little more; likewise a Palm high: The Cover (which goes shelving) two Palms long: In the midst thereof above is a handle, and within side a thin plate of Iron three fingers distant from the beginning of the Cover, to preserve the Tin from being burnt and marred by the flame of the Lamp. Below is another empty handle to put in a stick to hold it up on high. Within, in a Circle made on purpose in the bottom, is put an earthen Lamp with a great Week, and Oyl. With this kind of Lamp they also search bushes, hedges, and low trees, where they think Thrushes and other Birds pearch, and having discovered them, strike them down with an Instrument called *Ramata,* made like a Racket with a long handle, or if they be out of reach of that, shoot them with a Cross-bow.

IV. OF BAT-FOWLING

Bat-fowling is a taking by night of great and small Birds, that rest not on the ground, but pearch on shrubs, bushes, trees, &c. and is proper to woody and rough Countries.

First, one must carry a Vessel with fire (as in Low-belling) then others must have Poles bound with dry Wisps of Hay, Straw, pieces of Links, pitcht Hurds, or any other combustible matter that will make a blaze. Others must bear long Poles with rough and bushy tops. When you are come to the Birds haunts, kindle some of your fires, and with your Poles beat the bushes and trees: Which done the Birds (if any be) will rise, and fly to, and play about the Lights: It being their nature not to depart from them, but almost scorch their Wings in the same, so that they who have the bushy Poles may at their pleasure strike them down and take them.

Others carry with them a great Lime-bush made of the head of a Birch or Willow Tree, and pitching it down make their blazes close by it; and the birds will come and light upon it, and so be entangled.

In this Sport you must observe the directions given in Low-belling as to the choice of the night, and especially keeping silence, &c.

OF TAKING BIRDS WITH DAY-NETS

The time of the Year for these Nets is from *August* till *November:* Of the Day a little before Sun-rise, so as your Nets may be laid, and all your Implements in readiness to begin your work by peep of Sun. The milder the Air, and the clearer and brighter the Morning, the fitter is the season for this exercise. The best place is in Champain Countries, remote from any Town, Village, or common concourse of people, on short Barley stubbles, smooth green Layes, or level Meadows; if the place be not naturally even and plain where you pitch your Nets, you must make it so: That both lying and falling over they may couch so close to the ground that the shortest grass or stubble appearing through them, they may as it were lie hid and unperceived by the Birds, and that being covered they may not creep or flicker from under them.

Let your Nets be made of very fine Packthread, knit sure, the Mash not above an Inch square. Let them be about three fathoms long, and not above one deep, verged on each side with strong small Cords, the ends extended upon two small Poles as long as the Nets is broad, &c. in all things like the Net described § 1. save that that was to be but one single Net but here you must have two exactly of the same size and fashion, and placed at that distance, that when they are drawn the sides may just meet and touch one another. Your Nets being staked down with strong stakes, so that with any nimble twitch you

may cast them to and fro at pleasure; some twenty or thirty paces from the
Nets place your Giggs on the tops of long Poles, turned into the wind, so as
they may play and make a noise therein.   These Giggs are made of long
Goose-feathers in the manner of Shuttle-cocks, and with little tunnels of wood
running in broad and flat Swan-quills, made round like a small hoop, and so
with longer strings fastened to the Pole, will with any small wind twirl and
flicker in the Air after such a wanton manner that the Birds will come in great
flocks to wonder and play about the same.   After the placing of your Giggs,
you shall then place your Stale, which is a small stake of wood to prick down
fast in the earth, having in it a Morteise hole, in which a long slender piece of
wood, of about two foot, is so fastened that it may move up and down at pleasure, and to this longer stick you shall fasten a small Line, which running
through a hole in the stake aforesaid, and so coming up to the place where you
sit, you may be drawing the Line up and down to you (with your right hand)
raise and mount the longer stick from the ground, as oft as you shall find
occasion.

Now to this longer stick you shall fasten a live Lark, or Bunting, (for you
must be sure ever to preserve some alive for that purpose) or for want of such,
any other small Bird, which the Line making to flicker up and down by your
pulling, will entice the Larks to play about it, and swoop so near the ground,
that drawing your hand, you may cover them with your Nets at pleasure: Also
it will entice Hawks and any other Birds of prey to stoop and strike at the
same, so as you may with ease take them.   . . .

Of the Election and Training Up of a Setting Dog

Although the Water-Spaniel, Mungrel, shallow-slew'd Hound, Tumbler,
Lurcher, or small Bastard Mastiff may be brought to Set; yet none of them
is comparable to the true-bred Land-Spaniel, being of a size rather small than
gross, a strong and nimble ranger of a courageous fiery mettle, a quick sent,
delighting in toil and indefatigable, yet fearful of, and loving to his Master.
Of what colour he be it matters not much.

Having gotten you a Whelp of such a Breed, begin to handle and instruct
him at four to six months old at the furthest.

1. You must make him very loving to, and familiar with you, and fond of
you, so as to follow you up and down without taking notice of any man else,
by suffering no man to feed or cherish him but yourself.   You must also make
him stand in aw of, and fear you as well as love you, and that rather by a stern
countenance and sharp words than blows.

2. Then you must teach him to couch and lie down close to the ground,
first by laying him down on the ground, and saying to him, *Lie close*, or the
like, terrifying him with rough language when he doth any thing against your
command, and cherishing him, and giving him food when he doth as you bid

him. And thus by continued use and practicing the same thing, in a few days you shall bring him readily and presently to lie close on the ground, whenever you shall but say, *Couch, down, lie close,* or the like.

3. Next you shall teach him, being couched, to come creeping to you with his head and belly close to the ground so far, or so little way as you shall think good, by saying, *Come nearer,* or the like: First, till he understand your meaning by shewing him a piece of bread, or some other food to entice him. And if when he offers to come he either raise from the ground his fore or hinder parts, or so much as lift up his head, then you shall not only with your hand thrust down his body in such sort as you would have him keep it, but also chide and rate him so as to make him strive to perform your pleasure: And if that will not quicken him sufficiently, to the terrour of your voice add a sharp jerk or two with a Whip-cord lash. When he does your will either fully or in part according to his apprehension chearfully, then you must be sure to cherish him, and to feed him: And then renew his lesson again till he be perfect in it. In like manner you must make him stop and pause when you bid him.

4. Then you shall teach him to lead in a string, and follow you at your heels without straining his Collar, which you may easily do by practice, not striving too roughly with him.

5. When he is thus far taught, you may out into the field with him, and suffer him to range and hunt, yet at such command, that upon the first hem or warning of your voice he stop and look back upon you, and upon the second, that he forthwith either forbear to hunt further, or else come in to your foot, and walk by you. If in ranging you find he opens, you shall first chide him therefore, and if that prevail not, either bite him hard at the roots of his ears, or lash him with a sharp Whip-cord lash, till you have him so staunch, that he will hunt close and warily without once opening, either through wantonness, or the rising up of any small birds before him.

When you find that he is come upon the haunt of any Partridge, (which you shall know by his eagerness in hunting, and by a kind of wimpering and whining, as being greatly desirous to open, but for fear not daring) you shall then warn him to take heed by saying; *be wise,* or the like. But if notwithstanding he either rush in and so spring them, or else open or use any means by which the Partridge escapeth, you shall then correct him soundly, and cast him off again in another place where you are sure a Covey lies, and then as before give him warning. And if you see that through fear he standeth still and waveth his tail, looking forward as if he pointed at somewhat, be sure the Partridge is before him: Then make him lie close, and taking a large ring about him, look for the Partridge. When you have found them, if you see he hath set them too far off, you shall make him creep on his belly nearer, else let him lie close without stirring, and then drawing your Net take the *Partridge.* Encourage your Dog by giving him the heads, necks, and pinions of the Partridge, and also bread or other food. But if he chance by any rude-ness or want of taking heed to spring them again, you shall correct him as be-fore, and lead him home in your string, and tie him up that night, giving him nothing but a bit of bread and water, and the next day take him out, and do as before, but with somewhat more terrour and harshness, and doubtless the Dog will do according to your will: Which if he doth, you must by no means forget to bestow upon him all the cherishings of voice, hand, and foot.

It is a fault in a Dog to stand up right as it were looking over the Partridge when he sets them, and therefore you must chide him for it, not giving over till you make him lie close.

It is also a fault for him when you go in to the Covey, to spring up the Par-tridge into your Nets, to rush hastily after you, or spring them before you, for which you must correct him; and your self proceeding leisurely the next time, ever as you go speak to the Dog to lie close.

An Abridgment of Some Statutes Relating to the Preservation of Fowl

Now lest any one, either not legally qualified or licensed, or by taking Fowl at prohibited times, or by prohibited Engines, or by destroying of their Eggs, should through ignorance incur the danger of the Law, I have thought fit to subjoyn an Abridgement of such Statutes as relate to the preservation of Fowl, collected and sent me by my worthy Friend Mr. *Walter Ashmore.*

None to destroy or take away the Eggs of any Wild Fowl on pain of one years imprisonment; and to forfeit for every Egg of a Crane or Bustard so taken and destroyed 20 *d*. Of a Bittern, Heron, or Shoveler 8 *d*. Of a Mal-ard, Teal, or other Wild-fowl 1 *d*. to be divided between the King and the Prosecutor. And herein Justices of peace have power to hear, and enquire, and determine offences of this kind, as they use to do in cases of trespass. Yet this act not to extend to such as kill Crows, Choughs, Ravens, and Buz-zards.

A Hawk taken up shall be delivered to the Sheriff, who after Proclama-tion made in several Towns, (if challenged) shall deliver her to the right Owner. And if the Hawk were taken up by a mean man, and be not chal-lenged in four months, the Sheriff to have her, satisfying the Party for taking her: But if by a man of estate, who may conveniently keep a Hawk, the Sher-iff shall restore her to him again, he paying for the charge of keeping.

If any take away or conceal a Hawk he shall answer the value thereof to the Owner, and suffer two years imprisonment, and in case he be not able to answer the value, he shall remain in prison a longer time.

He that steals and carries away a Hawk, not observing the Ordinance of 34 *Ed.* 3. 22. shall be deemed a Felon.

None shall take Pheasants or Partridges with Engines in anothers ground without licence, in pain of ten pound to be divided between the Owner of the ground and the Prosecutor.

None shall take out of the Nest any Eggs of Falcon, Goshawk, Lanner, or Swan, in pain of a year and a days imprisonment, and to incur a Fine at the Kings pleasure, to be divided between the King and the Owner of the ground where the Eggs shall be so taken.

None shall bear any Hawk of English breed called a Nyesse, (Goshawk, Tarcel, Lanner, Lanneret, or Falcon) in pain to forfeit the same to the King.

He that brings a Nyesse Hawk from beyond the Seas shall have a Certificate under the Customers Seal where he lands, or if out of *Scotland,* then under the Seal of the Lord Warden or his Lieutenant, testifying she is a Foreign Hawk, upon the like pain of forfeiting the Hawk.

None shall take, kill, or fear away any of the said Hawks from their Coverts where they use to breed, in pain of ten pounds.

Every Freeman may have Eyries of Hawks within their own Woods which be within a Forest.

None shall kill or take Pheasants or Partridges by night, in pain of 20 *s.* a Pheasant, and 10 *s.* a Partridge, or one months imprisonment, and bond with Sureties not to offend again in the like kind.

Directions to recover the Forfeitures, *vid.* Statute.

None to hawk or hunt with Spaniels in standing Grain in pain of 40 *s.*

No person shall kill or take any Pheasant, Partridge, Pigeon, Duck, Heron, Hare, or other Game, or take or destroy the Eggs of Pheasants, Partridges, or Swans, in pain of 20 *s.* or imprisonment for every Fowl, Hare, or Egg, and to find Sureties in 20 *l.* not to offend in the like kind.

No person shall keep Dog or Net to take or kill any of the last mentioned Game, unless qualified as in the Act, in pain of 30 *s.*

No Person to buy or sell any Partridge or Pheasant upon pain to forfeit 20 *s.* for every Pheasant, and 10 *s.* for every Partridge.

No Person to be twice punished for one offence.

Persons are to be licensed in Sessions to kill Hawks meat, and to become bound in 20 *l.* not to kill any of the said Games, nor to shoot within 600 paces of a Heronry, within 100 paces of a Pigeon House, or in a Park, Forest, or Chase, whereof his Master is not Owner or Keeper.

Every person having hawked at, or destroyed any Pheasant or Partridge between the first of *July* and last of *August* shall forfeit 40 *s.* for every time so Hawking, and 20 *s.* for every Pheasant or Partridge so destroyed or taken.

This offence to be prosecuted within six months after it is committed.

Lords of Mannors and their Servants may take Pheasants or Partridges in their own grounds or Precincts in the day time between *Michaelmas* and *Christmas.*

Every person of a mean condition having killed or taken any Pheasant or

Partridge shall forfeit 20 *s.* for each one so killed, and shall become bound in 20 *l.* not to offend so again.

Constables and Headboroughs upon warrant to search houses, and seize Dogs or Nets, and destroy them at pleasure.

Lords of Mannors to appoint Game-keepers, who by a Warrant from a Justice may in the day-time take and seize all Guns, Bows, Grey-hounds, Setting-dogs, Lurchers, or other Dogs to kill Hares or Conies, Ferrets, Trammels, Low-bells, Hays, or other Nets, Hare-pipes, Snares, and other Engines for the taking and killing of Conies, Hares, Pheasants, Partridges, and other Game within the Precincts of such Mannor, as shall be used by any Person prohibited by that Act to keep or use the same.

Persons under the value of 100 *l. per annum;* or for term of life, or not having Leases for ninety nine years, or for a longer term of the value of 150 *l.* other than the Son and Heir apparent of an Esquire or other person of higher degree, and the Owners and Keepers of Forests, Parks, Chases, or Warrens, are not to have or keep for themselves or others any Guns, Bows, Greyhounds, Setting-dogs, Lurchers, Hays, Nets, Lowbells, Hare-pipes, Snares, or other Engine.

THE *Dodo*, CALLED BY CLUSIUS *Gallus gallinaceus peregrinus*, BY NUREMBERG *Cygnus eucullatus*, BY BONTIUS *Dronte*

This Exotic Bird, found by the *Hollanders* in the Island called *Cygnaea* or *Cerne* by the *Portugues*, *Mauritius* Island by the *Low Dutch*, of thirty miles compass, famous especially for black Ebony, did equal or exceed a *Swan* in bigness, but was of a far different shape: For its Head was great, covered as it were with a certain membrane resembling a hood: Beside, its Bill was not flat and broad, but thick and long; of a yellowish colour next the Head, the point being black: The upper Chap was hooked; in the nether had a bluish spot in the middle between the yellow and black part. They reported that it is covered with thin and short feathers, and wants Wings, instead whereof it hath only four or five long, black feathers; that the hinder part of the body is very fat and fleshy, wherein for the Tail were four or five small curled feathers, twirled up together, of an ash-colour. Its Legs are thick rather than long, whose upper part, as far as the knee, is covered with black feathers, the lower part, together with the Feet, of a yellowish colour: Its Feet divided into four toes, three (and those the longer) standing forward, the fourth and shortest backward; all furnished with black Claws. After I had composed and writ down the History of this Bird with as much diligence and faithfulness as I could, I hapned to see in the house of *Peter Pawius*, primary Professor of Physic in the University of *Leyden*, a Leg thereof cut off at the knee, lately brought over out of *Mauritius* his Island. It was not very long, from the knee to the bending of the foot being but little more than four inches; but of a great thickness, so that it was almost four inches in compass, and covered with thick-set scales, on the upper side broader, and of a yellowish colour, on the under [or backside of the Leg] lesser and dusky. The upper side of the Toes was also covered with broad scales, the under side wholly callous. The Toes were short for so thick a Leg: For the length of the greatest or middlemost Toe to the nail did not much exceed two inches, that of the other Toe next to it scarce came up to two inches: The back-toe fell something short of an inch and half: But the Claws of all were thick, hard, black, less than an inch long; but that of the back-toe longer than the rest, exceeding an inch. The Mariners in their dialect gave this bird the name of Walghvogel, that is, a nauseous, or yel-lowish bird: Partly because after long boyling its flesh became not tender, but continued hard, and of a difficult concoction; excepting the Breast and Gizzard, which they found to be of no bad relish; partly because they could easily get many *Turtle-Doves,* which were much more delicate and pleasant to the Palate. Wherefore it was no wonder that in comparison of those they despised this, and said they could well be content to be without it. Moreover they said, that they found certain stones in its Gizzard: *And no wonder, for all other birds as well as these swallow stones, to assist them in grinding their meat.* Thus far *Clusius.*

Bontius writes, that this Bird is for bigness of mean size, between an *Ostrich* and a *Turkey,* from which it partly differs in shape, and partly agrees with them, especially with the *African Ostriches,* if you consider the Rump quils, and feathers: So that it shews like a Pigmy among them, if you regard this shortness of its Legs. It hath a great, ill-favoured Head, covered with a kind of membrane resembling a hood: Great, black Eyes, a bending, prominent, fat Neck: An extraordinary long, strong, bluish white Bill, only the ends of each Mandible are of a different colour, that of the upper black, that of the nether yellowish, both sharp-pointed and crooked. It gapes huge wide, as being nat-urally very voracious. Its body is fat, round, covered with soft, grey feathers, after the manner of an *Ostriches:* In each side instead of hard Wing-feathers or quils, it is furnished with small soft-feathered Wings, of a yellowish ash colour; and behind the Rump, instead of a Tail, is adorned with five small curled feathers of the same colour. It hath yellow Legs, thick, but very short; four Toes in each foot, solid, long, as it were scaly, armed with strong, black Claws. It is a slow-pace and stupid bird, and which easily becomes a prey to the Fowlers. The flesh, especially of the Breast, is fat, esculent, and so copi-ous, that three or four *Dodos* will sometimes suffice to fill an hundred Sea-mens bellies. If they be old, or not well boyled, they are of difficult concoc-tion, and are salted and stored up for provision of victual. There are found in their stomachs stones of an ash-colour of divers figures and magnitude; yet not bred there as the common people and Seamen fancy, but swallowed by the Bird; as though by this mark also Nature would manifest that these Fowl are of the *Ostrich* kind in that they swallow any hard things, though they do not digest them. *Thus Bontius.*

We have seen this Bird dried, or its skin stuft in *Tradescants* Cabinet.

BIRTH OF MODERN SYSTEMATIC ZOOLOGY

RAY (Wray), John (British botanist and naturalist, 162?–1705). From *Synopsis methodica animalium quadrupedium et serpentini generis,* London, 1693; tr. of a fragment by T. S. Hall, for this volume.

A study of the works of the great animal systematists reveals that although they all wished to discover orderly relations among animals, some (Aristotle, Linnaeus, Ray) were more impressed by the *categorical* aspect of those relations, others (especially Lamarck and his successors) by their *serial* aspect. This is an important distinction because serial relations, much more than categorical ones, suggest the possibility of dynamic or causal relations between the phenomena in question. Had Ray's interest focused less on the categorical and more on the sequential he might have grasped more fully the dynamic connections between the animal groups. Actually, he never went further in his evolutionary thought than to admit the possibility of mutability of species.* Note the strictly dichotomic treatment brought over from the Greek schools even though Aristotle† had cited its inadequacies; also the all-important discussion concerning the definition of blood (see p. 31).

The animals with blood can be divided into those breathing by *lung* and those breathing by *gills.*

Those breathing by *lung* in turn into those having the heart characterized by two ventricles and, by *one.*

The warmer animals have the heart characterized by *two* ventricles; being either *viviparous,*—the *quadrupeds* among land forms and the cetaceans among water forms—or else *oviparous,*—as the birds.

Having the heart built of one ventricle are the oviparous quadrupeds and the genus of *serpents;* in this way must the genus of serpents be distinguished from that of oviparous quadrupeds.

All fishes except cetaceans breathe with gills; understand here the sanguineous fishes, for the name of fish is occasionally extended to aquatic animals without blood.

Animals without blood are either *greater,* or *less;* the greater are for the most part aquatic and have been conveniently subdivided by Aristotle into three genera, viz.: 1) the Mollia, or Mollusca; 2) the Crustacea; and 3) the Testacea. These, moreover [*i.e.,* the Testacea—Ed.] are either univalves, bivalves, or turbinates.

The *lesser* bloodless animals are called *insects.*

---

* The definitive work on Ray is C. E. Raven, *John Ray, naturalist, his life and works,* Cambridge University Press, 1942.

† *In De partibus animalium,* Book 1,

These subdivisions of animals appear to me as the exactest of all and as most conforming to nature.

The first subdivision of animals, and the most suitable of all, is into *those with* and *those without blood.*

I call 'blood' that ruddy liquor, life's seat and source, nourisher and carrier of the native heat, which, by a circular and undesisting motion of the two acting together, flows through the arteries and veins.

I am not unaware that some, even philosophers of great name who seem to have discernment above common, reject this distinction and contend that all animals possess blood in the sense that all possess a vital liquor of the same sort except that its color is not always red.

I, meanwhile, would not deny that some sort of vital fluid has been given to all animals, analogous to blood, equally governed by a circular motion, and distinguished by the same usefulness to the body; however, that should not be called blood which is to be deemed less noble and spirituous because it is deprived of the ruddy color with which, as though robed in purple, the blood of warmer and more perfect animals shines.

It must, moreover, not be pretended that any insect genus is truly sanguineous nor, truly, the earthworm, whose vital liquor is not so much analogous to blood but only tinctured with a purple color, while pure blood is undefiled. In truth one exception or even more does not overthrow or destroy a general law.

TAXONOMY APPLIED TO MAN

von LINNAEUS (Linné), Carl (Carolus) (Swedish systematic and general biologist, 1707–1778). From *Systema naturae*, Leyden, 1735; tr. by W. Turton (of *Last edition of the Systema Naturae of Linné by Gmelin*, Leipzig, 1788), London, 1806.

Linnaeus was not the first worker in the fields of taxonomy, nomenclature, reproductive plant physiology, and comparative anthropology, but the thesis could be maintained that his contributions did much to establish each of them in its modern form. His animal taxonomy as a whole is less original than his realization that taxonomy applies, among other animals, to man. Reprinted here are the first pages of the *Systema* (in its last edition).

Man, when he enters the world, is naturally led to enquire who he is; whence he comes; whither he is going; for what purpose he is created; and by whose benevolence he is preserved. He finds himself descended from the remotest creation; journeying to a life of perfection and happiness; and led by his endowments to a contemplation of the works of nature.

Like other animals who enjoy life, sensation, and perception; who seek for food, amusements, and rest, and who prepare habitations convenient for their kind, he is curious and inquisitive: but, above all other animals, he is noble in his nature, in as much as, by the powers of his mind, he is able to reason justly upon whatever discovers itself to his senses; and to look, with reverence and wonder, upon the works of Him who created all things.

That existence is surely contemptible, which regards only the gratification of instinctive wants, and the preservation of a body made to perish. It is therefore the business of a thinking being, to look forward to the purposes of all things; and to remember that the end of creation is, that God may be glorified in all his works.

Hence it is of importance that we should study the works of nature, than which, what can be more useful, what more interesting? For, however large a portion of them lies open to our present view; a still greater part is yet unknown and undiscovered.

All things are not within the immediate reach of human capacity. Many have been made known to us, of which those who went before us were ignorant; many we have heard of, but know not what they are; and many must remain for the diligence of future ages.

It is the exclusive property of man, to contemplate and to reason on the great book of nature. She gradually unfolds herself to him, who with patience and perseverance, will search into her misteries; and when the memory of the present and of past generations shall be entirely obliterated, he shall enjoy the high privilege of living in the minds of his successors, as he has been advanced in the dignity of his nature, by the labours of those who went before him.

The UNIVERSE comprehends whatever exists; whatever can come to our knowledge by the agency of our senses. The *Stars,* the *Elements,* and this our *Globe.*

The STARS are bodies remote, lucid, revolving in perpetual motion. They shine, either by their own proper lights, as the *Sun,* and the remoter *fixed Stars;* or are *Planets* receiving light from others. Of these the primary planets are solar; *Saturn, Jupiter, Mars,* the *Earth, Venus, Mercury,* and *Georgium Sidus:* the secondary are those subservient to, and rolling round the primary, as the *Moon* round the earth.

The ELEMENTS are bodies simple, constituting the atmosphere of, and probably filling the spaces between the stars.

Fire;	lucid,	resilient,	warm,	evolant,	vivifying.
Air;	transparent,	elastic,	dry,	encircling,	generating.
Water;	diaphanous,	fluid,	moist,	gliding,	conceiving.
Earth;	opaque,	fixed,	cold,	quiescent,	steril.

The EARTH is a planetary sphere, turning round its own axis, once in 24 hours, and round the sun once a year; surrounded by an *atmosphere* of elements, and covered by a stupendous crust of *natural bodies,* which are the objects of our studies. It is terraqueous; having the depressed parts covered with waters; the elevated parts gradually dilated into dry and habitable continents. The *land* is moistened by *vapours*, which rising from the waters, are collected into *clouds:* these are deposited upon the tops of mountains; form small *streams*, which unite into *rivulets*, and reunite into those ever-flowing *rivers*, which pervading the thirsty earth, and affording moisture to the productions growing for the support of her living inhabitants, are at last returned into their parent *sea*.

The study of natural history, simple, beautiful, and instructive, consists in the collection, arrangement, and exhibition of the various productions of the earth.

These are divided into the three grand kingdoms of nature, whose boundaries meet together in the Zoophytes.

MINERALS inhabit the interior parts of the earth in rude and shapeless masses; are generated by salts, mixed together promiscuously, and shaped fortuitously.

They are bodies *concrete*, without life or sensation.

VEGETABLES clothe the surface with verdure, imbibe nourishment through bibulous roots, breathe by quivering leaves, celebrate their nuptials in a genial metamorphosis, and continue their kind by the dispersion of seed within prescribed limits.

They are bodies *organized*, and have *life* and not sensation.

ANIMALS adorn the exterior parts of the earth, respire, and generate eggs; are impelled to action by hunger, congeneric affections, and pain; and by preying on other animals and vegetables, restrain within proper proportion the numbers of both.

They are bodies *organized*, and have *life, sensation,* and the power of locomotion.

MAN, the last and best of created works, formed after the image of his Maker, endowed with a portion of intellectual divinity, the governor and subjugator of all other beings, is, by his wisdom alone, able to form just conclusions from such things as present themselves to his senses, which can only consist of bodies merely natural. Hence the first step of wisdom is to know these bodies; and to be able, by those marks imprinted on them by nature, to distinguish them from each other, and to affix to every object its proper name.

These are the elements of all science; this is the great alphabet of nature: for if the name be lost, the knowledge of the object is lost also; and without these, the student will seek in vain for the means to investigate the hidden treasures of nature.

METHOD, the soul of Science, indicates that every natural body may, by inspection, be known by its own peculiar name; and this name points out whatever the industry of man has been able to discover concerning it: so that amidst the greatest apparent confusion, the greatest order is visible.

SYSTEM is conveniently divided into five branches, each subordinate to the other: *class, order, genus, species,* and *variety,* with their names and characters. For he must first know the name who is willing to investigate the object.

The science of nature supposes an exact knowledge of the nomenclature, and a systematic arrangement of all natural bodies. In this arrangement, the *classes* and *orders* are arbitrary; the *genera* and *species* are natural. All true knowledge refers to the species, all solid knowledge to the genus.

Of these three grand divisions the *animal* kingdom ranks highest in comparative estimation, next the *vegetable,* and the last and lowest is the *mineral* kingdom.

ANIMALS enjoy *sensation* by means of a living organization, animated by a medullary substance; *perception* by nerves; and *motion* by the exertion of the will.

They have *members* for the different purposes of life; *organs* for their different senses; and *faculties* or powers for the application of their different perceptions.

They all originate from an *egg.*

Their external and internal structure; their comparative anatomy, habits, instincts, and various relations to each other, are detailed in authors who professedly treat on these subjects.

The natural *division* of animals is into 6 *classes,* formed from their internal structure.

Heart with 2 auricles, 2 ventricles; blood warm, red.	viviparous.	MAMMALIA.	1.
	oviparous.	BIRDS.	2.
Heart with 1 auricle, 1 ventricle; blood cold, red.	lungs voluntary.	AMPHIBI	3.
	external gills.	FISHES.	4.
Heart with 1 auricle, ventricle o; sanies cold, white.	have antennae.	INSECTS.	5.
	——tentacula.	WORMS.	6.

I. MAMMALIA. *Lungs* respire alternately; *jaws* incumbent, covered; *teeth* usually within; *teats* lactiferous; *organs* of sense, tongue, nostrils, eyes, ears, and papillae of the skin; *covering,* hair, which is scanty in warm climates, and hardly any on aquatics; *supporters,* 4 feet, except in aquatics; and in most a *tail: walk* on the *earth,* and *speak.*

II. BIRDS. *Lungs* respire alternately; *jaws* incumbent, naked, extended, without teeth; *eggs* covered with a calcareous shell; *organs* of sense, tongue, nostrils, eyes, and ears without auricles; *covering,* incumbent, imbricate feathers; *supporters,* feet 2, wings 2; and a heart-shaped rump; *fly* in the *air,* and *sing.*

III. AMPHIBIA.   *Jaws* incumbent; *penis* (frequently) double; *eggs* (usually) membranaceous; *organs* of sense, tongue, nostrils, eyes, ears; *covering,* a naked skin; *supporters* various, in some 0; *creep* in *warm* places and *hiss.*

IV. FISHES.   *Jaws* incumbent; *penis* (usually) 0; *eggs* without white; *organs* of sense, tongue, nostrils ? eyes, ears; *covering,* imbricate scales; *supporters,* fins; *swim* in the *water,* and *smack.*

V. INSECTS.   *Spiracles,* lateral pores; *jaws,* lateral; *organs* of sense, tongue, eyes, antennae on the head, brain 0, ears 0, nostrils 0; *covering,* a bony coat of mail; *supporters,* feet, and in some, wings; *skip* on *dry* ground, and *buzz.*

VI. WORMS.   *Spiracles,* obscure; *jaws,* various; frequently *hermaphrodites; organs* of sense tentacula, (generally) eyes, brain 0, ears 0, nostrils 0; *covering,* calcareous or 0, except spines; *supporters,* feet 0, fins 0; *crawl* in *moist* places, and are *mute.*

CLASS I. MAMMALIA.

THESE suckle their young by means of lactiferous teats. In external and internal structure they resemble man: most of them are quadrupeds; and with man, their natural enemy, inhabit the surface of the earth. The largest, though fewest in number, inhabit the ocean.

They are distributed into 7 *Orders,* the characters of which are taken from the number, situation, and structure of the *teeth.*

I. PRIMATES.   *Fore-teeth* cutting, upper 4 parallel, (except in some species of bats which have 2 or 0); *tusks,* solitary, that is, one on each side, in each jaw; *teats* 2, pectoral; *feet,* 2 are hands; *nails,* (usually) flattened, oval; *food,* fruits, except a few who use animal food.

II. BRUTA.   *Fore-teeth* 0 in either jaw; *feet* with strong hoof-like nails; *motion,* slow; *food,* (mostly) masticated vegetables.

III. FERAE.   *Fore-teeth* conic, usually 6 in each jaw; *tusks* longer; *grinders* with conic projections; *feet* with claws; *claws* subulate; *food,* carcases and preying on other animals.

IV. GLIRES.   *Fore-teeth* cutting, 2 in each jaw; *tusks* o; *feet* with claws formed for running and bounding; *food,* bark, roots, vegetables, &c, which they gnaw.

V. PECORA.   *Fore-teeth,* upper o, lower cutting, many; *feet* hoofed, cloven; *food,* herbs which they pluck; *chew* the cud; *stomachs* 4, the *paunch* to macerate and ruminate the food, the *bonnet,* reticulate, to receive it, the *omasus,* or maniplies of numerous folds to digest it, and the *abomasus* or caille, fasciate, to give it acescency and prevent putrefaction.

VI. BELLUAE.   *Fore-teeth* obtuse; *feet* hoofed; *motion* heavy; *food* gathering vegetables.

VII. CETE.   *Fins* pectoral instead of feet; *tail* horizontal, flattened; *claws* o; *hair* o; *teeth,* in some cartilaginous, in some bony; *nostrils* o, instead of which is a fistulous opening in the anterior and upper part of the head; *food* molluscae and fish; *habitation,* the ocean.

These are necessarily arranged with the mammalia from their similarity of structure, though their habits and manners are like those of fish. *Heart* with 2 auricles, 2 ventricles; *blood* warm; *lungs* respiring alternately; *eyelids* moveable; *ears* hollow, receiving sound through the medium of the air; *vertebrae* of the neck 7; *lumbar* bones, and *coccyx*; *teats* lactiferous, with which they suckle their young.

CHARACTERS OF THE MAMMALIA.

I. PRIMATES. *Fore-teeth incisors,* 4; *tusk* 1.

1. **HOMO.**   Walks erect; body naked, except in a few places.
2. *Simia.*   Tusks distant from each other.
3. *Lemur.*   Fore-teeth, lower 6.
4. *Vespertilio.*   Fore-feet palmate, formed for flying.

II. BRUTA. *Fore-teeth* o, *in either jaw.*

10. *Rhinoceros.*   Horn on the middle of the forehead.
11. *Sukotyro.*   Horn on each side near the eyes.
12. *Elephas.*   Tusks and Grinders; nose elongated into a proboscis.
13. *Trichechus.*   Tusks upper; grinders rough bony; feet stretched backwards.
5. *Bradypus.*   Tusks o; anterior grinders longer; body hairy.
6. *Myrmecophaga.*   Teeth o; body hairy.
8. *Manis.*   Teeth o; body scaly.
9. *Dasypus.*   Grinders; Tusks o; body covered with a crustaceous shell.
7. *Platypus.*   Mouth like a duck's bill; feet palmate.

III. FERAE. *Fore-teeth conic,* (10, 6, 2.); *tusk* 1.

14. *Phoca.*	Fore-teeth upper 6, lower 4.
15. *Canis.*	Fore-teeth 6,6; intermediate upper ones lobate.
16. *Felis.*	Fore-teeth 6,6; lower ones equal; tongue aculeate.
17. *Viverra.*	Fore-teeth 6,6; intermediate lower ones shorter.
18. *Mustela.*	Fore-teeth 6,6; lower ones crowded; 2 alternate interior.
19. *Urfus.*	Fore-teeth 6,6; upper ones excavate; a crooked bone in the penis.
20. *Didelphis.*	Fore-teeth upper 10, lower 8.
21. *Talpa.*	Fore-teeth upper 6, lower 8.
22. *Sorex.*	Fore-teeth upper 2, lower 4.
23. *Erinaceus.*	Fore-teeth upper 2, lower 2.

IV. GLIRES. *Fore-teeth incisors,* 2; *tusks* 0.

24. *Hystrix.*	Body covered with spines.
25. *Cavia.*	Fore-teeth wedged; grinders 4 on each side; clavicle 0.
26. *Castor.*	Fore-teeth upper wedged; grinders 4 on each side; clavicle perfect.
27. *Mus.*	Fore-teeth upper wedged; grinders 3 on each side; clavicle perfect.
28. *Arctomys.*	Fore-teeth wedged; grinders upper 5, lower 4 on each side; clavicle perfect.
29. *Sciurus.*	Fore-teeth upper wedged, lower acute; grinders upper 5, lower 4 on each side; clavicle perfect; tail distichous; whiskers long.
30. *Myoxus.*	Whiskers long; tail round, thicker at the point.
31. *Dipus.*	Fore-feet short; hind-feet long.
32. *Lepus.*	Fore-teeth upper double.
33. *Hyrax.*	Fore-teeth upper broad; tail 0.

V. PECORA. *Upper fore-teeth* 0.

34. *Camelus.*	Horns 0; tusks many.
35. *Moschus.*	Horns 0; tusks solitary, upper ones projecting.
37. *Camelopardalis.*	Horns shortest; fore-feet longer than the hind.
36. *Cervus.*	Horns solid, branching, deciduous; tusks 0.
38. *Antilope.*	Horns solid, simple, persistent; tusks 0.
39. *Capra.*	Horns hollow, erect; tusks 0.
40. *Ovis.*	Horns hollow, reclined; tusks 0.
41. *Bos.*	Horns hollow, spread; tusks 0.

VI. BELLUAE. *Fore-teeth upper and lower.*

42. *Equus.*	Fore-teeth upper 6, lower 6.
43. *Hippopotamus.*	Fore-teeth upper 4, lower 4.
44. *Tapir.*	Fore-teeth upper 10, lower 10.
45. *Sus.*	Fore-teeth upper 4, lower 6.

VII CETE. *Teeth various; feet* 0.

46. *Monodon.*	Teeth in the upper jaw 2, protruding, bony.
47. *Balaena.*	Teeth in the upper jaw horny.
48. *Physeter.*	Teeth in the lower jaw only; bony.
49. *Delphinus.*	Teeth in both jaws; bony.

MAMMALIA.

ORDER I. PRIMATES.

Fore-teeth cutting; upper 4, parallel; teats 2 pectoral

1. HOMO.

Sapiens.   Diurnal; varying by education and situation.

2. Four-footed, mute, hairy.   *Wild man.*
3. Copper-coloured, choleric, erect.   *American.*
   Hair black, straight, thick; *nostrils* wide, *face* harsh; *beard* scanty;
   obstinate, content free. *Paints* himself with fine red lines. *Regulated*
   by customs.
4. Fair, sanguine, brawny.   *European.*
   Hair yellow, brown, flowing; *eyes* blue; *gentle,* acute, inventive. *Covered* with clo~e vestments. *Governed* by laws.
5. Sooty, melancholy, rigid.   *Asiatic.*
   Hair black; *eyes* dark; *severe,* haughty, covetous. *Covered* with loose
   garments. *Governed* by opinions.
6. Black, phlegmatic,.relaxed.   *African.*
   Hair black, frizzled; *skin* silky; *nose* flat; *lips* tumid; *crafty,* indolent,
   negligent. *Anoints* himself with grease. *Governed* by caprice.

Menstrosus   Varying by climate or **art.**
1. Small, active, timid.   *Mountaineer.*
2. Large, indolent.   *Patagonian.*
3. Less fertile.   *Hottentot.*
4. Beardless.   *American.*
5. Head conic.   *Chinese.*
6. Head flattened.   *Canadian.*

The anatomical, physiological, natural, moral, civil and social histories of
man, are best described by their respective writers.

BIOLOGICAL BASIS OF COMPARATIVE ANTHROPOLOGY

BLUMENBACH, Johann Friedrich (German anthropologist and anatomist, 1752–
1840). From *De generis humani varietate nativa,* Göttingen, 1775; tr. (of the 3d ed.,
Göttingen, 1795) by T. Bendyshe under the same title, in *The anthropological treatises
of J. F. B.,* London, 1865.

Linnaeus (*q.v.*) focused attention on the existence of a taxonomic problem as regards
man himself. The matter was crystallized with Blumenbach's establishment of his five
classical varieties derived from a study of many different traits. This is given in the following passage. Blumenbach likewise made contributions to comparative anatomy, and
wrote of them, and of his views on evolution and vitalism, in heavy periods, laced at
times with surprisingly raw allusions. Study of the following argument reveals that his
thinking was essentially biological, stemming directly from recent advances, some of
them his own, in the biological field (see Buffon).

Causes of degeneration.   Animal life supposes two faculties, depending
upon the vital forces as primary conditions and principles of all and singular its

functions; the one, namely, of so receiving the force of the stimuli which act upon the body that the parts are affected by it; the other of so reacting from this affection that the living motions of the body are in this way set in action and perfected. So there is no motion in the animal machine without a preliminary stimulus and a consequent reaction. These are the hinges on which all the physiology of the animal economy turns. And these are the fountains from which, just as the business itself of generation, so also the *causes of degeneration* flow; but in order to make this clear to those even who know but little of physiology, it will be as well to premise with a few words from that science.

Formative force. I have in another place professedly, and in a separate book devoted to this subject, endeavoured to show that the vulgar system of evolution, as it is called (according to which it is taught that no animal or plant is generated, but that all individual organic bodies were at the very earliest dawn of creation already formed in the shape of undeveloped germs and are now being only successively evolved), answers neither to the phenomena themselves of nature, nor to sound philosophic reasoning. But on the contrary, by properly joining together the two principles which explain the nature of organic bodies, that is the physico-mechanical with the teleological, we are conducted both by the phenomena of generation, and by sound reasoning, to lay down this proposition: That the genital liquid is only the shapeless material of organic bodies, composed of the innate matter of the inorganic kingdom, but differing in the force it shows, according to the phenomena; by which its first business is under certain circumstances of maturation, mixture, place, &c. to put on the form destined and determined by them; and afterwards through the perpetual function of nutrition to preserve it, and if by chance it should be mutilated, as far as lies in its power to restore it by reproduction.

Let me be allowed to distinguish this energy, so as to prevent its being confused with the other kinds of vital force, or with the vague and undefined words of the ancients, the plastic force, &c. by the name of the formative force (*nisus formativus*); by which name I wish to designate not so much the cause as some kind of perpetual and invariably consistent effect, deduced *à posteriori*, as they say, from the very constancy and universality of phenomena. Just in the same way as we use the name of attraction or gravity to denote certain forces, the *causes* of which however still remain hid, as they say, in Cimmerian darkness.

As then other vital forces, when they are excited by their appointed and proper stimuli, become active and ready for reaction, so also the formative force is excited by the stimuli which belong to it, that is, by the kindling of heat in the egg during the process of incubation. But as other vital forces, as contractility, irritability, &c. put themselves out only by the mode of motion, this, on the other hand, of which we are talking, manifests itself by increase, and by giving a determinate form to matter; by which it happens that every plant and every animal propagates its species in its offspring (either immediately, or gradually by the successive access and change of other stimuli, through metamorphosis).

Now the way in which the formative force may sometimes turn aside from its determined direction and plan is principally in three forms. First, by the production of monsters; then by hybrid generation through the mixture of the genital liquid of different species; finally, by degeneration into varieties, properly so called. The production of monsters, by which, whether through some disturbance and as it were mistake of the formative force, or even through accidental or adventitious circumstances, as by external pressure, &c. a structure manifestly faulty and unnaturally deformed is intruded upon organic bodies, has nothing to do with our present purpose. Nor is this the place to consider hybrids sprung from the commingling of the generation of different species, since by a most wise law of nature (by which the infinite confusion of specific forms is guarded against) hybrids of this kind, especially in the animal kingdom, scarcely ever occur except through the interference of man: and then they are almost invariably sterile, so as to be unable to propagate any further their new ambiguous shape sprung from anomalous venery.

Still, meanwhile, this subject we are now discussing may be illustrated by the history of hybrids sprung from different species; partly on account of their analogy with those hybrids which spring from different *varieties*, of which we shall speak by and by; partly, because, like everything else, they go as proofs to refute that theory about the evolution of pre-formed germs, and to display clearly the power and efficacy of the formative force; a consideration, which will escape no one who rightly appreciates those well-known and very remarkable experiments, in which, in the very rare instances of *prolific* hybrids, when their fecundation has been frequently repeated for many generations by the aid of the male seed of the same species, that new appearance of hybrid posterity has so sensibly deflected from the maternal form as more and more to pass into the paternal form of the other species, and so, finally, the former seems to become quite transmuted into the latter, by a sort of arbitrary metamorphosis.

But the mixture of specifically different generation, although it cannot overturn, or as it were suffocate, all the excitability of the formative force, still can impart to it a singular and anomalous direction. And so it happens that the continuous action, carried on for several series of generations of some peculiar stimuli in organic bodies, again has great influence in sensibly diverting the formative force from its accustomed path, which deflection is the most bountiful source of degeneration, and the mother of varieties properly so called. So now let us go to work and examine one by one the chief of these stimuli.

Climate. That the power of climate must be almost infinite, as on all organic bodies, so especially on warm-blooded animals, will quickly appear to any one who considers first, by how intimate and how constant a bond these animals are bound while alive to the action of the atmospheric air in which they dwell. Besides, how wonderfully this air (which was once held to be a simple element of itself) is made up of what they call multifarious elements, such as gasiform constituents, the accessories of light, heat, electricity, &c. Then of what different proportions of these matters does it not consist, and in consequence of this variety how different must be the atmospheric action on those we call animals! Especially when we throw in the consideration of so many other things, by whose accession climates differ so much, as the position of countries in respect of the zones of the globe, the elevation of the soil, mountains, the vicinity of the sea or lakes and rivers, the customary winds, and innumerable other things of this kind.

This air, then, which those we call animals suck in by breathing from the time of birth, modified so greatly by the variety of climates, is decomposed in their lungs as it were in a living laboratory. Part of what they inhale is distributed with the arterial blood over the whole body; but as a balance to another portion of this point, elements are liberated, which are partly deposited on the peripheral integuments of the body, and partly are carried back by the flow of venous blood to the respiratory organs; hence arise the various modifications of the blood itself, and the remarkable influxes of these humours, especially of fat, bile, &c. into the secretions. Hence finally the action of all these things as so many stimuli on a living solid, and hence the resulting reaction as well of this thus affected solid, as what especially belongs to our discussion, the direction and determination of the formative force. This great and perpetual influence of climate on the animal economy and the habit and conformation of the body, although there has been no time when it has not attracted the attention of good observers, has in our own time above all been illustrated and confirmed by the great advance that has been made in chemistry, and by a deeper study of physiology. Still it is always a difficult and arduous thing, in the discussion of these varieties, to settle what is to be attributed exclusively to climate, what rather to other causes of degeneration, and finally to the joint action of both. . . .

Some considerations to be observed in the examination of the causes of degeneration. Many of the causes of degeneration we have already spoken of are so very clear, and so placed beyond all possibility of doubt, that most phenomena of degeneration above enumerated may by an easy process be undoubtedly referred to them, as effects to their causes. But on the other hand even in, that very way there is frequently such a concurrence or such a conflicting opposition of many of them; such a diverse and multifarious proneness of organic bodies to degeneration, or reaction from it; and besides, these causes have such effects upon these bodies according as they act immediately (so to speak) or otherwise; and finally, such is the difference of these effects by which they are preserved unimpaired by a sort of tenacious constancy through long series of generations, or by some power of change withdraw themselves again in a short space of time, that in consequence of this diversified and various relation there is need of the greatest caution in the examination of varieties.

Let me then, if only for the benefit of the student, at the end of this discourse, before we pass to the varieties of men themselves, lay down some maxims of caution at least, as corollaries to be carefully borne in mind in the discussion we are entering upon:

1. The more causes of degeneration which act in conjunction, and the longer they act upon the same species of animals, the more palpably that species may fall off from its primeval conformation. Now no animal can be compared to man in this respect, for he is omnivorous, and dwells in every climate, and is far more domesticated and far more advanced from his first beginnings than any other animal; and so on him the united force of climate, diet, and mode of life must have acted for a very long time.

2. On the other hand an otherwise sufficiently powerful cause of degeneration may be changed and debilitated by the accession of other conditions, especially if they are as it were opposed to it. Hence everywhere in various regions of the terraqueous globe, even those which lie in the same geographical latitude, still a very different temperature of the air and an equally different and generally a contrary effect on the condition of animals may be observed, according as they differ in the circumstances of a higher or lower position, proximity to the sea, or marshes, or mountains, or woods, or of a cloudy or serene sky, or some peculiar character of soil, or other circumstances of that kind.

3. Sometimes a remarkable phenomenon of degeneration ought to be referred not so much to the immediate, as to the mediate, more remote, and at the first glance concealed influence of some cause. Hence the darker colour of peoples is not to be derived solely from the direct action of the sun upon the skin, but also from its more remote, as its powerful influence upon the functions of the liver.

4. Mutations which spring from the mediate influence of causes of this sort seem to strike root all the deeper, and so to be all the more tenaciously propagated to following generations. Hence, if I mistake not, we are to look for the reason why the brown colour of skin contracted in the torrid zone will last longer in another climate than the white colour of northern animals if they are transported towards the south.

5. Finally, the mediate influences of those sort of causes may lie hid and be at such a distance, that it may be impossible even to conjecture what they are, and hence we shall have to refer the enigmatical phenomena of degeneration to them, as to their fountains. Thus, without doubt, we must refer to mediate

causes of this kind, which still escape our observation, the racial and constant forms of skulls, the racial colour of eyes, &c.

On the Causes and Ways by Which Mankind Has Degenerated, as a Species.

Order of proceeding. Now let us come to the matter in hand, and let us apply what we have hitherto been demonstrating about the ways in and the causes by which animals in general degenerate, to the native variety of mankind, so as to enumerate one by one the modes of degenerating, and allot to each the particular cause to which it is to be referred. . . .

Racial form of skulls. That there is an intimate relation between the external face and its osseous substratum is so manifest, that even a blind man, if he has any idea of the vast difference by which the Mongolian face differs from the Ethiopian, can undoubtedly, by the mere touch, at once distinguish the skull of the Calmuck from that of the Negro. Nor would you persuade even the most ignorant person to bend over the head of one or other of them as he might over those after whose models the divine works of ancient Greece were sculptured. This, I say, is clear and evident so far as the general habit goes.

But it might have been expected that a more careful anatomical investigation of genuine skulls of different nations would throw a good deal of light upon the study of the variety of mankind; because when stripped of the soft and changeable parts they exhibit the firm and stable foundation of the head, and can be conveniently handled and examined, and considered under different aspects and compared together. It is clear from a comparison of this kind that the forms of skulls take all sorts of license in individuals, just as the colour of skins and other varieties of the same kind, one running as it were into the other by all sorts of shades, gradually and insensibly: but that still, in general, there is in them a constancy of characteristics which cannot be denied, and is indeed remarkable, which has a great deal to do with the racial habit, and which answers most accurately to the nations and their peculiar physiognomy. That constancy has induced some eminent anatomists from the time of Andr. Spigel to set up a certain rule of dimensions to which as to a scale the varieties of skulls might be referred and ranked; amongst which, above all others, the facial line of the ingenious Camper deserves special mention.

Facial line of Camper. He imagined, on placing a skull in profile, two right lines intersecting each other. The first was to be a horizontal line drawn through the external auditory meatus and the bottom of the nostrils. The second was to touch that part of the frontal bone above the nose, and then to be produced to the extreme alveolar limbus of the upper jaw. By the angle which the intersection of these two lines would make, this distinguished man thought that he could determine the difference of skulls as well in brute animals as in the different nations of mankind.

Remarks upon it. But, if I am correct, this rule contains more than one error. First: what indeed is plain from those varieties of the racial face I was speaking of, this universal facial line at the best can only be adapted to those varieties of mankind which differ from each other in the direction of the jaws, but by no means to those who, in exactly the contrary way, are more remarkable for their lateral differences.

Secondly: it very often happens that the skulls of the most different nations, who are separated as they say by the whole heaven from one another, have still one and the same direction of the facial line: and on the other hand many skulls of one and the same race, agreeing entirely with a common disposition, have a facial line as different as possible. We can form but a poor opinion of skulls when seen in profile alone, unless at the same time account be taken of their breadth. Thus as I now write I have before me a pair of skulls, viz.: an Ethiopian of Congo, and a Lithuanian of Sarmatia. Both have almost exactly the same facial line; yet their construction is as different as possible if you compare the narrow and, as it were, keeled head of the Ethiopian with the square head of the Sarmatian. On the other hand, I have two Ethiopian skulls in my possession, differing in the most astonishing manner from each other as to their facial line, yet in both, if looked at in front, the narrow and, as it were, squeezed-up skulls, the compressed forehead, &c. sufficiently testify to their Ethiopian origin.

Thirdly, and finally, Camper himself, in the plates appended to his work, has made such an arbitrary and uncertain use of his two normal lines, has so often varied the points of contact according to which he has drawn them, and upon which all their value and trustworthiness depends, as to make a tacit confession that he himself is uncertain, and hesitates in the application of them.

Vertical scale for defining the racial characters of skulls. The more my daily experience and, as it were, my familiarity with my collection of skulls of different nations increases, so much the more impossible do I find it to reduce these racial varieties—when such differences occur in the proportion and direction of the parts of the truly many-formed skull, all having more or less to do with the racial character—to the measurements and angles of any single scale. That view of the skull however seems to be preferable for the diagnosis which is our business that presents together at one glance the most and the principal parts best adapted for a comparison of racial characters. With this object I have found after many experiments that position answer best in which skulls are seen from above and from behind, placed in a row on the same plane, with the malar bones directed towards the same horizontal line jointly with the inferior maxillaries. Then all that most conduces to the racial character of skulls, whether it be the direction of the jaws, or the cheekbones, the breadth or narrowness of the skull, the advancing or receding outline of the forehead, &c. strikes the eye so distinctly at one glance, that it is not out of the way to call that view the vertical scale (*norma verticalis*). . . .

FIVE PRINCIPAL VARIETIES OF MANKIND, ONE SPECIES.

Innumerable varieties of mankind run into one another by insensible degrees. We have now completed a universal survey of the genuine varieties of mankind. And as, on the one hand, we have not found a single one which does not (as is shown in the last section but one) even among other warm-blooded animals, especially the domestic ones, very plainly, and in a very remarkable way, take place as it were under our eyes, and deduce its origin from manifest causes of degeneration; so, on the other hand (as is shown in the last section), no variety exists, whether of colour, countenance, or stature, &c., so singular as not to be connected with others of the same kind by such an imperceptible transition, that it is very clear they are all related, or only differ from each other in degree.

Five principal varieties of mankind may be reckoned. As, however, even among these arbitrary kinds of divisions, one is said to be better and preferable to another; after a long and attentive consideration, all mankind, as far as it is at present known to us, seems to me as if it may best, according to natural truth, be divided into the five following varieties; which may be designated and distinguished from each other by the names *Caucasian, Mongolian, Ethiopian, American,* and *Malay*. I have allotted the first place to the Caucasian, for the reasons given below, which make me esteem it the primeval one. This diverges in both directions into two, most remote and very different from each other; on the one side, namely, into the Ethiopian, and on the other into the Mongolian. The remaining two occupy the intermediate positions between that primeval one and these two extreme varieties; that is, the American between the Caucasian and Mongolian; the Malay between the same Caucasian and Ethiopian.

Characters and limits of these varieties. In the following notes and descriptions these five varieties must be generally defined. To this enumeration, however, I must prefix a double warning; first, that on account of the multifarious diversity of the characters, according to their degrees, one or two alone are not sufficient, but we must take several joined together; and then that this union of characters is not so constant but what it is liable to innumerable exceptions in all and singular of these varieties. Still this enumeration is so conceived as to give a sufficiently plain and perspicuous notion of them in general.

Caucasian variety. Colour white, cheeks rosy; hair brown or chestnut-coloured; head subglobular; face oval, straight, its parts moderately defined, forehead smooth, nose narrow, slightly hooked, mouth small. The primary teeth placed perpendicularly to each jaw; the lips (especially the lower one) moderately open, the chin full and rounded. In general, that kind of appearance which, according to our opinion of symmetry, we consider most handsome and becoming. To this first variety belong the inhabitants of Europe (except the Lapps and the remaining descendants of the Finns) and those of Eastern Asia, as far as the river Obi, the Caspian Sea and the Ganges; and lastly, those of Northern Africa.

Mongolian variety. Colour yellow; hair black, stiff, straight and scanty; head almost square; face broad, at the same time flat and depressed, the parts therefore less distinct, as it were running into one another; glabella flat, very broad; nose small, apish; cheeks usually globular, prominent outwardly; the opening of the eyelids narrow, linear; chin slightly prominent. This variety comprehends the remaining inhabitants of Asia (except the Malays on the extremity of the trans-Gangetic peninsula) and the Finnish populations of the cold part of Europe, the Lapps, &c. and the race of Esquimaux, so widely diffused over North America, from Behring's straits to the inhabited extremity of Greenland.

Ethiopian variety. Colour black; hair black and curly; head narrow, compressed at the sides; forehead knotty, uneven; malar bones protruding outwards; eyes very prominent; nose thick, mixed up as it were with the wide jaws; alveolar edge narrow, elongated in front; the upper primaries obliquely prominent; the lips (especially the upper) very puffy; chin retreating. Many are bandy-legged. To this variety belong all the Africans, except those of the north.

American variety. Copper-coloured; hair black, stiff, straight and scanty; forehead short; eyes set very deep; nose somewhat apish, but prominent; the face invariably broad, with cheeks prominent, but not flat or depressed; its parts, if seen in profile, very distinct, and as it were deeply chiselled; the shape of the forehead and head in many artificially distorted. This variety comprehends the inhabitants of America except the Esquimaux.

Malay variety. Tawny-coloured; hair black, soft, curly, thick and plentiful; head moderately narrowed; forehead slightly swelling; nose full, rather wide, as it were diffuse, end thick; mouth large, upper jaw somewhat prominent with the parts of the face when seen in profile, sufficiently prominent and distinct from each other. This last variety includes the islanders of the Pacific Ocean, together with the inhabitants of the Marianne, the Philippine, the Molucca and the Sunda Islands, and of the Malayan peninsula.

Divisions of the varieties of mankind by other authors. It seems but fair to give briefly the opinions of other authors also, who have divided mankind into varieties, so that the reader may compare them more easily together, and weigh them, and choose which of them he likes best. The first person, as far as I know, who made an attempt of this kind was a certain anonymous writer who towards the end of the last century divided mankind into four races; that is, first, one of all Europe, Lapland alone excepted, and Southern Asia, Northern Africa, and the whole of America; secondly, that of the rest of Africa; thirdly, that of the rest of Asia with the islands towards the east; fourthly, the Lapps. Leibnitz divided the men of our continent into four classes. Two extremes, the Laplanders and the Ethiopians; and as many intermediates, one eastern (Mongolian), one western (as the European).

Linnaeus, following common geography, divided men into (1) the red American, (2) the white European, (3) the dark Asiatic, and (4) the black Negro. Buffon distinguished six varieties of man: (1) Lapp or polar, (2) Tartar (by which name according to ordinary language he meant the Mongolian), (3) south Asian, (4) European, (5) Ethiopian, (6) American.

Amongst those who reckoned three primitive nations of mankind answering to the number of the sons of Noah, Governor Pownall is first entitled to praise, who, as far as I know, was also the first to pay attention to the racial form of skull as connected with this subject. He divided these stocks into white, red and black. In the middle one he comprised both the Mongolians and Americans, as agreeing, besides other characters, in the configuration of their skulls and the appearance of their hair. Abbé de la Croix divides man into white and black. The former again into white, properly so called, brown (*bruns*), yellow (*jaunâtres*), and olive-coloured.

Kant derives four varieties from dark-brown Autochthones: the white one of northern Europe, the copper-coloured American, the black one of Senegambia, the olive-coloured Indian. John Hunter reckons seven varieties: (1) of black men, that is, Ethiopians, Papuans, &c.; (2) the blackish inhabitants of Mauritania and the Cape of Good Hope; (3) the copper-coloured of eastern India; (4) the red Americans; (5) the tawny, as Tartars, Arabs, Persians, Chinese, &c.; (6) brownish, as the southern Europeans, Spaniards, &c., Turks, Abyssinians, Samoiedes and Lapps; (7) white, as the remaining Europeans, the Georgians, Mingrelians and Kabardinski.

Zimmermann is amongst those who place the aborigines of mankind in the elevated Scythico-Asiatic plain, near the sources of the Indus, Ganges and Obi rivers; and thence deduces the varieties of Europe (1), northern Asia, and the great part of North America (2), Arabia, India, and the Indian Archipelago (3), Asia to the north-east, China, Corea, &c. (4). He is of opinion that the Ethiopians deduce their origin from either the first or the third of these varieties.

Meiners refers all nations to two stocks: (1) handsome, (2) ugly; the first white, the latter dark. He includes in the handsome stock the Celts, Sarmatians, and oriental nations. The ugly stock embraces all the rest of mankind. Klügel distinguishes four stocks: (1) the primitive, autochthones of that elevated Asiatic plain we were speaking of, from which he derives the inhabitants of all the rest of Asia, the whole of Europe, the extreme north of America, and northern Africa; (2) the Negroes; (3) the Americans, except those of the extreme north; (4) the Islanders of the southern ocean. Metzger makes two principal varieties as extremes: (1) the white man native of Europe, of the northern parts of Asia, America and Africa; (2) the black, or Ethiopian, of the rest of Africa. The transition between the two is made by the rest of the Asiatics, the inhabitants of South America, and the Islanders of the southern ocean.

Notes on the five varieties of Mankind. But we must return to our pentad of the varieties of mankind. I have indicated separately all and each of the characters which I attribute to them in the sections above. Now, I will string together, at the end of my little work, as a finish, some scattered notes which belong to each of them in general.

Caucasian variety. I have taken the name of this variety from Mount Caucasus, both because its neighbourhood, and especially its southern slope, produces the most beautiful race of men, I mean the Georgian; and because all physiological reasons converge to this, that in that region, if anywhere, it seems we ought with the greatest probability to place the autochthones of mankind. For in the first place, that stock displays, as we have seen, the most beautiful form of the skull, from which, as from a mean and primeval type, the others diverge by most easy gradations on both sides to the two ultimate extremes (that is, on the one side the Mongolian, on the other the Ethiopian). Besides, it is white in colour, which we may fairly assume to have been the primitive colour of mankind, since, as we have shown above, it is very easy for that to degenerate into brown, but very much more difficult for dark to become white, when the secretion and precipitation of this carbonaceous pigment has once deeply struck root.

Mongolian variety. This is the same as what was formerly called, though in a vague and ambiguous way, the Tartar variety; which denomination has given rise to wonderful mistakes in the study of the varieties of mankind which we are now busy about. So that Buffon and his followers, seduced by that title, have erroneously transferred to the genuine Tartars, who beyond a doubt belong to our first variety, the racial characters of the Mongols, borrowed from ancient authors, who described them under the name of Tartars.

But the Tartars shade away through the Kirghis and the neighbouring races into the Mongols, in the same way as these may be said to pass through the Tibetans to the Indians, through the Esquimaux to the Americans, and also in a sort of way through the Philippine Islanders to the men of the Malay variety.

Ethiopian variety. This variety, principally because it is so different in colour from our own, has induced many to consider it, with the witty, but badly instructed in physiology, Voltaire, as a peculiar species of mankind. But it is not necessary for me to spend any time here upon refuting this opinion, when it has so clearly been shown above that there is no single character so peculiar and so universal among the Ethiopians, but what it may be observed on the one hand everywhere in other varieties of men, and on the other that many Negroes are seen to be without each. And besides there is no character which does not shade away by insensible gradation from this variety of mankind to its neighbours, which is clear to every one who has carefully considered the difference between a few stocks of this variety, such as the Foulahs, the Wolufs, and Mandingos, and how by these shades of difference they pass away into the Moors and Arabs.

The assertion that is made about the Ethiopians, that they come nearer the apes than other men, I willingly allow so far as this, that it is in the same way that the solid-hoofed variety of the domestic sow may be said to come nearer to the horse than other sows. But how little weight is for the most part to be attached to this sort of comparison is clear from this, that there is scarcely any other out of the principal varieties of mankind, of which one nation or other, and that too by careful observers, has not been compared, as far as the face goes, with the apes; as we find said in express words of the Lapps, the Esquimaux, the Caaiguas of South America, and the inhabitants of the Island Mallicollo.

American variety. It is astonishing and humiliating what quantities of fables were formerly spread about the racial characters of this variety. Some have denied beards to the men, others menstruation to the women. Some have attributed one and the same colour to each and all the Americans; others a perfectly similar countenance to all of them. It has been so clearly demonstrated now by the unanimous consent of accurate and truthful observers, that the Americans are not naturally beardless, that I am almost ashamed of the unnecessary trouble I formerly took to get together a heap of testimony, by which it is proved that not only throughout the whole of America, from the Esquimaux downwards to the inhabitants of Tierra del Fuego, are there groups of inhabitants who cherish a beard; but also it is quite undeniable as to the other beardless ones that they eradicate and pluck out their own by artifice and on purpose, in the same way as has been customary among so many other nations, the Mongolians for example, and the Malays. We all know that the beard of the Americans is thin and scanty, as is also the case with so many Mongolian nations. They ought therefore no more to be called beardless, than men with scanty hair to be called bald. Those therefore who thought the Americans were naturally beardless fell into the same error as that which induced the ancients to suppose and persuade others, that the birds of paradise, from whose corpses the feet are often cut off, were naturally destitute of feet.

The fabulous report that the American women have no menstruation, seems to have had its origin in this, that the Europeans when they discovered the new world, although they saw numbers of the female inhabitants almost entirely naked, never seem to have observed in them the stains of that excretion. For this it seems likely that there were two reasons; first, that amongst those nations of America, the women during menstruation are, by a fortunate prejudice, considered as poisonous, and are prohibited from social intercourse, and for so long enjoy a beneficial repose in the more secluded huts far from the view of men; secondly, because, as has been noticed, they are so commendably clean in their bodies, and the commissure of their legs so conduces to modesty, that no vestiges of the catamenia ever strike the eye.

As to the colour of the skin of this variety, on the one hand it has been observed above, that it is by no means so constant as not in many cases to shade away into black; and on the other, that it is easily seen, from the nature of the American climate, and the laws of degeneration when applied to the extremely probable origin of the Americans from northern Asia, why they are not liable to such great diversities of colour, as the other descendants of Asiatic autochthones, who peopled the ancient world. The same reason holds good as to the appearance of the Americans. Careful eye-witnesses long ago laughed at the foolish, or possibly facetious hyperbole of some, who asserted that the inhabitants of the new world were so exactly alike, that when a man had seen one, he could say that he had seen all, &c. It is, on the contrary, proved by the finished drawings of Americans by the best artists, and by the testimony of the most trustworthy eye-witnesses, that in this variety of mankind, as in others, countenances of all sorts occur; although in general that sort of racial conformation may be considered as properly belonging to them which we attributed to them above. It was justly observed by the first Europeans who visited the new continent, that the Americans came very near to the Mongolians, which adds fresh weight to the very probable opinion that the Americans came from northern Asia, and derived their origin from the Mongolian nation. It is probable that migrations of that kind took place at different times, after considerable intervals, according as various physical, geological, or political catastrophes gave occasion to them; and hence, if any place is allowed for conjecture in these investigations, the reason may probably be derived, why the Esquimaux have still much more of the Mongolian appearance about them than the rest of the Americans: partly, because the catastrophe which drove them from northern Asia must be much more recent, and so they are a much later arrival; and partly because the climate of the new country, which they now inhabit, is much more homogeneous with that of their original country. In fact, unless I am much mistaken, we must attribute to the same influence I mentioned above, which the climate has in preserving or restoring the racial appearance, the fact that the inhabitants of the cold southern extremity of South America, as the barbarous inhabitants of the Straits of Magellan, seem to come nearer, and as it were fall back, to the original Mongolian countenance.

The Malay variety. As the Americans in respect of racial appearance hold as it were a place between the medial variety of mankind, which we called the Caucasian, and one of the two extremes, that is the Mongolian; so the Malay variety makes the transition from that medial variety to the other extreme, namely, the Ethiopian. I wish to call it the Malay, because the majority of the men of this variety, especially those who inhabit the Indian islands close to the Malacca peninsula, as well as the Sandwich, the Society, and the Friendly Islanders, and also the Malambi of Madagascar down to the inhabitants of Easter Island, use the Malay idiom.

Meanwhile even these differ so much between themselves through various degrees of beauty and other corporeal attributes, that there are some who divide the Otaheitans themselves into two distinct races; the first paler in colour, of lofty stature, with face which can scarcely be distinguished from that of the European; the second, on the other hand, of moderate stature, colour and face little different from that of Mulattos, curly hair, &c. This last race then comes very near those men who inhabit the islands more to the south in the Pacific Ocean, of whom the inhabitants of the New Hebrides in particular come sensibly near the Papuans and New Hollanders, who finally on their part graduate away so insensibly towards the Ethiopian variety, that, if it was thought convenient, they might not unfairly be classed with them, in that distribution of the varieties we were talking about.

Conclusion. Thus too there is with this that insensible transition by which as we saw the other varieties also run together, and which, compared with what was discussed in the earlier sections of the book, about the causes and ways of degeneration, and the analogous phenomena of degeneration in the other domestic animals, brings us to that conclusion, which seems to flow spontaneously from physiological principles applied by the aid of critical zoology to the natural history of mankind; which is, *That no doubt can any longer remain but that we are with great probability right in referring all and singular as many varieties of man as are at present known to one and the same species.*

BEGINNINGS OF EXACT ANATOMETRY

CAMPER, Petrus (Peter) (Dutch anatomist and naturalist, 1722–1789). From *Verhandeling over het natuurlijk verschilder wezenstrekken in menschen van onderscheiden landaart en ouderdom, etc.,* Utrecht, 1791; tr. by T. Cogan in *The works of Professor Camper on the connexion between the science of anatomy and the arts of drawing, painting, statuary, &c., &c.,* London, 1794. (The following passage is from the edition of 1821.)

As Camper admits, he was not the first to interest himself in anthropometry, its origins being lost in antiquity. By making careful measurement, however, Camper discovered canons of proportion which were previously unknown and which led directly to the establishment of an important new science (see Blumenbach, Retzius).

. . . When I gave lectures in the public college at Amsterdam, as Professor of Anatomy, I found, by comparing bodies of various ages that were brought to me for dissection, that the oval was not calculated for the delineation of the features with any degree of accuracy or expedition. With this idea I sawed several heads, both of men and of animals, perpendicularly through the middle; and I was fully convinced that the ball of the head forming the cavity destined to contain the brains, was in general very uniform; but that the position of the upper and lower jaws was the manifest cause of the most striking differences. The same observation may be extended from quadrupeds down to the finny race: and it has suggested hints sufficiently numerous to form a separate Treatise.

The above examination has also enabled me to discover whence those changes arise which progressively take place in our features, from infancy to the most advanced age. But I still was unable to explain in what manner it was that the Greeks should have acquired, at a very remote period, that singular and dignified expression which they gave to their figures; and which I have never seen perfectly equalled. I perceived, moreover, that in the copies taken from these, the facial line did not differ from our own. . . .

Having contemplated the inhabitants of various nations with greater attention, I conceived that a striking difference was occasioned, not merely by the position of the inferior maxilla, but by the breadth of the face, and the quadrangular form of this maxilla. This idea was confirmed by contemplating a considerable collection which I afterwards made of heads, that acknowledged various countries for their parents; or of exact copies from them. Exclusive of several skulls of my countrymen, and of the adjacent nations, I possess two of English negroes (the one was a young person, the other advanced in years) —the head of a female Hottentot,—of an inhabitant of Mogul,—a Chinese, —a youth of Madagascar,—a Celebean,—and finally, the cranium of a Calmuck; that is, of eight different nations.

When I was at Oxford, in the year of 1786, I also took a sketch of the lower jaw of a native of Otaheite, that had been brought over by Captain King. I have never been able to obtain possession of the cranium of a native American, nor even of an Anglo-American, which has, however, some peculiarities that were pointed out to me by that celebrated artist Mr. West; of which, as he was born in Pennsylvania, he was the best qualified to judge. Their face is long and narrow; and the socket of the eye surrounds the ball in so close a manner, that no space is allowed for a large upper eye-lid; which is so graceful to the countenance of most Europeans.

When in addition to the skull of a negro, I had procured one of a Calmuck, and had placed that of an ape contiguous to them both, I observed that a line drawn along the forehead and the upper lip, indicated this difference in national physiognomy; and also pointed out the degree of similarity between a negroe and the ape. By sketching some of these features upon a horizontal plane, I obtained the lines which mark the countenance, with their different angles. When I made these lines to incline forwards, I obtained the face of an antique; backwards, of a negroe; still more backwards, the lines which mark an ape, a dog, a snipe, &c.—This discovery formed the basis of my edifice. . . .

PHYSIOLOGICAL OBSERVATIONS CONCERNING THE DIFFERENCE OF FACES IN PROFILE; FROM APES, OURANGS, NEGROES, AND OTHER CLASSES OF PEOPLE, UP TO THE ANTIQUE.

The assemblage of craniums, and profiles of two apes, a negro and a Calmuck, in the first plate, may perhaps excite surprise. The striking resemblance between the race of Monkies and of Blacks, particularly upon a superficial view, has induced some philosophers to conjecture that the race of blacks originated from the commerce of the whites with ourangs and pongos; or that these monsters, by gradual improvements, finally become men.

This is not the place to attempt a full confutation of so extravagant a notion. I must refer the reader to a physiological dissertation concerning the ourang-outang, published in the year 1782. I shall simply observe at present, that the whole generation of apes, from the largest to the smallest, are quadrupeds, not formed to walk erect; and that from the very construction of the larynx, they are incapable of speech. Further: They have a great similarity with the canine species, particularly respecting the organs of generation. The diversities observable in these parts, seem to mark the boundaries which the Creator has placed between the various classes of animals.

The proximity of the eyes to each other, the smallness and apparent flatness of the nose, and the projection of the upper lip, constitute the principal points of resemblance; and these are much exaggerated by our modern naturalists, by their heightened descriptions, and embellished plates; but they will immediately diminish in our estimation, if we give attention to the whole body, or minutely examine every part of the head. This will evidently appear by comparing together the different figures of the first plate.

All the figures . . . are sketched in profile. In this manner the differences may be more easily and accurately investigated. The bones of the cranium may also be the better contemplated as the basis of the features, which are immediately placed upon and under them.

In each of these figures the greatest accuracy and precision have been diligently studied. For example: An horizontal line has been drawn through the lower part of the nose (see Fig. 3*d*, N.) and the orifice of the ear C.; and the four skulls were arranged with care on the line A.B.; attention being also paid to the direction of the *jugale*, or cheek-bone Q. Figs. 3*c* and 3*d*.

In order to preserve the true form and relative situations of the parts, I

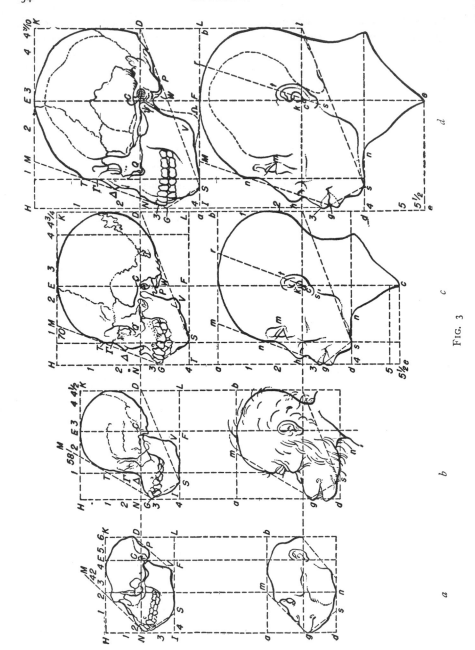

FIG. 3

did not view them from one fixed point, but my eye was always directed, in a right line, to the central point of the object, in the manner practised by masons and architects; avoiding the rules of perspective, by which particular parts are always distorted and misplaced. I viewed the object with only one eye.

To facilitate this business, I invented a machine sufficiently large to receive the largest skull. It consisted of an horizontal quadrangular table, upon which was placed a perpendicular frame, that was also quadrangular. In the laths which completed this frame a number of holes were bored parallel to each other; so that threads could be drawn through them, and be fastened in every direction required. By these I was able to make horizontal, perpendicular, or oblique lines at any convenient distance from each other.

The fore part of the square table is also divided into equal portions, by means of brass pegs, correspondent to the holes made in the upper part of the frame, that lines may also be drawn by means of threads obliquely downwards: thus may the true point of vision be obtained, by placing the eye in such a direction, that the oblique thread may perfectly coincide with the perpendicular one.

The table before me being elevated to such a height that my eye became parallel with the horizontal line A.B. I placed the skulls, by the side of each other, on the table behind the perpendicular threads of the frame. By extending the oblique threads in such a manner as to make them pass over the principal parts, and by means of the perpendicular lines, I was secure of all the points requisite to afford me an accurate drawing.

It was in this manner I discovered in all the figures, that the lines ND. and EF. intersect each other in C. before the aperture of the ear; and also the point of contact of the front teeth was at N. and of the occiput at D; by which the size of proportion of NC. to CD. that is, the relative distances from the extremity of the fore teeth to the aperture of the ear, and from thence to the extreme part of the occiput, became manifest.

The great utility of this method will fully appear hereafter. I shall only remark at present, that the point C. generally coincides in the human species with the line of gravity of the whole body (see Fig. 4, EF. or EF, e.) and thus in the centre of the head's motion: which is in the place of union of the condyles of the occiput with the first vertebrae of the neck. See PW. in Figs. 3c and 3d or W. in Fig. 4.

By means of the same instrument, the exact height of the heads could also be ascertained (see EF. in all the upper figures of Figs. 3 and 4) and also the proportionate size of EC. that is, of the head from the vertex or crown, to the aperture of the ear, compared with CF. or the distance from this aperture to the lower edge of the maxilla: likewise the proportions between HN. and NI. or the relative distances from the line of the vertex to that which passes

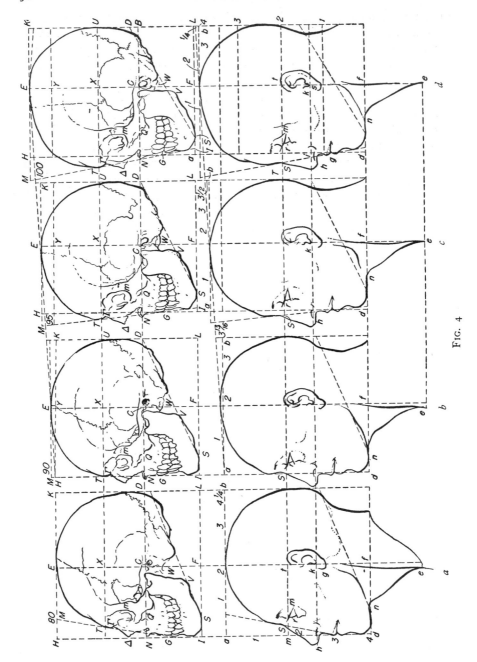

FIG. 4

under the nose; and from this to the lower edge of the maxilla. It also marks the squares H, I, L, K, in which these heads were delineated.

Further: As the closing of the teeth marks the mouth at G. I was able to draw an oblique line from G. to M. along the nasal bone △, and the forehead T. This, upon account of its great use in discriminating the differences of faces, may properly be termed the *linea facialis*, or the facial line.

Figure 3*a* represents the exact profile of a *Simia caudata*, or tailed ape. I do not recollect the particular species. It had a flat forehead, which was some-what elevated above the rim of the eye-sockets: It had five double teeth, and facculi; so that it was a native of Africa. The facial line MG. makes with AD. the angle MND.; which is equal to 42 degrees.

NC was to CD : : 8 : 2½ or : : 16 : 5.

EC : CF : : 7 : 7 . that is, EC = CF.

Or, in more familiar terms,

The distance from the mouth to the orifice of the ear, was compared with the distance of this orifice from the bottom of the lower jaw, as 8 is to 2½, or 16 to 5: and the distance from the vertex to the orifice of the ear, was pre-cisely equal to the distance of this, from the basis of the lower jaw.

Figure 3*b* is drawn from a small orang-outang, reduced to one fourth of its natural size. It is the same that I had delineated and described in a former Treatise. It was very young, and had not more than two double teeth.

The facial line MG. made with AB. or ND. an angle of 58 degrees: NC. compared with CD. was as 7 to 4.; and EC. compared with CF. nearly as six to four.

The high forehead of this animal gives it a greater resemblance to the hu-man species; and the sockets of the eyes are more elevated; which communi-cates a more animated appearance to the eyes themselves.

Edwards, who has but imperfectly delineated this species of ape, gives to the facial line an angle of 55 degrees. This small difference may be over-looked, as much greater are perceived in the human species.

The real pongo has been lately discovered in the Island of Borneo; and a description of it is given in the Batavian Transactions. This animal is, upon the whole, of a similar figure to the other; but it is about twice the size. I have in my possession the skull of one that was four feet five inches in height; whereas the smaller species seldom exceed two feet and a half. This however has less of the human form, as its forehead is flatter, the cheek-bones are broader, and the jaw-bone projects farther. The facial line makes with the horizon an angle of 47 degrees.

The cranium of the young negro, represented in Fig. 3*c*, immediately in-dicates the human countenance. He was changing his teeth; as may be known by the second grinder and a lower incisive tooth that were fallen out; and the succeeding teeth were advancing. He had only four teeth on each side. I dissected the body of this youth publicly at Amsterdam, in the year 1758.

The facial line MG. made an angle of 70 degrees with the horizontal line ND.

NC compared with CD was as 7¾ to 8, or as 31 to 32.

EC   :   CF   : : 8½ : 5, or as 17 to 10.

The projecting point of the jugal, or cheek-bone, Q. was in the centre between the mouth and the orifice of the ear; that is, NQ : QC : : 4 : 4. or NQ = QC. It is the projecting part Q. which gives the degree of flatness to the face. This is strongly marked on the medal of Bocchus, King of Mauritania.

Albert Dürer, having occasionally delineated a Moor, in his treatise on the changes of the facial line in different countenances, has made the facial line correspond with that of ours: its inclination being about 69 or 70 degrees.

The ancients seem to have paid great attention to the facial line. This is particularly observable in the *Recueil d'Antiq.* of Count Caeylus. In some of the plates, the head of a negro is represented upon an ornamental lamp, with singular accuracy.

Figure 3*d* represents the head of a Calmuck. As the teeth and under jaw were wanting, I have been obliged to supply the deficiency from the cranium of an aged negro, the size of which was nearly similar.

The facial line MG. made also an angle of 70 degrees with the horizontal line ND. NC : CD. was as 11 : : 7¼ or as 44 : 29. and EC : CF : : 10½ : 6. or 21 : 12. QC = 15. And thus NQ : QC : : 7 : 15. That is,

The distance from the extreme projection of the teeth to the orifice of the ear, compared with the distance of this from the extremity of the occiput, was 11 to 7¼, or as 44 to 29. The distance from the vertex to the orifice of the ear, compared with the distance of this orifice from the lower edge of the inferior maxilla, was 10½ to 6, or 21 to 12. The most projecting part of the jugal bone, from the orifice of the ear, was equal to 15; that is, the distance of the mouth from the process of the jugal bone, compared with the distance of this from the orifice of the ear, was 7 to 15.

From a large collection of European heads in my cabinet, I have selected the one represented in Fig. 4*a*. In this, as well as many other which I measured with care, the facial line MG. made an angle of 80 degrees with the horizontal line ND. or AB. The proportions were as follow:

NC was : CD : : 7½ : 7¾, or as 30 : 31.

EC : CF : : 9 : 5½. or as 18 : 11.

NQ : QC : : 3½ : 4. or as 7 : 8.

It follows from hence, that the angle of the facial line has in nature a *maximum* and a *minimum* from 70 to 80 degrees; which describe its greatest or smallest degree of elevation. When the *maximum* of 80 degrees is exceeded by the facial line, it is formed by the rules of art alone: and when it does not rise to 70 degrees, the face begins to resemble some species of monkies. This will be fully explained hereafter.

To proceed with as much perspicuity as possible, I shall place the facial line MG. erect in the perpendicular line HI. See Fig. 4*b*. The angle is now become 10 degrees larger, and the cavities of the eyes, cheek-bones, &c. are brought forwards and nearer to NM.

Imagine a cranium of a pliable consistence, and that the occiput could be pressed forwards and upwards; then must EC. or the distance from the aperture of the ear to the vertex increase, and again the space EY.; although the cavities of the eyes, and the eyes themselves, will still remain in the line TU.

The line SV. which marks the oblique direction of the lower jaw, rises also in the same proportion, until it approaches to D.; until it coincides with D. as in Fig. 4*c* or rises above it, as in Fig. 4*d*. On the other hand, the distance between TX.; that is, between the facial line and the perpendicular line that passes from the vertex by the orifice of the ear, gains as much as XU. has lost. The head becomes gradually narrower also in proportion as the facial line rises and inclines forwards into the 100th degree; which is the *maximum*, or utmost that the artificial line will permit. In this case the eyes, placed in the centre of their cavities, are exactly in the middle of the head, or at an equal distance from the vertex and the bottom of the chin. See Fig. 4*d*.

If the projecting part of the forehead be made to exceed the 100th degree, the head becomes misshapen, and assumes the appearance of the hydrocephalus, or watery head. It is very surprising that the artists of ancient Greece should have chosen precisely the *maximum*, while the best Roman artists have limited themselves to the 95th degree, which is not so pleasing; as the comparison of Figs. 4*c* and 4*d* will evince:

The two extremities therefore of the facial line are from 70 to 100 degrees, from the negro to the Grecian antique; make it under 70, and you describe an orang or an ape: lessen it still more, and you have the head of a dog. Increase the *minimum* and you form a fowl, a snipe, for example, the facial line of which is nearly parallel with the horizon; that is, both the maxillae will be lengthened, and the lower maxilla will gradually lose its angle CVS. No space is now left for teeth; which explains the reason why fowls are destitute of teeth.

I have sometimes amused myself with making these gradations upon a smaller scale, by sketching them on a long slip of paper; which exhibits a singular appearance. It is not necessary to give a specimen, as they can be easily made by every one skilled in drawing.

If attention be given to the angle MGS. which describes the angle formed by the facial line and the lower extremity of the chin (see Figs. 3*a*–3*d*) it will be immediately perceived that this becomes larger, *i.e.* more rectangular, in proportion as the facial line MG. ascends; it is therefore the largest in a European (as in Fig. 4*a*, GI.); and that it projects forwards with the facial line, which it always follows, as in Figs. 4*b*, 4*c*, 4*d*. In this situation the angle of the lower jaw becomes more erect, the distance from I. to F. becomes less, and V. is rounder. It is this which makes the maxillae of the antique heads rounder and more graceful; as will appear in Fig. 4*d*.

The eyes, which are placed nearly in a line with the upper edge of their sockets, gradually recede in an European and the antique; that is, Sr. or the distance from the eye to the ridge of the nose, gradually becomes greater. See the lower figures of Figs. 4*b*, 4*c*, 4*d*. This gives a certain elegance and dignity to the countenance of the antique, which cannot be otherwise acquired.

It is plain, if the cavity of the eye remains at the same distance from the perpendicular line IH. and the forehead be made to project forwards, that this depth or distance will increase according to the degree of projection. See the same figures.

If I am not deceived, the size of the mouth is in proportion to the distance of the *dentes canini*, or eye-teeth, in men and animals, with only a few exceptions. Or, to speak more properly, the angles terminate at the commencement of the first double tooth, or grinder. Many animals have not the eye-teeth.

In apes therefore, in the orang, and in the negro, the rim or angle of the mouth must be more distended than in an European, as the projection of the upper jaw enlarges the distance. For the same reason, the mouth of the antique will be the smallest.

The central line of the ear is, in all persons, somewhat inclined . . . It is never parallel with the facial line in white men. It is, however, in the negro; as is apparent from Fig. 3*c*.

I have placed the central line of the ear perpendicularly in Fig. 4, that the true distance of the eye from the ear may be more accurately ascertained.

Remarks Concerning Differences in the Facial Line, and the Changes Which Necessarily Arise from Thence.

In the preceding chapter, I have simply shewn the kind of angle which the oblique line MG. makes in all the figures of Figs. 3 and 4. Let us now pay attention to the triangle TGS. of Figs. 3*c* and 3*d* and it will appear that this triangle is not remarkably large in the European, represented in Fig. 4*a*. In Fig. 4*b* it is totally effaced: in Fig. 4*c* the angle becomes *minus*; and in Fig. 4*d* the *minus* is increased.

Let us now suppose that all these heads were of an equal size, and that the nose of each projected to an equal distance from the line or surface TS. (Figs. 3*c* and 3*d*) it is manifest that the nose of the negro and Calmuck will seem to be less, and, as it were, pressed inwards.

The nose of the European (Fig. 4*a*) will appear somewhat bent, and also to project farther than the upper lip. In the face of the antique (Fig. 4*d*) the nose will be nearly in a perpendicular line with the forehead, and project but a little from the lip.

The lower jaw, as well as the upper, is also much forwarder in the negro, Caffre, and Calmuck; and therefore it is that these people approach nearer to the figure of an ape than either the European or the antique. The lines mgs. are nearly the same with the lines MGS. Compare the lower sketches with the upper, in Figs. 3*b* and 3*c*.

In a Calmuck, the upper jaw is very flat before, because the cheek-bone Q. (Fig. 3*d*) being very large, nearly advances to the perpendicular line TT. that is, directly over the middle molaris or grinder. In the Chinese, Otaheites, and other orientals, the cheek-bone Q. corresponds with the division between the third and fourth grinder hindwards.

In the negro, CQ. is obviously shorter, and the line falls behind the third grinder. In the European, Q. is behind the fourth grinder; and in the antique head, it comes yet more forwards. Hence it follows that the features of antiques, those of Apollo, for example, must be flatter than ours; and, on the other hand, those of Asiatics and Africans still flatter; and those of the Calmucks the flattest of all.

The distance from N to G *i.e.* from the undermost part of the nose to the union of the upper and lower teeth, is greater in a Calmuck than in a negro; and in him greater than in us. On the contrary, NG. is very short in an Asiatic. The lips must necessarily be longer and thicker in proportion to this distance; and therefore is the upper lip the longest and thickest in a Calmuck, and the smallest in the antique.

If attention be paid to what may be called the Suspension of the face, *i.e.* the distance of PF. (Figs. 3*a*, 3*c*, and 3*d*) or the axis upon which the head moves from the line of the lower maxilla IL. in a negro or Calmuck (Figs. 3*c* and 3*d*) or the European (Fig. 4*a* W.) it will appear that the maxilla and the chin are deeper or lower in the two former than in the latter. The condyle also, or prominence on which the head turns, is in the same line, as the union of the teeth of the upper and lower jaw. See Fig. 3*c*, WG. Hence it follows that the neck of a Calmuck is shorter than that of an European: or rather, that it appears to be shorter, because the lower jaw, or chin, sinks so much lower. In proportion as the chin is lower, the condyle of the neck shorter, and the shoulders raised in consequence of the length of the clavicle (as is the case with the orang, and all deformed persons) will the head sink more upon the breast; and the stronger will be the resemblance to the people who are denominated Acephali; and who were said to exist in Guinea.

Again: As the *foramen magnum* of the occiput is not always placed at an equal distance from the perpendicular line KL. and as the condyles are placed

in an oblique direction before and on each side of the *foramen*, it follows, that
the centre of motion of the head will vary considerably in different people.
The line ND. extended from the extreme point of the mouth to that of the
occiput, may be compared to a lever, of which the centre of motion is in C.
Now, in proportion as the distance from N to C. is increased, will the face project forwards, and the neck will appear shorter.

The following appear to be the different proportions. In the Calmuck is
NC. or the distance from the extremity of the teeth to the orifice of the ear,
compared with CD. or the distance from this orifice to the extreme part of the
occiput, as $12\frac{1}{2}$ to 6: or nearly as 2 to 1.

In the negro is   NC : CD : : $7\frac{1}{2}$: $8\frac{1}{2}$: : 15 : 17.
In the European is NC : CD : : $7\frac{1}{2}$: $7\frac{1}{2}$: : 1 : 1.
In the antique is   NC : CD : : $7\frac{1}{2}$: $5\frac{1}{2}$: : 15 : 11.

The heads of the Calmucks must of consequence incline forwards, and sink
upon the shoulders.

The heads of negroes incline backwards, as the heaviest part is behind the
centre of motion.

The head of the orang-outang must be more forwards than that of the Calmuck, for the reasons given; and the head of the ape, the dog, horse, &c. still
more than either of these.

The heads of the Europeans remain in an equipoise; which gives them
something of an haughty mien.

In the antiques the gentle inclination of the head, particularly in the statues,
communicates the most state and dignity to the countenance.

Since I began to compose this Treatise, I have been able to procure the entire cranium of a Chinese, who died in the flower of his age. The facial line
was 75 degrees. The cavities of the eyes were in breadth, compared with
their height, as 12-8ths to 9-8ths $= 1\frac{1}{2}$ d. In the European they are equal.
It is not surprising, therefore, that the countenance of the Chinese should have
a melancholy aspect, and that the chinks or fissures formed by the upper and
lower eye-lids are naturally so long.

Their superior maxilla is narrow; that is, the space from N. to C. is very
small; so that they cannot have a large lip. However, the lower jaw is of a
more quadrangular form than in the European or the negro. In the Chinese
it makes an angle of 110 degrees; in the European, of 120; and in the negro,
125. Sees SVW. in Fig. 4*a* and Fig. 3*c*. The lower jaw of the Chinese has,
upon this account, something of the ape, and particularly of the orang, in its
form.

I took a sketch of the entire cranium of an Otaheite, who was brought into
Europe by Captain King, when I was at Oxford, in the year 1785, which
has a very great resemblance with that of the Chinese. The facial line was,
however, perpendicular; which may have been incidental. In the cranium of

an islander of the Celebese, are the same peculiarities as in that of the Chinese.

In the cranium of a man of the Celebese, and one of a Macassar, which I possess, there is a complete similarity; more than with that of a Moguller; which has, notwithstanding, much of the Asiatic in its form.

It is amusing to contemplate an arrangement of these, placed in a regular succession: apes, orangs, negroes, the skull of an Hottentot, Madagascar, Celebese, Chinese, Moguller, Calmuck, and divers Europeans. It was in this manner that I arranged them upon a shelf in my cabinet, in order that those differences might become more obvious which I have described in the preceding chapter.

To perceive at once the great utility of these principles, let any person sketch the profile of a negro, . . . resembling the one delineated in Fig. $3c$; the outlines of which are marked in the sixth plate by the letters KA, B, H, I, L, M, then draw the facial line of an European along the forehead FE, of 85 degrees; which will direct him to sketch from A to NE and O, and to terminate in I, and he will immediately have the face of an European.

Or let the face of an European be first sketched; and by inverting the mode, the physiognomy of a negro will be obtained.

By covering the dotted line ABH, with the tips of the fingers, the European face becomes more conspicuous: on the contrary, by covering NEO, the negro will more perfectly appear.

BASIC MORPHOLOGIC TYPES

von GOETHE, Johann Wolfgang (German naturalist, poet, and philosopher, 1749–1832). From *Erster entwurf eine allgemeinen zinleitung in die vergleichende anatomie ausgehend von der osteologie,** (= Heft 2 in) *Zur morphologie*, vol. 1, Stuttgart and Tübingen, 1817 (Heft 1) and 1820 (Heft 2); tr. by V. and M. Hamburger and T. S. Hall for this volume.

In the following passage Goethe develops the zoological implications of the "type concept" with which his name is commonly associated. His actual role in relation to this doctrine is subject to some doubts. For one thing, the scientist whose name is commonly associated with a scientific idea is rarely the one to whom it first occurred. The type concept, for example, bears unmistakable relations to the formal causes of the School of Athens. Furthermore, it is reasonably sure that Goethe's interest in the scientific implications of this doctrine was implanted in him by Herder in their Strasbourg days. Although he has been called an evolutionist, it seems safe to believe that the *Typus* never had for Goethe any specific material meaning. Perhaps its chief effect was an heuristic one, in stimulating the debate between the exponents of *Naturphilosophie* and those of *Materialism* for the latter of whom the Typus gradually assumed the role of a common ancestor. Thus Goethe's ideas, while used by many to combat the doctrine of descent, actually did much to establish it. He read Buffon and shows the latter's influence. ·The work of Lamarck was apparently unknown to him.

* Written Jena, 1795.

CONCERNING TYPES TO BE ESTABLISHED FOR THE FACILITATION
OF COMPARATIVE ANATOMY

The similarity of animals to each other and to man is obvious and already recognized in a general way. With regard to particulars, however, it is more difficult to observe, not always immediately demonstrable in detail, often misunderstood, sometimes even denied. Hence, the various opinions of observers are difficult to unify, since a norm against which to test the different parts is lacking, as is a sequence of principles which would also have to be acknowledged.

We have compared the animals to man and to each other, and in this way with much effort, made only detailed advances, but this accumulation of detail has made any kind of survey only more and more impossible. Since, by this method, one had to compare all animals with each and each with all, we realize at once the impossibility of ever achieving unification in this way.

For this reason, there is herewith proposed an anatomical Type—a general image in which the forms of all animals are potentially contained, and according to which the description of each animal would follow a certain order. As far as possible, the Type should be constructed with due regard to physiological considerations. It follows at once from the general idea of the Type that no single animal can be set up as such a canon of comparison; no single one can be the model for all.

Man in his high state of organic perfection must, because of that very perfection, not be set up as a measure for imperfect animals. Rather one should proceed as follows:

First, experience must teach us the parts common to all animals and wherein these parts differ. The Idea must govern the whole, must abstract the general image in a genetic way. Once such a Type has been set up, even though tentatively, then to test it we may employ the formerly customary modes of comparison.

We used to compare: animals with each other, animals with man, human races with each other, the two sexes, the main parts of the body (for instance upper and lower extremities), subordinate parts (for instance one vertebra with another).

All such comparisons can still be made after the Type is established; however, one will undertake them with better consequences and with a greater influence on the whole of science; one will judge what has been done heretofore and assign verified observations to their proper places.

Having established a Type, one precedes to the comparison in two ways: firstly, one describes individual animal species in terms of it. After this has been done, one need no longer compare animal with animal; one merely holds the descriptions against each other and the comparison takes care of itself.

One may also describe some one part throughout the principal genera, and in this manner an instructive comparison is perfectly effected. Both kinds of monographs ought to be as complete as possible to be fruitful; particularly in the latter, several observers might join forces. However, one should first of all agree on a general scheme whereafter the mechanical part of the labor could be expedited by a table on which each observer could base his work and so he would have worked for all, for science. As the matter now stands, it is sad to see that everyone must begin all over again.

General presentation of the Type

In the preceding, we have actually spoken only of mammalian anatomy from the comparative point of view and of possible ways of facilitating its study; now that we undertake the construction of the Type, however, we must look about us more widely in organic nature, because without such a survey we should be unable to establish a general image of the mammals, and because this image, if we consult all nature for its construction, can later be reversely modified in such a way that the images of imperfect organisms can be derived from it.

All fairly well developed creatures manifest in their outward structure three main divisions. Look, for example, at the highly developed insects! Their body consists of three parts which perform different vital functions and which, by their connection with each other and action upon each other, represent organic existence on a high level. These three divisions are: the head, middle, and hind regions; auxiliary organs are found attached to these in different ways.

The head is always anterior in position, is the meeting place of the different senses, and contains the ruling sensory instruments combined into one or several nerve nodes which we are accustomed to call the brain. The middle part contains the organs of the inner vital drive and of unceasing outward motion; the organs of the inner vital drive are less significant because in these creatures every part is apparently endowed with a life of its own. The hindmost division contains the organs of nutrition and reproduction as well as of cruder excrétion.

If the three designated divisions are separate and, as often happens, connected only by threadlike tubules, this indicates a perfect state. Therefore, the chief event in the progressive metamorphosis of caterpillar into insect is the progressive separating-off of systems which in the worm still lay concealed under the general covering and were in a largely ineffective, indeterminate state; but now that development has occurred, now that the final and most effective forces operate independently, free motion and action of the creature come into play and, as a result of manifold determination and separation of the organic systems, reproduction is possible.

In perfect animals, the head is more or less distinct from the second division, but by a continuation of the vertebral column, the third is combined with that anterior to it, and enveloped in a common cover. Dissection shows, however, that it is separated from the intermediate system, the chest, by a septum.

The head has the auxiliary organs needed for acquisition of food; they present themselves either as divided pincers or as a more or less connected pair of mandibles. In imperfect animals [*e.g.*, insects—Ed.], the middle region has a great variety of auxiliary organs: legs, wings, and wing coverings; in the perfect animals, (quadrupeds) this middle region has attached to it the middle auxiliary organs: arms or forelimbs. The hind region has no auxiliary organs in insects in their developed stage; however, perfect animals, where the two systems are approximated and crowded together, the most posterior auxiliary organs (called feet) are located at the hind end of the third section, and it is in this fashion that we shall find all mammals organized. Their last and hindermost part possesses, in greater or less degree, a continuation, the tail, which may be considered as suggesting the infinity of organic existences.

Application of the general representation of Type to specific cases

The parts of the animal, their respective forms, their relations, and their special character determine the vital needs of the individual; hence the definite, but limited, mode of life characteristic of each animal genus, and species. If according to the Type we have set up, so far only in a very general way, we examine the different parts of those most perfect of animals which we call mammals, then we shall find that, though the formative sphere of nature is indeed restricted, yet, because of the number of parts and their manifold modifiability, an infinite variety of form becomes possible.

When we know and study the parts accurately, we shall find that their manifoldness of form derives from the fact that this or that part is conceded dominance over the others.

For instance, in the giraffe, neck and extremities are favored at the expense of the body, whereas in the mole the reverse holds true.

From this point of view we are at once confronted by the law that no one thing can be added to any part unless something be deducted from another, and vice versa.

There exist limitations of animal nature within which the formative force apparently moves in a most admirable and almost arbitrary way yet without being in the least capable of breaking the circle or overstepping. The formative drive is here made ruler over a domain which, though limited, is nevertheless well ordered. The balance sheet over which its expenditures are to be distributed is outlined for it. It is free, to a certain degree, to decide how much to turn over to each item. If it wishes to bestow a relatively greater amount on any one part, it is not entirely hindered from doing so, but it is obliged to withdraw something from another immediately; in this way nature can never incur debt or become bankrupt.

With the help of this guide we wish to try to find our way through the labyrinth of animal formation. In the future we shall discover that it extends down even to the most formless organic natures. We wish to test it, first, on form, so that we may later on be able to apply it also to forces. We think, then, of the self-contained animal as a microcosm which exists for its own sake and entirely through itself. Thus, not only is each creature an end in itself; but also, since all its parts are in the most immediate interaction and are all related to each other and in this way continually renew the circle of life, each animal must be considered physiologically perfect. Although parts may appear to be useless viewed externally since the inner correlations in the nature of the animal molded them in this way without consideration of external relations. No part of it is useless when viewed in its internal relations nor, as is sometimes imagined, is it produced by the formative drive arbitrarily, as it were. Hence, in the future, one will not inquire of such parts, for example, as the canines of *Sus barbirussa, what do they serve for:* but, *where do they originate?* One will not contend that a bull has been endowed with horns "in order to" butt, but one will investigate *how* he could have horns with which to butt. The general Type which we shall first construct, on exploring into its parts, we shall find to be on the whole invariable. We shall find the highest class of animals, the mammals themselves, despite their most variable forms, still most conformative.

But now, though our concepts remain constant as regards that which is constant, we must at the same time learn to change our views and acquire manifold mobility as regards that which changes, so that we may be skilled in the pursuit of the Type in all its variability and so that the proteus may nowhere escape.

But if one inquires concerning the causes through which such a manifold capacity for being determined appears, the answer is: the animal is formed *by* circumstances *for* circumstances; hence its perfection, and hence its purposiveness viewed externally.

Now, in order to make apparent this idea of an economical give and take we adduce a few examples. The snake ranks quite high in organization; it has a definite head with perfect auxiliary organ, namely the lower mandible joined in front. However, its body is, as it were, infinite, and it can afford to be so because it has to invest neither matter nor energy in auxiliary organs. As soon as the latter appear, when for instance in lizards short arms and legs are produced, then the absolute length must at once contract and a shorter body occur. The long legs of the frog force the body of this creature into a very short form, and the misshapen toad is extended in width according to the same law.

The only remaining question is how far one wishes to pursue this principle, in cursory fashion, through the different natural classes, genera, and species, and make this idea generally evident and attractive by an appraisal of habitus and external character, so that our desire and courage may be stimulated to explore details with attention and pains.

But first the Type ought to be considered with reference to the way in which the different elementary natural forces affect it and also as to how it must submit, to a certain degree, to general external laws.

Water, decidedly, swells up the bodies which it surrounds and into which it penetrates to a greater or less degree. Thus the trunk of the fish, especially the flesh of it will become swollen up according to the laws of the element. Now according to the laws of the organic Type, this swelling up of the body must, no matter what other organs may arise, be followed by the contraction of the extremities or auxiliary organs.

Air by absorbing water, dries things out. Therefore the type which develops in air, will, the purer and the less humid the air is, become that much drier inside; and a more or less lean bird will originate, whose flesh and bony skeleton will be amply covered, and whose auxiliary organs will be sufficiently provided for, because enough material has been left over for the formative force. What is formed into flesh in the fish, is here left for the feathers. Thus the eagle is formed *by* air *for* air, *by* mountain height *for* mountain height. The swan, the duck, amphibious types, betray their inclination to water already by their shape.

How wonderfully the stork, the sandpiper, indicate both their proximity to water and their inclination to air. This merits our unceasing attention.

Thus we shall find the effects of climate, mountain height, warmth and cold, besides those of water and ordinary air, to exert a powerful influence in the formation of the mammals. Warmth and humidity cause swelling and produce seemingly inexplicable monsters even within the limits of the type, whereas heat and dryness produce the most perfect and highly developed creatures, however opposed those may be to man in nature and form; for example, lion and tiger. Thus the hot climate alone is capable of imparting some human likeness to even imperfectly organized beings, as occurs, for instance, in monkeys and parrots.

RISE OF THE TISSUE CONCEPT

BICHAT, Marie François Xavier (French anatomist and surgeon, 1771–1802). From *Anatomie générale appliquée à la physiologie et la médecine*, Paris, 1801; tr. by G. Hayward as *General anatomy applied to physiology and medicine*, Boston, 1822.

While Bichat's tissues (at first "textures") were more extensively described in the *Traité des membranes*, published in 1799, than in the *General Anatomy*, it was in the latter that he more explicitly stated the tissue *concept*. The tissues were not an accidental discovery but came into view as a new answer to an old question, viz.: Given, clearly defined, the really significant faculties of plants and animals, what material entities are responsible for them? Before Bichat, the answer to this question would have been "the organs." In the following passage, which bears strong affinities to Aristotle (*De partibus animalium*, Book I), we may learn exactly why Bichat felt that this old question must now receive a new answer.

[From the Preface]

The general doctrine of this work consists in analyzing with precision the properties of living bodies, in showing that every physiological phenomenon is ultimately referable to these properties considered in their natural state; that every pathological phenomenon derives from their augmentation, diminution, or alteration; that every therapeutic phenomenon has for its principle the restoration of the part to the natural type, from which it has been changed. . . .

OBSERVATIONS UPON THE ORGANIZATION OF ANIMALS [ENTIRE].

The properties, whose influence we have just analyzed, are not absolutely inherent in the particles of matter that are the seat of them. They disappear when these scattered particles have lost their organic arrangement. It is to this arrangement that they exclusively belong; let us treat of it here in a general way.

All animals are an assemblage of different organs, which, executing each a function, concur in their own manner, to the preservation of the whole. It is several separate machines in a general one, that constitutes the individual. Now these separate machines are themselves formed by many textures of a very different nature, and which really compose the elements of these organs. Chemistry has its simple bodies, which form, by the combinations of which they are susceptible, the compound bodies; such are caloric, light, hydrogen, oxygen, carbon, azote, phosphorus, &c. In the same way anatomy has its simple textures, which, by their combinations four with four, six with six, eight with eight, &c. make the organs. These textures are, 1st. the cellular; 2d. the nervous of animal life; 3d. the nervous of organic life; 4th. the arterial; 5th. the venous; 6th. the texture of the exhalants; 7th. that of the absorbents and their glands; 8th. the osseous; 9th. the medullary; 10th. the cartilaginous; 11th. the fibrous; 12th. the fibro-cartilaginous; 13th. the muscular of animal life; 14th. the muscular of organic life; 15th. the mucous; 16th. the serous; 17th. the synovial; 18th. the glandular; 19th. the dermoid; 20th. the epidermoid; 21st. the pilous.

These are the true organized elements of our bodies. Their nature is constantly the same, wherever they are met with. As in chemistry, the simple bodies do not alter, notwithstanding the different compound ones they form. The organized elements of man form the particular object of this work.

The idea of thus considering abstractedly the different simple textures of our bodies, is not the work of the imagination; it rests upon the most substantial foundation, and I think it will have a powerful influence upon physiology as well as practical medicine. Under whatever point of view we examine them, it will be found that they do not resemble each other; it is nature and not science that has drawn the lines of distinction between them.

1st. Their forms are everywhere different; here they are flat, there round. We see the simple textures arranged as membranes, canals, fibrous fasciae, &c. No one has the same external character with another, considered as to their attributes of thickness or size. These differences of form, however, can only be accidental, and the same texture is sometimes seen under many different appearances; for example, the nervous appears as a membrane in the retina, and as cords in the nerves. This has nothing to do with their nature; it is then from the organization and the properties, that the principal differences should be drawn.

2dly. There is no analogy in the organization of the simple textures. We shall see that this organization results from parts that are common to all, and from those that are peculiar to each; but the common parts are all differently arranged in each texture. Some unite in abundance the cellular texture, the blood vessels and the nerves; in others, one or two of these three commoner parts are scarcely evident or entirely wanting. Here there are only the exhalants and absorbents of nutrition; there the vessels are more numerous for other purposes. A capillary net-work, wonderfully multiplied, exists in certain textures, in others this net-work can hardly be demonstrated. As to the peculiar part, which essentially distinguishes the texture, the differences are striking. Colour, thickness, hardness, density, resistance, &c. nothing is similar. Mere inspection is sufficient to show a number of characteristic attributes of each, clearly different from the others. Here is a fibrous arrangement, there a granulated one; here it is lamellated, there circular. Notwithstanding these differences, authors are not agreed as to the limits of the different textures. I have had recourse, in order to leave no doubt upon this point, to the action of different re-agents. I have examined every texture, submitted them to the action of caloric, air, water, the acids, the alkalies, the neutral salts, &c. drying, putrefaction, maceration, boiling, &c. the products of many of these actions have altered in a different manner each kind of texture. Now it will be seen that the results have been almost all different, that in these various changes, each acts in a particular way, each gives results of its own, no one resembling another. There has been considerable inquiry to ascertain whether the arterial coats are fleshy, whether the veins are of an analogous nature, &c. By comparing the results of my experiments upon the different textures, the question is easily resolved. It would seem at first view that all these experiments upon the intimate texture of systems, answer but little purpose; I think however that they have effected an useful object, in fixing with precision the limits of each organized texture; for the nature of these textures being unknown, their difference can be ascertained only by the different results they furnish.

3dly. In giving to each system a different organic arrangement, nature has also endowed them with different properties. You will see in the subsequent part of this work, that what we call *texture* presents degrees infinitely varying, from the muscles, the skin, the cellular membrane, &c. which enjoy it in the highest degree, to the cartilages, the tendons, the bones, &c. which are almost destitute of it. Shall I speak of the vital properties? See the animal sensibility predominant in the nerves, contractility of the same kind particularly marked in the voluntary muscles, sensible organic contractility, forming the peculiar property of the involuntary, insensible contractility and sensibility of the same nature, which is not separated from it more than from the preceding, characterizing especially the glands, the skin, the serous surfaces, &c. &c. See each of these simple textures combining, in different degrees, more or less of these properties, and consequently living with more or less energy.

There is but little difference arising from the number of vital properties they have in common; when these properties exist in many, they take in each a peculiar and distinctive character. This character is chronic, if I may so express myself, in the bones, the cartilages, the tendons, &c.; it is acute in the muscles, the skin, the glands, &c.

Independently of this general difference, each texture has a particular kind of force, of sensibility, &c. Upon this principle rests the whole theory of secretion, of exhalation, of absorption, and of nutrition. The blood is a common reservoir, from which each texture chooses, that which is adapted to its sensibility, to appropriate and keep it, or afterwards reject it.

Much has been said since the time of Bordeu, of the peculiar life of each organ, which is nothing else than that particular character which distinguishes the combination of the vital properties of one organ, from those of another. Before these properties had been analyzed with exactness and precision, it was clearly impossible to form a correct idea of this peculiar life. From the account I have just given of it, it is evident that the greatest part of the organs being composed of very different simple textures, the idea of a peculiar life can only apply to these simple textures, and not to the organs themselves.

Some examples will render this point of doctrine which is important, more evident. The stomach is composed of the serous, organic muscular, mucous, and of almost all the common textures, as the arterial, the venous, &c. which we can consider separately. Now if you should attempt to describe in a general manner, the peculiar life of the stomach it is evidently impossible that you could give a very precise and exact idea of it. In fact the mucous surface is so different from the serous, and both so different from the muscular, that by associating them together, the whole would be confused. The same is true of the intestines, the bladder, the womb, &c.; if you do not distinguish what be-

longs to each of the textures that form the compound organs, the term peculiar life will offer nothing but vagueness and uncertainty. This is so true,
that oftentimes the same textures alternately belong or are foreign to their organs. The same portion of the peritoneum, for example, enters or does not
enter, into the structure of the gastric viscera, according to their fulness or
vacuity.

Shall I speak of the pectoral organs? What has the life of the fleshy texture of the heart in common with that of the membrane that surrounds it?
Is not the pleura independent of the pulmonary texture? Has this texture
nothing in common with the membrane that surrounds the bronchia? Is it
not the same with the brain in relation to its membranes, of the different parts
of the eye, the ear, &c.?

When we study a function, it is necessary carefully to consider in a general
manner, the compound organ that performs it; but when you wish to know
the properties and life of this organ, it is absolutely necessary to decompose it.
In the same way, if you would have only general notions of anatomy, you can
study each organ as a whole; but it is essential to separate the textures, if you
have a desire to analyze with accuracy its intimate structure.

NATURPHILOSOPHIE

OKEN, Lorenz (German natural philosopher, 1779–1851). From the preface to *Lehrbuch der naturphilosophie*, Jena, 1809; tr. (of the 3d ed., Zürich, 1843) by A. Tulk as
Elements of physiophilosophy, London, 1847.

Stemming largely from the Kantian reaction against mechanistic and positivistic science, there arose in eighteenth century Germany the subjectivistic school of so-called
Nature Philosophers (see Goethe). In this movement was attempted a new synthesis
of the mechanistic ideas introduced by Descartes with ideational "patterns" and final
causes reminiscent of the school of Athens. The distortive influence of such thinking upon the interpretation of biological facts culminates with the Physiophilosophy of
Lorenz Oken. The difficulties with which Oken wrestled tended to disappear with the
rise of the doctrine of descent. In the preface to the third edition (given below) he summarizes his philosophy and gives references leading to his original statement of each
main idea.

The first principles of the present work I laid down in my small pamphlet
entitled *Grundriss der Naturphilosophie, der Theorie der Sinne und der darauf
gegründeten Classification der Thiere*; Frankfurt bey Eichenberg, 1802, 8vo
(out of print). I still abide by the position there taken, namely, that the Animal Classes are virtually nothing else than a representation of the sense-organs,
and that they must be arranged in accordance with them. Thus, strictly
speaking, there are only 5 Animal Classes: *Dermatozoa*, or the Invertebrata;
Glossozoa, or the Fishes, as being those animals in whom a true tongue makes
for the first time its appearance; *Rhinozoa*, or the Reptiles, wherein the nose

opens for the first time into the mouth and inhales air; *Otozoa,* or the Birds, in which the ear for the first time opens externally; *Ophthalmozoa,* or the Thricozoa, in whom all the organs of sense are present and complete, the eyes being moveable and covered with two palpebrae or lids.   But since all vegetative systems are subordinated to the tegument or general sense of feeling, the Dermatozoa divide into just as many or corresponding divisions, which, on account of the quantity of their contents, may be for the sake of convenience also termed classes.   Thereby 9 classes of the inferior animals originate, but which, when taken together, have only the worth or value of a single class.   So much by way of explaining the apparent want of uniformity in the system.

I first advanced the doctrine, that all organic beings originate from and consist of *vesicles* or *cells,* in my book upon Generation. (*Die Zeugung.* Frankfurt bey Wesche, 1805, 8vo.) These vesicles, when singly detached and regarded in their original process of production, are the infusorial mass, or the protoplasma (*Ur-Schleim*) from whence all larger organisms fashion themselves or are evolved.   Their production is therefore nothing else than a regular agglomeration of Infusoria; not of course of species already elaborated or perfect, but of mucous vesicles or points in general, which first form themselves by their union or combination into particular species.   This doctrine concerning the primo-constituent parts of the organic mass is now generally admitted or recognised, and I need not, therefore, add anything by way of apology for it or defence.

In mine and Kieser's *Beyträgen zur vergleichenden Zoologie, Anatomie und Physiologie;* Frankfurt bey Wesche, 1806, 4to, I have shown that the intestines originate from the umbilical vesicle, and that this corresponds to the vitellus.   It is true *Friedrich Wolf* had already discovered it in the chick, but his was only a single instance, and completely forgotten.   I have also discovered it and without knowing anything about my being anticipated, since it was nowhere taught.   But I have elevated this structure to the light of a general law, and it is that unto which I may fairly lay claim.   In the same essay I have introduced into the Physiology the Corpora Wolfiana, or Primordial Kidneys, but, having failed to recognise their signification, any one who pleases may filch away the credit of their bare detection.

In my Essay: *Ueber die Bedeutung der Schädelknochen,* (Ein Programm beym Antritt der Professur an der Gesammt-Universität zu Jena; Jena gedruckt bey Gopfert, 1807, verlegt zu Frankfurt bey Wesche, 4to,) I have shown that the head is none other than a vertebral column, and that it consists of four vertebrae, which I have respectively named Auditory, Maxillary or Lingual, Ocular and Nasal vertebra; I have also pointed out that the maxillae are nothing else but repetitions of arms and feet, the teeth being their nails; all this is carried out more circumstantially and in detail in the *Isis,* 1817, S. 1204; 1818, S. 510; 1823. litt. *Anzeigen* S. 353 und 441.   This doctrine was at first scoffed at and repulsed; finally, when it began to force its way, several barefaced persons came forward, who would have made out if they could, that the discovery was achieved long ago. The reader will not omit to notice that the above essay appeared as my Antritts-Programm, or Inaugural discourse, upon being appointed Professor at Jena.

In my Essay entitled *Ueber das Universum als Fortsetzung des Sinnensystems;* Jena bey Frommann, 1808, 4to, I showed that the Organism is none other than a combination of all the Universe's activities within a single individual body. This doctrine has led me to the conviction that World and Organism are one in kind, and do not stand merely in harmony with each other. From hence was developed my Mineral, Vegetable and Animal system, as also my philosophical Anatomy and Physiology.

In my Essay entitled *Erste Ideen zur Theorie des Lichts, der Finsterniss, der Farben und der Wärme;* Jena bey Frommann, 1808, 4to, I pointed out, that the Light could be nothing but a polar tension of the aether, evoked by a central body in antagonism with the planets; and that the Heat were none other than the motion of this aether. This doctrine appears to be still in a state of fermentation.

In my Essay entitled *Grundzeichung des natürlichen Systems der Erze;* Jena bey Frommann, 1809, 4to, I arranged the Ores for the first time, *not* according to the Metals, but agreeably to their *combinations* with Oxygen, Acids, and Sulphur, and thus into *Oxyden, Halden, Glänzen,* and *Gediegenen.* This has imparted to the recent science of Mineralogy its present aspect or form.

In the first edition of my *Lehrbuch der Naturphilosophie,* 1810 and 1811, I sought to bring these different doctrines into mutual connexion, and to show, forsooth, that the Mineral, Vegetable, and Animal classes are not to be arbitrarily arranged in accordance with single or isolated characters, but to be based upon the cardinal organs or anatomical systems, from which a firmly established number of classes must of necessity result; moreover, that each of these classes commences or takes its starting-point from below, and consequently that all of them pass parallel to each other. This parallelism is now pretty generally adopted, at least in England and France, though with sundry modifications, which, from the principles being overlooked or neglected, are based at random, and are not therefore to be approved of. As in chemistry, where the combinations follow a definite numerical law, so also in Anatomy the organs, in Physiology the functions, and in Natural History the classes, families and even genera of Minerals, Plants and Animals, present a similar arithmetical ratio. The genera are indeed, on account of their great number and arbitrary erection to the rank whose title they bear, not to be circumscribed or limited in every case with due propriety, nor brought into their true scientific place in the system; it is nevertheless possible to render their parallelism with each other clear, and to prove that they by no means form a single ascending series. If once the genera of Minerals, Plants and Animals come to stand correctly opposite each other, a great advantage will accrue therefrom to the science of Materia Medica; for corresponding genera will act specifically upon each other.

These principles, which I have now carried out into detail, were retained in the *second*, and have been also in the *third* or present edition of the *Physiophilosophy*, the arrangement and serial disposition of the natural objects having, with my increase of knowledge and concomitant views of things, been amended, enlarged or diminished, as the case might require, especially in the Mineral, Vegetable and Animal systems. I am very well aware that there is many an object which does not stand in its right place; but where again is there a single system in which this is not still more strikingly the case? We have here dealt only with the restoration of the edifice, wherein, after years of long and oft-repeated attempts, the furniture may for the first time be properly distributed, without detriment to its general bearings or ground plan.

In my *Lehrbuch der Naturgeschichte,* the Mineralogical and Zoological portions of which are out of print, but the Botanical still to be had (Weimar, Industrie-Comptoir, 1826,), I have arranged for the first time the genera and species in accordance with the above principles, and stated everything of vital importance respecting these matters. This was the first attempt to frame a scientific Natural History, and one unto which I have remained true in my last work, the *Allgemeine Naturgeschichte*, the principles whereof I have sought to develop more distinctly and in detail in the work now before the reader.

Thus then have I prosecuted throughout a long series of years one kind of principle, and worked hard to perfectionate it upon all sides. Yet, notwithstanding my endeavour to amass the manifold stores of knowledge so requisite to an undertaking like this, I could not acquire within the vast circuit that appertains thereunto, many things which might be necessary unto a system extending into all matters of detail. This it is to be hoped the reader will acknowledge, and have forbearance for the errors, against which every one will stumble who has busied himself throughout life with a single branch of the natural sciences. Natural History is not a closed department of human knowledge, but presupposes numerous other sciences, such as Anatomy, Physiology, Chemistry and Physics, with even Medicine, Geography and History; so that one must be content with knowing only the main facts of the same, and relinquishing the Singular to its special science. The gaps and errors in Natural History can therefore be filled up or removed only by numerous writers and in the lapse of time.

MECKEL ON METHOD IN ANATOMY

MECKEL, Johann Friedrich (German anatomist, 1781–1833). From *Handbuch der menschlichen anatomie*, Halle and Berlin, 1816–20; tr. by A. S. Doane as *Manual of general, descriptive and pathological anatomy*, Philadelphia, 1832, from a French translation by Jourdain and Breschet.

Someone has said that Meckel "knew anatomy better than he understood it." Thus, the following "laws" of formation are admissible as such only by a rather broad definition of the term "law." Historically, however Meckel's search for underlying principles played an extremely influential role, for, when basic patterns appear in nature, science seeks their significance and their causes, and in this way discovery is stimulated. It was partly to provide a rational explanation of such empirical generalities as Meckel's laws that the great ideas of nineteenth-century biology (such as *cell theory, doctrine of descent,* and *biogenetic law*) developed and took hold. Important specific contributions of Meckel were the anatomy of the duckbill and cassowary, systematic and theoretical studies of malformations, etc. He later made even broader generalizations on the method of anatomy.*

The general laws of the organic form, and hence those which belong to man, are as follows: first, *the outline is not sharp and angular, but rounded.* This law is true, both in regard to the form of the whole body, as well as to that of each of its organs, and its smallest elements. The roundness of the form usually depends upon the fact, that all the solids are accompanied with fluids, for the first effect of solution is to smooth down the angles of solid bodies. We mention, as examples, the round form of the cavities of the body, of the viscera, vessels, nerves, muscles, bones, etc.

II. *The dimension of length exceeds the others.* This law is seen no less in the body as a whole, and in the external form, than in the internal form, or the texture of its parts. The whole length of the body, much exceeds its breadth and thickness. It is divided into three principal regions, the *head, trunk,* and the *limbs,* or *extremities.* Of these, the head alone is round, but it is only the upper and bulging extremity of the vertebral column, that is of the osseous base of the trunk, in which the dimension of length evidently predominates. This column is enveloped by lateral expansions, producing cavities designed to lodge the apparatus placed before them, but it is not entirely concealed. The excess of length is most manifest in the extremities, generally, and in their different parts. It is the same with all the particular systems. The dimension of length much exceeds the other two in the vascular and nervous systems. It is especially marked in the hair. The number of the long bones, muscles, and fibrous organs, is much greater than of those which are broad or thick. The intestinal canal, the trachoea, ureters, urethra, &c., are very narrow in proportion to their length.

---

* *System der vergleichenden anatomie*, Halle, 1821; see the foreword.

This rule applies exactly to the texture, since the fibrous is the most common of all, and every large fibre is divided into an endless multitude of others which gradually become smaller and smaller.

III. *The structure of the organism is radiated.* From the central parts, which are largest, originate others, which are smaller, which move in all directions, and in which the dimension of length especially predominates. Thus the extremities arise from the trunk: the long, and narrow ribs, from the spine; the nerves, from the brain, spinal marrow, and ganglions, the vessels from the heart. But, besides these grand centres, from which the rays commence, there is an infinite number of the second order, since each ray usually divides into several, which, in some systems, particularly the general, as that of the nerves and vessels, also, in *their turn* subdivide. The rays then *ramify*.

Another general law is, that the number of rays augments as they depart from the principal centre of radiation, and their volume diminishes in the same proportion. Thus, instead of one long bone, as in the arm and thigh, we find two, which are smaller, in the forearm and leg, twenty-six in the foot and twenty-seven in the hand, still smaller. The number of the muscles and tendons inserted into the bones, and of the ligaments, multiplies in the same manner, and they diminish in size in the same ratio. In the whole course of the nervous and vascular trunks, branches and twigs are constantly sent off in every direction, at different angles, and at certain distances from the principal trunks they divide into others still smaller, which are themselves again subdivided.

This division is true, not only in respect to length, and from without, inward: it also occurs in the thickness of the organs; for the muscles and nerves represent bundles composed of cords, which are, in their turn, formed of fibres and filaments.

IV. At the side of this *law of ramification* proceeds another, *the law of anastomosis.* These rays, it is true, subdivide a great many times, but the subordinate rays which result from this division, unite in different ways with each other, and with the principal ray. The same remark is true in regard to the continuity, that is, to the thickness of the organs; for these anastomoses take place from above, downwards, and also from within, outwards. The different trunks, branches, and twigs of the nerves and vessels, the different tendons of the same muscle, the simple fibres, their fasciculi, both large and small, in the nerves, or at least in many muscles, the fibres of the bones, and those of the fibrous organs, mutually anastomose together. We can, to a certain extent, mention here those bones placed side by side, which, in addition to the ligaments necessary to keep them in place, are connected by interosseous membranes.

V. *These rays are not straight, but usually more or less curved.* This law, which has been termed *the law of the spiral line,* is seen in the vertebral column, which describes several curves; it is confirmed also by most of the long bones. The cochlea, semicircular canals, several vessels, excretory ducts, and nerves, are also examples. Finally, this is sometimes seen very evidently in the double monsters: since, when two heads are placed side by side, or two bodies are united by a head, the direction of these two bodies or two heads is always different; so that the monster, considered as a whole, appears spiral.

VI. *The different organs are somewhat analogous.* It has already been stated that the texture of the most dissimilar organs can be reduced to two elements of form, which are generally united; we have there indicated the analogy which exists between the final structure of the organs, and to which many writers have wrongly given still more extent. As the structure of almost all the organs is radiated, it follows that the analogy of their external form is demonstrated. It exists even where there is no manifest radiation; since we there remark parts first dilated, and others which are contracted. Thus, the brain is joined to the spinal marrow, the vertebral column to the skull, the vascular trunks to the heart, the esophagus to the buccal cavity, the intestinal canal to the stomach, the trachea to the larynx, the cystic duct to the gall bladder, the ureter to the pelvis of the kidney, and the urethra to the bladder.

Another great analogy between the different systems is established by the circumstance that those which vary the most from each other are formed after the same type in the same parts of the body. Thus, the simple trunk of the arteries of the superior or inferior extremities is almost always divided into two different branches, at the place where the number of bones is doubled; in general, the divisions and unions correspond exactly in the different systems situated near each other. The number of the arteries destined for the fingers and toes is the same as that of these appendages. The nervous trunks and the vessels anastomose in the palms of the hands and soles of the feet. In the same manner, the tendons of the flexor and extensor muscles of the fingers and toes are united by mucous and tendinous slips. The different parts of the nervous system and of the vascular system proceed together.

The whole form of the body is repeated not only in those systems which are generally diffused in the entire economy, and which form a whole more or less continuous—as the cellular tissue and the nervous, vascular, osseous, and muscular systems—but also in each of the organs. We should refer to this the form already pointed out as belonging to so many systems, the peculiarity of which is an enlargement at one extremity and a continuation into a narrower process at the other. We must refer to this also the manner in which most of the organs, and principally the glands, receive their vessels. In fact, we always observe a considerable depression, a fissure, about the centre of these organs, through which the vessels enter and emerge exactly as in the fetus. Here the organ is open to a certain extent, and is much more so the nearer it is to its formation, exactly as in the fetus, which at first is open entirely before.

VII. *The body is formed symmetrically.* We find an analogy, and even a resemblance to a certain extent, not only between the different organs, but particularly also between their different regions. This analogy may be demonstrated, both in the breadth, and in the length, and thickness, or even between its right and left sides, between its upper and lower extremities, and between its anterior and posterior faces. We must here remark generally, that the *similitude is never perfect,* and that usually one extremity predominates, more or less, over the other. This is sometimes expressed by the greater volume, and sometimes by the greater development in the radiation of corresponding parts. Nor is the symmetry equally great in all directions, nor between the different corresponding regions. The most perfect is the lateral symmetry, or that of the two sides of the body; and the most imperfect, that of its anterior and posterior faces. The nervous, osseous, ligamentous, and muscular systems, and the genital apparatus, are the most symmetrical parts; we find less symmetry in the vascular system and in the thoracic and pelvic viscera, if we except the genital apparatus.

The *most perfect lateral symmetry* is in the external form, and on the surface of the body. Hence, it is better known there. In fact, the body seems to be composed of a right and a left half, since most of the organs are *double,* and those which are single, are placed more or less on the median line, so that a plane drawn from before backward, would divide them into two nearly equal parts, as they are formed of two lobes united and blended on the median line. These latter, when placed between cavities, form *septa;* and on the contrary, are called media of communication, *commissures,* when placed between two corresponding parts, which are otherwise separated.

Similar arrangements are found in all the systems, and we may say with justice that a commissure exists more or less perceptibly in all the body, although it is often interrupted; this, at the same time, forms a septum between the right and left sides. Thus, the falx cerebri descends from before backward, from the centre of the skull; and the internal ridges of the frontal and occipital bones correspond to it. Below it, is the corpus callosum which unites the two hemispheres of the cerebrum; below, is the septum lucidum, formed of two layers closely applied to each other, and which represent in the brain what the falx and spinous ridges have in the skull. The nasal cavity is divided into two parts by a partition, which is bony above and behind, and cartilaginous before; the former of these portions being formed by a part of the os ethmoides, and by a particular bone, the vomer. The frena of the lips in front, and the uvula behind, represent this septum in the mouth. In the chest, the internal parieties of the pleurae, which partly touch, and are partly separated by organs placed between them, form the anterior and posterior mediastina, and thus establish a line of demarkation between the two halves of the thoracic cavity. A longitudinal septum, generally perfect, exists between the right and left sides of the heart. This septum is only indicated in the abdomen, where the two halves seem blended together; the division has been destroyed, or its formation has been prevented, by the considerable mass of organs which are inclosed by this cavity. We trace it, however, forward and above, in the suspensory ligament of the liver which extends from the inferior face of this gland to the umbilicus, and below, in the analogous but less extensive fold of the peritonaeum, which reaches from the bladder to the umbilicus, covering the remains of the obliterated umbilical artery and urachus; and finally, behind, in the other fold of the peritonaeum, which goes from the anterior face of the lumbar vertebrae to the intestinal canal, and which is called the mesentery. On the median line of the penis in the male, and of the clitoris in the female, we find a perpendicular septum. The corpus spongiosum of the penis, and the septum and the raphe of the scrotum in the male, are situated exactly on the median line. The cellular tissue, which unites the skin to the subjacent parts, is thicker in all regions of the body on the anterior and superior, than on the posterior face. The vessels frequently anastomose together on the median line, as is seen in the coronary arteries of the lips, the sinuses of the medulla spinalis, and the cerebral arteries, which, supplying the two hemispheres of the brain, unite by numerous transverse branches. So, likewise, the two vertebral arteries unite on the median line, to produce the basilary, and the anterior and posterior spinal arteries descend along the spinal marrow. Several sinuses of the dura mater exist on the median line of the skull. The aorta, venae cavae, thoracic canal, the azygos vein, and partly even the esophagus, describe a curve, which corresponds very nearly to the median line of the thoracic and abdominal cavities.

The vertebral column, sternum, occiput, os frontis, os ethmoides, and os sphenoides, are those parts which are unmated and distinct, and serve to join the corresponding parts of the same system, as they are united with them, and wedged in by them. Those bones which meet on the median line, but remain always distinct, although united by an intermediate substance, as the ossa parietalia and ossa ilia, form the connecting link between them and those which do not touch in the least.

The brain and spinal marrow, the heart, womb, vagina, prostate gland, bladder, urethra, thyroid, and thymous glands, the intestinal canal, the trachea, larynx, and tongue, are unmated, but are formed of two similar portions, between which the median line passes, at least to a certain extent.

All the other organs are·mated, and are rarely united by their own proper substance. They are connected in various ways. Thus the kidneys lie on each side, and are united above by the blood-vessels, and below by the ureters, which go to the bladder. The lungs are joined above both by the trachea and pulmonary blood-vessels, while the extremities are perfectly insulated, or at least are united only at their upper ends, where but a small number of parts are found belonging to them in common.

VIII. *No organ possesses precisely the same qualities at all periods of the existence of the organism.* There are none which are alike at all periods of their existence. Each organ, and consequently the whole organism, passes through certain regular and normal stages. This very important law, called the *law of development,* gives rise to the following considerations:

1. There is for each organ, and for the whole organism, a period of imperfection, in which the whole development is not attained; this is called the period of *youth,* or *infancy;* a second called that of *mature age,* or *period of maturity, of perfection;* and a third, that of *old age,* or *of decline.* 2. *The resemblance is much greater between different organs, and the different regions of the body, the nearer each respective organ, and the whole organism, is to its origin; the more recent the organism, the more symmetrical it is.* . . . 3. *The color of the organs develops itself gradually.* . . . 4. *Every organ is softer and more fluid the nearer it is to its origin; it gradually acquires its normal degree of consistence and its cohesion increases till the end of life.* . . . 5. *This state of great fluidity is attended with a want of a determined texture during the first periods of existence.* . . . 6. *All the organs do not appear at the same time.* . . . 7. *These parts, which are but repetitions of other more perfect parts, and which especially correspond to them, are the last to appear.* . . . 8. *The external form develops itself much more rapidly than the texture and chemical composition of the organs.* . . . 9. *The organs arise almost entirely by separate parts, which gradually unite to form a whole.* . . . 10. *All the organs have not the same proportional volume at every epoch of life.* . . . 11. *The duration of the organs is not the same.* . . . 12. *Some systems pass through a greater variety of degrees than others: the history of their life is more complicated.* . . . 13. *In some parts we can always trace the primitive formation; in others we cannot, although we know not exactly the cause of this difference.* . . . 14. *The degrees of development through which man passes from birth to the period of perfect maturity correspond to constant formations in the animal kingdom.* . . . 15. *Man is distinguished from the other animals by the greater rapidity with which he passes through the inferior formations.* . . .

IX. *Although the form of the human organism varies at different periods of life, yet it presents certain peculiarities which distinguish it from all others, and characterize the human race as a separate species.* This species, however, is only one of numerous modifications of the primitive type which constitutes the base of all animal formations, so that its form necessarily resembles, in many respects, those of other animals, particularly those most allied to man. It is then almost incredible that, even recently, several writers would consider many of these conditions of the human form not as results of this law, but as proving positively that, after the original sin, man was even physically degraded from the great excellence he possessed in Paradise! They pretend that the traces of the intermaxillary bone prove that the cerebrum and cranium are diminished; they add, that the face is developed in the same proportion, that the plantaris muscle has attained at the same time the aponeurotic expansion of the sole of the foot, and that its actual rudimentary existence proved that men walked on all fours, etc. All these assertions are unfounded: all these phenomena demonstrate nothing, since it might be proved in the same manner, by the arrangements of some other part, that man, before the deluge, was a different animal from what he is now. The human structure has nothing to distinguish it entirely from that of animals: it ought then to have the same forms: those presented by it serve to remind us, here and there, of what is found in animals. But these marks, such for instance as the intermaxillary bone, are easily explained by the preceding law; they trace that series of degrees of the organization through which the embryo, but not the whole human race, always passes; or they are the vestiges of the primitive state when the human formation was depressed to the level of the animal formation. In order to give some probability to the opinion we oppose, it is necessary, *at least*, to compare the human skulls before the fall of man and the deluge with each other, and with those of the present time. Nor are there any facts to support a similar hypothesis, the partizans of which pretend, that as the human organism, in accordance with the preceding law, passes through different periods, so this is the case with the whole human family; and that certain races are now at a point formerly possessed by other races at present more elevated, and that these also are capable of gradual improvement. In refuting this hypothesis, we cannot deny but that the different classes of organisms are developed gradually, and in direct proportion to their greater or less degree of perfection.

X. Notwithstanding the peculiarities of conformation which prove that man, like every other organism, forms a separate species, daily observation demonstrates that, under any relation, all the individuals are not exactly alike. The principle difference, which extends to the whole species, is the distinction of this species into two sexes, *male* and *female*. 1st. *In size*. 2nd. *In the external form*. 3d. *In texture and physical qualities*. 4th. The *situation* of the parts is the same in both sexes, if we except those organs which relate to the generative functions.

Besides this grand fundamental difference, which divides the human species into two halves, there are others less striking which are common to both of these halves, which distinguish the species, not into sexes, but *races*.

Besides the differences of sexes and of races, the human formation presents others of a third kind. These latter are separate from the first, and are common to the whole species, and are found indiscriminately in both sexes and in all the races, although they may be more frequent in one sex, or race, than in another: they are called *abnormal formations*, or *deviations of formation*.

XI. The organic form every where presents traces of a formation in accordance with the purpose to be attained. It is impossible not to perceive that an intellectual power, whatever may be its relations to matter, has governed the formation of organized bodies. This is especially confirmed by those mechanical arrangements which we find in a multitude of places, and by the greater protection given to the organs essential to life. Among the phenomena of the first class, we shall mention the valves established in those vessels which have no immediate power of impulse, as the veins and the lymphatics, and the multiplicity of these valves, either at those points where the friction is greatest, as in the small veins and in the lymphatic vessels generally, or those where there is no mechanical impulse, as in the lymphatic system. On the contrary, there are no valves in those veins where the different trunks anastomose together.

As to the phenomena of the second kind, we see that the organs most essential to life, as the brain, spinal marrow, and lungs, are wholly or partially enclosed in large cavities, the skull, vertebral column, and the thorax, which are also particularly remarkable on account of their circular form. So, likewise, the veins are situated less deeply than the arteries.

The duplication of most of the organs deserves also to be regarded in this point of view, since it allows the continuance of the function, even when an organ, or a portion of it, is destroyed. The texture and external form of all the organs seem to harmonize with the final end of the organism, since most anomalies soon suspend its functions.

INTRODUCTION OF THE CELL THEORY

SCHWANN, Theodor (German zoologist and microscopist, 1810–1882). *Mikroskopische untersuchungen über die übereinsteimmung in der struktur und dem wachstum der tiere und pflanzen*, Berlin, 1839.

The next major step in the history of ideas concerning the organization of animal life was the publication of Schwann's general cell theory. For the essential passage, see p. 443.

ESTABLISHMENT OF THE UNICELLULAR CONDITION OF THE PROTOZOA

von SIEBOLD, Carl Theodor Ernst (German zoologist and parasitologist, 1808–1894). From *Lehrbuch der vergleichenden anatomie*, Berlin, 1848; tr. by W. I. Burnet as *Anatomy of the invertebrata*, Boston, 1854.

At the time this work appeared, there still existed an opposition of views on the nature of the infusoria. Were their complexities to be regarded as the organ systems of full-fledged organisms (see Ehrenberg), or as simple products existing by virtue of the inherent nature of the protoplasm, or sarcode, which they contained (Dujardin)? The establishment of the cell doctrine supplied the means of solving this problem, and Siebold at once postulated protozoa as unicellular organisms; his full answer to Ehrenberg is contained in the following.

CLASSIFICATION OF THE INVERTEBRATE ANIMALS.

The invertebrate animals are organized after various types, the limits of which are not always clearly defined. There is, therefore, a greater number of classes among them than among the vertebrates. But, as the details of their organization are yet but imperfectly known, they have not been satisfactorily classified in a natural manner.

There are among them many intermediate forms, which make it difficult to decide upon the exact limits of various groups.

The following division, however, from the lowest to the highest forms of organization, appears at present the best:

ANIMALIA EVERTEBRATA.

INVERTEBRATE ANIMALS.

Brain, spinal cord, and vertebral column, absent.

First Group.

Protozoa.

Animals in which the different systems of organs are not distinctly separated, and whose irregular form and simple organization is reducible to the type of a cell.

CLASS I.   INFUSORIA.
CLASS II.   RHIZOPODA. . . .

BOOK FIRST.

INFUSORIA AND RHIZOPODA.

CLASSIFICATION.

The Infusoria, using this word in a restricted sense, are far from being the highly-organized animals *Ehrenberg* has supposed. In the first place, on account of their more complicated structure, the Rotifera must be quite separated from them, as has already been done by *Wiegmann, Burmeister, R. Wagner, Milne Edwards, Rymer Jones,* and others. The same may be said of the so-called Polygastrica. In fact, a great number of the forms included under Closterina, Bacillaria, Volvocina, and others placed by *Ehrenberg* among the anenteric Polygastrica, belong, properly, to the vegetable kingdom. Indeed, this author has very arbitrarily taken for digestive, sexual, and nervous organs, the rigid vesicles, and the colored or colorless granular masses, which are met with in simple vegetable forms, but which are always absent in those low organisms of undoubtedly an animal nature. Cell-structure and free motion are the only two characteristics in common of the lowest animal and vegetable forms; and since *Schwann* has shown the uniformity of development and structure of animals and plants, it will not appear strange that the lowest conditions of each should resemble each other in their simple-cell nature. As to motion, the voluntary movements of Infusoria should be distinguished from those which are involuntary, of simple vegetable forms; a distinction not insisted upon until lately.. Thus, in watching carefully the motions of Vorticellina, Trachelina, Kolpodea, Oxytrichina, &c., one quickly perceives their voluntary character. The same is true of the power of contracting and expanding their bodies.

But in the motions of vegetable forms other conditions are perceived; and there is no appearance of volition in either change of place or form, their locomotion being accomplished either by means of cilia, or other physical causes not yet well understood. Cilia, therefore, belong to vegetable as well as to animal forms, and in this connection it is not a little remarkable that in animals they should be under the control of volition. With vegetable forms these organs are met with either in the shape of ciliated epithelium, as upon the spores of *Vaucheria,* or as long, waving filaments, as upon the earlier forms of many confervae, in which last can often be seen the so-called organization of *Ehrenberg's* Monadina and Volvocina. Until the fact that ciliated organs belong to both animals and vegetables was decided, the real place of many low organisms had to remain undetermined. However, notwithstanding their free motion from place to place by means of cilia, the vegetable nature of many organisms seemed clearly indicated by the rigid, non-contractile character of their forms. It is from a misapprehension of the true nature of these facts, that some modern naturalists have denied the existence of limits between the two kingdoms.

With Bacillareae and Diatomaceae, this question has another aspect. Many of these organisms have been taken for animals from their so-called voluntary movements, which truly entirely want the character of volition. In the movements of the rigid Diatomaceae, for instance, the whole plant has oscillatory motions like a magnetic needle, at the same time slightly changing its place forward and backward. When small floating particles come in contact with such an organism, they immediately assume the same motion. This may be well observed with the Oscillatoria. There are here, undoubtedly, no ciliary ·organs; in fact, they could not, if present, produce this kind of motion. According to *Ehrenberg,* the Naviculae can protrude ciliary locomotive organs through openings of their carapace; but this has not been observed by other naturalists.

The Rhizopoda, whose internal structure is as yet imperfectly known, are closely allied to the Infusoria. Like these last, their bodies are cellular, containing nuclear corpuscles, but no system of distinct organs. These two classes of Protozoa differ, however, in their external form, and the structure of their locomotive organs. The body of the Infusoria, notwithstanding its contractility, has a definite form, and moves chiefly by means of vibratile organs. That of the Rhizopoda, on the other hand, although equally contractile, has no definite form; their movements also are not due to ciliated organs, but to a change of the form of the body by various prolongations and digitations.

Owing to the present incomplete details upon the organization of these animals, little can here be said about them; and therefore, instead of devoting to them a separate chapter, it will be proper to treat of them with the Infusoria in general. . . .

CHAPTER I.

External Covering.

The Protozoa are surrounded by a very delicate cutaneous envelope, which is sometimes smooth, and sometimes covered with thickly-set cilia. Generally these cilia are arranged in longitudinal rows; but in *Actinophrys* they consist of long contractile filaments of a special nature.

CHAPTER II.

Muscular System and Locomotive Organs.

With the Protozoa a distinct muscular tissue cannot be made out, but the gelatinous substance of their body is throughout contractile.

It is only in the contractile peduncle of certain Vorticellina, that there can be perceived a distinct longitudinal muscle, which, assuming a spiral form, can contract suddenly like a spring.

The Vibratile Organs on the surface of Infusoria serve as organs of locomotion.

With many species they are found much developed at certain points, and are arranged in a remarkable order and manner.

With *Peridinium*, a crown of them encircles the body; with *Stylonychia*, they are quite long, and surround the flattened body like a fringe; while the Vorticellina have the anterior portion of their body surrounded by retractile cilia, arranged in a circular or spiral manner. In *Trichodina* there is, upon the ventral surface, besides a crown of these cilia upon the back, a very delicate ciliated membranous border, which is attached to a ring which is dentated, and composed of a compact homogeneous tissue. With *Trichodina pediculus* this border is whole and entire; but it is broken or ragged with *Trichodina mitra*.

By means of this organ these animals swim with facility, or invade with skill the arm-polyps and Planaria. With many Infusoria, the vibratile organs are situated at the anterior extremity of the body, as simple or double non-retractile filaments, which move in a manner to produce a vortical action of the water. But with others the locomotive organ is a long retractile proboscis. With the Oxytrichina and Euplota, there are fleshy movable points (uncini) upon the ventral surface, by which these animals move about as upon feet. During these movements with the Oxytrichina, the posterior portion of the body is supported by many setose and styloid processes, which point backward.

The singularly varied and branching locomotive organs of the Rhizopoda are short, and digitated with *Amoeba, Difflugia* and *Arcella.* But in the other genera they are elongated and filamentous.

CHAPTERS III. AND IV.

Nervous System and Organs of Sense.

Although the Infusoria clearly evince in their actions the existence of sensation and volition, and appear susceptible of sensitive impressions, yet no nervous tissue whatever has as yet been found in them. If *Ehrenberg* supposed the Polygastric Infusoria to possess a nervous system, he did so because, having decided that the red pigment points of these animals were eyes, he inferred that they necessarily had a nervous ganglion at their base.

With the naked Infusoria the sense of touch exists, undoubtedly, over the whole body. But beside this, it appears specially developed, in many species, in the long cilia forming vibratile circles, or in those movable foot-like and snout-like prolongations of the body. In the same manner, it is probable they have the sense of taste also; for they seem to exercise a choice in their food, although no gustatory organ has yet been found.

All species, whether they have red pigment points or not, seem affected by light. Without doubt, therefore, their vision consists simply in discriminating light from darkness, which is accomplished by the general surface of the body, and without the aid of a special optical organ.

The simple pigment point of many Infusoria, and which *Ehrenberg* has generally regarded as an eye, has no cornea, and contains no body capable of refracting light; there is, moreover, connected with it no nervous substance.

Ehrenberg attaches here too great an importance to the red color of the pigment, for the blue, violet and green pigments, seen in the eyes of insects and crustacea, show clearly that the red pigment is not essential to the eye.

CHAPTER V.

Digestive Apparatus.

The Infusoria are nourished, either by taking solid food into the interior of their body, or by absorbing by its entire surface nutritive fluids which occur in the media in which they live.

This last mode is illustrated in the Astoma, which have no distinct oral aperture or digestive apparatus. By the ingenious experiment first performed by *Gleichen,* of feeding these animals with colored liquids, no trace of these organs could be found.

Ehrenberg, who also had observed that they did not eat, regarded their internal vesicles as stomachal organs, which were in connection with the mouth by tubes. The correctness of this opinion, however, has not been verified. Indeed, the genus *Opalina* refutes it; here the species are quite large and visible to the naked eye, yet an oral aperture can be detected upon no part of their body, and never do they admit into its interior colored particles. Solid substances found in them cannot be regarded as food. That fluids are here introduced by surface-imbibition is shown by *Opalina ranarum;* this animal is found in bile in the rectum of frogs, and assumes a green color. When *Opalina* requiring only a certain quantity of liquid are placed in water, they quickly absorb it, become greatly swollen, and shortly after die. In such cases, the absorbed liquid is seen as clear, vesicular globules under the surface, and these globules have been taken by *Ehrenberg* as stomachal vesicles (VENTRICULI), and by *Dujardin* as VACUOLAE.

Those Infusoria which are nourished by solid food have a mouth at a certain place, and an oesophagus traversing the parenchyma of the body. Through this last the food is received, and is finally dissolved in the semi-liquid parenchyma of the body, without passing through stomachal or intestinal cavities. In many cases there is at the end of the body opposite the mouth an ANUS, through which the refuse material is expelled. But, when this is wanting, its function is often performed by the mouth. According to *Ehrenberg,* the *Infusoria polygastrica,* such as we have just been describing, differ from the *Infusoria rotatoria,* in having a great number of stomachs, which connect by hollow peduncles with the mouth in the division Anentera, and with the intestine in that of Enterodela. This organization, which, from its high authority, has generally been admitted by naturalists, is not, however, met with in any infusorium.

The vesicular cavities in the bodies of these animals, and which have been regarded by *Ehrenberg* as stomachal-pouches, never have a hollow peduncle, either connecting with the mouth (*Anentera*) or with the intestine (*Enterodela*). Indeed, it is doubtful if a digestive canal can be made out in these Infusoria.

The vesicular, irregular contracting cavities of their body contain a clear liquid, evidently the same as that in which they live, which, with the Astoma, has been absorbed through the surface of the body. But, with those having a mouth and oesophagus, it is received through them, and taken up by the yielding parenchyma of the body.

If the methods of feeding of *Gleichen* and *Ehrenberg* are employed, the colored particles are taken in by a vortical action of the water, caused by the cilia surrounding the mouth. This water, with its molecules, accumulates at the lower portion of the oesophagus, and so distends there the parenchyma as to cause the appearance of a vesicle. Thus situated, the whole has much the aspect of a pedunculated vesicle. But when, from contractions of the oesophagus; this water escapes into the parenchyma, it appears there as an unpedunculated globule, in which the colored particles still float. When the Stomatoda are full-fed in this manner, there appear many of these globules in various parts of the body; and thus substances previously ingested are taken up and disseminated throughout the body.

If the globules thus containing solid particles are closely aggregated, it sometimes happens that they fuse together; a fact which proves that they are not surrounded by a special membrane.

The solid particles of food of the Stomatoda, which are often the lower Algae, such as the Diatomaceae and Oscillatoria, and often other Infusoria, are sometimes deposited in the parenchyma without being surrounded by a vesicular liquid.

From observations made upon *Amoeba, Arcella* and *Difflugia,* it appears that the Rhizopoda ingest their food like the Stomatode Infusoria.

If the vesicular cavities containing the liquid and colorless food of the Stomatoda be examined under the microscope by a horizontal central incision, their contents appear colorless; but by changing the focus, viewing alternately the convex and concave surfaces of the vesicle, the points of junction between the colorless globules and the parenchyma appear colored pale-red. This appearance, due to an optical illusion, might easily deceive one into the opinion that the vesicles which are really colorless are colored.

From this it is probable that *Ehrenberg* has described *Bursaria vernalis* and *Trachelius meleagris* as having a red gastric juice.

The violet points which are found upon the back and neck of *Nassula elegans* and *Chilodon ornatus* are only collections of pigment granules, which, in the first case, are often absent, and in the second are often partially dissolved.

This last violet liquid has been regarded by *Ehrenberg* as a gastric juice resembling bile.

The solid particles of food, whether surrounded by the parenchyma or enclosed in a liquid vesicle, are moved hither and thither in the gelatinous tissue of the body, during the contracting and expanding movements of the animal. In some, the parenchyma with its contained food moves in a regularly circular manner, like the liquid contained in the articulated tubes of Chara. In *Loxodes bursaria* this circulation is remarkable, and of much physiological interest. Its cause is yet quite unknown, for in no case is it due to cilia, and it may be observed in individuals entirely at rest. *Ehrenberg,* therefore, is incorrect in regarding it as due solely to a contractile power of the parenchyma, displacing the molecules. Much less is his explanation satisfactory, since the digestive tube of an infusorium can be extended at the expense of its stomachal pouches, so as to fill the whole body, giving it the appearance of having a circulation of molecules throughout its entire extent.

The round or elongated oval mouth of Infusoria varies as to its position. Sometimes it is in front, sometimes behind; and in some cases, near the middle third of the body. Rarely naked, its borders are generally ciliated, and often its circumference is provided with a very remarkable ciliary apparatus. By the aid of this, these animals not only move about, but when quiet produce vortical actions of the water, which are felt at quite a distance; and all minute particles within its reach are quickly drawn towards its mouth, and then swallowed or rejected according to the option of the individual.

It is rare that this oral aperture is provided with a dental apparatus. The oral cavity, generally infundibuliform, extends into a longer or shorter, straight or curved oesophagus, which is lined throughout by a very delicate ciliated epithelium.

The anus, situated usually upon the dorsal surface of the posterior portion of the body, is sometimes, though rarely, indicated by a slight external projection.

CHAPTERS VI. AND VII.

CIRCULATORY AND RESPIRATORY SYSTEMS.

A vascular system entirely distinct by closed walls from the other organs is not found in the Protozoa. But with very many (with all the Stomatoda, without exception) there are contractile pulsatory cavities, the form, number and arrangement of which is quite varied.

They are situated in the denser and outer layers of the parenchyma of the body, and during the diastole they become swollen by a clear, transparent, colorless liquid, which, during the systole, entirely disappears.

These movements succeed each other at more or less regular intervals. When these cavities are numerous, a certain order in the succession and alternation of their contractions cannot always be observed. It is very probable that their liquid contained during the diastole is only the nutritive fluid of the parenchyma, and to which it returns during the systole. In this way it has a constant renewal, and all stagnation is prevented. This arrangement constitutes the *first appearance of a circulatory system,* and the *first attempt at a circulation of nutritive fluids.*

From an optical illusion similar to the one mentioned as belonging to the vacuolae the liquid of these pulsating cavities has a reddish hue.

A round, pulsating cavity is found in the genera *Vorticella, Epistylis, Loxodes,* and in the following species:—*Amoeba diffluens, Paramoecium kolpoda, Stylonychia mytilus, Euplotes patella, &c.* With *Actinophrys, Bursaria, Trichodina,* there are from one to two; with *Arcella vulgaris,* three to four; with *Nassula elegans,* there are four placed in a longitudinal line on the dorsal surface. With *Trachelius meleagris,* there is a series of eight to twelve upon the sides of the body, and with the various species of *Amphileptus* there are fifteen to sixteen arranged more or less regularly. With *Stentor,* there is a large cavity in the anterior portion of the body, and many similar cavities appear upon the sides, united sometimes into one long canal. A similar canal traverses the entire body of *Spirostomum ambiguum,* and *Opalina planariarum.* With *Paramoecium aurelia,* the two round cavities present a remarkable aspect, being surrounded by five or seven others, small and pyriform, the top of which being directed outward, the whole has a star-like appearance. During the pulsation, often the entire star disappears, sometimes only the two central cavities, and in some cases the rays only.

These cavities, entirely disappearing in the systole, reäppear in the diastole, and usually in the same place and with the same form and number. This would lead us to conclude that they are not simple excavations in parenchyma, but real vesicles or vessels, the walls of which are so excessively thin as to elude the highest microscopic power.

In some individuals, as, for instance, with *Trachelius lamella,* there appear, during the diastole, two or three small vesicles at the extremity of the body, which, after having increased in size, blend into one which is very large. These are probably only globules of nutritive fluid, separated from the parenchyma. Similar phenomena are observed in *Phialina vermicularis* and *Bursaria cordiformis.*

It sometimes happens with these animals that a forcible contraction of the whole body divides an elongated cavity into two spherical portions, as though it were a drop of oil. The observation of these phenomena would make it doubtful whether or not these cavities are true vesicles or vessels.

These cavities have been met with in only a few of the Astoma, and these are, *Cryptomonas ovata* and *Opalina planariarum.*

The Infusoria appear to respire solely by the skin. In those species whose bodies are covered with vibratile cilia this function is promoted by the vortical action of the water caused by these organs. In others, the contractile cavities just described are situated immediately under the skin, and the opinion may be entertained that the water so communicates with their liquid contents as to perform a respiratory function. In this respect *Actinophrys sol* is quite remarkable, for its contractile cavities are so superficial that when filled they raise the skin in the form of aqueous vesicles, which, however, are so elastic as entirely to disappear in the parenchyma. Here it is plain that a mutual relation between the external water and the contents of these cavities might easily take place.

CHAPTER VIII.

Organs of Secretion.

No special organ of secretion has been found in the Protozoa; their skin, however, has a power of secreting various materials, which in some species harden and form a carapace, or a head of a particular shape; while in others it serves to glue together foreign particles, forming a case, in which the animal retreats.

Among those having a carapace, may be mentioned *Vaginicola, Cothurnia,* and *Arcella.* This more or less hard envelope does not resist fire, and is probably of a corneous nature. In the Rhizopoda, however, it is usually calcareous, like the shells of Mollusca, and is not affected by heat. The *Difflugiae* carry about with them an envelope of this kind, composed of grains of sand.

CHAPTER IX.

ORGANS OF REPRODUCTION.

The Infusoria propagate by *fissuration* and *gemmation,* and never by eggs. They have therefore no proper sexual organs.

This fissuration occurs longitudinally with some, transversely with others, and in many of them by both at once. Gemmation, on the contrary, is very rare.

Nearly all the Infusoria and Rhizopoda have in their interior a nicely-defined body, a kind of a nucleus, which is quite different, in its compact texture, from the parenchyma by which it is surrounded. This nucleus, which, in different species, varies much in number and form, performs an essential part in the fissuration. For, every time the individual divides either longitudinally or transversely, this nucleus, which is usually situated in the middle, divides also. So that, in the end, each of the two new individuals has a nucleus. When an animal is about to undergo fissuration, there is generally first perceived a change in the nucleus. Thus, in *Paramoecium, Bursaria,* and *Chilodon,* the nucleus is sulcated longitudinally or transversely, or even entirely divided, before the surface of the body presents any constriction.

This nucleus, which is of a finely granular aspect and dense structure, retains perfectly its form when the animal is pressed between two plates of glass, and the other parts are spread out in various ways. By direct light its color appears pale yellow. It appears to lie very loosely in the parenchyma, and sometimes individuals may be observed turning their bodies around it as it rests motionless in the centre. From all this, it cannot be supposed that this nucleus attaches itself to other parts of the animal, and especially to the pulsatory cavities (*Vesiculae seminales* of *Ehrenberg*).

A simple, round, or oval nucleus is found in *Euglena, Actinophrys, Arcella, Amoeba, Bursaria, Paramoecium, Glaucoma, Nassula* and *Chilodon.* But there are two which are round, and placed one after the other in *Amphileptus anser* and *fasciola,* in *Trachelius meleagris,* and *Oxytricha pellionella.* With *Stylonychia mytilus,* there are four.

It is not rare that a variable number of these round nuclei, arranged in a row, traverse the body in a tortuous manner. This is so in *Stentor coeruleus* and *polymorphus,* in *Spirostomum ambiguum,* and in *Trachelius moniliger.* In many instances the nucleus has the form of an elongated band, which is slightly curved in *Vorticella convallaria, Epistylis leucoa, Prorodon niveus* and *Bursaria truncatella.* In *Stentor Roeselii,* it is spiral, and in *Euplotes patella* and *Trichodina mitra,* it is shaped like a horse-shoe. In *Loxodes bursaria,* it is kidney-form, and encloses in one of its extremities a small corpuscle (nucleolus).

The round nucleus of *Euglena viridis* has in its centre a transparent dot. In *Chilodon cucullulus,* the nucleolus has a similar dot, and thus the nucleus as a whole resembles a cell.

These nuclei, which make Infusoria resemble cells, deserve a special attention, since they do not die with the animal. Thus the nucleus of *Euglena viridis,* which, according to *Ehrenberg,* is globular when dying, and surrounded by a kind of cyst, remains unchanged a long time, or even increases in size, having no appearance of a dead body. It may be that the life of this animal, under these circumstances, is not finished, but only assumes another form.

CELLULAR BASIS OF HISTOLOGY

von KÖLLIKER, Rudolf Albert (Swiss histologist, 1817–1905). From *Handbuch der gewebelehre des menschen,* Leipzig, 1852; tr. by G. Busk and T. Huxley as *Manual of human histology,* London (for the Sydenham Society), 1853.

Kölliker's two major contributions were the placing of histology on a cellular basis and its application to embryology. This omits, of course, numerous valuable cytological contributions, among them the demonstration that sperm are cells. We have chosen for inclusion here Kölliker's explanation of the point of view from which histology may be considered a science and of the relations of this science to the cell theory of Schwann.

In characterising the present position of Histology and of its *objects,* we must by no means forget that, properly speaking, it considers only one of the three aspects which the elementary parts present to observation, namely, their form. Microscopical anatomy is concerned with the understanding of the microscopic forms, and with the laws of their structure and development, not with any general doctrine of the elementary parts. Composition and function are only involved, so far as they relate to the origin of forms and to their variety. Whatever else respecting the activity of the perfect elements and their chemical relations is to be found in Histology, is there either on practical grounds in order to give some useful application of the morphological conditions, or to complete them; or from its intimate alliance with the subject, it is added only because physiology proper does not afford a due place for the functions of the elementary parts.

If Histology is to attain the rank of a science, its first need is to have as broad and certain an objective basis as possible. To this end, the minuter structural characters of animal organisms are to be examined on all sides and not only in fully-formed structures, but in all the earlier period from their first development. When the morphological elements have been perfectly made out, the next object is to discover the laws according to which they arise, wherein one must not fail to have regard also to their relations of composition and function. In discovering these laws, here as in the experimental sciences generally, continual observation separates more and more, among the collective mass of scattered facts and observations, the occasional from the constant, the accidental from the essential, till at last a series of more and more general expressions of the facts arises,—from which, in the end, mathematical expressions or formulae proceed, and thus the laws are enunciated.

If we inquire how far Histology has satisfied these requirements, and what are its prospects in the immediate future, the answer must be a modest one. *Not only does it not possess a single law,* but the materials at hand from which such should be deduced, are as yet relatively so scanty, that not even any considerable number of general propositions appear well founded. Not to speak of a complete knowledge of the minuter structure of animals in general, we are not acquainted with the structure of a single creature throughout, not even of man, although he has been so frequently the object of investigation,— and therefore it has hitherto been impossible to bring the science essentially any nearer its goal. It would, however, be unjust to overlook and depreciate what we do possess; and it may at any rate be said, that we have acquired a rich store of facts and a few more trustworthy general propositions. To indicate only the more important of the former, it may be mentioned, that we have a very sufficient acquaintance with the perfect elementary parts of the higher animals and that we also understand their development, with the exception of the elastic tissue, and of the elements of the teeth and bones. The mode in which these are united into organs has been less examined, yet on this head also, much has been added of late, especially in man, whose individual organs with the exception of the nervous system, the higher organs of sense and a few glands (the liver, blood-vascular glands), have been almost exhaustively investigated. If the like progress continue to be made, the structure of the human body will in a few years be so clearly made out, that, except perhaps in the nervous system, nothing more of importance will remain to be done with our present modes of investigation. With comparative Histology it is otherwise; hardly commenced, not years but decades will be needed to carry out the necessary investigations. Whoever will do good work in this field must, by monographs of typical forms embracing their whole structure from the earliest periods of development, obtain a general view of all the divisions of the animal kingdom, and then, by the methods above described, strive to develop their laws.

As regards the general propositions of Histology, the science has made no important progress since Schwann, however much has been attained by the confirmation of the broad outlines of his doctrines. The position that all the higher animals at one time consist wholly of cells and develop from these their higher elementary parts, stands firm, though it must not be understood as if cells, or their derivatives, were the sole possible or existing elements of animals. In the same way, Schwann's conception of the genesis of cells, though considerably modified and extended, has not been essentially changed, since the cell nucleus still remains as the principal factor of cell development and of cell multiplication. Least advance has been made in the laws which regulate the origin of cells and of the higher elements, and our acquaintance with the elementary processes which take place during the formation of organs must be regarded as very slight. Yet the right track in clearing up these points has been entered upon; and a logical investigation of *the chemical relations of the elementary parts and of their molecular forces,* after the manner of Donders, Dubois, Ludwig, and others, combined with a more *profound microscopical examination* of them, such as has already taken place with regard to the muscles and nerves,—further, a histological treatment of embryology, such as has been attempted by Reichert, Vogt, and myself, will assuredly raise the veil, and bring us, step by step, nearer to the desired though perhaps never to be reached, end.

II. THE ACTIVITIES OF THE ANIMAL ORGANISM

Contributions to physiology

EARLY GROPINGS TOWARD AN IDEA OF METABOLISM

von HOHENHEIM, Aureolus Philippus Theophrastus Bombastus (Paracelsus) (Swiss physician, 1493–1541) ;* tr. by A. E. Waite from *The Hermetical and Alchemical Writings of Aureolus Philippus Theophrastus Bombast, of Hohenheim, called Paracelsus, the Great, etc.*, London, 1894; by permission of The de Laurence Company, Inc.

Paracelsus, the famous iconoclast of early sixteenth-century medicine, appears to gain occasional glimpses of fundamental physiological truth, usually through a haze of alchemical and astrological absurdities. The following excerpts show that he possessed the germ of a notion of metabolism and of the cycle of elements. It is said that before certain of his lectures he expressed his antitraditionalism by burning the works of Averroes and Galen; this symbolized not an awakening of empiricism in his point of view but rather his desire to return to scientific fantasies of the pre-Socratic era and, worse, of the Dark Ages.

Concerning the Three Prime Essences

Everything which is generated and produced of its elements is divided into three, namely, into Salt, Sulphur, Mercury. Out of these a conjunction takes place, which constitutes one body and an united essence. This does not concern the body in its outward aspect, but only the internal nature of the body.

Its operation is threefold. One of these is the operation of Salt. This works by purging, cleansing, balsaming, and by other ways, and rules over that which goes off in putrefaction. The second is the operation of Sulphur. Now, sulphur either governs the excess which arises from the two others, or it is dissolved. The third is of Mercury, and it removes that which changes into consumption. Learn the form which is peculiar to these three. One is liquor, and this is the form of mercury; one is oiliness, which is the form of sulphur; one is alcali, and this is from salt. Mercury is without sulphur and salt; sulphur is devoid of salt and mercury; salt is without mercury or sulphur. In this manner each persists in its own potency.

But concerning the operations which are observed to take place in complicated maladies, notice that the separation of things is not perfect, but two are conjoined in one, as in dropsy and other similar complaints. For those are mixed diseases which transcend their sap and tempered moisture. Thus, mercury and sulphur sometimes remove paralysis, because the bodily sulphur unites therewith, or because there is some lesion in the immediate neighbourhood.

---

* The questions as to when, where, and in what form the writings of Paracelsus first appeared are not as yet perfectly solved; see J. Ferguson, *Bibliographica Paracelsica*, Glasgow, 1877.

Observe, consequently, that every disease may exist in a double or triple form. This is the mixture, or complication, of disease. Hence the physician must consider, if he deals with a given simple, what is its grade in liquor, in oil, in salt, and how along with the disease it reaches the borders of the lesion. According to the grade, so must the liquor, salt, and sulphur be extracted and administered, as is required. The following short rule must be observed: Give one medicine to the lesion, another to the disease.

Salts purify, but after various manners, some by secession, and of these there are two kinds—one the salt of the thing, which digests things till they separate—the other the salt of Nature, which expels. Thus, without salt, no excretion can take place. Hence it follows that the salt of the vulgar assists the salts of Nature. Certain salts purge by means of vomiting. Salts of this kind are exceedingly gross, and, if they do not pass off in digestion, will produce strangulation in the stomach. Some salts purge by means of perspiration. Such is that most subtle salt which unites with the blood. Now, salts which produce evacuation and vomiting do not unite with the blood, and, consequently, produce no perspiration. Then it is the salt only which separates. Other salts purge through the urine, and urine itself is nothing but a superfluous salt, even as dung is superfluous sulphur. No liquor superfluously departs from the body, for the same remains within. Such are all the evacuations of the body, moisture expelled by salt through the nostrils, the ears, the eyes, and other ways. This is understood to take place by means of the Archeus from these evacuations. Now, as out of the Archeus a laxative salt comes forth, of which one kind purges the stomach because it proceeds from the stomach of the Archeus, so another purges the spleen because it comes from the spleen of the Archeus; and it is in like manner with the brain, the liver, the lungs, and other members, every member of the Archeus acting upon the corresponding member of the Microcosmus.

The species of salt are various. One is sweet as cassia, and this is a separated salt which is called antimony among minerals. Another is like vinegar, as sal gemmae; yet another is acid, as ginger. Another is bitter, as in rhubarb or colocynth. So, also, with alkali; there is some that is generated, as harmel; some extracted, as scammony; some coagulated, as absinth. In the same way, certain salts purge by perspiration alone, certain others by consuming alone, and so on. Wherever there is a peculiar savour, there is also a peculiar operation and expulsion. The operation is of two kinds—that which belongs to the thing and the extinct operation.

Concerning the Mass and the Matter out of Which Man Was Made.

It follows next in order to consider how it comes about that external causes are so powerful in man.

It must be realised, first of all, that God created all things in heaven and on earth—day and night, all elements, and all animals. When all these were created, God then made man. And here, on the subject of creation, two remarks have to be made. First, all things were made of nothing, by a word only, save man alone. God made man out of *something*, that is to say, from a mass, which was a body, a substance—a *something*. What it was—this mass —we will briefly enquire.

God took the body out of which He built up man from those things which He created from nothingness into something. That mass was the extract of all creatures in heaven and earth, just as if one should extract the soul or spirit, and should take that spirit or that body. For example, man consists of flesh and blood, and besides that of a soul, which is the man, much more subtle than the former. In this manner, from all creatures, all elements, all stars in heaven and earth, all properties, essences, and natures, that was extracted which was most subtle and most excellent in all, and this was united into one mass. From this mass man was afterwards made. Hence man is now a microcosm, or a little world, because he is an extract from all the stars and planets of the whole firmament, from the earth and the elements; and so he is their quintessence. The four elements are the universal world, and from these man is constituted. In number, therefore, he is fifth, that is, the fifth or quintessence, beyond the four elements out of which he has been extracted as a nucleus. But between the macrocosm and the microcosm this difference occurs, that the form, image, species, and substance of man are diverse therefrom. In man the earth is flesh, the water is blood, fire is the heat thereof, and air is the balsam. These properties have not been changed, but only the substance of the body. So man is man, not a world, yet made from the world, made in the likeness, not of the world, but of God. Yet man comprises in himself all the qualities of the world. Whence the Scripture rightly says we are dust and ashes, and into ashes we shall return; that is, although man, indeed, is made in the image of God, and has flesh and blood, and is not like the world, but more than the world, still, nevertheless, he is earth and dust and ashes. And he should lay this well to heart lest from his figure he should suffer himself to be led astray; but he should think what he has been, what he now is, and what hereafter he shall be.

Attend, therefore, to these examples. Since man is nothing else than what he was, and out of which he was made, let him not, even in imagination, be led astray. The knowledge of the fact tends to force upon him the confession that he is nothing but a mass drawn forth from the great universe. This being the case, he must know that he cannot be sustained and nourished therefrom. His body is from the world, and therefore must be fed and nourished by that world from which he has sprung. So it is that his food and his drink and all his aliment grow from the ground. The great universe contributes less to his food and nourishment. If man were not from the great world but from heaven, then he would take celestial bread from heaven along with the angels.

He has been taken from the earth and from the elements, and therefore must be nourished by these. Without the great world he could not live, but would be dead, and so he is like the dust and ashes of the great world. It is settled, then, that man is sustained from the four elements, and that he takes from the earth his food, from the water his drink, from the fire his heat, and from the air his breath. But these all make for the sustentation of the body only, of the flesh and the blood.

Now, man is not only flesh and blood, but there is within him the intellect which does not, like the complexion, come from the elements, but from the stars. And the condition of the stars is this, that all the wisdom, intelligence, industry of the animal, and all the arts peculiar to man are contained in them. From the stars man has these same things, and that is called the light of Nature; in fact, it is whatever man has found by the light of Nature. Let us illustrate our position by an example. The body of man takes its food from the earth, to which food it is destined by its conception and natural agreement. This is the reason why one person likes one kind of food, and another likes another, each deriving his pleasure from the earth. Animals do the same, hunting out the food and drink for their bodies which has been implanted in the earth. Now as there is in man a special faculty for sustaining his body, that is, his flesh and blood, so is it with his intellect. He ought equally to sustain that with its own familiar food and drink, though not from the elements, since the senses are not corporal but are of the spirit as the stars are of the spirit. He then attracts by the spirit of his star, in whom that spirit is conceived and born. For the spirit in man is nourished just as much as the body. This special feature was engrafted on man at his creation, that although he shares the divine image, still he is not nourished by divine food, but by elemental. He is divided into two parts; into an elemental body, that is, into flesh and blood, whence that body must be nourished; and into spirit, whence he is compelled to sustain his spirit from the spirit of the star. Man himself is dust and ashes of the earth. Such, then, is the condition of man, that, out of the great universe he needs both elements and stars, seeing that he himself is constituted in that way.

And now we must speak of the conception of man, how he is begotten and made. The first man was made from the mass, extracted from the machinery of the whole universe. Then there was built up from him a woman, who corresponds to him in his likeness to the universe. For the future, there proceeds from the man and the woman the generation of all children, of all men. Moreover, the hand of God made the first man after God's own image in a wonderful manner, but still composed of flesh and blood, that he may be very man. Afterwards the first man and his wife were subjected to Nature, and so far separated from the hand of God that man was no longer built up miraculously by God's hand, but by Nature. The generation of man, therefore, has

been entrusted to Nature and conferred on one mass from which he had proceeded. That mass in Nature is called semen. Most certain it is, however,
that a man and woman only cannot beget a man, but along with those two,
the elements also and the spirit of the stars. These four make up the man.
The semen is not in the man, save in so far as it enters into him elementarily.
When, in the act of conception, the elements do not operate, no body is begotten. Where the star does not operate, no spirit is produced. Whatever is
produced without the elements and the spirit of the stars is a monster, a mola,
an abortion contrary to Nature. As God took the mass and infused life into it,
so must the composition perpetually proceed from those four and from God, in
whose hand all things are placed. The body and the spirit must be there.
These two constituents make up the man—the human being, that is, the man
with the woman, and the semen, which comes from without, and is, as it were,
an aliment, something which the man has not within himself, but attracts from
without, just as though it were a potion. Such as the principle of food and
drink is, such is also that of the sperm, which the elements from without contribute to the body as a mass. The star, by means of its spirit, confers the
senses. The father and mother are the instruments of the externals by which
these are perfected. In order to make this intelligible, I will adduce an example: In the earth nothing grows unless the higher stars contribute their powers.
What are these powers? They are such that one cannot exist without the
other, but of necessity one must act in conjunction with the other. As those
without are, such are those within, so far as man is concerned. Hence it is inferred that the first man was miraculously made, and so existed as the work of
God. After that, man was subjugated to Nature, so that he should beget children in connection with her. Now, Nature means the external world in the
elements and in the stars. Now it is evident from this that those elements have
their prescribed course and mode of operation, just as the stars, too, have their
daily course. They proceed in their daily agreement, and at particular epochs
Nature puts forth new ones. Now, if this form of operation—if the father
and mother—with this concordance meet together for the work of conception, then the foetus is allotted the Nature of those from whom it is born,
namely, of the four parents—the father, the mother, the elements, the stars.
From the father and mother proceed a like image and essence of flesh and
blood. Besides this, from their imagination, which is the human star, there is
allotted the intellect, in proportion wherein the concordance and constellation
have exhibited themselves. So, too, from the elements there is allotted the
complexion and the quality of the nature. So, too, from the external stars
their intelligence. As these meet, the influence which is stronger than the
others, preponderates in the foetus, or else there is a mutual commingling of
all. Thus man becomes a microscosm. The father and mother are made
from the universe, and the universe is constantly contributing to the generation

of man. In this way, there is constituted a single body, but a double nature, a single spirit, but a twofold sense. At length the body returns to its primal body, and the senses to the primal sense. They die, pass away, and depart, never to return. The ashes cannot again be made wood, neither can man from that state in which he is ashes be brought back so as to be man again.

Now we have traced the generation of man to this point as a general and universal probation of the whole of astronomy, in order that it might be understood from thence why the astronomer studies and gets to know men by the stars, namely, because man is from the stars. As every son is known by his father, so is it here; and this science is very useful if a man knows who is from heaven, from the elements, from father and mother. The knowledge of the father and mother lies at the root. The knowledge of the elements pertains to medicine. The knowledge of the stars is astrological. There are many reasons why these cognitions are useful and good. Many men are mere brutes, and yet make themselves out angels. Many speak from their mother, calling themselves Samuels or Maccabees. Many in their earthly complexion fast and pray, and call themselves divine. Many handle those things which are not really what they are said to be. Anyone who is an astrologer knows what that spirit is which speaks and is seen. It is matter for regret that many hesitate between the two lights, culling and stealing from each in order to make themselves conspicuous. The spirits are known, indeed, to each, but in a different way, and this should not be so. But though things are thus, man is the work of God, but one only is His very son, that is, Adam. Others are sons of Nature, as Luke in his genealogy recounts of Joseph, that he was the Son of Helus, which Helus was the son of Mathat, which Mathat was the son of Levi, and so on back to Adam; yet there is no mention of the son of God. Thus man is a son in Nature, and does not desert his race, but follows the nature of his parents, the stars. Now, he who knows the father and mother of the stars and of the elements, and also the father and mother of the flesh and blood, he is in a position to discuss concerning that offspring, concerning its nature, essence, properties, in a word, concerning its whole condition. And as a physician compounds all simples into one, preparing a single remedy out of all, which cannot be made up without these numerous ingredients, so God performs His much more notable miracle by concocting man into one compound of all the elements and stars, so that man becomes heaven, firmament, elements, in a word, the nature of the whole universe, shut up and concealed in a slender body. And though God could have made man out of nothing by His one word "Fiat," He was pleased rather to build man up in Nature and to subject him to Nature as its son, but still so that He also subjected Nature to man, though still Nature was man's father. Hence it results that the astronomer knows man's conception by man's parentage. This is the reason why man can be healed by Nature through the agency of a physician, just as a father helps a son who has fallen into a pit. In this way Nature is subjected to man as to its own flesh and blood, its own son, its own fruit produced from itself; in the body of the elements wherein diseases exist; in the body of the spirit, where flourish the intelligence and reason; and the elements, indeed, by means of medicine, but the stars by their own knowledge and wisdom. Now, this wisdom in the sight of God is nothing; but the Divine wisdom is preeminent above all. So the names of wisdom differ. That wisdom which comes from Nature is called animal, because it is mortal. That which comes from God is named eternal, because it is free from mortality.

CIRCULATION POSTULATED

SERVETUS (Servet, Serveto), Miguel (Michael) (Spanish, later French, mystic and anatomist, 1511–1553). From *Christianismi restitutio*, Vienne, 1553; tr. by R. Willis in his *Servetus and Calvin*, London, 1877.

The circulation of the blood may be asserted to have been *postulated* by Servetus, *proved* by Harvey, and *seen* by Malpighi. Servetus was primarily a religious philosopher whose spiritual independence, especially his opposition to Calvinism, led to his death at the stake at the age of forty-two. The *Christianismi restitutio*, in which the following remarks are introduced incidentally to a transcendentalist speculation on the Holy Spirit, was published anonymously and secretly, and was later burned. It has been called* one of the rarest books in the world.

There is commonly said to be a threefold spirit in the body of man, derived from the substance of the three superior elements—a natural, a vital, and an animal spirit; there are, however, not really three, but only two distinct spirits. One of these, the first, characterised as *natural*, is communicated from the arteries to the veins by their anastomoses, and is primarily associated with the blood, the proper seat or home of which is the liver and veins. The second is the *vital* spirit, whose seat or dwelling-place is the heart and arteries. The third, the *animal* spirit, comparable to a ray of light, has its home in the brain and nerves. In each and all of these is the force—*energeia*—of the one spirit and light of God comprised. Now, that the natural spirit is imparted from the heart to the liver, and not from the liver to the heart, is proclaimed by the formation of man in the womb; for we see an artery associate with a vein sent from the mother through the navel of the foetus; and in the adult body we always find an artery and a vein conjoined. But it was truly into the heart of Adam that God breathed the breath of life or the soul. From the heart, therefore, it is that life is communicated to the liver; · for by the breathing into the mouth and nostrils it was that the soul was first truly imparted, the breath tending directly to the heart.

---

* By the translator; see also p. x in the editor's introduction to Harvey's *Motion of the heart and blood in animals*, tr. by R. Willis, New York, 1906, E. P. Dutton & Co.

The heart is the first organ that lives, and, situate in the middle of the body, is the source of its heat. From the liver the heart receives the liquor, the material as it were of life, and in turn gives life to the source of the supply. The material of life is therefore derived from the liver; but, elaborated as you shall hear, by a most admirable process, it comes to pass that the life itself is in the blood—yea that the blood is the life, as God himself declares (Genes. ix.; Levit. xvii.; Deut. xii.).

Rightly to understand the question here, the first thing to be considered is the substantial generation of the vital spirit—a compound of the inspired air with the most subtle portion of the blood. The vital spirit has, therefore, its source in the left ventricle of the heart, the lungs aiding most essentially in its production. It is a fine attenuated spirit, elaborated by the power of heat, of a crimson colour and fiery potency—the lucid vapour as it were of the blood, substantially composed of water, air, and fire; for it is engendered, as said, by the mingling of the inspired air with the more subtle portion of the blood which the right ventricle of the heart communicates to the left. This communication, however, does not take place through the septum, partition or midwall of the heart, as commonly believed, but by another admirable contrivance, the blood being transmitted from the pulmonary artery to the pulmonary vein, by a lengthened passage through the lungs, in the course of which it is elaborated and becomes of a crimson colour. Mingled with the inspired air in this passage, and freed from fuliginous vapours by the act of expiration, the mixture being now complete in every respect, and the blood become fit dwelling-place of the vital spirit, it is finally attracted by the diastole, and reaches the left ventricle of the heart.

Now that the communication and elaboration take place in the lungs in the manner described, we are assured by the conjunctions and communications of the pulmonary artery with the pulmonary vein. The great size of the pulmonary artery seems of itself to declare how the matter stands; for this vessel would neither have been of such a size as it is, nor would such a force of the purest blood have been sent through it to the lungs for their nutrition only; neither would the heart have supplied the lungs in such fashion, seeing as we do that the lungs in the foetus are nourished from another source—those membranes or valves of the heart not coming into play until the hour of birth, as Galen teaches. The blood must consequently be poured in such large measure at the moment of birth from the heart to the lungs for another purpose than the nourishment of these organs. Moreover, it is not simply air, but air mingled with blood that is returned from the lungs to the heart by the pulmonary vein.

It is in the lungs, consequently, that the mixture [of the inspired air with the blood] takes place, and it is in the lungs also, not in the heart, that the crimson colour of the blood is acquired. There is not indeed capacity or room enough in the left ventricle of the heart for so great and important an elaboration, neither does it seem competent to produce the crimson colour. To conclude, the septum or middle partition of the heart, seeing that it is without vessels and special properties, is not fitted to permit and accomplish the communication and elaboration in question, although it may be that some transudation takes place through it. It is by a mechanism similar to that by which the transfusion from the *vena portae* to the *vena cava* takes place in the liver, in respect of the blood, that the transfusion from the pulmonary artery to the pulmonary vein takes place in the lungs, in respect of the spirit.

The vital spirit (elaborated in the manner described) is at length transfused from the left ventricle of the heart to the arteries of the body at large, and in such a way that the more attenuated portion tends upwards, and undergoes further elaboration in the retiform plexus of vessels situated at the base of the brain, in which the *vital* begins to be changed into the *animal* spirit, reaching as it now does the proper seat of the rational soul. Here, still further sublimated and elaborated by the igneous power of the soul, the blood is distributed to those extremely minute vessels or capillary arteries composing the choroid plexus, which contain or are the seat of the soul itself. The arterial plexus penetrates even the most intimate part of the brain, its constituent vessels, interwoven in highly complex fashion, being distributed over the ventricles, and sent to the origins of the nerves which subserve the faculties of sensation and motion. Most wonderfully and delicately interwoven, these vessels, although spoken of as arteries, are really the terminations of arteries proceeding to the origins of nerves in the meninges. They are in truth a new kind of vessels; for, as in the transfusion from arteries to veins within the lungs we find a new kind of vessels proceeding from the arteries and veins, so, in the transfusion from arteries to nerves, is there a new kind of vessels produced from the arterial coats and the cerebral meninges.

VALVES DISCOVERED IN THE VEINS

FABRICIUS (Fabrizio, Fabrizzi) ab Aquapendente, Hieronymus (Girolamo) (Italian anatomist, 153?–1619). *De venarum ostiolis*, Padua, 1603; tr. by K. J. Franklin as *De venarum ostiolis 1603 of H. F. of A.*, Springfield, Ill., 1933; by courtesy of Charles C Thomas, publisher.

Of all the evidence adduced by Harvey for his doctrine of the circulation of the blood, none is more frequently emphasized by him than that of the *unidirectionality of its flow:* through the heart, through the lung, through the arteries and veins. It was his master, Fabricius, who *saw* the valves in the veins (see below) but it remained for Harvey himself to see them in their proper role, *i.e.*, as permitting the blood to flow in only one direction. Fabricius was last in line of the great Paduan anatomists (Vesalius, Colombus, Fallopius). For an embryological contribution by him, see p. 343.

Valves of veins is the name I give to some extremely delicate little membranes in the lumen of veins. They occur at intervals, singly or in pairs, especially in the limb veins. They open upwards in the direction of the main venous trunk, and are closed below, while, viewed from the outside, they resemble the swellings in the stem and small branches of plants.

My theory is that Nature has formed them to delay the blood to some extent, and to prevent the whole mass of it flooding into the feet, or hands and

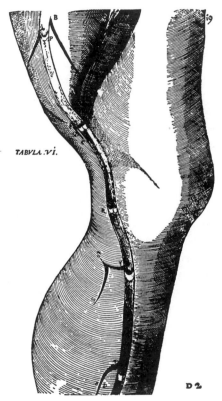

TABVLA.VI.

FIG. 5. *From text accompanying original figure:* "Explanation of Plate VI on Valves of Veins. Plate VI is continuous at B.C. with the preceding one, and shows a single vein only, B.P.Q.R.S.T.D. This vein is part of the smaller branch and comes from the bifurcation of the vein described in Plate V to constitute the Saphena. In it occurs at once a single valve P., which is placed fourth, at a distance of four fingers, among the valves described in the preceding Plate. Then along from this valve P. are seen twin-valves Q., having in front of them, a.β., a small obliquely coursing branch. After this, at an interval of four fingers and almost under the ham of the knee, twin-valves R. are again seen, but without any branch. Below these, at an interval of three fingers, is disclosed a single valve S., but it has in front of it a small oblique branch γ. δ. ε. And four fingers again from this valve S. is present another valve T., which also has in front of it a small branch ζ. η. Finally, at a distance of about five fingers, and in the following figure, continuous with the present plate at D, you will see a valve present, if you turn over the page."

fingers, and collecting there. Two evils are thus avoided, namely, under-nutrition of the upper parts of the limbs, and a permanently swollen condition of the hands and feet. Valves were made, therefore, to ensure a really fair general distribution of the blood for the nutrition of the various parts.

A discussion of these valves must be preceded by an expression of wonder at the way in which they have hitherto escaped the notice of Anatomists, both of our own and of earlier generations; so much so that not only have they never been mentioned, but no one even set eye on them till 1574, when to my great delight I saw them in the course of my dissection. And this despite the fact that anatomy has claimed many distinguished men among its followers, men, moreover, whose research was conducted with great care and attention to detail. But a certain amount of justification does exist for them in this case, for who would ever have thought that membranous valves could be found in the lumen of veins, especially as this lumen, designed for the passage of blood to the whole body, should be free for the free flow of the blood: just as in the case of the arteries, which are valveless, yet, in so far as they are channels for blood, are on the same footing as veins?

But a further justification can be advanced for the anatomists. All veins are not provided with valves. The vena cava, when it traverses the trunk of the body, the internal jugulars, and countless small superficial veins in like manner, are destitute of them. On the other hand, a reasonable charge may be made against the earlier workers. Either they neglected to investigate the function of the valves, a matter, one would think, of primary importance, or else they failed to see them in their actual demonstration of veins. For in the bare veins exposed to view, but still uninjured, the valves in a manner display themselves. Nay more, when assistants pass a ligature round the limbs preparatory to blood-letting, valves are quite obviously noticeable in the arms and legs of the living subject. And, indeed, at intervals along the course of the veins certain knotty swellings are visible from the outside; these are caused by the valves. In some people, in fact, such as porters and peasants, they appear to swell up like varices: but here I must correct myself. It must be clearly stated that actual varices are due entirely to the dilatation of the valves and veins by too long retention and thickening of the blood at the valves; since in the absence of valves the veins would be expected to swell up and dilate uniformly throughout their length, differing thus from varices. So that hereby another and that no mean function of the valves *may* come to light, namely, a strengthening action on the veins themselves. For as in cases of varix, with valvular incompetence or rupture as an expected finding, one always sees a greater or lesser degree of venous dilatation, one can doubtless say with safety that the Supreme Artificer made valves to prevent venous distension. Venous distension and dilatation would, moreover, have occurred readily since their coat is of membranous structure, single and delicate. And if they were to dilate, not only would the excessive accumulation of blood in them cause damage to themselves and the surrounding parts, and a swelling be caused, as is known to occur in cases of limb varix. There would also be a more or less defective nutrition of the parts above with the blood rushing in force, say, to a site of venous dilatation, and collected, as it were, in a pool. Arteries, on the other hand, had no need of valves, either to prevent distension—the thickness and strength of their coat suffices—or to delay the blood—an ebb and flow of blood goes on continuously within them. But let us, now, consider the number, shape, structure, site, distance, and other characteristics of valves. It was certainly necessary to make valves in the limb veins either of large or medium calibre—not the small ones—in order, no doubt, to slow the blood flow everywhere to an extent compatible with sufficient time being given for each small part to make use of the nourishment provided. Otherwise the whole mass of blood, owing to the slope of the limbs, would flood into their extremities, and collect there, causing a swelling of these lower parts, and wasting of the parts above. That the blood flow is slowed by the valves, evident even without this from their actual construction, can be tested by anyone either in the exposed veins of the cadaver, or in the living subject if he passes a ligature round the limbs as in blood-letting. For if one tries to exert pressure on the blood, or to push it along by rubbing from above downwards, one will clearly see it held up and delayed by the valves. This indeed was the way in which I was led to an observation of such nature. Small veins, however, had no need of valves, for two reasons. First, owing to their smallness, they held only a little blood and all that suffices for them: and secondly, it was sufficient for the nutriment to delay in the larger vessels as in a fountain-head, since by this means the small tributaries also would not lack what was necessary.

In the limbs, on the other hand, there was some need of valves. The legs and arms are very often engaged in local movement; this movement is at times vigorous and extremely powerful, and in consequence there is a very large and vigorous output of heat in them. There is no doubt that with this output of heat the blood would have flowed to the limbs and been drawn to them in such an amount that one of two things would have happened. The principal organs would have been robbed of their nutriment from the vena cava, or the limb vessels would have been in danger of rupture. Either alternative would have been fraught with very serious ill for the animal as a whole, since the principal organs, such as the liver, heart, lungs and brain, had constant need of a very plenteous blood-supply. It was for this reason, I imagine, that the vena cava, in its passage through the trunk, and likewise the jugulars, were made completely valveless. For the brain, heart, lungs, liver and kidneys, which are concerned with the welfare of the body as a whole, needed to be well-supplied with nutriment, and absence of even the briefest delay was essential, if lost substance was to be restored, and vital and animal spirits, by the agency of which animals continue to live, were to be generated. If, however, by chance you see valves at the beginning of the jugular veins in man, you may say they have been placed there to stop the blood rushing in spate to the brain, and collecting therein in undue amount, in the downward position of the head. For the reasons enumerated, then, valves were given to the medium and large sized veins though not to the small ones, in the limbs, and yet were not given to the trunk of the vena cava or to the jugulars. Though indeed valves have been put in very many places, where, for instance, smaller branches leave the main stem to continue in other directions, and this is a mark of rare wisdom, the object being, I imagine, that the blood may be delayed at that point where it needs distributing to other parts; whereas, without such an arrangement, it would doubtless have flowed in mass through the single wider and straighter venous channel. Valves are present, so to speak, as intelligent doorkeepers of the many parts, to prevent escape of the nutriment downwards, until the parts above have acquired their fitting share of it.

The shape of valves is such that they resemble the nail of the index or other three fingers. They open upwards in the direction of the main stem of the veins, while below and at the sides they are united to the vein-wall. Further, if one has seen the valves in the heart, or those of the great artery or arterial vein, one will have got a fairly accurate picture of the shape of all these venous valves, except that the cardiac ones are three-fold, while the venous ones occur singly or in pairs in their respective situations. In like manner the cardiac valves were made .very thick to meet the demands of diastole and systole, while the venous ones are extremely delicate. The valve membrane is thus delicate because it was unfitting for the blood's place to be largely occupied by membranes, and this would undoubtedly have occurred, had the valves been made too thick in this respect. But also deserving of comment is the fact that this extreme delicacy goes hand in hand with extreme toughness. This was done for the safety of the valves, and to obviate any risk of damage; that is to say, to prevent such valves being ruptured by a sudden strong rush of blood against them. The toughness and strength of these same valves are well exemplified by varices. For if they hold back and support for a long time, a tough, thick, inspissated blood such as the melancholic, one is forced to concede that these membranes are strong. Indeed, anyone, attempting even with some effort to push the blood down through the veins, would feel the resistance and power of the valves.

Double valves were put at intervals in individual places, not triple ones as in the heart, because in the heart it was more a question of preventing reflux of blood, in the veins, on the other hand, of delaying in some measure its passage. In the heart, again, a large diastole and systole were present, in the veins none. Finally, in the heart the openings are very great and channels very large in which valves are found; and for this reason triple valves have sometimes been seen in [the veins of] the ox, presumably on account of the great size of this animal.

If indeed double valves were adequate for veins, the provision of triple ones would have been pointless. But double ones were adequate because double ones were competent to slow somewhat the strong flow of blood, and yet in no wise to prevent its passage. For this reason, at intervals, not a few places are found with but a single valve in each, the object being, undoubtedly, to grade with extreme accuracy the slowing of the blood. Where need of delay was greater, two valves were put, where less, one only. But the situation of choice for a single valve is where nature is about to give off a somewhat smaller vessel obliquely from a larger branch. In such a case a valve is made like a floodgate in the larger branch, a little below the opening of the smaller one which is given off from it. Thus, doubtless, the descending blood is not only held back by the valve, but, striking on it, floods backwards, and, pooling, enters the mouth of the small vein.

The activity which Nature has here devised is strangely like that which artificial means have produced in the machinery of mills. Here engineers put certain hindrances in the water's way so that a large quantity of it may be kept back and accumulate for the use of the milling machinery. These hindrances are called in Latin *septa* (sluices) and *claustra* (dams), but in dialect *clausae* and *rostae*. Behind them collects in a suitable hollow a large head of water and finally all that is required. In like manner nature labours in the veins by means of valves, here singly, there in pairs, the veins themselves representing the channels for the streams.

Nor let anyone here be surprised that nature puts valves—frequently paired—in various places, where no branch is given off obliquely in the trunk of a vein, while nevertheless valves are required to hold back the blood somewhat and retain it. For valves are. placed in veins less with a view to causing a pooling and storing of blood before the oblique mouths of branches than with a view to checking it on its course and preventing the whole mass of it slipping headlong down and escaping. A row of many valves was needed, and individual ones contributing each a little, not only to delaying the hurrying blood as already described, but also everywhere to preventing distension of the veins.

Here again indeed one may fairly wonder at Nature's activity, for two valves are made very close to one another, and so large that they take and fill up more than half of the vein-lumen: the remaining half is free and completely valveless. So nothing was in a position to prevent the blood from being able to flow in mass through the free part of the vessel, and causing the inconveniences already described. Nature, therefore, besides providing very many valves at intervals of two, three, or four fingers along the length of a vessel, devised further an ingenious mechanism to slow the strong current of blood. More valves, placed along the same side or in a straight line, would have caused either no delay or else quite a negligible one, as the blood would have flowed straight through the free part of the vessel. On the other hand, had she filled the whole lumen in individual places with three or four valves, she would have completely stopped the passage of the blood. Nature has therefore so placed the valves that in every case the higher valves are on the opposite side of the vein to the valves immediately below them; not unlike the way in which in the vegetable kingdom flowers, leaves, and branches grow successively from opposite sides of the stem. In this way the lower valves always delay whatever slips past the upper ones, but meanwhile the passage of blood is not blocked.

Finally, a point in connection with valves needs investigation, namely, how it happens that in some people more frequent and more numerous valves are seen in both legs and arms, in others fewer; a fact which is very noticeable when attendants pass ligatures round the limbs in the living person for the purpose of blood-letting. It must be said that more are seen in such as have much, very thick melancholic blood, or alternatively very thin, bilious blood (in which cases there is over-functioning of the valves, either to delay the thin fluid blood in the one case, or, in the other, to prevent the thick blood from distending the vein). Or again more are seen in such as are of powerful build or inclined to flesh, and to that extent have more numerous veins, so that they need greater functioning of valves to provide blood for the oblique branches. Or have very wide vessels which demand many valves better to delay the current of blood and increase the strength of the veins. Or the parts receive long straight veins [and more valves are present] so that the length and straightness should not allow the blood to rush right along in a stream, undelayed. Or finally more valves occur if an animal is naturally rather agile in its movement. And such is the wisdom and ingenuity of Nature which by my own efforts I have discovered in this new field.

PROOF OF CIRCULATION

HARVEY, William (English physician and naturalist, 1578–1657). From *Prelectiones anatomiae per me Gulielmum Harveium Londinensem anatomie* (!) *et chirurgie* (!) *professorem*, a MS dated 1616, first published London, 1886, with an autotype of the original.

Harvey's own notes for his Lumleian lectures at the Royal College of Physicians exist in the form of a book. Acquired by Hans Sloane from an unknown source, this was seen in 1766, lost, rediscovered before 1877, and published by the College in 1886. The notes concern the anatomy and physiology of the body as a whole and are furnished with frequent allusions to both medical and general literature as well as with illustrative anecdote. The Latin style is disconnected, ungrammatical, abbreviated, and interspersed with English. After presenting extensive preliminary arguments, both traditional and empirical, the author, on p. 80, presents the following abstract of his theory of the circulation. To a remarkable degree these few lines express in condensed form the basic structure of the later great work on the motion of the heart and blood. They are given here in the original and in translation in so far as a translation is practicable.

> WH constat per fabricam cordis sanguinem
> per pulmones in aortem perpetuo
> transferri. as by two clacks of a
> water bellows to rayse water
> constat per ligaturam transitum sanguinis
> ab arterijs ad venes
> unde △ perpetuum sanguinis motum
> in circulo fieri pulsu cordis
> An? hoc gratia Nutritionis
> an magis Conservationis sanguinis
> at Membrorum per Infusionem calidem
> vicissimique sanguis Calefaciens
> membra frigifactum a Corde
> Calefit

By W. H. it has been shown from the structure of the heart that blood is carried across continuously through the lungs to the aorta as by two clacks of a water bellows to rayse water. Through [the use of] ligatures a passage of blood from the arteries across to the veins has been shown. Whence it has been proven that a perpetual motion of the blood in a circle is caused by the beat of the heart. [It is questionable] whether this [is] for the sake of nourishment or for greater preservation of blood or of the organs by [their receiving] a warm bath or whether, on the contrary, the blood, cooled while warming the organs, is [in this way] warmed by the heart.

THEORY OF THE CIRCULATION

HARVEY, William (English physician and naturalist, 1578–1657). From *Exercitatio anatomica de motu cordis et sanguinis in animalibus*, Frankfort on the Main, 1628; tr. by C. D. Leake as *An anatomical study of the movement of the heart and blood in animals*, Springfield, Ill., 1931; by courtesy of Charles C Thomas, publisher.

Harvey's *Exercitatio de motu cordis* is by general consensus one of the two or three principal writings in the whole history of physiology. It is especially admired not only for the amount of new knowledge it presents but also for its style, its epistemological integrity and interest, and its intensive portrayal of the author's great general culture and experimental imagination. The work is too well knit not to suffer by editorial deletion, yet a logically connected chain of basic propositions and proofs stands out with special clarity; these are presented here.

The Motions of the Heart as Observed in Animal Experiments

In the first place, when the chest of a living animal is opened, and the capsule surrounding the heart is cut away, one may see that the heart alternates in movement and rest. There is a time when it moves, and a time when it is quiet.

This is more easily seen in the hearts of cold-blooded animals, as toads, snakes, frogs, snails, shell-fish, crustaceans, and fish. It is also more apparent in other animals as the dog and pig, if one carefully observes the heart as it moves more slowly when about to die. The movements then become slower and weaker and the pauses longer, so that it is easy to see what the motion really is and how made. During a pause, the heart is soft, flaccid, exhausted, as in death.

Three significant features are to be noted in the motion and in the period of movement:

1. The heart is lifted, and rises up to the apex, so that it strikes the chest at that moment, and the beat may be felt on the outside.

2. It contracts all over, but particularly to the sides, so that it looks narrower and longer. An isolated eel's heart placed on a table or in the hand shows this well, but it may also be seen in the hearts of fishes and of cold-blooded animals in which the heart is conical or lengthened.

3. Grasping the heart in the hand, it feels harder when it moves. This hardness is due to tension, as when one grasps the fore-arm and feels its tendons become knotty when the fingers are moved.

4. An additional point may be noted in fishes and cold-blooded animals, as serpents and frogs. When the heart moves it is paler in color, but when it pauses it is of a deeper blood color.

From these facts it seems clear to me that the motion of the heart consists of a tightening all over, both contraction along the fibers, and constriction everywhere. In its movement it becomes erect, hard, and smaller. The motion is just the same as that of muscles when contracting along their tendons and fibers. The muscles in action become tense and tough, and lose their softness in becoming hard, while they thicken and stand out. The heart acts similarly.

From these points it is reasonable to conclude that the heart at the moment it acts, becomes constricted all over, thicker in its walls and smaller in its ventricles, in order to expel its content of blood. This is clear from the fourth observation above in which it was noted that the heart becomes pale when it squeezes the blood out during contraction, but when quiet in relaxation the deep blood red color returns as the ventricle fills again with blood. But no one need doubt further, for if the cavity of the ventricle be cut into, the blood contained therein will be forcibly squirted out when the heart is tense with each movement or beat.

The following things take place, then, simultaneously: the contraction of the heart; the beat at the apex against the chest, which may be felt outside; the thickening of the walls; and the forcible ejection of the blood it contains by the constriction of the ventricles.

So the opposite of the commonly received opinion seems true.   . . .

The Movement of the Arteries as Seen in Animal Experimentation

In connection with the movements of the heart one may observe these facts regarding the movements and pulses of the arteries:

1. At the instant the heart contracts, in systole, and strikes the breast, the arteries dilate, give a pulsation, and are distended. Also, when the right ventricle contracts and expels its content of blood, the pulmonary artery beats and is dilated along with the other arteries of the body.

2. When the left ventricle stops beating or contracting, the pulsations in the arteries cease, or the contractions being weak, the pulse in the arteries is scarcely perceptible. A similar cessation of the pulse in the pulmonary artery occurs when the right ventricle stops.

3. If any artery be cut or punctured, the blood spurts forcibly from the wound when the left ventricle contracts. Likewise, if the pulmonary artery is cut, blood vigorously squirts out when the right ventricle contracts.

In fishes, also, if the blood vessel leading from the heart to the gills is cut open, the blood will be seen to spurt out when the heart contracts.

Finally, in arteriotomy, the blood is seen squirted alternately far and near, the greater spurt coming with the distention of the artery, at the time the heart strikes the ribs. This is the moment the heart contracts and is in systole, and it is by this motion that the blood is ejected.

Contrary to the usual teaching, it is clear from the facts, that the diastole of the arteries corresponds to the systole of the heart, and that the arteries are filled and distended by the blood forced into them by the contraction of the ventricles. The arteries are distended because they are filled like sacs, not because they expand like bellows. All the arteries of the body pulsate because of the same cause, the contraction of the left ventricle. Likewise the pulmonary artery pulsates because of the contraction of the right ventricle.

To illustrate how the beat in the arteries is due to the impulse of blood from the left ventricle, one may blow into a glove, distending all the fingers at one and the same time, like the pulse. The pulse corresponds to the tension of the heart in frequency, rhythm, volume, and regularity.   . . .

In the more perfect warm-blooded adult animals, as man, the blood passes from the right ventricle of the heart through the pulmonary artery to the lungs, from there through the pulmonary vein into the left auricle, and then into the left ventricle.   First I shall show how this may be so, and then that it is so.

The Passage of Blood through the Substance of the Lungs from the Right Ventricle of the Heart to the Pulmonary Vein and Left Ventricle

That this may be so, and that there is nothing to keep it from being so, is evident when we consider how water filtering through the earth forms springs and rivers, or when we speculate on how sweat goes through the skin, or urine through the kidneys.   It is well known that those who use Spa waters, or those of *La Madonna* near Padua, or other acid waters which are drunk by the gallon, pass them all off in an hour or so by the bladder.   So much fluid must tarry a while in the digestive tract, it must pass through the liver (everyone agrees that the alimentary succus goes through this organ at least twice daily), through the veins, the substance of the kidneys, and through the ureters into the bladder.

I know there are those who deny that the whole mass of blood may pass through the lungs as the alimentary juices filter through the liver, saying it is impossible and unbelievable.   They are of that class of men, as I reply with the poet, who promptly agree or disagree, according to their whim, fearful when wanted, bold when there is no need.

The substance of the liver and also of the kidney is very dense, but that of the lung is much looser, and in comparison with the liver and kidney is spongy.

There is no propulsive force in the liver, but in the lung the blood is pushed along by the beat of the right ventricle of the heart, which must distend the vessels and pores of the lung.   Again, as Galen indicates (*De Usu Part., cap. 10*), the continual rising and falling of the lungs in respiration must open and close the vessels and porosities, as in a sponge or thing of similar structure when it is compressed and allowed to expand.   The liver, however, is quiet, it never seems to expand or contract.

No one denies that all the ingested nourishment may pass through the liver to the vena cava in man and all large animals.   If nutrition is to proceed, nutriment must reach the veins, and there appears to be no other way.   Why not hold the same reasoning for the passage of blood through the lungs of adults, and believe it to be true, with Columbus, that great anatomist, from the size and structure of the pulmonary vessels, and because the pulmonary vein and corresponding ventricle are always filled with blood, which must come from the veins and by no other route except through the lungs?   He and I consider it evident from dissections and other reasons given previously.

Those who will agree to nothing unless supported by authority, may learn that this truth may be confirmed by the words of Galen himself, that not only may blood be transmitted from the pulmonary artery to the pulmonary vein, then into the left ventricle, and from there to the arteries, but that this is accomplished by the continual beat of the heart and the motion of the lungs in breathing.

There are three sigmoid or semilunar valves at the opening of the pulmonary artery, which prevent blood forced into this pulmonary artery from flowing back into the heart. . . .

Galen proposes this argument to explain the passage of blood from the vena cava through the right ventricle to the lungs.   By merely changing the terms, we may apply it more properly to the transfer of blood from the veins through the heart to the arteries.   From the words of that great Prince of Physicians, Galen, it seems clear that blood filters through the lung from the pulmonary artery to the pulmonary vein as a result of the heart beat and the movement of the lungs and thorax.   (Consult Hofmann's excellent Commentary on Galen's 6th Book, *De Usu Part.*, which I saw after writing this.)   The heart, further, continually receives blood in its ventricles, as into a cistern, and expels it.   For this reason, it has four kinds of valves, two regulating inflow, and two outflow, so blood will not be inconveniently shifted back and forth like Euripus, neither flowing back into the part from which it should come, nor quitting that to which it should pass, lest the heart be wearied by vain labor and respiration be impeded.   Finally, our assertion is clearly apparent, that the blood continually flows from the right to the left ventricle, from the vena cava to the aorta, through the porosities of the lung. . . .

Amount of Blood Passing through the Heart from the Veins to the Arteries, and the Circular Motion of the Blood

So far we have considered the transfer of blood from the veins to the arteries, and the ways by which it is transmitted and distributed by the heart beat. There may be some who will agree with me on these points because of the authority of Galen or Columbus or the reasons of others.   What remains to be said on the quantity and source of this transferred blood, is, even if carefully reflected upon, so strange and undreamed of, that not only do I fear danger to myself from the malice of a few, but I dread lest I have all men as enemies, so much does habit or doctrine once absorbed, driving deeply its roots, become second nature, and so much does reverence for antiquity influence all men. But now the die is cast; my hope is in the love of truth and in the integrity of intelligence.

First I seriously considered in many investigations how much blood might be lost from cutting the arteries in animal experiments.   Then I reflected on the symmetry and size of the vessels entering and leaving the ventricles of the heart, for Nature, making nothing in vain, would not have given these vessels such relative greatness uselessly. Then I thought of the arrangement and structure of the valves and the rest of the heart. On these and other such matters I pondered often and deeply. For a long time I turned over in my mind such questions as, how much blood is transmitted, and how short a time does its passage take. Not deeming it possible for the digested food mass to furnish such an abundance of blood, without totally draining the veins or rupturing the arteries, unless it somehow got back to the veins from the arteries and returned to the right ventricle of the heart, I began to think there was a sort of motion as in a circle.

This I afterwards found true, that blood is pushed by the beat of the left ventricle and distributed through the arteries to the whole body, and back through the veins to the vena cava, and then returned to the right auricle, just as it is sent to the lungs through the pulmonary artery from the right ventricle and returned from the lungs through the pulmonary vein to the left ventricle, as previously described.

This motion may be called circular in the way that Aristotle says air and rain follow the circular motion of the stars. The moist earth warmed by the sun gives off vapors, which, rising, are condensed to fall again moistening the earth. By this means things grow. So also tempests and meteors originate by a circular approach and recession of the sun.

Thus it happens in the body by the movement of the blood, all parts are fed and warmed by the more perfect, more spiritous, hotter, and, I might say, more nutritive blood. But in these parts this blood is cooled, thickened, and loses its power, so that it returns to its source, the heart, the inner temple of the body, to recover its virtue.

Here it regains its natural heat and fluidity, its power and vitality, and filled with spirits, is distributed again. All this depends on the motion and beat of the heart.

So the heart is the center of life, the sun of the Microcosm, as the sun itself might be called the heart of the world. The blood is moved, invigorated, and kept from decaying by the power and pulse of the heart. It is that intimate shrine whose function is the nourishing and warming of the whole body, the basis and source of all life. But of these matters we may speculate more appropriately in considering the final causes of this motion.

The vessels for the conduction of blood are of two sorts, the vena cava type and the aortic type. These are to be classified, not on the basis of structure or make-up, as commonly thought with Aristotle, for in many animals, as I have said, the veins do not differ from the arteries in thickness of tunics, but on the basis of difference in function or use. Both veins and arteries were called veins by the ancients, and not unjustly, as Galen notes. The arteries are the vessels carrying blood from the heart to the body, the veins returning blood from the body to the heart, the one the way from the heart, the other toward the heart, the latter carrying imperfect blood unfit for nourishment, the former perfected, nutritious blood.

The Circulation of the Blood Is Proved by a Prime Consideration

If anyone says these are empty words, broad assertions without basis, or innovations without just cause, there are three points coming for proof, from which I believe the truth will necessarily follow, and be clearly evident.

First, blood is constantly being transmitted from the vena cava to the arteries by the heart beat in such amounts that it cannot be furnished by the food consumed, and in such a way that the total quantity must pass through the heart in a short time.

Second, blood is forced by the pulse in the arteries continually and steadily to every part of the body in a much greater amount than is needed for nutrition or than the whole mass of food could supply.

And likewise third, the veins continually return this blood from every part of the body to the heart.

These proved, I think it will be clear that the blood circulates, passing away from the heart to the extremities and then returning back to the heart, thus moving in a circle. . . .

The First Proposition, Concerning the Amount of Blood Passing from Veins to Arteries, During the Circulation of the Blood, Is Freed from Objections, and Confirmed by Experiments

Whether the matter be referred to calculation or to experiment and dissection, the important proposition has been established that blood is continually poured into the arteries in a greater amount than can be supplied by the food. Since it all flows past in so short a time, it must be made to flow in a circle.

Someone may say here that a great amount may flow out without any necessity for a circulation and that it all may come from the food. An example might be given in the rich milk supply of the mammae. A cow may give three or four, or even seven and more gallons of milk daily, and a mother two or three pints when nursing a baby or twins, all of which must obviously come from the food. It may be replied that the heart, by computation, does more in an hour or less.

Not yet persuaded, one may still insist that cutting an artery opens a very abnormal passage through which blood may forcibly pour, but that nothing like this happens in the intact body, with no outlet made. With the arteries filled, in their natural state, so large an amount cannot pass in so short a time as to make a return necessary. It may be replied that from the computation and reasons already given, the excess contained in the dilated heart in comparison with the constricted must be in general pumped out with each beat and this amount must be transmitted, as long as the body is intact and in a natural state. . . .

The Second Proposition Is Proven

Now, let an experiment be made on a man's arm, using a bandage as in blood-letting, or grasping tightly with the hand. The best subject is one who is lean, with large veins, warm after exercise when more blood is going to the extremities and the pulse is stronger, for then all will be more apparent.

Under these conditions, place on a ligature as tightly as the subject can stand. Then it may be observed that the artery does not pulsate beyond the bandage, in the wrist or elsewhere. Next, just above the ligature the artery is higher in diastole and beats more strongly, swelling near the ligature as if trying to break through and flood past the barrier. The artery at this place seems abnormally full. The hand, however, retains its natural color and appearance. In a little time it begins to cool a bit, but nothing is "drawn" into it.

After this bandage has been on for some time, loosen it to the medium tightness used, as I said, in blood-letting. You will see the whole hand at once become suffused and distended, and its veins become swollen and varicosed. After ten or fifteen beats of the artery you will see the hand become impacted and gorged with a great amount of blood "drawn" by this medium tight ligature, but without pain, heat, horror of a vacuum or any other cause so far proposed.

If one will place a finger on the artery as it beats at the edge of the bandage, the blood may be felt to flow under it at the moment of loosening. The subject, also, on whose arm the experiment is made, clearly feels, as the ligature is slackened, warmth and blood pulsing through, as though an obstacle has been removed. And he is conscious of it following the artery and diffusing through the hand, as it warms and swells.

In the case of the tight bandage, the artery is distended and pulsates above it, not below; in the mediumly tight one, however, the veins become turgid and the arteries shrink below the ligature, never above it. Indeed, in this case, unless you compress these swollen veins very strongly, you will scarcely be able to force any blood above the ligature or cause the veins there to be filled.

From these facts any careful observer may easily understand that blood enters a limb through the arteries. A tight bandage about them "draws" nothing, the hand keeps its color, nothing flows into it, neither is it distended. With a little slackening, as in a mediumly tight ligature, it is clear that the blood is instantly and strongly forced in, and the hand made to swell. When they pulsate, blood flows through them into the hand, as when a medium bandage is used, but otherwise not, with a tight ligature, except above it. Meanwhile, the veins being compressed, nothing can flow through them. This is indicated by the fact that they are much more swollen below the bandage than above it, or than is usual with it removed, and that while compressed they carry nothing under the ligature to the parts above. So it is clear that the bandage prevents the return of blood through the veins to the parts above it and keeps those below it engorged.

The arteries, however, for the simple reason that they are not blocked by the moderate ligature, carry blood beyond it from the inside of the body by the power and impulse of the heart. This is the difference between a tight and medium bandage, the former not only blocks the flow of blood in the veins but also in the arteries, the latter does not impede the pulsating force from spreading beyond the ligature and carrying blood to the extremities of the body.

One may reason as follows. Below a medium bandage we see the veins become swollen and gorged and the hand filled with blood. This must be caused by blood passing under the ligature either in veins, arteries or tiny pores. It cannot come through the veins, certainly not through invisible ducts, so it must flow through the arteries, according to what has been said. It obviously cannot flow through the veins since the blood cannot be squeezed back above the ligature unless it is completely loosened. Then we see the veins suddenly collapse, discharging themselves to the part above, the hand loses its flush, and the stagnant blood and swelling quickly fade away.

Further, he whose arm has been bound for some time with a medium bandage, and whose hand has been rendered somewhat swollen and cold, feels, as the ligature is loosened, something cold creeping up with the returning blood to the elbow or armpit. I think this cold blood returning to the heart, after removing the bandage in blood-letting, is a cause of fainting, which we sometimes see even in robust persons, usually when the ligature is removed, or, as is commonly said, when the blood turns.

Moreover, immediately on loosening a tight bandage to a medium one, we see the veins below it, but not the arteries, swollen with blood continually carried in by the arteries. This indicates that blood passes from arteries to veins, not the reverse, and that there is either an anastomosis of these vessels or pores in the flesh and solid parts permeable to blood. It also indicates that the veins inter-communicate, since, with a medium ligature above the elbow, they all swell up at the same time, and, if even a single venule be cut with a lancet, they all quickly shrink, giving up their blood to this one, and subside almost together. . . .

The Third Proposition Is Proven, and the Circulation of the Blood Is Demonstrated from it

So far we have considered the amount of blood flowing through the heart and lungs in the body cavity, and similarly from the arteries to the veins in the periphery. It remains for us to discuss how blood from the extremities gets back to the heart through the veins, and whether or not these are the only vessels serving this purpose. This done we may consider the three basic propositions proving the circulation of the blood so well established, so plain and obvious, as to force belief.

This proposition will be perfectly clear from a consideration of the valves found in the venous cavities, from their functions, and from experiments demonstrable with them.

The celebrated anatomist, Hieronymus Fabricius of Aquapendente, or, instead of him, Jacobus Sylvius, as Doctor Riolan wishes it, first described membranous valves in the veins, of sigmoid or semilunar shape, and being very delicate eminences on the inner lining of these vessels. They are placed differently in different individuals, but are attached to the sides of the veins, and they are directed upwards toward the main venous trunks. As there are usually two together, they face and touch each other, and their edges are so apt to join or close that they prevent anything from passing from the main trunks or larger veins to the smaller branches. They are so arranged that the horns of one set are opposite the hollow part of the preceding set, and so on alternately.

The discoverer of these valves and his followers did not rightly appreciate their function. It is not to prevent blood from falling by its weight into areas lower down, for there are some in the jugular vein which are directed downwards, and which prevent blood from being carried upwards. They are thus not always looking upwards, but more correctly, always towards the main venous trunks and the heart. Others as well as myself have sometimes found them in the milky veins and in the venous branches of the mesentery directed towards the vena cava and portal vein. To this may be added that there are none in the arteries, and that one may note that dogs, oxen, and all such animals have valves at the branches of the crural veins at the top of the sacrum, and in branches from the haunches, in which no such weight effect of an erect stature is to be feared.

Nor, as some say, are the valves in the jugular veins to prevent apoplexy, since the head is more likely to be influenced by what flows into it through the carotid arteries. Nor are they present to keep blood in the smaller branches, not permitting it to flow entirely into the larger more open trunks, for they are placed where there are no branches at all, although I confess they are more frequently seen where there are branchings. Nor are they present for slowing the flow of blood from the center of the body, for it seems likely it would flow slowly enough anyway, as it would then be passed from larger to smaller branches, become separated from the source and mass, and be moved from warmer to cooler places.

The valves are present solely that blood may not move from the larger veins into the smaller ones lest it rupture or varicose them, and that it may not advance from the center of the body into the periphery through them, but rather from the extremities to the center. This latter movement is facilitated by these delicate valves, the contrary completely prevented. They are so situated that what may pass the horns of a set above is checked by those below, for whatever may slip past the edges of one set is caught on the convexity of those beyond, so it may not pass farther.

I have often noticed in dissecting veins, that no matter how much care I take, it is impossible to pass a probe from the main venous trunks very far into the smaller branches on account of the valvular obstructions. On the contrary it is very easy to push it in the opposite direction, from the branches toward the larger trunks. In many places a pair of valves are so placed that when raised they join in the middle of the vein, and their edges are so nicely united that one cannot perceive any crack along their junction. On the other hand, they yield to a probe introduced from without inwards and are easily released in the manner of flood-gates opposing a river flow. So they intercept, and when tightly closed, completely prevent in many places a flow of blood back from the heart and vena cava. They are so constituted that they can never permit blood to move in the veins from the heart upwards to the head, downwards toward the feet, or sidewise to the arms. They oppose any movement of blood from the larger veins toward the smaller ones, but they favor and facilitate a free and open route starting from the small veins and ending in the larger ones.

This fact may be more clearly shown by tying off an arm of a subject as if for blood-letting (A, A, Fig. 6a). There will appear at intervals (especially in rustics) knots, or swellings, like nodules (B, C, D, E, F), not only where there is branching (E, F), but also where none occurs (C, D). These are caused by the valves, appearing thus on the surface of the hand and arm. If you will clear the blood away from a nodule or valve by pressing a thumb or finger below it (H, Fig. 6b), you will see that nothing can flow back, being entirely prevented by the valve, and that the part of the vein between the swelling and the finger (H, O, Fig. 6b), disappears, while above the swelling or valve it is well distended (O, G). Keeping the vein thus empty of blood, if you will press downwards against the valve, (O, Fig. 6c) by a finger of the other hand on the distended upper portion (K, Fig. 6c), you will note that nothing can be forced through the valve. The greater effort you make the more the vein is distended toward the valve, but you will observe that it stays empty below it (H, O, Fig. 6c).

From many such experiments it is evident that the function of the valves in the veins is the same as that of the three sigmoid valves placed at the opening of the aorta and pulmonary artery, to prevent, when they are tightly closed, the reflux of blood passing over them.

Further, with the arm bound as before and the veins swollen, if you will press on a vein a little below a swelling or valve (L, Fig. 6d) and then squeeze the blood upwards beyond the valve (N) with another finger (M), you will see that this part of the vein stays empty, and that no back flow can occur through the valve (as in H, O, Fig. 6b). But as soon as the finger (H) is

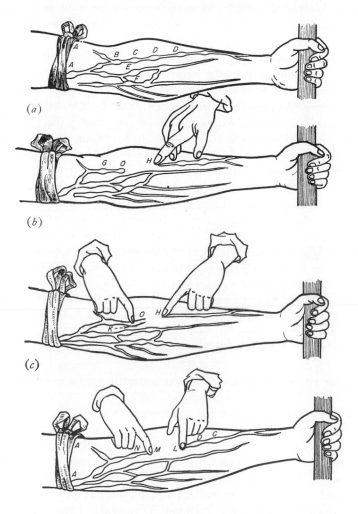

FIG. 6. Experiments on a bandaged arm in *An anatomical study of the movement of the heart and blood in animals.*

removed, the vein is filled from below (as in D, C, Fig. 6a). Thus it is clearly evident that blood moves through the veins toward the heart, from the periphery inwards, and not in the opposite direction. The valves in some places, either because they do not completely close, or because they occur singly, do not seem adequate to block a flow of blood from the center, but the majority certainly do. At any rate, wherever they seem poorly made, they appear to be compensated for in some way, by the greater frequency or better action of the succeeding valves. So, as the veins are the wide open passages for returning blood to the heart, they are adequately prevented from distributing it from the heart.

Above all, note this. With the arm of your subject bound, the veins distended, and the nodes or valves prominent, apply your thumb to a vein a little below a valve so as to stop the blood coming up from the hand, and then with your finger press the blood from that part of the vein up past the valve (L, N, Fig. $6d$), as was said before. Remove your thumb (L), and the vein at once fills up from below (as in D, C, Fig. $6a$). Again compress with your thumb, and squeeze the blood out in the same way as before (L, N, and H, O), and do this a thousand times as quickly as possible. By careful reckoning, of course, the quantity of blood forced up beyond the valve by a single compression may be estimated, and this multiplied by a thousand gives so much blood transmitted in this way through a single portion of the veins in a relatively short time, that without doubt you will be very easily convinced by the quickness of its passage of the circulation of the blood.

But you may say this experiment of mine violates natural conditions. Then if you will take as long a distance from the valve as possible, observing how quickly, on releasing your thumb, the blood wells up and fills the vein from below, I do not doubt but that you will be thoroughly convinced.

Conclusion of the Demonstration of the Circulation of the Blood

Briefly let me now sum up and propose generally my idea of the circulation of the blood.

It has been shown by reason and experiment that blood by the beat of the ventricles flows through the lungs and heart and is pumped to the whole body. There it passes through pores in the flesh into the veins through which it returns from the periphery everywhere to the center, from the smaller veins into the larger ones, finally coming to the vena cava and right auricle. This occurs in such an amount, with such an outflow through the arteries, and such a reflux through the veins, that it cannot be supplied by the food consumed. It is also much more than is needed for nutrition. It must therefore be concluded that the blood in the animal body moves around in a circle continuously, and that the action or function of the heart is to accomplish this by pumping. This is the only reason for the motion and beat of the heart.

DISCOVERY OF LYMPHATIC SYSTEM

BARTHOLIN (Bartholinus), Thomas (Danish physician, 1616–1680). From *De lacteis thoracicis in homine brutisque nuperrime observatis, historia anatomica publice proposita respondentem Michaele Lysero*, London, 1652; tr. as *The anatomical history of Thomas Bartholinus, doctor and king's professor: concerning the lacteal veins of the thorax, observed by him lately in man and beast. Publickly proposed to Michael Lyserus answering*, London, 1653.

The lacteals had been discovered by Aselli* and the thoracic ducts by Pecquet.† Bartholin was a member of "a family that long monopolized anatomical posts in Denmark" (his father and brother were famous physicians and anatomists). He demonstrated the interconnections of the system generally and, by anatomizing two recently well-fed and then strangled criminal prisoners, extended the observations from dog to man.

But *John Pecquet* of *Diep*, a Physician in the City of *Paris*, has to the eternal renown of his name, lately augmented and lengthned the bounds of the *Vene lacteae*, having set out new experiments in *Anatomie*, by which he has discovered the vessels of the *Chyle*, hitherto unknown, and did assert it at *Paris* in the year 1651 publickly, after three years observations: which notwithstanding in three dissections and lesse shall be shown to others. He first of all found out these in sheep, doggs, and other living creatures dissecting them four howers after repast. And first that a receptacle of the *Chyle*, did from the *Mesenterie* or these *milkie veins*, of *Asellius* already known, come out, into which by severall *milkie veins*, all the *Chylus* is drawn from the *Mesenterie*. Next, that from the same receptacle there did creep up through the *Thorax* to the *Subclavial veins*, other passages for the *Chyle*. He found out both of them by accident, being never heard of in so many ages, not thinking of any such thing, being only busied in finding out the motion of the *heart* in living dissections; for the bowels being taken out, and wiping off them the quantity of blood, happily there appeared to him this *milkie liquor*, and the white passages of the *Thorax*: after which he making further search, found the beginning of them about the *Diaphragma*, and the *Mesenterie*. Truly it is a wonderfull thing, that so apparent businesse has hitherto been hid from every ones eyes, even from theirs, who with vain arrogance boast that they have found out the utmost of *Anatomy*. But we shall easily deserve pardon, who daily importune Nature to find out some new thing, and yet ingenuously confesse that we shall never find it out. Whence we return thanks to *Pecquet*, that he rather ascribes our not finding out of these things to our infelicity, than to our carelessnesse. Being advertis'd by my Brother *Eras: Bartholinus* his Letter, a Physician, and eminent Mathematician, who at *Bloys* had made tryall of the

---

* Gasparo Aselli (1627), *De lactibus sive lacteis venis*, Basel, 1628.

† Jean Pecquet (1647), *Experimenta nova anatomica*, Amsterdam, 1661.

truth of this experiment, I did fellicitously enquire after *Pecquets* observations, and being set a-work by the care of this new businesse, so soone as I could get the Book, I did diligently search, together with *Michael Lyserus* my friend, and well skill'd in *Anatomie*, in a living Dog dissecting him four hours after he had eaten his belly full, and found most of these things which he had observ'd, and set down in his figure: But I did imagine, since in so great motion of the living creature, and his extreme struggling with his pain, the milkie vessels are sooner perish'd, that I might have more successe in the experiment, if either the creature some hours after his repast, might be strangled with a sudden noose, or if a man condemned, and being hang'd after he had eaten a good breakfast, might be immediately opend; because then the motion of the *Thorax* being hinder'd, and respiration being stopp'd, the humors which were distributed would stay in the same position, nor would so easily stir, by his pain in cutting.   Having resolv'd upon the way, the event was according to our wish; and having made many experiments in dogs, at last we got two mens bodies, by publick consent, and grant of the most gracious King *Frederick* the III, which had bin well fed, and were otherwise to have suffered upon the Wheel, and were to have been hang'd in Chains, upon which, both in the publik School of *Anatomie* by solemn demonstration, as also privately, we did so much the more diligently search, because we were the first that made this essay in a mans body.   The first corps we cut up, was of a fellow that had kill'd a child; the other was the body of a Thief; one of them was lean, and his bones scarce holding together: the tother was fat, and sound at all points.   Both of those bodies was fill'd sufficiently with meat, and wine, five hours before their breath was stop'd.   In the first, we saw the *Mesentericals, Lacteals,* and *Thoracicks,* full of *Chyle.*   In the other there was not so much as any appearance of *Chyle* found in the vessels; nevertheless we were not frustrate of our hopes, for in the *Mesenterie* the *Glandulae Lumbares* did appear, and in the *Thorax* the passages destin'd for the *Chyle:* which we took out with the same ease, as we had done formerly in the many experiments which we had made, in the dissection of many creatures, knowing the place and way how to doe it, to a hair.   In the first we shew'd the three *Lumbares Glandulae* full, in the place of *Pecquets Receptacle,* and outward insertion of the *Thoracicae Lacteae,* into the *Sinistra Axillaris,* clearly to be seen in three branches.   Our chiefest care was about the other, seeing the same *Glandules* in the inward or internal insertion of the *Lacteae Thoracicae,* which we viewed singularly well, together with the valves, by the help of a pair of bellows, and other assistants of the knife.   Let no ingenuous man doubt of our truth, and sincerity.   We never did cosen nor cheat the world.   Yet I know there will be some, whom either envy, or contempt, will carry out into prejudice.   But away with that generation of men; there belongs Assizes to the Law, let them try it if they can, if not, let them follow our foot-steps, and let them leave off to doe ill, and be ill reported of.

In the mean time, if they accuse me of a lye, I have testimonials by me; and witnesses both great and small, and very honorable Spectators of all sorts, who were present with us, both at our private and publick dissection: Especially the *Illustrious* and *Magnificent Nobleman, the Lord Christianus Thomaeus of Stougaard, Knight Baronet, Great Chancellour to the Kings Majestie, and a most bountiful Protectour of the Vniversity, and learned men,* and other worthy and Noble Persons, whom the newnesse of the businesse, and their accustomed favour to me, had drawn to this Spectacle of Nature; as likewise a great many most excellent Physicians, and most skilfull in *Anatomie, Iacobus Fabricius, Olaus Wormius, Simon Paul, Paulus Mothius, Henricus Fuiren, Iacobus Finokius,* and other of the Kings *Professors,* and a most honourable Assembly of choice Students. Whosoever will not give credit to so many, and to men of so great account, let them be of no credit for ever. But we did observe diverse frame of these new vessels in men, and beasts, according to the diversity of the *Species* and *Individuals,* as in the ensuing relations we shall more acurately set down, according to our accustomed Method: especially insisting upon men, because the first inventer of *Anatomie,* did only use dissection in beasts, and has adorn'd his observations with such obscure varnishes, both in his words and sentences, that the Reader having read him never so often, still remains doubtful. . . .

CHAPTER VIII

The Use of the New Glandules

Moreover because it is but an empty commendation unless this that we are about be profitable, I will shew the chiefest uses of the *receptacle,* or of the *Glandulae Lacteae,* as much as the noveltie of such a strange business will permit.

1. Man has receiv'd augmentation in the number of his parts, being hitherto either imperfect, or by us unperfectly known; Anatomie is made more perfect, and the functions are clearer.

2. The use of *Glandules* is more ready to receive *Chylus* out of the *milkie veins* of the *Mesenterie,* and somewhat too to prepare it by help of the neighbouring hot vessells the *arterie* and the *vein,* and being in a little while prepared to thrust it out to the *Thoraces* and other parts, sense teaches us that there is *Chylus* contain'd there, and it is known that it receivs it from the *Lacteal Mesaraick* by their compression, lastly the branching of the *Lactea* does assure us that it advances upwards, yea pressing the *Mesenterie,* and wounding or tying the *Lacteae's* in the *Thorax,* or thrusting them up with your finger, the *Chyle* either comes out that way or swels the place next to the *Mesenterie;* Besides, when there is none of this milkie humour, such wheish humour is found in them as is seen in the *Glandules.*

3. They squeeze out the *serum*, being separated from the *Chyle* in that light preparation, and endeavouring to expell, either into the *veins* hard by them, or into the emulgent *arteries* to which they send branches, or into the *Capsules* of the *Atrabilis* destined for melancholie, or lastly into the doubling of the *Peritoniū*, in which they abide.

CHAPTER IX

MANY CONVENIENCES BY THIS NEW INVENTION

And from hence, the reason of many diseases in the body may be easily given, which we were either forc'd to be ignorant of here before, or by tedious traversing to look for them.

The way is now clear, and the shortest cut found, by which drinkers who like *Promachus* wil drink foure gallons for his own share, or drink like *Bonesus* more than any man beside, who are not born to live but to drink, how they come to pisse this again so soon, and so abundantly.

For the ordinary way through the *Liver, Heart, Arteries, Emulgent veins, Ureters, Bladder,* is longer and though *Aquapend.* take a great deal of pains to find a shorter through the *Liver,* from the *Gastrick veins* of the *stomach,* and *Piso* and *Conringius* through the spleen; yet he finding the *Circulation,* missed of his aim, & they having found out ways in the conception of their own wit, cannot demonstrate them to sense.   . . .

How and by what near way many things swallow'd with meat are return'd with *urine,* any body may guesse. The story of the *Venetian* Virgin in *Alexander Benedictus Schenkius lib. 3. Obs. 9* in *Sanctorius l. 14. Meth. vit. Err. C. 11. Paraeus lib. 24. c. 19.* and by others too is affirmed, which had swallowed a hair bodkin four inches long, and after four years piss'd it out of her bladder, envelop'd with calculous matter. *Claudinus* in *Resp. Med. 40. Langius* in *Ep. p. 745, Sancto. 1. d.* to [do?—Ed.] torment themselves to find the wayes, but what ways soever they chuse, they will be very far from the purpose, and yet not free from hurt. But ours is very short & not dangerous. So judg likewise of the rest. That nothing is more ordinary than pissing of hair, *Tulpius* observ'd periodically in the Provost of *Hornes* Son, who every fourteenth day piss'd hairs a full inch long for four dayes together with a great deal of stopping in his water. *Nic. Flor.* saw some likewise of the same bignesse. *Schenkius* had bundles of hair which a woman piss'd. *Zactus* saw another that piss'd hairs a handbredth long, thick and hard like Hoggs bristles.

CHAPTER XII

The Insertion of the Milkie Veins of ·the Thorax into the Left Axillar

Now must we enquire after the insertion of the *Milkie veins* of the *Thorax*. From the third or fifth *vertebre* of the back, where they depart from the *spinal,* through the midst of which they crept all this way, they turn a little to the left and creep up under the *Oesophagus,* and the *Aorta,* and under the *Subclavial arterie,* and the *Glandules* of the *Thymus,* they goe forward to the *left clavicule.* Here *Pecquet* differs, who writes and cutts it as divided in two branches, from the third *vertebre* of the spinal, of which one on the left side goes to the *left clavicule* and that on the right side to the *right clavicule:* But the Graver seems to have added some things to this new invention, or else the conjecture of the inventor. For we could not observe that it was parted on both sides, neither in Brutes nor in Men, unlesse Nature deal otherwise with us in *Denmark,* than she does with those in *France.* But that it does always frō the third *vertebre* turn towards the left side, as we have set down in the Cut of a man. Putting in a bellows below, we could see no signe of a vessel on the right side, nor could our sight find out any thing like it. Nor do I think it can be otherwise, which I likewise found in man making a diligenter search by more eyes than mine own: Because at the rysing of the *spinal* near the throat on the right side of the *Axillar vein* arise immediately from the *Cava* near to the *Basis* of the *Heart,* but in the left it stretches out further, so that Nature seems to have design'd a more convenient place for this insertion. *Auzotius,* a man of profound learning, sayes that he saw and kept this vessell tyed at both ends, for many days, as he confessed to my brother *Erasmus Bartholinus* whom we praised before. But let me never be trusted by *Auzotius* or any body else, if ever I observ'd any insertion upon the right side. Nature sports her self many times, and did perchance in one corps reveal that to him, which a great many have denied to us. The solitary branch then on the left side stretches under the *Oesophagus, Thymus,* the *arterie* of the *Thorax* and *left clavicle* to the *left axillar vein,* and that sometimes with one branch only, sometimes with three, sometimes with a more numerous division through little holes, or one hole if there be one insertion, immediately where the outward *Jugularie vein* pours it self into the *Axillare vein.* And I have observ'd the entrie of it immediately under the outgoing of the outward *Jugularie vein,* and that in small *Glandules* it lay in the distance of the *Muscles* of the neck. Where there is but one insertion of a single branch, we have observed in men and in dogs that the hole is longer, and that there is a Miterlike *valve* of most thin contexture plac'd upon the hole in the concavitie of the *Axillar,* which hinders the regresse of the *Chyle* and of the blood downwards, and its ascent to the joynts, because it looks only towards the *heart*, and in the hinder part it is turn'd towards the *Axillar*.

We did not observe *Milkie veins* to passe any further, or reach to the head, because near to the *jugular* there stands a *valve*, or to the joynts, as well by reason of the *valve* of the *Arm-veins*, as for the continuall deflux of blood, whence by squeezing we came to know that no *Chyle* could come to those parts. It was a very singular thing which we saw in a dog, which we did cut up seven hours after his repast, certain wheyish vessels by his foremost feet, as in the *abdomen*, which went along by the *veins* of the feet on both sides of them, being transparent, like those in the *Thorax* and *Abdomen* observ'd by us, the ingresse & the egress of which we could not then perceive, but being intercepted by *ligature*, they swell'd upwards towards the joynts.

CHAPTER XIII

The Use of the Milkie Veins of the Thorax

The use of the *Milkie veins* of the *Thorax*, is the same with that of the *Meseraicks*, if you make a difference of the bound from which, and the bound to which. For the new ones carry the *Chylus* out of the new *Milkie glandules* or *Receptacle* of the *Mesenterie* to the *Subclavials*. The *Lacteae Mesaraicae* carry the *Chylus* from the *Intestines*, either to the *Liver*, or the *Receptacle*. It is manifest that it carries *Chylus*, because in the dissection it flows forth, and ligature shows it. Its insertion shows us likewise that it is infus'd into the *Subclavial veins*, whence also their flowes a milkie humour, and wind will likewise get through if you unty the string, so that sometimes the branches of the *Vena Cava* will look blew with it, and the Concavity of the *Thorax* look white. . . . It was hitherto thought a wonder, and related for a fable or a miracle which some Ecclesiastical stories do relate of *Saint Paul*, who was beheaded under the Empire of *Nero*, that a stream of milk flow'd out of his neck. This Relation had severall authors, but more especially the Acts of St. *Nereus*, and St. *Achilleus*, St. *Ambr*. Serm. *68*. *Chrysostom*, whose testimony *Baronius* brings too, Ann. Chr. *66* Sect. 11. and approves it. If the story be true, by finding the milkie *veins* of the *Thorax*, we shall easily find the way and manner of it. The Branches of the *axillars*, or rather the insertion of the *Lacteae* into the axillars was dilated, and did gape with doors wider than ordinary.

If you ask why the *Chylus* goes out of the *Subclavialls*, you shall have a certain answer, that it goes not to the head, nor to the joynts, for reasons before given, but that it is carried to the *heart* with the blood that runs down in Circulation, that it may be chang'd into blood, as it was partly the opinion of

Aristotle. Diverse experiments did evidence this to the first Inventor *Pecquet. cap. 2. fignata. I.* cutting out the *heart* of a live dog, and those things which did adhere to it, and wiping away the blood, there flowd a certain white liquor within the pipe of the *Vena Cava* about the place of the right ventricle. 2. He opened the *Vena Cava* from the *Diaphragma* to the throat, where this liquor did appear flowing, free from the mixture of blood. 3. There setled a white liquor within the *vein,* even from the *Subclavial* branches to the *Pericardium,* in all things like the milkie juyce of the *Mesentarie.* 4. We made another experiment often, at *Hasnia,* with bellowes, and a hollow pipe, which making a little hole, we put into the *Milkie vein* of the *Thorax* beneath the *Clavicules,* and sent in wind and water by blowing in't. They did rise up and swell, not only the vessells towards the *Axillar vein,* but that part of the *Axillar* which is towards the *heart,* and streight the *right ear* and the *right ventricle* of the *heart* itself, then the *lungs* by the *Vena Arteriosa* which was swell'd with water and blowing. Whosoever shall make tryall of this, I will engage he shall never fail, unless the *Tunicle* of the *veins* of the *Thorax,* which is very thin, being hurt, or falling flat, stop the hole which you make.

MECHANISTIC BASIS OF PHYSIOLOGY

DESCARTES, René (French philosopher and mathematician, 1596–1650). A fragment from *De homine figuris et latinitate donatis a Florentio Schuyl, etc.,* Leyden, 1662; tr. of this fragment by J. P. Mahaffy in his *Descartes,* Philadelphia, 1881.

In his dialogue called *Il saggiatore* Galileo states that all natural phenomena must ultimately be analyzed in terms of motion. The greatest early biological monument to the power of this approach is the work of Harvey, whose study on circulation is entitled *An exercise on the* motion *of the heart and of the blood.* Harvey's argument for the circulation in fact stems principally from his discovery that the motion of the heart causes that of the blood rather than vice versa as previously supposed. With Descartes, mechanism becomes a general physiological concept. He vivisects rabbits, dogs, eels, and fish, and anatomizes various animals and the human cadaver. While he especially develops the mechanistic approach in his posthumous treatise on *Man,* from which the following fragment is taken, he has already stated it with considerable explicitness and applied it to problems of physiological psychology in the *Passions of the soul* (see p. 273). For an example of the consequences of the doctrine see La Mettrie, p. 176. For another example of successful mechanistic physiology of the seventeenth century, see Borelli, p. 158.

I desire you next to consider that all the functions which I have attributed to this (animal) machine, such as digestion, the heating of the heart and arteries, nutrition and growth, breathing, waking and sleep, the perception of colours, sounds, tastes, heat, and other such qualities by the external senses, the impression of their ideas in the organ of *sensus communis* and of imagination, the retention or impression of these ideas in memory, the internal motions of appetites and passions; and finally, the external movements of all the limbs, which follow so suitably as well from the action of objects presented to sense as from the passions and impressions which are found in the memory, that they imitate as perfectly as is possible those of a real man,—I desire you to notice that these functions follow quite naturally in the machine from the arrangement of its organs exactly as those of a clock, or other automaton, from that of its weights and wheels; so that we must not conceive or explain them by any other vegetative or sensitive soul, or principle of motion and life, than its blood and its spirits agitated by the heat of the fire which burns continually in its heart, and which is of no other kind than all the fires which are contained in inanimate bodies.

EARLY STUDIES ON RESPIRATION

Lord BOYLE, Robert (English natural philosopher, 1627–1691). From *New experiments physicomechanical touching the spring of the air and its effects made for the most part in a new pneumatical engine*, Oxford, 1660.

These well-known researches of Boyle were prophetic in that they stated the classical issues of pneumatic chemistry and linked them intimately with the study of respiratory physiology. The settling of the same issues 100 years later by Scheele, Priestley, and Lavoisier would prove an important opening chapter in the history of modern chemistry. Boyle was fourteenth child of the Earl of Cork. He studied and wrote, usually with elegance and charm as well as scientific objectivity, about (among other subjects) the physical properties of air, the "chymical principles of mixed bodies," normal and abnormal physiology, heat and cold, blood transfusion, sun spots, buoyancy, sweetening of sea water, final causes, logic, profanity, duties of motherhood, and religious subjects.

Experiment XL.

It may seem well worth trying, whether or no in our exhausted glass the want of an ambient body, of the wonted thickness of air, would disable even light and little animals, as bees, and other winged insects, to fly. But though we easily foresaw how difficult it would be to make such an experiment, yet not to omit our endeavours, we procured a large flesh-fly, which we conveyed into a small receiver. We also another time shut into a great receiver a humming bee, that appeared strong and lively, though we had rather have made the trial with a butterfly, if the cold season would have permitted us to find any. The fly, after some exsuctions of the air, dropped down from the side of the glass whereon she was walking. But, that the experiment with the bee might be the more instructive, we conveyed in with her a bundle of flowers, which remained suspended by a string near the upper part of the receiver; and having provoked the bee, we excited her to fly up and down the capacity of the vessel, till at length, as we desired, she lighted upon the flowers: where upon we presently began to draw out the air, and observed, that though for some time the bee seemed to take no notice of it, yet within a while after she did not fly, but fall down from the flowers, without appearing to make any use of her wings to help herself. But whether this fall of the bee, and the other insect, proceeded from the medium's being too thin for them to fly in, or barely from the weakness, and as it were swooning of the animals themselves, you will easily gather from the following experiment.

Experiment XLI.

To satisfy ourselves in some measure about the account upon which respiration is so necessary to the animals that nature hath furnished with lungs, we took (being then unable to procure any other lively bird, small enough to be put into the receiver) a lark, one of whose wings had been broken by a shot of a man that we had sent to provide us some birds for our experiment; but notwithstanding this hurt, the lark was very lively, and did, being put into the receiver, divers times spring up in it to a good height. The vessel being hastily, but carefully closed, the pump was diligently plied, and the bird for a while appeared lively enough; but upon a greater exsuction of the air, she began manifestly to droop and appear sick, and very soon after was taken with as violent and irregular convulsions, as are wont to be observed, in poultry, when their heads are wrung off: for the bird threw herself over and over two or three times, and died with her breast upward, her head downwards, and her neck awry. And though upon the appearing of these convulsions, we turned the stop-cock, and let in the air upon her, yet it came too late; whereupon casting our eyes upon one of those accurate dials that go with a pendulum, and were of late ingeniously invented by the noble and learned *Hugenius,* we found that the whole tragedy had been concluded within ten minutes of an hour, part of which time had been employed in cementing the cover to the receiver. Soon after we got a hen sparrow, which being caught with bird-lime was not at all hurt; when we put her into the receiver, almost to the top of which she would briskly raise herself, the experiment being tried with this bird, as it was with the former, she seemed to be dead within seven minutes, one of which were employed in cementing the cover: but upon the speedy turning of the key, the fresh air flowing in, began slowly to revive her, so that after some pantings she opened her eyes, and regained her feet, and in about $\frac{1}{4}$ of an hour after, threatned to make an escape at the top of the glass, which had been unstopped to let in the fresh air upon her: but the receiver being closed the second time, she was killed with violent convulsions within five minutes from the beginning of the pumping.

A while after we put in a mouse, newly taken, in such a trap as had rather affrighted than hurt him; whilst he was leaping up very high in the receiver, we fastened the cover to it, expecting that an animal used to live in narrow holes with very little fresh air, would endure the want of it better than the lately mentioned birds: but though, for a while after the pump was set a work, he continued leaping up as before; yet, it was not long ere he began to appear sick and giddy, and to stagger: after which he fell down as dead, but without such violent convulsions as the bird died with. Whereupon, hastily turning the key, we let in some fresh air upon him, by which he recovered, after a while, his senses and his feet, but seemed to continue weak and sick: but at length, growing able to skip as formerly, the pump was plied again for eight minutes, about the middle of which space, if not before, a little air by a mischance got in at the stop cock; and about two minutes after that, the mouse divers times leaped up lively enough, though after about two minutes more he fell down quite dead, yet with convulsions far milder than those wherewith the two birds expired. This alacrity so little before his death, and his not dying sooner than at the end of the eighth minute, seemed ascribable to the air (how little soever) that slipt into the receiver. For the first time, those convulsions (that, if they had not been suddenly remedied, had immediately dispatched him) seized on him in six minutes after the pump began to be set a work. These experiments seemed the more strange, in regard that during a great part of those few minutes the engine could but inconsiderably rarefy the air (and that too, but by degrees) and at the end of them there remained in the receiver no inconsiderable quantity; as may appear by what we have formerly said of our not being able to draw down water in a tube, within much less than a foot of the bottom: with which we likewise considered, that by the exsuction of the air and interspersed vapours, there was left in the receiver a space some hundreds of times exceeding the bigness of the animal, to receive the fuliginous streams, from which expiration discharges the lungs; and which, in the other cases hitherto known, may be suspected, for want of room, to stifle those animals that are closely penned up in too narrow receptacles.

I forgot to mention, that having caused these three creatures to be opened, I could, in such small bodies, discover little of what we sought for, and what we might possibly have found in larger animals; for though the lungs of the birds appeared very red, and as it were inflamed, yet that colour being usual enough in the lungs of such winged creatures, deserves not so much our notice, as it doth, that in almost all the destructive experiments made in our engine, the animals appeared to die with violent convulsive motions: from which, whether physicians can gather any thing towards the discovery of the nature of convulsive distempers, I leave to them to consider.

Having proceeded thus far, though (as we have partly intimated already) there appeared not much cause to doubt, but that the death of the forementioned animals proceeded rather from the want of air, than that the air was overclogged by the steams of their bodies, exquisitely penned up in the glass; yet I, that love not to believe any thing upon conjectures, when by a not overdifficult experiment I can try whether it be true or no, thought it the safest way to obviate objections, and remove scruples, by shutting up another mouse as close as I could in the receiver; wherein it lived about three quarters of an hour, and might probably have done so much longer, had not a Virtuoso of quality, who in the mean while chanced to make me a visit, desired to see whether or no the mouse could be killed by the exsuction of the ambient air: whereupon we thought fit to open, for a little while, an intercourse betwixt the air in the receiver, and that without it, that the mouse might thereby (if it were needful for him) be refreshed; and yet we did this without uncementing the cover at the top, that it might not be objected, that perhaps the vessel was more closely stopped for the exsuction of the air than before.

The experiment had this event, that after the mouse had lived ten minutes (which we ascribed to this, that the pump, for want of having been lately oiled, could move but slowly, and could not by him that managed it be made to work as nimbly as it was wont) at the end of that time he died with convulsive fits, wherein he made two or three bounds into the air, before he fell down dead.

Nor was I content with this, but for your Lordship's farther satisfaction, and my own, I caused a mouse, that was very hungry, to be shut in all night, with a bed of paper for him to rest upon: and to be sure that the receiver was well closed, I caused some air to be drawn out of it, whereby, perceiving that there was no sensible leak, I presently readmitted the air at the stop-cock, lest the want of it should harm the little animal; and then I caused the engine to be kept all night by the fire-side, to keep him from being destroyed by the immoderate cold of the frosty night. And this care succeeded so well, that the next morning I found that the mouse was not only alive, but had devoured a good part of the cheese that had been put in with him. And having thus kept him alive full twelve hours, or better, we did, by sucking out part of the air, bring him to droop, and to appear swelled; and by letting in the air again, we soon reduced him to his former liveliness.

A Digression Containing Some Doubts Touching Respiration.

I fear your Lordship will now expect, that to these experiments I should add my reflections on them, and attempt, by their assistance, to resolve the difficulties that occur about respiration; since at the beginning I acknowledged a farther enquiry into the nature of that, to have been my design in the related trials. But I have yet, because of the inconvenient season of the year, made so few experiments, and have been so little satisfied by those I have been able to make, that they have hitherto made respiration appear to me rather a more, than a less mysterious thing, than it did before. But yet, since they have furnished me with some such new considerations, concerning the use of the air, as confirms me in my diffidence of the truth of what is commonly believed touching that matter; that I may not appear sullen or lazy, I am content not to decline employing a few hours in setting down my doubts, in presenting your Lordship some hints, and in considering whether the trials made in our en-

gine will at least assist us to discover wherein the deficiency lies that needs to
be supplied.   . . .

For first, many there are, who think the chief, if not sole use of respiration,
to be the cooling and tempering of that heat in the heart and blood, which
otherwise would be immoderate; and this opinion not only seems to be most
received amongst scholastic writers, but divers of the new philosophers, Cartesians and others, admitted with some variation; teaching, that the air is
necessary, by its coldness, to condense the blood that passeth out of the right
ventricle of the heart into the lungs, that thereby it may contain such a consistence as is requisite to make it fit fewel for the vital fire or flame, in the left
ventricle of the heart.   And this opinion seems favoured by this, that fishes,
and other cold creatures, whose hearts have but one cavity, are also unprovided
of lungs, and by some other considerations.   But though it need not be denied,
that the inspired air may sometimes be of use by refrigerating the heart, yet
(against the opinion that makes this refrigeration the most genuine and constant use of the air) it may be objected, that divers cold creatures (some of
which, as particularly frogs, live in the water) have yet need of respiration;
which seems not likely to be needed for refrigeration by them that are destitute
of any sensible heat, and besides, live in the cold water: that even decrepid old
men, whose natural heat is made very lanquid, and almost extinguished by
reason of age, have yet a necessity of frequent respiration: that a temperate air
is fittest for the generality of breathing creatures; and as an air too hot, so also
an air too cold may be inconvenient for them (especially if they be troubled
with an immoderate degree of the same quality which is predominant in the
air:) that in some diseases the natural heat is so weakened, that in case the
use of respiration were too cool, it would be more hurtful than beneficial to
breathe; and the suspending of the respiration may supply the place of those
very hot medicines that are wont to be employed in such distempers: that nature might much better have given the heart but a moderate heat, than such
an excessive one, as needs to be perpetually cooled, to keep it from growing
destructive; which the gentle, and not the burning heat of an animal's heart,
seems not intense enough so indispensably to require.   These, and other objections might be opposed, and pressed against the recited opinion; but we shall
not insist on them, but only add to them, that it appears not by our foregoing
experiments that in our exhausted receiver, where yet animals die so suddenly
for want of respiration, the ambient body is sensibly hotter than the common
air.

Other learned men there are, who will have the very substance of the air to
get in by the vessels of the lungs, to the left ventricle of the heart, not only to
temper its heat, but to provide for the generation of spirits.   And these alledge
for themselves the authority of the ancients, among whom *Hippocrates* seems
manifestly to favour their opinion; and both *Aristotle* and *Galen* do sometimes

(for methinks they speak doubtfully enough) appear inclinable to it. But for aught ever I could see in dissections, it is very difficult to make out, how the air is conveyed into the left ventricle of the heart, especially the systole and the diastole of the heart and lungs being very far from being synchronical: besides, that the spirits seeming to be but the most subtile and unctuous particles of the blood, appear to be of a very differing nature from that of the lean and incombustible corpuscles of air. Other objections against this opinion have been proposed, and pressed by that excellent Anatomist, and my industrious friend Dr. *Highmore,* to whom I shall therefore refer you.

Another opinion there is touching respiration, which makes the genuine use of it to be ventilation not of the heart, but of the blood, in its passage through the lungs; in which passage it is disburthened of those excrementitious steams preceding for the most part, from the superfluous serosities of the blood (we may add) and of the chyle too, which (by those new conduits of late very happily detected by the famous *Pecquet*) hath been newly mixed with it in the heart. And this opinion is that of the industrious *Moebius,* and is said to have been that of that excellent philosopher *Gassendus;* and hath been, in part, an opinion almost vulgar. But this hypothesis may be explicated two ways: for first, the necessity of the air in respiration may be supposed to proceed from hence; that as a flame cannot long burn in a narrow and close place, because the fulginous steams it incessantly throws out, cannot be long received into the ambient body; which, after a while, growing too full of them to admit any more, stifles the flame: so that the vital fire in the heart requires an ambient body of a yielding nature, to receive into it the superfluous serosities, and other recrements of the blood, whose seasonable expulsion is requisite to depurate the mass of blood and make it fit, both to circulate and to maintain the vital heat residing in the heart. The other way of explicating the above-mentioned hypothesis is, by supposing, that the air doth not only, as a receptacle, admit into its pores the excrementitious vapours of the blood, when they .are expelled through the wind-pipe, but doth also convey them out of the lungs, in regard that the inspired air reaching to all the ends of the *aspera arteria,* doth there associate itself with the exhalations of the circulating blood, and when it is exploded, carries them away with itself: as we see that winds speedily dry up the surfaces of wet bodies, not to say any thing of what we formerly observed touching our liquor, whose fumes were strangely elevated upon the ingress of the air.

Now, of these two ways of explicating the use of respiration, our engine affords us this objection against the first; that upon the exsuction of the air, the animal dies a great deal sooner than if it were left in the vessel; though by that exsuction, the ambient space is left much more free to receive the steams that are either breathed out of the lungs of the animal, or discharged by insensible transpiration through the pores of his skin.

But if the hypothesis proposed be taken in the other sense, it seems congruous enough to that grand observation, which partly the phaenomena of our engine, and partly the relations of travellers, have suggested to us; namely, that there is certain consistence of air requisite to respiration: so that if it be too thick, and already over-charged with vapours, it will be unfit to unite with, and carry off those of the blood, as water will dissolve and associate to itself but a certain proportion of saline corpuscles; and if it be too thin or rarefied, the number or size of the aërial particles is too small to be able to carry off the halituous excrements of the blood, in such plenty as is requisite.

Now that air too much thickened (and as it were clogged) with steams, is unfit for respiration, may appear by what is wont to happen in the lead mines of *Devonshire* and for aught I know, in those too of other countries, (though I have seen mines where no such thing was complained of) for I have been informed by more than one credible person (and particularly by an ingenious man that hath often, for curiosity, digged in those mines and been employed about them) that there often riseth damps (as retaining the German word by which we call them) which doth so thicken the air, that, unless the workmen speedily make signs to them that are above, they would (which also sometimes happens) be presently stifled for want of breath: and though their companions do make haste to draw them up, yet frequently, by that time they come to the free air, they are, as it were, in a swoon, and are a good while before they come to themselves again. And that this swooning seems not to proceed from any arsenical or poisonous exhalation contained in the damp, as from its overmuch condensing the air, seems probable from hence; that the same damps oftentimes leisurely extinguish the flames of their candles or lamps; and from hence also that it appears (by many relations of authentical authors) that in those cellars where great store of new wine is set to work, men have been suffocated by the too great plenty of the steams exhaling from the must, and too much thickning the air: as may be gathered from the custom that is now used in some hot countries, where those that have occasion to go into such cellars, carry with them a quantity of well-kindled coals, which they hold near their faces; whereby it comes to pass, that the fire discussing the fumes, and rarefying the air, reduceth the ambient body to a consistence fit for respiration.

We will add (by way of confirmation) the following experiment: in such a small receiver, as those wherein we killed divers birds, we carefully closed up one, who, though for a quarter of an hour he seemed not much prejudiced by the closeness of his prison, afterwards began first to pant very vehemently, and keep his bill very open, and then to appear very sick; and last of all, after some long and violent strainings, to cast up some little matter out of his stomach; which he did several times, till growing so sick that he staggered and gasped, as being just ready to die. We perceived, that within about three quarters of an hour from the time that he was put in, he had so thickened and tainted the air with the steams of his body, that it was become altogether unfit for the use of respiration: which he will not much wonder at, who hath taken notice in *Sanctorius* his *Statica Medicina,* how much that part of our aliments which goeth off by insensible transpiration, exceeds in weight all the visible and grosser excrements both solid and liquid.

That (on the other side) an air too much dilated is not serviceable for the ends of respiration, the hasty death of the animal we killed in our exhausted receiver seems sufficiently to manifest. And it may not irrationally be doubted, whether or no, if a man were raised to the very top of the atmosphere, he would be able to live many minutes, and would not quickly die for want of such air as we are wont to breathe here below. And that this conjecture may not appear extravagant, I shall, on this occasion, subjoin a memorable relation that I have met with in the learned *Josephus Acosta,* who tells us, that when he himself passed the high mountains of *Peru* (which they call *Pariacaca*) to which, he says, that the *Alps* themselves seemed to them but as ordinary houses in regard of high towers, he and his companions were surprized with such extreme pangs of straining and vomiting (not without casting up blood too) and with so violent a distemper, that he concludes he should undoubtedly have died, but that this lasted not above three or four hours, before they came into a more convenient and natural temperature of air: to which our learned author adds an inference, which being the principal thing I designed in mentioning the narrative, I shall set down in his own words: *I therefore* (says he) *persuade myself, that the element of the air is there so subtile and delicate, as it is not proportionable with the breathing of man, which requires a more gross and temperate air; and I believe it is the cause that doth so much alter the stomach, and trouble all the disposition.* Thus far our author, whose words I mention, that we may guess, by what happens somewhat near the confines of the atmosphere (though probably far from the surface of it) what would happen beyond the atmosphere. That, which some of those that treat of the height of mountains, relate out of *Aristotle,* namely, that those that ascend to the top of the mountain *Olympus,* could not keep themselves alive, without carrying with them wet spunges, by whose assistance they could respire in that air, otherwise too thin for respiration (that relation, I say, concerning this mountain) would much confirm what hath been newly recited out of *Acosta,* if we had sufficient reason to believe it. But I confess I am very diffident of the truth of it; partly, because when I passed the *Alps,* I took notice of no notable change betwixt the consistence of the air at the top and the bottom of the mountain; partly, because in a punctual relation made by an English gentleman, of his ascension to the top of the pike of *Tenariff* (which is by great odds higher than *Olympus*) I find no mention of any such difficulty of breathing; and partly also, because the same author tells us out of *Aristotle,* that upon the top of *Olympus* there is no motion of the air, insomuch that letters traced upon the dust, have been, after many years, found legible and not discomposed; whereas that inquisitive *Busbequius* (who was ambassador from the German to the Turkish emperor) in one of his eloquent Epistles, tells us, upon his own knowledge, that Olympus *may be seen from* Constantinople, *blanched with perpetual snow;* which seems to argue, that the top of that, as well as of divers other tall hills, is not above that region of the air wherein meteors are formed. Though otherwise, in that memorable narrative which *David Fraelichius* made of his ascent to the top of the prodigiously high Hungarian mountain *Carpathus,* he tells us, *that when having passed through very thick clouds, he came to the very top of the hill, he found the air so calm and subtile, that not a hair of his head moved, whereas, in the lower stages of the mountain, he felt a vehement wind.* But this might well be casual, as was his, having clear air where he was, though there were clouds, not only beneath him, but above him.

But, though what hath been hitherto discoursed, incline us to look upon the ventilation and the depuration of the blood, as one of the principal and constant uses of respiration; yet methinks it may be suspected that the air doth something more than barely help to carry off what is thrown out of the blood, in its passage through the lungs, from the right ventricle of the heart to the left. For we see, in phlegmatic constitutions and diseases, that the blood will circulate tolerably well, notwithstanding its being excessively serous: and in asthmatical persons, we often see that though the lungs be very much stuffed with tough phlegm, yet the patient may live some months, if not some years. So that it seems scarce probable, that either the want of throwing out the superfluous serum of the blood for a few moments, or the detaining it, during so short a while, in the lungs, should be able to kill a perfectly sound and lively animal: I say, for a few moments, because, that having divers times tried the experiment of killing birds in a small receiver, we commonly found, that within half a minute of an hour, or thereabout, the bird would be surprised by mortal convulsions, and within about a minute more would be stark dead, beyond the recovery of the air, though never so hastily let in. Which sort of experiments seem so strange, that we were obliged to make it several times, which gained it the advantage of having persons of differing qualities, professions and sexes (as not only ladies and lords, but doctors and mathematicians) to witness it. And to satisfy your Lordship that it was not the narrowness of the vessel, but the sudden exsuction of the air that dispatched these creatures so soon; we will add, that we once inclosed one of these birds in one of these small receivers, where, for a while, he was so little sensible of his imprisonment, that he eat very cheerfully certain seeds that were conveyed in with him, and not only lived ten minutes, but had probably lived much longer, had not a great person, that was spectator of some of these experiments, rescued him from the prosecution of the trial. Another bird being within about half a minute cast into violent convulsions, and reduced into a sprawling condition, upon the exsuction of the air, by the pity of some fair ladies, related to your Lordship, who made me hastily let in some air at the stop-cock, the gasping animal was presently recovered, and in a condition to enjoy the benefit of the ladies compassion. And another time also, being resolved not to be interrupted in our experiment, we did at night shut up a bird in one of our small receivers, and observed that for a good while he so little felt the alteration of the air, that he fell asleep with his head under his wing; and though he afterwards awaked sick, yet he continued upon his legs between forty minutes and three quarters of an hour: after which, seeming ready to expire, we took him out, and soon found him able to make use of the liberty we gave him for a compensation of his sufferings.

If to the foregoing instances of the sudden destruction of animals, by the removal of the ambient air, we should now annex some, that we think fitter to reserve till anon; perhaps your Lordship would suspect, with me, that there is some use of the air which we do not yet so well understand, that makes it so continually needful to the life of the animals. *Paracelsus*, indeed, tells us, *that as the stomach concocts meat, and makes part of it useful to the body, rejecting the other part; so the lungs consume part of the air, and proscribe the rest.* So that, according to our Hermetic philosopher (as his followers would have him styled) it seems we must suppose, that there is in the air a little vital quintessence (if I may so call it) which serves to the refreshment and restauration of our vital spirits, for which use the grosser and incomparably greater part of the air being unserviceable, it need not seem strange, that an animal stands in need of almost incessantly drawing in fresh air. But though this opinion is not (as some of the same author) absurd, yet besides that it should not be barely asserted, but explicated and proved; and besides that some objections may be framed against it, out of what hath been already argued against the transmutation of air into vital spirits: besides these things, it seems not probable, that the bare want of the generation of the wonted quantity of vital spirits, for less than one minute, should, within that time, be able to kill a lively animal, without the help of any external violence at all.

But yet, on occasion of this opinion of *Paracelsus*, perhaps it will not be impertinent if, before I proceed, I acquaint your Lordship with a conceit of that deservedly famous Mechanician and Chymist, *Cornelius Drebell*, who, among other strange things that he performed, is affirmed, by more than a few credible persons, to have contrived, for the late learned King *James*, a vessel to go under water; of which, trial was made in the *Thames*, with admired success, the vessel carrying twelve rowers, besides passengers; one of which is yet alive, and related it to an excellent Mathematician that informed me of it. Now that for which I mention this story is, that having had the curiosity and opportunity to make particular enquiries among the relations of *Drebell*, and especially of an ingenious Physician that married his daughter, concerning the grounds upon which he conceived it feasible to make men unaccustomed to continue so long under water without suffocation, or (as the lately mentioned person that went in the vessel affirms) without inconvenience; I was answered, that *Drebell* conceived, that it is not the whole body of the air, but a certain quintessence (as Chymists speak) or spirituous part of it, that makes it fit for respiration; which being spent, the remaining grosser body, or carcase, if I may so call it, of the air, is unable to cherish the vital flame residing in the heart: so that, for aught I could gather, besides the mechanical contrivance of his vessel, he had a chymical liquor, which he accounted the chief secret of his submarine navigation. For when, from time to time, he perceived that the finer and purer part of the air was consumed, or overclogged by the respiration and steams of those that went in his ship, he would, by unstopping a vessel full of this liquor, speedily restore to the troubled air such a proportion of vital parts, as would make it again, for a good while, fit for respiration, and it gave us also occasion to suspect, that if insects have no lungs, nor any part analogous thereunto, the ambient air affects them, and relieves them at the pores of their skin; it not being irrational to extend to these creatures that of *Hippocrates,* who saith, that a living body is throughout perspirable; or, to use his expression, εἰσπνόιε ἐχπνὸν, disposed to admit and part with what is spirituous. Which may be somewhat illustrated by what we have elsewhere noted, that the moister parts of the air readily insinuate themselves into, and recede from the pores of the beards of wild oats, and those of divers other wild plants; which almost continually wreath and unwreath themselves according to, even, the light variations of the temperature of the ambient air.

This circumstance of our experiment we particularly took notice of, that when at any time, upon the ingress of the air, the bee began to recover, the first sign of life she gave, was a vehement panting, which appeared near the tail; which we therefore mention, because we have observed the like in bees drowned in water, when they first come to be revived by a convenient heat: as if the air were in the one case as proper to set the spirits and alimental juice moving, as heat is in the other; and this may, perchance, deserve a farther consideration.

We may add, that we scarce ever saw any thing that seemed so much as this experiment to manifest, that even living creatures (man always excepted) are a kind of curious engines, framed and contrived by nature (or rather the author of it) much more skillfully than our gross tools and imperfect wits can reach to. For in our present instance we see animals, vivid and perfectly sound, deprived immediately of motion, and any discernable signs of life, and reduced to a condition that differs from death, but in that it is not absolutely irrecoverable. This (I say) we see performed without any, so much as the least external violence offered to the engine; unless it be such as is offered to a wind-mill, when the wind ceasing to blow on the sails, all the several parts remain moveless and useless, till a new breath put them into motion again.

And this was farther very notable in this experiment; that whereas it is known that bees and flies will not only walk, but fly for a great while, after their heads are off; and sometimes one half of the body will, for divers hours, walk up and down, when it is severed from the other: yet, upon the exsuction of the air, not only the progressive motion of the whole body, but the very motions of the limbs do forthwith cease; as if the presence of the air were more necessary to these animals, than the presence of their own heads.

But, it seems, that in these insects, that fluid body (whether it be juice or flame) wherein life chiefly resides, is nothing near so easy dissipable as in the perfect animals. For whereas we have above recited, that the birds we conveyed into our small receiver were within two minutes brought to be past, we were unable (though by tiring him that pumped) to kill our insects by the exsuction of the air: for though, as long as the pump was kept moving, they continued immovable, yet, when he desisted from pumping, the air that pressed in at the unperceived leaks did, though slowly, restore them to the free exercise of the functions of life.

But, my Lord, I grow troublesome, and therefore shall pass on to other experiments: yet without despairing of your pardon for having entertained you so long about the use of respiration, because it is a subject of that difficulty to be explained, and yet of that importance to human life, that I shall not regret the trouble my experiments have cost me, if they be found in any degree serviceable to the purposes to which they were designed. And though I despair not but that hereafter our engine may furnish us with divers phaenomena useful to illustrate the doctrine of respiration; yet having not, as yet, had the opportunity to make other trials, of various kinds, that I judge requisite for my information, I must confess to your Lordship, that in what I have hitherto said, I pretend not so much to establish or overthrow this or that hypothesis, as to lay together divers of the particulars, because I could add many others, but that I want time, and fear that I should need your Lordship's pardon for having been so prolix in writing; and that of Physicians (which perhaps I shall more easily obtain) for having invaded anatomy, a discipline which they challenge to themselves, and indeed have been the almost sole improvers of. Without denying then, that the inspired and exspired air may be sometimes very useful, by condensing and cooling the blood that passeth through the lungs; I hold that the depuration of the blood in that passage, is not only one of the ordinary, but one of the principal uses of respiration. But I am apt also to suspect, that the air doth something else in respiration, which hath not yet been sufficiently explained; and therefore, till I have examined the matter more deliberately, I shall not scruple to answer the questions that may be asked me, touching the genuine use of respiration, in the excellent words employed by the acute St.

Austin, to one who asked him hard questions: *Mallem quidem* (says he) *eorum quae à me quaesivisti, habere scientiam quam ignorantiam: sed quia id nondum potui, magis eligo cautam ignorantiam consiteri, quam falsam scientiam profiteri.*

DIRECT OBSERVATION OF CAPILLARY CIRCULATION

MALPIGHI, Marcello (Italian physician and microscopist, 1628–1694). From *De pulmonibus epistola altera,* Bononia, 1661; tr. by T. S. Hall for this volume.

As the first to employ the microscope for exhaustive systematic observation of animals and plants, Malpighi may be considered the principal founder of the science of microanatomy. He here first describes, in his usual difficult style, the connections between arteries and veins. The demand for these had existed since Harvey's proof of the circulation published in the year of Malpighi's birth (see Harvey, Fabricius, Servetus).

To eyes otherwise attentive solely to structure and composition [*i.e.,* of the lung—Ed.], microscopic observation reveals things even more remarkable. For, provided the heart still beats, contrary movements of the blood in the veins are (admittedly with difficulty) to be seen. By this is manifestly revealed the blood's circulation, which is also, and even more happily, to be discerned in the mesentery and other major veins contained in the abdomen. In this way the blood pours floodlike into the smallest openings, through the arteries, and by one or another of the branches crossing through or terminating here, into each little compartment. And, by being so much subdivided, it loses its red color, and, conducted roundaboutly, is everywhere distributed until it reaches the compartmental walls and angles, and the reabsorbing branches of the veins.

The power of the unaided eye could not be extended further in the dissected, animate living being, whence I had been led to believe that the blood's substance was emptied into a vacant space, and, by some open pathway, was recollected by the peculiar structure of the compartmental walls, the possibility of which was attested by the blood's motion (twisting, and pouring out in all directions) and its coming together (all toward one side). My belief was shaken, however, by dried frog's lung which keeps its blood-redness even in the smallest little *vessels* (for that such was their nature was to be grasped later) where, by means of a more perfect glass, there met the eye not a skin fashioned of specks, as in what we call 'Sagrino,' but actually small vessels attached to each other like rings. And so extensive is the subdivision of these little vessels, issuing thus from artery and vein, that the orderly arrangement proper to vessels is no longer maintained, but instead there comes into view a net, constituted by the extensions of the two sorts of vessels. Not only does the net occupy the whole empty region, but it extends also to the compartmental walls, and is augmented by the excurrent vessels,—just as I was able to observe, also, (admittedly with difficulty) in the elongated and equally membranous and transparent lung of the tortoise. Here it appeared that the subdivided blood, by this impetus, ran through twisted vessels and was not poured out into empty spaces, but was driven through little tubes and dispersed by the frequent turnings of the vessels. Nor is it uncustomary for nature to join together the open ends of the vessels, for she does the same thing in the intestines and other body regions, which will not seem more remarkable than that she should join upper termini of veins to lower ones in a visible anastomosis, as the most learned Fallopius has very well shown.

TRANSFUSIONS ACCOMPLISHED

LOWER, Richard (English physician, 1631–1691). From *Tractatus de corde (item de motu & colore sanguinis, et chyli in eum transitu, etc.*), Amsterdam, 1771; tr. by M. W. Hollingsworth in his *Blood transfusion by Richard Lower in 1665,* in *Annals of Medical History,* vol. 10, No. 3, p. 213, 1928; by permission of Paul B. Hoeber, Inc.

It is probable that mutual blood transfusions were performed in 1492 between Pope Innocent VIII and three healthy boys, an experiment culminating in the deaths of all concerned including the Pope.* Richard Lower is usually accredited with the first published account of a successful transfusion.†

London, June 26, 1666.

Honorable Sir:

At a meeting of the Royal Society at Gresham College last Wednesday I heard from Dr. Wallis that you had acquitted yourself successfully of that difficult experiment of transfusing blood from one animal into another [he having been present at your experiment]. I decided the operation to be worthy of being made known to that honorable body and accordingly suggested they call upon Dr. Wallis for a description of it. His reply added not a little to your reputation among us. He was of the opinion that you should furnish a written report, rather than that he should tell about it, particularly as the request had come so unexpectedly. I then personally mentioned the fact that a little while

---

* For details, see B. J. Ficarra, *Evolution of blood transfusion,* in *Annals of Medical History,* series 3, vol. 4, No. 4, page 302, 1942.

† See *Diary of Samuel Pepys,* November 14, 1666. "To the Pope's Head, where all the Houblons were, and Dr. Croone. Dr. Croone told me, that, at the meeting at Gresham College to-night, which it seems, they now have every Wednesday again, there was a pretty experiment of the blood of one dog let out, till he died, into the body of another on one side, while all his own ran out on the other side. The first died upon the place, and the other very well, and likely to do well. This did give occasion to many pretty wishes, as of the blood of a Quaker to be let into an Archbishop, and such like; but, as Dr. Croone says, may, if it takes, be of mighty use to a man's health, for the amending of bad blood by borrowing from a better body."

ago you had promised to acquaint me with the details of the experiment [when convenient], but that I realized you were more familiar with the subject and consequently in a better position than I to intelligently describe the operation before such an august assemblage. Therefore, I earnestly request you to consider it worth your while to explain the entire technique of the experiment, in which you succeeded so admirably. Now I urge you to be very exact in this because several clever men who are right up to the mark, critical enough but not too credulous, will pass severe judgment on this lofty subject. They might consider it spoken rashly when, on being questioned by the Royal Society, I incidentally mentioned a few months previously you had already attempted this experiment at Oxford before this although at that time the experiment did not succeed in all respects because some of the apparatus was not suitable. Nevertheless, do not disappoint me in not releasing the story at an early date. I am called away at this moment but I entreat you to favor me by not making my request one in vain. I will not regret causing you this inconvenience if this dignified convention recognizes you on a favorable occasion. There are many men among them who feel under obligation to you and who would like to cultivate your friendship. As for the rest,

Yours respectfully,
Robert Boyle.

To my honorable friend,
D. Richard Lower, M.D.,
Oxford.

Oxford, July 6, 1666.

Dear friend:

I received your letter and in accordance with your request I herewith briefly explain the whole method of blood transfusion. Select either a dog or whatever animal you wish to use as a donor to supply blood to another animal of like or diverse species, expose a cervical artery, separate it from the eighth nerve and denude it almost a finger's length. The cephalic end is ligated firmly with a cord as it will not be necessary to unfasten this during the whole operation. Another ligature is placed around the same vessel a half-finger's length below the first and secured by a slip-knot that can be tightened or loosened at will. After the ligatures have been arranged in this manner, that portion of the artery lying between the ligatures is slit with a scalpel and in the incision a reed is inserted towards the heart, with one end projecting free like a little wooden staff. Another ligature is now placed around the vessel enclosing the reed and drawn tight.

The jugular vein of the animal that is to receive the blood is denuded a half-finger's length and at both extremities of the denuded portion ligatures are so placed that it is possible to tighten or loosen the knots at will. Another pair of ligatures is then placed in position between the first pair. That portion of the denuded vein lying between the two middle ligatures is now incised lengthwise and the two reeds inserted, one towards the heart to conduct the blood from the donor and the other towards the head from which the animal is to pour forth its blood into a saucer. In the mean time, it may be necessary to stop up the ends of the reeds with wooden stopples until ready, or to ligate the vein above. The two dogs are now pinioned in juxtaposition so that the reed of one is approximated to that of the other. It will now be found necessary to supply an intermediate coupling to direct the blood from one reed to the other.

On completion of this arrangement of apparatus, the reed leading downward into the jugular vein of the second dog is opened and the one projecting from the cervical artery of the first dog also opened and the two ends connected. Afterwards, the farther knots are to be loosened and immediately the blood will flow through the reeds as rapidly as through an unbroken artery, in fact, just as in an anastomosis. As soon as possible the neck of the recipient is constricted with a torniquet as is customary in phlebotomy, or at least the vein on the opposite side of the neck is compressed with the fingers. The reed in the upper end of the jugular is immediately unstoppered so that as the lower end is being filled with blood from the donor, blood, which always divides up into dark red clots as human blood does, may be released from the recipient into a saucer from time to time, until the donor, midst wailing, fainting, and terminal spasms, will give up the ghost with his vital fluid.

After this tragedy is enacted, both reeds are removed from the surviving animal and the slip knots drawn tight where the vein is cut open. This causes practically no inconvenience to a dog as it has adequate anastomoses of the jugulars around the larynx which suffice to carry the blood from the head. The rent in the skin is sutured, the chains removed and the dog permitted to jump down from the table. After shaking itself as though just awakening from sleep, it will be more lively than its partner from whom life and strength departs with its blood.

I have only one admonition, my dear friend, and that is that the reeds be secured in the vessels with tight ligatures before they are approximated; otherwise they will be loosened by the struggles of the animals and the whole thing will be to do over again. As I have become more experienced, I have lately constructed a silver cannula which does not tear the vessel where it is inserted and which has a projecting ring or ridge running around the outside near the end which facilitates securing it inside the vessel. It is more convenient and affords less danger of tearing or obstructing the vessel, particularly when the animals toss and twist about, to procure a section from the cervical artery of a cow or horse and insert smaller cannulas in each end which may be used to connect the cannula of the recipient with that of the donor. This intermediate section from an artery has afforded us better success not only because of its elasticity which allows it to yield to the movements of the struggling animals but also because in case the tube becomes stopped up one can strip down the tube with the fingers, imparting a squeezing motion and start the blood flowing again. This letter furnishes you with the data for the Honorable Society and I trust it will not at any time be found wanting in any respect.

Yours most respectfully,
Richard Lower.

Written to the most honorable
Robert Boyle, D.D., London.

Although transfusion had been perfected by me toward the end of February, 1665, and Boyle's letter to me under date of the sixth of the following June and my reply to same were ordered set up in type for the Transactions of the Philosophical Society the following December, no mention of this was made on the part of Dionys during the whole year afterwards; and furthermore, let him acknowledge [since, as he says, no thought concerning this in any way had entered his mind during the previous ten years] that he had learned of transfusion and the manner of performing it from the Philosophical Journal of England. I will leave it to anyone to decide whether the invention of this experiment should be accredited to him.

But when a man's mind has reached such a state of depravity that nothing pleases him he does not himself originate, or nothing is well planned he does not plan himself, he will not be a worry to me. In any case I am not concerned that this idea ever occurred to anyone else, for abundant testimony of famous men is quoted above that it was perfected by me. Meanwhile there is no doubt that anyone, whoever he may be, who might perform this operation skillfully just at this time, will become famous.

One would hardly consider human blood less compatible with itself than with that of other animals, as recently proved by numerous experiments in France. We have lately proved this in the case of A. C., an amiable man, in whom we have kept back impending insanity by transfusing several ounces of sheep's blood into his forearm at different times before meetings of the Royal Society; this was done without any great inconvenience to him. This therapy was repeated several times in this case and was conducive to a more sound mind, unless his mental state, rather than his physical condition, should prove our hopes illfounded.

It is not assumed from this that the blood of all animals is compatible, nor is it conceded that the remedy is so useful that an ill-advised or inopportune administration of it will not incur ill repute. I will briefly list a few of the indications for transfusion. In those whose blood is exceedingly putrid and foul, or in whom it becomes completely saturated with extraneous ferments and poisons, or in whom their tissues are polluted and weakened, especially from scurvy, syphilis, leprosy, poisoning, or other chronic illnesses where the tissues are wasted everywhere, much improvement from transfusion is not to be expected.

Without doubt blood becomes impure by absorbing ferments during its repeated circulation through vitiated organs, the ferments of which destroy it, and thus it becomes contaminated with any particular diseased condition present. If such blood is used on a healthy animal and allowed to circulate through its body the animal will contract the disease and quickly sicken, in the same way that wine will become odorous from mouldy containers and soon spoil.

Unless a clean dish is used, whatever is poured into it sours.

When a vein is severed by accident or inflicted wounds, or when simple hemorrhage results in such a great loss of blood that succor from some source is demanded in the emergency, there is no doubt but that the blood of an animal could be allowed to slip in and substitute in its place to advantage. But in the case of the arthritics and the maniacs whose robust bodies, sound organs and dense brains have not yet become vitiated, and in whom the blood has not become infected with putrid poisons, possibly no less benefit may be expected from the new blood transfusion than from the older purgative treatment.

In order that men may acquire greater confidence in the application of this famous experiment and its usage become known, it has not seemed to me to be out of place to recommend it to physicians the world over as a treatment to be used whenever the occasion presented itself.

At least it is a comfort to our Nation and a credit to our fame that Harvey became preeminent by first demonstrating that blood circulated in the vessels inside the body. That this circulation could be extended outside the body was first discovered by me.

INTERPRETATION OF KIDNEY FUNCTION

MALPIGHI, Marcello (Italian microscopical anatomist, 1628–1694). From *De viscerum structura*, Amsterdam, 1669; tr. by J. M. Hayman as *Malpighi's Concerning the structure of the kidneys*, in *Annals of Medical History*, vol. 7, No. 3, p. 242, New York, 1925; by permission of Paul B. Hoeber, Inc., publisher.

Malpighi's study of the kidney was in essence an empirical extension of Aristotle's statement that when blood "percolates through the kidneys the excretion which results collects into the middle of the kidneys where the hollow is in most cases."[*] Drawing clear distinctions between observation and speculation and recognizing the inadequacy of either used alone, Malpighi admits that he has not obtained full proof for his extension of Aristotle's idea. Time has rewarded his precautions by upholding the view which,

---

[*] *De partibus animalium* 671 b 25.

with every effort at objectivity, he advanced. Other studies by this pioneer microanatomist of Pisa and Bologna were made on the lung (see this volume), the brain[†] (with many misinterpretations), the chick embryo,[‡] the silkworm,[§] plants.[||]

Concerning the Kidneys
INTRODUCTION

For a long time the kidneys have been the subject of varying opinions, some even having regarded them as superfluous and unnecessary, a thought which is certainly not a tribute to Nature. More recently, however, because of their wonderful structure, and because of the very necessary function attributed to them, they have attained a place among the important parts of the body. So many different views regarding their composition are held by anatomists that there is little agreement.

The ancients conceived of a sieve which provided a means for separating the urine. Many have been satisfied simply with the name "parenchyma." In the meantime, the idea of fibers for drawing out the fluid pleased some, and this idea was strengthened by the similar structure of the heart. Among subsequent writers the existence of fibers in the kidney appeared doubtful and unlikely, whereupon they announced that when the substance of the kidney was cut, certain little canals were to be seen. Later some have contended that the substance of the kidney is complex, still more recently it has been stated in a very elaborate work that the substance of the kidney consists of a single fibrous substance, permeated by little canals. This was determined from cut sections in which, everything but vessels having been excluded, it was evident that the body of the kidney consists of nothing but a collection of little canals or channels which increase in size uninterruptedly from the external surface toward the center.

The fact that the human mind has pondered these and similar ideas about the kidneys through the ages stimulated me to further investigation, or at least to the confirmation of the statements of others. Study with me, then, a few things in the spirit of truth alone, so that we may establish the manner of Nature's operations in the individual viscera as I have revealed it in the liver and other organs. For this essay which I plan will perhaps shed light upon the structure of the kidney. Do not stop to question whether these ideas are new or old, but ask, more properly, whether they harmonize with Nature. And be assured of this one thing, that I never reached my idea of the structure of the kidney by the aid of books, but by the long, patient, and varied use

---

[†] *Epistolae anatomicae* (with Fracassati), Amsterdam, 1662.

[‡] *Anatome plantarum cum appentice observationes de ovo incubato continente*, London, 1672.

[§] *Dissertatio epistolica de bombyce*, London, 1669.

[||] See third footnote above.

of the microscope. I have gotten the rest by the deductions of reason, slowly, and with an open mind, as is my custom.

THE GLANDS IN THE INTERIOR OF THE KIDNEY AND THEIR CONNECTION WITH THE VESSELS

Since in the previous section we have mentioned the glands that have been discovered in the kidney, which, as will be shown below, contribute a special service in the excretion of the urine, it is now proper to pause briefly to consider their structure. These glands, situated in the outer part of the kidney, are almost innumerable, and probably, as I think, correspond in number to the urinary vessels by which the mass of the kidney is formed. The number of the urinary vessels in each fasciculus, by which the small lobules which have been described are formed, exceeds forty.

As to the structure of the glands, a distinct outline cannot be obtained on account of their minute size and translucency. They appear, however, spherical, precisely like fish eggs: and when a dark fluid is perfused through the arteries they grow dark, and one would say that all around them are the extreme ends of the blood vessels, which run along like creeping tendrils, so that they appear as it were crowned, with this reservation, however, that the part which is fastened to the branch of the artery grows black; the rest retains its former color.

The glands are connected with the branches of the arteries in the following manner. They spring from the deeper branches of the arteries and in some instances from the superficial branches which are bent inward and continued into intricate branchlets. That they are attached directly to these vessels is clearly demonstrated by the perfusion of colored fluid through the renal artery, for the glands and the connected arteries are stained with the same color, so that the eye very easily perceives their connection.

The glands are also connected with the veins, which follow the ramifications of the arteries; for when the veins are filled with ink, although the glands may not be filled with the same fluid, nevertheless the color seems to be worked in toward them, so that nothing intervenes between the glands and the ends of the veins. It is probable that the liquid injected through the veins by force, having overcome their valves, clings to the mouths of the glands and is shut out from their different passages.

To these arguments it may be added that the glands are at first white and almost translucent, then red, which certainly happens because of the blood of the arteries which is poured out. Moreover, it is known from the usual custom of Nature that the radicles of the veins take their origin from the same place in which the terminal arteries end, whence, although the senses do not perceive the connection, reason, nevertheless, sufficiently prevails.

About the nerves there will certainly be no difficulty, for it is commonly observed that these are carried throughout the interior of the kidney, whence it is probable that they are led to the glands along with the other vessels, as is seen in like conditions in other organs.

Another question can be asked, namely, whether a portion of the ureter reaches these glands which lie in the blood path; for the pelvis, as will be shown below, embraces in its own curve the veins and the arteries even to their capillary branchings. So we are permitted to think that it may be bound to the glands by its own fibrils.

There remains another vessel of excretion of the urine, of whose branches, as we have hinted above, the outer part of the kidney is chiefly composed. I worked a long time in order that I might subject to the eye this evident connection which reason sufficiently attests. For I have never been able to observe liquids perfused through the arteries penetrating the urinary vessels, even though they fill the glands, and the same is true when the veins are perfused. Liquid perfused by the same method through the ureter stains some divisions of the pelvis with its color, but is not able to blacken these excretory vessels of the urine which some call fibers, and thence the glands. So that in spite of many attempts (but in vain) I could not demonstrate the connection of the glands and the urinary vessels. Recently, however, a method occurred to me for attempting it in a living animal, which I have more often carried out in a dog. I tied the renal veins with a ligature and also, at the same time, the ureter; so that when the animal had lived a long time, I obtained a kidney greatly swollen from the blood which was driven in. When this was cut longitudinally through the back, it distinctly showed the branches of the urinary vessels, or fibers, together with the glands, between which, in some places, where the cortex of the kidney was more distended by the intercepted blood, and a fasciculus of the urinary vessels in turn more opened out, I seemed to see a certain connection and continuation, although not such as satisfied the senses in all particulars. Reason, however, can bring assistance. For if in the liver, in the brain, and in other glands, it is the invariable rule that each single acinus, or globule, of the gland throws out its own excretory duct besides the arteries and veins, the same will have to be said about glandular bodies of this sort. Since, moreover, these urinary vessels, or fibers, are elongated bodies, contiguous surely with the glands, and from their nature must be excretory vessels of the humors (for when compressed they pour out urine) it seems most probable that they have an intimate connection with the glands. If, indeed, as is probably agreed by all, the material of the urine is derived from the arteries, and since it is clear from the evidence above that the ends of the arteries lie open in these very numerous glands, and since the urine is eliminated into the pelvis through the fibers of the kidney as though through its own peculiar excretory vessels, there must of necessity be granted a continuity and communication between them, for otherwise no secreted fluid would be strained out from the arteries into the pelvis.

THE FUNCTION OF THE KIDNEYS

The constant trickle of urine from the kidneys through the ureters, thence carried to the bladder, and voided at a fixed interval, is sufficient indication of their function. But by what means this is accomplished is most obscure. It is reasonable to assume that this is wholly the result of the work of the glands: but since the minute and simple structure of the openings within the glands escapes us, we can only postulate some things in order to give a satisfactorily probable answer to this question. It is obvious that this mechanism accomplishes the work of separation of the urine by its internal arrangement. But whether this arrangement is similar to those devices which we make use of here and there for human needs, and in imitation of which we build rough contrivances, is doubtful. For although similar sponge-like bodies, structures with sieve-like fistulae, may be encountered, it is difficult to determine to which of these the structure of the kidneys is similar in all respects. And since the manifestation of Nature's working is most varied, we may discover mechanisms which are unknown to us and whose operations we cannot understand.

FIRST PUBLISHED ACCOUNT OF SPERM

van LEEUWENHOEK (Leeuwenhoeck), Antony (Dutch microscopist, 1632–1723). A communication to the Secretary of the Royal Society, in *Philosophical Transactions of the Royal Society*, London, 1679; tr. by C. R. Dobell in F. J. Cole, *Early theories of sexual generation*, Oxford, 1930; by arrangement with the Clarendon Press.

The reader will note that Leeuwenhoek's attention was directed to the sperm by another. The letter was communicated in 1677. For details as to who actually first *saw and described sperm*, Cole may be consulted. Authorities agree that claimants to priority (*e.g.*, Hartsoeker) got the idea of looking for sperm from Leeuwenhoek himself.

THE OBSERVATIONS OF MR. ANTONY LEEUWENHOEK, ON ANIMALCULES ENGENDERED IN THE SEMEN [LETTER NO. 22].

A LETTER FROM THE OBSERVER TO THE RIGHT HONOURABLE THE VISCOUNT BROUNCKER; WRITTEN IN LATIN, AND DATED NOVEMBER, 1677; WHICH THE EDITOR* CONSIDERED SHOULD BE PUBLISHED IN THE VERY WORDS IN WHICH IT WAS SENT.

After the distinguished Professor of Medicine Craanen had himself many times honoured me with a visit, he besought me, in a letter, to demonstrate some of my observations to his kinsman Mr. Ham. On the second occasion

---

* Nehimiah Grew.

when this Mr. Ham visited me [in August, 1677], he brought with him, in a small glass phial, the spontaneously discharged semen of a man who had lain with an unclean woman and was suffering from gonorrhoea; saying that, after a very few minutes (when the matter had become so far liquefied that it could be introduced into a small glass tube) he had seen living animalcules in it which he believed to have arisen by some sort of putrefaction. He judged these animalcules to possess tails, and not to remain alive above twenty-four hours. He also reported that he had noticed that the animalcules were dead after the patient had taken turpentine.

In the presence of Mr. Ham, I examined some of this matter which I had introduced into a glass tube, and saw some living creatures in it: but when I examined the same matter more carefully by myself, I observed that they were dead after the lapse of two or three hours.

I have divers times examined the same matter (human semen) from a healthy man (not from a sick man, nor spoiled by keeping for a long time, and not liquefied after the lapse of some minutes; but immediately after ejaculation, before six beats of the pulse had intervened): and I have seen so great a number of living creatures in it, that sometimes more than a thousand were moving about in an amount of material the size of a grain of sand. I saw this vast number of living animalcules not all through the semen, but only in the liquid matter which seemed adhering to the surface of the thicker part. In the thicker matter of the semen, however, the animalcules lay apparently motionless. And I conceived the reason of this to be, that the thicker matter consisted of so many coherent particles that the animalcules could not move in it. These animalcules were smaller than the corpuscles which impart a red colour to the blood; so that I judge a million of them would not equal in size a large grain of sand. Their bodies were rounded, but blunt in front and running to a point behind, and furnished with a long thin tail, about five or six times as long as the body, and very transparent, and with the thickness of about one twenty-fifth that of the body; so that I can best liken them in form to a small earth-nut with a long tail.[1] The animalcules moved forward with a snake like motion of the tail, as eels do when swimming in water: and in the somewhat thicker matter, they lashed their tails some eight or ten times in advancing a hair's breadth. I have sometimes fancied that I could even discern different parts in the bodies of these animalcules: but forasmuch as I have not always been able to do so, I will say no more. Among these animalcules there

---

[1] Leeuwenhoek is here comparing the spermatozoa with the 'nuts' of the plants which form our common 'Earth-nuts' or 'Pig-nuts' (*Bunium flexuosum*). *Bunium* has a tuber-like swelling on its root, and when this is dug up it bears an associated rootlet—hence mimicking the form of the spermatozoon with its head and tail. This comparison is not only apt in itself, but characteristic of Leeuwenhoek's simple manner of expressing himself.

were some still smaller particles, to which I can ascribe nothing but a globular form.

I remember that some three or four years ago I examined seminal fluid at the request of the late Mr. Oldenburg, Secretary of the Royal Society. Looking into the matter I find that he wrote asking me to do so from London, on the 24th of April, 1674: and among other things, he besought me also to examine saliva, chyle, sweat, &c.: but at that time I took the animalcules just described for globules. Yet as I felt averse from making further inquiries, and still more so from writing about them, I did nothing more at that time. What I here describe was not obtained by any sinful contrivance on my part, but the observations were made upon the excess with which Nature provided me in my conjugal relations. And if your Lordship should consider such matters either disgusting, or likely to seem offensive to the learned, I earnestly beg that they be regarded as private, and either published or suppressed as your Lordship's judgment dictates.

I have already many times observed with wonder the parts themselves whereof the denser substance of the semen is mainly made up. They consist of all manner of great and small vessels, so various and so numerous that I misdoubt me not that they be nerves, arteries, and veins. Nay, I have indeed observed these vessels in such great numbers, that I believe I have seen more in a single drop of semen than an anatomist would meet with in a whole day's dissection of any object. And when I saw them, I felt convinced that, in no full-grown human body, are there any vessels which may not be found likewise in sound semen.

Once I fancied I saw a certain form, about the size of a sand grain, which I could compare with some inward part of our body. When this matter had been exposed to the air for some moments, the mass of vessels aforesaid was turned into a watery substance mingled with large oily globules, such as I have formerly described as lying among the vessels of the spinal marrow. On seeing these oily globules, I conceived that the vessels might perhaps serve for the conveyance of the animal spirits, and that they are composed of such a soft substance in order that, as the humour or animal spirits continually flowed through them, they might thereby become consolidated into oily globules of sundry sizes—especially when they are exposed to the air.

Moreover, when this matter had stood a little while, there appeared therein some three-sided bodies terminating at either end in a point, and of the length of the smallest grains of sand, though some may have been a bit bigger. And these were furthermore as bright and clear as if they had been crystals.

THE PHYSICS OF FLIGHT

BORELLI, Giovanni Alfonso (Italian physiologist, 1608–1679). From *De motu animalium*, Rome, 1680; tr. as *The flight of birds*, London, 1911; by permission of the Secretary of the Royal Aeronautical Society.

Borelli's *De motu* constitutes, in more ways than one, the establishment of physiology in what was to become its modern form, namely, the effort to describe organic function in mechanical and physicochemical terms. As such it constitutes the empirical equivalent of the more speculative approach of Descartes (*q.v.*). It is interesting that, in direct opposition to Borelli's pessimism concerning the possibility of flight by men, his precisely reasoned analysis was exactly what was needed before men could ever undertake flight on other than a trial-and-error basis. Borelli studied, taught, practiced medicine at Pisa, Messina, Florence, Rome; knew Galileo and wished to reconstitute medicine in terms of Galilean physics; taught Malpighi.

Wherein are set forth the reasons for the immense power of the wings.

Such excessive power of the pectoral muscles of Birds seems to arise, firstly, from their large size and from the more compact and stronger organic structure of the fibres of the pectoral muscles; for these fibres are thicker and closer, forming a dense and compact fleshy structure, whereas the muscles of the legs are formed of meagre, spare flesh. By reason hereof the former can be extended more forcefully and vehemently, so that the former are able to exert more power than the latter.

Secondly, the action of the wings is increased by the decrease in resistance, for the body of a Bird is disproportionately lighter than that of man or of any quadruped; that is, the weight of a Bird is in smaller proportion to the weight of the latter animals than its mass to theirs. This is evident since the bones of a Bird are porous, hollowed out to extreme thinness like the roots of the feathers, and the shoulder-bones, ribs, and wing-bones are of little substance; the breast and abdomen contain large cavities filled with air; while the feathers and the down are of exceeding lightness. Hence the power of the wings is increased in duplicate ratio: firstly, by the increase in the force of the muscles, and secondly by the decrease of the weight to be supported.

This downward pull is diminished the more as its downward movement is retarded by the spread of the wings and of the tail; hence the force of the wings can the more readily effect the leaps through the air, as the resistance of the downward pull of the Bird itself is diminished.

Thirdly, in leaping from the earth the projectile momentum is immediately extinguished so soon as the feet come into contact with the earth again; whence it follows that the momentum must forthwith be renewed. On the contrary, when a Bird is flying through the air, the projectile force is not extinguished by the fluid air, wherefore it assists the succeeding leaps which are made by the beating of the wings.

Fourthly, in effecting separate leaps from the earth, the soles of the feet come into contact with the ground not without experiencing hurt and painful injury, whence arise fatigue and weakness. But no such hurt results from leaping through the air; wherefore, since the motive force is not weakened to the same extent, longer, more powerful, and more lasting leaps may be made through the air. The various causes set forth above render the process abundantly clear.

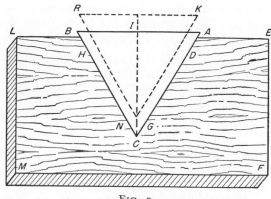

FIG. 7.

In what manner an oblique transverse force may propel straightly a body unaffected by the motion.

It is taught by the science of mechanics that the action of the wedge ABC, through which two parts EFG and LMN of the same body must be separated from one another, amounts to the forcing of the two resisting bodies DE and HM over the inclined surfaces CA and CB of the wedge, along which they seek to ascend when the wedge is driven in a direction from I to C. And the same transverse motion over the inclined faces CA and CB must take place when the two adjacent bodies DF and HM are forced towards each other; for in this case the smooth wedge ABC seeks to escape in the opposite direction and to recoil from C to I, being expelled through the pressure of the collateral bodies, in the same manner as the smooth pips of a fruit may be projected to a long distance by being compressed between one's fingers. And this propulsion is made with the same force and momentum as that wherewith the bodies DF and HM compress the inclined faces CA and CB: the expelling force having the same proportion to their absolute force as the heights AI, BI of the planes to the lengths AC, BC of their inclination.

If a Bird suspended in the air strike with its outspread wings the undisturbed air, with a motion perpendicular to the horizon, it will fly with a transverse movement parallel to the horizon.

Let the Bird RS be suspended in the air with its wings BEA and BCF expanded and its belly downwards, and the under surfaces of the wings BEA and BCF strike against the wind perpendicularly to the horizon with such force as to prevent the bird from falling, then I hold that it will be impelled horizontally from S towards R. And this happens because the two osseous rods (*virgae*) BC and BE by muscular strength and on account of their hardness are able to resist the pressure of the wind, and, moreover, to retain their shape, but the afterparts of any kind of wing yield to the air pressure, as the flexible feathers are able to move about the wing bones (*manubria*) or their boney axes BC and BE; and so it follows that the ends A and F of the feathers close in towards one another, by which means the wings assume the form

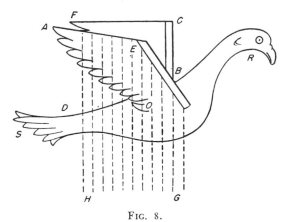

FIG. 8.

of a wedge with its apex towards AF. But as the surfaces of the wedge are compressed on all sides by the ascending air, the wedge is of necessity squeezed and driven towards its base CBE. And as the said wedge formed by the wings cannot move forward without taking with it, since it is attached thereto, the body of the Bird RS, which is swimming in the air and can therefore be moved freely from its position, for this reason it is able to give room to the incoming air in the place of the air driven out; and therefore the bird moves with a horizontal motion towards R.

Let us now take the case of undisturbed opposing air which is struck by the flexible portions of the wings with a movement perpendicular to the horizon. Since the sails and flexible portions of the wings assume the shape of a wedge, with the apex towards the tail, when acted upon by the force and compression of the air, whether the wings strike the quiescent air beneath, or whether the air rushes up against the outstretched wings with their rigid wing-bones; in both cases the flexible feathers of each wing yield to the pressure and close in towards one another. Therefore, of necessity, as will be presently shown, the bird will be moved forwards towards R.

Wherein is explained the way in which the horizontal flight of Birds is effected.

To have brought about flight, it is evident that Nature impelled birds upward and held them suspended in the air, and afterwards they were enabled by horizontal movements to be carried about. The first step could not have been accomplished except by successive leaps; next the heavy bird was carried up and its descent prevented by the beating of its wings, and then, as the downward pull of its weight is perpendicular to the horizon, beats with the flat face of its wings would be made by striking the air in the same perpendicular direction; and in this fashion has Nature brought about the suspension of the Birds in the air.

Concerning the second and transverse motion of Birds, some people do blunder strangely, for they think that it ought to be done as in Ships, which, by the exertion of a horizontal force towards the stern, through the means of oars, the while floating on the quiet and therefore resisting water beneath, recoil at the contrary motion, and so are moved forward. In the same way they affirm that the wings are flapped with a horizontal movement towards the tail and so strike against the undisturbed air, the resistance of which occasions, by the reflex action, their forward motion.

But this is repellent to the evidence of the senses and of reason, for we never see the larger Birds, such as Swans, Geese, and the like, while flying, to flap their wings toward the tail with a horizontal motion, but always to incline them downwards, describing circles set perpendicularly to the horizon. Moreover, in Ships, the horizontal motion of the oars can be easily accomplished and a perpendicular stroke upon the water would be useless and unnecessary, as there is no need to prevent their descent when they are sustained by the weight and density of the water. But in the case of Birds, it would be foolish to make such a horizontal motion, which would rather hinder flight as the speedy downfall of the heavy Bird would result from it; wherefore, the Bird must be sustained by continual vibrations of the wings perpendicularly to the horizon.

Wherefore Nature was compelled to use, with remarkable shrewdness, a movement which both sustained the Bird and propelled it horizontally. . . .

How Birds, without flapping their wings, can sometimes rise in the air for a short time not only horizontally, but also obliquely upward.

It is clear from what has been said that the projectile force is communicated to a Bird's body by the flapping of the wings in the same way as motion is given to a Ship by the strokes of the oars, which motion is of a constant nature.

Suppose, however, that the action of the oars stops, nevertheless the Ship proceeds upon its way until its movement is arrested by external forces.

Therefore both Bird and Ship from the motion imparted to them have the same properties as an arrow and other projectiles; and just as in a Ship in motion, if its axis is deflected from a straight course by the strength of the helm, then this same motion comes into play on the altered course, and the voyage is continued; so also in the Bird A, moving horizontally along the straight line ABC, as often as its axis is directed upward through BD by the force of its tail acting as a helm, of necessity its impetus follows an upward movement through the parabolical curve BEF, but it is true that such ascent stops suddenly, the natural gravity of a Bird producing this effect and tending to bring

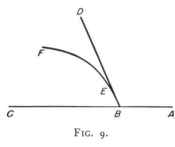

it down. While the force of gravity is less than the velocity, the Bird rises upwards through BE, and when at F the forces equalise, the Bird is seen to float at that point for a little while, moving with expanded wings, almost in the same plane parallel to the horizon; for a bird cannot remain entirely motionless at the same point in the air, therefore upward flight cannot be

FIG. 9.

made exactly perpendicularly to the horizon, but always obliquely through the line of a parabola, as projectiles move.

Therefore, after this rise is made, either the bird continues in a horizontal course for a short time because the equalisation of the forces soon ceases, or as the projectile force is spent it descends at a constantly increasing speed until brought up by external forces. Hence the necessity arises for renewing the impulse through the air by fresh strokes of the wings. . . .

It is impossible that men should be able to fly craftily by their own strength.

Three principal points ought to be considered in flying: firstly, the motive power by which the body of the Animal may be sustained through the air; secondly, the suitable instruments, which are wings; thirdly, the resistance of the Animal's heavy body.

The degree of motive power is known by the strength and quantity of the muscles, which are designed to bend the arms or to flap the wings. And because the motive force in Birds' wings is apparently ten-thousand times greater than the resistance of their weight, and as Nature has endowed Birds with so great an excess of motive power, the Bird largely increases the strength of its pectoral muscles and skilfully decreases the weight of its body, as we have hinted above.

When, therefore, it is asked whether men may be able to fly by their own strength, it must be seen whether the motive power of the pectoral muscles (the strength of which is indicated and measured by their size), is proportionately great, as it is evident that it must exceed the resistance of the weight of the whole human body ten-thousand times, together with the weight of enormous wings which should be attached to the arms. And it is clear that the motive power of the pectoral muscles in men is much less than is necessary for flight, for in Birds the bulk and weight of the muscles for flapping the wings are not less than a sixth part of the entire weight of the body. Therefore, it would be necessary that the pectoral muscles of a man should weigh more than a sixth part of the entire weight of his body; so also the arms, by flapping with the wings attached, should be able to exert a power ten-thousand times greater than the weight of the human body itself. But they are far below such excess, for the aforesaid pectoral muscles do not equal a hundredth part of the entire weight of a man. Wherefore either the strength of the muscles ought to be increased or the weight of the human body must be decreased, so that the same proportion obtains in it as exists in Birds.

Hence it is deduced, that the Icarian invention is entirely mythical because impossible; for it is not possible either to increase a man's pectoral muscles or to diminish the weight of the human body; and whatever apparatus is used, although it is possible to increase the momentum, the velocity or the power employed can never equal the resistance; and therefore wing flapping by the contraction of muscles cannot give out enough power to carry up the heavy body of a man.

There only remains the diminution of the weight of the human body, not in itself, for this is impossible, its mechanism must remain intact, but especially and respectively to the aerial fluid in the same way as a strip of lead can float on water if a certain amount of cork be attached to it which causes the entire mass of lead and cork to float, being of like weight to the amount of water which it displaces, according to the law of Archimedes. And this device Nature uses in fishes. She places in their bellies a sack full of air by means of which they are able to maintain their equilibrium, so that they can remain in the same place as if they were part of the water itself.

By this same device some have lately persuaded themselves that the weight of the human body is able to be brought into equilibrium with the air, that is to say by the use of a large vessel, either a vacuum or very nearly so, of so great a size that it is possible to sustain a human body in the air together with the vessel.

But we easily perceive this to be a vain hope as it is necessary to construct the vessel of some hard metal such as brass or copper, and squeeze out and take away all the air from its interior, and it must also be of so great a size that when in the air it displaces a quantity of air of the same weight as itself, together with the man fastened to it; wherefore it would have to occupy a space of more than 22,000 cubic feet; moreover, the plates composing the sphere must be reduced to an extraordinary thinness. Furthermore, so thin a vessel of this size could not be constructed, or, if constructed, preserved intact, nor could it be exhausted by any pump, much less by mercury, of which so large a quantity is not to be found in the world, nor could be extracted from the earth, and if such a great vacuum were made the thin brass vessel could not resist the strong pressure of the air, which would break or crush it. I pass over the fact that so great a machine of the same weight as the air would not be able to keep itself in exact equilibrium with the air, and therefore would incontinently rise to the highest confines of the air like clouds, or would fall to the ground.

Again, such a large mass could not be moved in flight on account of the resistance of the air; in the same way feathers and soap bubbles can be moved only with difficulty through the air, even when they are blown by a light breeze, just as clouds, poised in the air, are driven by the wind.

At this point we cease to wonder that Nature, who is accustomed everywhere to imitate others' advantages, makes the swimming of fishes in water so easy and the flying of Birds through the air so difficult, for we see whereas fishes can remain in the midst of water, being of their own accord and without effort held up and poised, and can very easily descend and ascend, and are only moved by the strength of muscles placed transversely and obliquely to the direction of motion; on the other hand, Birds are not able to float in the air but owe their sustentation to the continual exertion of strength and a projectile force, not external, but natural and intrinsic, by contracting their pectoral muscles by which they make a series of bounds through the air; and this requires enormous strength, as they are not going upon feet supported on solid ground, but on wings supported by very fluid and greatly agitated air.

EARLY STUDIES ON THE MECHANICS OF CIRCULATION

HALES, Stephen (English clergyman, botanist, and physiologist, 1677–1761). From *Statical essays: containing Haemastaticks; or An account of some hydraulick and hydrostatical experiments made on the blood and blood-vessels of animals, etc.*, London, 1733.

Hale's admirably conceived experiments on plant and animal physiology stand in sharp relief against the taxonomic and philosophical productions which chiefly characterized the biology of this period. Along with the researches of Réaumur, Spallanzani, and relatively few others, they constitute a connecting link between the brilliant physiological studies of the seventeenth century (Harvey, Borelli, *et al.*) and the important expansion which would occur in the nineteenth century. An unassuming country clergyman, better known for his botanical than his animal experiments, Hales found time for important practical researches (*e.g.*, ventilation of prisons) and philanthropies. We reprint here the Preface in which Hales outlines his attitude toward research, and one of the more famous experiments.

THE PREFACE.

What I had at first intended only as additional Observations and Experiments to the first Volume, is now grown into the Size of another Volume, so fruitful are the Works of the great Author of Nature in rewarding by farther Discoveries, the Researches of those *who have Pleasure therein:* We can never indeed want Matter for new Experiments; and tho' the History of Nature as recorded from almost innumerable Experiments, which have been made within the compass of a Century, be very large, yet the Properties of Bodies are so various, and the different Ways by which they may be examined so infinite, that 'tis no wonder that we are as yet got little farther than to the Surface of Things: Yet ought we not to be discouraged, for tho' we can never hope to attain to the compleat Knowledge of the Texture, or constituent Frame and Nature of Bodies, yet may we reasonably expect by this Method of Experiments, to make farther and farther Advances abundantly sufficient to reward our Pains.

And tho' this Method be tedious, yet our Abilities can proceed no faster; for as the learned Author of the *Procedure of human Understanding* observes, *pag.* 205, 206. "All the real true Knowledge we have of Nature is entirely *experimental,* insomuch that, how strange soever the Assertion seems, we may lay this down as the first fundamental unerring Rule in Physicks, *That it is not within the compass of human Understanding to assign a purely speculative Reason for any one Phenomenon in Nature.*" So that in natural Philosophy, we cannot depend on any meer Speculations of the Mind; we can only with the Mathematicians, reason with any tolerable Certainty from proper *Data,* such as arise from the united Testimony of many good and credible Experiments.

Yet it seems not unreasonable on the other hand, tho' not far to indulge, yet to carry our Reasonings a little farther than the plain Evidence of Experiments will warrant; for since at the utmost Boundaries of those Things which we clearly know, there is a kind of Twilight cast from what we know, on the adjoining Borders of *Terra incognita,* it seems therefore reasonable in some degree to indulge Conjecture there; otherwise we should make but very slow Advances in future Discoveries, either by Experiments or Reasoning: For new Experiments and Discoveries do usually owe their first Rise only to lucky Guesses and probable Conjectures, and even Disappointments in these Conjectures, do often lead to the Thing sought for: Thus by observing the Errors and Defects of a first Experiment in any Researches, we are sometimes carried to such fundamental Experiments, as lead to a large Series of many other useful Experiments and important Discoveries.

If therefore some may be apt to think that I have sometimes too far indulged

Conjecture, in the Inferences I have drawn from the Events of some Experiments; they ought to consider that it is from these kind of Conjectures that fresh Discoveries first take their Rise; for tho' some of them may prove false, yet they often lead to farther and new Discoveries. It is by the like Conjectures that I have been led on, Step by Step, thro' this long and laborious Series of Experiments; in any of which I did not certainly know what the Event would be, till I had made the Trial, which Trial often led on to more Conjectures and farther Experiments.

In which Method we may be continually making farther and farther Advances in the Knowledge of Nature, in proportion to the Number of Observations which we have: But as we can never hope to be furnished with a sufficient Number of these, to let us into a thorough knowledge of the great and intricate Scheme of Nature, so it would be but dry Work to be ever laying Foundations, but never attempting to build on them. We must be content in this our infant State of Knowledge, while we *know in part* only, to imitate Children, who for want of better Skill and Abilities, and of more proper Materials, amuse themselves with slight Buildings. The farther Advances we make in the Knowledge of Nature, the more probable and the nearer to Truth will our Conjectures approach: so that succeeding Generations, who shall have the Benefit and Advantage both of their own Observations, and those of preceding Generations, may then make considerable Advances, when *many shall run to and fro, and Knowledge shall be increased,* Dan. xii. 4. In the mean time, it would but ill become us in this our State of Uncertainty, to treat the Errors and Mistakes of others with Scorn and Contempt, when we cannot but be conscious, that we ourselves see Things *but as thro' a Glass darkly,* and are very far from any Pretensions to Infallibility.

As it has ever been of great Importance to the Welfare of Mankind, to make the best Researches they can into the Nature of our Bodies, so have many eminent Persons, from time to time, made considerable Discoveries therein. And as the animal Fluids move by Hydraulick and Hydrostatical Laws, so I have here made some enquiry into the Nature of their Motions by a suitable Series of Experiments. The disagreeableness of the Work did long discourage me from engaging in it; but I was on the other hand spurred on by the hopes that we might thereby get some farther Insight into the animal Oeconomy. We find here a large Field for Experiments, which may be multiplied and varied many Ways, of which I have here only given a few Specimens. As these Experiments do obviously and clearly give an Account of some *Phaenomena,* so they may possibly be of Service in the Hands of those who are well skilled in the animal Oeconomy and the History of Diseases, to explain many other of the innumerable Variety of Cases, which occur in so complicated a Subject as an animal Body is.

In which we are assured that all Things are wisely adjusted in Number,

Weight and Measure, yet with such complex Circumstances as require many *Data* from Experiments, whereon to found just Calculations: But tho' many of the following Calculations are founded only on such inaccurate Mensurations as the Nature of the Subject would allow of, yet may we thence fairly draw many rational Deductions in relation to the animal Oeconomy.

In which there is so just a Symmetry of Parts, such innumerable Beauties and Harmony in the uniform Frame and Texture of so vast a Variety of solid and fluid Parts, as must ever afford Room for farther Discoveries to the diligent Enquirer; and thereby yield fresh Instances to illustrate the Wisdom of the divine Architect, the Traces of which are so plain to be seen in every Thing, that the *Psalmist* had good reason to call him a *Fool* who could be so senseless as to *say in his Heart that there is no God;* whose masterly Hand is so evident in every Part of Nature, that if there be any who pretend they cannot see it, it can be no Breach of Charity to say that they are wilfully blind, and therefore Lyars.

In the Treatise on the *Calculus,* I have endeavoured by a great Variety of Experiments, to enquire into the Nature of that most formidable Concretion; but tho' I have not thence been able, to find out either a sure Preventive, or a safe Dissolvent, yet I am not without hopes that these Researches, as they were sincerely intended, so they will prove of considerable Benefit to Mankind in many Cases, by shewing not only the Nature, but also the Causes which are most apt, either to promote or retard the Progress of this Secret and terrible Intruder; not that it is to be expected, that all should find Benefit thereby, tho' haply some may. The Instrument described at the End of this Treatise, will, I doubt not, be very beneficial in instantly relieving many, who might otherwise suffer great Torture for several Days, and some of them lose their Lives by a dangerous Incision.

The Appendix contains several miscellaneous Observations and Experiments, some of which relate to Subjects in *Vegetable Staticks,* but the greatest Part to the *Analysis* of the Air in the first Volume.

I have added a general Index of the Matters contained in these two Volumes. But as this Index is adapted to the Second Edition of the first Volume; so there must be an Allowance for a difference of eight Pages less, from Page 75 of the first Edition of the first Volume; which difference in the numbering of the Pages of the two Editions of the first Volume, continually decreases, so as to be none in the last Page of each Edition.

Experiment I.

In *December* I caused a Mare to be tied down alive on her Back, she was fourteen Hands high, and about fourteen Years of Age, had a *Fistula* on her Withers, was neither very lean, nor yet lusty: Having laid open the left crural Artery about three Inches from her Belly, I inserted into it a brass Pipe whose bore was one sixth of an Inch in Diameter; and to that, by means of another brass Pipe which was fitly adapted to it, I fixed a glass Tube, of nearly the same Diameter, which was nine Feet in Length: Then untying the Ligature on the Artery, the Blood rose in the Tube eight Feet three Inches perpendicular above the Level of the left Ventricle of the Heart: But it did not attain to its full Height at once; it rushed up about half way in an Instant, and afterwards gradually at each Pulse twelve, eight, six, four, two, and sometimes one Inch: When it was at its full Height, it would rise and fall at and after each Pulse two, three, or four Inches; and sometimes it would fall twelve or fourteen Inches, and have there for a time the same Vibrations up and down at and after each Pulse, as it had, when it was at its full Height; to which it would rise again, after forty or fifty Pulses.

2. The Pulse of a Horse that is well, and not terrified, nor in any Pain, is about thirty six Beats in a Minute, which is nearly half as fast as the Pulse of a Man in Health: This Mare's Pulse beat about fifty five times in a Minute, and sometimes sixty or a hundred she being in pain.

3. Then I took away the glass Tube and let the Blood from the Artery mount up in the open Air, when the greatest Height of its Jet was not above two Feet.

4. I measured the Blood as it run out of the Artery, and after each Quart of Blood was run out, I refixed the glass Tube to the Artery, to see how much the Force of the Blood was abated; this I repeated to the eighth Quart, and then its Force being much abated, I applied the glass Tube after each Pint had flowed out: The Result of each Trial was as is set down in the following Table, in which are noted the greatest Heights it reached after every Evacuation: It was usually about a Minute before it rose to these several Heights, and did not rise gradually, but would stand during several Pulses much lower, than what it would at length reach to; so that I often thought it had done rising, when on a sudden it would rise for some time four, eight, twelve or sixteen Inches higher, where it would stay for some time, and then on a sudden fall four, eight, twelve or sixteen Inches.

There was about a Quart lost in making the several Trials, so there flowed out in all seventeen Quarts, and half a Pint after the last Trial, when she expired. This whole Quantity of Blood was equal to 1185.3 cubick Inches.

5. We may observe from this Table, that the Decrease of the Force of the Blood in the Arteries, was not proportioned to the several Quantities of Blood which were evacuated; for at the eighth Trial, when seven Quarts were drawn off, the Height of the Blood was four Feet eight Inches, after which it decreased in the five following Trials to three Feet odd Inches, sometimes a little lower and then a few Inches higher. But at the fourteenth Trial, after ten Quarts and a Pint had been drawn off, it rose again up to four Feet 3 + ½ Inches, and it came nearly to the same Height again at the twentieth Trial when thirteen Quarts and a Pint had been drawn off.

The several Trials.	The Quantities of Blood let out in Wine Measure.		The several Heights of the Blood after these evacuations.	
	Quarts	Pints	Feet	Inches
1	0	5 Ounces*	8	3
2	1	0	7	8
3	2		7	2
4	3		6	6½
5	4		6	10½
6	5		6	½
7	6		5	5½
8	7		4	8
9†	8		3	3
10	8	1	3	7½
11	9	0	3	10
12	9	1	3	6½
13	10	0	3	9½
14	10	1	4	3½
15	11	0	3	8
16	11	1	3	10½
17	12	0	3	9
18	12	1	3	7½
19	13	0	3	2
20	13	1	4	½
21	14	0	3	9
22	14	1	3	3
23	15	0	3	4½
24	15	1	3	1
25	16	0	2	4

* These 5 Ounces lost in preparing the Artery.

† By this time there is a Pint lost in making the several Trials, which is not allowed for in this Table.

6. This disproportionate Inequality in the several Heights was principally owing to her violent straining to get loose, which made the Blood in the fourteenth Trial rise higher than it had done in several of the preceding ones.

7. About the twentieth Trial she grew very faint and uneasy, and breathed quick: The violent straining to get loose, did by the acting of most of her muscles, especially the abdominal, impell the Blood from all Parts to the *Vena Cava*, and consequently there was a greater Supply for the Heart, which must therefore throw out more at each Pulsation, and thereby increase the Force of the Blood in the Arteries.

8. For the same Reason too, it would be somewhat increased in Height upon deep sighing, because the Lungs being then put into greater Motion and more dilated, the Blood passed more freely and in greater Quantity to the left Auricle and thence to the Ventricle.

9. This plainly shews how sighing increases the Force of the Blood, and consequently proportionately chears and relieves Nature, when oppressed by its too slow Motion, which is the Case of those who are dejected and sad.

10. Hence also we see evidently, that the Blood moves fastest and most freely thro' the Lungs when they are in a dilated State: For which Reason

Animals when they are near expiring, do usually breathe quick, the Lungs then labouring to heave fast, that the languid Blood may thereby have a freer Course thro' them, to supply the then almost bloodless Pulsations of the Heart, as was we see the Case of this Mare when her Blood was near exhausted.

11. When between fourteen and fifteen Quarts of Blood had been evacuated, and thereby the Force of that which remained in the Vessels greatly decreased, then the Mare fell into cold clammy Sweats, such as frequently attend dying Persons; which shews to how low a State the vital Force of the Blood is at that Time reduced: Whence we see, that these faint Sweats are not occasioned by a greater protrusive Force of the Blood at that Time, but rather by a general Relaxation of the Pores, as well as of all other Parts of the Body. And it seems hence probable, that the Vigour of the Blood in the Arteries is much abated, when Persons who are not in a dying State, have colliquative Sweats, as in violent Colic Pains, Fear, *&c.*

12. Upon opening the Mare's Body, I found little or no Blood in the *Aorta,* about an Ounce in the left Ventricle, but none in the right; the *Vena Porta* and *Cava* were full: She bled two or three Ounces, but very slowly, and not without pressing the jugular Vein, which was opened as soon as she expired.

13. There might be about two Quarts and three Quarters of Blood left in the large Veins, which, with what was drawn out at the Artery, makes up twenty Quarts, equal to 1154 cubick Inches, or forty four Pounds; which at a low Estimation, may be reckoned the Quantity of current Blood in a Horse; there is doubtless considerably more, but it is not easy to determine how much.

14. As this Experiment shews how much the Force of the Blood in the Arteries is abated by different Degrees of Evacuation; so it may be of use to direct what Quantity to let out at a Time in bleeding: For whatever the real Quantity of the circulating Blood be, it is certain that the Estimate of what can be with safety let out at once, must be taken from the Proportion which that bears to the whole Quantity of Blood, which will flow out of the Vein or Artery of the Animal till it dies.

15. We see also from this Experiment, the Reasonableness of the Practice of bleeding at several distant Times, where it is requisite to take away a great Quantity of Blood, and not to do it all at once, which would too much weaken the Force of the Blood. For since it was found by several Instances in this Experiment, that when the Force of the Blood was much depressed by Evacuations, it would be considerably raised again by the Action of the Muscles, out of whose very fine and long capillary Vessels it moves but slowly, as also by the Motion of all Parts of the Mare; so the case is doubtless the same when the Vigour of the Blood is in any Degree rebated in the large Vessels, by Blood-letting, that Vigour will in some measure be in a little Time restored again, not only by the Action of the several Parts of his Body, whereby the Blood would have time to flow in from all Parts, to supply the most evacu-

ated Vessels, whereby there would be a just proportionate evacuation of all
Parts; but also because the Vessels themselves would thereby have time to
contract themselves in some Proportion to the Degree of their Evacuation.

REPRODUCTIVE PHYSIOLOGY OF INVERTEBRATES

SWAMMERDAM, Jan (Dutch naturalist and microscopist, 1637–1680). From *Biblia
natura* first published (tr. by H. D. Gaubius) as *Bijbel de natuure, of, historie der
insecten, tot zeekere zoorten gebracht, etc.*, Leyden, 1737 (posth.); tr. by T. Flloyd (revision by J. Hill) as *The book of nature or, the history of insects reduced to distinct
classes, confirmed by particular instances, etc.*, London, 1758.

Swammerdam's contributions are significant in point both of the method employed,
namely monographic analysis of certain single species, and of the knowledge which the
method produced. The detailed organization which he discovered in small-scaled invertebrates raised the whole question of the lower size limit for highly complicated creatures (see Ehrenberg, Dujardin, Siebold) as well as that of preformation vs. epigenesis
(see Harvey, Roux, Driesch). We have selected, for illustration of his approach, a set
of observations on the morphology, reproductive physiology, and natural history of reproduction in the snail.

Though the Snail was reckoned by the Jews among unclean animals, which
they were forbid to use as food, they did not scruple the application of it to
other purposes. The royal psalmist borrows a moral simile from it, and prays,
that the wicked may "consume away like a Snail"; and, however impure and
slimy, it must notwithstanding claim the consideration of those, who are desirous of being acquainted with the wonderful works of the creation. . . .

Of the Manner in Which Snails Mutually Perform the Business of Coition.

Having hitherto, in part, shewn the method whereby Snails generate, I shall
now give you a full description thereof; since it is a matter very worthy of notice that, an hermaphrodite little creature should have need of a companion for
the purpose of generation.

The Snails gather together for some days before their coition, and lye quiet
near each other, eating very little in the mean time; but they settle their bodies
in such a posture, that the neck and head are placed upright. Thus, whilst the
shell of each rests upon the earth, with its double head, the Snails are raised
upwards, and they support themselves erect, by the extreme ends of the fringes
and verge of their bodies, in the same manner as it is said Serpents engender,
that is, in an erect situation, and twisted about each other.

At that time the verge, or its aperture, is continually open to take in the air;
but the opening of the genitals in the neck is sometimes observed to be alternately open and contracted. This action is performed in the same manner as

the agitation of the outer parts of the vulva in Dogs and Hens, when they desire coition.   The Snails being thus animated, softly approach each other, and
apply their bodies one to the other, as smoothly as the palms and fingers of
both hands can be grasped together: and by this means, not only their bodies,
but their necks and heads, are raised up and pressed close to each other.   Then
are seen the most wonderful motions of their heads and eight horns, which
surpass all imagination; like Turtles, they are continually observed to kiss each
other, and to join lip to lip.   The horns are affected with such various motions,
that one can scarce think, how they can possibly have so many and such different muscles.   Another circumstance that deserves notice is, that when they
touch each other in the least with their horns, they immediately draw them in,
or move them up or down again, or sideways; and these motions are often
repeated.

   These motions sometimes continue for three days, during which time the
Snails turn in and out, and join together their genitals, so that the penis and
uterus, of each, are sometimes seen to hang entirely out of the body.   But since
I cannot. observe that Snails have organs proper for seeing and discovering
things near and at hand, but only for remote objects; hence it is, that for want
of this knowledge, I cannot observe their coition but by chance.   For though
they very often shoot their genitals, like an arrow out of a bow, yet the coition
is scarce performed once in three times.   This mistake seems to happen the oftener, because every snail carries its penis in the right side of its neck: it must
therefore happen before they have turned their respective heads cross-wise towards each other, that they often attempt a coition unsuccessfully.   But they
have leisure enough to repeat this business, since they feel for a long time the
incentives of their venery, though they have already gone through the business
of coition ten or twelve times before.   Nay, I have known some of them indulge their venereal desires three weeks afterwards, and that they repeated
them again in six weeks after that.

   But when they associate in coition, as they should do, each of them stretches
its penis, together with the orifice of the uterus, entirely out of the body,
which is not done by erection only, as it is in quadrupedes, but principally by
turning the inward parts out, as happens in the penis of Drakes.   The first
thing that opens in the Snail is the aperture, that lies in the neck between the
upper and lower horns; then the inward parts of generation are observed to
come out like two apertures; so that by this means the lower horn is pushed
out of its place; afterwards these parts are very suddenly thrown out of the
body; yet so as that the aperture of the uterus appears first, and immediately
after the penis, the thicker part of which turns out first, and afterwards the
sharp part.   After this begins the coition, and the two lower horns are then so
far thrust out of their places, that they touch and press each other.   These
parts are afterwards remarkably swelled by the humours that flow towards

them, so that they resemble the clammy white of the boiled egg of a Lapwing, which, being mixed as it were with a transparent whiteness, makes a very agreeable sight, resembling an agat. For the appendage of the penis is observed to run so far, and by the clearness of those parts, is so evidently seen, that its motion is also obvious to the eye.

After coition the parts before-mentioned may be still perceived for a quarter of an hour hanging out of the body, that is until their swelling has fallen, and then it is surprising to see what wonderful motions the penis has. But if any one should in the mean time handle these parts, the Snail endeavours to draw them in by force, but however much it stretches all its nerves, yet it cannot by any force bring them in, unless they first become flaccid. A certain lympid moisture then distils from these parts, which soon coagulates, and becomes tenacious and firm in the air. The upper horns being always bent like a circle in the venereal act, are observed not to move much, unless that they are sometimes drawn in, and again stretched out. After all is finished, the little creature, having wantonly consumed the strength of life, becomes dull and heavy; and thence calmly retiring into its shell, rests quietly without much creeping, until the furious lust of generation gathers new strength, and effaces the memory of the uneasiness suffered after the former coition.

In a Snail, dissected a little time after the act of generation, I observed that the penis was smaller, but the uterus a little more expanded and glutinous in its cavity. The ovary was manifestly swelled, and was longer, thicker, and larger, so that it now seemed full of sperm like milk. Afterwards, however, I saw it much larger, and filled with more glutinous moisture. But in five weeks after coition, the ovary became yellow, and like real glue; yet the eggs were very soft like slime, and were scarce visible. When I afterwards viewed the ovary in such a boiled Snail, every thing was callous, and as far as I could discern, an infinite number of little eggs presented themselves in both the Snails, which had copulated with each other. The uterus itself was likewise at that time much more expanded, and became, as it were glandulous; so that when thrown into the water, it swelled very much. When the eggs were held a little while in the hand, the fingers stuck together. Therefore the ovary, the longer it passed after the time of coition, became more tenacious, compact and yellow; for all the eggs of the Snails are covered with very clammy membranes, and are, at length, perfected in the uterus. They cannot remain in the ovary, since this is placed between the spiral part of the shell, and throws its eggs into the cavity of the uterus. Some Snails lay their eggs up and down on the ground, others tie them all together like a chain. I have seen a little chain of eggs of this kind, which the vulgar thought dropped down from heaven, and therefore immediately framed a great many superstitious fables concerning it: so far is ignorance the mother of error. The testicles, after coition, are found deprived of their sperm. The blind appendage of the uterus likewise, in the beginning, though not very much, became afterwards contracted, and had thrown off its bone. The common tube between the penis and uterus suffered no change. The vas deferens was more dilated, and in it, as I have said, I found the little bone. Hence it is probable, that this little bone, at the time of coition, carries some of the spermatic humour through the upper tube of the vas deferens into the uterus; whilst the penis, in the mean time, throws its sperm from the inward part into it. To conclude, the little part in the form of a chain had undergone no change, unless that on the side near the spiral convolution of the liver, the vessels that it distributed there, were here and there very unequally dilated and filled with a calcarious humour; except that some little round whitish membranes, which appear perforated in the middle, and marked with a black spot, were here and there observed fluctuating in its hollow canals, which resemble the leaves of trees.

PARTHENOGENESIS PROVEN

BONNET, Charles (Swiss naturalist, 1720–1793). From *Traité d'insectologie ou observations sur les pucerons*, Paris, 1745; tr. by T. S. Hall for this volume.

Bonnet's outstanding entomological work grew progressively more speculative as, with an aggravating eye affliction, observation became difficult. Of wealthy parentage, he had strong predilections toward a rather dogmatic form of Calvinistic thought. Quite aside from its excellence as a specimen of scientific pains and objectivity, where shall we find a more charming treatment of scientific subject matter than the following? See also his less convincing statement on preformation (p. 377).

Is there, then, no copulation among plant lice? So as to have something more than conjecture on this matter, M. de Réaumur proposed an experiment which he tried 4 or 5 times without success: this was to take a plant louse on emergence from its mother's body and rear it in such a way that it could not have commerce with any insect of its species. "If a plant louse which would thus have been raised by itself," said M. de Réaumur, "produced plant lice, it must either have done so without copulating or else have copulated within its mother's body."

Stimulated by the invitation of M. de Réaumur, I undertook in 1740 to try the experiment on the plant louse of the spindle tree.

FIRST OBSERVATION. *First experiment on an aphid of the spindle tree to decide whether plant lice reproduce without copulating.* Several methods of raising a plant louse in isolation present themselves. Here is that which I decided upon. In a flower pot filled with ordinary earth, I pushed down as far as its neck a vial full of water. Into this vial I introduced the cut end of a small branch of the spindle tree on which I left only 5 or 6 leaves after examining them on both sides with the utmost care. I then placed on one of these leaves a plant louse whose mother lacked wings and had just given it birth while I looked on. I finally covered the little branch with a glass vase whose rim was applied exactly against the surface of the earth in the flower pot; from which I was surer of my prisoner's conduct than Acrisius was of Danae's, although she was locked up by his order in a bronze tower.[1]

It was on May 20th, at 5 P.M. that my plant louse was committed, from birth, to the confinement which I have just described. I took care from then on to keep an exact diary of its life. In this record, I noted its least movement; not a step it took was a matter of indifference to me. Not only did I watch it every day, starting usually at 4 to 5 in the morning and scarcely ever stopping before 9 to 10 at night, but I likewise observed it several times in the same hour and always with an eyeglass to render the observation more exact and to inform myself of the most secret action of our little solitary prisoner. But if this application cost me some trouble and kept me somewhat confined, I received in return something for which I could congratulate myself that I had gone through with it.

My louse changed its skin four times; on the evening of the 23rd; at 2 P.M. on the 26th; at 7 A.M. on the 29th; and at 7 P.M. on the 31st.

It was immediately after ridding itself of its skin that my louse was working to turn over. With its two hind legs, as with two arms, it embraced its cast skin, trying to raise it so as to disengage the barbs attaching it to the leaf or twig on which it moults. It repeated its efforts in a different direction. Little by little it managed to disengage one of its legs and finally all the others. As soon as the cast skin was no longer attached, the louse pushed it up into the air and abandoned it. All this was a somewhat rough task for a louse whose legs had not had time to strengthen themselves.

Perhaps I may be accused of childishness if I record the misgivings which my louse caused me at this last moult. Although it was always shut up in such a way as to leave no fear that some other insect might slip into its solitude, I found it so swollen and so shiny that it seemed to be in the same condition as lice which are nourishing worms inside themselves. What added to my fear and increased my chagrin was that it appeared to be motionless. Unfortunately I had only the light of a candle by which to observe it. Having finally realized that it was only changing its skin I was somewhat reassured; but I was still not without misgivings. It had been lying on its side, and soon was on its back so that its belly was entirely visible. I saw it move its legs, which up to that time it had held to its chest, as nymphs do, it shook them several times as if it desired to use them to change its position; but weak as they were, having just cast their old skin, they did not seem strong enough to acquit themselves of the task. In this position and on an almost vertical leaf, the louse was held only by its cast, to the end of which its body was still fastened.

---

[1] Yet became the mother of Perseus.—Ed.

It was thus exposed to the possibility of a fatal fall. This crisis disquieted me so that I regained my peace of mind only as, little by little, it righted itself.

But it is time to come to the most interesting moment in the life of our little hermit. Happily delivered from four maladies which might have carried it off, it finally arrived at the stage to which I had hoped by my pains to bring it. It had become a full-grown plant louse. On June the 1st, around 7 o'clock in the evening, I saw, to my great happiness, that it was about to give birth; after which I felt it mandatory to refer to it as a she-louse. Beginning on that day, and until the 21st inclusive, she had 95 young, all born alive and most of them arriving while I looked on.

Finally, to finish the story of our louse, I have only to say that, being obliged to absent myself all the 25th until 5 o'clock the next morning, I was chagrinned on my return, not to find her where I had left her, nor anywhere in the neighborhood which I searched in vain. Inasmuch as, since she had begun to give birth, I had not considered it necessary to keep her completely confined, she no doubt took advantage of this fact to depart and to finish her days in other parts. One will easily judge that I was not insensible to this loss. I had seen her born, I had followed her carefully for a month, and I would have enjoyed continuing to observe her with the same care until death.

MECHANISM IN THE EIGHTEENTH CENTURY

de LA METTRIE, Julien Òffray (Offroy) (French physician and speculative physiologist, 1709–1751). From *L'homme machine*, Leyden, 1748; tr. by G. C. Bussey as *Man a machine*, Chicago, 1912.

Mettrie's materialism is an offshoot of the philosophy of Descartes but goes beyond it to a far-reaching agnosticism. This brought him into difficulties with ecclesiastical authority. Unlike Servetus and Bruno who died in flames for their beliefs, Mettrie fled to Holland, and later Prussia, to die eventually, it is said, of overeating. There have been many to reply to what was deemed a scurrilous assault on the dignity of man.[*]

It is not enough for a wise man to study nature and truth; he should dare state truth for the benefit of the few who are willing and able to think. As for the rest, who are voluntarily slaves of prejudice, they can no more attain truth, than frogs can fly.

I reduce to two the systems of philosophy which deal with man's soul. The first and older system is materialism; the second is spiritualism.

The metaphysicians who have hinted that matter may well be endowed with the faculty of thought have perhaps not reasoned ill. For there is in this case a certain advantage in their inadequate way of expressing their meaning. In

[*] E. Luzak, *Man more than a machine*, London, 1752; J. Needham, *Man not a machine*, London, 1927. See also L. C. Rosenfeld, *From beast machine to man machine*, New York, 1941.

truth, to ask whether matter can think, without considering it otherwise than in itself, is like asking whether matter can tell time. It may be foreseen that we shall avoid this reef upon which Locke had the bad luck to make shipwreck.

The Leibnizians with their monads have set up an unintelligible hypothesis. They have rather spiritualized matter than materialized the soul. How can we define a being whose nature is absolutely unknown to us?

Descartes and all the Cartesians, among whom the followers of Malebranche have long been numbered, have made the same mistake. They have taken for granted two distinct substances in man, as if they had seen them, and positively counted them.

The wisest men have declared that the soul can not know itself save by the light of faith. However, as reasonable beings they have thought that they could reserve for themselves the right of examining what the Bible means by the word "spirit," which it uses in speaking of the human soul. And if in their investigation, they do not agree with the theologians on this point, are the theologians more in agreement among themselves on all other points?

Here is the result in a few words, of all their reflections. If there is a God, He is the Author of nature as well as of revelation. He has given us the one to explain the other, and reason to make them agree.

To distrust the knowledge that can be drawn from the study of animated bodies, is to regard nature and revelation as two contraries which destroy each the other, and consequently to dare uphold the absurd doctrine, that God contradicts Himself in His various works and deceives us. . . .

Experience and observation should therefore be our only guides here. Both are to be found throughout the records of the physicians who were philosophers, and not in the works of the philosophers who were not physicians. The former have traveled through and illuminated the labyrinth of man; they alone have laid bare to us those springs [of life] hidden under the external integument which conceals so many wonders from our eyes. They alone, tranquilly contemplating our soul, have surprised it, a thousand times, both in its wretchedness and in its glory, and they have no more despised it in the first estate, than they have admired it in the second. Thus, to repeat, only the physicians have a right to speak on this subject. What could the others, especially the theologians, have to say? Is it not ridiculous to hear them shamelessly coming to conclusions about a subject concerning which they have had no means of knowing anything, and from which on the contrary they have been completely turned aside by obscure studies that have led them to a thousand prejudiced opinions,—in a word, to fanaticism, which adds yet more to their ignorance of the mechanism of the body?

But even though we have chosen the best guides, we shall still find many thorns and stumbling blocks in the way.

Man is so complicated a machine that it is impossible to get a clear idea of the machine beforehand, and hence impossible to define it. For this reason, all the investigations have been vain, which the greatest philosophers have made *à priori*, that is to say, in so far as they use, as it were, the wings of the spirit. Thus it is only *à posteriori* or by trying to disentangle the soul from the organs of the body, so to speak, that one can reach the highest probability concerning man's own nature, even though one can not discover with certainty what his nature is.

Let us then take in our hands the staff of experience, paying no heed to the accounts of all the idle theories of philosophers. To be blind and to think that one can do without this staff is the worst kind of blindness. How truly a contemporary writer says that only vanity fails to gather from secondary causes the same lessons as from primary causes! One can and one even ought to admire all these fine geniuses in their most useless works, such men as Descartes, Malebranche, Leibniz, Wolff and the rest, but what profit, I ask, has any one gained from their profound meditations, and from all their works? Let us start out then to discover not what has been thought, but what must be thought for the sake of repose in life.

There are as many different minds, different characters, and different customs, as there are different temperaments. Even Galen knew this truth which Descartes carried so far as to claim that medicine alone can change minds and morals, along with bodies. . . . It is true that melancholy, bile, phlegm, blood etc.—according to the nature, the abundance, and the different combination of these humors—make each man different from another. . . .

The transition from animals to man is not violent, as true philosophers will admit. What was man before the invention of words and the knowledge of language? An animal of his own species with much less instinct than the others. In those days, he did not consider himself king over the other animals, nor was he distinguished from the ape, and from the rest, except as the ape itself differs from the other animals, i.e., by a more intelligent face. Reduced to the bare intuitive knowledge of the Leibnizians he saw only shapes and colors, without being able to distinguish between them: the same, old as young, child at all ages, he lisped out his sensations and his needs, as a dog that is hungry or tired of sleeping, asks for something to eat, or for a walk. . . .

I do not mean to call in question the existence of a supreme being; on the contrary it seems to me that the greatest degree of probability is in favor of this belief. But since the existence of this being goes no further than that of any other toward proving the need of worship, it is a theoretic truth with very little practical value. Therefore, since we may say, after such long experience, that religion does not imply exact honesty, we are authorized by the same reasons to think that atheism does not exclude it.

Furthermore, who can be sure that the reason for man's existence is not simply the fact that he exists? Perhaps he was thrown by chance on some spot on the earth's surface, nobody knows how nor why, but simply that he must live and die, like the mushrooms which appear from day to day, or like those flowers which border the ditches and cover the walls.

Let us not lose ourselves in the infinite, for we are not made to have the least idea thereof, and are absolutely unable to get back to the origin of things. Besides it does not matter for our peace of mind, whether matter be eternal or have been created, whether there be or be not a God. How foolish to torment ourselves so much about things which we can not know, and which would not make us any happier even were we to gain knowledge about them!

But, some will say, read all such works as those of Fénelon, of Nieuwentyt, of Abadie, of Derham, of Rais, and the rest. Well! what will they teach me or rather what have they taught me? They are only tiresome repetitions of zealous writers, one of whom adds to the other only verbiage, more likely to strengthen than to undermine the foundations of atheism. The number of the evidences drawn from the spectacle of nature does not give these evidences any more force. Either the mere structure of a finger, of an ear, of an eye, a single observation of Malpighi proves all, and doubtless much better than Descartes and Malebranche proved it, or all the other evidences prove nothing. Deists, and even Christians, should therefore be content to point out that throughout the animal kingdom the same aims are pursued and accomplished by an infinite number of different mechanisms, all of them however exactly geometrical. For what stronger weapons could there be with which to overthrow atheists? It is true that if my reason does not deceive me, man and the whole universe seem to have been designed for this unity of aim. . . .

But since all the faculties of the soul depend to such a degree on the proper organization of the brain and of the whole body, that apparently they are but this organization itself, the soul is clearly an enlightened machine. For finally, even if man alone had received a share of natural law, would he be any less a machine for that? A few more wheels, a few more springs than in the most perfect animals, the brain proportionally nearer the heart and for this very reason receiving more blood—any one of a number of unknown causes might always produce this delicate conscience so easily wounded, this remorse which is no more foreign to matter than to thought, and in a word all the differences that are supposed to exist here. Could the organism then suffice for everything? Once more, yes; since thought visibly develops with our organs, why should not the matter of which they are composed be susceptible of remorse also, when once it has acquired, with time, the faculty of feeling?

The soul is therefore but an empty word, of which no one has any idea, and which an enlightened man should use only to signify the part in us that thinks. Given the least principle of motion, animated bodies will have all that is necessary for moving, feeling, thinking, repenting, or in a word for conducting themselves in the physical realm, and in the moral realm which depends upon it.

Yet we take nothing for granted; those who perhaps think that all the difficulties have not yet been removed shall now read of experiments that will completely satisfy them.

1. The flesh of all animals palpitates after death. This palpitation continues longer, the more cold blooded the animal is and the less it perspires. Tortoises, lizards, serpents, etc. are evidence of this.

2. Muscles separated from the body contract when they are stimulated.

3. The intestines keep up their peristaltic or vermicular motion for a long time.

4. According to Cowper, a simple injection of hot water reanimates the heart and the muscles.

5. A frog's heart moves for an hour or more after it has been removed from the body, especially when exposed to the sun or better still when placed on a hot table or chair. If this movement seems totally lost, one has only to stimulate the heart, and that hollow muscle beats again. Harvey made this same observation on toads.

6. Bacon of Verulam in his treatise "Sylva Sylvarum" cites the case of a man convicted of treason, who was opened alive, and whose heart thrown into hot water leaped several times, each time less high, to the perpendicular height of two feet.

7. Take a tiny chicken still in the egg, cut out the heart and you will observe the same phenomena as before, under almost the same conditions. The warmth of the breath alone reanimates an animal about to perish in the air pump.

The same experiments, which we owe to Boyle and to Stenon, are made on pigeons, dogs, and rabbits. Pieces of their hearts beat as their whole hearts would. The same movements can be seen in paws that have been cut off from moles.

8. The caterpillar, the worm, the spider, the fly, the eel—all exhibit the same phenomena; and in hot water, because of the fire it contains, the movement of the detached parts increases.

9. A drunken soldier cut off with one stroke of his sabre an Indian rooster's head. The animal remained standing, then walked, and ran: happening to run against a wall, it turned around, beat its wings still running, and finally fell down. As it lay on the ground, all the muscles of this rooster kept on moving. That is what I saw myself, and almost the same phenomena can easily be observed in kittens or puppies with their heads cut off.

10. Polyps do more than move after they have been cut in pieces. In a week they regenerate to form as many animals as there are pieces. I am sorry that these facts speak against the naturalists' system of generation; or rather I

am very glad of it, for let this discovery teach us never to reach a general conclusion even on the ground of all known (and most decisive) experiments.

Here we have many more facts than are needed to prove, in an incontestable way, that each tiny fibre or part of an organized body moves by a principle which belongs to it. Its activity, unlike voluntary motions, does not depend in any way on the nerves, since the movements in question occur in parts of the body which have no connection with the circulation. But if this force is manifested even in sections of fibres the heart, which is a composite of peculiarly connected fibres, must possess the same property. I did not need Bacon's story to persuade me of this. It was easy for me to come to this conclusion, both from the perfect analogy of the structure of the human heart with that of animals, and also from the very bulk of the human heart, in which this movement escapes our eyes only because it is smothered, and finally because in corpses all the organs are cold and lifeless. If executed criminals were dissected while their bodies are still warm, we should probably see in their hearts the same movements that are observed in the face-muscles of those that have been beheaded.

The motive principle of the whole body, and even of its parts cut in pieces, is such that it produces not irregular movements, as some have thought, but very regular ones, in warm blooded and perfect animals as well as in cold and imperfect ones. No resource therefore remains open to our adversaries but to deny thousands and thousands of facts which every man can easily verify.

If now any one ask me where is this innate force in our bodies, I answer that it very clearly resides in what the ancients called the parenchyma, that is to say, in the very substance of the organs not including the veins, the arteries, the nerves, in a word, that it resides in the organization of the whole body, and that consequently each organ contains within itself forces more or less active according to the need of them.

Let us then conclude boldly that man is a machine, and that in the whole universe there is but a single substance differently modified. This is no hypothesis set forth by dint of a number of postulates and assumptions; it is not the work of prejudice, nor even of my reason alone; I should have disdained a guide which I think to be so untrustworthy, had not my senses, bearing a torch, so to speak, induced me to follow reason by lighting the way themselves. Experience has thus spoken to me in behalf of reason; and in this way I have combined the two.

But it must have been noticed that I have not allowed myself even the most vigorous and immediately deduced reasoning, except as a result of a multitude of observations which no scholar will contest; and furthermore, I recognize only scholars as judges of the conclusions which I draw from the observations; and I hereby challenge every prejudiced man who is neither anatomist, nor acquainted with the only philosophy which can here be considered, that of the

human body. Against so strong and solid an oak, what could the weak reeds of
theology, of metaphysicis, and of the schools, avail,—childish arms, like our
parlor foils, that may well afford the pleasure of fencing, but can never wound
an adversary. Need I say that I refer to the empty and trivial notions, to the
pitiable and trite arguments that will be urged (as long as the shadow of prejudice or of superstition remains on earth) for the supposed incompatibility of two
substances which meet and move each other unceasingly? Such is my system,
or rather the truth, unless I am much deceived. It is short and simple. Dispute it now who will.

SOLUTION OF FOOD IN THE STOMACH

de RÉAUMUR, René Antoine Ferchault (French scientist, 1683–1757). From *Sur la
digestion des oiseaux*, in *Histoire de l'Académie Royale des Sciences*, Paris, publ. 1756 for
the year 1752; tr. by J. F. Fulton in *Selected readings in the history of physiology*,
Springfield, Ill., 1930; courtesy of Charles C Thomas, publisher.

In this classical experiment on digestion, clear distinction is drawn between the problem of what happens in the stomach and the problem of what causes it to happen. A new
branch of comparative physiology is thus initiated. Other subjects investigated by
Réaumur were forestry,* industrial engineering and chemistry of steel and iron,† ther-
mometry,‡ entomology,§ animal electricity,‖ etc.

It will be easily seen that sponge is the most convenient substance to use in
these experiments, for birds of prey do not normally eat it and therefore, judging by previous observations, we may conclude that they cannot digest it. I
had no doubt as to the success of the experiment which I was about to undertake, and I accordingly put several small pieces of sponge into a tube, without
filling it too full; numerous small holes were then made in the tube and it was
swallowed by the bird and brought up as usual. Before the pieces of sponge
were placed in the tube they weighed only 13 grains, but when I took them
out their weight was 63 grains; they therefore absorbed 50 grains of fluid,
most of which I was easily able to squeeze out into a vessel prepared for the
purpose.

This experiment alone proves that a fairly large quantity of juice can be obtained quite easily. Two or three of these tubes containing sponge, administered in the course of the day would yield double or treble the above amount,

---

* *Réflexions sur l'état des bois du royaume, Histoire de l'Académie Royale des Sci-
ences*, p. 289, 1721.
† *L'art de convertir le fer forgé en acier*, Paris, 1722.
‡ *Explication des principes établis par M. de Réaumur pour la construction des ther-
momètres*, Paris, 17??.
§ *Mémoires pour servir à l'histoire des insectes*, Paris, 1734.
‖ *Des effets que produit le poisson appellé en françois "torpille,"* Amsterdam, 1719.

i.e. 100 or 150 grains of fluid, while for a small outlay two or three birds could be kept for a week or two and thus 200, 300, or even 450 grains of juice obtained daily. A still greater quantity of this fluid might be provided without any expense by those who keep numbers of these birds for the sport at which they excel by their swift flight and great strength. If it were feared that the sharp edges of the tin tube might injure either the bird's stomach, or the passage by which it enters the stomach and is ejected from it,—though it never seemed to do my buzzard any harm,—then let a leaden tube with smooth rounded edges be substituted. Such a tube would be no more harmful than the lead pills given to birds of prey as a medicine, and would, indeed, take their place. If the amount of fluid obtainable from a buzzard, or from any bird of prey of that size, is still insufficient, one could get much larger quantities, which would be enough for a great many experiments, from larger birds such as vultures or eagles. My buzzard died before the series of experiments which I had intended to perform upon it were completed, and I blame my negligence in not replacing it by another buzzard or similar bird of prey. However, I shall make amends for this and try some further experiments which seem to me desirable: I shall now indicate which of these I consider to be the most important, in order to encourage other physicians to attempt them if they have the opportunity.

Before my buzzard died I had only twice obtained, by means of sponges, this juice which can dissolve bones and meat. When I squeezed the fluid out of the sponges into a dish it was quite unlike the clear liquid which is got by different distillations, for it was thick and cloudy and a muddy yellowish white in colour. I am not sure whether its natural colour and transparency had been altered, but further observations will elucidate this point. In the first experiments which I made I did not take the precaution of thoroughly washing the sponges, so that if there were any sediment or other matter in them it would change the consistency of the fluid and make it cloudy.

Apart from anything which the fluid might absorb from the sponges, there may be another reason for its impurity. If, before entering the tube, it came in contact with fragments of meat in the stomach it would not fail to act upon them to a certain extent, and some part of this meat would, when digested and reduced to pulp, almost certainly be mixed with the juice. Therefore, although the juice contained in the sponges could dissolve meat it must not be supposed that it is pure. To obtain fluid which is absolutely pure,—or at least much purer than that which we have just mentioned,—it is only necessary to make sure that the bird swallows the tube containing the sponge on an empty stomach, and that it does not have any food while it retains it. This will not be such a hardship as one might imagine, for nature enables these birds to endure very long fasts; they are not always successful in their search for prey and often go for days without catching, and consequently without eating, anything.

While M. le Commandeur Godeheu was at Malta he received a live vulture (which he intended for my laboratory, where it now is) from Tripoli, on which he performed an experiment which proves that such birds do not need daily nourishment: in order to dissect it more easily after death he wished the bird to become very thin and accordingly he gave it no food at all; the bird withstood this severe fast for 17 days.

When I put some of the juice from the buzzard's stomach on my tongue, it tasted salt rather than bitter, although, on the contrary, the bones which had been reduced to a jelly by a similar fluid, and their remains, on which the fluid had acted, had not a salt but a bitter taste.

When blue [litmus?] paper was moistened with the fluid it became red.

One of the first experiments that ought to be tried with this fluid,—both because it would be most interesting and because it would prove that it is this juice which reduces meat and bones to pulp,—would be to make it dissolve meat in a vessel just as it dissolves it in the stomach. Actual digestion of aliments taking place under such abnormal conditions would be a most singular and interesting phenomenon. Although my efforts in this direction were not successful I shall recount them. I only made two attempts and they would need to be repeated with precautions which I did not take; indeed, they would probably have to be made many times before all the necessary measures were discovered.

EARLY EXPERIMENTAL STUDY ON COAGULATION

HEWSON, William (English physiologist, physician and surgeon, 1739–1774). From *An inquiry into the properties of the blood*, London, 1772.*

Before Hewson, coagulation was thought due to cooling, quiescence, rouleau formation. Hewson's studies redirected the approach along modern lines. Hewson died when thirty-five years old of a wound infection received during a dissection. Another of his discoveries was the lymphatics in cold-blooded vertebrates.†

OF THE SEPARATION OF THE SERUM; THE COLOUR OF THE CRASSAMENTUM; AND OF THE CAUSES OF THE COAGULATION OF THE BLOOD.

When fresh blood is received into a basin, and suffered to rest, in a few minutes it jellies or coagulates, and soon after separates into two parts, distinguished by the names of crassamentum and serum. These two parts differ in their proportions in different constitutions: in a strong person the crassamen-

---

* Republished in *The Works of William Hewson*, London (for the Sydenham Society), 1846.

† *Ibid.*, pp. 119*ff.*

tum is in greater proportion to the serum than in a weak one; and the same difference is found to take place in diseases; thence is deduced the general conclusion, that the less the quantity of serum is in proportion to the crassamentum, bleeding, diluting liquors, and a low diet, are the more necessary; whilst in some dropsies, and other diseases where the serum is in a great, and the crassamentum in a small proportion, bleeding and diluting would be highly improper.

As it is therefore supposed useful to attend to the proportions of these parts in many disorders, and even to take indications of cure from them, it has been an object with those who have made experiments on the blood, to determine the circumstances on which its more perfect separation into these two parts depends; it being obvious, that till this be done, our inferences from their proportions will be liable to considerable fallacies. Two of the latest writers on this subject agree, that if the blood, after being taken from a vein, be set in a cold place, it will not easily separate, and that a moderate warmth is necessary: this is a fact that is evinced by daily experience. They likewise say, that the heat should be less than that of the animal, or than 98° of Fahrenheit's thermometer; and that, if fresh blood be received into a cup, and that cup put into water heated to 98°, it will not separate; nay, they even say, that it will not coagulate; but this, I am persuaded from experiments, is ill founded.

A tin vessel, containing water, was placed upon a lamp, which kept the water in a heat that varies between 100° and 105°. In this water was placed a phial, containing blood that instant taken from the arm of a person in health; the phial was previously warmed, then filled, and corked to exclude air. In the same water was placed a teacup half full of blood, just taken from the same person; a third portion of the blood was then received from the same vein into a basin, and was set upon a table, the heat of the atmosphere being at 67°. Now, according to their opinion, the two former should neither have coagulated nor separated, when that in the basin began to separate; but, on the contrary, they were all three found to coagulate nearly in the same time; and those in the warm water not only did separate as well as the other, but even sooner.

The same experiment was repeated on the blood of a person that laboured under the acute rheumatism, whilst the heat of the atmosphere was no higher than 55°, and that of the warm water was 108°; and the result of this experiment was not only a confirmation of what was observed in the first, but it even showed, that this degree of heat was so far from lessening, that it increased the disposition to coagulate; for the blood in the cup and in the phial was not only congealed, but the separation was much advanced before the whole of the blood in the basin was coagulated. Thence I am led to conclude, that the separation of the blood in a given time, is in proportion as the heat in which it stands is nearer to the animal heat. or 98°; or greater in that heat than in any of a less degree. And I am confirmed in this inference by experiments hereafter to be related, where the blood in the living animal, whilst at rest, was found both to coagulate and to separate.

It is well known, that the crassamentum consists of two parts, of which one gives it solidity, and is by some called the fibrous part of the blood, or the gluten, but by others with more propriety termed the coagulable lymph; and of another, which gives the red colour to the blood, and is called the red globules. These two parts can be separated by washing the crassamentum in water, the red particles dissolving in the water, whilst the coagulable lymph remains solid. That it is the coagulable lymph, which, by its becoming solid, gives firmness to the crassamentum, is proved by agitating fresh blood with a stick, so as to collect this substance on the stick, in which case the rest of the blood remains fluid.[1]

The surface of the crassamentum, when not covered with a size, is in general of a more florid red than the blood was when first taken from the vein, whilst its bottom is of a dark colour, or blackish. This floridness of the surface is justly attributed by some of the more accurate observers to the air with which it is in contact; for, if the crassamentum be inverted, the colours are changed, at least that which is now become the upper surface assumes a more florid redness. This difference of colour, others have endeavoured to explain from the different proportions of the red particles, or globules as they are called, which, say they, being in a greater proportion at the bottom of the crassamentum, make it appear black; but, if inverted, the globules then settle from the surface which is now uppermost, and that becomes redder. But this, I think, is not probable; for the lymph in the crassamentum is so firmly coagulated, as to make it too dense to allow of bodies even heavier than the red particles to gravitate through it; for example, gold. That air has the power of changing the colour of the blood, has been long known; and the following experiment shows it very satisfactorily, and hardly leaves room to refer the appearance to another cause. . . .

As the subject seemed to me of importance, I have endeavoured to ascertain the circumstance to which this coagulation is owing by several experiments, in each of which the blood was generally exposed to but one of the suspected causes at a time. Thus, in order to see whether the blood's coagulation out of the body was owing to its being at rest, I made the following experiment:

---

[1] It may be proper to mention here, that till of late the coagulable lymph has been confounded with the serum of the blood, which contains a substance that is likewise coagulable. But in these sheets, by the *lymph*, is always meant that part of the blood which jellies, or becomes solid spontaneously when blood is received into a basin, which the coagulable matter that is dissolved in the serum does not; but agrees more with the white of an egg, in remaining fluid when exposed to the air, and coagulating when exposed to heat, or when mixed with ardent spirits, or some other chemical substances.

Having laid bare the jugular vein of a living dog, I made a ligature upon it in two places, so that the blood was at rest between the ligatures; then covering the vein with the skin, to prevent its cooling, I left it in this situation. From several experiments made in this way, I found in general, that after being at rest for ten minutes, the blood continued fluid; nay, that after being at rest for three hours and a quarter, above two thirds of it were still fluid, though it coagulated afterwards. Now the blood, when taken from a vein of the same animal, was completely jellied in about seven minutes. The coagulation therefore of the blood in the basin, and of that which is merely at rest, are so different, that rest alone cannot be supposed to be the cause of the coagulation out of the body.

To see the effects of cold on the blood, I made this experiment:

I killed a rabbit, and immediately cut out one of its jugular veins, proper ligatures being previously made upon it; I then threw the vein into a solution of sal ammoniac and snow, in which the mercury stood at the 14th degree of Fahrenheit's thermometer. As soon as the blood was frozen and converted into ice, I took the vein out again, and put it into lukewarm water till it thawed and became soft; I then opened the vein, received the blood into a teacup, and observed that it was perfectly fluid, and in a few minutes it jellied or coagulated as blood usually does. Now, as in this experiment the blood was frozen and thawed again without being coagulated, it is evident that the coagulation of the blood out of the body is not solely owing to cold any more than it is to rest.

Next, to see the effects of air upon the blood, I tried as follows:

Having laid bare the jugular vein of a living rabbit, I tied it up in three places, and then opened it between two of the ligatures, and emptied that part of its blood. I next blew warm air into the empty vein, and put another ligature upon it, and letting it rest till I thought the air had acquired the same degree of heat as the blood, I then removed the intermediate ligature, and mixed the air with the blood. The air immediately made the blood florid, where it was in contact with it, as could be seen through the coats of the vein. In a quarter of an hour I opened the vein, and found the blood entirely coagulated; and as the blood could not in this time have been completely congealed by rest alone, the air was probably the cause of its coagulation.

From comparing these experiments, may we not venture to conclude, that the air is a strong coagulant of the blood, and that to this its coagulation when taken from the veins is chiefly owing, and not to cold, nor to rest?

But although it appears from these experiments that the coagulation of the blood in the basin is owing to the air alone; for cold has no such effect, nor has rest in a sufficient degree, because the coagulation of the blood in the basin takes place in a few minutes, whilst that which is merely at rest in the veins is not completely coagulated in three hours or more. Yet the blood is in time completely coagulated merely by its being at rest in the veins; but then in this case it coagulates in a different manner from what it does in the basin; and as it probably is in this way that the blood is coagulated in the body, I have been more particularly attentive to it, and have endeavoured to determine by experiment how it takes place. With this view I have several times repeated Experiment the Fourth, which was made with a view to determine whether the blood would coagulate by rest. In the first trial, the vein was not opened till the end of three hours and a quarter; and just before it was opened I had observed through its coats, that the upper part of the blood was transparent, owing to the separation of the lymph. On letting out this blood, it seemed to me entirely fluid: a part indeed had been lost, but the greatest part was collected in the cup, and which afterwards coagulated as blood commonly does when exposed to the air. From this experiment I imagined that the whole had been fluid; but from others made since, I am persuaded that the part which was lost had been coagulated; for, from a variety of trials, I now find, that though the whole of the blood is not congealed in this time by rest alone, yet a part of it is. But as it would be trespassing too much on the reader's time to relate every experiment I have been obliged to make for this purpose, I shall only mention the general result of the whole.

After fixing a dog down to a table, and tying up his jugular veins, I have in general found that on opening them at the end of ten minutes, the blood was still entirely fluid, or without any appearance of coagulation.[2] If they were opened at the end of fifteen minutes, at first sight it also appeared quite fluid; but on a careful examination I found sometimes one, and sometimes two or three small particles about the size of a pin's head, which were coagulated parts of the blood. When opened later than this period, a larger and larger coagulum was observed; but so very slowly does this coagulation proceed, that in an experiment where I had the curiosity to compare more exactly the clotted part with the unclotted, I found, after the vein had been tied two hours and a quarter, that the coagulum weighed only two grains; whilst the rest of the blood, which was fluid, on being suffered to congeal, weighed eleven grains. I can advance nothing farther in this part of my subject with precision.

---

[2] I say *in general* it was fluid at the end of ten minutes; but I must likewise mention that in one dog I found two very small particles of beginning coagulation, even at this period; yet in another I could not observe any such appearance, even at the end of fifteen minutes.

RESPIRATORY INTERDEPENDENCE OF ANIMALS AND PLANTS

PRIESTLEY, Joseph (English teacher, naturalist, and divine, 1733–1804). From *Experiments and observations on different kinds of air*, Birmingham, 1774–1777.

At the time when Priestley discovered the "purification" of air and water by vegetation, he thought that the burning of objects in air or foods in the body produced a substance called phlogiston, a notion inherited from Stahl. Enough phlogiston rendered air unfit to support combustion, or life. Purification of air, therefore, would be the removal from it of phlogiston. This, he believed, was the role of vegetation. Even his own great subsequent achievement, oxygen from mercuric oxide, did not suggest that a similar phenomenon might explain the "purifying" action of plants. Priestley was born poor, studied both for commerce and the ministry, acquiring proficiency in the French, German, Italian, Latin, Greek, Hebrew, Chaldean, Syriac, and Arabic tongues. His scientific work, never systematic, competed for his attention with energetic preaching of various nonconformist doctrines and with his interest in political liberalism. Persecuted for these, he went to America.

OF THE RESTORATION OF AIR INFECTED WITH ANIMAL RESPIRATION, OR PUTREFACTION, BY VEGETATION.

That candles will burn only a certain time, in a given quantity of air is a fact not better known, than it is that animals can live only a certain time in it; but the cause of the death of the animal is not better known than that of the extinction of flame in the same circumstances; and when once any quantity of air has been rendered noxious by animals breathing in it as long as they could, I do not know that any methods have been discovered of rendering it fit for breathing again. It is evident, however, that there must be some provision in nature for this purpose, as well as for that of rendering the air fit for sustaining flame; for without it the whole mass of the atmosphere would, in time, become unfit for the purpose of animal life; and yet there is no reason to think that it is, at present, at all less fit for respiration than it has ever been. I flatter myself, however, that I have hit upon one of the methods employed by nature for this great purpose. How many others there may be, I cannot tell.

When animals die upon being put into air in which other animals have died, after breathing in it as long as they could, it is plain that the cause of their death is not the want of any *pabulum vitæ*, which has been supposed to be contained in the air, but on account of the air being impregnated with something stimulating to their lungs; for they almost always die in convulsions, and are sometimes affected so suddenly, that they are irrecoverable after a single inspiration, though they may be withdrawn immediately, and every method has been taken to bring them to life again. They are affected in the same manner, when they are killed in any other kind of noxious air that I have tried, viz.

fixed air[1], inflammable air[2], air filled with the fumes of sulphur, infected with putrid matter, in which a mixture of iron filings and sulphur has stood, or in which charcoal has been burned, or metals calcined, or in nitrous air, &c.

As it is known that *convulsions* weaken, and exhaust the vital powers, much more than the most vigorous *voluntary* action of the muscles, perhaps these universal convulsions may exhaust the whole of what we may call the *vis vitae* at once; at least that the lungs may be rendered absolutely incapable of action, till the animal be suffocated, or be irrecoverable for want of respiration.

If a mouse (which is an animal that I have commonly made use of for the purpose of these experiments) can stand the first shock of this stimulus, or has been habituated to it by degrees, it will live a considerable time in air in which other mice will die instantaneously. I have frequently found that when a number of mice have been confined in a given quantity of air, less than half the time that they have actually lived in it, a fresh mouse being introduced to them has been instantly thrown into convulsions, and died. It is evident therefore, that if the experiment of the Black Hole, at Calcutta, were to be repeated, a man would stand the better chance of surviving it, who should enter at the first, than at the last hour.

I have also observed, that young mice will always live much longer than old ones, or than those which are full grown, when they are confined in the same quantity of air. I have sometimes known a young mouse to live six hours in the same circumstances in which an old mouse has not lived one. On these accounts, experiments with mice, and, for the same reason, no doubt, with other animals also, have a considerable degree of uncertainty attending them; and therefore, it is necessary to repeat them frequently, before the result can be absolutely depended upon. But every person of feeling will rejoice with me in the discovery of *nitrous air*, which supersedes many experiments with the respiration of animals; being a much more accurate test of the purity of air.

The discovery of the provision in nature for restoring air, which has been injured by the respiration of animals, having long appeared to me to be one of the most important problems in natural philosophy, I have tried a great variety of schemes in order to affect it. In these, my guide has generally been to consider the influences to which the atmosphere is, in fact, exposed; and, as some of my unsuccessful trials may be of use to those who are disposed to take pains in the farther investigation of this subject, I shall mention the principal of them.

The noxious effluvium with which air is loaded by animal respiration, is not absorbed by standing, without agitation, in fresh or salt water. I have kept it many months in fresh water, when, instead of being meliorated, it has seemed to become even more deadly, so as to require more time to restore it, by the methods which will be explained hereafter, than air which has been lately

---

[1] Carbon dioxide.
[2] Hydrogen.

made noxious. I have even spent several hours in pouring this air from one glass vessel into another, in water, sometimes as cold, and sometimes as warm, as my hands could bear it, and have sometimes also wiped the vessels many times, during the course of the experiment, in order to take off that part of the noxious matter, which might adhere to the glass vessels, and which evidently gave them an offensive smell; but all these methods were generally without any sensible effect. The *motion*, also, which the air received in these circumstances, it is very evident, was of no use for this purpose. I had not then thought of the simple, but most effectual method of agitating air in water, by putting it into a tall jar and shaking it with my hand.

This kind of air is not restored by being exposed to the *light*, or any other influence to which it is exposed, when confined in a thin phial, in the open air, for some months.

Among other experiments, I tried a great variety of different *effluvia*, which are continually exhaling into the air, especially of those substances which are known to resist putrefaction; but I could not by these means effect any melioration of the noxious quality of this kind of air.

Having read, in the Memoirs of the Imperial Society, of a plague not affecting a particular village, in which there was a large sulphur-work, I immediately fumigated a quantity of this kind of air; or (which will hereafter appear to be the very same thing) air tainted with putrefaction, with the fumes of burning sulphur, but without any effect.

I once imagined, that the *nitrous acid* in the air might be the general restorative which I was in quest of; and the conjecture was favoured, by finding that candles would burn in air extracted from saltpetre. (This was the first instance of my finding dephlogisticated air, but without knowing it to be at all different from common air). I therefore spent a good deal of time in attempting, by a burning glass, and other means, to impregnate this noxious air with some effluvium of saltpetre, and, with the same view introduced into it the fumes of the smoking spirit of nitre; but both these methods were altogether ineffectual.

In order to try the effect of *heat*, I put a quantity of air, in which mice had died, into a bladder, tied to the end of the stem of a tobacco pipe, at the other end of which was another bladder, out of which the air was carefully pressed. I then put the middle part of the stem into a chafing-dish of hot coals, strongly urged with a pair of bellows; and, pressing the bladders alternately, I made the air pass several times through the heated part of the pipe. I have also made this kind of air very hot, standing in water before the fire. But neither of these methods were of any use.

Rarefaction and *condensation* by instruments were also tried, but in vain.

Thinking it possible that the *earth* might imbibe the noxious quality of the air, and thence supply the roots of plants with such putrescent matter as is known to be nutritive to them, I kept a quantity of air in which mice had died, in a phial, one half of which was filled with fine garden-mould; but, though it stood two months in these circumstances, it was not the better for it.

I once imagined that, since several kinds of air cannot be long separated from common air, by being confined in bladders, in bottles well corked, or even closed with ground stoppers, the affinity between this noxious air and the common air might be so great, that they would mix through a body of water interposed between them; the water continually receiving from the one, and giving to the other, especially as water receives some kind of impregnation from, I believe, every kind of air to which it is contiguous; but I have seen no reason to conclude, that a mixture of any kind of air with the common air can be produced in this manner.

I have kept air in which mice have died, air in which candles have burned out, and inflammable air, separated from the common air, by the slightest partition of water that I could well make, so that it might not evaporate in a day or two, if I should happen not to attend to them; but I found no change in them after a month or six weeks. The inflammable air was still inflammable, mice died instantly in the air in which other mice had died before, and candles would not burn where they had burned out before.

Since air tainted with animal or vegetable putrefaction is the same thing with air rendered noxious by animal respiration, I shall now recite the observations which I have made upon this kind of air, before I treat of the method of restoring them.

That these two kinds of air are, in fact, the same thing, I conclude from their having several remarkable common properties, and from their differing in nothing that I have been able to observe. They equally extinguish flame, they are equally noxious to animals, they are equally, and in the same way, offensive to the smell, they equally precipitate lime in lime water, and they are restored by the same means.

Since air which has passed through the lungs is the same thing with air tainted with animal putrefaction, it is probable that one use of the lungs is to carry off a putrid effluvium, without which, perhaps, a living body might putrefy as soon as a dead one.

Insects of various kinds live perfectly well in air tainted with animal or vegetable putrefaction, when a single inspiration of it would have instantly killed any other animal. I have frequently tried the experiment with flies and butterflies. The *aphides* also will thrive as well upon plants growing in this kind of air, as in the open air. I have been frequently obliged to take plants out of the putrid air in which they were growing, on purpose to brush away the swarms of these insects which infected them; and yet so effectually did some of them conceal themselves, and so fast did they multiply, in these circumstances, that I could seldom keep the plants quite clear of them.

When air has been freshly and strongly tainted with putrefaction, so as to smell through the water, sprigs of mint have presently died, upon being put into it, their leaves turning black; but if they do not die presently, they thrive in a most surprising manner. In no other circumstances have I ever seen vegetation so vigorous as in this kind of air, which is immediately fatal to animal life. Though these plants have been crowded in jars filled with this air, every leaf has been full of life; fresh shoots have branched out in various directions, and have grown much faster than other similar plants, growing in the same exposure in common air.

This observation led me to conclude, that plants, instead of affecting the air in the same manner with animal respiration, reverse the effects of breathing, and tend to keep the atmosphere sweet and wholesome, when it is become noxious, in consequence of animals either living and breathing, or dying and putrefying in it.

In order to ascertain this, I took a quantity of air, made thoroughly noxious, by mice breathing and dying in it, and divided it into two parts; one of which I put into a phial immersed in water; and to the other (which was contained in a glass jar, standing in water) I put a sprig of mint. This was about the beginning of August, 1771, and after eight or nine days, I found that a mouse lived perfectly well in that part of the air, in which the sprig of mint had grown, but died the moment it was put into the other part of the same original quantity of air; and which I had kept in the very same exposure, but without any plant growing in it.

This experiment I have several times repeated; sometimes using air in which animals had breathed and died; and at other times using air tainted with vegetable or animal putrefaction; and generally with the same success.

Once, I let a mouse live and die in a quantity of air which had been noxious, but which had been restored by this process, and it lived nearly as long as I conjectured it might have done in an equal quantity of fresh air; but this is so exceedingly various, that it is not easy to form any judgment from it; and in this case the symptom of *difficult respiration* seemed to begin earlier than it would have done in common air.

Since the plants that I made use of manifestly grow and thrive in putrid air; since putrid matter is well known to afford proper nourishment for the roots of plants; and since it is likewise certain that they receive nourishment by their leaves as well as by their roots, it seems to be exceedingly probable, that the putrid effluvium is in some measure extracted from the air, by means of the leaves of plants, and therefore that they render the remainder more fit for respiration.

Towards the end of the year some experiments of this kind did not answer so well as they had done before, and I had instances of the relapsing of this restored air to its former noxious state. I therefore suspended my judgment concerning the efficacy of plants to restore this kind of noxious air, till I should have an opportunity of repeating my experiments, and giving more attention to them. Accordingly I resumed the experiments in the summer of the year 1772, when I presently had the most indisputable proof of the restoration of putrid air by vegetation; and as the fact is of some importance, and the subsequent variation in the state of this kind of air is a little remarkable, I think it necessary to relate some of the facts pretty circumstantially.

The air, on which I made the first experiments, was rendered exceedingly noxious by mice dying in it on the 20th of June. Into a jar nearly filled with one part of this air, I put a sprig of mint, while I kept another part of it in a phial, in the same exposure; and on the 27th of the same month, and not before, I made a trial of them, by introducing a mouse into a glass vessel, containing two ounce measures and a half, filled with each kind of air; and I noted the following facts.

When the vessel was filled with the air in which the mint had grown, a very large mouse lived five minutes in it, before it began to shew any sign of uneasiness. I then took it out, and found it to be as strong and vigorous as when it was first put in; whereas in that air which had been kept in the phial only, without a plant growing in it, a younger mouse continued not longer than two or three seconds, and was taken out quite dead. It never breathed after, and was immediately motionless. After half an hour, in which time the larger mouse (which I had kept alive, that the experiment might be made on both the kinds of air with the very same animal) would have been sufficiently recruited, supposing it to have received any injury by the former experiment, was put into the same vessel of air; but though it was withdrawn again, after being in it hardly one second, it was recovered with difficulty, not being able to stir from the place for near a minute. After two days, I put the same mouse into an equal quantity of common air, and observed that it continued seven minutes without any sign of uneasiness; and being very uneasy after three minutes longer I took it out. Upon the whole, I concluded that the restored air wanted about one fourth of being as wholesome as common air. The same thing also appeared when I applied the test of nitrous air.

In the seven days, in which the mint was growing in this jar of noxious air, three old shoots had extended themselves about three inches, and several new ones had made their appearance in the same time. Dr. Franklin and Sir John Pringle happened to be with me, when the plant had been three or four days in this state, and took notice of its vigorous vegetation, and remarkably healthy appearance in that confinement.

On the 30th of the same month, a mouse lived fourteen minutes, breathing naturally all the time, and without appearing to be much uneasy, till the last two minutes, in the vessel containing two ounce measures and a half of air which had been rendered noxious by mice breathing in it almost a year before, and which I found to be most highly noxious on the 19th of this month, a plant having grown in it, but not exceedingly well, these eleven days: on which account I had deferred making the trial so long. The restored air was affected by a mixture of nitrous air, almost as much as common air.

That plants are capable of perfectly restoring air injured by respiration, may, I think, be inferred with certainty from the perfect restoration, by this means, of air which had passed through my lungs, so that a candle would burn in it again, though it had extinguished flame before, and a part of the same original quantity of air still continued to do so. Of this one instance occurred in the year 1771, a sprig of mint having grown in a jar of this kind of air, from the 25th of July to the 17th of August following; and another trial I made, with the same success, the 7th of July, 1772, the plant having grown in it from the 29th of June preceding. In this case also I found that the effect was not owing to any virtue in the leaves of mint; for I kept them constantly changed in a quantity of this kind of air, for a considerable time, without making any sensible alteration in it.

These proofs of a partial restoration of air by plants in a state of vegetation, though in a confined and unnatural situation, cannot but render it highly probable, that the injury which is continually done to the atmosphere by the respiration of such a number of animals, and the putrefaction of such masses of both vegetable and animal matter, is, in part at least, repaired by the vegetable creation. And, notwithstanding the prodigious mass of air that is corrupted daily by the above-mentioned causes; yet, if we consider the immense profusion of vegetables upon the face of the earth, growing in places suited to their nature, and consequently at full liberty to exert all their powers, both inhaling and exhaling, it can hardly be thought, but that it may be a sufficient counterbalance to it, and that the remedy is adequate to the evil.

Dr. Franklin, who, as I have already observed, saw some of my plants in a very flourishing state, in highly noxious air, was pleased to express very great satisfaction with the result of the experiments. In his answer to the letter in which I informed him of it, he says,

"That the vegetable creation should restore the air which is spoiled by the animal part of it, looks like a rational system, and seems to be of a piece with the rest. Thus fire purifies water all the world over. It purifies it by distillation, when it raises it in vapours, and lets it fall in rain; and farther still by filtration, when, keeping it fluid, it suffers that rain to percolate the earth. We knew before that putrid animal substances were converted into sweet vegetables, when mixed with the earth, and applied as manure; and now, it seems, that the same putrid substances, mixed with the air, have a similar effect. The strong thriving state of your mint in putrid air seems to shew that the air is mended by taking something from it, and not by adding to it." He adds, "I hope this will give some check to the rage of destroying trees that grow near houses, which has accompanied our late improvements in gardening, from an opinion of their being unwholesome. I am certain, from long observation, that there is nothing unhealthy in the air of woods; for we Americans have every where our country habitations in the midst of woods, and no people on earth enjoy better health, or are more prolific."

May not plants also restore air diminished by putrefaction, by absorbing part of the phlogiston with which it is loaded? The greater part of a dry plant, as well as of a dry animal substance, consists of inflammable air, or something that is capable of being converted into inflammable air; and it seems to be as probable that this phlogistic matter may have been imbibed by the roots and leaves of plants, and afterwards incorporated into their substance, as that it is altogether produced by the power of vegetation. May not this phlogistic matter be even the most essential part of the food and support of both vegetable and animal bodies?

Having discovered that vegetation restores, to a considerable degree of purity, air that had been injured by respiration or putrefaction, I conjectured that the phlogistic matter, absorbed by the water, might be imbibed by plants, as well as form other combinations with substances under the water. A curious fact, which has since been communicated to me, very much favours this supposition.

Mr. Garrick was so obliging as to give me the first intimation of it, and Mr. Walker, the ingenious author of a late English Dictionary, from whom he received the account, was pleased to take some pains in making farther inquiries into it for my use. He informed me that Mr. Bremner, who keeps a music-shop opposite to Somerset-house, was at Harwich, waiting for the packet; and observed that a reservoir at the principal inn was very foul on the sides. This made him ask the innkeeper why he did not clean it out; who immediately answered, that he had done so once, but would not any more; for that after cleansing the reservoir, the water which was caught in it grew fetid, and unfit for use; and that it did not recover its sweetness till the sides and bottom of the reservoir grew very foul again. Mr. Walker questioned Mr. Bremner, whether there were any vegetables growing at the sides and bottom of it; but of this he could not be positive. However, as he said it was covered with a *green substance*, which is known to be vegetable matter (and indeed nothing else could well adhere to the *sides*, as well as to the bottom of the reservoir) I think it will be deemed probable, that it was this vegetating matter that preserved the water sweet, imbibing the phlogistic matter that was discharged in its tendency to putrefaction.

I shall be happy, if the mention of this fact should excite an attention to things of this nature. Trifling as they seem to be, they have, in a philosophical view, the greatest dignity and importance; serving to explain some of the most striking phenomena in nature, respecting the general plan and constitution of the system, and the relation that one part of it bears to another.

RESPIRATION AS UTILIZATION OF OXYGEN

LAVOISIER, Antoine-Laurent (French naturalist, 1743–1794). From *Expériences sur la respiration des animaux, et sur les changemens qui arrivent à l'air en passant par leur poumon*, in *Histoire de l'Académie Royale des Sciences* for 1777, p. 185, 1780, tr. by T. Henry as *Experiments on the respiration of animals and on the changes effected on the air passing through their lungs* in his *Essays, on the effects produced by various processes on atmospheric air, etc.*, Warrington, 1783.*

It was chiefly by his overthrow of the phlogiston doctrine (see Priestley) that Lavoisier was able to take this important forward step in our knowledge of respiration. His clear-cut establishment of oxygen as an element, and his corresponding reform of the nomenclature, produced profound effects on the development of chemistry. Although Lavoisier had a philanthropic and liberal record as public servant, he was executed for his affluence as a farmer-general. Part of his prosperity had resulted from his own progressive studies in agricultural chemistry.

Of all the phenomena of the animal economy, none is more striking, none more worthy the attention of philosophers and physiologists than those which accompany respiration. Little as our acquaintance is with the object of this singular function, we are satisfied that it is essential to life, and that it cannot be suspended for any time without exposing the animal to the danger of immediate death. . . .

The experiments of some philosophers, and especially those of Messrs. Hales and Cigna, had begun to afford some light on this important object; and, Dr. Priestley has lately published a treatise, in which he has greatly extended the bounds of our knowledge; and has endeavoured to prove, by a number of very ingenious, delicate, and novel experiments, that the respiration of animals has the property of phlogisticating air, in a similar manner to what is effected by the calcination of metals and many other chemical processes; and that the air ceases not to be respirable, till the instant when it becomes surcharged, or at least saturated, with phlogiston.

However probable the theory of this celebrated philosopher may, at first sight, appear; however numerous and well conducted may be the experiments by which he endeavours to support it, I must confess I have found it so contradictory to a great number of phenomena, that I could not but entertain some doubts of it. I have accordingly proceeded on a different plan, and have found myself led irresistibly, by the consequences of my experiments, to very different conclusions.

Now air which has served for the calcination of metals, is, as we have already seen, nothing but the mephitic residuum of atmospheric air, the highly respirable part of which has combined with the mercury, during the calcination: and the air which has served the purposes of respiration, when deprived

---

* The portion reprinted here is the same as that which appears in John F. Fulton, *Selected readings in the history of physiology*, pp. 122–126, Springfield, Ill., 1930.

of the fixed air[1], is exactly the same; and, in fact, having combined, with the latter residuum, about $\frac{1}{4}$ of its bulk of dephlogisticated air, extracted from the calx of mercury, I re-established it in its former state, and rendered it equally fit for respiration, combustion, etc., as common air, by the same method as that I pursued with air vitiated by the calcination of mercury.

The result of these experiments is, that to restore air that has been vitiated by respiration, to the state of common respirable air, two effects must be produced: 1st. to deprive it of the fixed air it contains, by means of quicklime or caustic alkali: 2dly. to restore to it a quantity of highly respirable or dephlogisticated air, equal to that which it has lost. Respiration, therefore, acts inversely as these two effects, and I find myself in this respect led to two consequences equally probable, and between which my present experience does not enable me to pronounce. . . .

The first of these opinions is supported by an experiment which I have already communicated to the academy. For I have shewn in a memoir, read at our public Easter meeting, 1775, that dephlogisticated air may be wholly converted into fixed air by an addition of powdered charcoal; and, in other memoirs, I have proved that this conversion may be effected by several other methods: it is possible, therefore, that respiration may possess the same property, and that dephlogisticated air, when taken into the lungs, is thrown out again as fixed air. . . . Does it not then follow, from all these facts, that this pure species of air has the property of combining with the blood and that this combination constitutes its red color. But whichever of these two opinions we embrace, whether that the respirable portion of the air combines with the blood, or that it is changed into fixed air in passing through the lungs; or lastly, as I am inclined to believe, that both these effects take place in the act of respiration, we may from facts alone, consider as proved:

1st. That respiration acts only on the portion of pure or dephlogisticated air, contained in the atmosphere; that the residuum or mephitic part is a merely passive medium which enters into the lungs, and departs from them in nearly the same state, without change or alteration.

2dly. That the calcination of metals, in a given quantity of atmospheric air, is effected, as I have already often declared, only in proportion as the dephlogisticated air, which it contains, has been drained and combined with the metal.

3dly. That, in like manner, if an animal be confined in a given quantity of air, it will perish as soon as it has absorbed, or converted into fixed air, the major part of the respirable portion of air, and the remainder is reduced to a mephitic state.

---

[1] Carbon dioxide.

4thly. That the species of mephitic air, which remains after the calcination of metals, is in no wise different, according to all the experiments I have made, from that remaining after the respiration of animals; provided always, that the latter residuum has been freed from its fixed air: that these two residuums may be substituted for each other in every experiment, and that they may each be restored to the state of atmospheric air, by a quantity of dephlogisticated air, equal to that of which they had been deprived. A new proof of this last fact is, that if the portion of this highly respirable air, contained in a given quantity of the atmospheric, be increased or diminished, in such proportion will be the quantity of metal which we shall be capable of calcining in it, and, to a certain point, the time which animals will be capable of living in it.

ANIMAL ELECTRICITY

GALVANI, Luigi (Aloisio, Aloysio) (Italian natural scientist, 1737–1798). From *De viribus electricitatis in motu musculari commentarius etc.*, Mutinae, 1791; tr. by J. F. Fulton as *Concerning electrical forces in muscular movement* in his *Readings in the history of physiology*, Springfield, Ill., 1930; courtesy of Charles C Thomas, publisher.

The difficulties in tracing Galvani's central thought have been cited by Dr. Fulton in the introductory note* to his translation. Probably Galvani's work is built upon the notion that if muscles can respond to an external electrical stimulus (which he demonstrated) it may be postulated that the normal internal stimulus is also of an electrical nature. That tissues can generate electrical stimuli was likewise proved by Galvani (although the authorship of the latter experiments was contested).†

Concerning Electrical Forces in Muscular Movement, First Part

I dissected and prepared a frog . . . and placed it with everything else at hand on a table on which was an electric machine, but the frog was completely removed from its conductor and by a considerable interval. One of those who were helping me accidently lightly touched with the point of his scalpel the inner nerve of this frog's leg, and suddenly all of the leg muscles appeared to become so contracted that they seemed to have fallen into fairly violent tonic convulsion. Another who was helping us in producing the electricity, seemed to observe that the phenomenon occurred while the spark was obtained from the conductor of the machine. He, marvelling at the novelty of the occurrence, at once drew my attention to this since I was at the time completely occupied with other things and was absorbed in my own thoughts. I was fired at once with an extraordinary eagerness and desire to perform the same experiment and to bring into the light of day the hidden secrets of the

---

* See p. 209.
† See E. Du Bois-Reymond, *Untersuchungen über thierische elektricität*, Berlin, 1848.

phenomenon.   Therefore I myself placed the point of the scalpel on one or other of the leg nerves while one of those who were present produced a spark. The whole phenomenon was by this same means repeated, and, at the same time as the sparks were obtained, there were produced wonderfully strong contractions in each muscle of the joints just as if an animal in tetanus had been used.

But fearing that those very movements arose rather from the contact of the point which accidentally acted as a stimulus than from the spark, I tested the same nerves with the point of my scalpel by the same method in the other frogs, and indeed with greater pressure, without, however, any spark on that occasion being produced by anyone; but there seemed to be no movements at all.   As a result of this I reasoned to myself that perhaps in order for the phenomenon to be established there were needed at the same time both the contact of some body and the application of a spark.   Therefore I again placed the edge of the scalpel on the nerves and held it there motionless, both when a spark was produced and when the machine was absolutely still.   But the phenomenon appeared only so long as the spark lasted.

We repeated the experiment always using the same scalpel, but to our amazement when the spark was produced sometimes the movements previously described took place, at other times were absent.

Stirred by the novelty of the occurrence, we began to test it by different experimental methods, always, however, using the same scalpel so that we might, if it were possible, discover the reasons for the unexpected difference.   This new task was not empty of result, for we discovered that the whole matter was to be attributed to the question of which different part of the scalpel we were holding with our fingers: the scalpel had a bone handle and, if this handle were held in the hand when a spark was produced, no movements resulted, but they did result when the fingers were placed either on the metal blade or on the iron nails securing the blade of the scalpel.

Therefore since the rather dry bone produced an electric force of its own, and the metal blade and iron nails a conducting or "anelectric" force, as they say, we suspected that perhaps it happened that when we held the bone handle with our fingers all approach to the electric current, entering by some means into the frog, was stopped, but that the current was released when we took hold of the blade or the nails setting up the same current.

Therefore in order to put the matter beyond all doubt, we used instead of the scalpel first a thin glass tube carefully polished so that all moisture and specks of dust were removed, and secondly one of iron.   We not only touched the leg nerves with the glass but as it were rubbed them when the spark was produced, but all our efforts were in vain as no phenomenon resulted although innumerable and stronger sparks were obtained from the conductor of the machine and at a very small distance from the animal.   But the phenomenon did result when the iron tube was applied to the same nerves and when quite small sparks were obtained.

[Later he found] that prepared frogs suspended by brass hooks through the spinal marrow from an iron lattice round a hanging garden of our house, exhibited convulsions not only during thunderstorms but occasionally also in fair weather. [Pursuing this experiment he later observed] when I pressed the brass hook, fixed in the spinal marrow, against the iron plate, behold! the same contractions, the same movements as before. I tried other metals with the same result, except that the amount of contraction depended on the metals used.

RISE OF MODERN PHYSIOLOGY OF DIGESTION

BEAUMONT, William (American surgeon and physiologist, 1785–1853). *A case of wounded stomach*, by Joseph Lovell [*sic*] Surgeon-General, U.S. Army, in *Medical Recorder of Original Papers and Intelligence in Medicine and Surgery*, vol. 3, p. 14, 1825; and *Further experiments in the case of Alexis San Martin, who was wounded in the stomach by a load of duck-shot, detailed in the Recorder for Jan. 1825, ibid.*, vol. 9, p. 94, 1826.

Beaumont's famous experiments on digestion are best known through his definitive account of them published in 1833. Actually, they were first announced (or some of them) in two issues of the Philadelphia *Recorder*. These papers are reprinted here Substitution of Lovell's for Beaumont's name as author of the first paper was an editorial error. The experiments not only greatly accelerated the establishment of modern digestive physiology but also gave rise to many household practices as regards diet, thorough mastication, use of condiments, etc.

ART. III. A CASE OF WOUNDED STOMACH. BY JOSEPH LOVELL, SURGEON-GENERAL, U.S. ARMY.

Alexis San Martin, a Canadian lad, 18 years of age, of good constitution, robust and healthy, was accidentally wounded by the discharge of a musket, on the 6th of June, 1822.

The charge, consisting of powder and duck shot, was received in his left side, at a distance of not more than one yard from the muzzle of the piece. It entered posteriorly, and in an oblique direction forward and inward; carrying away the integuments and muscles to the size of a man's hand; fracturing and entirely blowing off the anterior half of the sixth rib; fracturing the fifth; lacerating the lower portion of the left lobe of the lungs, and the diaphragm; and perforating the stomach. The whole contents of the musket, with fragments of clothing and pieces of the fractured ribs, were driven into the muscles and the cavity of the chest.

I saw him in 25 or 30 minutes after the accident, and, on examination, found a portion of the lung, as large as a turkey's egg, protruding through the external wound, lacerated and burnt; and immediately below this another protrusion, which, on further inspection, proved to be a portion of the stomach, lacerated through all its coats, and pouring out the food he had taken at breakfast, through an orifice large enough to admit my fore finger.

In attempting to return the protruded portion of the lung, I was prevented by a sharp point of the fractured rib, over which it had caught by its membranes; but, by raising it with my finger, and clipping off the point of the rib, I was able to return it into its proper cavity; though it could not be retained there on account of the incessant efforts to cough. The projecting portion of the stomach was nearly as large as that of the lung; and it passed through the lacerated diaphragm and the external wound, mingling the food with the bloody mucus blown from the lung.

After cleansing the wound of the charge and other extraneous matter, and replacing the stomach and lung, as far as practicable, I applied the carbonated fermenting poultice; keeping the surrounding parts constantly embrocated with the "lotio ammoniae muriat. cum aceto," and giving, internally, the "aq. ammon. acetat. cum camphor." in liberal quantities.

Under this treatment a strong reaction took place in about 24 hours, accompanied with high arterial excitement, fever, marked symptoms of inflammation of the lining membranes of the chest and abdomen, great difficulty of breathing, and distressing cough. He was bled to the amount of 18 to 20 ounces, and took a cathartic. The bleeding reduced the arterial action and gave relief; but the cathartic had no effect, as it escaped from the stomach through the wound.

On the 5th, a partial sloughing of the integuments and muscles took place. Some of the protruded portions of the lung and the lacerated parts of the stomach also sloughed; and left the perforation in the stomach plainly to be seen, and large enough to admit the whole length of my fore finger into its cavity; and also a passage into the chest half as large as my fist, exposing to view a part of the lung, and admitting the free escape of air and bloody mucus at every respiration.

A violent fever continued for ten days, running into a typhoid type; and the wound became very foetid. On the 11th, a more extensive sloughing took place; the febrile symptoms subsided; and the whole surface of the wound assumed a healthy and granulating appearance. For 17 days all that entered his stomach by the oesophagus soon passed out through the wound; and the only means of sustaining him was by nutricious injections per anum, until compresses and adhesive straps could be applied to retain his food. During this period no alvine evacuations could be obtained, although cathartic enemata and various other means were adopted to promote them.

In a few days after the firm dressings were applied, and the contents of the stomach retained, the bowels became gradually excited, and, with the aid of cathartic injections, a very foetid, hard, black stool was produced, followed by several similar ones, when his bowels became quite regular, and have so continued.

The cataplasms were applied until the sloughing was completed, and the granulating process fully established; and were afterwards occasionally resorted to when the wound became ill-conditioned. The aq. ammon. acet. cum camphor. was also continued for several weeks, in proportion to the febrile symptoms, and the fetid condition of the wound. No sickness nor unusual irritation of the stomach, nor even nausea, was manifested during the whole time; and, after the fourth week, the appetite became good, digestion regular, the alvine evacuations natural, and all the functions of the system perfect and healthy.

By the adhesion of the sides of the protruded portion of the stomach to the pleura costalis and the external wound, a free exit was afforded to its contents, and thereby effusion into the abdominal cavity prevented. Cicatrization and contraction of the external wound, commenced in the fifth week; the stomach became more firmly attached to the pleura and intercostals, by its external coat, but showed not the least disposition to close its orifice by granulations, which terminated as if at a natural boundary, and left the perforation resembling, in all but a sphincter, the natural anus with a slight prolapsus. Whenever the wound was dressed, the contents of the stomach would flow out in proportion to the quantity recently taken; if it happened to be empty, or nearly so, a partial inversion would take place, unless prevented by the application of the finger; and frequently, in consequence of the derangement of the bandages, the inverted part would be found of the size of a hen's egg. No difficulty, however, occurred, in reducing it by gentle pressure with the finger, or a sponge wet with cold water, neither of which produced the least pain.

In the seventh week, exfoliation of the ribs, and a separation of their cartilaginous ends, began to take place. The 6th rib was denuded of its periosteum, for about two inches from the fractured end; so that I was obliged to amputate it about 3 or 4 inches from the articulation with the spine; which I did by dissecting back the muscles, securing the intercostal artery, and sawing it off with a fine saw, introduced between it and the 5th rib, without injury to the neighbouring parts. Healthy granulations soon appeared, and formed soundly over the amputated end. About half of the inferior edge of the 5th rib also exfoliated and separated from its cartilage. After the removal of these pieces of bone, I attempted to contract the wound and close the perforation of the stomach, by gradually drawing the edges together with adhesive straps, laid on in a radiated form. The circumference of the external wound was at least 12 inches; and the orifice of the stomach nearly in the center, two inches below the left nipple, and in a line drawn from this to the point of the left ilium. To retain his food and drinks, I kept a plug and compress of lint, fitted to the shape and size of the hole, confined there by the adhesive straps. After trying every means in my power, for 8 or 10 months, to close the orifice, by exciting adhesive inflammation in the lips of the wound, without the least appearance of success, I gave it up as impracticable in any other way than that of incising them, and bringing them together by suture; an operation to which the patient would not submit.

By the sloughing of the injured portion of the lung, a cavity was left, as large as a common-sized tea-cup, which continued a copious discharge of pus for three months, when it became filled with healthy granulations, firmly adhering to the pleura, and soundly cicatrized over that part of the wound.

Four months after the injury was received, an abscess formed about two inches below the wound, nearly over the cartilaginous ends of the 1st and 2d false ribs, very painful and extremely sore, and producing violent symptomatic fever. On the application of emollient poultices, it pointed externally. It was then laid open to the extent of 3 inches, and several shot and pieces of wad extracted; after which, a gum elastic bougie could be introduced 4 or 5 inches in the longitudinal direction of the ribs, towards the spine. Great pain and soreness extended from the opening of the abscess, along the track of the cartilaginous ends of the false ribs, to the spine, with a copious discharge from the sinus. In 5 or 6 days, there came away a cartilage one inch in length. In 6 or 7 days more, another, an inch and a half long, and in about the same length of time a third, two inches long; and they continued to come away every 5 or 6 days, until five were discharged at the same opening, the last 3 inches in length. They were all entire, and evidently separated from the false ribs. The discharge, pain, and irritation, during the 4 or 5 weeks these cartilages were working out, greatly reduced the strength of the patient, produced a general febrile habit, and stopped the healing process in the original wound.

Directly after the passage of the last cartilages, inflammation began over the lower end of the sternum, which, by the usual applications, terminated in a few days, in a large abscess; and from which, by laying it open two inches, I extracted another cartilage 3 inches in length. The inflammation then abated; and in a day or two another short piece came away, and the discharge subsided.

To support the strength under all these debilitating accidents, I administered wine with the diluted muriatic acid and 30 or 40 drops of the tincture of assafoetida, three times a day; which seemed to have the desired effect, and very much improved the condition of the wound.

On the 3d of January, 1823, I extracted another cartilage from the opening over the sternum, an inch and a half long; and on the fourth another, two and a half inches in length, an inch broad at one end, and narrowing to less than half an inch at the other; which must have been the ensiform cartilage of the sternum. After this the sinus closed, and there was no return of in-flammation.

On the 6th of June, a year from the time of the accident, the injured parts were all sound and firmly cicatrized, with the exception of the stomach, which continued in much the same condition as it was six weeks after the wound was received. The perforation was about the size of a shilling piece, with its edges firmly attached to the pleura and intercostals, and the food and drinks con-tinually exuded, unless prevented by the plug, compress, and bandage.

The lad is now (Sept. 1824) in perfect health, and says he feels no incon-venience from the wound, except the trouble of dressing it. He eats as heart-ily, and digests as perfectly, as he ever did; is strong and able bodied, perform-ing any kind of labour, from that of a house servant to chopping wood or mowing in the field. He has been in my service since April, 1823, during which period he has not had a day's sickness, sufficient to disqualify him for his ordinary duties. He will drink a quart of water or eat a dish of soup, and then by removing the compress can immediately throw it out through the wound. On removing the dressings, I frequently find the stomach inverted to the size, and about the shape, of a half blown damask rose; yet he complains of no pain, and it will return itself, or is easily reduced by gentle pressure. When he lies upon the opposite side, I can look directly into the cavity of the stom-ach, observe its motion, and almost see the process of digestion—I can pour in water with a funnel, or put in food with a spoon, and draw them out again with a syphon. I have frequently suspended flesh, raw and roasted, and other substances in the hole, to ascertain the length of time required to digest each; and at one time used a plug of raw beef, instead of lint, to stop the orifice, and found that in less than five hours it was completely digested off, as smooth and even as if it had been cut with a knife.

This case affords a most excellent opportunity of experimenting upon the gastric fluids, and the process of digestion. It would give no pain, nor cause the least uneasiness, to extract a gill of fluid every two or three days; for it frequently flows out spontaneously in considerable quantities; and one might introduce various digestible substances into the stomach, and easily examine them during the whole process of digestion. I may, therefore, be able here-after to give some interesting experiments on these subjects.

[Beginning of second paper]

EXPERIMENT 1ST.

On the 1st of August 1825, at 12 o'clock, M. I introduced through the perforation between the ribs, into the stomach of *Alexis San Martin*, the fol-lowing articles of diet, suspended by a silk string, and fastened at proper dis-tances from each other, so as to pass in without giving pain—to wit: a piece of high seasoned *alamode beef*, a piece of raw salted *lean beef*, a piece of raw salted *fat pork*, a piece a raw lean *fresh beef*, a piece of boiled *corned beef*, a piece of *stale bread*, and a bunch of raw cabbage, each piece containing about two drachms. The lad continued his usual domestic employments about the house.

1 o'clock P.M.—withdrew and examined them—found the *cabbage* and *bread* about half digested; the pieces of meat unchanged. Returned them to the stomach.

2 o'clock P.M.—withdrew them—found the *cabbage, bread, pork* and boiled beef all cleanly digested and gone from the string; the other pieces of. meat but very little affected. Returned them into the stomach again.

3 o'clock P.M.—withdrew them—found the *alamode* partly digested, and the raw beef slightly macerated on the surface, but its general texture firm and entire. The fluids of the stomach smell and taste slightly rancid. Lad complains of some pain and uneasiness at the breast. Returned them again.

5 o'clock P.M.—The lad complains of considerable distress at the stomach, general debility and lassitude, and some pain in the head. Withdrew them— found all the remaining pieces of meat much in the same condition as when last drawn out; the fluids more rancid and sharp. The boy complaining considerably, I did not return them any more.

EXPERIMENT 2D.

11 o'clock A.M. August 7th—After having kept the boy fasting seventeen hours, I introduced the glass tube taken from the plate of a thermometer (Kendall's) through the perforation into the stomach, nearly the whole length of the stem, to ascertain the degrees of natural warmth of the system; in five minutes, or less, the mercury rose to 100°, and then remained stationary, (which I determined by marking the heighth of the mercury upon the glass with ink, as it stood in the stomach, and then withdrawing and placing it upon the graduated plate again.)

I then introduced a gum-elastic syphon, and drew off an ounce of pure gastric liquor, unmixed with any other matter, into a three-ounce phial; took a piece of *corned boiled beef*, large as my little finger, and put it into the liquor in the phial; corked it tight and placed it in a saucepan filled with water, raised precisely to 100°, and kept it at that point by placing it in a gentle and nicely regulated sandbath. In forty minutes digestion had distinctly commenced over the surface of the meat; at fifty the fluid became quite opake and cloudy, the external texture began to separate and become loose; at sixty minutes chyme began to form; at one o'clock P.M. (digestion having progressed with the same regularity as in the last half hour) the cellular texture seemed to be entirely destroyed, leaving the muscular fibres loose and unconnected, floating about in small fine shreds, very tender and soft.

At 3 o'clock P.M. the muscular fibres had diminished one half from one o'clock.

At 5 o'clock P.M. nearly all digested, a few fibres only remaining.

At 7 o'clock P.M. its texture was completely broken down, and very few of the small particles only floating in the fluid.

At 9 o'clock P.M. every particle of the meat completely and perfectly digested.

The gastric fluid, which, when taken from the stomach, was clear and almost transparent as water, was at this time about the colour of whey, and after standing at rest a few minutes, a fine sediment, the colour of meat, precipitated to the bottom of the phial.

EXPERIMENT 3D.

11 o'clock A.M. August 7th, (the same time that I commenced the second experiment,) I suspended a piece of meat exactly similar to that put into the phial, into the stomach, through the wound.

12 o'clock M—Withdrew and found it just about as much affected by digestion as that in the phial; little or no difference in their appearance. Returned it again into the stomach.

1 o'clock P.M.—Withdrew the *string*, but found the meat completely digested and gone. The effect of the gastric fluid upon the piece of meat suspended in the stomach was exactly similar to that in the phial, only more rapid after the first hour, and sooner completed. Digestion commenced and was confined to the surface entirely in both situations. Agitation accelerated the solution in the phial, as it removed the coat that was digested upon the surface and enveloped the remainder of the meat, and gave the gastric fluid access to the undigested portions.

EXPERIMENT 4TH.

August 8th—9 o'clock A.M.—Drew off 1½ ounces of gastric liquor into a three-ounce phial, and suspended two pieces of boiled chicken, from the breast and back, and placed it in the same situation and temperature as in the second experiment, observing the same regularity and minuteness.

Digestion commenced and progressed much the same as in the foregoing experiments, only rather slower, the fowl appearing to be more difficult of digestion than the flesh, the texture of the chicken being firmer than the beef, the gastric juice seemed not to insinuate itself into the interstices of the muscular fibres as it did into the beef, but operated entirely upon the surface, dissolving it as a piece of gumarabic wastes in the mouth, until the last particle was digested.

The colour of the fluid, after digesting the chicken-meat, was of a grayish-white, and more resembling milky fluid than *whey*, as in the experiment of beef in the phial, and the sediment was lighter coloured; but in other respects every way similar. The contents of both phials, kept perfectly tight, remained free from any foetor, acidity, or offensive smell or taste, from the time it was taken out of the stomach (7th and 8th of August) to the 6th of September, at which time that containing the boiled beef became very offensive and putrid, while that containing the chicken was perfectly bland and sweet. Both were kept in exactly similar situations.

The foregoing are given, as nearly as I have ability to observe and language to express, exactly as they occurred, and such as they be, I submit them for consideration, etc.

The man absconded to Canada, and it is to be regretted that the experiments have, on that account, been suspended.

BIOCHEMICAL NATURE OF NUTRITION

von LIEBIG, (Freiherr) Justus (German physiological chemist, 1803–1873). From *Die thier-chemie; oder, die organische chemie in ihrer anwendung auf physiologie und pathologie*, Braunschweig, 1842; tr. as *Animal chemistry or organic chemistry in its application to physiology and pathology*, Cambridge, 1842.*

In order to assess validly Liebig's contribution to physiological chemistry, we must remember that scientific progress depends both upon the totality of amassed information and upon the general points of view governing the thinking of scientists at the period in question. Thus, chemical physiologists in the nineteenth century came to regard animal chemistry as consisting of the actions (independent and interdependent) of three great groups of organic compounds: proteins, fats, and, as they would later be called, carbohydrates. This inflection to physiological science was given by Justus von Liebig (see below). Other contributions of Liebig were: isomerism of organic acids; important works (with Wöhler) on benzaldehyde, amygdalin, and urea; coating of mirrors; artificial fertilizers; and others. Liebig founded several journals and influenced chemical education more than any other single individual in the nineteenth century.

If we hold, that increase of mass in the animal body, the development of its organs, and the supply of waste,—that all this is dependent on the blood, that is, on the ingredients of the blood, then only those substances can properly be called nutritious, or considered as food, which are capable of conversion into blood. To determine, therefore, what substances are capable of affording nourishment, it is only necessary to ascertain the composition of the food, and to compare it with that of the ingredients of the blood.

Two substances require especial consideration as the chief ingredients of the blood; one of these separates immediately from the blood when withdrawn from the circulation. It is well known that in this case blood coagulates, and separates into a yellowish liquid, the *serum* of the blood, and a gelatinous mass,

* The English translation did not await the appearance of the German but was "Edited from the Author's Manuscript by William Gregory, M.D., F.R.S.E."

which adheres to a rod or stick, in soft, elastic fibres, when coagulating blood is briskly stirred. This is the *fibrine* of the blood, which is identical in all its properties with muscular fibre, when the latter is purified from all foreign matters.

The second principal ingredient of the blood is contained in the serum, and gives to this liquid all the properties of the white of eggs, with which it is identical. When heated, it coagulates into a white elastic mass, and the coagulating substance is called *albumen*.

Fibrine and albumen, the chief ingredients of blood, contain, in all, seven chemical elements, among which nitrogen, phosphorus, and sulphur are found. They contain also the earth of bones. The serum retains in solution sea salt and other salts of potash and soda, in which the acids are carbonic, phosphoric, and sulphuric acids. The globules of the blood contain fibrine and albumen, along with a red coloring matter, in which iron is a constant element. Besides these, the blood contains certain fatty bodies in small quantity, which differ from ordinary fats in several of their properties.

Chemical analysis has led to the remarkable result, that fibrine and albumen contain the same organic elements united in the same proportion, so that two analyses, the one of fibrine and the other of albumen, do not differ more than two analyses of fibrine or two of albumen respectively do, in the composition of 100 parts.

In these two ingredients of blood the particles are arranged in a different order, as is shown by the difference of their external properties; but in chemical composition, in the ultimate proportion of the organic elements, they are identical.

This conclusion has lately been beautifully confirmed by a distinguished physiologist (Dénis), who has suceeded in converting fibrine into albumen, that is, in giving it the solubility, and coagulability by heat, which characterize the white of egg.

Fibrine and albumen, besides having the same composition, agree also in this, that both dissolve in concentrated muriatic acid, yielding a solution of an intense purple color. This solution, whether made with fibrine or albumen, has the very same reactions with all substances yet tried.

Both albumen and fibrine, in the process of nutrition, are capable of being converted into muscular fibre, and muscular fibre is capable of being reconverted into blood. These facts have long been established by physiologists, and chemistry has merely proved, that these metamorphoses can be accomplished under the influence of a certain force, without the aid of a third substance, or of its elements, and without the addition of any foreign element, or the separation of any element previously present in these substances.

If we now compare the composition of all organized parts with that of fibrine and albumen, the following relations present themselves:—

All parts of the animal body which have a decided shape, which form parts of organs, contain nitrogen. No part of an organ which possesses motion and life is destitute of nitrogen; all of them contain likewise carbon and the elements of water; the latter, however, in no case in the proportion to form water.

The chief ingredients of the blood contain nearly 17 per cent. of nitrogen, and no part of an organ contains less than 17 per cent. of nitrogen.

The most convincing experiments and observations have proved, that the animal body is absolutely incapable of producing an elementary body, such as carbon or nitrogen, out of substances which do not contain it; and it obviously follows, that all kinds of food fit for the production either of blood, or of cellular tissue, membranes, skin, hair, muscular fibre, &c., must contain a certain amount of nitrogen, because that element is essential to the composition of the above-named organs; because the organs cannot create it from the other elements presented to them; and, finally, because no nitrogen is absorbed from the atmosphere in the vital process.

The substance of the brain and nerves contains a large quantity of albumen, and, in addition to this, two peculiar fatty acids, distinguished from other fats by containing phosphorus (phosphoric acid?). One of these contains nitrogen (Frémy).

Finally, water and common fat are those ingredients of the body which are destitute of nitrogen. Both are amorphous or unorganized, and only so far take part in the vital process as that their presence is required for the due performance of the vital functions. The inorganic constituents of the body are, iron, lime, magnesia, common salt, and the alkalies.

The nutritive process in the carnivora is seen in its simplest form. This class of animals lives on the blood and flesh of the graminivora; but this blood and flesh is, in all its properties, identical with their own. Neither chemical nor physiological differences can be discovered.

The nutriment of carnivorous animals is derived originally from blood; in their stomach it becomes dissolved, and capable of reaching all other parts of the body; in its passage it is again converted into blood, and from this blood are reproduced all those parts of their organization which have undergone change or metamorphosis.

With the exception of hoofs, hair, feathers, and the earth of bones, every part of the food of carnivorous animals is capable of assimilation.

In a chemical sense, therefore, it may be said that a carnivorous animal, in supporting the vital process, consumes itself. That which serves for its nutrition is identical with those parts of its organization which are to be renewed.

The process of nutrition in graminivorous animals appears at first sight altogether different. Their digestive organs are less simple, and their food consists of vegetables, the great mass of which contains but little nitrogen.

From what substances, it may be asked, is the blood formed, by means of which their organs are developed? This question may be answered with certainty.

Chemical researches have shown, that all such parts of vegetables as can afford nutriment to animals contain certain constituents which are rich in nitrogen; and the most ordinary experience proves, that animals require for their support and nutrition less of these parts of plants in proportion as they abound in the nitrogenized constituents. Animals cannot be fed on matters destitute of these nitrogenized constituents.

These important products of vegetation are especially abundant in the seeds of the different kinds of grain, and of pease, beans, and lentils; in the roots and the juices of what are commonly called vegetables. They exist, however, in all plants, without exception, and in every part of plants in larger or smaller quantity.

These nitrogenized forms of nutriment in the vegetable kingdom may be reduced to three substances, which are easily distinguished by their external characters. Two of them are soluble in water, the third is insoluble.

When the newly-expressed juices of vegetables are allowed to stand, a separation takes place in a few minutes. A gelatinous precipitate, commonly of a green tinge, is deposited, and this, when acted on by liquids which remove the coloring matter, leaves a grayish white substance, well known to druggists as the deposit from vegetable juices. This is one of the nitrogenized compounds which serves for the nutrition of animals, and has been named *vegetable fibrine*. The juice of grapes is especially rich in this constituent, but it is most abundant in the seeds of wheat, and of the cerealia generally. It may be obtained from wheat flour by a mechanical operation, and in a state of tolerable purity; it is then called *gluten*, but the glutinous property belongs, not to vegetable fibrine, but to a foreign substance, present in small quantity, which is not found in the other cerealia.

The method by which it is obtained sufficiently proves that it is insoluble in water; although we cannot doubt that it was originally dissolved in the vegetable juice, from which it afterwards separated, exactly as fibrine does from blood.

The second nitrogenized compound remains dissolved in the juice after the separation of the fibrine. It does not separate from the juice at the ordinary temperature, but is instantly coagulated when the liquid containing it is heated to the boiling point.

When the clarified juice of nutritious vegetables, such as cauliflower, asparagus, mangel-wurtzel, or turnips, is made to boil, a coagulum is formed, which it is absolutely impossible to distinguish from the substance which separates as a coagulum, when the serum of blood or the white of an egg, diluted with water, are heated to the boiling point. This is *vegetable albumen*. It is found in the greatest abundance in certain seeds, in nuts, almonds, and others, in which the starch of the gramineae is replaced by oil.

The third nitrogenized constituent of the vegetable food of animals is *vegetable caseine*. It is chiefly found in the seeds of pease, beans, lentils, and similar leguminous seeds. Like vegetable albumen, it is soluble in water, but differs from it in this, that its solution is not coagulated by heat. When the solution is heated or evaporated, a skin forms on its surface, and the addition of an acid causes a coagulum, just as in animal milk.

These three nitrogenized compounds, vegetable fibrine, albumen, and caseine, are the true nitrogenized constituents of the food of graminivorous animals; all other nitrogenized compounds, occurring in plants, are either rejected by animals, as in the case of the characteristic principles of poisonous and medicinal plants, or else they occur in the food in such very small proportion, that they cannot possibly contribute to the increase of mass in the animal body.

The chemical analysis of these three substances has led to the very interesting result that they contain the same organic elements, united in the same proportion by weight; and, what is still more remarkable, that they are identical in composition with the chief constituents of blood, animal fibrine, and albumen. They all three dissolve in concentrated muriatic acid with the same deep purple color, and even in their physical characters, animal fibrine and albumen are in no respect different from vegetable fibrine and albumen. It is especially to be noticed, that, by the phrase, identity of composition, we do not here imply mere similarity, but that even in regard to the presence and relative amount of sulphur, phosphorus, and phosphate of lime, no difference can be observed.

How beautifully and admirably simple, with the aid of these discoveries, appears the process of nutrition in animals, the formation of their organs, in which vitality chiefly resides! Those vegetable principles, which in animals are used to form blood, contain the chief constituents of blood, fibrine, and albumen, ready formed, as far as regards their composition. All plants, besides, contain a certain quantity of iron, which reappears in the coloring matter of the blood. Vegetable fibrine and animal fibrine, vegetable albumen and animal albumen, hardly differ, even in form; if these principles be wanting in the food, the nutrition of the animal is arrested; and when they are present, the graminivorous animal obtains in its food the very same principles on the presence of which the nutrition of the carnivora entirely depends.

Vegetables produce in their organism the blood of all animals, for the carnivora, in consuming the blood and flesh of the graminivora, consume, strictly speaking, only the vegetable principles which have served for the nutrition of the latter. Vegetable fibrine and albumen take the same form in the stomach of the graminivorous animal as animal fibrine and albumen do in that of the carnivorous animal.

From what has been said it follows, that the development of the animal organism and its growth are dependent on the reception of certain principles identical with the chief constituents of blood.

In this sense we may say, that the animal organism gives to blood only its form; that it is incapable of creating blood out of other substances which do not already contain the chief constituents of that fluid. We cannot, indeed, maintain, that the animal organism has no power to form other compounds, for we know that it is capable of producing an extensive series of compounds, differing in composition from the chief constituents of blood; but these last, which form the starting point of the series, it cannot produce.

The animal organism is a higher kind of vegetable, the development of which begins with those substances, with the production of which the life of an ordinary vegetable ends. As soon as the latter has borne seed, it dies, or a period of its life comes to a termination.

In that endless series of compounds, which begins with carbonic acid, ammonia, and water, the sources of the nutrition of vegetables, and includes the most complex constituents of the animal brain, there is no blank, no interruption. The first substance capable of affording nutriment to animals is the last product of the creative energy of vegetables.

The substance of cellular tissue and of membranes, of the brain and nerves, these the vegetable cannot produce.

The seemingly miraculous in the productive agency of vegetables disappears in a great degree, when we reflect that the production of the constituents of blood cannot appear more surprising than the occurrence of the fat of beef and mutton in cocoa beans, of human fat in olive oil, of the principal ingredient of butter in palm oil, and of horse fat and train oil in certain oily seeds.

While the preceding considerations leave little or no doubt as to the way in which the increase of mass in an animal, that is, its growth, is carried on, there is yet to be resolved a most important question, namely, that of the function performed in the animal system by substances containing no nitrogen, such as sugar, starch, gum, pectine, &c.

The most extensive class of animals, the graminivora, cannot live without these substances; their food must contain a certain amount of one or more of them, and if these compounds are not supplied, death quickly ensues.

This important inquiry extends also to the constituents of the food of carnivorous animals in the earliest periods of life; for this food also contains substances, which are not necessary for their support in the adult state.

The nutrition of the young of carnivora is obviously accomplished by means similar to those by which the graminivora are nourished; their development is dependent on the supply of a fluid, which the body of the mother secretes in the shape of milk.

Milk contains only one nitrogenized constituent, known under the name of *caseine*; besides this, its chief ingredients are butter (fat), and sugar of milk.

The blood of the young animal, its muscular fibre, cellular tissue, nervous matter, and bones, must have derived their origin from the nitrogenized constituent of milk, the caseine; for butter and sugar of milk contain no nitrogen.

Now, the analysis of caseine has led to the result, which, after the details given in the last section, can hardly excite surprise, that this substance also is identical in composition with the chief constituents of blood, fibrine and albumen. Nay, more, a comparison of its properties with those of vegetable caseine has shown that these two substances are identical in all their properties; insomuch, that certain plants, such as peas, beans, and lentils, are capable of producing the same substance which is formed from the blood of the mother, and employed in yielding the blood of the young animal.

The young animal, therefore, receives, in the form of caseine, which is distinguished from fibrine and albumen by its great solubility, and by not coagulating when heated, the chief constituent of the mother's blood. To convert caseine into blood no foreign substance is required, and in the conversion of the mother's blood into caseine, no elements of the constituents of the blood have been separated. When chemically examined, caseine is found to contain a much larger proportion of the earth of bones than blood does, and that in a very soluble form, capable of reaching every part of the body. Thus, even in the earliest period of its life, the development of the organs, in which vitality resides, is, in the carnivorous animal, dependent on the supply of a substance, identical in organic composition with the chief constituents of its blood.

What, then, is the use of the butter and the sugar of milk? How does it happen that these substances are indispensable to life?

Butter and sugar of milk contain no fixed bases, no soda or potash. Sugar of milk has a composition closely allied to that of the other kinds of sugar, of starch, and of gum; all of them contain carbon and the elements of water, the latter precisely in the proportion to form water.

There is added, therefore, by means of these compounds, to the nitrogenized constituents of food, a certain amount of carbon, or, as in the case of butter, of carbon and hydrogen; that is, an excess of elements, which cannot possibly be employed in the production of blood, because the nitrogenized substances contained in the food already contain exactly the amount of carbon which is required for the production of fibrine and albumen.

The following considerations will show that hardly a doubt can be entertained, that this excess of carbon alone, or of carbon and hydrogen, is expended in the production of animal heat, and serves to protect the organism from the action of the atmospheric oxygen.

The excrements of a buzzard which had been fed with beef, when taken out of the rectum, consisted, according to L. Gmelin and Tiedemann, of urate of ammonia. In like manner, the faeces in lions and tigers are scanty and dry, consisting chiefly of bone earth, with mere traces of compounds containing carbon; but their urine contains, not urate of ammonia, but urea, a compound in which carbon and nitrogen are to each other in the same ratio as in *neutral* carbonate of ammonia.

Assuming that their food (flesh, &c.) contains carbon and nitrogen in the ratio of eight equivalents to one, we find these elements in their urine in the ratio of one equivalent to one; a smaller proportion of carbon, therefore, than in serpents, in which respiration is so much less active.

The whole of the carbon and hydrogen which the food of these animals contained, beyond the amount which we find in their excrements, has disappeared, in the process of respiration, as carbonic acid and water.

In the preceding pages it has been assumed that the elements of the food are converted by the oxygen absorbed in the lungs into oxidized products; the carbon into carbonic acid, the hydrogen into water, and the nitrogen into a compound containing the same elements as carbonate of ammonia.

This is only true in appearance; the body, no doubt, after a certain time, acquires its original weight. The amount of carbon, and of the other elements, is not found to be increased—exactly as much carbon, hydrogen, and nitrogen has been given out as was supplied in the food; but nothing is more certain than that the carbon, hydrogen, and nitrogen given out, although equal in amount to what is supplied in that form, do not directly proceed from the food.

It would be utterly irrationable [!] to suppose that the necessity of taking food, or the satisfying the appetite, had no other object than the production of urea, uric acid, carbonic acid, and other excrementitious matters—of substances which the system expels, and consequently applies to no useful purpose in the economy.

In the adult animal, the food serves to restore the waste of matter; certain parts of its organs have lost the state of vitality, have been expelled from the substance of the organs, and have been metamorphosed into new combinations, which are amorphous and unorganized.

The food of the carnivora is at once converted into blood; out of the newly formed blood those parts of organs which have undergone metamorphoses are reproduced. The carbon and nitrogen of the food thus become constituent parts of organs.

The flesh and blood consumed as food yield their carbon for the support of the respiratory process, while its nitrogen appears as uric acid, ammonia, or urea. But previously to these final changes, the dead flesh and blood become living flesh and blood, and it is, strictly speaking, the carbon of the compounds formed in the metamorphoses of living tissues that serves for the production of animal heat.

According to what has been laid down in the preceding pages, the substances of which the food of man is composed may be divided into two classes; into *nitrogenized* and *non-nitrogenized*. The former are capable of conversion into blood; the latter incapable of this transformation.

Out of those substances which are adapted to the formation of blood are formed all the organized tissues. The other class of substances, in the normal state of health, serve to support the process of respiration. The former may be called the *plastic elements of nutrition*; the latter, *elements of respiration*. Among the former we reckon—

Vegetable fibrine.
Vegetable albumen.
Vegetable caseine.
Animal flesh.
Animal blood.

Among the elements of respiration in our food, are—

Fat.   Pectine.
Starch.   Bassorine.
Gum.   Wine.
Cane sugar.   Beer.
Grape sugar.   Spirits.
Sugar of milk.

FORMATION OF SUGAR IN THE LIVER

BERNARD, Claude (French physiologist, 1813–1878). *De l'origine du sucre dans l'économie animale*, in *Archives Générales de Médecine*, vol. 18, p. 303, 1848; tr. as *The Origin of Sugar in the Animal Body*, in *Medical Classics*, vol. 3, No. 5, p. 567, 1939; by permission of The Williams and Wilkins Company.

This famous paper on liver function, presented here in its entirety, was reported orally to the *Société Biologique* in October of the publication year. Subsequent studies of liver function in carbohydrate metabolism have made sugar formation "independent of a sweet or starchy alimentation" unacceptable, except with important qualifications.

Sugar is spread with profusion throughout the vegetable kingdom, but it also exists in animals. The vegetables are unable to find it all prepared in the earth, and it is evident that they form it in their organs. With animals, is it the same, or is it that the sugar which is encountered in their bodies is furnished exclusively by the sweet and starchy vegetables which serve them for food? Such is an important question, that for a long time has occupied physiologists and chemists, and which we shall attempt to solve experimentally.

As a food, sugar is a neutral substance which is consumed by man and the animals in different states. The sugars which may be introduced habitually into the alimentary tract are: 1st, cane sugar, otherwise called "sugar of the first species", which is found in sugar-cane, the beet, the carrot, etc. 2nd, grape sugar, or "sugar of the second species", which exists in grapes, the sweet fruits, etc. Starch, which constitutes a very abundant alimentary material, may be compared to the sugars, because as the result of digestive phenomena, it is changed in the intestinal canal into sugar of the second species, 3rd, milk sugar, which exists in the milk of animals, etc.

This is not at all the place to trace the distinctive characters of the different sugars, nor to determine what are the changes and transformations which they must undergo to become qualified for the ulterior phenomena of nutrition. I merely state that certain foods being susceptible of providing considerable quantities of sugary material, they have been considered as the sole source whence comes the sugar which is encountered in the blood or fluids of animals. It is, indeed, at this explanation that we have stopped, in the ideas ruling at present on nutrition. It is admitted today that sugar does not exist in the blood of animals, except under the condition that they had previously eaten substances which contained it or were capable of producing it. But, on the one hand, the acquired chemical facts inform us that there is only starch, among the foods, which may transform itself into sugar, and on the other hand, connecting this question with the ingenious idea that animals do not create any immediate principle and do nothing but destroy that which is provided by the vegetable kingdom, one feels sufficiently authorized to refuse in the most explicit manner to the animal organism the faculty of making sugar, and he has recognized in it only the faculty of destroying sugar and of making it disappear. The facts contained in this work, the details of which will follow, show us that physiology is opposed to this point of view.

First series of experiments. It has been observed that during the digestion of a sweet or starchy food, the blood of man and the animals contains sugar, and the fact has been relied on to conclude that the sugar is furnished by the aliments. The experimental result, taken as an isolated fact, is exact, but the experiment is incomplete, and consequently the conclusion is faulty, as will be seen.

First experiment. On a lively rabbit, healthy and of medium size, having eaten some bran and carrots, I also ingested into the stomach, with the aid of a sound, 30 gm. of starch mixed with ¼ liter of boiled water which had been cooled. Five hours later, the rabbit was killed by a blow on the nape of the neck; immediately I opened the breast and collected about 30 gm. of blood which ran out in dividing the cavities of the heart.

After one hour, the blood was well coagulated. I then examined the clear alkaline serum which was separated from the curd, and in it I confirmed the presence of sugar in the most positive manner.[1]

The stomach and the intestine contain sugar deriving from the carrots and

---

[1] The procedure used to determine the sugar will be given in detail in the 3rd series of experiments.

the transformation of the starch. The stomach, with acid reaction, contained unchanged starch. The urine was cloudy, alkaline, and did not contain sugar.

Second experiment. An adult dog, healthy, fasting for 24 hours, ate without difficulty 300 gms. of fresh starch paste, bought at the grocer's. Five hours later, the dog was killed: I immediately opened the chest, and collected the blood in the cavities of the heart. After three quarters of an hour, coagulation having taken place, I confirmed the presence of sugar in the clear alkaline serum, which had separated from the blood clot.

The stomach, with acid reaction, still contained starch which was unchanged and bare traces of sugar. In the intestine, which offered an alkaline reaction, all the starch had been converted, and there sugar was found in great quantity. The urine did not contain sugar.

Third experiment. An adult and healthy female dog made a big meal on cooked head of mutton, bought from the dealer in entrails, and in addition, some bones of fowl. Seven hours later, the animal was killed. The chest having been opened at once, I collected the blood which ran out of the incision of the heart. After one and a half hours, I found the blood coagulated, and an opalescent, milky, alkaline serum had separated. I examined it, and I established in an unequivocal manner the presence of sugar. The animal was in full intestinal digestion. The material found in the stomach and small intestine had an acid reaction and did not contain the least trace of sugar. The urine, with acid reaction, did not contain any sugar either.

Fourth experiment. An adult and healthy dog was left without food. After two days of complete abstinence from solid and liquid foods, the animal was killed. The chest opened immediately, I collected the blood from the cavities of the heart, and one hour later it had separated into clot and a clear, non-milky, alkaline serum; I examined it with the reagent, and established the presence of sugar with the greatest evidence. The stomach and small intestine, absolutely empty and retracted on themselves, consequently did not contain sugar. In the large intestine there was a little fecal matter, hard and black. The acid urine did not contain sugar.

The experiments reported above have been reproduced a large number of times with similar results. The general fact which derives from them is readily grasped: it is, that sugar constantly exists in the blood of animals with all alimentary regimens and also with abstinence. It was a mistake, therefore, to rely on the presence of sugar in the blood during the digestion of starchy materials to conclude that it came from the food; because if, for the animals which were the subjects of the 1st and 2nd experiments, the sugar found in the alimentary canal could account for that which was in the blood, it is evident that this reason is no longer valid for the animal of the 3rd experiment, which had eaten only meat, and in which the absence of sugary material in the digestive tract had been established. For the animal in the 4th experiment, fasting for two days and having the alimentary tract empty, the matter becomes still more difficult to explain.

This example could be chosen from among many others to show how very easy error becomes in physiology, when one does not free himself of preconceived ideas and when one does not make comparative experiments. Indeed, if one had had less confidence in the theory which he wished to support, he would not have contented himself with examining the blood of animals digesting starchy or sweet foodstuffs; he would have thought of examining the blood comparatively with other alimentations, and without doubt he would have conducted, as I have, a research to find the source of sugar in animals which eat neither sweetened material nor starch. The account of this research will be the subject of the second series of experiments.

Second series of experiments. Whence comes the sugar which exists in the blood of animals which are fed on meat or even are subjected to abstinence? That is the interesting question for the solution of which we must now institute new experiments. It was quite presumable that the sugar was not manufactured in the heart, where we found it, but that it had been quite simply transported from some point in the organism. After a few gropings which I think it unnecessary to report here, I was led to seek the source of the sugar among the glandular organs of the abdomen, and here is how I experimented.

First experiment. An adult and healthy dog, having made a big meal on bones and bits of cooked meat, was killed seven hours later. I opened the abdomen at once, and I found the phenomena which accompany digestion when it is in full activity; that is, a turgescent state of all the organs of the lower abdomen, among which the circulation was going forward very actively, and in addition, the repletion of the chyliferous vessels and of the thoracic duct by a white, milky chyle, quite homogeneous.

I collected: 1) the blood which flowed from the incision made in the trunk of the portal vein near the point where the splenic vein enters; 2) I obtained some chyle by opening the thoracic duct; 3) I took some blood from the cavities of the heart. Next, I separated with care the contents of the stomach and the small intestine, and I looked for the presence of sugar in all the products.

1. The alimentary material contained in the stomach and in the small intestine showed an acid reaction, and with the reagents did not give the slightest trace of sugar.

2. The pinky-white chyle extracted from the thoracic duct permitted the separation of a milky, alkaline serum, in which I established the absence of sugar.

3. The blood of the portal vein having coagulated, there separated off an opalescent, slightly milky, alkaline serum, in which I verified the presence of a very large quantity of sugar.

4. The blood taken from the right ventricle of the heart soon coagulated and gave an alkaline and milky serum, in which the reagents demonstrated much sugar, but in lesser abundance than in the blood of the portal vein.

Second experiment. An adult and healthy dog was killed on the third day of an absolute abstinence. I opened the abdomen at once, and I found the phenomena which accompany the inactivity of the digestive organs, namely, a state of pallor and of anemia of the organs of the lower abdomen, and in addition, the vacuity and retraction of the stomach and intestines. The chyliferous vessels and the thoracic duct contained transparent lymph with a very slight opaline reflection. I collected separately:

1) Blood from the trunk of the portal vein. I confirmed very clearly in the limpid alkaline serum which separated off, the presence of sugar, although it was in lesser abundance than in the preceding experiment.

2) Blood from the right ventricle. In its clear, alkaline serum the presence of sugar was indubitable.

3) Lymph from the thoracic duct, in which I found not the least trace of sugar.

I repeated these experiments several times, under similar circumstances, with identical results and without succeeding in understanding how the blood of the portal vein could contain so much sugar when the intestines did not contain any. Reflecting that it must be that the sugar came from some nearby organ, considering that the walls of the portal vein probably did not have the property of secreting it, I made the following experiment:

Third experiment. Having killed as rapidly as possible, that is, in a few seconds, by section of the medulla oblongata, a dog in the process of digestion of food free from sugar and starch, I opened the abdominal cavity immediately, then with the greatest celerity possible I placed ligatures: 1) on the venous branches which emanated from the small intestine, not far from the intestine; 2) on the splenic vein, a few centimeters from the spleen; 3) on the venous branches emanating from the pancreas; 4) on the trunk of the portal vein, before its entry into the liver. Then incising the different veins behind the ligatures which I had placed, or said in another way, between the ligature and the organ, I was able to collect separately the blood coming from the small intestine, from the spleen, from the pancreas, and that from the liver. 1) In the blood of the intestinal veins, just as in the material contained in the intestine, I found sugar. 2) The blood coming from the spleen no longer contained any trace of sugar; 3) in the blood of the pancreatic veins, I did not find any either; 4) finally, in the blood which flowed back in abundance from the hepatic veins, after the opening of the trunk of the portal vein above the ligature, it was not without astonishment that I found great quantities of sugar, although the tissue of the spleen, the pancreas, the mesenteric ganglia of the same animal, equally washed and examined with care, did not indicate any trace of sugar with the reagents.

Accordingly, it became evident that it was the liver from which the sugar proceeded.

But how, one will say, is the sugar encountered in the blood of the portal vein and in the hepatic veins, for in supposing that it was formed in the tissue of the liver, the current of the blood should carry in the direction of the suprahepatic veins, towards the heart, and prevent it from flowing back by the hepatic veins into the portal vein. That remark would be correct if it concerned the general circulation, where one does not see the blood traverse, by a retrograde movement, a capillary tissue which it has already passed over in a progressive movement. But for the liver, this is not so, and the reflux of the blood of the suprahepatic veins into the portal vein is a very easy matter. As for the cause which, in my experiments, has determined the reflux of sugary blood from the liver into the portal vein, it is very easy to understand. Indeed, the circulation of the blood in the portal vein, in the physiological state, is produced especially by the pressure exerted on the viscera by the abdominal walls. It therefore results that, the trunk and the branches of the portal vein being naturally compressed, when one goes to open the abdomen, this compression ceases due to the issue of the abdominal viscera. If one adds that, by this hernia of the organs, the vascular branches find themselves pulled and lengthened, one will see that the opening of the abdomen must make a sort of depletion throughout the entire extent of the portal vein, and particularly in the large trunks: this sort of vacuum aspirates the blood of the liver and of the other organs with even greater facility, since there are no valves to hinder the retrograde flow of the blood. I shall not insist any longer on this point, to which I shall have occasion to return in other circumstances. I only indicate that in our experiments, the presence of sugar in the portal vein should be regarded as accidental; because I have been able to avoid it after knowing the cause, by placing a ligature on the portal vein at its entry into the liver, before performing the debridement and eventration of the animal. From whence it follows that in the physiologic state, there is no sugar in the blood which enters the liver.

As a résumé for this second series of experiments, we have learned that sugar exists in large quantity in the liver; that this sugar dissolves or pours itself out into the blood which passes through the liver and thus is carried by the suprahepatic veins and the inferior vena cava into the right heart, where it is constantly found.

Third series of experiments. The facts reported above have led us to find a source of sugar in animals. This discovery seems too important for us not to encircle it with all the guarantees possible. We shall indicate, therefore, the procedures which we have used for the investigation of sugar, so that everyone may be in a position to repeat the experiments, if he wishes to do so.

1. Investigation of sugar in the liver. It is sufficient to take a certain quantity of liver-tissue, to pulverize it in a mortar or otherwise, after which one boils it for a few moments in a small quantity of water, then filters it to obtain the liquid of the decoction. This decoctum, which ordinarily presents an opaline appearance, possessed all the characters of a sweetened liquid. 1st, it turns brown when one boils it with potash, and it reduces, in similar circumstances, the double tartrate of potash and of copper. If one adds brewer's yeast with a suitable temperature, at the end of a very short time the fermentation is established and goes forward very actively. It can be confirmed that carbonic acid is given off, and when the fermentation is complete, if the liquid is distilled, one obtains alcohol, which if sufficiently concentrated by several distillations, will burn and is recognized by all its characters.

The considerable proportions of sugar which are shown in the liver by the above-indicated reactions, lead one to believe that he might succeed, by taking a sufficient quantity of the organ, in extracting from it the sugar in its natural form. The most simple procedure would consist of taking decoctions or macerations of liver which are sufficiently concentrated, treating them with alcohol to separate the albuminoid materials, then rapidly evaporating over a gentle heat to the syrupy consistency suitable to obtain crystallization. By working in this manner or by other analogous means, it has been possible to obtain the concentration of the sweetened liquor, but the crystallization has never been effected. That is due to the fact that the tissue of the liver, besides great quantities of fatty and albuminous materials, contains enormous quantities of salts, particularly sodium chloride. If, by water at first and then by sufficiently rectified alcohol, one frees himself of the first substances, it becomes extremely difficult to obtain the separation of the salts, which by remaining in the sugar solution, prevent the crystallization of the sugar and form a veritable molasses. It would perhaps be of great interest for chemists to be able to separate and analyze the sugar in the liver; but, from my point of view, the matter is not indispensable, because the aggregate of characteristics which we have given, especially the fermentation and the formation of carbonic acid and of alcohol, seems to me more than enough to establish the existence of sugar in the liver.

When one has devoted himself to the investigation of sugar in the liver, and when he sees this substance exists there in such abundance, that nothing is more simple and easy than to confirm its presence by fermentation, he is surprised that this fact has remained so long unknown. The liver is indeed a very common affair, and it suffices to take a piece of this organ from the butcher's shop to see all that we have said above. However, there is one thing which should strike one: that is the extreme bitterness of the bile and the particular sweet flavor of the tissue of the liver. It is evident that the bitterness of the bile contained in the hepatic ducts is tempered or masked by the sugar of the liver, and with accuracy one may say that in this organ honey finds itself beside gall.

The test for sugar in the blood is made very simply. After the blood has been extracted from the heart of the vessels, I let it coagulate; taking, in a tube which is closed at one end, a part of the serum which has separated, I add to it about one sixth in volume of double tartrate of copper and of potash; then, by boiling the mixture, there is a reduction of the copper salt proportional to the quantity of sugar contained in the serum. This procedure, very simple and very rapid, shows the least trace of sugar. When one is working with comparative experiments, he may, if necessary, rely on this characteristic. Meanwhile, if he wished to have more security, he would add brewer's yeast to the serum, and would collect the gas in an appropriate apparatus. If the quantity of sugar in the serum was not sufficiently great to give fairly clear fermentation products, he would coagulate a sufficient quantity of the serum with alcohol, and then treat the alcoholic solution, which had been filtered and properly concentrated. There is a point which should never be lost to view when one tests for sugar in the blood, and that is that it is quickly and spontaneously destroyed, so that it is necessary to act on the serum as quickly as possible, and immediately after its separation. If one wishes to prevent the destruction of the sugar, he needs only to coagulate the blood on its leaving the vessels, by alcohol or by lead acetate; then the sugar material would be conserved perfectly intact in the solution in alcohol or lead acetate.

We must now express ourselves on the kind of sugar which is found in the liver and blood. In recalling the reactions which it gave, it could be concluded that it was neither milk sugar nor cane sugar. It is not cane sugar because it turns brown with potash and reduces the salts of copper; it is not milk sugar, because it ferments with great rapidity. There remains grape sugar or glucose, of which liver sugar has shown·the chemical characters, although at the same time it differs from the physiological point of view. Later, in a work which will follow this, in applying myself to the mechanism by which sugar is destroyed in the blood, I shall show that the sugar of diabetes, which has been considered as chemically identical with grape sugar (glucose), differs notably from it by certain physiological characters. Thus, I may say by anticipation, the sugar which is found in the liver is the sugar of diabetes.

Fourth series of experiments. We know now that the sugar which is concentrated in the bodies of animals is found especially concentrated in their liver. But whence does it come definitely? In this regard, two suppositions may be made: either it results directly from a particular transformation of certain elements of the liver, or one may admit that the sugar is only deposited or accumulated in the organ as the result of exterior alimentations. Indeed, the animals fed on meat or put on abstinence perhaps had eaten, one might say, on the preceding days of bread or sugar, and as these substances, absorbed especially by the portal vein, must of necessity traverse the tissue of the liver, one could admit, I say, that the liver retains a part of this sugar. One could even add, to corroborate this point of view, that it is already known that the liver has the property of retaining in this manner arsenic and certain other metallic poisons, etc. Without denying that in certain cases the liver may play the rôle of condensing organ, I may say that the following experiments are not favorable to this manner of thinking.

First experiment. An adult and healthy dog was put on abstinence of solid and liquid foods for 8 days; after this time, the animal was fed for 11 days, abundantly and exclusively with cooked meat (mutton head). The 19th day of its sequestration, the animal was killed during full digestion. The blood contained much sugar, and the tissue of the liver furnished quantities equally as great as in our first experiments.

I repeated this experiment three times in the same manner, with similar results.

These experiments no longer permit, it seems to me, of thinking that the liver only retains the sugar of foods, for after 19 days its elimination certainly would have been effected, as one may convince himself by the facts which follow.

Second experiment. On an adult and vigorous rabbit, in the middle of digestion of grass and carrots, I cut the two pneumogastrics in the middle region of the neck. Seventeen hours later the animal was found dead and still warm. I performed its autopsy with care, and I did not find a trace of sugar either in the blood or in the liver. The bile, which is habitually alkaline in these animals, was very clearly acid and greenish.

Third experiment. On an adult and vigorous dog, during digestion, I cut the two pneumogastric nerves in the middle region of the neck. The dog died the second day, and its liver nor blood, examined immediately after the death, showed no presence of sugar with the reagents. The bile contained in the gallbladder was equally acid.

The effect of this section of the pneumogastric nerves on the function of the liver, if the result is maintained in repeating the experiment, seems to me to be excessively singular. The result would be indeed, that this formation of sugar in the liver, which is evidently a chemical fact, is found to be directly connected to the influence of the nervous system. On the other hand, these experiments prove that the elimination of the sugar previously contained in the liver must have occurred very rapidly, because one no longer finds it with a sensitive method, even when the animals have it in the stomach (experiment 2).

If it were necessary to demonstrate by new arguments that the formation of sugar in the liver is independent of foodstuffs, I would say that I had established, with young calves taken at the slaughtering-houses, that sugar exists in the liver in very large proportion during the intra-uterine life. However, it is not until the fourth or fifth month of the intra-uterine life that this presence of sugar begins to manifest itself in the liver, and the proportion of the principle increases as one approaches the birth.

From all this, I believe it is possible to conclude that sugar forms in the liver, and that this organ is at the same time the *seat* and the *source* of the sugar material in animals.

CONCLUSIONS AND REFLECTIONS

The conclusions which seem to me to derive from the facts contained in this memorandum are:

1. That in the physiologic state, there exists constantly and normally the sugar of diabetes in the blood of the heart[2] and in the liver of man and animals.

2. That the formation of this sugar takes place in the liver, and that it is independent of a sweet or starchy alimentation.

3. That this formation of sugar in the liver begins to operate in the animal before birth, and consequently before the direct ingestion of foodstuffs.

4. That this production of sugary material, which would be one of the functions of the liver, appears to be bound to the integrity of the pneumogastric nerves.

It is evident that before these facts, that law, that animals do not create any immediate principle, but only destroy those furnished them by the vegetables, must cease to be true, because animals in the physiologic state are able, like the vegetables, to create and destroy sugar.

Although the animal organism produces sugar without starch, something which the known chemical means do not permit us to do, I would not conclude therefore that the importance of chemical knowledge should be diminished in the study of the phenomena of life. I am, on the contrary, one of those who appreciate the more all the progress that organic chemistry has caused physiology to make. Only, I believe, as I have already had occasion to state,[3] that in order to avoid error and to render all the service of which it is capable, chemistry should never go adventuring alone in the examination of animal functions; I believe that it alone can solve, in many cases, the difficul-

---

[2] We shall see later, in speaking of the destruction of sugar, that it is able to disappear before arriving in the superficial veins of the body whence blood is usually drawn.

[3] *Experiments with the different chemical manifestations of substances introduced into the body.* Archives génér. de méd., 4 s., Vol. 16, 1848.

ties which deter physiology, but it cannot precede the latter, and I think, finally, that in no case may chemistry believe itself authorized to restrain the resources of nature, which we know not, to the limits of the facts or procedures which constitute our knowledge of the laboratory.

The question of the origin of sugar in the animals, which we have just examined in this work, is still far from being known to us in all its elements. Indeed, if we already possess very positive results, there are, in another direction, facts to elucidate. We should indicate these facts, in order to point them out for study and to show the extent of our subject, which we have only skirted in this first work.

According to what we have said about the existence of sugar in the liver, it should not be believed that in going into an amphitheater and in taking the liver of a cadaver, one would surely find sugar in it. There are, indeed, a large number of diseases in which the sugar disappears and is no longer found in the liver after death. In diabetics, it is known that the sugar disappears from the urine in the last stage of life; it disappears equally from the liver, because the liver of a diabetic which I had occasion to examine in this connection contained no sugar. I have investigated the sugar in the cadavers of 18 subjects who died of different diseases: there were some who showed different proportions of sugar, and there were others who did not contain a trace. My observations on this point are not numerous enough so that I may decide if there are diseases in which the sugar constantly disappears, while it persists in others. In animals weakened by a very long abstinence, which have become sick or have died of disease the sugar often diminishes considerably, and even disappears completely. All the livers of butchered animals, however, should contain much sugar, if they were killed under proper circumstances. The livers obtained at the butcher's have always given me large quantities of sugar. Finally, there is a question which we should examine with care: it is to know whether sugar exists in the same proportion in all classes of animals, taken in conditions as similar as possible. I may already affirm that there seem to be differences in this regard: 1) in birds (chicken, pigeon) the proportion of sugar is quite considerable; 2) in mammals (dog, rabbit, pig, beef, calf, horse) the proportion of sugar is also quite considerable; 3) in reptiles (frog, lizard) the quantity of sugar found in the liver is very small; 4) in fish, in the ray and the eel, the livers of which I examined in as fresh a state as possible, I did not find the least trace of sugar. Whence comes this disappearance of sugar in certain cold-blooded animals? Is that related to the lesser energy of the respiratory phenomena, which, as we shall see later on, are in very intimate relation with the formation of sugar in the liver?

ANALYSIS OF ADRENAL FUNCTION

ADDISON, Thomas (English physician, 1793–1860). *On the constitution and local effects of disease of the supra-renal capsules*, London, 1855.*

In a note published in 1849, Addison tells us that he "could not help entertaining a very strong impression that these hitherto mysterious bodies—the supra-renal capsules—may be either directly or indirectly concerned in sanguification; and that a diseased condition of them, functional or structural, may interfere with the proper elaboration of the body generally, or of the red particles more especially."† It was only after accumulation of further cases that he published the following classical description, which uses a biological as much as a medical approach. Through its influence on Brown-Séquard and others, this paper stimulated investigation of glandular action and hence helped to establish the modern physiology of secretion. Addison was Senior Physician to Guy's Hospital, London.

To The Right Honourable Lord Hawke, as a tribute of respect, and in grateful acknowledgment of a long, cordial, and most disinterested friendship, this little work is dedicated by His Lordship's obliged friend and humble servant,

THOMAS ADDISON.

PREFACE

If Pathology be to disease what Physiology is to health, it appears reasonable to conclude, that in any given structure or organ, the laws of the former will be as fixed and significant as those of the latter; and that the peculiar characters of any structure or organ may be as certainly recognized in the phenomena of disease as in the phenomena of health. When investigating the pathology of the lungs, I was led, by the results of inflammation affecting the lung-tissue, to infer, contrary to general belief, that the lining of the air-cells was not identical and continuous with that of the bronchi; and microscopic investigation has since demonstrated in a very striking manner the correctness of that inference,—an inference, be it observed, drawn entirely from the indications furnished by pathology. Although Pathology therefore, as a branch of medical science, is necessarily founded on Physiology, questions may nevertheless arise regarding the true character of a structure or organ, to which occasionally the pathologist may be able to return a more satisfactory and decisive reply than the physiologist,—these two branches of medical knowledge being thus found mutually to advance and illustrate each other. Indeed, as re-

---

* Unabbreviated except for the case histories which follow the passage reprinted and which are in several instances reports by other workers.

† *Anemia: disease of the supra-renal capsules*, in *The London Medical Gazette*, n.s. **vol. 8, p. 517.**

gards the functions of individual organs, the mutual aids of these two branches of knowledge are probably much more nearly balanced than many may be disposed to admit; for in estimating them, we are very apt to forget how large an amount of our present physiological knowledge, respecting the functions of these organs, has been the immediate result of casual observations made on the effects of disease. Most of the important organs of the body, however, are so amenable to direct observation and experiment, that in respect to them the modern physiologist may fairly lay claim to a large preponderance of importance, not only in establishing the solid foundation, but in raising and greatly strengthening the superstructure of a rational pathology. There are still, however, certain organs of the body, the actual functions and influence of which have hitherto entirely eluded the researches and bid defiance to the united efforts of both physiologist and pathologist. Of these not the least remarkable are the "Supra-Renal Capsules,"—the *Atrabiliary* Capsules of Caspar Batholinus; and it is as a first and feeble step towards an inquiry into the functions and influence of these organs, suggested by Pathology, that I now put forth the following pages.

T. A.

24 New Street, Spring Gardens,
    May 21, 1855.

It will hardly be disputed that at the present moment, the functions of the supra-renal capsules, and the influence they exercise in the general economy, are almost or altogether unknown. The large supply of blood which they receive from three separate sources; their numerous nerves, derived immediately from the semilunar ganglia and solar plexus; their early development in the foetus; their unimpaired integrity to the latest period of life; and their peculiar gland-like structure; all point to the performance of some important office: nevertheless, beyond an ill-defined impression, founded on a consideration of their ultimate organization, that, in common with the spleen, thymus and thyroid body, they in some way or other minister to the elaboration of the blood, I am not aware that any modern authority has ventured to assign to them any special function or influence whatever.

To the physiologist and to the scientific anatomist, therefore, they continue to be objects of deep interest, and doubtless both the physiologist and anatomist will be inclined to welcome, and regard with indulgence, the smallest contribution calculated to open out any new source of inquiry respecting them. But if the obscurity, which at present so entirely conceals from us the uses of these organs, justify the feeblest attempt to add to our scanty stock of knowledge, it is not less true, on the other hand, that any one presuming to make such an attempt, ought to take care that he do not, by hasty pretensions, or by partial and prejudiced observation, or by an over-statement of facts, incur the just rebuke of those possessing a sounder and more dispassionate judgement than himself. Under the influence of these considerations I have for a considerable period withheld, and now venture to publish, the few facts bearing upon the subject that have fallen within my own knowledge; believing as I now do, that these concurring facts, in relation to each other, are not merely casual co-incidences, but are such as admit of a fair and logical inference—an inference, that where these concurrent facts are observed, we may pronounce with con-siderable confidence, the existence of diseased supra-renal capsules.

As a preface to my subject, it may not be altogether without interest or un-profitable, to give a brief narrative of the circumstances and observations by which I have been led to my present convictions.

For a long period I had from time to time met with a very remarkable form of general anaemia, occurring without any discoverable cause whatever; cases in which there had been no previous loss of blood, no exhausting diar-rhoea, no chlorosis, no purpura, no renal, splenic, miasmatic, glandular, stru-mous, or malignant disease. Accordingly, in speaking of this form of anaemia in clinical lecture, I, perhaps with little propriety, applied to it the term "idio-pathic," to distinguish it from cases in which there existed more or less evi-dence of some of the usual causes or concomitants of the anaemic state.

The disease presented in every instance the same general character, pursued a similar course, and, with scarcely a single exception, was followed, after a variable period, by the same fatal result. It occurs in both sexes, generally, but not exclusively, beyond the middle period of life, and so far as I at present know, chiefly in persons of a somewhat large and bulky frame, and with a strongly-marked tendency to the formation of fat. It makes its approach in so slow and insidious a manner, that the patient can hardly fix a date to his earliest feeling of that languor, which is shortly to become so extreme. The countenance gets pale, the whites of the eyes become pearly, the general frame flabby rather than wasted; the pulse perhaps large, but remarkably soft and compressible, and occasionally with a slight jerk, especially under the slightest excitement; there is an increasing indisposition to exertion, with an uncom-fortable feeling of faintness or breathlessness on attempting it; the heart is readily made to palpitate; the whole surface of the body presents a blanched, smooth and waxy appearance; the lips, gums and tongue seem bloodless; the flabbiness of the solids increases; the appetite fails; extreme languor and faint-ness supervene, breathlessness and palpitations being produced by the most trifling exertion or emotion; some slight oedema is probably perceived about the ankles; the debility becomes extreme, the patient can no longer rise from his bed, the mind occasionally wanders, he falls into a prostrate and half-torpid state, and at length expires: nevertheless to the very last, and after a sickness of perhaps several months' duration, the bulkiness of the general frame and the amount of obesity often present a most striking contrast to the failure and exhaustion observable in every other respect.

With, perhaps, a single exception, the disease, in my own experience, resisted all remedial efforts, and sooner or later terminated fatally. On examining the bodies of such patients after death, I have failed to discover any organic lesion that could properly or reasonably be assigned as an adequate cause of such serious consequences; nevertheless, from the disease having uniformly occurred in fat people, I was naturally led to entertain a suspicion that some form of fatty degeneration might have a share at least in its production; and I may observe, that in the case last examined, the heart had undergone such a change, and that a portion of the semilunar ganglion and solar plexus, on being subjected to microscopic examination, was pronounced by Mr. Quekett to have passed into a corresponding condition. Whether any, or all, of these morbid changes are essentially concerned, as I believe they are, in giving rise to this very remarkable disease, future observation will probably decide.

The cases having occurred prior to the publication of Dr. Bennett's interesting essay on "Leucocythaemia," it was not determined by microscopic examination whether there did, or did not, exist an excess of white corpuscles in the blood of such patients.

It was whilst seeking in vain to throw some additional light upon this form of anaemia, that I stumbled upon the curious facts, which it is my more immediate object now to make known to the Profession; and however unimportant or unsatisfactory they may at first sight appear, I cannot but indulge the hope, that by attracting the attention and enlisting the coöperation of the Profession at large, they may lead to the subject being properly examined and sifted, and the inquiry so extended, as to suggest, at least, some interesting physiological speculations, if not still more important practical indications.

The leading and characteristic features of the morbid state to which I would direct attention, are, anaemia, general languor and debility, remarkable feebleness of the heart's action, irritability of the stomach, and a peculiar change of colour in the skin, occurring in connexion with a diseased condition of the "supra-renal capsules."

As has been observed in other forms of anaemic disease, this singular disorder usually commences in such a manner, that the individual has considerable difficulty in assigning the number of weeks or even months that have elapsed since he first experienced indications of failing health and strength; the rapidity, however, with which the morbid change takes place, varies in different instances. In some cases that rapidity is very great, a few weeks proving sufficient to break up the powers of the constitution, or even to destroy life; the result, I believe, being determined by the extent, and by the more or less speedy development, of the organic lesion. The patient, in most of the cases I have seen, has been observed gradually to fall off in general health; he becomes languid and weak, indisposed to either bodily or mental exertion; the appetite is impaired or entirely lost; the whites of the eyes become pearly; the pulse small and feeble, or perhaps somewhat large, but excessively soft and compressible; the body wastes, without, however, presenting the dry and shrivelled skin, and extreme emaciation, usually attendant on protracted malignant disease; slight pain or uneasiness is from time to time referred to the region of the stomach, and there is occasionally actual vomiting, which in one instance was both urgent and distressing; and it is by no means uncommon for the patient to manifest indications of disturbed cerebral circulation. Notwithstanding these unequivocal signs of feeble circulation, anaemia, and general prostration, neither the most diligent inquiry, nor the most careful physical examination, tends to throw the slightest gleam of light upon the precise nature of the patient's malady: nor do we succeed in fixing upon any special lesion as the cause of this gradual and extraordinary constitutional change. We may indeed suspect some malignant or strumous disease; we may be led to inquire into the condition of the so-called blood-making organs; but we discover no proof of organic change anywhere,—no enlargement of spleen, thyroid, thymus or lymphatic glands,—no evidence of renal disease, of purpura, of previous exhausting diarrhoea, or ague, or any long-continued exposure to miasmatic influences: but with a more or less manifestation of the symptoms already enumerated, we discover a most remarkable, and, so far as I know, characteristic discoloration taking place in the skin,—sufficiently marked indeed as generally to have attracted the attention of the patient himself, or of the patient's friends. This discoloration pervades the whole surface of the body, but is commonly most strongly manifested on the face, neck, superior extremities, penis and scrotum, and in the flexures of the axillae and around the navel. It may be said to present a dingy or smoky appearance, or various tints or shades of deep amber or chestnut-brown; and in one instance the skin was so universally and so deeply darkened, that, but for the features, the patient might have been mistaken for a mulatto.

In some cases this discoloration occurs in patches, or perhaps rather certain parts are so much darker than others, as to impart to the surface a mottled or somewhat checkered appearance; and in one instance there were, in the midst of this dark mottling, certain insular portions of the integument presenting a blanched or morbidly white appearance, either in consequence of these portions having remained altogether unaffected by the disease, and thereby contrasting strongly with the surrounding skin, or, as I believe, from an actual defect of colouring matter in these parts. Indeed, as will appear in the subsequent cases, this irregular distribution of pigment-cells is by no means limited to the integument, but is occasionally also made manifest on some of the internal structures. We have seen it in the form of small black spots, beneath the peritoneum of the mesentery and omentum—a form which in one instance presented itself on the skin of the abdomen.

This singular discoloration usually increases with the advance of the disease; the anaemia, languor, failure of appetite, and feebleness of the heart, become aggravated; a darkish streak usually appears upon the commissure of the lips; the body wastes, but without the extreme emaciation and dry harsh condition of the surface so commonly observed in ordinary malignant diseases; the pulse becomes smaller and weaker, and without any special complaint of pain or uneasiness, the patient at length gradually sinks and expires. In one case, which may be said to have been acute in its development as well as rapid in its course, and in which both capsules were found universally diseased after death, the mottled or checkered discoloration was very manifest, the anaemic condition strongly marked, and the sickness and vomiting urgent; but the pulse, instead of being small and feeble as usual, was large, soft, extremely compressible, and jerking on the slightest exertion or emotion, and the patient speedily died.

My experience, though necessarily limited, leads to a belief that the disease is by no means of very rare occurrence, and that were we better acquainted with its symptoms and progress, we should probably succeed in detecting many cases, which, in the present state of our knowledge, may be entirely overlooked or misunderstood; and, I think, I may with some confidence affirm, that although partial disease of the capsules may give rise to symptoms, and to a condition of the general system, extremely equivocal and inconclusive, yet that a more extensive lesion will be found to produce a state, which may not only create a suspicion, but be pronounced with some confidence to arise from the lesion in question. When the lesion is acute and rapid, I believe the anaemia, prostration, and peculiar condition of the skin will present a corresponding character, and that whether acute or chronic, provided the lesion involve the entire structure of both organs, death will inevitably be the consequence.

If this statement be correct, and I quite believe it to be so, the chief difficulty that remains to be surmounted by further experience in this, I fear, irremediable disease, is a correct and certain diagnosis;—how we may at the earliest possible period detect the existence of this form of anaemia, and how it is to be distinguished from other forms of anaemic disorder. As I have already observed, the great distinctive mark of this form of anaemia is the singular dingy or dark discoloration of the skin; nevertheless at a very early period of the disorder, and when the capsules are less extensively diseased, the discoloration may, doubtless, be so slight and equivocal as to render the source of the anaemic condition uncertain. Our doubts, in such cases, will have reference chiefly to the sallow anaemic conditions resulting from miasmatic poisoning or malignant visceral disease; but a searching inquiry into the history of the case, and a careful examination of the several parts or organs usually involved in anaemic disease, will furnish a considerable amount of at least negative evidence; and when we fail to discover any of the other well-known sources of that condition, when the attendant symptoms resemble those numerated as accompanying disease of the capsules, and when to all this is superadded a dark, dingy or smoky-looking discoloration of the integument, we shall be justified at least in entertaining a strong suspicion in some instances,—a suspicion almost amounting to certainty in others. It must, however, be observed, that every tinge of yellow, or mere sallowness, throws a still greater doubt over the true nature of the case, and that the more decidedly the discoloration partakes of the character described, the stronger ought to be our impression as to the capsular origin of the disorder.

The morbid appearances discovered after death will be described with the cases in which they occurred; but I may remark that a recent dissection (March 1855) has shown that even malignant disease may exist in both capsules, without giving rise to any marked discoloration of the skin; but, in the case alluded to, the deposit in each capsule was exceedingly minute, and could not have seriously interfered with the functions of the organs: extensive and fatal malignant disease had, however, affected other parts. It may be observed in conclusion, that on subjecting the blood of a patient, who recently died from a well-marked attack of this singular disease, to microscopic examination, a considerable excess of white corpuscles was found to be present.

RECOGNITION OF THE COLLOID CONDITION

GRAHAM, Thomas (British chemist and physicist, 1805–1869). From *Liquid diffusion applied to analysis*, in *Philosophical Transactions of the Royal Society*, vol. 151, p. 183, London, 1861.

During the nineteenth century, the interest of physiologists turned successively from organs to tissues (see Bichat), from tissues to cells (see Schwann), and from cells to protoplasm (see Schultze). The study of protoplasm then resolved itself largely into an effort to account for the behavior of this substance (temporarily believed to be uniform for the cells of all species) in terms of its physical and chemical properties on two levels: the level of "true solutes" or "crystalloids," and the level of colloids. This was a distinction whose significance, and, importantly enough, whose limitations, were pointed out in the following classical paper. It was not unnatural that Graham should be found measuring the diffusion of substances against obstacles, since it was this method which he had already used (namely, the opposition of plaster of paris to gaseous diffusion) to develop his famous law.*

The property of volatility, possessed in various degrees by so many substances, affords invaluable means of separation, as is seen in the ever-recurring processes of evaporation and distillation. So similar in character to volatility

---

* The rate of diffusion is inversely proportional to the square root of the density.

is the Diffusive power possessed by all liquid substances, that we may fairly reckon upon a class of analogous analytical resources to arise from it. The range also in the degree of diffusive mobility exhibited by different substances appears to be as wide as the scale of vapour tensions. Thus hydrate of potash may be said to possess double the velocity of diffusion of sulphate of potash, and sulphate of potash again double the velocity of sugar, alcohol, and sulphate of magnesia. But the substances named belong all, as regards diffusion, to the more "volatile" class. The comparatively "fixed" class, as regards diffusion, is represented by a different order of chemical substances, marked out by the absence of the power to crystallize, which are slow in the extreme. Among the latter are hydrated silicic acid, hydrated alumina, and other metallic peroxides of the aluminous class, when they exist in the soluble form; with starch, dextrin and the gums, caramel, tannin, albumen, gelatine, vegetable and animal extractive matters. Low diffusibility is not the only property which the bodies last enumerated possess in common. They are distinguished by the gelatinous character of their hydrates. Although often largely soluble in water, they are held in solution by a most feeble force. They appear singularly inert in the capacity of acids and bases, and in all the ordinary chemical relations. But, on the other hand, their peculiar physical aggregation with the chemical indifference referred to, appears to be required in substances that can intervene in the organic processes of life. The plastic elements of the animal body are found in this class. As gelatine appears to be its type, it is proposed to designate substances of the class as *colloids,* and to speak of their peculiar form of aggregation as the *colloidal condition of matter.* Opposed to the colloidal is the crystalline condition. Substances affecting the latter form will be classed as *crystalloids.* The distinction is no doubt one of intimate molecular constitution.

Although chemically inert in the ordinary sense, colloids possess a compensating activity of their own arising out of their physical properties. While the rigidity of the crystalline structure shuts out external impressions, the softness of the gelatinous colloid partakes of fluidity, and enables the colloid to become a medium for liquid diffusion, like water itself. The same penetrability appears to take the form of cementation in such colloids as can exist at a high temperature. Hence a wide sensibility on the part of colloids to external agents. Another and eminently characteristic quality of colloids, is their mutability. Their existence is a continued metastasis. A colloid may be compared in this respect to water while existing liquid at a temperature under its usual freezing-point, or to a supersaturated saline solution. Fluid colloids appear to have always a *pectous*[1] modification; and they often pass under the slightest

---

[1] Πηκτὸς, *curdled.* As fibrin, casein, albumen. But certain liquid colloid substances are capable of forming a jelly and yet still remain liquefiable by heat and soluble in water. Such is gelatine itself, which is not pectous in the condition of animal jelly; but may be so as it exists in the gelatiferous tissues.

influences from the first into the second condition. The solution of hydrated silicic acid, for instance, is easily obtained in a state of purity, but it cannot be preserved. It may remain fluid for days or weeks in a sealed tube, but is sure to gelatinize and become insoluble at last. Nor does the change of this colloid appear to stop at that point. For the mineral forms of silicic acid, deposited from water, such as flint, are often found to have passed, during the geological ages of their existence, from the vitreous or colloidal into the crystalline condition (H. Rose). The colloidal is, in fact, a dynamical state of matter; the crystalloidal being the statical condition. The colloid possesses ENERGIA. It may be looked upon as the probable primary source of the force appearing in the phenomena of vitality. To the gradual manner in which colloidal changes take place (for they always demand time as an element), may the characteristic protraction of chemico-organic changes also be referred.

A simple and easily applicable mode of effecting a diffusive separation is to place the mixed substance under a column of water, contained in a cylindrical glass jar of 5 or 6 inches in depth. The mixed solution may be conducted to the bottom of the jar by the use of a fine pipette, without the occurrence of any sensible intermixture. The spontaneous diffusion, which immediately commences, is allowed to go on for a period of several days. It is then interrupted by siphoning off the water from the surface in successive strata, from the top to the bottom of the column. A species of cohobation has been the consequence of unequal diffusion, the most rapidly diffusive substance being isolated more and more as it ascended. The higher the water column, sufficient time being always given to enable the most diffusive substance to appear at the summit, the more completely does a portion of that substance free itself from such other less diffusive substances as were originally associated with it. A marked effect is produced even where the difference in diffusibility is by no means considerable, such as the separation of chloride of potassium from chloride of sodium, of which the relative diffusibilities are as 1 to 0·841. Supposing a third metal of the potassium group to exist, standing above potassium in diffusibility as potassium stands above sodium, it may be safely predicated that the new metal would admit of being separated from the other two metals by an application of the jar-diffusion above described.

A certain property of colloid substances comes into play most opportunely in assisting diffusive separations. The jelly of starch, that of animal mucus, of pectin, of the vegetable gelose of PAYEN, and other solid colloidal hydrates, all of which are, strictly speaking, insoluble in cold water, are themselves permeable when in mass, as water is, by the more highly diffusive class of substances. But such jellies greatly resist the passage of the less diffusive substances, and cut off entirely other colloid substances like themselves that may be in solution. They resemble animal membrane in this respect. A mere film of the jelly has the separating effect. Take for illustration the following simple experiment.

A sheet of very thin and well-sized letter paper, of French manufacture, having no porosity, was first thoroughly wetted and then laid upon the surface of water contained in a small basin of less diameter than the width of the paper, and the latter depressed in the centre so as to form a tray or cavity capable of holding a liquid. The liquid placed upon the paper was a mixed solution of cane-sugar and gum-arabic, containing 5 per cent. of each substance. The pure water below and the mixed solution above were therefore separated only by the thickness of the wet sized paper. After twenty-four hours the upper liquid appeared to have increased sensibly in volume, through the agency of osmose. The water below was found now to contain three-fourths of the whole sugar, in a condition so pure as to crystallize when the liquid was evaporated on a water-bath. Indeed the liquid of the basin was only in the slightest degree disturbed by subacetate of lead, showing the absence of all but a trace of gum. Paper of the description used is sized by means of starch. The film of gelatinous starch in the wetted paper has presented no obstacle to the passage of the crystalloid sugar, but has resisted the passage of the colloid gum. I may state at once what I believe to be the mode in which this takes place.

The sized paper has no power to act as a filter. It is mechanically impenetrable, and denies a passage to the mixed fluid as a whole. Molecules only permeate this septum, and not masses. The molecules also are moved by the force of diffusion. But the water of the gelatinous starch is not directly available as a medium for the diffusion of either the sugar or gum, being in a state of true chemical combination, feeble although the union of water with starch may be. The hydrated compound itself is solid, and also insoluble. Sugar, however, with all other crystalloids, can separate water, molecule after molecule, from any hydrated colloid, such as starch. The sugar thus obtains the liquid medium required for diffusion, and makes its way through the gelatinous septum. Gum, on the other hand, possessing as a colloid an affinity for water of the most feeble description, is unable to separate that liquid from the gelatinous starch, and so fails to open the door for its own passage outwards by diffusion.

The separation described is somewhat analogous to that observed in a soap-bubble inflated with a gaseous mixture composed of carbonic acid and hydrogen. Neither gas, as such, can penetrate the water-film. But the carbonic acid, being soluble in water, is condensed and dissolved by the water-film, and so is enabled to pass outwards and reach the atmosphere; while hydrogen, being insoluble in water, or nearly so, is retained behind within the vesicle.

It may perhaps be allowed to me to apply the convenient term *dialysis* to the method of separation by diffusion through a septum of gelatinous matter. The most suitable of all substances for the dialytic septum appears to be the commercial material known as vegetable parchment or parchment-paper, which was first produced by M. GAINE, and is now successfully manufactured by Messrs. De la Rue. This is unsized paper, altered by a short immersion in sulphuric acid, or in chloride of zinc, as proposed by Mr. T. Taylor. Paper so metamorphosed acquires considerable tenacity, as is well known; and when wetted it expands and becomes translucent, evidently admitting of hydration. A slip of 25 inches in length was elongated 1 inch in pure water, and 1·2 inch in water containing one per cent. of carbonate of potash. In the wetted state parchment-paper can easily be applied to a light hoop of wood, or better, to a hoop made of sheet gutta percha, 2 inches in depth and 8 or 10 inches in diameter, so as to form a vessel like a sieve in form (Fig. 10). The disc of parchment-paper used should exceed in diameter the hoop to be covered by 3 or 4 inches, so as to rise well round the hoop. It may be bound to the hoop by string, or by an elastic band, but should not be firmly secured. The parchment-paper must not be porous. Its soundness will be ascertained by sponging the upper surface with pure water, and then observing that no wet spots show themselves on the opposite side. Such defects may be remedied by applying liquid albumen, and then coagulating the same by heat. Mr. De la Rue recommends the use of albumen in cementing parchment-paper,

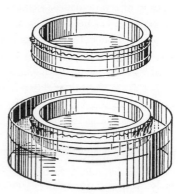

FIG. 10. A hoop dialyser.

which thus may be formed into cells and bags very useful in dialytic experiments. The mixed fluid to be dialysed is poured into the hoop upon the surface of the parchment-paper to a small depth only, such as half an inch. The vessel described (*dialyser*) is then floated in a basin containing a considerable volume of water, in order to induce the egress of the diffusive constituents of the mixture. Half a litre of urine, dialysed for twenty-four hours, gave its crystalloidal constituents to the external water. The latter, evaporated by a water-bath, yielded a white saline mass. From this mass urea was extracted by alcohol in so pure a condition as to appear in crystalline tufts upon the evaporation of the alcohol.

I may be allowed to advert again to the radical distinction assumed in this paper to exist between colloids and crystalloids in their intimate molecular constitution. Every physical and chemical property is characteristically modified in each class. They appear like different worlds of matter, and give occasion to a corresponding division of chemical science. The distinction between these kinds of matter is that subsisting between the material of a mineral and the material of an organized mass.

The colloidal character is not obliterated by liquefaction, and is therefore more than a modification of the physical condition of solid. Some colloids are soluble in water, as gelatine and gum-arabic; and some are insoluble, like gum-tragacanth. Some colloids, again, form solid compounds with water, as gelatine and gum-tragacanth, while others, like tannin, do not. In such points the colloids exhibit as great a diversity of property as the crystalloids. A certain parallelism is maintained between the two classes, notwithstanding their differences.

The phenomena of the solution of a salt or crystalloid probably all appear in the solution of a colloid, but greatly reduced in degree. The process becomes slow; time, indeed, appearing essential to all colloidal changes. The change of temperature, usually occurring in the act of solution, becomes barely perceptible. The liquid is always sensibly gummy or viscous when concentrated. The colloid, although often dissolved in a large proportion by its solvent, is held in solution by a singularly feeble force. Hence colloids are generally displaced and precipitated by the addition to their solution of any substance from the other class. Of all the properties of liquid colloids, their slow diffusion in water, and their arrest by colloidal septa, are the most serviceable in distinguishing them from crystalloids. Colloids have feeble chemical reactions, but they exhibit at the same time a very general sensibility to liquid reagents, as has already been explained.

While soluble crystalloids are always highly sapid, soluble colloids are singularly insipid. It may be questioned whether a colloid, when tasted, ever reaches the sentient extremities of the nerves of the palate, as the latter are probably protected by a colloidal membrane, impermeable to soluble substances of the same physical constitution.

It has been observed that vegetable gum is not digested in the stomach. The coats of that organ dialyse the soluble food, absorbing crystalloids and rejecting all colloids. This action appears to be aided by the thick coating of mucus which usually lines the stomach.

The secretion of free hydrochloric acid during digestion—at times most abundant—appears to depend upon processes of which no distinct conception has been formed. But certain colloidal decompositions are equally inexplicable upon ordinary chemical views. To facilitate the separation of hydrochloric acid from the perchloride of iron, for instance, that salt is first rendered basic by the addition of peroxide of iron. The comparatively stable perchloride of iron is transformed, by such treatment, into a feebly-constituted colloidal hydrochlorate. The latter compound breaks up under the purely physical agency of diffusion, and divides on the dialyser into colloidal peroxide of iron and free hydrochloric acid. The super-induction of the colloidal condition may possibly form a stage in many analogous organic decompositions.

A tendency to spontaneous change, which is observed occasionally in crystalloids, appears to be general in the other class. The fluid colloid becomes pectous and insoluble by contact with certain other substances, without combining with these substances, and often under the influence of time alone. The pectizing substance appears to hasten merely an impending change. Even while fluid a colloid may alter sensibly, from colourless becoming opalescent; and while pectous the degree of hydration may become reduced from internal change. The gradual progress of alteration in the colloid effected by the agency of time, is an investigation yet to be entered upon.

The equivalent of a colloid appears to be always high, although the ratio between the elements of the substance may be simple. Gummic acid, for instance, may be represented by $C_{12}H_{11}O_{11}$, but judging from the small proportions of lime and potash which suffice to neutralize this acid, the true numbers of its formula must be several times greater. It is difficult to avoid associating the inertness of colloids with their high equivalents, particularly where the high number appears to be attained by the repetition of a smaller number. The inquiry suggests itself whether the colloid molecule may not be constituted by the grouping together of a number of smaller crystalloid molecules, and whether the basis of colloidality may not really be this composite character of the molecule.

With silicic acid, which can exist in combination both as a crystalloid and colloid, we have two series of compounds, silicates and cosilicates, the acid of the latter appearing to have an equivalent much greater (thirty-six times greater in one salt) than the acid of the former. The apparently small proportion of acid in a variety of metallic salts, such as certain red salts of iron, is accounted for by the high colloidal equivalent of their bases. The effect of such an insoluble colloid as prussian blue in carrying down small proportions of the precipitating salts, may admit of a similar explanation.

Gelatine appears to hold an important place as a colloidal base. This base unites with colloidal acids, giving a class of stable compounds, of which tannogelatine only appears to be hitherto known. Gelatine is precipitated entirely by a solution of metaphosphoric acid added drop by drop, 100 parts of gelatine uniting with 3·6 parts of the acid. The compound formed is a semitransparent, soft, elastic, and stringy solid mass, presenting a startling resemblance to animal fibrin. It will be an interesting inquiry whether metaphosphoric acid is a colloid, and enters into the compound described in that character, or is a crystalloid, as the small proportion and low equivalent of the acid would suggest. Gelatine is also precipitated by carbolic acid.

The hardness of the crystalloid, with its crystalline planes and angles, is replaced in the colloid by a degree of softness, with a more or less rounded outline. The water of crystallization is represented by the water of gelatination. The water in gelatinous hydrates is aptly described by M. CHEVREUL as retained by "capillary affinity," that is, by an attraction partaking both of the physical and chemical character. While it is here admitted that chemical affinity of the lowest degree may shade into capillary attraction, it is believed that the character of gelatinous hydration is as truly chemical as that of crystalline hydration. Combination of a colloid with water is feeble, it is true, but so is combination in general with the colloid. Notwithstanding this, anhydrous colloids can decompose certain crystalloid hydrates. The water in alcohol of greater strength than corresponds with the density 0·926, which represents the definite hydrate $C_4H_6O_2+6HO$, is certainly in a state of chemical union. But alcohol so high as 0·906, contained in a close vessel, is concentrated in a notable degree by contact with dry mucus, gelatine, and gum, and sensibly even by dry parchment-paper. Dilute alcohol divided from the air of the atmosphere by a dry septum of mucus, gelatine, or gum, is also concentrated by evaporation, as in the well-known bladder experiment of SÖMMERING. The selective power is here apparent of the colloid for water, that fluid being separated from alcohol, and travelling through the colloidal septum by combination with successive molecules of the latter, till the outer surface is reached and evaporation takes place. The penetration in this manner of a colloid by a foreign substance may be taken as an illustration of the phenomena of cementation. Iron and other substances which soften under heat, may be supposed to assume at the same time a colloidal constitution. So it may be supposed does silica when fused into a glass by heat, and every other vitreous substance.

Gelatinous hydrates always exhibit a certain tendency to aggregation, as is seen in the jelly of hydrated silicic acid and of alumina. With some the jelly is also adhesive, as in glue and mucus. But unless they be soluble in water, gelatinous hydrates, when once formed, are not in general adhesive. Separated masses do not reunite when brought into contact. This want of adhesiveness is very remarkable in the gelose of PAYEN, which resembles gelatine so closely in other respects. Layers of a gelose solution, allowed to cool and gelatinize in succession in a diffusion-jar, do not adhere together.

Ice itself presents colloidal characters at or near its melting-point, paradoxical although the statement may appear. When ice is formed at temperatures a few degrees under 0°C., it has a well-marked crystalline structure, as is seen in water frozen from a state of vapour, in the form of flakes of snow and hoar-frost, or in water frozen from dilute sulphuric acid, as observed by Mr. FARADAY. But ice formed in contact with water at 0°, is a plain homogeneous mass with a vitreous fracture, exhibiting no facets or angles. This must appear singular when it is considered how favourable to crystallization are the circumstances in which a sheet of ice is slowly produced in the freezing of a lake or river. The continued extrication of latent heat by ice as it is cooled a few degrees below 0°C., observed by M. PERSON, appears also to indicate a molecular change subsequent to the first freezing. Further, ice, although exhibiting none of the viscous softness of pitch, has the elasticity and tendency to rend seen in colloids. In the properties last mentioned, ice presents a distant analogy to gum incompletely dried, to glue, or any other firm jelly. Ice further appears to be of the class of adhesive colloids. The redintegration (regelation of FARADAY) of masses of melting ice, when placed in contact, has much of a colloid character. A colloidal view of the plasticity of ice demonstrated in the glacier movement will readily develope itself.

A similar extreme departure from its normal condition appears to be presented by a colloid holding so high a place in its class as albumen. In the so-called blood-crystals of FUNKE, a soft and gelatinous albuminoid body is seen to assume a crystalline contour. Can any facts more strikingly illustrate the maxim that in nature there are no abrupt transitions, and that distinctions of class are never absolute?

LIFE EXISTING INDEPENDENTLY IN AN "INTERNAL ENVIRONMENT"

BERNARD, Claude (French physician, philosopher, and physiologist, 1813–1878). From *Les trois formes de la vie*, in *Leçons sur les phénomènes de la vie communs aux animaux et aux végétaux*, Paris, 1878–1879; tr. by T. S. Hall for this volume.

In this lecture, the great French physiologist applies the term *milieu interieur* to the plasma of the blood and lymph and remarks that its existence endows the "life" of the cells with considerable independence of the world outside. While this raises a problem of definitions, in that the individual becomes an organism independent of its body fluids, it has furnished a remarkably fruitful platform from which to comprehend the compensatory regulations of warm-blooded animals.

Constant or Free Life

. . . Constant, or free, life is the third[1] form of existence and pertains only to those animals which are highest in organization. With these animals, life is never, under any circumstance, found to be suspended. It pursues a course which is constant and apparently indifferent to alterations in the cosmic environment and changes in the material conditions surrounding the animal. Organs, mechanisms, tissues function in an apparently stable manner, without evincing such considerable variations as appear among animals with variable life. This comes about because, actually, the internal environment which surrounds organs, tissues, and tissue elements does not change; atmospheric variations are checked by it, so that, it may truthfully be said, the physical conditions of the environment are, for the higher animal, constant. It is enveloped in an internal environment which acts for it as an atmosphere of its own in the midst of an ever changing outer cosmic environment. The higher organism has, in effect, been placed in a hot house. Here it is beyond the reach of the perpetual changes of the cosmic environment. It is not bound up in them; it is free and independent.

---

[1] The other two forms of life, in addition to *la vie constante*, are *la vie oscillante* (as in cold-blooded animals) and *la vie latente* (as in spores and encysted microbes.)

I believe I was the first to insist upon this idea that there are for the animal really two environments: an external environment in which the organism is situated, and an internal environment in which the tissue elements live. Thus life goes on for it not in the external environment,—(atmospheric air for the aërial animal, fresh or salt water for the aquatic one),—but in the liquid internal environment formed by the circulating organic liquid which surrounds and bathes all the anatomical elements of the tissues. It is the lymph, or the plasma, the liquid part of the blood, which, among the higher animals, penetrates the tissues, and makes up the ensemble of all interstitial liquids, and is the outward expression of all local nutritional activity and is the source and common ground of every fundamental exchange. A complex organism must be considered as a collection of simple beings,—*viz.*, the anatomical elements which live in the liquid internal environment.

The invariability of the internal environment is the essential condition of free independent life: the mechanism which permits this constancy is precisely that which insures the maintenance in the internal environment of all conditions necessary to the life of the elements. This makes it clear that there can be no question of free, independent life for simple beings whose constituent elements are in direct contact with the cosmic environment, but that this form of life is, on the contrary, the exclusive possession of beings which have achieved the highest complexity and organic differentiation.

Stability of the environment implies a perfection of the organism such that external variations shall be at every instant compensated and brought into equilibrium. Consequently, far from being indifferent to the external world, the higher animal is, on the contrary, narrowly and wisely attuned to it in such a way that, from the continual and delicate compensation, established as if by the most sensitive balance, equilibrium results.

Those conditions which are necessary to the life of the elements and which, for the functioning of free existence, must be mobilized and kept constant in the internal environment are already known to us: water, oxygen, heat, chemical substances, or reserves.

These conditions are the same as those needed for life by simple creatures; except that, with the perfected animal, manifesting independent life, the nervous system is called upon to preserve harmony among them.

1. WATER

This element, qualitatively and quantitatively, is an indispensable constituent of the milieu in which the living elements move and act. Among animals with free life, there must exist an ensemble of arrangements for regulating losses and acquisitions in such a way as to maintain a requisite quantity of water in the internal environment. With the lower organisms, variations in water compatible with life are quantitatively more narrow; but, on the other hand, the organism is here without influence in the regulation of them. This is why it is bound up in the vicissitudes of climate, torpid with latent life in time of drought, reanimated in time of moisture.

For man in particular, and for higher animals in general, loss of water occurs: with secretion; with the urine and sweat, especially; to a less degree with respiration (which includes a noticeable quantity of water vapor); and, finally, with cutaneous perspiration.

As for gains, they occur through the ingestion of liquids, or of foods which contain water, or even, for some animals, through cutaneous absorption. In any case, it is very likely that the entire amount of water in the organism comes from outside by one or the other of these routes. We have not succeeded in demonstrating that the animal organism really produces water; the contrary opinion seems almost certain.

It is the nervous system, as we have said, which makes up the machinery of compensation between acquisitions and losses. The sensation of thirst which is under the dependence of this system always makes itself felt at times when the proportion of liquid diminishes in the body after some such condition as hemorrhage or abundant sweating; the animal finds itself compelled to repair, by ingestion of liquids, the losses which it has sustained. But this very same ingestion is also regulated, in the sense that it will not augment, beyond a certain degree, the quantity of water which exists in the blood; the urinary excretions and others eliminate the surplus somewhat like an overflow device. The mechanisms which cause the quantity of water to vary and which reestablish it are very numerous. They set in operation the large number of secretory devices as well as those of respiration, ingestion, and circulation, which transports the ingested, absorbed liquids. These mechanisms are varied but concur in the same result: namely, the presence of water in the internal environment in a set proportion, a necessary condition for the maintenance of free life.

It is not only for water that these compensatory mechanisms exist. We recognize them equally well for the majority of mineral or organic substances contained in solution in the blood. We know that the blood does not take on appreciable quantities of sodium chloride, for example; the excess, beyond a certain limit, is carried off with the urine. I have also discovered the same thing as regards sugar which, remaining normal in the blood, is, beyond a certain amount, rejected by the urine.

2. Heat

We know that there exists for each organism, elementary or complex, limits of external temperature between which their functioning is possible, and a middle-point which corresponds to the maximum of vital energy. Moreover, this is true not only for creatures which have achieved maturity but also for the egg and embryo. All these beings manifest variable life, but for the higher, so-called warm-blooded animals, the temperature compatible with manifestations of life is narrowly fixed. This fixed temperature maintains itself in the internal environment, despite the most extreme climatic variations, and assures continuity and independence of existence. There is, in a word, among animals possessing constant and free life, a heat producing property which does not exist among animals with oscillatory life.

For this function there exists a group of mechanisms governed by the nervous system. There are thermic nerves, and vasomotor nerves which I have demonstrated, whose activity produces either elevation or depression of temperature depending upon circumstances.

Heat production is due, in the living world as in the inorganic world, to chemical phenomena. Such is the great law made known to us by Lavoisier and Laplace. It is in the chemical activity of the tissues that the higher organism discovers a source for that heat which it keeps, in its internal environment, at a practically fixed degree, 38 to 40 degrees for mammals, 45 to 47 degrees for birds. Calorific regulation is accomplished as I have said by means of two sorts of nerves: (1) The nerves which I have called thermic, which belong to the greater sympathetic system and which in some way bridle the chemico-thermal activities for which the living tissues are the seat. When these nerves act, they diminish interstitial combustions and lower the temperature; when their influence is weakened, by the suppression of their action, or, by the antagonism of other nervous influences, then the combustions are heightened and the temperature of the internal environment rises considerably; (2) The vasomotor nerves which, accelerating circulation at the periphery of the body or in the central organs, likewise participate in the mechanism of equilibration of animal heat.

I will add only this last point. When one attenuates considerably the action of the cerebrospinal system by permitting that of the greater sympathetic (the thermic nerve) to persist in full activity one witnesses a considerable lowering of the temperature, and the warm-blooded animal finds itself, as it were, transformed into a cold-blooded one. This is the experiment which I carried out on rabbits, by cutting their spinal cord between the seventh cervical vertebra and the first dorsal one. When, on the contrary, one destroys the greater sympathetic, leaving intact the cerebrospinal system, one witnesses a rise in temperature, at first locally, then generally; this is the experiment which I carried out on horses by cutting the great sympathetic, especially when they were previously in poor condition. Under these conditions a veritable fever supervenes. I have developed elsewhere at length an account of these mechanisms. I wish here to recall them only in order to establish that the calorific function proper to warm-blooded animals is due to a perfecting of the nervous mechanism, which, by incessant compensation, maintains a practically fixed temperature in the internal environment in the midst of which live those organic elements to which we must always very definitely refer all manifestations of life.

3. Oxygen

The manifestations of life demand for their production the intervention of air, or better, of its active element, oxygen, in a soluble form and in such condition as to facilitate its arrival within the actual organism. Moreover, it is necessary, up to a certain point, that this oxygen exist in fixed proportions in the internal environment: amounts too small and too great are equally incompatible with vital function.

Hence it is necessary that, among animals with constant life, appropriate mechanisms regulate the quantity of gas which is received into the internal environment and keep it practically constant. Thus, among advanced animals, the penetration of oxygen into the blood depends upon respiratory movements and upon the quantity of this gas which exists in the surrounding medium. On the other hand, the quantity of oxygen which is found in the air results, as we learn from physics, from the percentage composition of the atmosphere, as well as its pressure. One can thus understand how an animal can live in an environment poor in oxygen, provided the accrued pressure manages to compensate for the oxygen decrease, and, conversely, how the same animal can live in an environment richer in oxygen than ordinary air is, provided a lowering of pressure compensates for the increment. We have here an important general proposition which results from the work of M. Paul Bert. In this case, one notes, variations in the environment are compensated and are equilibrated *by themselves*, without the intervention of the animal. As the pressure augments or diminishes, if the percentage composition diminishes or augments in inverse ratio, the animal finds in the environment exactly the same quantity of oxygen, and its life activities go forward under the same conditions.

But it is possible for there to be in the animal itself mechanisms to establish the compensation when it is not done outside and to assure the penetration into the internal environment of the amount of oxygen necessary for vital function. We are here referring to the quantitative variations which may be experienced by the hemoglobin, a substance actively absorbent of oxygen,—variations still little known but which certainly also play their part.

All these mechanisms like the preceding ones are effective only within rather restricted limits; they are distorted and rendered powerless in extreme conditions. They are regulated by the nervous system. When air is rarefied due to any cause, such as an ascension in a balloon or in the mountains, the respiratory movements become more ample and frequent, and compensation is established. Nevertheless the mammals, man included, cannot long sustain this compensatory struggle when the rarefaction is exaggerated,—when, for example, they find themselves carried to altitudes higher than 5000 meters.

It is unnecessary here to enter into the particular details which the question involves. It is sufficient for us merely to pose the question. We will simply point out a case which M. Campana has made known. It relates to birds of high flight, such as the birds of prey, and, especially, the condors which rise to heights of 7000 to 8000 meters. Here, in an atmosphere which would prove fatal for a mammal, they remain, moving about for a long time. The principles previously stated permitted the prediction that the *internal* respiratory environment of these animals would be found to escape, by means of an appropriate mechanism, the lowered pressure of the *external* environment; in other words, that the oxygen contained in their arterial blood must not be variable at these great heights. Now, there exist in these birds of prey, enormous pneumatic sacs attached to the wings and functioning only when the wings are moved. If the wings are lifted, the sacs fill with external air. If lowered, they push this air into the pulmonary parenchyma. The result of which is that when, and to the extent that, the air is rarefied, the work of the bird's wing is not only supporting it but involuntarily increasing the supplementary supply of oxygen which enters the lung. The compensation for the rarefaction of external air by the increase in the quantity of air inspired is thus assured, as is also, in this way, the constancy of the respiratory environment requisite to the bird.

These examples, which we could multiply, show us that all vital mechanisms, however they may vary, have but a single end, namely that of preserving the unity of conditions of life in the internal environment.

4. Reserves

Finally, the animal must have, for the maintenance of life, reserve materials which assure a constancy of the composition of its internal environment. The higher organisms secure with their diet the material contents of their internal environment but since they cannot secure an identical or exclusive diet, there must be in them certain mechanisms which draw from these *variable* nutriments *similar* materials and which regulate the proportion in which each of them must enter the blood.

I have shown, and we will see later, that nutrition is not *direct*, as accepted chemical theories teach us, but that, on the contrary, it is indirect and is accomplished through reserves. This fundamental law is a consequence of the variety of the diet, as opposed to the constancy of the (internal) environment. Briefly, one does not live upon his actual food substance, but only upon those which one has previously eaten, modified, and somehow brought into existence by the act of assimilation. The same is true of respiratory combustion. It is by no means direct, as we shall show later.

There are, then, reserves, prepared from nutrients and expended at each instant in greater or less proportion. In this way, vital activities lead to the de-

struction of provisions which have their origin, to be sure, in the first place outside the body, but which have been elaborated under the influence of the tissues of the organism, and which, poured into the blood, assure the constancy
of its physicochemical constitution.

When the mechanisms of nutrition are disturbed, and when the animal finds
itself unable to provide these reserves, since it does naught but consume those
which it had previously accumulated, it moves toward a destruction which can
only end in the impossibility of life, namely in death. It will then be of no
use for it to eat. It still will not nourish itself, will not assimilate, will perish.

Something similar happens in the case of an animal in a state of fever. It
uses food without reconstructing it and this, if it persists to the point of an entire exhaustion of materials accumulated by previous nutrition, is fatal.

Thus edible substances penetrating the organism, animal or vegetable, are
not directly and immediately used for nutrition. The nutritive phenomenon
accomplishes itself at two different times; and these are separated from each
other by a longer or shorter period whose duration is a function of a large
number of circumstances. Nutrition is preceded by a special process of work-
ing-over, which ends, either in the animal or the vegetable, with the placing
of reserves in storage. This fact permits us to understand how a being continues to live often for a very long time without taking nourishment: it lives
on the reserves accumulated in its own substance; it consumes itself.

These reserves, in different animals and vegetables, vary greatly, according to the creatures which one is studying, and according to different circumstances. This is not the place to analyze so vast a subject; we desired merely
to show that the formation of reserves is not only the general law of all forms
of life, but also it constitutes an active and indispensable mechanism for the
maintenance of constant and free existence, independent of the variations of
the surrounding cosmic environment.

Conclusion

We have successively examined the three general forms in which life
appears:—latent life, oscillatory life, constant life,—in order to see whether,
in any of them, we would find an interior vital principle capable of causing
manifestations of life, independent of the exterior physicochemical conditions.
The conclusion to which we find ourselves led is easy to discern. We see
that in latent life the being is dominated by exterior physicochemical conditions
to such a point that every vital manifestation can be stopped. In oscillating
life, although the living creature is not absolutely submitted to these conditions,
yet it remains so bound up in them that it undergoes all their variations. In
constant life, the creature appears to be free, and the vital manifestations seem
to be effective and controlled by an internal vital principle, entirely free from
the influence of external physicochemical conditions. This appearance is an

illusion. Quite to the contrary, it is exactly in the mechanism of constant or free life that these narrow relationships are particularly evident. We cannot therefore admit in living organisms a free vital principle struggling against the influence of physical conditions. The opposite has been proved, and thus all of the contrary conceptions of the vitalists are seen to be overthrown.

PHYSIOLOGICAL ACTION OF A HORMONE DEMONSTRATED

OLIVER, George (1841–1915) and SHARPEY–SCHAEFER, Sir Edward Albert (1850–1935) (British physiologists). *On the physiological action of extract of the suprarenal capsules*, in *Proceedings of the Physiological Society* (London) (with the *Journal of Physiology*, vol. 16, p. i following p. 318, 1894); by permission of the editors of the *Journal*.

The following paper is usually considered the first experimental demonstration of the specific physiological action of a substance produced by an endocrine gland.

The suprarenal capsules yield to water (cold or hot), to alcohol or to glycerine a substance which exerts a most powerful action upon the blood vessels, upon the heart, and upon the skeletal muscles. These effects have been investigated upon the dog, cat, rabbit and frog. In the frog the solutions were injected into the dorsal lymph-sac, in the rabbit subcutaneously and into a vein, in the other animals into a vein. The alcohol extracts were first dried and the residue extracted with normal saline; the watery decoctions were made with normal saline, and the glycerine extracts were largely diluted with the same previous to injection. The doses employed have varied from a mere trace up to an amount of extract equivalent to 3 grains (0.2 gramme) of the fresh gland; in one or two instances we have given larger doses with the object of obtaining if possible a lethal result. The extracts used have been made from the suprarenals of the calf, sheep and dog. Exactly similar effects have been obtained in each case. Except in the case of the frog we have not obtained any marked effect from subcutaneous injections of comparatively small doses. We are, however, able to confirm the statement of Foa and Pellacani that hypodermic injections of the aqueous extract produce death in 24 hours in the rabbit; but a large dose, equal to 50 grs. of the gland, was necessary. Even after intravenous injection the symptoms entirely pass off after a few minutes, showing that the poison must be rapidly eliminated.

The effect upon the blood vessels is to cause extreme contraction of the arteries, so that the blood-pressure is enormously raised. This is most evident when the vagi are cut in order to obviate the inhibitory action upon the heart which otherwise occurs; it is also seen after section of the cervical cord. The blood-pressure may rise from 2 to 4 times above normal. This extreme contraction of the vessels is evidenced by the plethysmograph; section of the nerves going to the limb produces no difference in the result.[1] The effect is therefore peripheral. This can also be shown in the frog with its nerve-centres destroyed, and through the blood vessels of which normal saline is allowed to circulate; if only a small quantity of suprarenal extract is added to the saline the flow almost entirely ceases.

The time which elapses between the injection into a vein and the first effect upon the blood vessels is in the dog from 25 to 30 seconds. But if an experiment has been conducted for some three hours or more, the animal during the whole of this time having been under the influence of morphia and curare, the contraction of the vessels of a limb is preceded by a preliminary expansion, the cause of which is as yet not clear. It is however a constant phenomenon.

Traube's curves are abolished during the greater part of the time that the substance is producing its effect upon the vessels. The effect on the heart, as long as one vagus remains uncut, is to produce powerful inhibition. During this time the auricles may come to a complete standstill, the ventricles continuing to beat with a slow independent rhythm. But if both vagi are cut the heart-beat becomes greatly accelerated (to twice its former rate) and also augmented; the augmentation showing itself most markedly upon the auricular tracing. The latency of this augmented action of the heart is less than that of the vascular contraction. So far as we are able to judge from one or two experiments which were kindly made for us by Dr. Signey Ringer, the direct action of the drug upon the frog ventricle as recorded in a Roy tonometer does not appear to be marked. When added to the circulating fluid used for perfusion the extract produced only a slight increase in the strength of the beat, and even this was not maintained.

The effect upon the skeletal muscles has been investigated in the frog. The movements of a frog to which a hypodermic injection of extract of suprarenal capsule (equal to 1 or 2 grains of the fresh gland) have been given soon become slow, and after about half-an-hour the reflexes are very faint and almost abolished; the animal soon appears completely paralysed. The muscles however still contract on being stimulated, either directly or through the motor nerves, but the contractions are modified, the relaxation period being greatly prolonged, as with veratria poisoning. The period of latent stimulation is not greatly, if at all, lengthened. The fatigue curves were rapidly developed. The effect is not at all comparable to that produced by curare.

We have noticed a slight effect to be produced upon the respiration, which may become shallower; but in the doses we have used the result was very

---

[1] This is the rule, but we have had one notable exception in which in the earlier part of the experiment the limb, the nerves of which were cut, showed a large expansion. After a short time, however, this passed off and the limbs gave similar curves.

slight when compared with the prodigious effects upon the heart and blood vessels which were obtained.

Although poisonous effects resulting from hypodermic or intravenous injections of extract of suprarenal capsules have been mentioned by more than one previous observer, we have not been able to find any record of experiments tending to show exactly what organs were affected, nor do such symptoms as have been described by others agree in most respects with the changes we have here recorded. As to the chemical nature of the active substance we are unable at present to make any positive statement, since the process of isolating such a body in a fairly pure condition is necessarily slow and difficult. We are, however, able to make the negative affirmation that it is not neurin, although neurin has by more than one observer been conjectured to be the active agent in such extract. For the symptoms produced by neurin do not in any way agree with those produced by the active substance in our extract; this we are able to affirm, both from the careful work regarding the physiological action of neurin which has been done by Cervello, and also by determining for ourselves by the same methods as we have employed in investigating the suprarenal extract what is the action of neurin upon the several organs. Suffice it to say that the most prominent effect produced by neurin is paralysis of the respiration. The effects upon the heart and vessels are relatively slight, and, such as they are, are quite different from suprarenal extract. We have also investigated the action of the salts of neurine (phosphate, hydrochlorate) and have obtained exactly the same results as with the base itself. We are unable to confirm the statement of Marino-Zuco that the addition of either mineral acid or a base to suprarenal extract abolishes its toxic properties (which he ascribes to the presence of phosphate or glycerophosphate of neurine).

We are indebted to Mr. Moore for the preparations of neurine which we have used, and for much other help in the chemical part of the investigation, and to Messrs. Willows, Francis and Butler for a large supply of glycerine extract and alcohol extract of calf suprarenals, which they have been at much trouble to prepare specially for this work. We have also tested the effect of a clear watery extract obtained from Paris (in a sealed tube) and furnished to us by the kindness of Dr. Hale White.

The physiological effect of this solution, as to the mode of preparation of which we are wholly ignorant, proved to be precisely the same as that of our own extracts.

The above communication was illustrated by two experiments, as well as by numerous tracings and photographs of tracings.

BIOCHEMICAL DEFINITION OF LIFE

VERWORN, Max (German physiologist, 1862–1921). From *Allgemeine physiologie; ein grundriss der lehre vom leben*, Jena, 1895; tr. (of the 2d ed., Jena, 1897) by F. S. Lee as *General physiology; an outline of the science of life*, New York, 1899.

Through the whole history of biological thought, one encounters the premise that for an object to manifest life, a special material composition is necessary (see Schultze, Dujardin). While denouncing the notion of living particles (whose prototypes go back at least to the "vegetative seeds" of Anaximines), Verworn introduces *chemical* particles whose special *chemical* actions are, in his view, what we really mean by *life*. Verworn's paper is outstanding as illustrative of the perennial dependence of speculative biology upon the contemporary status of physical and chemical knowledge. It adumbrates recent studies locating metabolic systems in cellular granules.

Psycho-monism

When the history of the problems that have kept the human intellect busy during the long course of its evolution is studied, it is found that many problems that perplexed the ancients have continued unchanged and unsolved down to the present day; others have been solved; while still others that have been prominent even for centuries have afterwards disappeared without finding a solution. The ancient question of the squaring of the circle, over which many a brain has puzzled in vain, that of perpetual motion, which since early times has been prominent in physics, and many others, have quite disappeared, although no one has ever squared the circle and no one has constructed a machine for perpetual motion. If it be asked how it happens that this is so, the answer is, because it is recognised that the basis of these supposed problems is false, and they are, therefore, insoluble. If the attempt be made to divide all the numbers of a series by 2 without a remainder, it is found impossible to do it. So it is with the above problems, which for centuries have harassed one generation of thinkers after another.

So it is also with the attempted explanation of psychical by physical events. It still engages unremittingly the attention of those who are not pleased with having limitations to their conception of the world, yet no one, however earnest his thought, comes nearer a solution. Only gradually will the conviction force its way, that this problem, like those above mentioned, will always resist solution because the question is falsely put.

That the attempted explanation is wrong is at once clear from the preceding considerations. It was found that the sole reality that we are able to discover in the world is mind. The idea of the physical world is only a product of the mind, and with the alteration of an old sentence of the sensualists, it can be said: *Nihil est in universo, quod non antea fuerit in intellectu*. But this idea is not the whole of mind, for we have many mental constituents, such as the simple sensations of pain and of pleasure, that are not ideas of bodies. The task of psychology, *i.e.*, the investigation of mind, consists in the analysis of all mental constituents. By investigating the contents of mind, by decomposing the higher psychical phenomena, the more complex groups and series of ideas, into their simple constituents, psychology arrives, finally, at the most primitive psychical phenomena, the psychical elements, and in the same degree discovers the laws of the arrangement of these elements into the higher groups and series of ideas. Just as in mathematics the endless variety of numbers is formed according to laws out of the numerical unit, so psychology reduces the endless variety of psychical phenomena to their formation, according to laws, out of the psychical elements. But the idea of matter, or, better, of an atom, is not a psychical element, it is a great complex of highly developed ideas. An atom is nothing but a thing possessing all the properties of a body, such as hardness, impenetrability, form, and extension, all of which presuppose very complex psychical processes. The endeavour of natural science to reduce the phenomena of the physical world to the mechanics of atoms is justifiable; it is an endeavour to derive the phenomena of large bodies from the properties of their material parts. But the attempt to reduce to the motions of atoms all psychical phenomena, not only ideas of the physical world but others, such as simple sensations, is precisely as absurd as the endeavour to reduce all numbers in the numerical series to 2 instead of to the numerical unit, for the complex notion of the atom is not a unit, not a psychical element. Herein lies the fallacy of the problem, and hence, as the history of human thought has shown so strikingly, all attempts to explain the psychical by the physical must fail.

The actual problem is precisely the reverse. It consists not in explaining psychical by physical phenomena, but rather in reducing to its psychical elements physical, like all other psychical, phenomena.

In natural science the view is frequently met with, that knowledge of the world falls into two sharply separated categories, namely, metaphysics and science. Metaphysics is left to philosophy, and science is limited to the investigation of the physical world. But the fact is often overlooked or intentionally neglected, that every process of knowledge, including scientific knowledge, is merely a psychical event, that science also deals with "metaphysics," as in accordance with an ancient and unfortunate manner of expression it is customary to term it, and even that science cannot exist without metaphysics. This fact cannot be banished by the well-known method of the ostrich.

It thus appears to be a contradiction to contrast nature ($\phi \acute{\upsilon} \sigma \iota s$) with something "beyond" nature ($\mu \epsilon \tau \grave{\alpha} \ \tau \grave{\eta} \nu \ \phi \acute{\upsilon} \sigma \iota \nu$). There is but *one* world, whether this be termed nature, mind, reality, or anything else. It follows also that there is but one kind of knowledge, and not two. As soon, therefore, as the question arises of principles and bases of knowledge, all artificial boundaries disappear. We should not be deceived. The goal toward which the human mind is striving in its theoretical investigations is not simply a knowledge of the lifeless physical world or of living bodies, or of this or that psychical phenomenon, but it is a knowledge of the world. A division of labour among investigators, however, should not only not be deprecated, but, because of the excessive multitude of phenomena, should be encouraged; nevertheless, the purely supplementary purpose of such a division must be kept in mind, and the artificial boundaries between the various fields of labour must not be confounded with the natural boundaries. A severe blow will be inflicted upon the coming centuries, if the gulf between philosophy and science widens constantly from both sides; if, upon the one side, confused speculation, and, upon the other, narrow specialisation constantly prevail and prevent a mutual approach toward a beneficent common labouring-ground. Science cannot make salutary advances without a philosophical working-plan, and we see in the history of science that great discoveries have been made, not by restricted specialisation, but by investigators working philosophically, *i.e.*, systematically, methodically, and cognisant of their aim. But philosophy can obtain really important results just as little by a purely speculative method, by not keeping close to established facts and not submitting its speculations to the severe criticism of experience. The history of science proves that true advance comes only by thoughtful investigation. The above theoretical considerations regarding knowledge ought to give us a basis for investigation such as every thinking investigator sooner or later must make for himself, and upon which he must build broadly and freely in order that his labours may be fruitful.

The most important result afforded by the above considerations is the monistic standpoint, in accordance with which the world appears as unitary, and the dualism of the physical world and mind as an illusion. The fact, which appears so remarkable from other standpoints, is, therefore, not surprising, that the laws that control the physical world and those that control mental phenomena are completely identical. This appears necessarily so when we find that the phenomena of the physical world are arranged according to space, time and causality, and when we recognise therein the logical principles of our own thought; the laws of the physical world are the laws according to which our own psychical phenomena occur, because the physical world is only our own idea. All science, therefore, is in this sense psychology.

We will now summarise our considerations regarding investigation. We started out with the question, whether there are impassable limits to a knowledge of the world. If we understand by knowledge the reduction of phenomena to the motions or the mechanics of atoms, limits do, indeed, exist. For not only is the atom, and hence matter, yet to be explained, but, as du Bois-Reymond's clever undertaking has shown very clearly, it is impossible to reduce psychical phenomena to the mechanics of atoms. If, however, we conceive knowledge in a more general and the only justified sense, namely, the reduction of phenomena to the elements of reality, we find that no limits exist, for the sole reality is our mind and all phenomena are only its contents; explanation, therefore, consists simply in the reduction of all psychical phenomena to their elements. In this sense, all science, and in general all knowledge, is in the end psychology. We thus come to the only consistent standpoint, namely, monism, the unitary view of the world, which seeks to derive all phenomena from a single cause. From this standpoint we see why we meet with limits when we define knowledge to be the reduction of phenomena to the mechanics of atoms. An atom is not an element of reality but a complex idea, hence all phenomena are not reducible to atoms; just as in a series of numbers the element of which is the unit 1, all are reducible to the common unit but not to a number more complicated than 1, *e.g.*, 2. It is thus evident that a limit can no more exist to the investigation of physical, than to that of psychical phenomena; for, since bodies, in other words, atoms or matter, are only ideas, in other words, psychical phenomena, they may be reduced to the same psychical elements as ideas.

THE CELL AS AN ELEMENTARY ORGANISM

When the organic world inhabiting the surface of the earth is examined, it is found that living substance does not form a single coherent mass, but that it is divided into separate organic individuals. It is not wholly easy to define the *conception of the organic individual;* yet many investigators, in recent times particularly Haeckel ('66), have endeavoured to give it a generally valid form. It arose in early times by a process of abstraction from ideas of man and the higher animals, which appear as unitary living beings independent of one another. But, as with all such early conceptions which spring from a limited circle of experiences and later come to cover a larger circle, the conception of the individual in its original form has become too narrow and requires an extension.

The original idea upon which the conception of individuality was based, was that of indivisibility. According to this an individual was a unitary whole, which was incapable of division without losing its characteristic properties. So long as none but men, vertebrates and perhaps insects were in mind this definition held good, for a man, a vertebrate or an insect cannot be divided into several independent individuals. But difficulties appear when we descend lower in the animal series or attempt to apply the conception to plants.

In fresh-water ponds and lakes there exists a peculiar representative of the great group of *Cnidaria,* the fresh-water polyp *Hydra.* This small animal, about one centimetre long, with its slender, tube-like body bearing several long thread-like tentacles that serve for catching prey began to attract the attention of observers soon after the discovery of the microscope. It was found that this remarkable creature could be divided by a cross-cut into two halves, each one of which could transform itself again into a complete, but correspondingly smaller individual. The anterior half, bearing the tentacles, simply closes up the wound and attaches itself again at its posterior end, while from the posterior half new tentacles soon sprout out from the edges of the wound, and in a short time both pieces have become complete Hydras. The halves can be divided still further, and the animal can even be cut into a large number of small pieces, each one of which can transform itself into a complete individual. The unitary individual has thus been divided into two or even several individuals. If, therefore, indivisibility alone be the standard of individuality, *Hydra* is not an individual, for it can be divided without the loss, by the pieces, of the characteristics of the original animal; and the same is true of every tree and every shrub.

The criterion of the individual is, therefore, not to be found in indivisibility, but rather in undividedness or unity. So long as *Hydra* was undivided, it was an individual, a whole, a unit. By the division, however, the original individual came to an end and from it two new units arose which, so long as they are not further cut into pieces, represent complete individuals. Hence the fact of unity alone is decisive in defining the conception of individuality, if the latter is to be stated in such general terms that it holds good for all special cases. An organic individual would accordingly be merely a unitary mass of living substance.

But in this very general form the definition is too broad. According to it a small particle of living substance, cut off from the living cell under the microscope, would be an individual. Such a particle, however, cannot be so considered when it is seen how every minute mass of living substance, which has not the value of the cell, sooner or later invariably perishes. The capability of self-preservation may, therefore, be added to the conception of the individual and the latter may be defined as follows: *An organic individual is a unitary mass of living substance which under definite external vital conditions is capable of self-preservation.*

This definition applies to all single, free-living organisms which are spatially separated from one another and are not artificially divided, in other words to all organisms in the form in which they occur in nature. But it includes more than single organisms; it includes groups of organisms, each one of which is separated from the others by space, but which together form a unit. An example of this is a community of ants. The community represents a single individual in so far as it is a unitary whole in which the single parts work together like the parts of an organism. But it consists of many single individuals, males, females, workers, and soldiers. It is thus seen that individuality may be of very different grades. It seems advantageous to distinguish the grades of individuality by terming the more comprehensive form an individual of a higher order, and the forms composing it individuals of a lower order. The condition in the coral-stem is like the relation between the ant-community and the individual ants. The coral-stem is an individual of a higher order, the single coral-polyp an individual of a lower order. The·sole difference between this case and that of the community of ants is that here the individuals of the lower order are in physical connection with one another.

It will be advantageous to look about the organic world and see what different grades of individuality are to be found. The *community*, the *colony*, is evidently the highest grade, for a sum of communities is not a new and higher unit. The next lower stage in the community is the *person*. The coral-colony can be regarded in a certain sense as a person which consists of single organs; this relation, however, is clearer in another group of *Coelenterata*, the *Siphonophora*. The *Siphonophora* represent persons which consist of a number of variously developed *organs*. Some of these organs are for purposes of movement, others for nutrition, others for reproduction, others for protection of the whole body, and all are grouped in regular order about a longitudinal axis. But all the organs are single individuals, for the embryology of the *Siphonophora* shows that they all arise from morphologically homologous parts by budding; and that in certain cases single individuals, as *e.g.*, the swimming-bells, can separate themselves from the stem and lead an independent existence as medusae. It is seen, therefore, that the person of the *Siphonophora* can be considered as a colony of single organs, and that the stage of individuality of the person includes the lower stages of individuality of the organs. Careful dissection of an organ, *e.g.*, a human arm, shows that it is composed of various constituents, which are termed *tissues*. The arm contains muscle-tissue, nerve-tissue, bone-tissue, etc.; the characteristic of the organ is its composition out of one or more tissues. The next lower stage of individuality, therefore, is the *tissue*. Certain organisms consist of but a single tissue, in which all the constituents are alike. Such free-living tissues are widely represented among the *Algae*. *Eudorina elegans*, *e.g.*, is a small transparent ball of jelly, in which many spherical particles lie embedded, which upon close examination prove to be bits of living substance separated from one another. These single minute particles of living substance are termed *cells*. In this particular case each cell has two delicate flagella, by the movement of which the whole mulberry-mass of jelly is driven about in the water. Every such flagellate cell is an independent individual, and continues to live when separated from the ball of jelly, which happens, *e.g.*, spontaneously in reproduction. It is seen, therefore, that the tissue contains within itself the single cell. The tissue is a colony of cells. In the cell the lowest stage of individuality has been reached. The cell is, indeed, composed of various constituents, of a soft ground-substance, the protoplasm, and a more solid cell-nucleus embedded in it; but in no case can these two constituents be separated without the death of both. Many experiments have shown that protoplasm is incapable of self-preservation without the cell-nucleus, and the nucleus similarly incapable without the protoplasm. Hence, according to the above definition of individuality, neither of the two represents an individual. In all nature no organism is known which represents a lower stage of individuality than the cell. As Brücke says, the cell is the "elementary organism."

Apparently in contradiction with this idea is the fact, recently established by many experiments, that under certain conditions the cell can be artificially divided into pieces which continue to live and even reproduce. If, *e.g.*, a free-living infusorian cell, such as the delicate *Stentor Roeselii*, which lives in fresh water and is especially adapted for this experiment, be divided into two parts in such a manner that each possesses a piece of the long rod-like nucleus, the same phenomenon appears as in *Hydra*: the two pieces regrow into small complete Stentors and continue to live in all respects normally. In such an experiment the cell, an individual of the lowest order, has become divided into two individuals, and can even be divided into more, if the operation be performed so that each piece possesses some protoplasm as well as a piece of the nucleus. This fact is of fundamental importance, and we shall have occasion to recall it frequently. In the present case it stands only apparently in contradiction with the idea of the cell as the elementary individual; for by the cutting operation there are obtained, not new stages of individuality, but complete Stentors, *i.e.*, individuals of the value of a cell. In all such divisions of cells, wherever protoplasm and nucleus are present in the pieces, the latter have the value of cells; in the process we do not go below the cell. If, however, the cut be made so that one piece contains protoplasm and nucleus, and the other only protoplasm, the former continues to live and represents a complete cell, while the latter, possessing no longer the individuality of a cell, invariably perishes. In every case, therefore, the cell remains the elementary organism.

If the above considerations be summarised, it is found that five stages of individuality can be distinguished in the organic world, and can be characterised as follows:

1. Individuals of the first order are *cells*. They represent elementary organisms that are not composed of lower units capable of life. An example is the unicellular, ciliate infusorian *Stentor*.

2. Individuals of the second order are *tissues*. The tissues are associations of individuals of the first order, each one of which is like the others. An example is the flagellated spherical alga, *Eudorina*.

3. Individuals of the third order are *organs*. The organs are associations of various kinds of individuals of the second order. An example is *Hydra*, the body of which consists of only two layers of tissues.

4. Individuals of the fourth order are *persons*. The persons are associations of various individuals of the third order. An example is man, whose body consists of various organs united.

5. Individuals of the fifth order are *communities*. The communities are associations of individuals of the fourth order. Examples are communities of ants and bees.

This scheme requires one more remark. It shows that every individual of a higher order consists of an assemblage of individuals of the next lower order, but the constituents of an individual of the higher order are not always real individuals, *i.e.*, they are capable of self-preservation when living in union with, but not when separated from, their fellows; in other words, they are only virtual individuals. A person or individual of the fourth order, for example a man, consists of single organs, which are equal to individuals of the third order. These organs, however, are virtual, not real, individuals, for they perish when separated from their fellows. It is the same with individuals of all orders. *E.g.*, the cell of an animal tissue, if separated from its fellows, is in itself incapable of life; in the tissue, therefore, it is only a virtual individual. In other cases, however, the constituents of an individual of a higher order, when separated from their fellows, can become real individuals of the next lower order, as is shown, *e.g.*, by *Eudorina*, in which the single cells when separated are in themselves capable of life.

From these considerations the important facts follow that in the end all living individuals of whatever order either are composed of cells as the elementary structural components or are themselves free-living cells. The cell must, therefore, be the seat of those events the expression of which is life.

The Mechanism of Life

The principle which the early civilised races with their mythical ideas poetically personified and represented as the cause of all life in the world, lies at the foundation of all vital phenomena according to the scientific knowledge of to-day. Among most people this principle has found expression in its original form in the allegory of the shifting contest between two hostile forces. These forces are life and death, which the ancient Egyptian personified in the forms of Horus and Typhon; bloom and decay, which the German clothed in the legends of Baldur and Loki; Ahriman struggling with Ormuzd, by which the Persian represented the interchange of the good and the evil in life; God striving with the Devil, in which the Christian of the middle ages perceived the all-creating positive element in its opposition to the all-destroying, "ever-denying spirit"; and, finally, they are recognised in the ever-alternating processes of becoming and passing away, of building up and breaking down, which control every living being and every vital event.

We have already recognised in the continual construction and destruction of living substance or, in brief, in unbroken metabolism, the real vital process, upon which the physical phenomena of life are based. We have become acquainted with these phenomena, have investigated the conditions under which they make their appearance, and have determined the changes that they experience under external influences. We must now endeavour to construct a bridge between the vital phenomena and the vital process, and, so far as the present condition of our knowledge allows, derive the former mechanically from the latter; the investigation of the mechanism of life forms the nucleus of the science that deals with the physical phenomena of life.

The Vital Process

As previous treatment of this subject has shown, our knowledge of the individual events in the metabolism of living substance is unfortunately thus far very meagre. Investigation of the mechanism of the physical phenomena of life is necessarily still far from complete, and progress can be made only slowly. An essential advance in this direction can be expected only from the detailed study of the processes in the cell, for the cell is the place where the vital process itself has its seat, and where all vital phenomena occur in their simplest form. Not until the physiology of organs, which is able to explain only the gross performances of the complex cell-community, develops into cell-physiology, can we hope essentially to enlarge our knowledge of the more delicate mechanism of life. Thus far only the first steps have been taken in this direction.

If, therefore, we attempt to form, so far as possible upon the basis of our present knowledge, a picture of the vital process in living substance, it can be only a sketch in which the most general elements are indicated in gross outline. Notwithstanding this, some kind of a picture of the vital process is necessary for further systematic investigation.

BIOGENS

It has been seen in a previous chapter that, in general, the characteristic of living organisms in comparison with those dead or apparently dead consists in their metabolism, the expression of which constitutes the vital phenomena. It is necessary to go a step beyond this general fact.

It will be recalled that in the determination of the chemical compounds that constitute living substance investigation deals exclusively with the dead cell. For the completion of a picture of living substance two questions now remain to be answered, viz.: first, do the chemical compounds which are found in the dead cell occur as such in the living cell? and, second, are there in the living cell still other compounds which are not present in the dead cell, which, in other words, are bound up inseparably with the life of the cell?

The first of these questions is relatively easy of answer. A careful comparison especially of the solid bodies that may be found as reserve-substances for a time unchanged in the living cell, with the corresponding substances of the dead cell shows that there occur in the living cell proteids, carbohydrates and fats, in other words, the three chief groups of organic compounds, and like-wise the products of their decomposition; in brief, there occur all the essential substances that are found in the dead cell.

There remains only the question whether, in addition, compounds exist in the living substance which are destroyed at death and hence are not to be found in the dead cell. A comparison of the chemical behaviour of living and dead cell-substance forces us to assume the existence of such compounds. Physiolog-ical chemistry has shown that between the two kinds of substance very essential chemical differences exist, which prove that living substance experiences in dy-ing pronounced chemical changes. A wide-spread difference between the two consists in their reaction. The reaction of living substance is almost without exception alkaline or neutral and with death changes usually to acid. Fur-ther, certain proteids that are in solution in living cell-substance, as, *e.g.,* the myosin of muscle, experience very remarkable changes. In death they coagu-late and pass into the solid state, which is very unfit for further chemical trans-formations. Physiological chemistry has shown similar changes in death in great number. All these facts prove that in the death of living cell-substance certain chemical compounds undergo transformations; hence substances exist in it which are not to be found in dead cell-substance.

The fact that these chemical compounds are only present in the living sub-stance and are decomposed with death necessitates the conclusion that the vital process is associated very closely with their existence. At all events an impor-tant property belonging to them is their great inclination toward transforma-tion, which is for life an indispensable element. When it is borne in mind how few causes are able to produce death, how almost all chemical substances that are at all soluble in water enter into chemical relations with living cell-sub-stance, while dead cell-substance usually behaves wholly indifferently to the same influences, it must be said that the substances that distinguish living from dead cell-substance possess a very loose constitution.

This conclusion is still more obvious when the fact of metabolism is con-sidered. Metabolism shows that the living cell-substance is being continually broken down and reformed, this process being made possible by the continual giving-off and taking-in of material. In contrast to this, under favourable conditions, dead cell-substance is capable of preservation for an extraordinarily long time without its excreting more than a trace of the material that living cell-substance gives off continually. Hence, in contrast to the former, the lat-ter must be distinguished by the possession of complexes of atoms that have very great tendency toward chemical transformations and are continually un-dergoing self-decomposition. The great lability of these complexes depends upon the fact that their transformation can be considerably augmented by slight influences from the outside, as the excitation of metabolism by stimuli clearly shows. Since, however, metabolism constitutes the real vital process, it is seen at once that life depends directly upon the existence of these labile complexes of atoms. We are, therefore, justified in examining these significant substances more in detail and investigating their nature somewhat further.

In searching after them we can best start from the decomposition-products excreted in metabolism. It is here found that among other substances, such as carbonic acid, water and lactic acid, which contain only the elements carbon, hydrogen and oxygen, compounds also occur that contain nitrogen. The non-nitrogenous decomposition-products may possibly be derived from the decomposition of carbohydrates, fats, etc.; but those containing nitrogen can come only from the transformation of proteids or their derivatives, for these are the sole bodies containing nitrogen that are present in all living substance. This important fact directs attention first to the proteids.

That this is the right path becomes at once clear when the facts concerning the proteids are recalled that have been mentioned in the course of the previous considerations. These facts show without doubt that the proteids stand at the centre of all organic life.

It is an important fact that in all cases where large quantities of reserve-substances, such as fat, starch, and glycogen, are not accumulated in cells, the proteids constitute by far the largest part of the organic compounds of living substance. This proves that they must play a significant *rôle* in the life of the cell. The dominant position of the proteids among the chemical compounds of living substance, however, is at once attested by the fact that they are the only substances that can be found in every cell without exception. It is a further fact that of all the more important substances in the cell the proteids and their compounds present the highest complexity in chemical composition, they comprise the largest number and variety of atoms in their molecules. The known chemical relations of the non-nitrogenous organic substances, especially the carbohydrates and fats, to the proteids are in harmony with this dominant position of the latter in living substance; for, so far as their history is known, those substances either are consumed in building up the proteid molecule, or are derived from the transformations of the latter. The former is, of course, shown most clearly by plants, in which all organic compounds are manufactured synthetically out of simpler inorganic substances. In the cells of the green plant occurs the synthesis of the first organic product, starch, out of carbonic acid and water. This carbohydrate constitutes the organic basis from which the proteid molecule is developed synthetically in a complex and still partly unknown manner with the help of nitrogenous and sulphur-containing salts taken from the earth. Regarding fat, it is known that it can serve for the construction of carbohydrate by transformations in the plant; the carbohydrate then gives off in turn the material for the formation of proteid, for in the seeds of *Paeonia*, which are filled with fatty oils, all oil disappears, e.g., after long exposure to the air, and starch appears in its place. It is thus seen most clearly in the plant how different substances serve for the construction of the proteid molecule; but the animal demonstrates best the fact that the most important non-nitrogenous groups of atoms in living substance, especially carbohydrates and fats, can be derived from the decomposition of the proteid molecule. Thus, the fact that fat can be derived from proteid has been demonstated by Leo in his experiments on phosphorus poisoning in frogs, and by Franz Hofmann in his experiments on the nutrition of the larvae of flies with blood freed from fat. Further, Claude Bernard and recently Mering have proved upon dogs whose bodies were freed from glycogen by fasting, that after the feeding of proteid glycogen is again manufactured in great quantity, in other words, that this carbohydrate can be derived from the transformation of proteid. Finally, Gaglio has established the fact that the lactic acid in the body is derived from the transformation of the proteid molecule, since the quantity of it in the blood is dependent solely upon the quantity of proteid that is eaten. Regarding the nitrogenous excretory products of the body, it is evident that they can be derived only from the transformation of proteids and their compounds, since no other nitrogenous bodies are present among the essential organic compounds of living substance. But the most striking proof of the fact that all substances, both non-nitrogenous and nitrogenous, that are essential to the life of the cell, can be derived by chemical transformation from proteids, is afforded by one of the most significant facts of physiology, namely, the possibility that carnivora are capable of maintaining their life upon pure proteid and, as Pflüger has recently shown, possess great capacity for doing work. Nothing demonstrates better than this fact the controlling position of the proteid molecule in the vital process.

Hence, not only does it follow from the fact of metabolism that very labile complexes of atoms exist in living substance, with the presence of which life is inseparably associated, but it is the proteids whose presence constitutes the general, essential condition and focus of life. If we endeavour to harmonize these two facts, the unavoidable necessity arises of assuming in living cell-substance, besides the known proteids that occur also in dead substance, certain other proteids or compounds of proteids, that are present in life only and terminate life with their decomposition.

Dead proteid, as it is found in the dead egg of the fowl, or as it is stored in quantity in living egg-cells in the form of vitellins, is able to exist for an extraordinarily long time without undergoing the slightest decomposition, if protected from bacteria. Certain proteids or proteid compounds of living substance, however, are continually undergoing spontaneous decomposition, even when the living substance is under wholly normal conditions, and, as is shown by the products that are given off, the slightest action of stimuli increases the decomposition. A long time ago Pflüger, as has been seen elsewhere, called attention to this important difference between the proteid in dead and that in living cell-substance in his valuable work upon oxidation in living substance, and distinguished clearly between living proteid and dead proteid. The fundamental difference between the two consists in the fact that the atoms of the dead proteid molecule are in a condition of stable equilibrium, while the living proteid molecule possesses a very labile constitution.

Pflüger's assumption of living proteid, which distinguishes living cell-substance from dead and in the loose constitution of which lies the essence of life, is necessitated. But this substance must be of essentially different composition from dead proteid, although, as follows from the character of its decomposition-products, certain characteristic atomic groups of the proteids are contained in it. The great lability that distinguishes it from other proteids, can be conditioned only by an essentially different constitution. Further, critics will rightly object to the terming of this hypothetical compound a "living proteid molecule," for there is a certain contradiction in calling a molecule living. The word "living" can be applied only to something that exhibits vital phenomena. Hence, the expression "living substance" is well justified, for vital phenomena may be observed in living substance as a whole. But a molecule cannot exhibit vital phenomena, at least as long as it exists as such; for if any changes appear in it it is no longer the original molecule; and, if it continues unchanged, vital phenomena are not present in it. The latter, which are based upon chemical processes, can be associated only with the construction or the destruction of the molecule in question; and thus the application of another name to the compound that is at the focus of life is doubly justified. In order to distinguish this body, therefore, from dead proteid and to indicate its high significance in the occurrence of vital phenomena, it appears fitting to replace the term "living proteid" with that of *biogen*. The expressions "plasma molecule," "plasson molecule," "plastidule," etc., which Elsberg and Haeckel have employed, and the conceptions of which are comprised approximately in the expression "biogen molecule," are less fitting in so far as they easily give the impression that protoplasm is a chemically unitary body, which consists of wholly similar molecules; such a view must be expressly rejected. Protoplasm is a morphological, not a chemical conception.

Extremely little is known concerning biogens, and this fact should not be concealed. Since the constitution of the proteids themselves, *i.e.*, substances that can be investigated chemically at any moment, is not at all known, it is readily understood that we possess much less knowledge concerning the biogens, the composition of which can only be inferred from their decomposition-products. It can be maintained of them only that they are extraordinarily labile, and this property gives to them a certain similarity to explosive bodies. Pflüger has employed certain facts in a most ingenious manner for the purpose of obtaining conclusions regarding certain characteristics of biogens, which make intelligible the great lability of the biogen molecule in comparison with the molecule of dead proteid.

The starting-point of Pflüger's discussion is a comparison of the decomposition-products that arise spontaneously and continually in the oxidation of living proteid, such as in respiration, with those that are obtained by the artificial oxidation of dead proteid. This demonstrates the important fact that the non-nitrogenous decomposition-products in the two cases agree essentially, while the nitrogenous products possess not the slightest similarity. "It follows from this that, as regards its hydrocarbon radicals, living proteid is not essentially different from the proteid of food." The important difference between the two consists rather in the arrangement of the nitrogenous groups of atoms. If, however, the nitrogenous decomposition-products of living proteid be examined, such as urea, uric acid, creatin, etc., as well as the nuclein bases, adenin, hypoxanthin, guanin and xanthin, it is found that, in contrast to the nitrogenous products that appear in the oxidation of dead proteid, some can be artificially prepared from cyanogen compounds, while others contain cyanogen (CN) as a radical. Hence it is highly probable that the carbon and the nitrogen are combined in the biogen molecule into cyanogen, a radical that is wanting in dead proteids.

Thus there is presented a very fundamental difference in the constitution of biogens and that of dead proteids; this explains also the great lability of the biogen molecule, for cyanogen is a radical that contains a great quantity of internal energy, all its compounds possessing strong inclination toward decomposition. This fact enables us to understand the process of respiration, for when in the biogen molecule two atoms of oxygen come into the vicinity of the very labile cyanogen radical, by reason of the active intramolecular vibrations of the carbon and nitrogen atoms in cyanogen the carbon atom will unite with the oxygen to form the very stable molecule of carbonic acid. In fact, cyanogen is very easily combustible, and in its combustion yields carbonic acid. Thus, Pflüger believes that the continual taking-in of oxygen and giving-out of carbonic acid on the part of living substance depends upon the presence of the cyanogen radical, and that the intramolecular oxygen is the essential condition of the tendency of living substance to decompose.

In these considerations we find a basis for an idea of the manner in which the formation of a biogen molecule takes place in an animal cell out of the ingested food. By the co-operation of the biogens already present, the atoms of the dead proteid molecule introduced in the food undergo in the cell a rearrangement, in such a manner that an atom of nitrogen always unites with an atom of carbon to form the cyanogen radical with the loss of water. The changes that necessarily appear at the same time in the other groups of the proteid molecule are for the present wholly unknown, but, if we may judge from the essential agreement in the non-nitrogenous decomposition-products of the living and of the dead proteid, they do not appear to be of fundamental importance. By the intramolecular addition of inspired oxygen the biogen molecule finally arrives at the maximum of its power of decomposition, so that only very slight impulses are required to bring about the union of the atoms of oxygen with the carbon in the cyanogen. The material of the non-nitrogenous groups of atoms afforded by the explosive decomposition of the biogen molecule can easily be regenerated by the residue of the biogen molecule from the carbohydrates and fats that are present in the living substance and contain such groups; in fact, it has been seen that these substances are consumed in the building-up of proteid. "Probably this is the essential significance of these satellites of the proteid molecule," as Pflüger very fittingly terms the carbohydrates and fats. If, finally, the living substance dies, the labile cyanogen-like compound of nitrogen passes over again into the more stable condition of the ammonia radical with the absorption of water, the nitrogen uniting with the hydrogen of the water. Thus we have again the stable compounds of dead proteid, such as serve for food. These are, in brief, some of the essential features of the abbreviated path followed by the food in the construction of the biogen molecule in the animal cell. The much longer path, which in the plant cell leads from the ingestion of the simplest inorganic compounds through the synthesis of the first carbohydrate and on to the construction of the biogens, is for the present much more obscure.

Notwithstanding the facts that the views here developed have been confirmed by experiment only in part, and that they contain many large gaps, which can be filled only slowly, they afford at least a basis for an understanding of the fundamental processes in living substance. The metabolism of living substance, upon which all life is based, is conditioned by the existence of certain very labile compounds, which stand next to the proteids and on account of their elementary significance in life are best termed biogens. To a certain degree the biogens are continually undergoing spontaneous decomposition, just as is the case with other organic bodies, *e.g.*, prussic acid. But this decomposition is much more extensive, if even slight external stimuli act upon the living substance. We must imagine that by reason of the extremely active intramolecular vibration of the atoms, which is the cause of the labile condition, certain atoms, partly spontaneously and partly as a result of external commotions, come under the influence of others for which they possess greater affinity than for their original neighbours, and in this manner more stable groupings of atoms arise as independent compounds. In this respect the biogens can be compared to explosive substances, the atoms of which possess likewise very labile equilibrium and which upon receiving violent shocks explode, *i.e.*, rearrange their atoms into more stable compounds; *e.g.*, nitroglycerine or trinitrate of glyceryl, which is employed for making dynamite, is decomposed by mechanical impulses or electric shocks into water, carbonic acid, nitrogen and oxygen: $2C_3H_5(ONO_2)_3 = 5H_2O + 6CO_2 + 6N + O$. But, in contrast to other explosive bodies, we must evidently ascribe to the biogens the peculiarity that in decomposition the whole molecule is not destroyed, but that certain groups of atoms, which are formed by rearrangement, are split off, while the residue is again built up into a complete biogen molecule at the expense of the materials found in its vicinity, just as in the manufacture of concentrated sulphuric acid the nitrous acid formed from nitric acid by the withdrawal of oxygen is rebuilt into nitric acid with the aid of the oxygen of the air. The substances still present in the living substance in addition to the biogens are merely "satellites" of the biogen molecule, and either serve for its construction or are derived from its transformations. Thus far no substances have been made known in living matter, which can stand in any nearer or more remote relations to the biogens. Nevertheless, from the variety in the decomposition-products that are excreted by different kinds of cells in metabolism, it must be concluded with great probability that biogen molecules have not in all cells exactly the same chemical composition, but that there are various biogen bodies, and even that the biogens not only of different cells, but of the various differentiations of the same cell, such as exoplasm, myoids or contractile fibres, muscle-fibrillae, cilia, etc., have different constitutions, although they agree in essential structure. The biogens, therefore, are the real bearers of life. Their continual decomposition and reformation constitutes the life-process, which is expressed in the manifold vital phenomena.

BIOCHEMICAL CHARACTER OF DIETARY DEFICIENCY

HOPKINS, Frederick Gowland (English biochemist, 1861–   ). From *The analyst and the medical man,* in *The Analyst,* vol. 31, No. 369, p. 385, 1906.

In this paper the eminent British physiologist and biochemist, who also made historic studies of muscular chemistry and cellular respiration,* reports in a preliminary way the first biochemical investigation of dietary deficiency.

While upon the business of prophecy, I am tempted to put another series of prognostications before you, the credibility of which is at the present time, perhaps, more obvious to the physiological chemist than to anybody else. I pass from pathology to an aspect of dietetics. This is a subject in which the medical man is the recognised authority, charged with instruction of the public, but for a scientific knowledge of which he depends largely on the chemical physiologist and the analyst.

Putting on one side the aspect of affairs which especially concerns this Society—the maintenance of purity and freedom from adulteration—and leaving out questions such as digestibility and the like, the chief practical points which have hitherto been considered in relation to the daily rations of mankind are the total energy value requisite for maintenance, the optimum ratio of fats and

* See his *On the autoxidisable substance of the cell* [glutathione—Ed.], in *Biochemical Journal,* vol. 15, pp. 286*ff,* 1921.

carbohydrates, and the optimum supply of protein. Now, these questions have recently received fresh attention, and experimental work has been done lately yielding, as you know, somewhat startling results, tending at first sight to modify our views concerning maximal, minimal, and optimum dietaries. But I am not going to discuss the work of Atwater or Chittenden, proposing rather to put before you very briefly facts of another sort, less known and seemingly academic. I believe, however, that my theme, which is that of the influence of minimal qualitative variations in dietaries, will one day become recognised as of great practical importance.

Physiological chemistry, chiefly owing to the work of Emil Fischer, has recently gained the knowledge that individual proteins, and among them those which contribute to human dietaries, may each bear a special chemical stamp; that a given protein may differ so widely from another protein as to have, quite possibly, a different nutritive value. I will illustrate this, first of all, by a somewhat extreme case. A protein, zein, forming no inconsiderable proportion of the total nitrogenous constituents of maize, is entirely deficient in at least one characteristic molecular grouping. It yields on digestion no tryptophane, the product which represents the indol group present in the molecule of most typical proteins.

In mentioning tryptophane, I cannot deny myself a moment's harmless gibe at your expense. The well-known colour reaction which you have used for so many years as a test for formaldehyde in milk is really a reaction due to this indol group of the casein. Now, as it was a similar colour reaction which led some of us at Cambridge to separate the tryptophane of protein for the first time, I have felt that some of you, being authorities on food-stuffs, ought with proper enterprise to have anticipated us in this not unimportant discovery.

Recently we have fed animals with this indol-free maize protein in such a way that it formed the only supply of protein, though associated with abundant fat and carbohydrate and suitable salts. The diet wholly failed to maintain tissue growth in young animals, which, however, grew at once when their zein was replaced by pure casein. When tryptophane was added to the zein diet, there was still inability to maintain tissue growth, doubtless because the zein has other deficiencies as a protein. But now an interesting fact came to light. The animals which received the missing indol derivative in addition to the zein did not grow, in fact, continued to lose weight daily, and were afterwards in much better health than, and long outlived, those which had the zein alone. These experiments seem to show two important facts: First, that in an extreme case a particular protein may wholly fail to support life, just as is the case with gelatin; and next, that a group in the protein molecule may serve some purpose in the body other than that of forming tissue or supplying energy. The usual discussions about food-stuffs attribute to them these two functions only— repair of the tissues and energy supply. But the body has other and more subtle needs equally urgent. Here, there, or elsewhere in the organs must appear special, indispensable, active substances which the tissues can only make from special precursors in the diet.

The indol grouping in the protein molecule serves some such special purpose, quite distinct from its necessary function in tissue repair. This matter of qualitative differences in proteins may be of no small significance in dietaries. It may account for what I believe is proved by experience—that rice may serve the races which rely upon it as an almost exclusive source of protein, while wheat is only suitable for races that take a much more varied dietary. It may explain many variations in nutritive values which at present we feel and recognise only vaguely. In the future the analyst will be asked to do more than determine the total protein of a food-stuff; he must essay the more difficult task of a discriminative analysis.

But, further, no animal can live upon a mixture of pure protein, fat, and carbohydrate, and even when the necessary inorganic material is carefully supplied the animal still cannot flourish. The animal body is adjusted to live either upon plant tissues or the tissues of other animals, and these contain countless substances other than the proteins, carbohydrates, and fats.

Physiological evolution, I believe, has made some of these well-nigh as essential as are the basal constituents of diet. Lecithin, for instance, has been repeatedly shown to have a marked influence upon nutrition, and this just happens to be something already familiar, and a substance that happens to have been tried. The field is almost unexplored; only is it certain that there are many minor factors in all diets of which the body takes account.

In diseases such as rickets, and particularly in scurvy, we have had for long years knowledge of a dietetic factor; but though we know how to benefit these conditions empirically, the real errors in the diet are to this day quite obscure. They are, however, certainly of the kind which comprises these minimal qualitative factors that I am considering.

Scurvy and rickets are conditions so severe that they force themselves upon our attention; but many other nutritive errors affect the health of individuals to a degree most important to themselves, and some of them depend upon unsuspected dietetic factors.

I can do no more than hint at these matters, but I can assert that later developments of the science of dietetics will deal with factors highly complex and at present unknown.

But am I at present justified in troubling you, as practical men, with such matters—you who are interested in professional chemistry, and not in what is still more or less academic physiology?

I have been led to do so from two considerations. First, it is abundantly clear that the foundation of future progress in chemical pathology and dietetics on the lines I have been indicating calls for large efforts in purely analytical chemistry—efforts which have been too long delayed. And the delay has arisen from a circumstance of no small interest and importance.

The scientific chemist—unlike his predecessors, the pioneers of sixty or seventy years ago—has long ceased to be much interested in the animal or the plant. Further, the triumph of synthetic work in advancing theory has led the pure chemist away from the especial difficulties of analytical work. His extraordinary developed technique concerns itself only secondarily and imperfectly with analytical studies of the kind still necessary in physiological problems. I mean the endeavours to identify and separate unknown substances, with unknown properties, present in complex mixtures. Only now and again has he made special efforts in this direction, such as that with which Fischer started his work upon proteins. Such work really requires special instincts, and the pure chemist has largely lost them. He is but a poor analyst, as the physiological explorer finds on turning to him for help. I feel that this help, so far as the immediate future is concerned, will have to come from the pupils primarily trained in your own laboratories, where the analytical instinct is developed. Some of your students, it is to be hoped, will have their attention turned in this direction, and to at least a few there may ultimately come opportunities for research; for research, in all callings, even that of the academic teacher, is only to be snatched from leisure. There are the beginnings just now of a renewed interest in biology on the part of all chemists. May the analyst feel this too. It is not only the manufacturer and the sanitary authority that require his help.

In the second place, I am not afraid to assert that progress in dietetics, no less than in chemical pathology, is about to react largely on professional chemical practice. Fresh problems and new ideas will unfailingly extend the field of professional operations.

All progress of the kind I have been hinting at cannot fail to be of the greatest importance to the doctor; and if I may seem to have maligned him in previous paragraphs, I know well how ready and able he is to make use of all knowledge that he believes to yield advantage to his patients.

I see abundant reasons for believing that in the near future events will march to the consummation of mutual appreciation and helpfulness, and to the disappearance of all misunderstanding, in the relations between analyst and medical man.

III. THE BASIS OF ANIMAL BEHAVIOR

Contributions to physiological and general animal psychology

MECHANISTIC BASIS OF PHYSIOLOGICAL PSYCHOLOGY

DESCARTES, René (French philosopher and mathematician, 1596–1650). From *Les passions de l'âme*, Amsterdam, 1649; tr. by E. S. Haldane and G. R. T. Ross as *Passions of the soul* in their *The philosophical works of Descartes*, Cambridge, 1911–1912; (Cambridge University Press, London); used by permission of The Macmillan Company, publisher.

See the note accompanying the fragment on p. 133.

The Passions of the Soul
article XXX.

That the soul is united to all the portions of the body conjointly.

But in order to understand all these things more perfectly, we must know that the soul is really joined to the whole body, and that we cannot, properly speaking, say that it exists in any one of its parts to the exclusion of the others, because it is one and in some manner indivisible, owing to the disposition of its organs, which are so related to one another that when any one of them is removed, that renders the whole body defective; and because it is of a nature which has no relation to extension, nor dimensions, nor other properties of the matter of which the body is composed, but only to the whole conglomerate of its organs, as appears from the fact that we could not in any way conceive of the half or the third of a soul, nor of the space it occupies, and because it does not become smaller owing to the cutting off of some portion of the body, but separates itself from it entirely when the union of its assembled organs is dissolved.

article XXXI.

That there is a small gland in the brain in which the soul exercises its functions more particularly than in the other parts.

It is likewise necessary to know that although the soul is joined to the whole body, there is yet in that a certain part in which it exercises its functions more particularly than in all the others; and it is usually believed that this part is the brain, or possibly the heart: the brain, because it is with it that the organs of sense are connected, and the heart because it is apparently in it that we experience the passions. But, in examining the matter with care, it seems as though I had clearly ascertained that the part of the body in which the soul exercises its functions immediately is in nowise the heart, nor the whole of the brain, but merely the most inward of all its parts, to wit, a certain very small gland which is situated in the middle of its substance and so suspended above the duct whereby the animal spirits in its anterior cavities have communication with those in the posterior, that the slightest movements which take place in it may alter very greatly the course of these spirits; and reciprocally that the smallest changes which occur in the course of the spirits may do much to change the movements of this gland.

ARTICLE XXXII.

How we know that this gland is the main seat of the soul.

The reason which persuades me that the soul cannot have any other seat in all the body than this gland wherein to exercise its functions immediately, is that I reflect that the other parts of our brain are all of them double, just as we have two eyes, two hands, two ears, and finally all the organs of our outside senses are double; and inasmuch as we have but one solitary and simple thought of one particular thing at one and the same moment, it must necessarily be the case that there must somewhere be a place where the two images which come to us by the two eyes, where the two other impressions which proceed from a single object by means of the double organs of the other senses, can unite before arriving at the soul, in order that they may not represent to it two objects instead of one.   And it is easy to apprehend how these images or other impressions might unite in this gland by the intermission of the spirits which fill the cavities of the brain; but there is no other place in the body where they can be thus united unless they are so in this gland.

ARTICLE XXXIII.

That the seat of the passions is not in the heart.

As to the opinion of those who think that the soul receives its passions in the heart, it is not of much consideration, for it is only founded on the fact that the passions cause us to feel some change taking place there; and it is easy to see that this change is not felt in the heart excepting through the medium of a small nerve which descends from the brain towards it, just as pain is felt as in the foot by means of the nerves of the foot, and the stars are perceived as in the heavens by means of their light and of the optic nerves; so that it is not more necessary that our soul should exercise its functions immediately in the heart, in order to feel its passions there, than it is necessary for the soul to be in the heavens in order to see the stars there.

ARTICLE XXXIV.

How the soul and the body act on one another.

Let us then conceive here that the soul has its principal seat in the little gland which exists in the middle of the brain, from whence it radiates forth through all the remainder of the body by means of the animal spirits, nerves, and even the blood, which, participating in the impressions of the spirits, can carry them by the arteries into all the members. And recollecting what has been said above about the machine of our body, i.e. that the little filaments of our nerves are so distributed in all its parts, that on the occasion of the diverse movements which are there excited by sensible objects, they open in diverse ways the pores of the brain, which causes the animal spirits contained in these cavities to enter in diverse ways into the muscles, by which means they can move the members in all the different ways in which they are capable of being moved; and also that all the other causes which are capable of moving the spirits in diverse ways suffice to conduct them into diverse muscles; let us here add that the small gland which is the main seat of the soul is so suspended between the cavities which contain the spirits that it can be moved by them in as many different ways as there are sensible diversities in the object, but that it may also be moved in diverse ways by the soul, whose nature is such that it receives in itself as many diverse impressions, that is to say, that it possesses as many diverse perceptions as there are diverse movements in this gland. Reciprocally, likewise, the machine of the body is so formed that from the simple fact that this gland is diversely moved by the soul, or by such other cause, whatever it is, it thrusts the spirits which surround it towards the pores of the brain, which conduct them by the nerves into the muscles, by which means it causes them to move the limbs.

ARTICLE XXXV.

Example of the mode in which the impressions of the objects unite in the gland which is in the middle of the brain.

Thus, for example, if we see some animal approach us, the light reflected from its body depicts two images of it, one in each of our eyes, and these two images form two others, by means of the optic nerves, in the interior surface of the brain which faces its cavities; then from there, by means of the animal spirits with which its cavities are filled, these images so radiate towards the little gland which is surrounded by these spirits, that the movement which forms each point of one of the images tends towards the same point of the gland towards which tends the movement which forms the point of the other image, which represents the same part of this animal. By this means the two images which are in the brain form but one upon the gland, which, acting immediately upon the soul, causes it to see the form of this animal.

ARTICLE XXXVI.

Example of the way in which the passions are excited in the soul.

And, besides that, if this figure is very strange and frightful—that is, if it has a close relationship with the things which have been formerly hurtful to the body, that excites the passion of apprehension in the soul and then that of courage, or else that of fear and consternation according to the particular temperament of the body or the strength of the soul, and according as we have to begin with been secured by defence or by flight against the hurtful things to which the present impression is related. For in certain persons that disposes the brain in such a way that the spirits reflected from the image thus formed on the gland, proceed thence to take their places partly in the nerves which serve to turn the back and dispose the legs for flight, and partly in those which so increase or diminish the orifices of the heart, or at least which so agitate the other parts from whence the blood is sent to it, that this blood being there rarefied in a different manner from usual, sends to the brain the spirits which are adapted for the maintenance and strengthening of the passion of fear, i.e. which are adapted to the holding open, or at least reopening, of the pores of the brain which conduct them into the same nerves. For from the fact alone that these spirits enter into these pores, they excite a particular movement in this gland which is instituted by nature in order to cause the soul to be sensible of this passion; and because these pores are principally in relation with the little nerves which serve to contract or enlarge the orifices of the heart, that causes the soul to be sensible of it for the most part as in the heart.

ARTICLE XXXVII.

How it seems as though they are all caused by some movement of the spirits.

And because the same occurs in all the other passions, to wit, that they are principally caused by the spirits which are contained in the cavities of the brain, inasmuch as they take their course towards the nerves which serve to enlarge or contract the orifices of the heart, or to drive in various ways to it the blood which is in the other parts, or, in whatever other fashion it may be, to carry on the same passion, we may from this clearly understand why I have placed in my definition of them above, that they are caused by some particular movement of the animal spirits.

Example of the movements of the body which accompany the passions and do not depend on the soul.

For the rest, in the same way as the course which these spirits take towards the nerves of the heart suffices to give the movement to the gland by which fear is placed in the soul, so, too, by the simple fact that certain spirits at the same time proceed towards the nerves which serve to move the legs in order to take flight, they cause another movement in the same gland, by means of which the soul is sensible of and perceives this flight, which in this way may be excited in the body by the disposition of the organs alone, and without the soul's contributing thereto.

EARLY CONCEPT OF CEREBRAL LOCALIZATION

SWEDENBORG, Emanuel (Swedish natural scientist and religious founder, 1688–1772). From *Concerning the cortical and medullary substances of the brain and concerning the blood vessels within the cerebrum*, 1738?; tr. from the Latin MS by A. Acton, Philadelphia, 1938; by permission of the Swedenborg Scientific Association, publisher.

Swedenborg appears to have grasped reasonably clearly the concept of physical continuity between certain parts of the brain and certain parts of the body. Since the work did not appear until 200 years after its production, it does not lie in the main line of the history of neurology. By the time its contents became generally known, clearer ideas on the same subject were already in existence. Also possibly unduly neglected was Swedenborg's insight into the problem of epigenesis. Elevated to the nobility for his scientific work, Swedenborg underwent conversion to mysticism and died in poverty.*

The cortical substance, in respect to the rest of the brain, appears as something infinite in respect to the finite; or as indefinites and inconstancies in respect to integers or constancies—as may be seen in the infinite calculus, and it may also be demonstrated by reason. Hence a conclusion can be made, to some extent, as to what appears to be the relation of the one to the other. As concerns the origin of these cerebella or little brains, it is the general opinion of our authors that they have their origin close to vessels which are not subject to the keenness of ocular sight. With respect to the abundance of the cortical substances, it is hardly to be doubted but that they occupy every corner. As regards their size, according to the calculation made by LEEUWENHOEK, "not even the sixty-fourth part of a myriadth, that is, of a millionth, of any substance equal in size to a small grain [of sand], could enter into the minute vessels which are seen distributed like a network among the pellucid globules,

---

* A discussion of Swedenborg's work on the brain is included in the introduction to the book from which this excerpt is copied.

and which are affixed to them." This infinity, however, has respect only to their number. But as regards their faculties, which are nourished by means of the blood of the body, which is a compound liquor ascending to the head, these are exalted to higher powers. Again there is no room for doubt since the art of optics has detected the exaltation of these glands from the one degree to the other. Thus we tread the whole way with confidence.

But it must be noted that from the minute stamens of the cerebral arteries where these terminate as ultimate threads, glands come into existence; that, on their other side, these glands send out the fibrils of the medullary substance to which they have given birth; and that by their mediation, the arteries seem to be continued. For, as testified by the common opinion of all our authors, the glands occupy the middle place; wherever the one is met with, there, on the other side, appears also the other. Thus in the cortex exists the artery, and in the medulla the fibres; and there is no fibre that does not possess its own spherule, nor any spherule that does not have its own fibre. The parts of the cortex tumesce when the purer spherules tumesce, and they are equally distended by a regular animation. Moreover, according to BOER-HAAVE, since fibres are appended to them, therefore a subtle humor is passed through them. Hence, according to VIEUSSENS, "If they are presented to the eyes quite broken up, they nowhere appear smooth and round, but are everywhere shortened and, as it were, contracted inward." But more concerning the medullary fibre presently.

Therefore, this cortical substance faces two ways. On the one side it looks to the arteries that come to it, and on the other, to the medullary fibres that go out from it—but differently in the one case than in the other. Consequently, it appears to be a uniting medium.

Therefore nothing can exist in the whole body and its artery and blood, nor anything in the medullas and their juice, of which the cortical substance is not at once rendered conscious, because mediately present in the cortex itself, the tunic whereof has been so formed as to become a meninx connected with the artery, and from which is produced the fibre with the nerves and fascicles. Consequently, this noble substance is the centre and, as it were, the meeting place of all contingencies that come up from the body or terminate in the body.

The cortical substance is, therefore, the seat wherein sensation finally ceases. . . .

From these transactions of the learned it is quite clear that the cortex or ash of the cerebrum, cerebellum and both medullas, is the substance from which all things in the animal body derive their principles; for there the meninges and arteries go off into their last subtilty, and there the fibres and nerves begin from their first subtilty; and these comprise everything that enters into the texture of the body. Therefore, he who would labor to know from causes the physical, economic, and psychological state of his body, that is, who would labor to search this out all the way to its causes, must make his start from this substance, or must by all means draw thither the thread of his labors. . . .

The copious medullary or fibrillar substance of the cerebrum is expended for the most part on the members of the chemical laboratory of its spirits,—of which members we shall speak in detail below; nor is it aught but a small part that passes off by the annular protuberance, and, on the opposite side, above the region of the testes and the pineal gland. But though only a small part,— gathered, however, from the whole cerebrum,—yet it salutes, touches, and decussates with all the fibres that are begotten in the medullas oblongata and spinalis; so that not one of them does anything whatever without the consciousness of the cerebrum; or, there is not one of the fibres which move the muscles, that does not depend on the will of the cerebrum. That nevertheless the cerebrum with its own fibres emitted through the two medullas finally passes out of the vertebral sheath, to wit., when it has thus effected these wonderful contacts and connections with all the intermediates, will come to be confirmed in our third Transaction. The same applies likewise to the cerebellum and its medullary fibres, but especially to its third process or restiform body.

Therefore, in a machine that is thus continued and determined by fibres, no part can in any way be touched in the extremity of the body where the fibres have suitably unfolded, without at once passing on the sensation to the cerebrum. And according to the order of nature, this sensation cannot stop midway but must go on to its last where also is the first. Nor does it stop here, since there is something ulterior or prior; but by formed paths of determinations it runs on in a moment and almost in an instant to its last or first substance; that is, all the way to the centre. Thus when it comes to the spherules of the cortical substance, it instantaneously proceeds therefrom to degrees of a superior or prior order; nor does it stop, except in the purest and most simple organ; and this to the end that it may communicate immediately with the soul which is thus rendered conscious of it. The reason is, because it is the soul that has regard to all things and is regarded by all; and for the sake of which they all exist.

The mode whereby modification or sensation passes over from the organs of the one degree to those of the other, and this without stopping except in the first and last, can in some measure be concluded from the mode whereby it is transmitted from the external organs to the cerebrum itself, or from the cerebrum to the cortical substance. And yet that progress hardly allows of being set forth in the customary formulas and signs; for when nature ascends or descends from one degree to another, then, with the change of clothing and form which she puts on in the superior degree, and which to an indefinite extent, is more perfectly qualified, she changes also the words expressive of the same. Thus sight, or an image like to the visual, when it goes further is called an idea and thought; and this again takes its essence from something still further, and from the rational soul. In a word, the same words do not come into use in one degree as obtain in another, even though, the one can be compared with the other by analogy. Hence for the expression of qualities belonging to the former degree, and also of the modes whereby is effected the passage from the one to the other and so on to the purest of all, a different Ontology must be used. The same is confirmed also by all the qualities, met with in describing the substances of ascending and descending degrees. Thus the modification existing in the air is called in the ear Sound, and the perception thereof Hearing; while modification existing in the ether, that is, in an aura of a superior degree, is called Sight; and yet the latter is circumstanced in like manner as the former, but in an indefinitely more perfect way. Moreover, modification of the auras becomes sensation in the organs of animals because a third element is added, which is due to the soul, namely, living, perceiving, understanding, etc. But of these matters we shall speak elsewhere.

This seems to be the reason why the parts of the cortical substance cannot be suitably called either *Vessels* or *Glands*. They cease to be vessels since they no longer carry blood but an essence purer than blood and prior thereto. Nor is the term tunic or cavity competent to them as it is to vessels; for the cortical spherule is surrounded by a most delicate pia mater, and pierced by a most subtle pore; moreover, it no longer secretes liquor but an essence suitable to the spirits; nor is it an excretory tube but a fibre. Hence by analogy it can be compared with whatever in the body receives the common mode of producing effects—here, doing this in principles.

Nor does a trace of this word remain when ascent is made from this degree of the organs to one still higher, although there remains an idea composed of things indefinitely many. Still less is there a trace when ascent is made to the supreme degree where is left nothing that can be expressed. The words which we have learned are properly suited to organs of the inferior degree and to the actions and affections of the body. Hence it is not possible to signify in particular the qualities and faculties of interior degrees except by signs such as can be deduced by an analytical way from those things that preserve an analogy; or, to use the manner of speech customary with authors, which are said to be *emulous of* or *to emulate* them.

But to resume. Since the scarcely visible spherules of the cortical substance, that is, the spherules of the third degree, conceive and produce their own fibres, it follows that the spherules of the superior substance, or those of the second degree, also do likewise. For the entire surface of a single spherule goes off into the entire surface of the fibril extended therefrom.

But the inmost organic substance, or last organism, does not seem to put forth any fibrils,—and this for various reasons, to wit.: Because it is the last thing, which, when conceived, at once flows into the recipient and ambient fibres; and being dispersed and disseminated into the whole corporeal system by means of the fibres and blood, is called *animal spirit;* from which, and also from the veriest unique and universal substance, is composed all else that is subject to the soul and its government. Hence, also, since the soul is in these, the soul is everywhere in its body. Again, the unconnected modes of the other parts do not allow that it shall fix its roots in fibres as do all other substances; and if it be so fixed, besides the fact that it can then no longer enter into the fluids of the animal, the larger compound cannot be inspired by it, as by the first, to becoming like it or a likeness of it. But on these matters the reader may consult our third Transaction.

It does not seem possible that this truly vital substance or animal spirit conceived within the organs of the second degree, can be for long contained in the fibres of those organs without danger of its flying away. Hence it is consociated with a purest elementary fluid elicited from the blood in the spherules of the third degree, that is, in the spherules proper of the cortical substance. This elementary fluid, consociated with the former at the very fountain, falls into the fibres, and is called the *nervous juice.* This is confirmed by the primitive birth and production of these fibres, which at once interweave, join together, and contract; also by an examination of the members of the cerebrum and of its whole chemical laboratory of the spirits,—of which members we shall treat in detail; and also by innumerable other effects which cannot be explained at this threshold of our work.

Consequently there are as many degrees of fibres as there are of organs, excepting one. The first, is that which is born and continued from the spherules of the second degree; the second from those of the third degree, that is, from the spherules of the cortical substance; the third from the spherules of the organs of the fourth degree, that is, from the brain as a whole,—which in producing itself into a medulla oblongata and spinalis is emulous of a grand fibre.

The manner in which each fibre is produced by its organ can in some measure be perceived if we suppose that in each organ the surface, contextured of an indefinite number of very minute threads, that is, the whole structure of the expanse of the spherule (with the exception of those ends which express a dew or exhalation to be mingled with the spirits in the organs of the third degree) descends for the formation of this its production and appendix; and if we further suppose, what will however be confirmed in the following Transaction, that each spherule makes perpetual alternations of animation, that is, of expansion and constriction, whence, in the most perfect spherules there must necessarily exist a mode of spiral or perpetuo-spherical contortion and retorsion; thus there results the surface of a most minute fibre convoluted and woven of threads which flow into a spiral form; and by such revolution it can give to its spirits and nervous juice the most suitable outlet. This is

confirmed by all that has hitherto been observed respecting the nerves; also
by all the anatomical effects that appear in the animal body as the products of
these fibres; and by innumerable other signs, of which we shall speak in particular; moreover, Leeuwenhoek says that he has seen the least parts of the
blood gyrating around their centre.   Thus the mesh of the fibre is not unlike
a mesh of stamens; and the whole fabric is as it were the work of the weaver
Athene newborn from the brain.

IRRITABILITY RECOGNIZED AND DEFINED

von HALLER, Albrecht (Swiss botanist, anatomist, physiologist, physician, and author,
1708–1777). From *Primae lineae physiologiae in usum praelectionum academicarum,*
Göttingen, 1747, *auctae et emendatae,* Göttingen, 1751; tr. (from the 3d Latin ed., Edinburgh, 1767) as *First lines of physiology,* Troy, N.Y., 1803.

To a remarkable degree Haller's *Primae lineae* gave to the nerve-muscle problem the
form along which it has developed down to the present time.   Irritability is recognized,
and the problem is raised as to the nature of the dependence of muscle action on nerve
transmission.   Haller's ideas on these subjects were based on actual experiment.   His
other contributions included important medical and physiological studies especially on
respiration,[*] a definitive work on the Swiss flora,[†] poetical writings, and three didactic
political novels.[‡]

The structure of the ultimate fibre, considered as the elements of a muscle,
when investigated by the microscope in man and other animals, has always
appeared similar to the structure of the larger fibres; and except very minute
filaments, connected by cellular substance, nothing upon which we can rely
has been observed.   There is no series of vesicles or chain of rhombs.   Are
these fibres hollow?   Are they continuous with the arteries?   Does the difference betwixt muscular and tendinous fibres consist in the latter being rendered solid by being compressed and having their fluids expelled?   That the
blood is not concerned, is proved by the slenderness of the fibres, which are
smaller than the blood globules, by the whiteness of the muscles, after the
blood is washed from them, and by physiological reasons.   And, in general,
more strength may be expected from a solid fibre.

A muscle is endowed at least with a three-fold power.   First, the dead one,
common to it with other animal fibres.   Then another, which we have called
the vis insita, possessing different phenomena.   For, in the first place, it is
peculiar to life, and to the first hours after death, and it disappears much

---

* *De respiratione experimenta anatomica,* Göttingen, 1746–1747; see also M. Foster,
Lectures on the history of physiology, Lecture IX, Cambridge, 1901.

† *Historia stirpium indigenarum Helvetiae inchoata,* Bern, 1678.

‡ *Usong,* Bern, 1771; *Alfred,* Bern, 1773; *Fabius und Cato,* Göttingen and Bern,
1774.

sooner than the dead one. Again, in most cases, its action consists in alternate oscillations; so that moving to and fro, at one moment it contracts itself towards the middle; and at the next, extends itself from the middle towards the extremities, and so on successively for several times. Moreover, it is manifest, quick, and performs very considerable motions; the dead force, only such as are small and scarcely apparent. It is excited both by the touch of a sharp instrument, and in the hollow muscles by inflated air, by water, and every kind of acrimony, but more powerfully than by any other stimulus by electricity. Lastly, it is peculiar to the muscular fibre, and in no other part of the human body is it found possessed of the qualities above mentioned. But its phenomena deserve to be more particularly explained.

It is natural to every muscle to shorten itself, by retracting its extremities towards its belly or middle. In order to discover the moving power from the fabric which we have described, it will be of use to consider the phenomena of muscular contraction. Every muscle when in action becomes shorter and thicker. This contraction of its length is various; less in some, more in others; and in particular instances very considerable, for example, in some of the sphincters, iris, diaphragm and intercostals, insomuch that it appears that the length of a muscle may be contracted much more than one third, which computation was derived from an erroneous hypothesis.

The intestines are exceedingly tenacious of their vis insita; they continue to contract, after they are taken out of the body, and even after they are cold. The heart is even more tenacious than these, if you consider all things; as is most evident in the chick, and in cold blooded animals. Different muscles are most readily excited, by different stimuli; as the bladder by urine, the heart by the blood, and the intestines by air. Though their nerves are removed, or their connexion with the brain cut off, muscles lose but little of their irritable nature. It appears also, that this irritable disposition is very widely extended through the animal fibre, from the examples of polypi and other insects, which have neither brain nor nerves, and yet are exceedingly impatient of any stimulus; and from the analogy of plants, of which very many flowers and leaves open or contract, according to the various degrees of heat and cold, some even so quickly, that they are nothing inferior in this respect to animals. This power is totally different from any other known property of matter, and is new. It does not depend either upon gravity, or attraction, or elasticity, for it is inherent in soft fibres, and is destroyed, when they become indurated.

But that a cause of motion is conveyed through the nerves into the muscles, is certain from the observations, already noticed. For the nerve alone possesses feeling; alone conveys the dictates of the mind; and neither retains any influence over, nor receives any perceptions from any part, whose nerve is either tied or cut, or which has no nerve. On irritating the nerve or spinal marrow, even in a dead animal, the muscle or muscles, which have nervous branches from those parts, are most violently convulsed. When the nerve of any muscle is cut or tied, or the part of the spinal marrow, or brain, from whence the nerve has its origin, is compressed, the muscle becomes paralytic and feeble, and cannot by any power be recalled into action similar to the vital one. But if the compression be removed from the nerve, the muscle recovers the power by which it is put into action. When the nerve is irritated below the place where it is cut, the muscle to which that nerve belongs is contracted. Numerous experiments have been made, to prove this, especially on the phrenic and recurrent nerves.

This power is not the same with the vis insita. The former is adventitious to the muscle; whereas the latter is inherent in it. The former ceases along with life; whereas the latter, according to certain experiments, subsists long after it. The former is suppressed, by tying a ligature upon the nerve, by injuring the brain, or by the exhibition of opium. The latter is not affected by these circumstances, but continues after the nerve is tied or cut, and even in the intestines, though taken out of the body; it also exists in animals destitute of brain: parts of the body possess motion, which are destitute of sensation, while others possess sensation, which are destitute of motion. The will excites and removes the nervous action, but has no power over the vis insita.

In muscular action, whether proceeding from the vis insita, or from the nervous power, the fibres are contracted towards the middle of its belly, and expand outwards: they are varied by transverse wrinkles, and the whole muscle becomes shorter, and draws its extremities towards its centre, and therefore carries towards each other those parts with which it is connected, in the reciprocal proportion of their firmness. Muscles, during their contraction, swell, and at the same time become hard, and, as it were, increase their circumference every where. I have never observed them to turn pale. Whether, on the whole, they are increased in bulk, and acquire more in breadth than they lose in length, is difficult to be known. They draw after them the passive tendons, which of themselves are neither moveable or irritable. The whole of a muscle may be moved at once, or only a part of it: if one extremity is fixed to an immoveable part, that only is moved, which is capable of yielding.

Do the arteries contribute in any way to muscular motion, as indicated by the paralysis of the lower extremities, produced by tying the aorta? Not at all, unless by preserving the integrity of the muscles, and mutual relation of the parts, by secreting vapour and fat, and by nourishing them. For by dividing or tying its artery, a muscle does not become paralysed, unless after a considerable time, when the muscles begin to be destroyed by gangrene. The irritation of the artery has no effect on the muscles. Moreover, it is impracticable to explain the motion of peculiar muscles from a cause, which, proceeding from the heart, operates with equal force on all parts of the body.

Lastly, the influence of the will is confined to the nerves, and does not extend to the arteries or other solid parts of the body.

But the manner in which the nerves excite motion in the muscles, is so obscure that we may almost despair of discovering it. And we do not even attempt to investigate the vis insita, which seems to be an increased attraction of the elementary parts of the fibre, by which they mutually approach each other, and accumulate contortions in the middle of the fibre. This force of attraction, which is implanted by nature in the moving fibre, is excited and increased by stimuli. The rest is mere hypothesis. As to nervous vesicles swelling by a quicker influx of the nervous fluid, they are inconsistent with anatomical truth, which demonstrates the fibres to be cylindrical, and in no part vesicular; and likewise with the celerity with which muscular motion is performed, and with the bulk of a muscle being rather diminished than increased during its action. The chains and rhombs of the inflated fibres are in the same manner repugnant to anatomical inspection, and to the celerity; they would also occasion an immense waste of power, and render the muscle but little shorter. The nerves want that irritable nature which is observed in the muscular fibre; and besides, it is by no means demonstrable, that the fibres, so numerous, can arise from nerves, so few and distributed in a different direction, almost transversely with respect to the muscular fibres. The idea of nerves being disposed round arterial fibres, compressing them by their elasticity, is founded upon a false structure of the fibre, which is gratuitously assumed to be filled with blood, and supposes nerves, where cellular fibres only can be demonstrated. Moreover, the phenomena of animals, which have neither brain or nerves, and are yet very capable of motion, demonstrate the fabric of the muscles to be sufficient for their motion, even without nerves. Blood globules, filled with air, and the explanations derived therefrom, suppose a false nature of that fluid; namely, that elastic air exists, where it does not. The animal spirits are not of the nature of electricity.

If we may add any thing to the phenomena, we may suppose the nervous liquor to be of a stimulating nature, forcing the elementary particles of the muscular fibre to approach nearer to each other. The motive cause which occasions the influx of the spirits into the muscle, so as to excite it into action, seems not to be the soul, but a law established by the Creator. For animals, newly born, or newly transformed, without any attempt, or exercise, know how to perform compound motions, very difficult to be defined by calculation. But the soul learns those things, which it performs, slowly, imperfectly, and experimentally. Muscles, therefore, contract, which in a given time receive more of the nervous fluid, whether that be occasioned by the will, or by some irritating cause arising in the brain, or applied to the nerve.

Though the soul may be supposed to act in nervous motions, it cannot be admitted in those arising from the vis insita. The heart and intestines, also some organs of the venereal appetite, are governed by the vis insita, and by stimuli. These powers do not arise from the will; nor are they lessened, or excited, or suppressed, or changed by it. No custom or art can subject these organs of inherent motion to the will, or cause a satellite of voluntary motion to forget to obey the commands of the soul. It is so certain, that motion is produced by the body alone, that we cannot even suspect any motion to arise from a spiritual cause, except that which the will seems to excite in animals; and, even in the very organs of animal volition, a stimulus will occasion the most excessive actions, in direct opposition to the will.

There seems to be this difference between the muscles obeying the will, and those which are governed by the vis insita, that the latter are more irritable, and are very easily excited into action by a gentle stimulus; as, for instance, the heart and intestines; which organs are most manifestly, and greatly, and constantly, irritable. On the other hand, the muscles which obey the will, are less easily, and less durably irritable. Hence, they either need the agency of the will, or of a powerful stimulus; by which, indeed, even these may be excited to action, independent of the will. Thus, it happens, that, in apoplexy, the muscles which obey the will, being deprived of all influx from the brain, languish, and become paralytic; while the vital muscles, having no occasion for the operation of the brain, continue to be excited into contraction by their stimuli; the heart by the blood, and the intestines by the air and aliments.

The strength of this action is very considerable in all persons, but more especially in madmen, and in some strong men; since frequently, with a few muscles only, they will raise a weight, much greater than that of the whole human body. But even in healthy people, very slender muscles have elevated 200 or 300 pounds. The muscles of the back will even sustain 3000. Notwithstanding this, much the greater part of the force or power exerted by a muscle, is always lost, without producing any visible effect. For all muscles are inserted nearer the fulcrum, than the weights are appended; and therefore their action is lessened, in proportion as their lever is shorter than that of the weight. Moreover, most of the muscles are inserted into the bones, especially in the limbs, at very acute angles; whence, again, the effect which a muscle exerts in action, is proportionably less than the effort which it exerts, as the sine of the angle intercepted betwixt the bone and the muscle, is less than the whole sine. Again, the half of every muscular effort is lost, because it may be considered as an elevating cord, drawing an opposite weight to its fixed point. Besides, many of the muscles are seated in the angle between two bones, arising from the one, and moving the other; and therefore, on that bone being moved, they are bent, and, like inflected cords, require a new force to extend them. Many of them pass over several joints, each of which they bend in some degree, so that only a small part of their effort remains to bend their proper joint. The fleshy fibres themselves of the muscles very often form angles with their common tendon, whence a great part of their force is again lost, and only that proportion of the whole remains, which is as the sine of the angle of their insertion to the whole. Finally, the muscles move their opposed weights with very great velocity and ease, so that they not only overcome the equilibrium, but likewise add a considerable excess of velocity.

RECOGNITION OF REFLEX ACTION

WHYTT,* Robert (Scotch physician and physiologist, 1714–1766). From *An essay on the vital and other involuntary motions of animals*, Edinburgh, 1751.

Distinctions between voluntary and involuntary actions had been drawn by members of the Greek schools, but it was only with Descartes, and more explicitly Whytt, that reflexes took their place as actual physiological entities. Whytt's interest in involuntary action was, however, subordinate to a more central concern, namely, his desire to destroy the supposed distinction between the sentient and rational consciousness. His complete works (London, 1765) include studies of sensation, circulation, cure of stone, and the causes of psychiatric disturbances.† Twice married, Whytt had sixteen children.

Many Philosophers have supposed two distinct principles in man; one of which has been called the *anima*, or soul; the other, the *animus*, or mind: by the former, they understood the principle of life and sense influencing the vital motions; and by the latter, the seat of reason or intelligence. According to them, we have the *anima*, or vital and sentient soul, in common with the brutes; but *animus*, or *mens*, which is of a more exalted nature, is proper to rational creatures alone.[1]

Some modern Materialists have imagined the *anima* to be no other than a more subtile kind of matter lodged, chiefly, in the brain and nerves, and circulating with the grosser fluids. But such spirits, or subtile matter, can no more be acknowledged the vital principle or source of animal life, than the blood from which they are derived; and with still less reason can this material *anima* be supposed endued with sense, since matter, of itself, and unactuated by any higher principle, is equally as incapable of sense or perception, pleasure or pain, as it is of self-motion. Indeed, a few authors have run even such lengths, as to suppose the very *animus*, or rational soul itself, material: but surely the powers and faculties of the mind are not to be found in matter, or in any of those principles, or elements, whereof either the antients or moderns

---

* Pronounced as if *white*.

† See L. Carmichael, *Robert Whytt, a contribution to the history of physiological psychology*, in *Psychological Review*, vol. 34, p. 287, 1927.

[1] *Indulsit communis conditor illis*
Tantum animas, nobis, animum quoque.

Sat. 15. lin. 148. & 149.

have imagined it to consist: fire itself, the most subtile and active among these, being as incapable of thought and reflexion, as water or earth; the most sluggish: and in what manner self-motion; sense or reason can possibly result from the figure, connexion, situation or arrangement of the various parts of the body, (without supposing a mind) is a point which the abettors of Materialism, to their confusion, will never be able to clear up.

As I cannot therefore agree with those, who, in ascribing all our powers to mere matter, seem willing to deprive us wholly of mind; so neither, at the same time, do I see any reason for multiplying principles of this kind in man: and, therefore, I am inclined to think the *anima* and *animus*, as they have been termed, or the sentient and rational soul, to be only one and the same principle acting in different capacities. Nay, *Epicurus* himself, according to *Lucretius*, did not look upon these two as separate beings, but regarded the mind as a kind of *mouvement* produced by the *anima* or soul.[2]

That the involuntary motions in man are not owing to a principle distinct from the rational mind, seems evident, from the muscles and organs, whose action has been generally ascribed to the *anima*, being, in many cases, subject to the power of the *animus* or rational principle; as well as, on the other hand, from the motions of the voluntary muscles often becoming involuntary, or independent upon the will. Thus the diaphragm, whose motions in the hiccup are altogether involuntary, and in ordinary respiration go on without our consciousness of them, is nevertheless subject to the immediate influence and direction of the mind; since its motions in breathing can, by an effort of the will, either be augmented or lessened, retarded or accelerated.—The evacuation of the *intestinum rectum* and urinary bladder, which, when the *stimulus* is gentle, is in part voluntary, become altogether involuntary and convulsive, when the irritation is greater.—The eye-lids, which the mind seems to have a full power over, move, commonly, not only without our attention but, in some cases, even against every effort of the will to the contrary.—The action of the *acceleratores urinae* is voluntary in expelling the last drops of urine; but in expelling the *semen*, it is involuntary.—The contraction of the pupil, which, in order to distinct vision, is voluntary, becomes altogether involuntary when owing to the light. In short, there is not a voluntary muscle in the body, whose motion does not become involuntary, as often as it is either directly, or from its consent with some neighbouring part, affected by any considerable *stimulus:* if the irritation be very gentle, we still retain a greater or less power over the muscle; but when it becomes stronger, we lose all this power.

Further, in man the sentient and rational principle must be acknowledged to be one; since we are all conscious that what feels, reasons, and exerts itself

---

[2] *Nunc* animum *atque* animam *dico conjuncta teneri*
  Inter se, atque unam naturam *conficere ex se.*
                                        LUCRET. lib. 3. vers. 136 & 137.

in moving the body, is one and the same, and not distinct beings. It is the mind, therefore, that feels, thinks, remembers and reasons; which, though one principle, is nevertheless possessed of these different powers, and acts in these different capacities: nay, since memory is as widely different from the present perception of ideas, or the exertion of the will in order to action, as sense is from reason, it might with equal propriety be maintained, that we are endued with four souls, namely, with a rational, a reminiscent, an active, and a sentient one, as that we have two. In brutes of the lowest kind there is evidently a sentient principle; but it seems to be wholly devoid of reason or intelligence: in those, however, of a higher class, we can perceive faint traces of something like what we call reason and reflexion in man. Why, therefore, may not the human mind, which enjoys all the powers belonging to the souls of the inferior creatures, and has also reason superadded to those powers, be allowed sometimes to act as a sentient, and at other times as a rational being, *i.e.* in different capacities?

But, if any one yet contends, that the sentient principle, governing the vital motions, is different from the rational, I shall not think it much worth while to dispute the matter with him: since whatever is advanced, in the present Essay, upon the subject of the involuntary motions of animals, will hold equally true, whether the sentient and rational soul be supposed distinct, or otherwise.

However, although we conceive it to be the most probable opinion, that the sentient and rational principle in man are one and the same; yet we think it a very clear point, that the mind does not, as Dr *Stahl* and others would persuade us, preside over, regulate, and continue the vital motions, or, upon extraordinary occasions, exert its power in redoubling them, from any rational views, or from a consciousness that the body's welfare demands her care in these particulars: for infants, ideots, and brutes of the lowest kind, (which last are certainly destitute of reason), perform these motions in as perfect a manner as the wisest Philosopher; and the mind, when life is endangered by the too violent circulation of the blood, neither does, nor can moderate the heart's motion. If the contraction of the heart were owing to any previous deduction of reason, or conviction of its being necessary to the continuance of health of life, the mind ought to have a power of restraining the uniform motions of its auricles and ventricles, or of repeating them at shorter or longer intervals, notwithstanding their having become, like those of the eyes, in a manner necessary through long habit: for though we cannot, indeed, move our eyes in every different direction, yet we can restrain or vary their uniform motions as we please.

Further, if there were any exercise of reason necessary to the continuance of the vital motions, the mind certainly ought to be conscious of this: since, in every ratiocination respecting action, there must first be a comparison of

things, and then, in consequence of this comparison, a preference or election:
but, I believe, few Philosophers will be found hardy enough to maintain, that
the mind can compare two, or more ideas, and thence form certain conclusions
and determinations, without being so much as conscious, in any degree, of what
it has been all the while employed about: for though, when we are solicitously
engaged in any action, deeply involved in any thought, or strongly hurried
away by any passion, we may often be unconscious of the impressions made by
material causes on the organs of sense; yet we cannot but be sensible of the
ideas formed within us by the internal operation of our minds, because their
very existence depends upon our being conscious of them, and is at an end, as
soon as either we attend not to, or forget them: to say therefore that such
ideas may be formed and exist in the mind without consciousness, is, in effect,
to say that they may, and may not exist at the same time; than which nothing
can be more absurd.

'Vital' Motions

Further, the motions excited by any pain, or irritation, are so instantaneous,
that there can be no time for the exercise of reason, or a comparison of ideas
in order to their performance; but they seem to follow as a necessary and
immediate consequence of the disagreeable perception.   And as the DEITY
seems to have implanted in our minds a kind of SENSE respecting *Morals*,
whence we approve of some actions, and disapprove of others, almost instantly,
and without any previous reasoning about their fitness or unfitness; a FACULTY
of singular use, if not absolutely necessary for securing the interests of virtue
among such creatures as men! so, methinks, the analogy will appear very
easy and natural, if we suppose our minds so formed and connected with our
bodies, as that, in consequence of a *stimulus* affecting any organ, or of an
uneasy perception in it, they shall immediately excite such motions in this or
that organ, or part of the body, as may be most proper to remove the irritating cause; and this, without any previous rational conviction of such motions
being necessary or conductive to this end.   Hence, men do not eat, drink, or
propagate their kind, from deliberate views of preserving themselves or their
species, but merely in consequence of the uneasy sensations of hunger,
thirst, *&c.*

The mind, therefore, in producing the vital and other involuntary motions,
does not act as a rational, but as a sentient principle; which, without reasoning
upon this matter, is as necessarily determined by an ungrateful sensation or
stimulus affecting the organs, to exert its power, in bringing about these
motions, as is a balance, while, from mechanical laws, it preponderates to that
side where the greatest weight prevails.

The general and wise intention of all the involuntary motions, is the removal of every thing that irritates, disturbs, or hurts the body: hence, those

violent motions of the heart, in the beginning of fevers, small-pox, measles, &c. when frequently the blood, from its being affected by the mixture of some peculiar *miasma*, acts as a stronger *stimulus* than usual upon this organ. Nevertheless, as, in many instances, the very best things may, by excess, become hurtful; so this endeavour to free the body, or any of its parts, from what is noxious, is unhappily, sometimes, so strong and vehement, as to threaten the entire destruction of the animal fabric. But, in the main, this FACULTY must be confessed highly useful and beneficial; since, without it, we should constantly have cherished in our bodies the lurking principles of diseases, slowly indeed and by imperceptible degrees, but not less surely, ruining our health and constitutions.

Upon the whole, there seems to be in man one sentient and intelligent PRINCIPLE, which is equally the source of life, sense and motion, as of reason; and which, from the law of its union with the body, exerts more or less of its power and influence, as the different circumstances of the several organs actuated by it may require. That this principle operates upon the body, by the intervention of something in the brain or nerves, is, I think, likewise probable; though, as to its particular nature, I presume not to allow myself in any uncertain conjectures; but, perhaps, by means of this connecting *medium*, the various impressions, made on the several parts of the body either by external or internal causes, are transmitted to, and perceived by the mind; in consequence of which it may determine the nervous influence variously into different organs, and so become the cause of all the vital and involuntary motions, as well as of the animal and voluntary. It seems to act necessarily, and as a sentient principle only, when its power is exerted in causing the former; but, in producing the latter, it acts freely, and both as a sentient and rational agent.

BEGINNINGS OF MODERN THEORY OF VISION AND COLOR VISION

NEWTON, Isaac (English philosopher and mathematician, 1642–1727) and YOUNG, Thomas (English physicist, 1773–1829). *Hypothesis III* in Young's *On the theory of light and colors*, in *Philosophical Transactions of the Royal Society*, p. 12, London, 1802.

Taking up the problem of color vision at the point where Newton had left it 140 years before, Young initiates the period of active investigation of this subject. Following the principle of parsimony, he postulates the fewest practicable number of kinds of color receptors: one for each of the three primary colors. Although Helmholtz, Hering, Edridge-Green, and Ladd-Franklin were later stimulated to study this problem, physiological approaches to its experimental investigation proved exceedingly difficult and real progress awaited the arrival of biochemical approaches introduced only after another hundred years or more.

The Sensation of different Colours depends on the different frequency of Vibrations excited by Light in the Retina.

Passages from Newton

"The objector's hypothesis, as to the fundamental part of it, is not against me. That fundamental supposition is, that the parts of bodies, when briskly agitated, do excite vibrations in the ether, which are propagated every way from those bodies in straight lines, and cause a sensation of light by beating and dashing against the bottom of the eye, something after the manner that vibrations in the air cause a sensation of sound by beating against the organs of hearing. Now, the most free and natural application of this hypothesis to the solution of phenomena I take to be this—that the agitated parts of bodies, according to their several sizes, figures, and motions, do excite vibrations in the ether of various depths or bignesses, which, being promiscuously propagated through that medium to our eyes, effect in us a sensation of light of a white colour; but if by any means those of unequal bignesses be separated from one another, the largest beget a sensation of a red colour, the least or shortest of a deep violet, and the intermediate ones of intermediate colours, much after the manner that bodies, according to their several sizes, shapes, and motions, excite vibrations in the air of various bignesses, which, according to those bignesses, make several tones in sound: that the largest vibrations are best able to overcome the resistance of a refracting superficies, and so to break through it with least refraction; whence the vibrations of several bignesses, that is the rays of several colours, which are blended together in light, must be parted from one another by refraction, and so cause the phenomena of prisms and other refracting substances; and that it depends on the thickness of a thin transparent plate or bubble, whether a vibration shall be reflected at its further superficies, or transmitted; so that, according to the number of vibrations, interceding the two superficies, they may be reflected or transmitted for many successive thicknesses. And, since the vibrations which make blue and violet are supposed shorter than those which make red and yellow, they must be reflected at a less thickness of the plate; which is sufficient to explicate all the ordinary phenomena of those plates or bubbles and also of all natural bodies, whose parts are like so many fragments of such plates. These seem to be the most plain, genuine, and necessary conditions of this hypothesis; and they agree so justly with my theory, that, if the animadversor think fit to apply them, he need not, on that account, apprehend a divorce from it; but yet, how he will defend it from other difficulties I know not." (*Phil. Trans.* Vol. VII, p. 5088. Abr. Vol. I, p. 145. Nov. 1672.)

(Further) Passages from Newton

[Three are quoted by Young of which only the second and third are pertinent.—Ed.]

"To explain colours, I suppose, that as bodies of various sizes, densities, or sensations, do by percussion or other action excite sounds of various tones, and consequently vibrations in the air of different bigness; so the rays of light, by infringing on the stiff refracting superficies, excite vibrations in the ether, of various bigness; the biggest, strongest, or most potent rays, the largest vibrations; and others shorter, according to their bigness, strength, or power: and therefore the ends of the capillamenta of the optic nerve, which pave or face the retina, being such refracting superficies, when the rays impinge upon them, they must there excite these vibrations, which vibrations (like those of sound in a trunk or trumpet) will run along the aqueous pores or crystalline pith of the capillamenta, through the optic nerves into the sensorium; and there, I suppose, affect the sense with various colors, according to their bigness and mixture; the biggest with the strongest colors, reds and yellows; the least with the weakest, blues and violets; the middle with green, and a confusion of all with white—much after the manner that, in the sense of hearing, nature makes use of aerial vibrations of several bignesses to generate sounds of divers tones, for the analogy of nature is to be observed." (*Birch*, Vol. III, p. 62. Dec. 1675).

"Considering the lastingness of the motions excited in the bottom of the eye by light, are they not of a vibrating nature? Do not the most refrangible rays excite the shortest vibrations, the least refrangible the largest? May not the harmony and discord of colors arise from the proportions of the vibrations propagated through the fibres of the optic nerve into the brain, as the harmony and discord of sounds arise from the proportions of the vibrations of the air?" (*Optics*, Qu. 16, 13, 14).

Scholium. Since, for the reason here assigned by Newton, it is probable that the motion of the retina is rather of a vibratory than of an undulatory nature, the frequency of the vibrations must be dependent on the constitution of this substance. Now, as it is almost impossible to conceive each sensitive point of the retina to contain an infinite number of particles, each capable of vibrating in perfect unison with every possible undulation, it becomes necessary to suppose the number limited, for instance, to the three principal colours, red, yellow, and blue, of which the undulations are related in magnitude nearly as the numbers 8, 7, and 6; and that each of the particles is capable of being put in motion less or more forcibly by undulations differing less or more from a perfect unison; for instance, the undulations of green light being nearly in the ratio of $6\frac{1}{2}$ will affect equally the particles in unison with yellow and blue, and produce the same effect as light composed of those two species, and each sensitive filament of the nerve may consist of three portions, one for each principal colour.

FUNCTIONS OF THE DORSAL AND VENTRAL ROOTS OF THE SPINAL NERVES

BELL, Sir Charles (Scotch physiologist, physician and surgeon, 1774–1842). *Idea of a new anatomy of the brain*, London, 1811.

Controversy has raged from the start as to how clearly Bell really grasped the functional difference between the dorsal and ventral roots of the spinal nerves. His own later claim that he was perfectly clear on this question Fulton* has reluctantly labeled "dishonest." The paper is important, however, because of the presence in it, in varying stages of differentiation, of this and many other important neurological concepts (see Swedenborg). Bell's life is summarized in *Selections from the writings of Sir Charles Bell with biography and bibliography*, in *Medical Classics*, vol. 1, No. 2, p. 81, 1936.

The want of any consistent history of the Brain and Nerves, and the dull unmeaning manner which is in use of demonstrating the brain, may authorize any novelty in the manner of treating the subject.

I have found some of my friends so mistaken in their conception of the object of the demonstrations which I have delivered in my lectures, that I wish to vindicate myself at all hazards. They would have it that I am in search of the seat of the soul; but I wish only to investigate the structure of the brain, as we examine the structure of the eye and ear.

It is not more presumptuous to follow the tracts of nervous matter in the brain, and to attempt to discover the course of sensation, than it is to trace the rays of light through the humours of the eye, and to say, that the retina is the seat of vision. Why are we to close the investigation with the discovery of the external organ?

It would have been easy to have given this Essay an imposing splendour, by illustrations and engravings of the parts, but I submit it as a sketch to those who are well able to judge of it in this shape.

The prevailing doctrine of the anatomical schools is, that the whole brain is a common sensorium; that the extremities of the nerves are organized, so that each is fitted to receive a peculiar impression; or that they are distinguished from each other only by delicacy of structure, and by a corresponding delicacy of sensation; that the nerve of the eye, for example, differs from the nerves of touch only in the degree of its sensibility.

It is imagined that impressions, thus differing in kind, are carried along the nerves to the sensorium, and presented to the mind; and that the mind, by

---

* *Selected readings in the history of physiology*, p. 252, Springfield, Ill., 1930.

the same nerves which receive sensation, sends out the mandate of the will to the moving parts of the body.

It is further imagined, that there is a set of nerves, called vital nerves, which are less strictly connected with the sensorium, or which have upon them knots, cutting off the course of sensation, and thereby excluding the vital motions from the government of the will.

This appears sufficiently simple and consistent, until we begin to examine anatomically the structure of the brain, and the course of the nerves,—then all is confusion: the divisions and subdivisions of the brain, the circuitous course of nerves, their intricate connections, their separation and re-union, are puzzling in the last degree, and are indeed considered as things inscrutable. Thus it is, that he who knows the parts the best, is most in a maze, and he who knows least of anatomy, sees least inconsistency in the commonly received opinion.

In opposition of these opinions, I have to offer reasons for believing, That the cerebrum and cerebellum are different in function as in form; That the parts of the cerebrum have different functions; and that the nerves which we trace in the body are not single nerves possessing various powers, but bundles of different nerves, whose filaments are united for the convenience of distribution, but which are distinct in office, as they are in origin from the brain:

That the external organs of the senses have the matter of the nerves adapted to receive certain impressions, while the corresponding organs of the brain are put in activity by the external excitement: That the idea or perception is according to the part of the brain to which the nerve is attached, and that each organ has a certain limited number of changes to be wrought upon it by the external impression:

That the nerves of sense, the nerves of motion, and the vital nerves, are distinct through their whole course, though they seem sometimes united in one bundle; and that they depend for their attributes on the organs of the brain to which they are severally attached.

The view which I have to present, will serve to shew why there are divisions, and many distinct parts in the brain: why some nerves are simple in their origin and distribution, and others intricate beyond description. It will explain the apparently accidental connection between the twigs of nerves. It will do away the difficulty of conceiving how sensation and volition should be the operation of the same nerve at the same moment. It will shew how a nerve may lose one property, and retain another; and it will give an interest to the labours of the anatomist in tracing the nerves.

IDEA, &c.

When in contemplating the structure of the eye we say, how admirably it is adapted to the laws of light! we use language which implies a partial, and

consequently an erroneous view. And the philosopher takes not a more enlarged survey of nature when he declares how curiously the laws of light
are adapted to the constitution of the eye.

This creation, of which we are a part, has not been formed in parts. The
organ of vision, and the matter or influence carried to the organ, and the qualities of bodies with which we are acquainted through it, are parts of a system
great beyond our imperfect comprehension, formed as it should seem at once
in wisdom; not pieced together like the work of human ingenuity.

When this whole was created, (of which the remote planetary system, as
well as our bodies, and the objects more familiar to our observation, are but
parts,) the mind was placed in a body not merely suited to its residence, but
in circumstances to be moved by the materials around it; and the capacities
of the mind, and the powers of the organs, which are as a medium betwixt
the mind and the external world, have an original constitution framed in relation to the qualities of things.

It is admitted that neither bodies nor the images of bodies enter the brain.
It is indeed impossible to believe that colour can be conveyed along a nerve; or
the vibration in which we suppose found to consist can be retained in the brain:
but we can conceive, and have reason to believe, that an impression is made
upon the organs of the outward senses when we see, or hear, or taste.

In this inquiry it is most essential to observe, that while each organ of sense
is provided with a capacity of receiving certain changes to be played upon it, as
it were, yet each is utterly incapable of receiving the impressions destined for
another organ of sensation.

It is also very remarkable that an impression made on two different nerves
of sense, though with the same instrument, will produce two distinct sensations; and the ideas resulting will only have relation to the organ affected.

As the announcing of these facts forms a natural introduction to the Anatomy of the Brain, which I am about to deliver, I shall state them more fully.

There are four kinds of Papillae on the tongue, but with two of those only
we have to do at present. Of these, the Papillae of one kind form the seat of
the sense of taste; the other Papillae (more numerous and smaller) resemble
the extremities of the nerves in the common skin, and are the organs of touch
in the tongue. When I take a sharp steel point, and touch one of *these*
Papillae, I feel the sharpness. The sense of *touch* informs me of the shape of
the instrument. When I touch a Papilla of taste, I have no sensation similar
to the former. I do not know that a point touches the tongue, but I am sensible of a metallic taste, and the sensation passes backward on the tongue.

In the operation of couching the cataract, the pain of piercing the retina
with a needle is not so great as that which proceeds from a grain of sand under
the eyelid. And although the derangement of the stomach sometimes marks

the injury of an organ so delicate, yet the pain is occasioned by piercing the outward coat, not by the affection of the expanded nerve of vision.

If the sensation of light were conveyed to us by the retina, the organ of vision, in consequence of that organ being as much more sensible than the surface of the body as the impression of light is more delicate than that pressure which gives us the sense of touch; what would be the feelings of a man subjected to an operation in which a needle were pushed through the nerve. Life could not bear so great a pain.

But there is an occurrence during this operation on the eye, which will direct us to the truth: when the needle pierces the eye, the patient has the sensation of a spark of fire before the eye.

This fact is corroborated by experiments made on the eye. When the eyeball is pressed on the side, we perceive various coloured light. Indeed the mere effect of a blow on the head might inform us, that sensation depends on the exercise of the organ affected, not on the impression conveyed to the external organ; for by the vibration caused by the blow, the ears ring, and the eye flashes light, while there is neither light nor sound present.

It may be said, that there is here no proof of the sensation being in the brain more than in the external organ of sense. But when the nerve of a stump is touched, the pain is as if in the amputated extremity. If it be still said that this is no proper example of a peculiar sense existing without its external organ, I offer the following example: Quando penis glandem exedat ulcus, et nihil nisi granulatio maneat, ad extremam tamen nervi pudicae partem ubi terminatur sensus supersunt, et exquisitissima sensus gratificatio.

If light, pressure, galvanism, or electricity produce vision, we must conclude that the idea in the mind is the result of an action excited in the eye or in the brain, not of any thing received, though caused by an impression from without. The operations of the mind are confined not by the limited nature of things created, but by the limited number of our organs of sense. By induction we know that things exist which yet are not brought under the operation of the senses. When we have never known the operation of one of the organs of the five senses, we can never know the ideas pertaining to that sense; and what would be the effect on our minds, even constituted as they now are, with a superadded organ of sense, no man can distinctly imagine.

As we are parts of the creation, so God has bound us to the material world by this law of our nature, that it shall require excitement from without, and an operation produced by the action of things external to rouse our faculties: But that once brought into activity, the organs can be put in exercise by the mind, and be made to minister to the memory and imagination, and all the faculties of the soul.

I shall hereafter shew, that the operations of the mind are seated in the great mass of the cerebrum, while the parts of the brain to which the nerves of sense tend, strictly form the seat of the sensation, being the internal organs of sense. These organs are operated upon in two directions. They receive the impression from without, as from the eye and ear: and as their action influences the operations of the brain producing perception, so are they brought into action and suffer changes similar to that which they experience from external pressure by the operation of the will; or, as I am now treating of the subject anatomically by the operation of the great mass of the brain upon them.

In all regulated actions of the muscles we must acknowledge that they are influenced through the same nerves, by the same operation of the sensorium. Now the operations of the body are as nice and curious, and as perfectly regulated before Reason has sway, as they are at any time after, when the muscular frame might be supposed to be under the guidance of sense and reason. Instinctive motions are the operations of the same organs, the brain and nerves and muscles, which minister to reason and volition in our mature years. When the young of any animal turns to the nipple, directed by the sense of smelling, the same operations are performed, and through the same means, as afterwards when we make an effort to avoid what is noxious, or desire and move towards what is agreeable.

The operations of the brain may be said to be three-fold: 1. The frame of the body is endowed with the characters of life, and the vital parts held together as one system through the operation of the brain and nerves; and the secret operations of the vital organs suffer the controul of the brain, though we are unconscious of the thousand delicate operations which are every instant going on in the body. 2. In the second place, the instinctive motions which precede the developement of the intellectual faculties are performed through the brain and nerves. 3. In the last place, the operation of the senses in rouzing the faculties of the mind, and the exercise of the mind over the moving parts of the body, is through the brain and nerves. The first of these is perfect in nature, and independent of the mind. The second is a prescribed and limited operation of the instrument of thought and agency. The last begins by imperceptible degrees, and has no limit in extent and variety. It is that to which all the rest is subservient, the end being the calling into activity and the sustaining of an intellectual being.

Thus we see that in as far as is necessary to the great system, the operation of the brain, nerves, and muscles are perfect from the beginning; and we are naturally moved to ask, Might not the operations of the mind have been thus perfect and spontaneous from the beginning as well as slowly excited into action by outward impressions? Then man would have been an insulated being, not only cut off from the inanimate world around him, but from his fellows; he would have been an individual, not a part of a whole. That he may have a motive and a spring to action, and suffer pain and pleasure, and become an intelligent being, answerable for his actions,—sensation is made to result from external impression, and reason and passion to come from the experience of good and evil; first as they are in reference to his corporeal frame, and finally as they belong to the intellectual privations and enjoyments.

The brain is a mass of soft matter, in part of a white colour, and generally striated; in part of a grey or cineritious colour, having no fibrous appearance. It has grand divisions and subdivisions: and as the forms exist before the solid bone incloses the brain; and as the distinctions of parts are equally observable in animals whose brain is surrounded with fluid, they evidently are not accidental, but are a consequence of internal structure; or in other words they have a correspondence with distinctions in the uses of the parts of the brain.

On examining the grand divisions of the brain we are forced to admit that there are four brains. For the brain is divided longitudinally by a deep fissure; and the line of distinction can even be traced where the sides are united in substance. Whatever we observe on one side has a corresponding part on the other; and an exact resemblance and symmetry is preserved in all the lateral divisions of the brain. And so, if we take the proof of anatomy, we must admit that as the nerves are double, and the organs of sense double, so is the brain double; and every sensation conveyed to the brain is conveyed to the two lateral parts; and the operations performed must be done in both lateral portions at the same moment.

I speak of the lateral divisions of the brain being distinct brains combined in function, in order the more strongly to mark the distinction betwixt the anterior and posterior grand divisions. Betwixt the lateral parts there is a strict resemblance in form and substance: each principal part is united by transverse tracts of medullary matter; and there is every provision for their acting with perfect sympathy. On the contrary, the *cerebrum*, the anterior grand division, and the *cerebellum* the posterior grand division, have slight and indirect connection. In form and division of parts, and arrangement of white and grey matter, there is no resemblance. There is here nothing of that symmetry and correspondence of parts which is so remarkable betwixt the right and left portions.

I have found evidence that the vascular system of the cerebellum may be affected independently of the vessels of the cerebrum. I have seen the whole surface of the cerebellum studded with spots of extravasated blood as small as pin heads, so as to be quite red, while no mark of disease was upon the surface of the cerebrum. The action of vessels it is needless to say is under the influence of the parts to which they go; and in this we have a proof of a distinct state of activity in the cerebrum and cerebellum.

From these facts, were there no others, we are entitled to conclude, that in the operations excited in the brain there cannot be such sympathy or corresponding movement in the cerebrum and cerebellum as there is betwixt the lateral portions of the cerebrum; that the anterior and posterior grand divisions of the brain perform distinct offices.

In examining this subject further, we find, when we compare the relative magnitude of the cerebrum to the other parts of the brain in man and in brutes, that in the latter the cerebrum is much smaller, having nothing of the relative magnitude and importance which in man it bears to the other parts of the nervous system; signifying that the cerebrum is the seat of those qualities of mind which distinguish man. We may observe also that the posterior grand division, or *cerebellum* remains more permanent in form: while the cerebrum changes in conformity to the organs of sense, or the endowments of the different classes of animals. In the inferior animals, for example, where there are two external organs of the same sense, there is to be found two distinct corresponding portions of cerebrum, while the cerebellum corresponds with the frame of the body.

In thinking of this subject, it is natural to expect that we should be able to put the matter to proof by experiment. But how is this to be accomplished, since any experiment direct upon the brain itself must be difficult, if not impossible?—I took this view of the subject. The *medulla spinalis* has a central division, and also a distinction into anterior and posterior fasciculi, corresponding with the anterior and posterior portions of the brain. Further we can trace down the crura of the *cerebrum* into the anterior fasciculus of the spinal marrow, and the crura of the *cerebellum* into the posterior fasciculus. I thought that here I might have an opportunity of touching the *cerebellum*, as it were, through the posterior portion of the spinal marrow, and the cerebrum by the anterior portion. To this end I made experiments which, though they were not conclusive, encouraged me in the view I had taken.

I found that injury done to the anterior portion of the spinal marrow, convulsed the animal more certainly than injury done to the posterior portion; but I found it difficult to make the experiment without injuring both portions.

Next considering that the spinal nerves have a double root, and being of opinion that the properties of the nerves are derived from their connections with the parts of the brain, I thought that I had an opportunity of putting my opinion to the test of experiment, and of proving at the same time that nerves of different endowments were in the same cord, and held together by the same sheath.

On laying bare the roots of the spinal nerves, I found that I could cut across the posterior fasciculus of nerves, which took its origin from the posterior portion of the spinal marrow without convulsing the muscles of the back; but that on touching the anterior fasciculus with the point of the knife, the muscles of the back were immediately convulsed.

Such were my reasons for concluding that the cerebrum and the cerebellum were parts distinct in function, and that every nerve possessing a double function obtained that by having a double root. I now saw the meaning of the double connection of the nerves with the spinal marrow; and also the cause of that seeming intricacy in the connections of nerves throughout their course, which were not double at their origins.

The spinal nerves being double, and having their roots in the spinal marrow, of which a portion comes from the cerebrum and a portion from the cerebellum, they convey the attributes of both grand divisions of the brain to every part; and therefore the distribution of such nerves is simple, one nerve supplying its destined part. But the nerves which come directly from the brain, come from parts of the brain which vary in operation; and in order to bestow different qualities on the parts to which the nerves are distributed, two or more nerves must be united in their course or at their final destination. Hence it is that the 1st nerve must have branches of the 5th united with it: hence the *portio dura* of the 7th pervades every where the bones of the cranium to unite with the extended branches of the 5th: hence the union of the 3d and 5th in the orbit: hence the 9th and 5th are both sent to the tongue: hence it is, in short, that no part is sufficiently supplied by one single nerve, unless that nerve be a nerve of the spinal marrow, and have a double root, a connection (however remotely) with both the cerebrum and cerebellum. Such nerves as are single in their origin from the spinal marrow will be found either to unite in their course with some other nerve, or to be such as are acknowledged to be peculiar in their operation.

The 8th nerve is from the portion of the *medulla oblongata*[1] which belongs to the cerebellum: the 9th nerve comes from the portion which belongs to the cerebrum. The first is a nerve of the class called Vital nerves, controuling secretly the operation of the body; the last is the Motor nerve of the tongue, and is an instrument of volition. Now the connections formed by the 8th nerve in its course to the viscera are endless; it seems no where sufficient for the entire purpose of a nerve; for every where it is accompanied by others, and the 9th passes to the tongue, which is already profusely supplied by the 5th.

Understanding the origin of the nerves in the brain to be the source of their powers, we look upon the connections formed betwixt distant nerves, and upon the combination of nerves in their passage, with some interest; but without this the whole is an unmeaning tissue. Seeing the seeming irregularity in one subject, we say it is accident; but finding that the connections never vary, we say only that it is strange, until we come to understand the necessity of nerves being combined in order to bestow distinct qualities on the parts to which they are sent.

The *cerebellum* when compared with the *cerebrum* is simple in its form. It has no internal tubercles or masses of cineritious matter in it. The medullary matter comes down from the cineritious cortex, and forms the *crus;* and

---

[1] The medulla oblongata is only the commencement of the spinal marrow.

the *crus* runs into union with the same process from the cerebrum; and they together form the *medulla spinalis,* and are continued down into the spinal marrow; and these crura or processes afford double origin to the double nerves of the spine. The nerves proceeding from the Crus Cerebelli go every where (in seeming union with those from the Crus Cerebri); they unite the body together, and controul the actions of the bodily frame; and especially govern the operation of the viscera necessary to the continuance of life.

In all animals having a nervous system, the *cerebellum* is apparent, even though there be no *cerebrum.* The cerebrum is seen in such tribes of animals as have organs of sense, and it is seen to be near the eyes, or principal organ of sense; and sometimes it is quite separate from the *cerebellum.*

The cerebrum I consider as the grand organ by which the mind is united to the body. Into it all the nerves from the external organs of the senses enter; and from it all the nerves which are agents of the will pass out.

If this be not at once obvious, it proceeds only from the circumstance that the nerves take their origin from the different parts of the brain; and while those nerves are considered as simple cords, this circumstance stands opposed to the conclusion which otherways would be drawn. A nerve having several roots, implies that it propagates its sensation to the brain generally. But when we find that the several roots are distinct in their endowments, and are in respect to office distinct nerves; then the conclusion is unavoidable, that the portions of the brain are distinct organs of different functions.

To arrive at any understanding of the internal parts of the cerebrum, we must keep in view the relation of the nerves, and must class and distinguish the nerves, and follow them into its substance. If all ideas originate in the mind from external impulse, how can we better investigate the structure of the brain than by following the nerves, which are the means of communication betwixt the brain and the outward organs of the senses?

The nerves of sense, the olfactory, the optic, the auditory, and the gustatory nerve, are traced backwards into certain tubercles or convex bodies in the base of the brain. And I may say, that the nerves of sense either form tubercles before entering the brain, or they enter into those convexities in the base of the *cerebrum.* These convexities are the constituent parts of the cerebrum, and are in all animals necessary parts of the organs of sense: for as certainly as we discover an animal to have an external organ of sense, we find also a medullary tubercle; whilst the superiority of animals in intelligence is shewn by the greater magnitude of the hemispheres or upper part of the cerebrum.

The convex bodies which are seated in the lower part of the cerebrum, and into which the nerves of sense enter, have extensive connexion with the hemispheres on their upper part. From the medullary matter of the hemispheres, again, there pass down, converging to the crura, Striae, which is the medullary matter taking upon it the character of a nerve; for from the Crura Cerebri, or its prolongation in the anterior Fasciculi of the spinal marrow, go off the nerves of motion.

But with these nerves of motion which are passing outward there are nerves going inwards; nerves from the surfaces of the body; nerves of touch; and nerves of peculiar sensibility, having their seat in the body or viscera. It is not improbable that the tracts of cineritious matter which we observe in the course of the medullary matter of the brain, are the seat of such peculiar sensibilities; the organs of certain powers which seem resident in the body.

As we proceed further in the investigation of the function of the brain, the discussion becomes more hypothetical. But surely physiologists have been mistaken in supposing it necessary to prove sensibility in those parts of the brain which they are to suppose the seat of the intellectual operations. We are not to expect the same phenomena to result from the cutting or tearing of the brain as from the injury to the nerves. The function of the one is to transmit sensation; the other has a higher operation. The nature of the organs of sense is different; the sensibilities of the parts of the body are very various. If the needle piercing the retina during the operation of couching gives no remarkable pain, except in touching the common coats of the eye, ought we to imagine that the seat of the higher operations of the mind should, when injured, exhibit the same effects with the irritation of a nerve? So far therefore from thinking the parts of the brain which are insensible, to be parts inferior (as every part has its use), I should even from this be led to imagine that they had a higher office. And if there be certain parts of the brain which are insensible, and other parts which being injured shake the animal with convulsions exhibiting phenomena similar to those of a wounded nerve, it seems to follow that the latter parts which are endowed with sensibility like the nerves are similar to them in function and use; while the parts of the brain which possess no such sensibility are different in function and organization from the nerves, and have a distinct and higher operation to perform.

If in examining the apparent structure of the brain, we find a part consisting of white medullar Striae and fasciculated like a nerve, we should conclude that as the use of a nerve is to transmit sensation, not to perform any more peculiar function, such tracts of matter are media of communication, connecting the parts of the brain; rather than the brain itself performing the more peculiar functions. On the other hand, if masses are found in the brain unlike the matter of the nerve, and which yet occupy a place guarded as an organ of importance, we may presume that such parts have a use different from that of merely conveying sensation; we may rather look upon such parts as the seat of the higher powers.

Again, if those parts of the brain which are directly connected with the nerves, and which resemble them in structure, give pain when injured, and occasion convulsion to the animal as the nerves do when they are injured; and if on the contrary such parts as are more remote from the nerves, and of a different structure, produce no such effect when injured, we may conclude, that the office of the latter parts is more allied to the intellectual operations, less to mere sensation.

I have found at different times all the internal parts of the brain diseased without loss of sense; but I have never seen disease general on the surfaces of the hemispheres without derangement or oppression of the mind during the patient's life. In the case of derangement of mind, falling into lethargy and stupidity, I have constantly found the surface of the hemispheres dry and preternaturally firm, the membrane separating from it with unusual facility.

If I be correct in this view of the subject, then the experiments which have been made upon the brain tend to confirm the conclusions which I should be inclined to draw from strict anatomy; viz. that the cineritious and superficial parts of the brain are the seat of the intellectual functions. For it is found that the surface of the brain is totally insensible, but that the deep and medullary part being wounded the animal is convulsed and pained.

At first it is difficult to comprehend, how the part to which every sensation is referred, and by means of which we become acquainted with the various sensations, can itself be insensible; but the consideration of the wide difference of function betwixt a part destined to receive impressions, and a part which is the seat of intellect, reconciles us to the phenomenon. It would be rather strange to find, that there were no distinction exhibited in experiments on parts evidently so different in function as the organs of the senses, the nerves, and the brain. Whether there be a difference in the matter of the nervous system, or a distinction in organization, is of little importance to our enquiries, when it is proved that their essential properties are different, though their union and co-operation be necessary to the completion of their function—the developement of the faculties by impulse from external matter.

All ideas originate in the brain: the operation producing them is the remote effect of an agitation or impression on the extremities of the nerves of sense; directly they are consequences of a change or operation in the proper organ of the sense which constitutes a part of the brain, and over these organs, once brought into action by external impulse, the mind has influence. It is provided, that the extremities of the nerves of the senses shall be susceptible each of certain qualities in matter; and betwixt the impression of the outward sense, as it may be called, and the exercise of the internal organ, there is established a connection by which the ideas excited have a permanent correspondence with the qualities of bodies which surround us.

From the cineritious matter, which is chiefly external, and forming the surface of the cerebrum; and from the grand center of medullary matter of the cerebrum, what are called the *crura* descend. These are fasciculated processes

of the cerebrum, from which go off the nerves of motion, the nerves governing the muscular frame. Through the nerves of sense, the *sensorium* receives
impressions, but the will is expressed through the medium of the nerves of
motion. The secret operations of the bodily frame, and the connections which
unite the parts of the body into a system, are through the cerebellum and
nerves proceeding from it.

SPINAL NERVE ROOTS

MAGENDIE, François (French experimental surgeon, 1783–1855). *Expériences sur
les fonctions des racines des nerfs qui naissent de la moelle épinière*, in *Journal de Physiologie Expérimentale et Pathologique*, vol. 4, 1822; tr. by A. Walker as *Experiments
upon the functions of the roots of the nerves which arise from the spinal marrow* in his
Documents and dates of modern discoveries in the nervous system, London, 1839.

For the question of precedence on assigning separate functions to the dorsal and ventral roots of the spinal nerves, see Bell (pp. 294–305) and Walker. It is sometimess said
that Magendie's chief contribution to science was the body of experimental techniques,
criticized at the time for their drastic treatment of the living organism, which he transmitted to his more profoundly intellectual pupil, Bernard (pp. 216, 241, 518).

"Next considering that the spinal nerves have a double root, and being of
opinion that the properties of the nerves are derived from their connections
with the parts of the brain, I thought that I had an opportunity of putting my
opinion to the test of experiment, and of proving at the same time that nerves
of different endowments were in the same cord and held by the same sheath.

"On laying bare the roots of the spinal nerves, I found that I could cut
across the fasciculus of nerves, which took its origin from the posterior portion
of the spinal marrow, without convulsing the muscles of the back; but that
on touching the anterior fasciculus with the point of a knife, the muscles of the
back were immediately convulsed."

It is seen by this citation of a work which I could not know, since it had
not been published, that *Mr. Bell, conducted by his ingenious ideas on the
nervous system, was very near discovering the functions of the spinal roots;*
AT THE SAME TIME, THE FACT THAT THE ANTERIOR ARE DESTINED TO MO-
TION, WHILST THE POSTERIOR BELONG MORE ESPECIALLY TO SENSATION, AP-
PEARS TO HAVE ESCAPED HIM: IT IS THEN TO THE ESTABLISHMENT OF THIS
FACT IN A POSITIVE MANNER THAT I MUST LIMIT MY PRETENSIONS. . . .

The facts which I announced in the preceding number are too important
to be passed over without my seeking to throw light upon them by new researches.

I at first wished to ascertain if it might not be possible to cut the anterior
and posterior roots of the spinal nerves without opening the great canal of the
vertebral dura-mater; because, by exposing the spinal marrow to the air and to

a cold temperature, the nervous action is sensibly weakened, and consequently the results sought for are obtained in a manner but little apparent.

The anatomical position of the parts did not render the thing impossible; for each bundle of spinal roots goes for some time in a particular canal before uniting and confounding itself with the other bundle. I found indeed that with the help of scissors blunt at the points, a sufficient quantity of the plates and lateral parts of the vertebrae may be taken away to expose the ganglion of each lumbar pair; and then with a small stylet there is not much difficulty in separating the canal which contains the posterior roots, and the section becomes easy. This mode of making the experiment gave me the same results as those I had previously observed; but as the experiment is much longer and more laborious than the preceding one in which the great canal of the spinal dura-mater is opened, I do not think that this mode of making the experiment should be followed in preference to the first.

I afterwards wished to submit to more particular proof the results of which I have previously spoken. Every one knows that nux vomica determines both in man and animals, general and very violent tetanic convulsions. I was curious to ascertain if these convulsions would still take place in a member in which the nerves of motion had been cut, and if they would appear to be as strong as usual, a section of the nerves of sensation having been made. The result accorded entirely with the preceding; that is to say in an animal in which the posterior roots were cut, the tetanus was complete and as intense as if the spinal nerves had been untouched: on the contrary, in an animal in which I had cut the nerves of motion of one of the posterior members, the members remained supple and immovable at the time when, under the influence of the poison, all the other muscles of the body suffered the most violent tetanic convulsions.

On directly irritating the nerves of sensation, or the posterior spinal roots, would contractions be produced? Would a direct irritation of the nerves of motion excite pain? These were the questions which I asked myself and which experience alone could resolve.

With this view, I began to examine THE POSTERIOR ROOTS or the nerves of sensation. The following are the results of my observations: *in pinching, pulling, pricking these roots, the animal gives signs of pain;* BUT IT IS NOT TO BE COMPARED IN INTENSITY WITH THAT WHICH OCCURS IF THE SPINAL MARROW BE ONLY SLIGHTLY TOUCHED AT THE PART WHERE THESE ROOTS ARISE. *Nearly every time that these posterior roots are thus excited, contractions are produced in the muscles to which the nerves are distributed;* THESE CONTRACTIONS ARE HOWEVER BUT SLIGHTLY MARKED, AND IN-FINITELY WEAKER THAN IF THE SPINAL MARROW ITSELF BE TOUCHED. IF ONE OF THE POSTERIOR BUNDLES OF ROOTS BE CUT AT ONCE, A GENERAL MOVEMENT IS PRODUCED IN THE MEMBER TO WHICH THE BUNDLE GOES.

I have repeated the same experiments upon THE ANTERIOR BUNDLES, and I have obtained analogous results, but in an inverse sense; for THE CONTRAC-TIONS *excited by the pinching, pricking,* &c. *are* EXTREMELY STRONG AND EVEN CONVULSIVE, *whilst the* SIGNS OF SENSIBILITY *are* SCARCELY VISIBLE. These facts then are confirmative of those already announced; only they seem to establish that SENSATION DOES NOT BELONG EXCLUSIVELY TO THE POS-TERIOR ROOTS, ANY MORE THAN MOTION TO THE ANTERIOR.

Nevertheless a difficulty might arise. *When, in the preceding experiments, the roots were cut, they were continuous with the spinal marrow: might not the disturbance communicated to the latter have been the real origin, either of the contractions or of the pain felt by the animals? To remove this doubt,* I REPEATED THE EXPERIMENTS, AFTER HAVING SEPARATED THE ROOTS FROM THE SPINAL MARROW; AND I OUGHT TO SAY THAT, WITH THE EX-CEPTION OF TWO ANIMALS IN WHICH I SAW CONTRACTIONS UPON PINCH-ING AND PULLING THE ANTERIOR AND POSTERIOR BUNDLES, IN ALL THE REST I DID NOT OBSERVE ANY SENSIBLE EFFECT FROM THE IRRITATION OF THE ANTERIOR OR POSTERIOR ROOTS THUS SEPARATED FROM THE SPINAL MARROW.

I had still to make *another kind of experiment on the spinal roots; that of galvanism.* By its means, I accordingly excited these parts, first leaving them in their ordinary state, and afterwards cutting them at their spinal extremities to place them upon an isolating body. In these various cases, I OBTAINED CONTRACTIONS FROM EACH SORT OF ROOTS; but those which followed the excitation of the anterior roots were in general much stronger and more com-plete than those which took place when the electric current operated upon the posterior. The same phenomena took place either by applying the zinc or copper pole to the nerve.

It nows remains for me to give an account of my researches to endeavour to follow motion and sensation distinctly beyond the roots of the nerves, that is to say, into the spinal marrow; this is the subject of my present occupation.

Before finishing this article, I ought to give some further explanations as to the novelty of the results which I have announced.

When I wrote the note contained in the preceding number, I believed I was the first who had thought of dividing the roots of the spinal nerves; but I was soon undeceived by a small work by Mr. Shaw, which this young and laborious practitioner had the politeness to send me as soon as he had received the number of my journal. It is said in that work that Mr. Charles Bell made this section *thirteen*[1] *years ago,* and that he had discovered that the section of

---

[1] Thirteen years before 1822, the date of this Paper, would make the date of printing the "Idea of a NEW ANATOMY OF THE BRAIN," 1809—*just as early* as Mr. Walker's "NEW ANATOMY AND PHYSIOLOGY OF THE BRAIN ["]! yet Sir C. Bell now states that his NEW ANATOMY was really two years later than Mr. Walker's NEW ANATOMY! The coincidence in the two Titles is certainly strange.

the posterior roots did not prevent the continuance of motion. Mr. Shaw
adds that Mr. Charles Bell had stated this result in a small pamphlet printed
solely for the use of his friends, but not for publication. I immediately asked
Mr. Shaw to have the kindness to send me if possible the pamphlet of Mr.
Charles Bell, in order that I might render him all the justice that was his due.
A few days afterwards I received it from Mr. Shaw.

This phamphlet is entitled:

Idea of a NEW ANATOMY OF THE BRAIN, *submitted for the Observations
of his Friends, by* CHARLES BELL, F.R.S.E. It is very curious, inasmuch as
there is to be found in it the germ of the recent discoveries of the author in
the nervous system. At page 22, the passage indicated by Shaw is to be found:
I shall transcribe the whole of it.

SPECIFIC NERVE ENERGIES

MÜLLER, Johannes Peter (German physiologist and pedagogue, 1801–1858). From
Handbuch der physiologie des menschen für vorlesungen, Coblenz, 1834; tr. by
W. Baly as *Elements of physiology*, London, 1827.

Although speculation on the nature of sensation is as old as objective science,* Müller's investigations are usually regarded as beginning modern sensory physiology. Previously it was believed that each different environmental stimulus affected sensory nerves
in a manner peculiar to that stimulus. Müller notices, however, that a given nerve tends
always to produce the same type of sensation regardless of the initiating stimulus. He
thus postulates that sensory distinctions result from physiological peculiarities of the different kinds of sensory nerves.

In addition to possessing scientific significance, Müller's ideas were of central importance in the history of metaphysical thought. Scientists had realized since the time of
Heraclitus that appearances were deceptive but ordinarily hoped to be able to correct
their sensory illusions through experiment and so obtain an idea of the "reality" behind
common experience. By emphasizing the limiting influence of our sensory equipment in
our attempt to frame ideas concerning "real" things, Müller's theories led to the concept
of science as conventional interpretation which was to be developed by Vaihinger† and
others.

OF THE PECULIAR PROPERTIES OF INDIVIDUAL NERVES

Of the Nerves of Special Sense.

The nerves have always been regarded as conductors, through the medium
of which we are made conscious of external impressions. Thus the nerves of
the senses have been looked upon as mere passive conductors, through which
the impressions made by the properties of bodies were supposed to be transmitted unchanged to the sensorium. More recently, physiologists have begun

---

* See G. M. Stratten, *Theophrastus and the Greek physiological psychology before
Aristotle*, London, 1911.

† In *Die philosophie des als ob*, Berlin, 1911.

to analyse these opinions. If the nerves are mere passive conductors of the impressions of light, sonorous vibrations, and odours, how does it happen that the nerve which perceives odours is sensible to this kind of impressions only, and to no others, while by another nerve odours are not perceived; that the nerve which is sensible to the matter of light, or the luminous oscillations, is insensible to the vibrations of sonorous bodies; that the auditory nerve is not sensible to light, nor the nerve of taste to odours; while, to the common sensitive nerve, the vibrations of bodies give the sensation, not of sound, but merely of tremours? These considerations have induced physiologists to ascribe to the individual nerves of the senses a special sensibility to certain impressions, by which they are supposed to be rendered conductors of certain qualities of bodies, and not of others.

This last theory, of which ten or twenty years since no one doubted the correctness, on being subjected to a comparison with facts, was found unsatisfactory. For the same stimulus, for example, electricity, may act simultaneously on all the organs of sense,—all are sensible to its action; but the nerve of each sense is affected in a different way,—becomes the seat of a different sensation: in one, the sensation of light is produced; in another, that of sound; in a third, taste; while, in a fourth, pain and the sensation of a shock are felt. Mechanical irritation excites in one nerve a luminous spectrum; in another, a humming sound; in a third, pain. An increase of the stimulus of the blood causes in one organ spontaneous sensations of light; in another, sound; in a third, itching, pain, &c. A consideration of such facts could not but lead to the inference that the special susceptibility of nerves for certain impressions is not a satisfactory theory, and that the nerves of the senses are not mere passive conductors, but that each peculiar nerve of sense has special powers or qualities which the exciting causes merely render manifest.

Sensation, therefore, consists in the communication to the sensorium, not of the quality or state of the external body, but of the condition of the nerves themselves, excited by the external cause.—We do not feel the knife which gives us pain, but the painful state of our nerves produced by it. The probably mechanical oscillation of light is itself not luminous; even if it could itself act on the sensorium, it would be perceived merely as an oscillation; it is only by affecting the optic nerve that it gives rise to the sensation of light. Sound has no existence but in the excitement of a quality of the auditory nerve; the nerve of touch perceives the vibration of the apparently sonorous body as a sensation of tremour. We communicate, therefore, with the external world merely by virtue of the states which external influences excite in our nerves.

By the knowledge of the fact just announced, we are led not only to recognise the peculiar qualities of the different nerves of sensation, in addition to their general distinction from the motor nerves; but we are also enabled to banish for ever from the doctrines of physiology a number of erroneous notions regarding the supposed power of the nerves to perform the functions of each other. It has long been known that blind persons cannot recognise colours with their fingers, *as colours:* but we perceive now why it is impossible for them to do so. However acute the sense of touch in the finger of the blind may be rendered by practice, it can still be but the one sense proper to the nerves of the fingers,—*touch.* . . .

Sensation consists in the sensorium receiving through the medium of the nerves, and as the result of the action of an external cause, a knowledge of certain qualities or conditions, not of external bodies, but of the nerves of sense themselves; and these qualities of the nerves of sense are in all different, the nerve of each sense having its own peculiar quality or energy.

The special susceptibility of the different nerves of sense for certain influences,—as of the optic nerve for light, of the auditory nerve for vibrations, and so on,—was formerly attributed to these nerves having each a specific irritability. But this hypothesis is evidently insufficient to explain all the facts. The nerves of the senses have assuredly a specific irritability for certain influences; for many stimuli, which exert a violent action upon one organ of sense, have little or no effect upon another: for example, light, or vibrations so infinitely rapid as those of light, act only on the nerves of vision and common sensation; slower vibrations, on the nerves of hearing and common sensation, but not upon those of vision; odorous substances only upon the olfactory nerves. The external stimuli must therefore be adapted to the organ of sense —must be "homogeneous": thus light is the stimulus adapted to the nerve of vision; while vibrations of less rapidity, which act upon the auditory nerve, are not adapted to the optic nerve, or are indifferent to it; for, if the eye be touched with a tuning-fork while vibrating, a sensation of tremours is excited in the conjunctiva, but no sensation of light. We have seen, however, that one and the same stimulus, as electricity, will produce different sensations in the different nerves of the senses; all the nerves are susceptible of its action, but the sensations in all are different. The same is the case with other stimuli, as chemical and mechanical influences. The hypothesis of a specific irritability of the nerves of the senses for certain stimuli, is therefore insufficient; and we are compelled to ascribe, with Aristotle, peculiar energies to each nerve,— energies which are vital qualities of the nerve, just as contractility is the vital property of muscle. The truth of this has been rendered more and more evident in recent times by the investigation of the so-called "subjective" phenomena of the senses by Elliot, Darwin, Ritter, Goethe, Purkinje, and Hjort. Those phenomena of the senses, namely, are now styled "subjective," which are produced, not by the usual stimulus adapted to the particular nerve of sense, but by others which do not usually act upon it. These important phenomena were long spoken of as "illusions of the senses," and have been regarded in an erroneous point of view, while they are really true actions of the senses, and must be studied as fundamental phenomena in investigations into their nature.

The sensation of sound, therefore, is the peculiar "energy" or "quality" of the auditory nerve; the sensation of light and colours that of the optic nerve; and so of the other nerves of sense. An exact analysis of what takes place in the production of a sensation would of itself have led to this conclusion. The sensations of heat and cold, for example, make us acquainted with the existence of the imponderable matter of caloric, or of peculiar vibrations in the vicinity of our nerves of feeling. But the nature of this caloric cannot be elucidated by sensation, which is in reality merely a particular state of our nerves; it must be learnt by the study of the physical properties of this agent, namely, of the laws of its radiation, its development from the latent state; its property of combining with and producing expansion of other bodies, &c. All this again, however, does not explain the peculiarity of the sensation of warmth as a condition of the nerves. The simple fact devoid of the theory is this, that warmth, as a sensation, is produced whenever the matter of caloric acts upon the nerves of feeling; and that cold, as a sensation, results from this matter of caloric being abstracted from a nerve of feeling.

So, also, the sensation of sound is produced when a certain number of impulses or vibrations are imparted, within a certain time, to the auditory nerve: but sound, as we perceive it, is a very different thing from a succession of vibrations. The vibrations of a tuning-fork, which to the ear give the impression of sound, produce in a nerve of feeling or touch the sensation of tickling; something besides the vibrations must consequently be necessary for the production of the sensation of sound, and that something is possessed by the auditory nerve alone. Vision is to be regarded in the same manner. A difference in the intensity of the action of the imponderable agent, light, causes an inequality of sensation at different parts of the retina: whether this action consists in impulses or undulations, (the undulation theory,) or in an infinitely rapid current of imponderable matter, (the emanation theory,) is a question here of no importance. The sensation of moderate light is produced where the action of the imponderable agent on the retina is not intense; of bright light where its action is stronger, and of darkness or shade where the imponderable agent does not fall; and thus results a luminous image of determinate form according to the distribution of the parts of the retina differently acted on. Colour is also a property of the optic nerve; and, when excited by external light, arises from the peculiarity of the so-called coloured rays, or of the oscillations necessary for the production of the impression of colour,—a peculiarity, the nature of which is not at present known. The nerves of taste and smell are capable of being excited to an infinite variety of sensations by external causes; but each taste is due to a determinate condition of the nerve excited by the ex-

ternal cause; and it is ridiculous to say that the property of acidity is communicated to the sensorium by the nerve of taste, while the acid acts equally
upon the nerves of feeling, though it excites there no sensation of taste.

The essential nature of these conditions of the nerves, by virtue of which
they see light and hear sound,—the essential nature of sound as a property
of the auditory nerve, and of light as a property of the optic nerve, of taste, of
smell, and of feeling,—remains, like the ultimate causes of natural phenomena
generally, a problem incapable of solution.   Respecting the nature of the
sensation of the color "blue," for example, we can reason no farther; it is
one of the many facts which mark the limits of our powers of mind.   It
would not advance the question to suppose, the peculiar sensations of the different senses excited by one and the same cause, to result from the propagation
of vibrations of the nervous principle of different rapidity to the sensorium.
Such an hypothesis, if at all tenable, would find its first application in accounting for the different sensations of which a single sense is susceptible; for example, in explaining how the sensorium receives the different impressions of
blue, red, and yellow, or of an acute and a grave tone, or of painful and
pleasurable sensations, or of the sensations of heat and cold, or of the tastes of
bitter, sweet, and acid.   It is only with this application that the hypothesis is
worthy of regard; tones of different degrees of acuteness are certainly produced by vibrations of sonorous bodies of different degrees of rapidity; and
a slight contact of a solid body, which singly excites in a nerve of common
sensation merely the simple sensation of touch, produces in the same nerve
when repeated rapidly, as the vibrations of a sonorous body, the feeling of
tickling; so that possibly a pleasurable sensation, even when it arises from
internal causes independently of external influences, is due to the rapidity of
the vibrations of the nervous principle on the nerves of feeling.

It was perhaps from an obscure acquaintance with the phenomena of the
sensation of light from internal causes, that even the older philosophers derived their imperfect idea of the essential part which the eye itself plays in the
sensations of light and colour.   Such an idea can evidently be traced in
Plato's doctrine of vision in the Timaeus.[1] . . .

Aristotle's treatise on dreams contains views in themselves more correct,
and stated in a more scientific form.   His explanation of spectral appearances
as the result of internal actions of the sense of vision, is quite on a level with the
present state of science.   He adduces indeed the observation since made by
Spinoza, that images seen during sleep can still be perceived in the organs of
vision after waking; and the varying colours of the ocular spectra produced by
gazing at the sun were well known to him.

In the present more perfect state of the different branches of natural science, which are studied separately, and in part independently of each other, it
still remains a task, well deserving the labour it would cost, to test the theories

---

[1] Timaeus, 45–47.

of fundamental phenomena, more especially of those which interest different sciences, such as the actions of light upon organic beings. But this would be a task of extreme difficulty, requiring for its proper performance a critical examination of the various facts.

During recent years, philosophy has done little in this field of inquiry. The manifestation of different objects to each other cannot express the nature of light; that it renders objects visible to us depends merely on our having an organ of vision with vital properties. And in this way many other agents have the same power of rendering objects manifest; were we endowed with as delicate an organic re-agent for electricity as for light, electricity would have the same influence as light in rendering manifest the corporeal world.

From the foregoing considerations we have learnt most clearly that the nerves of the senses are not mere conductors of the properties of bodies to our sensorium, and that we are made acquainted with external objects merely by virtue of certain properties of our nerves, and of their faculty of being affected in a greater or less degree by external bodies. Even the sensation of touch in our hands makes us acquainted, not absolutely with the state of the surfaces of the body touched, but with changes produced in the parts of our body affected by the act of touch. By imagination and reason a mere sensation is interpreted as something quite different.

The accuracy of our discrimination by means of the senses depends on the different manner in which the conditions of our nerves are affected by different bodies; but the preceding considerations show us the impossibility that our senses can ever reveal to us the true nature and essence of the material world. In our intercourse with external nature it is always our own sensations that we become acquainted with, and from them we form conceptions of the properties of external objects, which may be relatively correct; but we can never submit the nature of the objects themselves to that immediate perception to which the states of the different parts of our own body are subjected in the sensorium.

THE SPEED OF THE NERVE IMPULSE

von HELMHOLTZ, Hermann (German physicist and physiologist, 1821–1894). From *Messungen ueber den zeitlichen verlauf der zuckung animalischer muskeln und die fortpflanzungsgeschwindigkeit der reizung in den nerven*, in *Archiv fuer anatomie, physiologie, und wissenschaftliche medicin herausgegeben von Dr. Johannes Mueller*, p. 276, 1850; and from *Messungen ueber fortpflanzungsgeschwindigkeit der reizen in den nerven*, *ibid.*, p. 199, 1852; both tr. by M. and V. Hamburger and T. S. Hall for this volume.

The electrical sign of nervous conduction had been described by Du Bois-Reymond.* Here Helmholtz measures its speed of propagation. He also made basic major contributions in electricity, mechanics, theory of sensation; and his famous essay *Die Erhaltung der Kraft* proved basic to subsequent developments in energy theory.

* *Gesammelte Abhandlungen*, 1877, Bd. ii. S. 319.

The duration of the twitch of an animal muscle is ordinarily only a small fraction of a second, except for a longer lasting, weak after-effect. Since our senses are not capable of immediate perception of single time elements of such short duration, we must use more artificial methods to observe and measure them. Two of these especially are to be considered here. In the first, the events whose time intervals one wishes to find out are recorded by a suitable mechanism on a surface which moves with even speed. The time intervals appear on it as proportional space differences and can be measured by the latter. Ludwig has already used this method for physiological purposes in order to show the fluctuations of blood pressure in the arteries and of atmospheric pressure in the pleural cavity. The other, essentially different, method of measuring time is the one proposed by Pouillet. The duration is here measured by the effect which a force of known intensity has produced during this interval. Pouillet has a galvanic current act on a resting magnet. The beginning and end of the current correspond exactly to the beginning and end of the interval to be measured; the magnitude of arc of the excursions which the magnet performs is, then, proportional to the duration to be measured. . . .

The foundation of Pouillet's method for measuring small time intervals is as follows: the time during which a galvanic current of known intensity from a coil has affected a magnet can be calculated exactly from its changed movement. Up to the present, one cannot anticipate a lower limit of time divisions measurable in this way, since one can increase at will the intensity of the acting current and the magnitude of its effect on the magnet by increasing the electromotor cells and the windings on the coil. But a limitation is imposed in the application of this procedure; namely, one must know how to cause the beginning and end of the supposed current, which from now on we shall call the time-measuring one, to coincide exactly with the beginning and end of the mechanical process the duration of which is to be measured. In the experiments to be described here the time-measuring current started at the moment when an instantaneous electric shock passed through the muscle or its nerve, and stopped when the circuit within which it circulated was interrupted by the contraction of the muscle. At the same time one could determine exactly the tension which the muscle had to develop in order to be able to separate the conductive metals from each other. The duration of the time-measuring current to be calculated is therefore identical with the time which elapses between the stimulation of the muscle, or of its nerve, and the moment at which its tension reaches a certain magnitude. . . .

[There follows a detailed description of an apparatus in which the nerve-muscle preparation is suspended and weighed down by a weight, with slight stretch being accomplished by overweight. Electrodes for stimulation by either galvanic or induction current could be placed at any point on the nerve. The

time-measuring current is interrupted when the tension of the muscle, following stimulation, exceeds the sum of weight and overweight. This is measurable.—Ed.]

In making measurements of the time which elapses between the stimulation
of the nerve and the lifting of the overweight by the muscle, one finds that
the time depends upon the point on the nerve at which one applies the electrical shock; the time is the longer, the longer the portion of the nerve between the stimulated point and the muscle. The experiment . . . can be
repeated any number of times, by placing two of the four conducting wires,
about two to three lines apart, on the nerve close to where it enters the
muscle, and the two others, just as far apart, on the pelvic part of the nerve.
I found it to be of advantage to move this second place not quite to the tran-
sected end of the sciatic plexus, but approximately to the place where the
strands of this plexus combine to form the trunk of the sciatic nerve, because
the extreme cut ends become inefficient relatively fast. Depending on whether
one connects the first or the second pair of the leads with the induction coil,
either the nerve point closer to the muscle, or the more distant one, will be
affected by the current. Comparative measurements, which incidentally are
carried out like those previously discussed, prove that the deflections of the
magnet by the time-measuring current are on the average from 5–7 dial
parts larger when the more distant point of the nerve is stimulated than the one
closer to the muscle.

Apparently this difference cannot be caused by any of the formerly discussed sources of error, which are based on the mechanical and electrical
occurrences in our measuring procedure, because all of these affect the experiments involving stimulation of the distant or near nerve point equally.
Rather occurrences inside of the nerve itself must be the cause. . . .

We must . . . make sure that the intensity of stimulation is the same at
both places. If this is so, then experiments show that whatever places on the
nerve are stimulated, corresponding energy stages will follow each other at
like time intervals, but the time between each of these energy stages (and the
stimulation) is larger by a definite amount, as the stimulated spot is further
away from the muscle. Therefore, if we express by curves the rise and fall
of energy for two different nerve points, then the curve corresponding to the
stimulation of the more distant point is congruent with the other, but between
its starting point and the point corresponding to the moment of stimulation,
there lies a larger part of the abscissa. From the nature of the time lapse
which the muscle exhibits following stimulation we can draw conclusions concerning the course of the corresponding processes in the nerve which are
mostly still unknown. . . . Now, since duration and strength of the stimulating electric current are exactly the same in both stimulated places the re-
tardation of the effect must be due to the fact that a certain time elapses until

it has spread from the more distant spot to the muscle. These experiments, therefore, enable us to find out the rate of propagation of the impulse[1] in the motor nerves of the frog, provided that we understand by impulse[2] those processes in the nerve, which develop in it as a result of an external *stimulus.*[2]

As long as the physiologists thought that nerve action could be ascribed to the propagation of an imponderable or psychic principle, it would have appeared incredible that the speed of this current should be measurable within the short distances of the animal body. At present we know from the investigations on the electromotor properties of nerves by Du Bois-Reymond, that the activity by which the conduction of an impulse is mediated is at least closely associated with, perhaps even essentially caused by, a changed arrangement of their material molecules. Accordingly, the conduction in the nerve would belong to the group of propagated molecular effects of ponderable bodies, to which, *e.g.*, belongs sound conduction in air and in elastic substances or the discharge of a tube filled with an explosive mixture. Under these circumstances, it is no longer surprising to see that the rate of conduction is not only measurable but as we shall see, even very moderate. Incidentally, the impossibility of observing time intervals of this kind in the daily perceptions of our own body, or in physiological experiments on muscle twitches must not surprise us, since the intervals which we may be sure that we observe between sensations involving the nerve fibers of our different sense organs are not much smaller than a second. One will recall that the most experienced astronomers differ by a full second in the comparative observation of visual and acoustic perceptions. . . .

[There follows a description of the measures taken to eliminate differences in the degree of contraction when the near vs. the far end of the nerve is stimulated. It is reported that the far end of the nerve is the first to show a decrease of sensitivity following repeated stimulation.—Ed.]

From the greater number of my experimental series, all of which gave the same result, with more or less exactness, I shall present herewith those which seem to be the most reliable on account of their extent or the correspondence of their single observations. For stimulation, we have invariably used currents which brought about maximum excitation. This was controlled by simultaneously observed elevations expressed in millimeters.

The series are arranged according to different plans. In some of them, all observations are made with the same or only two different overweights, in order to get as extensive figures as possible for the calculation of the essential time interval. For these I have calculated the means of the time-lapse be-

---

[1] Helmholtz uses "Reizung" three times here, the distinction between stimulus and impulse being recognized in fact but not in words.

[2] See footnote 1.

tween stimulation and muscle reaction for both points on the nerve, the difference between these means which corresponds to the rate of conduction in the nerve, and finally, in order to evaluate their exactness, the probable errors of all these values according to the rules of probability.

In other experimental series, the overweights have been exchanged as often as possible in order to prove that the delay is the same, for different degrees of muscle energy, provided one stimulates from the more distant point on the nerve, but the form of the energy increase is not altered. Obviously, the few experiments made with each overweight cannot furnish such exact values for differences due to nerve conduction as would longer series; therefore the individual means for these differences often vary considerably. However, the larger and smaller values are distributed entirely irregularly, and those for different overweights do not differ more from each other than those for the same overweight in successive observations. It follows that the magnitude of the difference does not depend noticeably on the amount of overweight, as is so definitely the case when the deflections of the magnet increase by decrease of the stimulation.

Finally, the rate of propagation of the nerve impulse was calculated after each experimental series. To do this, one must know the length of the traversed nerve piece, that is, the distance between the terminals at the two stimulated nerve places closest to the muscle. Unfortunately this length is very uncertain on account of the great extensibility of the nerve. If the nerve is not stretched, its fibers are bent in an undulating fashion; in order to measure its length I have always stretched it to such an extent that the transverse satin-like striations of its surface disappeared, on the assumption that the fibers would then run approximately straight. But a few millimeters are then always left to one's own discretion. Incidentally, it would not yet pay to devise an improved measuring technique since the inaccuracies of the time measurements are relatively much greater than those of the length measurements. Therefore it is not surprising that the established values of the rate of conduction still differ considerably from each other. . . .

[The following is part of one of the twenty experimental series reported in the paper—Ed.]

SERIES X.

Done on December 29 with the muscles of a frog kept for four months. Through the more distant point on the nerve is sent a stronger current, generated with the coils touching each other, and through the nearer point a weaker current with a distance between the coils of 2½ cm. After each two observations the muscle is reset.

A. Right muscle—nerve length 40 mm., deflection before 116.09, after 112.45, mean 114.27.

MEASUREMENTS OF THE TIME-LAPSE OF THE TWITCH OF ANIMAL * MUSCLES AND OF THE RATE OF PROPAGATION OF THE NERVE IMPULSE

Number	Overweight	Lift	Difference of deflection on stimulation of	
			further	nearer
			nervepoint	
1	20 gr.	1.19	100.09	
2		1.22	96.15	
3		1.22		93.92
4		1.15		97.19
5		1.10	97.70	
6		1.10	104.33	
7		1.17		93.87
8		1.12		92.27
9		1.15	106.43	
10		1.15	101.74	
11		1.12		98.00
12		1.17		98.60
13		1.12	96.81	
14		1.10	103.99	
Mean			100.98	95.64
Probable error of the mean			±0.86	±0.66
The same of the single observation			±2.42	±1.61
Duration of the time in sec. from stimulation to lifting			0.02437	0.02307
Probable error of the same			±0.00020	±0.00016
Time difference due to propagation			0.00130 ± 0.00027	
Rate of propagation			30.8 ± 6.4†	

* *I.e.*, skeletal.

† *I.e.*, meters/sec.

The values found for the rate of propagation between 11 and 21 °C. are therefore:

a) from series IX, X, and XI.

$$24.6 \pm 2.0$$
$$30.8 \pm 6.4$$
$$32.0 \pm 9.7$$
$$31.4 \pm 7.1$$
$$38.4 \pm 10.6$$

From these one finds by the method of least squares as most probable mean: 26.4 [meters/sec.].

b) from series XII, XIII, and XIV.

$$29.1$$
$$25.1$$
$$\underline{26.9}$$

Mean 27.0

Finally, I summarize the results of the present investigations:

1) If animal (skeletal) muscle, or its nerve, is stimulated by a momentary

electric shock, a short time passes during which its elastic tension does not change noticeably; then it gradually rises to a maximum, and just as gradually falls again. The contraction of animal muscle differs from that which occurs in organic (visceral), nonrythmically reacting muscle, after a relatively short stimulation, only in that its single phases pass much more rapidly.

If two different points of a motor nerve are stimulated by a momentary stimulus and if the magnitude of the stimulation is the same for both, then the time-lapse of the subsequent muscle twitch, is also the same; however, if the more distant point on the nerve has been stimulated, all of the muscle twitch stages occur later by an equal amount. From this, we conclude that the conduction of the nerve impulse to the muscle requires a measurable time. . . . [In addition to Pouillet's method, the graphic method was used. The results of these experiments are reported in the second paper, of which the essential portion follows.—Ed.]

In the first series of my investigations on the time relations of muscle and nerve activity I have proved by the electromagnetic method of measuring time, that the mechanical reactions of the muscle, following a nerve stimulation, set in later if the excitation has to pass a longer portion of the nerve before getting to the muscle. The method mentioned offers, in fact, the best guarantee where safe execution of exact measurements is desired, but it has the great disadvantage of yielding the said result only after extensive and tedious series of experiments, which on account of their long duration require an especially favorable condition of the frog preparation. The other graphic method of measuring time, the application of which has been mentioned before, is essentially one in which the muscle during twitching records the magnitudes of its contractions on a moving surface; this promised a much simpler and easier demonstration of the rate of propagation in the nerves, and, since this seemed to me sufficiently important, I undertook to follow up the matter in this way, and I was perfectly successful.

The procedure of the experiments, I have already briefly indicated in the previous paper. A pen which is raised by the twitching muscle draws a curve on a surface moving with uniform speed, the vertical coordinates of the curve are proportionate to the contractions of the muscle, the horizontal ones proportionate to the time. As a starting point of this curve we shall fix that point which corresponds to the moment of stimulation of the muscle or its nerve. Now, if we arrange for two curves to be drawn in succession, and if we take care that at the moment of stimulation the pen occupies always exactly the same point on the surface, then both curves will have the same starting point, and from the congruence or noncongruence of their individual parts one can observe whether or not the different stages of the mechanical muscle response have occurred, in both instances, at the same or a later time after stimulation. . . .

If the animal parts are rather vigorous and fresh, then the shapes of the double curves are all alike, at whatever nerve spot one may start the stimulation. Then, each drawing consists of two curves of congruent shape which are shifted in a horizontal direction with respect to each other by a certain amount as in Fig. 11, such that the curve which has been drawn upon stimulation of the nearer nerve spot, is also nearer to the starting point of stimulation. The curve *adefg* corresponds to the stimulation of the nearer nerve point, *aδεφg* to the one of the more distant nerve point. . . .

When we look at the double curve Fig. 11 it is evident that both of the muscle twitches recorded have been entirely identical as to strength, duration, and course of the different stages of contraction except that the one has

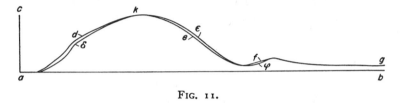

FIG. 11.

started later after stimulation than the other one. Now, since the arrangement of the apparatus and the mechanical forces of the muscle have been exactly the same, the delay of the reaction in one instance can only have been derived from the longer time of propagation in the nerve. . . .

It is the great advantage of the described method that one can recognize immediately in each single drawing from the shape of the two curves whether the muscle has worked uniformly in both instances, whereas this fact could be deduced by the electromagnetic method of time measurement only from a long series of single experiments. As to the absolute value of the rate of propagation, the horizontal distances of the two curves cannot be measured with great accuracy; nevertheless, the values of that rate are about the same as in the former method. For instance, the horizontal distance in Fig. 11 is about 1 mm, the circumference of the cylinder corresponding to $\frac{1}{6}$ sec, is 85.7 mm, therefore the length of the abscissae is 514.2 mm per second. One mm corresponds therefore to $1/514.2$ sec. The length of the nerve involved in propagation was 53 mm, from which follows a rate of propagation of 27.25 m per sec. The most probable value from previous experiments was 26.4 m.[3]

---

[3] The modern figures for rate in frog gastrocnemius and sciatic nerve is 30–43 meters/sec. for coarse fibers, 16 meters/sec. for thinnest fibers.

ON THE INSTINCTS OF INSECTS

FABRE, Jean Henri Casimir (French entomologist, teacher, and writer, 1823–1915). From *Hautes théories* (= chap. ix in) *Souvenirs entomologiques Études sur l'instinct et les moeurs des insectes*, Paris, 1879; tr. by A. Teixiera de Mattos as *Advanced theories* (= chap. vii in) *The hunting wasps*, New York, 1916; by arrangement with Dodd, Mead & Company, Inc., publisher.

In 1879, Fabre published the first of 11 volumes of his entomological memoirs. The method, one of painstaking interpretive observation was not new (see Aristotle on the bee in the *Historia*), but many of the observations were. By the time this work appeared, Fabre had already published some three dozen volumes of elementary school arithmetic, chemistry, astronomy, physics, botany, agriculture, etc., as well as a number of scientific contributions. Perhaps the most provocative of his discussions were those concerning the attempt to interpret insect behavior in anthropomorphic terms. The following, one of his earliest definite discussions of this problem, is illustrative.

This is the place to interpolate a certain passage from Lacordaire's *Introduction to Entomology* against which I am eager to protest. Here it is:

"Darwin [*i.e.*, Erasmus D.—Ed.], who wrote a book on purpose to prove the identity of the intellectual principle actuating men and animals, was walking one day in his garden when he saw on the path a Sphex who had just possessed herself of a Fly almost as large as herself. He saw her cut off the victim's head and abdomen with her mandibles, keeping only the thorax, to which the wings remained attached, after which she flew away; but a breath of wind, striking the Fly's wings, made the Sphex spin round and prevented her progress; hereupon she alighted again on the path, cut off one of the Fly's wings and then the other and, after thus destroying the cause of her difficulties, resumed her flight with what remained of her prey. This fact carries with it manifest signs of reasoning-power. Instinct might have led this Sphex to cut off her victim's wings before carrying it to her nest, as do some species of the same genus; but here there was a sequence of ideas and results from those ideas, which are quite inexplicable unless we allow the intervention of reason."

This little story, which so lightly grants reason to an insect, lacks I will not say truth, but even mere likelihood, not in the act itself, which I accept without reserve, but in the motives for the act. Darwin saw what he tells us; only he was mistaken as to the heroine of the drama, the drama itself, and its significance. He was profoundly mistaken and I will prove it.

First of all, the old English scientist was bound to know enough about the creatures to which he gives these high dignities to call things by their right names. Let us therefore take the word Sphex in its strict scientific meaning. Under this assumption, by what strange aberration was this English Sphex, if any such there be, choosing a Fly for her prey, when her kinswomen hunt such different game, Orthoptera? Even admitting what I consider to be inadmissible, a Fly to form the quarry of a Sphex, other difficulties come crowding up. It is now duly proved that the Burrowing Wasps do not take dead bodies to their larvae, but a victim merely numbed, paralysed. Then what is the meaning of this prey of which the Sphex cuts off the head, the abdomen, the wings? The stump carried away is no more than a fragment of a corpse, which would infect the cell with its rottenness, without being of any use to the larva, whose hatching is not due for some days yet. It is as clear as daylight: when making his observation, Darwin did not have before him a Sphex in the strict sense of the word. Then what did he see?

The term Fly, by which the captured prey is designated, is a very elastic word, which can be applied to the immense order of Diptera and which therefore leaves us undecided among thousands of species. The expression Sphex is most likely also employed in an equally indefinite sense. At the end of the eighteenth century, when Darwin's book appeared, this expression was used to denote not only the Sphegidae proper, but particularly the Crabronidae. Now, among the latter, some, when storing provisions for their larvae, hunt Diptera, Flies, the prey required by the unknown Hymenopteron of the English naturalist. Then was Darwin's Sphex a Crabro? No; for these Dipteron-hunters, like the hunters of any other prey, want game that keeps fresh, motionless but half-alive, for the fortnight or three weeks required for the hatching of the eggs and the complete development of the larvae. All these little ogres need meat killed that day and not gone bad or even a little high. This is a rule to which I know of no exception. The word Sphex cannot be accepted therefore, even with its old meaning.

Instead of a precise fact, really worthy of science, we have a riddle to read. Let us continue to examine the riddle. Different species of the Crabro family are so like the Social Wasps in size, in shape and in their black-and-yellow livery as to deceive any eye unversed in the delicate distinctions of entomology. To any one who has not made a special study of such subjects a Crabro is a Common Wasp. May it not have happened that the English observer, looking at things from a height and thinking unworthy of strict investigation the tiny fact which nevertheless was to corroborate his transcendental theories and help to bestow reason upon an animal, made a mistake in his turn, but one in the other direction and quite pardonable, by taking a Wasp for a Crabro? I would almost dare swear so; and here are my reasons.

Wasps, if not always, at least often bring up their family on animal food; but, instead of accumulating a provision of game in each cell beforehand, they distribute the food to the larvae, one by one and several times a day; they feed them with their mouths, as the father and mother feed young birds with their beaks. And the mouthful consists of a fine mash of chewed insects, ground between the mandibles of the Wasp nurse. The favourite insects for the preparation of this infants' food are Diptera, especially Common Flies; when fresh meat can be had, it is a windfall eagerly turned to account. Who has seen Wasps boldly enter our kitchens or pounce upon the meat hanging in the butchers' shops, to cut off a scrap that suits them and carry it away forthwith, as *spolia opima* for the use of the grubs? When the half-closed shutters admit a streak of sunlight to the floor of a room, where the Housefly is taking a luxurious nap or polishing her wings, who has not seen the Wasp rush in, swoop down upon the Fly, crush her in her mandibles and make off with the booty? Once again, a morsel reserved for the carnivorous nurselings.

The prey is dismembered now on the spot where captured, now on the way, now at the nest. The wings, which possess no nutritive value, are cut off and rejected; the legs, which are poor in juices, are also sometimes disdained. There remains a mutilated corpse, head, thorax, abdomen, united or separated, which the Wasp chews and rechews to reduce it to the pap beloved of the larvae. I have tried to take the place of the nurses in this method of rearing grubs on Fly-soup. The subject of my experiment was a nest of *Polistes gallica,* the Wasp who fastens her little rosette of brown-paper cells to the roots of a shrub. My kitchen-table was a flat piece of marble on which I crushed the Fly-pap after cleaning the heads of game, that is to say, after removing the parts that were too tough, the wings and legs; lastly, the feeding-spoon was a fine straw, at the tip of which the dish was served, from cell to cell, to each nurseling, which opened its mandibles just as the young birds in the nest might do. I used to go to work in exactly the same way and succeeded no better when bringing up broods of Sparrows, that joy of my childhood. All went well as long as my patience did not fail me, tried as it was by the cares of so finikin and absorbing an education.

The obscurity of the enigma gives way to the full light of truth thanks to the following observation, made with all the deliberateness which strict precision calls for. In the early days of October, two large clumps of asters in blossom outside the door of my study became the meeting-place of a host of insects, among which the Hive-bee and an Eristalis-fly (*Eristalis tenax*) predominate. A gentle murmur rose from them, like that of which Virgil sings:

Saepe levi somnum suadebit inire susurro.[1]

But, where the poet finds but an incitement of the delights of sleep, the naturalist beholds a subject for study: all this small folk making holiday on the last flowers of the year will perhaps furnish him with some fresh data. Behold me then on observation-duty before the two clumps with their thousands of lilac petals.

The air is absolutely still, the sun blazing, the atmosphere heavy: signs of

---

[1] The busy bees, with a soft murmuring strain,
 Invite to gentle sleep the labouring swain.

　　　　　　　　　　　Pastorals, i.e., Dryden's translation

an approaching storm, but conditions eminently favourable to the work of the Hymenoptera, who seem to foresee to-morrow's rain and redouble their activity to improve the opportunity. And so the Bees plunder eagerly, while the Eristalis fly clumsily from flower to flower. At times, the peaceable multitude, filling its crop with nectar, is disturbed by the sudden invasion of the Wasp, a ravening insect attracted hither by prey, not honey.

Equally ardent in carnage, but very unequal in strength, two species divide the hunting between them: the Common Wasp (*Vespa vulgaris*), who catches Eristalis, and the Hornet (*Vespa Crabro*), who preys on Hive-bees. The methods are the same in either case. Both bandits explore the expanse of flowers with an impetuous flight, going backwards and forwards in a thousand directions, and then make a sudden rush for the coveted prey, which is on its guard and flies away while the kidnapper's impetus brings her up with a bump against the deserted flower. Then the pursuit continues in the air, as though a Sparrow-hawk were chasing a Lark. But the Bee and the Eristalis, by taking brisk turns, soon baffle the attempts of the Wasp, who resumes her evolutions above the clustering blossoms. At last, sooner or later, some quarry less quick at flight is captured. Forthwith, the Common Wasp drops on to the lawn with her Eristalis; I also instantly lie on the ground, quietly removing with my hands the dead leaves and bits of grass that might interfere with my view; and I witness the following tragedy, if I have taken proper precautions not to scare the huntress.

First, there is a wild struggle in the tangle of the grass between the Wasp and the Eristalis, who is bigger than her assailant. The Fly is unarmed, but powerful; a shrill buzz of her wings tells of her desperate resistance. The Wasp carries a dagger; but she does not understand the methodical use of it, is unacquainted with the vulnerable points so well-known to the marauders who need a prey that keeps fresh for long. What her nurselings want is a mess of Flies that moment reduced to pulp; and, so long as this is achieved, the Wasp cares little how the game is killed. The sting therefore is used blindly, without any method. We see it pointed indifferently at the victim's back, sides, head, thorax or belly, according to the chances of the scuffle. The Hunting Wasp paralysing her victim acts like a surgeon who directs his scalpel with a skilled hand; the Social Wasp killing her prey behaves like a common assassin who stabs at random. For this reason, the Eristalis' resistance is prolonged; and her death is the result of scissor-cuts rather than dagger-thrusts. When the victim is duly garrotted, motionless between its ravisher's legs, the head falls under a snap of the mandibles; then the wings are cut off at their juncture with the shoulder; the legs follow, severed one by one; lastly, the belly is flung aside, but emptied of the entrails, which the Wasp appears to add to the one favoured portion. This choice morsel is solely the thorax, which is richer in lean meat than the rest of the Eristalis' body. Without further delay, the Wasp flies off with it, carrying it in her legs. On reaching the nest, she will make it into potted Fly and serve it in mouthfuls to the larvae.

The Hornet who has caught a Bee acts in much the same manner; but, in the case of an assailant of her dimensions, the struggle cannot last long, notwithstanding the victim's sting. The Hornet may prepare her dish on the very flower where the capture was effected, or more often on some twig of an adjacent shrub. The Bee's crop is first ripped open and the honey that runs out of it lapped up. The prize is thus a twofold one: a drop of honey for the huntress to feast upon and the Bee herself for the larvae. Sometimes the wings are removed and also the abdomen; but generally the Hornet is satisfied with reducing the Bee to a shapeless mass, which she carries off without disdaining anything. Those parts which have no nutritive value, especially the wings, will be rejected on arriving at the nest. Lastly, she sometimes prepares the mash in the actual hunting-field, that is to say, she crushes the Bee between her mandibles after removing the wings, the legs and at times the abdomen as well.

Here then, in all its details, is the incident observed by Darwin. A Wasp (*Vespa vulgaris*) catches a big Fly (*Eristalis tenax*); she cut off the victim's head, wings, abdomen and legs with her mandibles and keeps only the thorax, which she carries off flying. But here there is not the least breath of wind to explain the carving-process; besides, the thing happens in a perfect shelter, in the thick tangle of the grass. The butcher rejects such parts of her prey as she considers valueless to her larvae; and that is all about it.

In short, the heroine of Darwin's story is certainly a Wasp. Then what becomes of that rational calculation on the part of the insect which the better to contend with the wind, cuts off its prey's abdomen, head and wings and keeps only the thorax? It becomes a most simple incident, leading to none of the mighty consequences which the writer seeks to deduce from it: the very trivial incident of a Wasp who begins to carve up her prey on the spot and keeps only the stump, the one part which she considers fit for her larvae. Far from seeing the least sign of reason in this, I look upon it as a mere act of instinct, one so elementary that it is really not worth expatiating upon.

To disparage man and exalt animals in order to establish a point of contact, followed by a point of union, has been and still is the general tendency of the "advanced theories" in fashion in our day. Ah, how often are these "sublime theories," that morbid craze of the time, based upon "proofs" which, if subjected to the light of experiment, would lead to as ridiculous results as the learned Erasmus Darwin's Sphex!

CONDITIONED REFLEXES

PAVLOV, Ivan Petrovitch (several spellings) (Russian physiologist, 1849–1936). From *Lectures on conditioned reflexes*, New York, 1928; by arrangement with International Publishers, Inc.

The outstanding interpretive and heuristic importance of Pavlov's neurophysiological studies needs no emphasis. His interpretation of behavior may be considered an extension into modern biology of the mechanistic thinking of Descartes. He directs it here especially against the anthropomorphic predilections of the "zoopsychologists." In 1904, Pavlov was awarded the Nobel prize in recognition of his studies on digestion, especially the nervous control of digestive secretion.*

Experimental Psychology and Psycho-pathology in Animals

(Read before the International Congress of Medicine, Madrid, April, 1903.)

Observing the normal activity of these (salivary) glands, it is impossible not to be struck with the high degree in which they are adapted to their work. Give the animal some dry, hard food, and there is a great flow of saliva, but with watery food there is much less. Now it is clear that for the chemical testing of the food and for mixing it and preparing it as a bolus capable of being swallowed, water is necessary. This water is supplied by the salivary glands. From the mucous salivary glands there flows for every kind of food, saliva rich in mucin. This facilitates the passage of the food through the oesophagus. Upon all strongly irritant chemical substances, as acids and salts, there is also a free flow of saliva, varying to a certain degree with the strength of the stimulus. The purpose of this is to dilute or neutralise the irritant, and to cleanse the mouth. This we know from every-day experience. This saliva contains much water and little mucin. For, what could be the use of mucin here?

If you put some quartz pebbles into a dog's mouth he moves them around, or may try to chew them, but finally drops them. There is no flow of saliva, or at most only two or three drops. Indeed, what purpose could saliva serve here? The stones are easily ejected and nothing remains in the mouth. But, if you throw some sand in the dog's mouth (the same stones but pulverised), there is an abundant flow of saliva. It is apparent that without fluid in the mouth, the sand could neither be ejected nor passed on to the stomach. We see here facts which are exact and constant, and which really seem to imply intelligence. The entire mechanism of this intelligence is plain. On the one hand physiology has known for a long time of the centrifugal nerves to the salivary glands which may cause either water or organic material to pass into the saliva. On the other hand, in certain regions, the lining of the oral

---

* See his *The work of the digestive glands*, London, 1902.

cavity acts as a receptor for mechanical, chemical and thermal stimuli. These different stimuli may be further subdivided: the chemical, for example, into salts, acids, etc. There is reason for assuming the same in regard to mechanical irritants. From these special regions of the oral cavity the specific centripetal nerves take their origin.

All these reactions of adaptation depend upon a simple reflex act which has its beginning in certain external conditions, affecting only certain kinds of centripetal nerve endings. From here the excitation runs along a certain nerve path to the centres whence it is conducted to the salivary glands, calling out their specific function. . . .

All the foregoing substances, which when placed in the mouth influence specifically the salivary glands, act exactly the same upon these glands, at least in a qualitative way, when they are a certain distance from the dog. Dry food, even from a distance, produces much saliva; moist food, only a little. To the stimulation by food at a distance, there flows into the mouth from the mucous glands a thick, lubricating saliva. Inedible substances also produce a secretion from all the glands, but the secretion from the mucous glands is watery and contains only a small amount of mucin. The pebbles when shown to the dog have no effect on the glands, but the sand provokes an abundant flow of saliva. The above facts were partly discovered, partly systematised by Dr. Wolfson in my laboratory. The dog sees, hears, and sniffs all these things, directs his attention to them, tries to obtain them if they are eatable or agreeable, but turns away from them and evades their introduction into the mouth if they are undesired or disagreeable. Every one would say that this is a psychical reaction of the animal, a psychical excitation of the salivary glands. . . .

In our "psychical" experiments on the salivary glands (we shall provisionally use the word "psychical"), at first we honestly endeavored to explain our results by fancying the subjective condition of the animal. But nothing came of it except unsuccessful controversies, and individual, personal, incoordinated opinions. We had no alternative but to place the investigation on a purely objective basis. The first and most important task before us, then, is to abandon entirely the natural inclination to transpose our own subjective condition upon the mechanism of the reaction of the experimental animal, and instead, to concentrate our whole attention upon the investigation of the correlation between the external phenomena and the reaction of the organism, which in our case is the salivary secretion. Reality must decide whether the elaboration of these new phenomena is possible in this direction. I dare to think that the following account will convince you, even as I am convinced, that in the given cases there opens before us an unlimited territory for successful research in a second immense part of the physiology of the nervous system as a system which establishes the relation, not between the individual parts of the organism with which we previously dealt, but between the organism and the surrounding world. Unfortunately, up to the present time, the influence of the environment on the nervous system has been explained for the most part subjectively, and this comprises the whole contents of the contemporary physiology of the sense organs.

In our psychical experiments we have before us definite, external objects, exciting the animal and calling forth in it a definite reaction—the secretion of the salivary glands. The effect of these objects, as has been shown, is essentially the same as in the physiological experiments in which they come in contact with the tongue and palate, as in eating. This is nothing more than a further adaptation, i.e., that the object influences the salivary glands if it is brought even *near* the mouth. . . .

It is not difficult to recognise in the first psychical experiments certain important conditions which insure constant results and guarantee the success of the experiment. You stimulate an animal (i.e., his salivary glands) by food from a distance; the success of the experiment depends exactly upon whether the animal has been prepared by a previous period of fasting. In a hungry dog we get a positive result, but, on the contrary, in even the most avaricious and greedy beast we fail to get a response to food at a distance if he has just satiated himself. Thinking physiologically we can say that we have a different excitability of the salivary centre—in the one case greatly increased, in the other decreased. One may rightly suppose that just as the carbonic acid of the blood determines the energy of the respiratory centre, the composition of the blood in the fasting or fed animal likewise regulates the threshold of excitability of the salivary centre, as noted in our experiment. From the subjective point of view this change in excitability could be designated as attention. With an empty stomach the sight of food causes the mouth to "water"; in a satiated animal this reaction is very weak or may be entirely lacking.

Let us go further. If you only show the dog food, or some undesired substance, and repeat this several times, at each repetition you get a weaker result, and finally no reaction whatever. But there is a sure method of restoring the lost reaction: this is by giving the dog some food or by putting any undesired substance into the mouth. This provokes, of course, the usual strong reflex, and the object is again effective from a distance. It is immaterial for our result whether food is given or the undesired substance is put into the mouth. For instance, if meat powder, having been repeatedly brought before the dog, fails to produce a flow of saliva, we may again make it active by either giving it to the dog to eat (after showing it), or by putting an undesired substance into his mouth, e.g., acid. Owing to the direct reflex, the irritability of the salivary centre has been increased, and now the weak stimulus—the object at a distance—becomes strong enough to produce its effect. Does it not happen the same with us when, having no desire for food, an

appetite comes as we begin to eat, or also when we have experienced shortly
before some unpleasant emotion (anger, etc.)?

Here is another series of constantly recurring facts. The object acts upon
the salivary glands at a distance not only as a complex of all its properties, but
through each of its individual properties. You can bring near the dog your
hand having the odour of the meat powder, and that will be enough to produce
a flow of saliva. In the same manner the sight of the food from a further
distance, and consequently only its optical effect, can also provoke the reaction
of the salivary glands. But the combined action of all these properties always
gives at once the larger and more significant effect, i.e., the sum of the stimuli
acts more strongly than they do separately.

The object acts from a distance upon the salivary glands not only through
its inherent properties but also through accidental qualities accompanying the
object. For example, if we colour the acid black, then water to which we
add a black colour will affect the salivary glands from a distance. But these
accidental properties of the substance become endowed with the quality of
stimulating the salivary glands from a distance only if the object with the new
property has been introduced into the mouth at least once. The black coloured
water acts on the salivary glands only in case the black coloured acid has been
previously put into the mouth. To this group of conditioned properties belong stimuli of the olfactory nerves. The experiments of Snarsky in our
laboratory showed that there exist simple physiological reflexes from the nasal
cavity acting on the salivary glands, and that they are conducted only through
the trigeminal nerve; for example, ammonia, oil or mustard, etc., always produce a constant action in the curarised animal. This action fails, however, if
the trigeminal nerves are cut. Odours without local irritating effects have no
influence on the salivary glands. If you bring before a dog with a salivary
fistula oil of anise for the first time, there is no secretion of saliva. If, however, simultaneously with the odour of anise you touch the oral cavity with
this oil (producing a strong local reaction), there will afterwards be a secretion of saliva from only the smell of the oil of anise.

If you combine food with an undesired object, or even with the qualities of
this object—for instance, if you show the dog meat moistened with acid—notwithstanding the fact that the dog approaches the meat, you note a secretion
from the parotid gland (there is no secretion from this gland with pure
meat), i.e., a reaction to an undesired object. And further, if the effect of
the undesired object at a distance, owing to its repetition, is diminished, combining it with food which attracts the animal always strengthens the reaction. . . .

All the above facts lead, on the one hand, to important and interesting conclusions about the processes in the central nervous system, and, on the other
hand, to the possibility of a more detailed and successful analysis. Let us

now consider some of our facts physiologically, beginning with the cardinal
ones. If a given object—food or a chemical—is brought in contact with the
special oral surface, and stimulates it by virtue of those of its properties upon
which the work of the salivary glands is especially directed, then it happens
that at the same time other properties of the object, unessential for the activity of these glands, or the whole medium in which the object appears, stimulate simultaneously other sensory body surfaces. Now these latter stimuli
become evidently connected with the nervous centre of the salivary glands,
whither (to this centre) is conducted through a fixed centripetal nervous path
also the stimulation of the essential properties of the object. It can be assumed
that in such a case the salivary centre acts in the central nervous system as a
point of attraction for the impulses proceeding from the other sensory body
surfaces. Thus from the other excited body regions, paths are opened up to
the salivary centre. But this connection of the centre with accidental path-
ways is very unstable and may of itself disappear. In order to preserve the
strength of this connection it is necessary to repeat time and again the stimulation through the essential properties of the object simultaneously with the unessential. There is established in this way a temporary relation between the
activity of a certain organ and the phenomena of the external world. This
temporary relation and its law (reinforcement by repetition and weakening if
not repeated) play an important role in the welfare and integrity of the organism; by means of it the fineness of the adaptation between the activity of
the organism and the environment becomes more perfect. Both parts of this
law are of equal value. If the temporary relations to some object are of
great significance for the organism, it is also of the highest importance that
these relations should be abandoned as soon as they cease to correspond to
reality. Otherwise the relations of the animal, instead of being delicately
adapted, would be chaotic.

Physiology and Psychology in the Study of the Higher Nervous Activity of Animals

(Read before the Philosophical Society, Petrograd, November 24, 1916.)

. . . Here you see a diagram of our animal. On it are two black spots,
one on the front leg, one on the thigh of the hind leg. These are the places
where we attached the apparatus for mechanical stimulation of the skin. We
proceeded as follows. After we have started mechanical irritation of these
places with the pricking apparatus, then acid is poured into the mouth of the
dog. The secretion of saliva produced by the acid is, of course, a simple inborn reflex. This was repeated several times, yesterday, to-day, and day
after day . . . After a number of experiments a state of affairs results in
which we get a flow of saliva when we begin only to irritate that spot of the

skin; it is just as if we had poured acid into the dog's mouth, though in reality no acid is given.

Now I come to the discussion of our fact, and will do it physiologically and then as far as I can possibly psychologically, as a zoö-psychologist would do it. I can not guarantee that I shall use the correct phrases, because I am out of practice in these expressions, but I shall approximate to those I have heard from others. The facts are these. I apply lightly the mechanical irritation of the skin and then give the acid. Saliva is secreted—the simple reflex. When this has been repeated several times, then only the mechanical irritation of the skin is necessary to call out the flow of saliva. Our explanation was that a new reflex was formed, a new nerve path was made between the skin and the salivary glands. The zoö-psychologist, who wants to penetrate into the dog's soul, says that the dog directed his attention and remembered that when he felt the irritation of the skin at a certain place he would receive the acid and, therefore, when there was only irritation of the skin, he imagined the acid was coming, and he reacted correspondingly—saliva flowed, etc. Let it be so. But let us proceed further. We shall perform another experiment. We had elaborated a reflex and every time it gave perfectly accurate results. Now I start the mechanical irritation and receive as formerly a complete motor and secretory reaction, but this time I do not give the acid. One or two minutes pass and I repeat the experiment. Now the action already is less, the motor reaction is not so marked and there is not so much saliva. Again the acid is not given. We allow two or three minutes to elapse and repeat the mechanical irritation. The resulting reaction is still less. When we have done this four or five times, the reaction is entirely absent; there is no movement and no secretion of saliva. Here you have a clear, absolutely exact fact.

But here is the difference between the physiologist and the zoö-psychologist. I say that there develops our well-known inhibition. This I base on the fact that if I now interrupt the experiment and wait two hours, then the mechanical irritation again has its action on the salivary glands. For me as a physiologist this is perfectly clear. It is known that all processes in the nervous system in the course of time and with the cessation of the active causes become obliterated. The zoö-psychologist is also not at a loss for an explanation, and he says that the dog noticed that now after the mechanical stimulation acid was not given, and therefore after four or five such skin irritations he ceases to react.

So far there is no difference between us. You can agree with one as well as the other. But we shall proceed to more complicated experiments. Now you are aware that when the zoö-psychologist and the physiologist vie with each other to see whose explanations are correct, and more appropriate, then we must be well acquainted with the conditions which the facts are to explain. The prerequisite is, as you know, that the explanation should account for all that really occurs. The facts must all be explained without changing the point of view. This is the first requirement, and the second is even more obligatory. This is that from the given explanation it should be possible to foretell the explained phenomena under consideration. He who can say what will happen is right compared with him who can not give any kind of prediction. The failure of the latter here will mean his bankruptcy.

I shall complicate my experiment as follows. I have a dog in which our reflex has been elaborated at several places, let us say three. After the mechanical stimulation of each of these places there appears the same acid reaction, measured by a definite flow of saliva. This is the simplest way to measure the reaction; the measurement of the motor component would be more difficult. The motor and the salivary reactions go together, they are parallel. They are the components of a single complicated reflex. Now we have several skin reflexes formed. They are all equal, they act with absolute exactness, they give the same number of divisions of the tube used to measure the salivary secretion, for example, 30 divisions for one half-minute stimulation. I stimulate the place on the front leg in the way I have just said, i.e., I do not combine it with the influence of the acid, and so after about five or six times the mechanical irritation does not show any action. To the physiologist this means that I have obtained a complete inhibition of the reflex. When this has happened to the place on the front leg, I can stimulate another spot on the hind leg. And there developed such phenomena. If now I take the mechanical stimulation on the thigh—just as I did on the front leg, where I got zero— so that there is no interval between the end of that stimulation and the beginning of this, then at the new place I obtain a full action, 30 divisions on our tube, and the dog behaves as if this were the first application of the stimulus. Saliva flows freely, the motor reaction occurs, the dog acting as if he were rejecting acid from the mouth with the tongue, although there is no acid present—in short, the whole reaction appears. If in the next experiment I try the effect of the irritation on the front leg until again there is no secretion (by repeating the mechanical stimulation without giving acid), and then irritate the place on the hind leg, not after zero seconds but after five seconds, then I receive not 30 divisions from the new place but only 20. The reflex has become weaker. The next time I use an interval of fifteen seconds, and I get a slight action from the new place,—5 divisions. Finally if I stimulate after twenty seconds there is no action whatever. If I go further and employ a great interval, thirty seconds, then again I get an action from this place. With an interval of about fifty seconds, there is considerable secretion, 25 divisions, and with an interval of sixty seconds we see the full reaction. On the same place, on the shoulder, after we obtained zero result, if the irritation is repeated with an interval of five, ten, fifteen minutes, then we get zero (I do not know if I have made this clear to you). What does this mean?

I invite the zoö-psychologists to give their explanation of these data. More than once I have questioned intelligent people, having a scientific education—doctors, etc., about these same facts, and asked them for an explanation of the phenomena. The majority of the naïve zoö-psychologists gave explanations, but each one his own, and different from the others. In general the result was disastrous. They examined the facts as much as possible, but there was no way of making the various interpretations agree. Why is it that on the shoulder, when the experiment was so conducted that we got zero, the apparatus produced no further action, but here at the other place we obtain now a full action, now nothing, in a fine dependence upon different intervals of time between the stimuli?

I came here to get an answer to this question from the point of view of the zoö-psychologists.

Now I shall tell you what we think. Our explanation is purely physiological, purely objective, purely *spatial*. It is obvious that in our case the skin is a projection of the brain mass. The different points of the skin are a projection of the points of the brain. When at a certain point of the brain, through the corresponding skin area on the shoulder, I evoke a definite nervous process, then it does not remain there, but makes a considerable excursion. It first *irradiates* over the brain mass, and then returns, *concentrating* at its point of origin. Both of these movements naturally require time. Having produced inhibition at the point of the brain corresponding to the shoulder, when I stimulated another place (the thigh) I found the inhibition had not yet spread this far. After twenty seconds it had gotten here; and in twenty seconds, though not before, complete inhibition occurred at this point. The concentration required forty seconds, and after sixty seconds from the end of the zero irritation on the shoulder, we already had a complete restoration of the reflex, on the second spot (the thigh). But on the primary place (the shoulder) the reflex was not yet restored even after five to ten or fifteen minutes.

This is my interpretation, the interpretation of a physiologist. I have had no difficulty in explaining these facts. For me it fits in perfectly with other facts in the physiology of the nervous process.

Now, gentlemen, we shall test the truth of this explanation. I have a means of verifying it. If actually we have a movement, then consequently in all the intervening points we should be able to predict the effect, judging by the fact that this movement occurs in two directions. I take only one intermediate point. What is to be expected at this place? In proportion to its proximity to that area where I produce the inhibition it will be inhibited. Consequently in it the zero effect appears sooner and lasts longer—while the inhibition passes further and then recedes. At this spot the return to the normal irritability occurs later. Thus it came to pass in the actual experiment. Here at the middle point after an interval of zero seconds, there were not 30 but 20 divisions. Then the zero effect appeared already after ten seconds, when the full inhibition had reached here, and this effect remained for a long time, both while the inhibition was spreading further, and also when it was contracting, and passing in the opposite direction. It is clear why on the shoulder the normal reactivity returned after one minute, but here only after two minutes.

This is one of the most astonishing facts that I have seen in the laboratory. In the depth of the brain mass there occurs a special process, and its movement can be mathematically foretold.

So here, gentlemen, is the complexity of our experiment, and its relation to the physiologist. I do not know how the zoö-psychologists will answer me, how they will consider these facts, but answer them they must. If, indeed, they refuse to give an explanation, then with full justice I can say that their point of view is in general unscientific, and unsuitable for accurate investigation.

IV. THE ORIGIN AND DEVELOPMENT OF THE INDIVIDUAL

REVIVAL OF DESCRIPTIVE EMBRYOLOGY

COITER, Volcher (Dutch-born, later German, physician, anatomist, and embryologist, 1543–1576). From *De ovorum gallinaceorum generationis primo exordio progressuque et pulli gallinacei creationis ordine*, in *Externarum et internarum principalium humani corporis tabulae*, Nüremberg, 1572; tr. by H. B. Adelmann as *The "De Ovorum" etc., of Volcher Coiter*, in *Annals of Medical History*, n.s. vol. 5, No. 4, p. 327, 1933; with the permission of Paul B. Hoerber, Inc., publisher.

With unimportant exceptions the following is the first published description of directly observed development of the chick after Aristotle. Coiter studied under Aldrovandi, Fallopio, Eustachi, and Rondelet, and knew Fabricius. He published important descriptions of the passages of the head, the origins of the spinal nerves, the living heart, and especially the skeletons of vertebrates.* Adelmann's translation is provided with invaluable historical and explanatory notes, to which the reader is referred. Coiter practiced medicine in Nüremberg and died when thirty-three, reputedly of leprosy.

As for myself, in May 1564 at Bologna, stimulated by Doctor Ulysses Aldrovandus (that most distinguished professor of ordinary philosophy, a man expert in the knowledge of the various sciences and arts as well as of philosophy and especially natural philosophy, my encourager and teacher to be cherished with perpetual regard), and encouraged by other teachers and students, I had selected two clucking hens ready for incubation. Under each of them I had placed twenty-three eggs and in the presence of those men, I opened one on each day, in order that we might determine these two things especially, to wit —the origin of the veins and what is first formed in the animal.

In an egg of the first day I saw that the yolk had acquired a white circle[1] which was not very large. In its middle there was a point or disc[1] of the same color. From the circle two similar circles[1] flowed away. One of them was thicker and longer than the other. The yolk was more fluid than the yolk of a fresh egg.

On the second day I saw two very thin pellicles or membranes, one[2] of which adhered immediately to the shell; the other,[3] which surrounded the substance of the egg, was separated a little from the shell, for a small space at the end of the egg was empty. The yolk had risen toward the pointed end of the egg as Aristotle says happens on the third day. The middle part of the yolk was seen to be clearer than the remaining portion.

---

* In *Diversorum animalium sceletorum explicationes iconibus artificiosis et genuinis illustrae*, Nüremberg, 1575.

[1] Reference to area pellucida, area opaca, and possibly the primitive streak.

[2] Shell membrane.

[3] Blastoderm.

In the middle I saw something resembling semen. The point and the circle were found under the membrane surrounding the substance of the egg. They were bestrewn with certain sanguineous filaments. The white, like the yolk, was a little more fluid than usual. The white glass-like substance,[4] which is commonly held to be the semen of the cock, was a little harder.

No change in the albumen was detected on the third day; but the sanguineous point or globule[5] previously found in the yolk, but now found rather in the albumen, manifestly pulsated[5] and, as we were able to judge by the color, gave rise to one venous trunk which, after dividing into two, gave off many small branches which encircled the pulsating point.[5] These small branches were supported by a very thin membrane which represented the secundine both in function and substance. Hence three membranes were found here; the first is assigned to the shell,[6] the second to the whole substance of the egg[7] and the third to the secundine.[8]

On the fourth day I opened the egg from the blunter end and there occurred first a space so empty that it easily admitted the first joint of the finger. The aqueous white, on account of its tenuity, flowed off. The globule and the other structures were found to be larger. Moreover, the globule pulsated; the vessels did not. On one side, three translucid, glass-like globules[8a] were found joined to one another. On the other side there were two branches or vessels, not unlike arteries; which, however, did not pulsate. When the albumen had been poured off, the rather liquid yolk came into view. It was besprinkled here and there indiscriminately with blood.

On the fifth day we observed that the second membrane, surrounding the whole substance of the egg and perfused with many veins, was so free from the shell-membrane and so strong that, without injury, it could be removed together with the substance of the egg. When this membrane was opened, I saw that the sanguineous, pulsating globule had sunk more deeply than usual. I took an egg of the fifth day from each hen. There appeared in one of them merely the still formless pulsating globule which was, however, surrounded by blood vessels, as has been stated. The lateral globules tended to be black in color. Between them there were still smaller connected globules, which represented the brain after a fashion. The third globule was unchanged except in size. In the albumen dissimilar parts were seen. Some were thin and approached glossy whiteness, others were denser. The yolk was easily mixed with the albumen. In the other egg the head of the chick clearly appeared

---

[4] Chalazae.
[5] Beating heart and area vasculosa.
[6] Shell membrane.
[7] Blastoderm.
[8] Allanto-chorion.
[8a] To be eyes and brain.

very large with respect to the size of the body. On both sides of it there was a blackish eye, which was found to be very clear in the middle. Between the eyes there was a third globule.[9] The rest of the rather long body hung down from the head. The heart or pulsating disc was situated not far from the head. From it the veins were seen to arise. I was able to find no trace of the liver, so confused were the viscera.

On the sixth day, in addition to the other parts which were all larger, we saw from the blunter end, after the [removal] of the membranes or pellicles, the albumen first, for the yolk remained in the bottom of the egg. The living, animate chick, imperfectly delineated in all its parts, floated in the albumen. It had been granted a head which was very large in proportion to the body. There was nothing more perfect or larger than the eyes. For, although I had some difficulty, I was able to distinguish from one another all the tunics and humors in the eyes. Hence from the fact that in birds the eyes are developed first, Lactantius Firmianus decided, not without reason, that the head begins to develop first in birds. Perhaps the eyes, just as very ingenious and elaborate instruments, require a long time for completion. Wherefore, since birds remain in the egg for a short time, the creator of things commences to weave this fabric first with the eyes. Furthermore, superiorly between the eyes there was a rather large, transparent globule which has been seen to represent the brain. Below the eyes there was something resembling a beak.

On the seventh day everything could be seen more clearly; first, the three pellicles, especially the third called the secundine[9a] which supports the veins and surrounds all the fluid. When the yellowish white fluid had been poured off, the small, rudely-formed animal appeared. Its head was larger and more perfect than the other parts of the body. Both the cranium and the beak were observed to be more perfect. The rest of the body and its members were formless. I was able to distinguish almost none of the viscera both because they were included in the thorax and because internally they had not yet attained their true and manifest form, for they were seen to be in disorder everywhere.

On the eighth day I found that all parts including the legs and wings were larger and better formed, so that the embryo already exhibited the likeness of a chick. It moved around in the albumen with more vigorous motion. In addition I found rather conspicuous veins entering the umbilicus of the chick.

On the ninth day I observed that the secundine had increased with respect to the number and size of its veins. In short, it was more distended with blood than usual; some very thin blood had even transuded. After the secundine, the chick was visible floating in the white with half of the yolk on each side as if the yolk had been cut into two equal parts. The umbilical vessels

---

[9] Mid-brain.

[9a] Allanto-chorion.

were much larger. After the chick had been taken out, I removed the proper membrane immediately surrounding the chick so that it might not perish in the liquid. The chick was now perfect in all parts and it was observed to have large pores everywhere, from which the primordia of feathers were about to emerge. The head was largest of all. The eyes were very perfect indeed and provided with lids. I saw three transparent globules in the head, the combination producing the form of the avian brain. All parts below the neck had attained their complete and natural form. A pulsation appeared externally in the thorax in the region of the heart. The heart had attained its form and was whitish in color. Moreover it beat for a long time outside the chick. The rather pale liver, the stomach, the intestines, the ribs and all the other viscera were found to be very soft and flaccid. There were two albuminous substances and each was surrounded by its own individual membrane. One portion was transparent and contained the chick swimming in it.[10] The other, viscous, dense and pale,[11] the excrementitious portion of the albumen, as it were, remained in the bottom of the egg.

An egg of the tenth day had a secundine somewhat stronger than usual and veins bedewed with blood on the outside. The fetus, which lay upon the yolk as on a couch, was perfect in all its parts. The legs, the back and the buttocks were covered with the primordia of feathers. The body or belly of the chick was now larger than the head. Aristotle states the contrary. In the eyes I saw all the tunics, all the humors, and the external covers or lids as well as the membrane with which they wink. The brain had its convolutions and gyri. The heart, indeed, had its natural form and was white in color like the spermatic parts. The lung and the ribs were formed. In the lower part of the belly were the liver which tended from white to reddish in color, the stomach and the intestines which were very white. Two umbilical vessels emerged from the belly of the fetus next the anus and proceeded to the yolk. When the chick had been removed, we noticed that the greater part of the albumen had been used up and what remained was a dense excrement. Little of the yolk, however, had disappeared.

On the eleventh day we found no difference except in size. The head had grown more evident and the feathers had sprouted a little. The fetus, resting upon the yolk, floated in a white and tenuous albumen. The dense and viscid albumen adhered below the yolk.

On the twelfth day, in addition to other things we examined the membranes and humors. There were four membranes. The first[12] was fused with the shell for the sake of strength, for so strongly does it hold the very fragile shell together that a fresh, even a raw egg, cannot be broken by even the

---

[10] Allanto-chorion.

[11] Yolk-sac overlying albumen remnant.

[12] Probably shell membrane.

strongest man if the pointed ends are compressed by both hands. The second[13] membrane surrounds the entire substance of the egg. It is depressed and contracted for the diminution and contraction of the egg. In this way it prevents the mingling of parts of the egg which could easily be brought about by shaking since everything becomes more liquid during incubation. The third[14] membrane affords a support for the veins and arteries, for it is full of arteries and veins. Moistened externally by tenuous blood, it surrounds the whole egg and performs the function of the secundine. The third membrane is followed by the membrane[15] proper to the chick, which contains not a little water in addition to the chick. It seemed to me that this tunic or membrane is dilated next to the umbilicus and that it invests the umbilical vessels. As for the chick, it was larger and more solid and was covered with larger feathers. Something white adhered loosely to the end of the upper and lower bill.[16] Women say that this prevents the picking of grain and for this reason they remove the white substance as soon as the chicks have hatched.

On the thirteenth day we saw that the substance of the egg was diminished in quantity. The chick was larger and more completely feathered. Besides these things there was nothing extraordinary.

On the fourteenth day the chick had grown in proportion to the degree to which the albumen had disappeared. It was entirely covered with small feathers. The transparent, dense and viscous fluid, smaller in quantity than usual, was contained in its own membrane just as the yolk was contained in a pellicle full of vessels. The umbilical vessels seemed to me to be five in number. One vein ran to the third membrane or secundine to which it supplied many branches, but two other veins were distributed throughout the tunic investing the yolk. In addition to these veins, I found among the umbilical vessels two small arteries which, coiled like worms, extended to the yolk. I was unable to observe their attachment.

On the fifteenth day the vessels were scattered everywhere through the secundine. They had tunics so thin that they were injured by slight contact with the knife. Very red, thin blood flowed out. These vessels or veins proceeded from one rather thick trunk, which extended for a long distance through the middle of the chorion. When the secundine was cut open, no little watery serum flowed out. Then there came into view the vesicle or membrane[17] in which the fetus lay floating in water, and likewise the yolk, covered by its own membrane.[18] This membrane was fused to the secundine[14] just as

---

[13] Chorion.
[14] Allanto-chorion.
[15] Amnion.
[16] Egg tooth.
[17] Amnion.
[18] Vitelline membrane.

in the human fetus the amnion is fused to the chorion. The vitellus was clearly divided into two parts which cohered for half a finger's breadth, but for both parts there was one membrane which arose from the umbilicus or skin of the chick. The rest corresponded to what has already been described.

On the sixteenth day I saw that the dense and viscous portion of the albumen had been used up almost everywhere and that the branch of the pulsating artery[19] was distributed throughout the membrane surrounding the yolk. The membrane of the chick and of the yolk were the same as formerly. Other things which should have grown were observed to be larger; those which should have diminished in size were seen to be smaller. Among other things, the belly was much more swollen and thicker than usual. Within the belly, the viscera had grown very much. In the stomach I found a chyle-like substance, the major portion of which, however, was in the end of the esophagus next to the stomach. Two worm-like vessels which resembled intestines rather than arteries, hung down outside the belly.

On the seventeenth day I saw an aqueous substance, scanty in amount both between the third and fourth membranes and between the fourth and the chick. And then I saw emerge from the umbilicus one artery,[20] three veins[20] and an outgrowth from the intestines to the yolk. Two of the veins entered the mesenteric veins. The third went to the vena cava. The large artery, passing by the stomach opened into the heart as I saw at times. I saw the intestines hanging out of the umbilicus for some distance and passages extending from the intestines to the yolk of the egg. Very hard, shell-like feces were observed between the fourth membrane[20a] and the chick. In the stomach some chyle, mixed with water was found. No little yolk had been drawn into the belly. Its disposition and location corresponded to the description of Aristotle.

On the eighteenth day I noticed some true excrement from the chick between the fourth membrane and the chick. The fourth membrane surrounded the yolk and the chick. The chick was completely feathered and pot-bellied. When the membrane belonging to the chick was separated I found that the membrane proper to the yolk was filled with many veins and that in its middle portion it had been pulled into the belly together with the yolk, now drawn out in length. The viscous white substance which has been mentioned was entirely used up. Within the chick, the stomach, which had grown very much, contained a little aqueous fluid and a great deal of a white, cheesy substance which we thought was chyle. Some such substance, indeed, flowed from the intestines.

On the nineteenth day the head of the chick was bent down and concealed under the right wing and leg. The aqueous fluid had diminished everywhere.

---

[19] Vitelline artery.

[20] Vitelline blood vessels.

[20a] Amnion.

Between the fourth tunic and the chick there were encountered true feces as well as a watery fluid with which the whole chick was wet. The membrane of the yolk was clearly seen to arise from the skin of the chick. The yolk was seen, by the contraction of this membrane, to be drawn inside and absorbed; indeed it had now been almost absorbed. In addition to this improper membrane, the yolk had also a proper one filled with many veins and arteries. We saw in the belly the things which we had seen before but they were more perfect. I found no yolk in the intestines (as Albertus Magnus would have it). The chick already had testes and a comb, for it was a cock. Furthermore, we heard peeping. When the opening was dilated with a knife, the yolk emerged invested with its own membrane and drawn out in length. There was a smaller amount of yolk than at the beginning. It was joined to the intestines by means of a canal[21] prolonged from them. The umbilical artery, which was mentioned in an egg of the seventeenth day was observed to pulsate much more clearly than before, and to open into the heart. Two veins joined branches of the portal vein far from the liver. From the stomach, when cut open, there flowed out chyle which was much yellower than usual, and a little water. Water flowed from the crop and greenish feces from the rectum. The color of the liver tended to be yellow rather than reddish. The lung was reddish.

On the twentieth day the chick was fully formed. Within the shell were found four membranes[22] or tunics attached to one another at one place. The inner two of these membranes contained many veins and arteries. As for the chick, we saw three umbilical vessels inside. Two of these seemed to be arteries, one a vein. The umbilical opening was closed, but it could easily be opened when a probe was introduced.

EARLIEST ILLUSTRATED EMBRYOLOGICAL TREATISE

FABRICIUS ab Aquapendente, Hieronymous (Italian anatomist, 153?–1619). From *De formatione ovi et pulli tractatus accuratissimus*, Padua, 1621 (posth.); tr. by H. B. Adelmann as *The formation of the egg and of the chick* in his *The embryological treatises of Hieronymous Fabricius of Aquapendente*, Ithaca, 1942; with the permission of Cornell University Press, and of the translator.

Fabricius was successor of Vesalius and immediate follower after Fallopius in the famous anatomical professorship of Padua. The facsimile edition with translation, from which, with permission of author and publisher, the following passage was excerpted, is exemplary as a specimen both of bookmaking and of research in the historical development of biology. From the standpoint of the history of biological ideas, the papers are especially interesting as specimens of the difficulty experienced by earlier Renaissance thinkers in detaching themselves from their scholastic predilections.

---

[21] Urachus.

[22] Shell membrane, chorion, allantois, amnion.

THE ACTION OF THE EGG, THAT IS, THE GENERATION OF THE CHICK

We must now inquire how the generation of the chick proceeds from the egg, setting out from that principle of Aristotle and Galen which everyone admits—namely, that everything made in this world is clearly observed to be produced by these three factors: agents, instruments, and material.

As, therefore, in works of art the coppersmith himself is the agent, the hammer and anvil the instruments, and the copper itself the material, and the result or final cause is, for example, a kettle or trumpet, so too in the works of Nature, there will be needed an agent, instruments, and material. But it is well to know that in the products of art the agent and his instrument are separate, as the smith and his hammer, the painter and his brush; but in the works of Nature the two are combined as one. Thus the liver is both the agent and the instrument for the production of blood, the stomach the agent and instrument for chylification, and the same is true of every part of the body. Hence Aristotle has justly said that it is not easy to distinguish the efficient causes from the instruments. Galen gives the reason for this in his book, *On the formation of the fetus*, namely, that in productions of art the agent acts from without, whereas in the works of Nature the efficient cause has been imparted to the instruments and has permeated all the organs. Therefore, in the generation of the chick only an agent and material will be needed and the discussion must turn upon these two things especially. Aristotle, in explaining these two, says that the male contributes the form and the principle of motion [efficient cause], but the female the substance or material, and he cites the comparable action of these two causes in milk. For in the coagulation of milk, the substance is the milk itself, but the rennet possesses the thickening and curdling principle.

But since not only the generation of the chick, but also its growth and nutrition take place in the egg, therefore not only the agent and the material, but also the aliment in the egg must be investigated, because as soon as the generation of the chick begins, its nourishment and growth immediately begin too. Hence Aristotle says that Nature places in the egg at once both the material for the living fetus and enough food for its growth.

Moreover, we gather that both material and aliment for the chick are contained in the egg from Hippocrates' argument that when a bird has been hatched there is no fluid worth mentioning left in the eggshell. In discussing the generation of the chick in the egg, therefore, we must consider three factors: the agent, the material, and the aliment. . . .

Now with regard to the first difficulty, which pertains to the material and aliment of the chick, since Hippocrates, Anaxagoras, Alcmaeon, Menander, and the ancients on the one hand, and Aristotle and Pliny on the other, are at odds with one another, I do not therefore see how this controversy between them can be resolved or composed. On that account, in order that you, Gentlemen, may judge it, I will state my own opinion, which I am very ready to change if necessary.

In the first place, I agree with both, that is, with Hippocrates, that the chick is nourished by the albumen, and with Aristotle, that it is nourished by the yolk. I disagree with both, that is, with Hippocrates, that the chick is produced from the yolk, and with Aristotle, that it is formed from the albumen. In short, I think that both the yolk and the albumen are merely food for the chick and not the material at all, an opinion which, as you see, partly agrees and partly disagrees with the authors I have mentioned. Although, as I have said, Aristotle gives only one reason to confirm his opinion, and Hippocrates has said that his was proved by experience, I shall nevertheless prove this opinion of mine by three arguments based upon reason.

First, that neither fluid of the egg is the material of the chick is demonstrated in this way: What has to be the material from which the chick is formed and engendered must be consumed as the generation of the chick is consummated and completed. This is my major premise, which is proved, according to Galen and Averroes, in this way: When the seed of an animal has been introduced into the uterus, or the seed of a plant has been placed in the earth, we see that it transforms very gradually into the particles of the body, nor do the parts of the seed cease to change and transform until all parts of the body are perfectly finished from it.

This is my minor premise: Neither the yolk nor the albumen is consumed while the generation of the chick is being completed. This is an established fact, since when the generation of the chick has been completed, yolk and albumen still remain up to the end, that is, until the chick hatches. Therefore the yolk and albumen cannot be the material from which the chick is formed.

I prove next that the yolk and the white are the food of the chick by this reasoning: Food must be supplied to the chick not only while the fetus is shut up in the egg, but also after the chick has hatched out, because the whole course of our life is bound up with and attended by nutrition. But outside the egg, the chick is nourished by mouth and by food from without. Therefore, while the animal remains in the egg it will be nourished by the contents of the egg, and this nourishment will also be stored in the egg until the chick hatches. Now, in fact, the yolk and albumen are stored in the egg up to the time the chick emerges, and therefore they will be the chick's food. Nor can they be the material for generation, because generation, as I have said, is completely finished within a few days, and after a few days the generative faculty ceases to act, and lies idle. On the other hand, the albumen and yolk are conserved and do not appear to be consumed and transformed.

All of these facts are still more fully substantiated by another argument taken from the vessels which branch and run through the yolk and albumen.[1] Since these are numerous and are given off from the chick to the membranes of both the albumen and the yolk,[2] and since they gradually diminish and consume the substance of both until they have been almost entirely used up and absorbed, and then the chick is hatched, they therefore show very clearly that the yolk and albumen are not the material of the chick, neither both of them together nor one more than the other, because both are very gradually diminished at a certain rate.

But if one of these were the material for the generation of the chick, in the first place it would be consumed when the generation of the chick has been completed, just as I have said; and in the second place the yolk and albumen would not need vessels for the generation of the chick because vessels are instruments of nutrition rather than of generation, although just as soon as generation has been completed, the vessels are at hand to deliver and supply nutriment. . . .

But, you will say, suppose the white and the yellow in the egg are the aliments of the chick, what then must be established as the material for the chick, since it has already been stated that the semen is not present in the egg? You may discover this material by induction, from a complete enumeration of the parts of the egg. There remain in the egg the shell, two membranes, and the chalazae; no one would maintain that the membranes and the shell of the egg are the material of the chick; therefore, the chalazae alone will be suitable material for it. But these too present difficulties. First, indeed, the chalazae seem merely to play the role of ligaments in the egg, since it is clearly apparent that the yolk is attached to the albumen and the membranes by means of them. Second, if the chalazae were the material of the chick, they would be present only in the blunt end of the egg where the chick is engendered. But chalazae are also found in the sharp end of the egg; therefore, the chick cannot be incorporated from the chalazae as the material. Fourth and fifth, the chalaza which is present in the blunt end of the egg is a structure so small and insignificant that it could in no way be material sufficient for the formation of so many organs and the large mass of the chick. Finally, opposed to the chalazae is the authority of Aristotle who writes that they contribute nothing to the generation of animals.

Accordingly, from what material the chick is engendered, is a very abstruse and recondite problem. Nevertheless, it is my opinion that the chick is incorporated from the chalazae as the material. This is proved, first, by the fact that only three of the structures constituting the egg are suitable for the generation of the chick—the albumen, the yolk, and the chalazae. The albumen and yolk are the nourishment for the whole chick as I have already proved;

---

[1] Vitelline and allantoic vessels.
[2] Allanto-chorion, vitelline membrane.

therefore the chalazae alone will be the material from which the chick is produced. Moreover, among the parts of the egg, the chalazae are structures *sui generis,* and distinct from the white and yolk in the properties of their substance. They differ from the yellow, as is apparent, and from the white too, for they are round, spherical, and nodular little bodies, whiter than the albumen, and covered with a bright sheen, like hail. Therefore, if they are different from the yolk and the albumen, they likewise serve a purpose different and distinct from both; and this purpose cannot correctly be established as other than the one I have proposed.

Further, the chalazae are situated at that place in the egg where the chick is engendered; therefore, the chick is formed from them. For if you break into a cooked egg at the blunt end as far as the chalazae, you will observe that they are related to that cavity beneath which the chick is formed, and especially where its head lies. Besides, if you inspect a chick at its very commencement, three or four days after conception, you will observe four things: First, you will notice the large head, entirely white and almost transparent, and in it the orbit, defined by a ring-like black line, and in the middle of it, a round white pupil. Second, the spine continuous with this head will be quite distinctly visible. It is also white, viscous, and almost transparent, so that you cannot conceive that these structures, the head and spine, have been formed from any material other than the chalaza. For this entire mass consisting of the head and spine resembles exactly the substance of the chalaza, and just as the chalaza is an elongated rather than a round body, so too is the body of the chick when it is first formed. Third, you will observe a redness or a red body situated anteriorly and downward beneath the head, which from its location is doubtless the heart and liver. Fourth, you will notice two veins[2a] which will extend to the albumen and to the yolk. Their trunks will be adjacent, but branches will be distributed to both the white and the yolk. I have had all this painted, which has, indeed, been done, but the artist was unable to reproduce the transparency. Yet whoever has seen the chalaza and such a conception will believe that he has seen what has reference to the body.

Immediately convincing, also, is the generation of frogs, which begins with black animalcules, here commonly called *ranabottoli* (tadpoles), of which nothing is to be seen but the head and tail, that is, the head and spine, without a trace of fore or hind limbs. However, when the tadpoles have grown larger with the passage of time, the black color fades and the true color of frogs appears, and also at the same time the fore and hind limbs gradually grow out. These are at first very small and imperfect, but they subsequently become perfect and complete.

Furthermore, no chalazae are to be seen when the chick has once been completely formed and fully shaped; but when a residue of parts still remains to be

[2a] Vitelline and allantoic vessels.

formed, such as the wings and legs, which are made last, remnants of the chalazae are also left over; therefore, chicks are formed from the chalazae. And another reason: If there are only three swellings in the chalazae, these nodes seem, properly, to correspond to the three cavities, the head, the thorax, and the abdomen, or to the three principal parts, the brain, the heart, and the liver. Now suppose five nodes are counted, they will correspond not only to the former structures, but also to the wings and legs. But if four nodes are never observed in the chalazae, this too will be a clear sign, plainly showing that the nodes of the chalazae correspond in number to the principal parts of the chick.

In addition to all these arguments there is yet another drawn from analogy, for just as a viviparous animal is incorporated from a small quantity of seminal matter, while the material supplied for food and nourishment is very abundant, so the little chalaza will suffice for the generation of the chick, and all the other structures contained in the egg will be merely food for it. Thus, plants spring from tiny little mustard seeds, and great trees, to be sure, from seeds of the oak, the apple, or the pear; that is, they are nourished and increased in size from an abundant supply of aliment. Hence we need not be surprised if Nature placed both the white and the yolk in the egg for nourishment only, but dedicated the little chalazae to incorporating the chick. . . .

I have already explained above that even though the semen of the cock lies outside the egg, its most important and particular use is to fecundate the uterus and the egg to prevent sterility. It is reasonable to believe that the semen is supplied in great abundance in order that the cock may be equal to the many matings he performs in a few hours, nay even in one hour. And therefore, as is clear from what I have stated before, a very large vessel has properly been constituted and formed to receive and carry this great quantity of semen. Further, the semen is very white, like milk. Aristotle writes that the prolific semen of all birds is white, and that of other animals as well,[3] and I, moreover, may add that all seed both of animals and plants is white, one differing from another only in its degree of whiteness. Hence Aristotle[4] says that cartilaginous fishes emit a milky fluid. He also says,[5] in stating the reason why all semen is white, that it is because the semen is foam, and foam is white because of the admixture of abundant spirit or air; and for this reason the semen is also very light in order that it may not sink down when emitted but be preserved.

If, however, all semen is white, it is therefore cold, since I have repeatedly said that everything white in the body of an animal is cold. This, however, is to be understood only of the corporeal substance of the semen. For since the semen is foamy and also spirituous and airy, by the same token it possesses

---

[3] *De gen. an.* II, 2, 736a10.

[4] *Hist. an.*, V, 5, 540b32.

[5] *De gen. an.* II, 2, 736a10–20.

much innate heat. Indeed, this heat is lodged in a cold substance, because if it had been placed in a warm one it would easily be dissipated and pass away before the animal was formed from it, and this the coldness prevents. The adhesive quality of the semen likewise contributes to this end and also makes it possible for the semen when introduced into the uterus to adhere there, while its softness enables it to be fashioned into a body more easily.

Now it is worthy of belief that Nature has prepared suitable restraints and obstacles in order that a proper amount of semen may be emitted at a single coitus and that there may yet be enough for all the matings, that is to say, that valves or a narrowing at the lowest part of the vessel, or vessels coiled like tendrils[6] have been used for this purpose. Finally, although a very small amount of semen is emitted, its force and power are nevertheless very great, not to say divine, for there is nothing found in the realm of Nature which of itself reveals so many and such important faculties as the semen.

THE USES OF THE CHALAZAE AND OF THE SMALL SCAR-LIKE DISC IN THE YOLK

There is, however, nothing for me to say about the chalazae and the remains of the little scar appearing on the surface of the yolk, and, as it were, adnate to it, since I have previously discussed the chalazae in detail in the appropriate context; unless, perhaps, I add this, that in a cooked egg, the chalazae are so contracted upon themselves that they resemble the form of a conception or of a chick just formed and. engendered. Moreover, there is also nothing for me to say about the remains of the pedicle, which resembles a scar, since it is now of no use, but is merely a trace of the separation of the stalk [from the yolk]. . . .

THE USES OF THE CAVITY WHICH LIES IN THE BLUNT END OF BOTH THE GRAVID AND NON-GRAVID EGG

There is a cavity which appears in all eggs, those which contain a chick as well as those which do not, but especially in those possessing one. It lies in the blunter portion of the egg, where the head and beak are situated, and it is observed to be the larger the nearer the chick is to its birth or hatching. Further, it is not seen exactly at the end of the egg but somewhat toward the side; it is not level and plane but slanting, and it is larger where the beak of the chick lies. . . .

The third state of the cavity exists when the chick needs a greater quantity of air to cool it on account of its further growth and the increased heat. At this time respiration is already required, and the chick draws air by the breathing which is evoked, but which is quite weak and very gentle. In so doing, the chick draws air through the pores of the second membrane itself rather

---

[6] Probably ductus epididymidis.

than directly from the cavity, because a membrane fences off the cavity, as it were, and the chick is enveloped by the membrane,[7] but it is reasonable too to believe that it also draws some air from the space which intervenes between the chick and the membrane.

The fourth use of the cavity occurs when the chick has already grown to such an extent that it requires more powerful respiration, in which air is no longer drawn from the cavity through the pores of the second membrane but directly from the cavity itself without the barrier of the membrane. At this time the chick, as if impelled by the necessity for cooling, punctures and breaks the membrane with its beak, and when the cavity has thus been laid open, the chick carries on its breathing by means of the free air contained in it. This air is then far more abundant because for the same reasons the cavity has grown larger, and has become so capacious that it is almost half as large as the egg.

The cavity is last used when the chick is at length so large and perfect that it can be nourished by mouth, and for that reason is close to the time of hatching. It therefore no longer needs the respiration which is afforded by the cavity and the air contained in it, nor that food which is supplied from within by the humors of the egg, but it does need both air and food coming from the outside. Moreover, it needs the outside air first, and sooner than food, since some food still remains within. At this juncture, unable to break the hard shell because of the softness of its beak and the distance of the shell from the beak drawn under the wing, the chick now signals its mother that the shell must be broken; and it does this by peeping. For at this time the chick is so strong, and the cavity has become so large and the air contained in it so abundant that respiration has at length grown vigorous enough for the chick to exhale forcibly and call out.

While the cavity is indeed quite small in the beginning, afterwards it dries out very gradually and grows larger until it becomes very large; then because of the large amount of air in it, the chick is able to exhale forcibly and cheep, and it clearly produces the sound which is natural for the chick at birth, and perhaps has something of the nature of a petition. This can be heard even by a bystander. Furthermore, Pliny[8] and Aristotle[9] give the same testimony, that if the egg is moved on the twentieth day, the voice of the living chick may already be heard within the shell, for the chick peeps a little. (So says Aristotle.)

As soon as the brooding hen hears the chick's voice, she breaks the shell with her beak, as if she recognized the necessity for breaking it so that the chick may at once enjoy the fresh air for its survival; or if you prefer, you may say that the mother hen is touched by a desire to look upon the chick, her beloved

---

[7] Probably allanto-chorion.

[8] *Nat. Hist.*, X, 53(74).

[9] *Hist. an.*, VI, 3, 561*b*26.

child.   This is accomplished without difficulty because the shell has become more fragile and easily broken in the region near the cavity, which has been destitute of humors for a long time, having been dried out by the contained air and by the heat.

The peeping, therefore, is the first and most important sign of the chick's desire to be liberated and of its need of fresh air.   This the mother hen perceives so exactly that if by chance she recognizes that the peeping of the chick is within and below, she then turns the egg up with her feet, so that she may break the shell only at the place whence the voice comes without any injury to the chick.

Hippocrates[10] in his *De natura pueri* adds another indication of the chick's desire to leave the egg—namely, that when the chick lacks nourishment, it moves about vigorously seeking more abundant food, and the membranes around it are broken.   When the mother feels the chick moving vigorously she hatches it by breaking the shell.   This occurs on the twentieth day.

It is now clear, therefore, how many uses proceed at all times from the cavity described, and how important they are, especially for both the generation of the chick and its preservation.

Women who understand all these matters best because they are well versed in the procedure for setting eggs, do not set them for brooding except after an interval of six or seven days from the time of laying.   At this time, although the egg was previously full, or provided with too small a cavity, it is already dried out and evaporated by the surrounding air and thus acquires and forms a larger cavity in that region.   This is confirmed by Pliny,[11] who states that it is most advantageous to set eggs within seven days after they have been laid; he thinks that older or fresher eggs are infertile, and therefore when a fresh egg is set for incubation the conception turns out unsuccessfully.

EPIGENETIC THEORY OF DEVELOPMENT

HARVEY, William (English physician and naturalist, 1578–1657).   *Exercitatio XXXXV* in *Exercitationes de generatione animalium quibus accedunt quaedam de partv, de membranis ac humoribus vteri, et de conceptione,* London, 1651; tr. by R. Willis as *What is the material of the chick and how is it formed in the egg?* (= *Exercise the forty-fifth* in) *Anatomical exercises on the generation 6f animals, etc.,* in his *The works of W. H., M.D., physician to the king, professor of anatomy and surgery to the college of physicians,* London (for the Sydenham Society), 1847.

Although Harvey's intellectual and scientific genius is still somewhat apparent in this work, the results can only be regarded as distinctly less satisfactory than those in his great study on the heart and blood.   He is often, on the grounds of the following, said to have revived epigenesis, an Aristotelian tradition.   This is true, but it is to be noted that he

---

[10] *De natura pueri,* cap. 30; Littré, vol. 7, p. 537.

[11] *Nat. Hist.,* X, 54(75).   Misquoted: ten days not seven.

contrasts epigenesis not with preformation but with metamorphosis in which he includes not only the sudden transformation of undifferentiated precursors into organized beings, as in insects, but also spontaneous generation.

Since, then, we are of opinion, that for the acquisition of truth, we cannot rely on the theories of others, whether these rest on mere assertions, or even may have been confirmed by plausible arguments, except there be added thereto a diligent course of observation; we propose to show, by clearly-arranged remarks derived from the book of nature, what is the material foetus, and in what manner it thence takes its origin.   We have seen that one thing is made out of another (tanquam ex materia) in two ways, and this as well in works of art, as in those of nature, and more particularly in the generation of animals.

One of these ways; viz., when the object is made out of something pre-existing, is exemplified by the formation of a bed out of wood, or a statue from stone; in which case, the whole material of the future piece of work has already been in existence, before it is finished into form, or any part of the work is yet begun; the second method is, when the material is both made and brought into form at the same time.   Just then as the works of art are accomplished in two manners, one, in which the workman cuts the material already prepared, divides it, and rejects what is superfluous, till he leaves it in the desired shape (as is the custom of the statuary); the other, as when the potter educes a form out of clay by the addition of parts, or increasing its mass, and giving it a figure, at the same time that he provides the material, which he prepares, adapts, and applies to his work; (and in this point of view, the form may be said rather to have been *made* than *educed;*) so exactly is it with regard to the generation of animals.

Some, out of a material previously concocted, and that has already attained its bulk, receive their forms and transfigurations; and all their parts are fashioned simultaneously, each with its distinctive characteristic, by the process called metamorphosis, and in this way a perfect animal is at once born; on the other hand, there are some in which one part is made before another, and then from the same material, afterwards receive at once nutrition, bulk, and form: that is to say, they have some parts made before, some after others, and these are at the same time increased in size and altered in form.   The structure of these animals commences from some one part as its nucleus and origin, by the instrumentality of which the rest of the limbs are joined on, and this we say takes place by the method of epigenesis, namely, by degrees, part after part; and this is, in preference to the other mode, generation properly so called.

In the former of the ways mentioned, the generation of insects is effected where by metamorphosis a worm is born from an egg; or out of a putrescent material, the drying of a moist substance or the moistening of a dry one, rudiments are created, from which, as from a caterpillar grown to its full size, or from an aurelia, springs a butterfly or fly already of a proper size, which never attains to any larger growth after it is first born; this is called metamorphosis. But the more perfect animals with red blood are made by epigenesis, or the superaddition of parts. In the former, chance or hazard seems the principal promoter of generation, and there, the form is due to the potency of a pre-existing material; and the first cause of generation is 'matter' rather than 'an external efficient'; whence it happens too that these animals are less perfect, less preservative of their own races, and less abiding, than the red-blooded terrestrial or aquatic animals, which owe their immortality to one constant source, viz. the perpetuation of the same species; of this circumstance we assign the first cause to nature and the vegetative faculty.

Some animals then are born of their own accord, concocted out of matter spontaneously, or by chance, as Aristotle seems to assert, when he speaks of animals whose matter is capable of receiving an impulse from itself, viz. the same impulse given by hazard, as is attributable to the seed, in the generation of other animals. And the same thing happens in art, as in the generation of animals. Some things, which are the result of art, are so likewise of chance, as good health; others always owe their existence to art; for instance, a house. Bees, wasps, butterflies, and whatever is generated from caterpillars by metamorphosis, are said to have sprung from chance, and therefore to be not preservative of their own race; the contrary is the case with the lion and the cock; they owe their existence as it were to nature or an operative faculty of a divine quality, and require for their propagation an identity of species, rather than any supply of fitting material.

In the generation by metamorphosis forms are created as if by the impression of a seal, or, as if they were adjusted in a mould; in truth the whole material is transformed. But an animal which is created by epigenesis attracts, prepares, elaborates, and makes use of the material, all at the same time; the processes of formation and growth are simultaneous. In the former the plastic force cuts up, and distributes, and reduces into limbs the same homogeneous material; and makes out of a homogeneous material organs which are dissimilar. But in the latter, while it creates in succession parts which are differently and variously distributed, it requires and makes a material which is also various in its nature, and variously distributed, and such as is now adapted to the formation of one part, now of another; on which account we believe the perfect hen's-egg to be constituted of various parts.

Now it appears clear from my history, that the generation of the chick from the egg is the result of epigenesis, rather than of metamorphosis, and that all its parts are not fashioned simultaneously, but emerge in their due succession and order; it appears, too, that its form proceeds simultaneously with its growth, and its growth with its form; also that the generation of some parts supervenes on others previously existing, from which they become distinct; lastly, that its origin, growth, and consummation are brought about by the method of nutrition; and that at length the foetus is thus produced. For the formative faculty of the chick rather acquires and prepares its own material for itself than only finds it when prepared, and the chick seems to be formed and to receive its growth from no other than itself. And, as all things receive their growth from the same power by which they are created, so likewise should we believe, that by the same power by which the chick is preserved, and caused to grow from the commencement, (whether that may have been the soul or a faculty of the soul,) by that power, I say, is it also created. For the same efficient and conservative faculty is found in the egg as in the chick; and of the same material of which it constitutes the first particle of the chick, out of the very same does it nourish, increase, and superadd all the other parts. Lastly, in generation by metamorphosis the whole is distributed and separated *into* parts; but in that by epigenesis the whole is put together *out of* parts in a certain order, and constituted *from* them.

Wherefore Fabricius was in error when he looked for the material of the chick, (as a distinct part of the egg, from which its body was formed,) as if the chick were created by metamorphosis, or a transformation of the material in mass; and as if all, or at least the principal parts of the body sprang from the same material, and, to use his own words, were incorporated simultaneously. (He is, therefore, of course opposed to the notion) of the chick being formed by epigenesis, in which a certain order is observed according to the dignity and the use of parts, where at first a small foundation is, as it were, laid, which, in the course of growth, has at one and the same time distinct structures formed and its figure established, and acquires an additional birth of parts afterwards, each in its own order; in the same way, for instance, as the bud bursting from the top of the acorn, in the course of its growth, has its parts separately taking the form of root, wood, pith, bark, boughs, branches, leaves, flowers, and fruit, until at length out comes a perfect tree; just so is it with the creation of the chick in the egg: the little cicatrix, or small spot, the foundation of the future structure, grows into the eye and is at the same time separated into the colliquament; in the centre of which the punctum sanguineum pulsans commences its being, together with the ramification of the veins; to these is presently added the nebula, and the first concretion of the future body; this also, in proportion as its bulk increases, is gradually divided and distinguished into parts, which however do not all emerge at the same time, but one after the other, and each in its proper order. To conclude, then: in the generation of those animals which are created by epigenesis, and are formed in parts, (as the chick in the egg,) we need not seek one material for the incorporation of the foetus, another for its commencing nutrition and growth; for it receives such nutrition and growth from the same material out of which it is made; and, vice versa, the chick in the egg is constituted out of the materials of its nutrition and growth. And an animal which is capable of nutrition is of the same potency as one which is augmentative, as we shall afterwards show; and they differ only, as Aristotle says, in their distinctness of being; in all other respects they are alike. For, in so far as anything is convertible into a substance, it is nutritious, and under certain conditions it is augmentative: in virtue of its repairing a loss of substance, it is called nutriment, in virtue of its being added, where there is no such loss of substance, it is called increment. Now the material of the chick, in the processes of generation, nutrition, and augmentation is equally to be considered as aliment and increment. We say simply that anything is generated, when no part of it has pre-existed; we speak of its being nourished and growing when it has already existed. The part of the foetus which is first formed is said to be begotten or born; all substitutions or additions are called adnascent or aggenerate. In all there is the same transmutation or generation from the same to the same; as concerns a part, this is performed by the process of nutrition and augmentation, but as regards the whole, by simple generation; in other respects the same processes occur equally. For from the same source from which the material first takes its existence, from that source also does it gain nutriment and increase. Moreover, from what we shall presently say, it will be made clear that all the parts of the body are nourished by a common nutritious juice; for, as all plants arise from one and the same common nutriment, (whether it be dew or a moisture from the earth,) altered and concocted in a diversity of manners, by which they are also nourished and grow; so likewise to identical fluids of the egg, namely, the albumen and the yelk, do the whole chick and each of its parts owe their birth and growth.

We will explain, also, what are the animals whose generation takes place by metamorphosis, and of what kind is the pre-existent material of insects which take their origin from a worm or a caterpillar; a material from which, by transmutation alone, all their parts are simultaneously constituted and embodied, and a perfect animal is born; likewise, to what animals any constant order in the successive generation of their parts attaches, as is the case with such as are at first born in an imperfect condition, and afterwards grow to maturity and perfection; and this happens to all those that are born from an egg. As in these the processes of growth and formation are carried on at the same time, and a separation and distinction of parts takes place in a regularly observed order, so in their case is there no immediate pre-existing material present, for the incorporation of the foetus, (such as the mixture of the semina of the male and female is generally thought to be, or the menstrual blood, or some very small portion of the egg,) but as soon as ever the material is created and prepared, so soon are growth and form commenced; the nutriment is immediately accompanied by the presence of that which it has to feed.

And this kind of generation is the result of epigenesis as the man proceeds from the boy; the edifice of the body, to wit, is raised on the punctum saliens as a foundation; as a ship is made from a keel, and as a potter makes a vessel, as the carpenter forms a footstool out of a piece of wood, or a statuary his statue from a ᵇlock of marble. For out of the same material from which the first part of the chick or its smallest particle springs, from the very same is the whole chick born; whence the first little drop of blood, thence also proceeds its whole mass by means of generation in the egg; nor is there any difference between the elements which constitute and form the limbs or organs of the body, and those out of which all their similar parts, to wit, the skin, the flesh, veins, membranes, nerves, cartilages, and bones, derive their origin. For the part which was at first soft and fleshy, afterwards, in the course of its growth, and without any change in the matter of nutrition, becomes a nerve, a ligament, a tendon; what was a simple membrane becomes an investing tunic; what had been cartilage is afterwards found to be a spinous process of bone, all variously diversified out of the same similar material. For a similar organic body (which the vulgar believe to consist of the elements) is not created out of elements at first existing separately, and then put together, united, and altered; nor is it put together out of constituent parts; but, from a transmutation of it when in a mixed state, another compound is created: to take an instance, from the colliquament the blood is formed, from the blood the structure of the body arises, which appears to be homogeneous in the beginning, and resembles the spermatic jelly; but from this the parts are at first delineated by an obscure division, and afterwards become separate and distinct organs.

Those parts, I say, are not made similar by any successive union of dissimilar and heterogeneous elements, but spring out of a similar material through the process of generation, have their different elements assigned to them by the same process, and are made dissimilar. Just as if the whole chick was created by a command to this effect, of the Divine Architect: "let there be a similar colourless mass, and let it be divided into parts and made to increase, and in the meantime, while it is growing, let there be a separation and delineation of parts; and let this part be harder, and denser, and more glistening, that be softer and more coloured," and it was so. Now it is in this very manner that the structure of the chick in the egg goes on day by day; all its parts are formed, nourished, and augmented out of the same material. First, from the spine arise the sides, and the bones are distinguishable from the flesh by minute lines of extreme whiteness; in the head three bullae are perceived, full of crystalline fluid, which correspond to the brain, the cerebellum, and one eye, easily observable by a black speck; the substance which at first appears a milky coagulum, afterwards gradually becomes cartilaginous, has spinous processes attached to it, and ends in being completely osseous; what was at first of a mucous nature and colourless, is converted at length into red flesh and parenchyma;

what was at one time limpid and perfectly pure water, presently assumes the form of brain, cerebellum, and eyes. For there is a greater and more divine mystery in the generation of animals, than the simple collecting together, alteration, and composition of a whole out of parts would seem to imply; inasmuch as there the whole has a separate constitution and existence before its parts, the mixture before the elements. But of this more at another time, when we come to specify the causes of these things.

THE QUEST FOR THE MAMMALIAN EGG

de (DE) GRAAF, Regnier (Reinier) (Dutch physician and anatomist, 1641–1673). *De mulierum organis generationi inservientibus (tractatus novus demonstrans tam homines et animalia caetera omnia quae vivipara dicuntur, haud minus quam ovipara ab ova originem ducere)*, Leyden, 1672; tr. by G. W. Corner of chap. xii entire as *On the female testes or ovaries*, in *Essays in biology in honor of Herbert M. Evans*, Berkeley and Los Angeles, 1943; with the permission of the Director of the University of California Press, publisher.

From a misinterpreted observation, de Graaf draws a true and important inference. The misinterpretation was the mistaking of follicle for egg. The true inference was that mammalian "testes" produce ova. This work stimulated the long search for the egg itself* culminating in its discovery by von Baer† in 1827. De Graaf also cannulated the pancreas.‡ His unquestioned genius as a biologist was hampered in its expression by his profession of Catholicism in an officially protestant country; he died when but thirty-two years old.

The testes of women differ much from those of the male as to position, form, size, substance, integuments, and function, as we are about to describe.

Thus, they have not an external position as in men, but are located in the lowest portion of the abdominal cavity about two finger breadths from each side of the fundus of the uterus, to which they are attached by a strong ligament which is called "Vas Deferens" by many anatomists, because they believed that semen was transferred through it from the testes to the uterus; on the other sides they are firmly attached to the peritoneum about the region of the iliac bone by the spermatic vessels, which supply them, and by the membranes with which the spermatic vessels are involved; so that the testes, fixed on each side, as if suspended, reach about the same level as the fundus of the uterus in the non-pregnant; in the pregnant, however, although they follow the fundus of the uterus to some extent, they do not rise to an equal degree, and thus, the more the fundus of the uterus rises, the farther they are from it, always keeping a lower position.

---

* See A. W. Meyer, *The rise of embryology*, p. 98.
† E. von Baer, *De ovi mammalium et hominis genesi*, Leipzig, 1827.
‡ *De succi pancreatici natura et usu*, in *Opera omnia*, Leiden, 1677.

The testicles are not suspended by any cremaster muscle, although some state this opinion, following Soranus.

They are located in the interior cavity of the abdomen, in order that they may be nearer the uterus and serve the better and more easily their intended purpose, which will be fully demonstrated below.

The testes of women, since they are broad and flattened on their anterior and posterior sides, differ much from those of men, for in their lower part they have a semi-oval bulge, while in the upper part, which the blood vessels enter, they appear more flat than humped, so that the testicles when separated from the blood vessels and the ligaments present a somewhat flattened semi-oval form.

The surface is more uneven than in males, because on account of the contents it projects unequally here and there, and displays certain small fissures in different places from time to time, due to depression or retraction of its coverings. Moreover, their size varies not a little with age, for in developing girls and [in women] in the flower of their life they weigh almost one and a half drachms, so that they attain a size about half that of the male testis, although in proportion they are wider and more succulent. In the old and decrepit, they are smaller, firmer, and more dried up, and slowly wither more and more, but never disappear completely; we have observed that the smallest testicles of old women weigh one scruple. In newborn and young infants they usually weigh from five grains to half a scruple; and therefore are smaller in these than in the very old, although most anatomists say they are larger in infants and gradually diminish with the thymus gland. In exceptional cases, however, the testicles grow to a remarkable size and contain within them such a quantity of fluid that they are dropsical: of which condition Schenck gives many examples in his observations, and Riolan and others, as well.

Moreover, the coverings of these testicles differ much from the male, for the latter are enveloped by many tunics, so that although hanging freely, they are protected from all injury; but the former are not altogether deprived of such protection, since they are invested by a tunic peculiar to themselves, called *dartos* by Galen; which although it is only moderately tough, is not easily removed from the substance of the testes, for it adheres to them as if continuous with their substance.

A membrane arising from the peritoneum covers the upper part of the [male] testicles and the blood vessels supplying them, and the same is generally considered to be true also of the female testicles. Some, however, who neither by boiling [the ovary] nor by any other artifice have been able to distinguish this *membrana propria* from the peritoneum, by a different appearance, agree with us that the [female] testes are covered with a single membrane originating from the peritoneum, and that it seems thicker [than the

covering of the male testis] because it is so much more firmly bound and united
to the [subjacent] parenchyma that one can scarcely see how it may be separated from it or divided into many membranes; but since this is not a matter
of great importance, let us leave the decision as to the number of these coverings free to everyone.

When the covering of the testes is removed, their whitish substance is revealed, which in every way differs from the substance of the male testicles,
for the former, excluding from consideration certain membranes and nutritive
vessels, are composed of seminal vessels which if mutually joined to each other,
would exceed twenty or even forty ells in length; the testicles of women are
not composed of similar vessels and no one, diligent as he may be, can in the
least separate them [into vessels].

Their internal substance is chiefly composed of many membranes and fibrils,
loosely bound to one another, in the interstices of which are found many
bodies which are either normal or abnormal. The normal structures, regularly found in the membranous substance of the testicles just described, are
vesicles full of liquor, nerves, and nutritive vessels, which [that is, the blood
vessels] run to the testes in almost the same way as in males, as we have already described, and course throughout the whole of their substance, and
enter the vesicles, within whose tunics many branches end after free division,
in just the same way as we have seen happening in the ovaries of fowls composed of clustered egg yolks.

As to whether the lymphatics found in the testes enter their substance, we
have not yet determined with sufficient clearness to venture an assertion, although this seems probable.

Those structures, which though normal, are only at certain times found
in the testes of women, are globular bodies in the form of conglomerate
glandulae which are composed of many particles, extending from the center
to the circumference in straight rows, and are enveloped by a special membrane. We assert that these globules do not exist at all times in the testicles
of females; on the contrary, they are only detected in them after coitus, [be-
ing] one or more in number, according as the animal brings forth one or more
foetuses from that congress. Nor are these always of the same nature in all
animals, or in the same kind of animal; for in cows they exhibit a yellow
color, in sheep red, in others ashen; because a few days after coitus they are
composed of a thinner substance and contain in their interior a limpid liquor
enclosed in a membrane, which when ejected with the membrane leaves only a
small space within the body which gradually disappears, so that in the latter
months of gestation they seem to be composed of a solid substance; but when
the foetus is delivered these globular bodies again diminish and finally disappear.

Finally, the abnormal objects sometimes found in the testes of women are
hydatids, calculi, steatomata, and other similar things.

From what has just been said, everyone will readily gather that it is the vesicles or their contents solely, which the nerves, arteries, veins, integuments, and the other structures normally observed in the testes are designed to serve.

These vesicles have been described under various names by Vesalius, Fallopius, Volcher Coiter, Laurentius, À Castro, Riolan, Bartholin, Wharton, Dom. de Marchettis, and others, whose accounts it would be too tedious to repeat here in full; it will not, however, be amiss to quote two of them at the present time, in order that the truth may be confirmed by their words. Fallopius says in his anatomical observations: "I have seen in them indeed certain vesicles, as it were, swollen out with a watery humor, in some yellow, in others transparent"; À Castro, also (lib. 1, cap. 4: "De Natura Mulierum"), says: "The testes have within them, besides the vessels, certain cavities full of a thin and watery humor which is like whey or white of egg." Some call these vesicles hydatids, but the celebrated Dr. van Horne in his *Prodromum* preferred to call them *ova*, a term which since it seems to me more convenient than the others, we shall in the future use, and we shall call these vesicles *ova* as does that distinguished man, on account of the exact similitude which they exhibit to the eggs contained in the ovaries of birds; for these, while they are still small contain nothing but a thin liquor like albumen. That albumen is actually contained in the ova of women will be beautifully demonstrated if they are boiled, for the liquor contained in the ova of the testicles acquires upon cooking the same color, the same taste and consistence as the albumen contained in the eggs of birds.

It is of no importance that the ova of women are not, like those of fowls, enveloped in a hard shell, for the latter are incubated outside the body in order to hatch the chickens, but the former remain within the female body during development, and are protected as thoroughly from all external injuries by the uterus as by a shell.

But before we proceed farther in their description it must be determined whether they are found in animals of all kinds and in what way they differ from hydatids.

We may assert confidently that eggs are found in all kinds of animals, since they may be observed not only in birds, in fishes, both oviparous and viviparous, but very clearly also in quadrupeds and even in man himself. Since it is known to everyone that eggs are found in birds and fishes, this needs no investigation; but also in rabbits, hares, dogs, swine, sheep, cows, and other animals which we have dissected, those structures similar to vesicles exhibit themselves to the eyes of the dissectors like the germs of eggs in birds. Occurring in the superficial part of the testicles, they push up the common tunic, and sometimes shine through it, as if their exit from the testis is impending.

These ova differ much in animals of various kinds, for we have observed that in rabbits and hares they scarcely exceed the size of rape seeds; in swine and sheep they reach the size of a pea or larger; in cows they often exceed the size of a cherry.

It must be noted, however, that in these animals, besides the large ova, lesser ones are found, of which some are so minute that they may scarcely be seen, for age and sexual intercourse cause great changes in the ova. In younger animals they are smallest, in older animals they become greater, and after coitus they are changed into the globules formerly described, of which one or more are formed, according to the number of embryos produced by the animal. These ova are so plentiful that we have sometimes seen twenty or more, filled with very clear liquor, in one testicle. Believing that these conditions are found in all animals dissected as yet by us, we asked the eminent Dr. N. Steno if he would deign to communicate to us what he had observed of the female testes in various other animals, of which we had not sufficient specimens, or which we had had no opportunity to dissect. He granted our request and generously informed us that in fallow deer, guinea pigs, badgers, red deer, wolves, asses, even in mules, and in other animals he had found ova of diverse sizes. These observations, combined with our own, more than sufficiently confirm the finding of ova in the females of all species. If any one inquires why they are present in the aged and in mules, which are incapable of reproduction, we say only that they are no more serviceable than the uterus, [male] testes and other reproductive organs customarily found in these as well as in fertile animals; many reasons may be given for their sterility, as for instance an improper conformation of their organs, insufficiency of the material of the ova for conception, or many other of the possible causes for sterility.

Since we have said above that we have sometimes found in the substance of the testicles or in their membranes vesicles of another kind very similar to the ova, it becomes important to cite in this place the chief differences between them and the ova.

Vesicles of the other kind, called *hydatids,* are usually formed with a double tunic. The interior layer, although very thin, is by no means difficult to separate from the exterior and the liquid contents is not easily coagulated by boiling. On the contrary, the common coats of the ova are separated from each other with great difficulty and their liquor is coagulated by boiling; hence, whenever we have found in testes which have been boiled some vesicles filled with hardened substance and others with a liquid humor, we have considered the former ova, the latter hydatids. It must be added that the hydatids now and then are suspended from the membranes of the testicles as if by a peduncle, which as yet we have never found to be the case with true ova.

These ova arise and are developed in the testes in exactly the same way as the eggs in the ovaries of birds, inasmuch as the blood flowing to the testes through the nutritive arteries deposits in their membranous substance materials suitable for the formation and nourishment of the ova, and the residual humors are carried back to the heart through the nutritive veins or lymphatic vessels. After the ova acquire their normal size they become invested with numerous tunics or follicles and in these immediately after sexual intercourse a kind of glandular substance grows up, of which the contents of the globules just described is composed. We shall attempt below to explain the purpose for which nature has so arranged [them].

Thus, the general function of the female testicles is to generate the ova, to nourish them, and to bring them to maturity, so that they serve the same purpose in women as the ovaries of birds. Hence, they should rather be called ovaries than testes because they show no similarity, either in form or contents, with the male testes properly so called. On this account, many have considered these bodies useless, but this is incorrect, because they are indispensable for reproduction. This is proved by the remarkable convolutions of the nutritive vessels about them, and is confirmed by the castration of females, which is invariably accompanied by sterility. Varro writes that spayed cows will conceive, if they copulate immediately; a thing which is no doubt true of males, in which the seminal vesicles are still filled with spermatic fluid [after castration], but not of females, in which such vesicles are not present. Hoffmann's assertion appears incorrect, a fact which we should amply demonstrate, if it had not already been done by Wharton. How the ova are fertilized and proceed to the uterus will be explained in the following chapters.

EARLY EXPERIMENTAL STUDIES
OF "SPONTANEOUS GENERATION"

REDI, Francesco (Florentine physician and naturalist, 1621–1697). From *Esperienze intorno alla generazione degl'insetti, fatte da Francesco Redi e da lui scritte in una lettera all'illustrisimo Signor Carlo Dati*, Florence, 1688; tr. by M. Bigelow as *Experiments on the generation of insects by Francesco Redi*, Chicago, 1909.

Philosophically committed to accept the evidence of observation and experiment, Redi found it impossible to remain outside the conflict, then raging, between the Aristotelians of the Jesuit schools and the new disciples of empirical objectivity. Living in time to witness the persecutions of Galileo, however, Redi contrived to steer a safer course: experimenting, practicing medicine, writing verse, and traveling as a favorite with the court of his Medicean patrons. His thesis here is *omne vivum* not *e vivo* (at least not from a parent of the same species) but *ex ovo*. This is clear from a comparison of the first and third excerpts.

But that great philosopher of our time, the immortal William Harvey, also held that all living things *are born from seed* as from an egg, be it the seed of animals of the same species or elsewhere derived; thus he says, "Because this is common to all living creatures, viz: that they derive their origin either from semen or eggs whether this semen have proceeded from others of the same kind, or have come by chance from something else. For what some-times happens in art occasionally occurs in nature also; those things, namely, take place by chance or accident which otherwise are brought about by art; of this health (according to Aristotle) is an illustration. And the thing is not different as respects generation (in so far as it is from seed) in certain animals; their semina are either present by accident, or they proceed from an univocal agent of the same kind. For even in fortuitous semina there is an inherent motive principle of generation, which procreates from itself and of itself, and this is the same as that which is found in the semina of congenera-tive animals—a power, to wit, of forming a living creature." . . .

Although content to be corrected by any one wiser than myself, if I should make erroneous statements, I shall express my belief that the Earth, after having brought forth the first plants and animals at the beginning by order of the Supreme and Omnipotent Creator, has never since produced any kinds of plants or animals, either perfect or imperfect, and everything which we know in past or present times that she has produced, came solely from the true seeds of the plants and animals themselves, which thus, through means of their own, preserve their species. And, although it be a matter of daily observation that infinite numbers of worms are produced in dead bodies and decayed plants, I feel, I say, inclined to believe that these worms are all generated by insemination and that the putrefied matter in which they are found has no other office than that of serving as a place, or suitable nest where animals deposit their eggs at the breeding season, and in which they also find nourishment; otherwise, I assert that nothing is ever generated therein. And, in order, Signor Carlo, to demonstrate to you the truth of what I say, I will describe to you some of those insects, which being most common, are best known to us.

In being thus, as I have said, the dictum of ancients and moderns, and the popular belief, that the putrescence of a dead body, or the filth of any sort of decayed matter engenders worms; and being desirous of tracing the truth in the case, I made the following experiment:

At the beginning of June I ordered to be killed three snakes, the kind called eels of Aesculapius. As soon as they were dead, I placed them in an open box to decay. Not long afterwards I saw that they were covered with worms of a conical shape and apparently without legs. These worms were intent on devouring the meat, increasing meanwhile in size, and from day to day I observed that they likewise increased in number; but, although of the same shape, they differed in size, having been born on different days. But all, little and big, after having consumed the meat, leaving only the bones intact, escaped from a small aperture in the closed box, and I was unable to discover their hiding place. Being curious, therefore, to know their fate, I again pre-pared three of the same snakes, which in three days were covered with small worms. These increased daily in number and size remaining alike in form, though not in color. Of these, the largest were white outside, and the smallest ones, pink. When the meat was all consumed, the worms eagerly sought an exit, but I had closed every aperture. On the nineteenth day of the same month some of the worms ceased all movements, as if they were asleep, and appeared to shrink and gradually to assume a shape like an egg. On the twentieth day all the worms had assumed the egg shape, and had taken on a golden white color, turning to red, which in some darkened, becoming almost black. At this point the red, as well as the black ones, changed from soft to hard, resembling somewhat those chrysalides formed by caterpillars, silkworms, and similar insects. My curiosity being thus aroused, I noticed that there was some difference in shape between the red and the black eggs (pupae), though it was clear that all were formed alike of many rings joined together; nevertheless, these rings were more sharply outlined, and more apparent in the black than in the red, which last were almost smooth and without a slight depression at one end, like that in a lemon picked from its stalk, which further distinguished the black egg-like balls. I placed these balls separately in glass vessels, well covered with paper, and at the end of eight days, every shell of the red balls was broken, and from each came forth a fly of gray color, torpid and dull, misshapen as if half finished, with closed wings; but after a few minutes they commenced to unfold and to expand in exact proportion to the tiny body, which also in the meantime had acquired symmetry in all its parts. Then the whole creature, as if made anew, having lost its gray color, took on a most brilliant and vivid green; and the whole body had expanded and grown so that it seemed incredible that it could ever have been contained in the small shell. Though the red eggs (pupae) brought forth green flies at the end of eight days, the black ones labored fourteen days to produce certain large black flies striped with white, having a hairy abdomen, of the kind that we see daily buzzing about the butchers' stalls.

Having considered these things, I began to believe that all worms found in meat were derived directly from the droppings of flies, and not from the putrefaction of the meat, and I was still more confirmed in this belief by having observed that, before the meat grew wormy, flies had hovered over it, of the same kind as those that later bred in it. Belief would be vain without the confirmation of experiment, hence in the middle of July I put a snake, some fish, some eels of the Arno, and a slice of milk-fed veal in four large, wide-mouthed flasks; having well closed and sealed them, I then filled the same number of flasks in the same way, only leaving these open. It was not long before the meat and the fish, in these second vessels, became wormy and flies were seen entering and leaving at will; but in the closed flasks I did not see a worm, though many days had passed since the dead flesh had been put in them. Outside on the paper cover there was now and then a deposit, or a maggot that eagerly sought some crevice by which to enter and obtain nourishment. Meanwhile the different things placed in the flasks had become putrid. . .

Not content with these experiments, I tried many others at different seasons, using different vessels. In order to leave nothing undone, I even had pieces of meat put under ground, but though remaining buried for weeks, they never bred worms, as was always the case when flies had been allowed to light on the meat. One day a large number of worms, which had bred in some buffalo-meat, were killed by my order; having placed part in a closed dish, and part in an open one, nothing appeared in the first dish, but in the second worms had hatched, which changing as usual into egg-shaped balls (pupae), finally became flies of the common kind. In the same experiment tried with dead flies, I never saw anything breed in the closed vessel.

Hence I might conjecture that Father Kircher, though a man worthy of esteem, was led into erroneous statements in the twelfth book of *"The Subterranean World,"* where he describes the experiment of breeding flies in the dead bodies of the same. "The dead flies", says the good man, "should be besprinkled and soaked with honey-water, and then placed on a copper-plate exposed to the tepid heat of ashes; afterward very minute worms, only visible through the microscope, will appear, which little by little grow wings on the back and assume the shape of very small flies, that slowly attain perfect size." I believe, however, that the aforesaid honey-water only serves to attract the living flies to breed in the corpses of their comrades and to drop their eggs therein. . . .

Leaving this long digression and returning to my argument, it is necessary to tell you that although I thought I had proved that the flesh of dead animals could not engender worms unless the semina of live ones were deposited therein, still, to remove all doubt, as the trial had been made with closed vessels into which the air could not penetrate or circulate, I wished to attempt a new experiment by putting meat and fish in a large vase closed only with a fine Naples veil, that allowed the air to enter. For further protection against flies, I placed the vessel in a frame covered with the same net. I never saw any worms in the meat, though many were to be seen moving about on the net-covered frame. These, attracted by the odor of the meat, succeeded at last in penetrating the fine meshes and would have entered the vase had I not speedily removed them. It was interesting, in the meanwhile, to notice the number of flies buzzing about which, every now and then, would light on the outside net and deposit worms there. I noted that some left six or seven at a time there and others dropped them in the air before reaching the net. Perhaps these were of the same breed mentioned by Scaliger, in whose hand, by a lucky accident, a large fly deposited some small worms, whence he drew the conclusion that all flies bring forth live worms directly and not eggs. But what I have already said on the subject proves how much this learned man was in error. It is true that some kinds of flies bring forth live worms and some others eggs, as I have proved by experiment.

. . . I hope it will not tire you if I describe another little animal mentioned by Father Kircher, supposed to breed in rotten cane and straw. While with the Court this year at Artiminio, for the hunt, I saw on the broom in the woods an infinite number of queer little creatures, called by the peasants "cavallucci" (mantis). I found these to be of two sorts; some were green with two white parallel lines running the length of the body, and the others were rusty red, like the stalks of broom; both kinds have little horns, with many articulations. They move slowly and solemnly. They have six legs, and every leg has three joints; the two forelegs arise just under the part to which the head is atttached. The head is very small, less than a grain of wheat; the eyes are hard and upturned, and smaller than a poppy-seed, and are red. All the space between the last pair of legs and the tip of the tail is composed and marked by ten rings, incisions, or knots; and from the last of these knots, two very fine spurs protruded. The whole body is not longer than the width of five fingers, and is of the same size from head to tail; though some are larger in the abdomen, these are always females, and are large-bellied in proportion to the number of eggs they carry. The males, as well as the females, cast their skin whole, like snakes, spiders, and other insects, the skin being nothing more than a fine white tunic shaped like the body. When these little animals were brought to me I was so fortunate as to have with me Signor Nicholas Steno, of Denmark, a famous anatomist, as you know, and a man of gracious and amiable manners, who is being royally entertained at this Court by His Highness, the Grand Duke. It occurred to both of us to examine the entrails and the internal structure of these little creatures as far as their minuteness would allow. We saw that a canal, starting from the mouth and extending through the body to an aperture near the last joint of the tail, performs the functions of the oesophagus, stomach, and intestine; and around this little canal we found a confused mass of many and divers filaments that are, perhaps, veins and arteries. From the middle of the body to the end of the tail we saw a large number of eggs, bound together and enclosed in a sac hardly discernible on account of its thinness. These eggs were not larger than millet grains, and some were soft and tender; others were hard. The soft ones seemed yellowish and almost transparent; but the hard ones, though yellow inside, had a black shell; and taking them all together, black and yellow, in a single animal we counted up to seventy. While we were thus engaged we observed that notwithstanding the fact that we had torn the entrails out of some of these animals, they continued to live and to move in the same way as do disembowelled reptiles. Whereupon we cut off the heads of others, and the head lived for a short time without the body; but the headless body was extremely lively and groped about as if in full possession of its parts. Then, just for a joke, and to amuse the company at the Villa, we resolved to graft the head on the body again, which we succeeded in doing with the same ease as the enchanter, Orrilo, who put his dismembered body together, and of whom the epic poet of Ferrara thus sings:

> Oft times asunder his stout limbs they hewed,
> Nor could they thus succeed the man to kill,
> For hand or leg cut off, with ready skill,
> The bold enchanter, in its place, renewed.
> His head does Grifon cleave, now, to the chin,
> Now Aquilante's sword sinks to his chest;
> He laughs at this blow as at all the rest.
> In helpless rage they curse this Satan's kin.
> Hast seen the slippery silver break and fall,
> Called mercury by some old alchemist?
> How fast it runs in many a shining ball,
> Then joins in one, and not a drop is miss'd!
> Thus Orrilo does seek his severed head,
> Nor till he finds it, leaves off stumbling round;
> Then seizing it by the long locks of red,
> Or by the nose, straight on his neck, 'tis bound.
> Now Grifon hurls him, with a mighty hand,
> Into the stream, but this does nought avail;
> Swimming below, without an ache or ail,
> Orrilo crosses to the other strand.

Thus our little animals, with their grafted heads, lived, not only all that day, but for five continuous days, to the great surprise of all who were not in the secret. And in that state they not only dropped their excrement, but even laid their eggs. Hence, an overhasty writer would have had many eyewitnesses to vouch for the truth of this experiment, but in asserting the restoration of the heads as genuine, he would be writing sheer nonsense, for the heads adhered to the trunks by means of a green, viscous fluid, that oozed from the bodies, and on drying, caused the parts to join firmly together; but though the bodies lived, the heads gave no sign of life; the trunks, with or without the heads, continued to live for five days. In case you should have the curiosity to see how these little creatures look, without seeking them in Kircher or Johnston, I send them to you, here enclosed, drawn from life, together with a drawing of one of their eggs, enlarged by means of a very fine microscope. You will notice that one end is oval and the other has a raised border, which resembles one of those divided, wooden eggs, that we use for boxes, and that screw together in the middle. . . .

Who knows? Perhaps many of the fruits of trees are produced with a secondary, rather than a primary purpose, not as pre-eminent in themselves, but as objects of utility, destined as a matrix for the generation of these worms, which remain in them for a determined length of time, and thence come forth to enjoy the sunshine.

I think that my idea will not seem paradoxical to you, if you consider the great variety of growths, such as glands, galls, knobs, warts, etc., produced by oaks, holm oaks, live oaks, and other acorn-bearing trees. In the hairy tufts of the oak, and in the woody tufts of the ilex, and in the galls of the holm oak leaf, it is manifest that the first and principal intention of Nature is to create therein a winged animal, for there is an egg in the interior of the gall, and this egg enlarges and matures in proportion to the development of the gall, and finally gives birth to the worm, which, when the gall has reached maturity, becomes a fly, that breaking the egg and commencing to gnaw the gall, makes a narrow and always round road from the center to the circumference, and abandoning its native prison, escapes and flies boldly away in search of food. . . .

It would be superfluous to make any further remarks on this subject, as some parts are not entirely new to you, such as my experiments made at Artiminio when the Court stayed there last year to enjoy the delightful sports of the chase. So I will keep silence in good faith, begging the continuation of your interest in another work which I am preparing to publish, i.e., a history of divers fruits and animals generated by oaks and other trees.

PREFORMATION THEORY

SWAMMERDAM, Jan (Dutch naturalist and microscopist, 1637–1680). From *Biblia naturae*, first published (tr. by H. D. Gaubius) as *Bijbel de natuure, of, historie der insecten, tot zeekere zoorten gebracht: etc.*, Leyden, 1737 (posth.); tr. by T. Flloyd (revision by J. Hill as *The book of nature or, the history of insects reduced to distinct classes, confirmed by particular instances, etc.*, London, 1758.

When one considers the wealth of small-scale organization which Swammerdam's microscope revealed to him, it is not strange that he should have been attracted by the notion of a minute preformed individual. His arguments for preformation were sufficiently persuasive to supplant, at least briefly, Harvey's ideas to the contrary. The epigenetic interpretation was revived shortly after by C. F. Wolff (see Wolff, Bonnet).

After an attentive examination of the nature and fabrick of the least and largest animals, I cannot but allow the less an equal, or perhaps superior degree in dignity. Whoever duly considers the conduct and instinct of the one, with the manners and actions of the other, must acknowledge all are under the direction and controul of a supreme and singular intelligence; which, as in the largest, it extends beyond the limits of our comprehension, escapes our researches in the smallest. If, while we dissect with care the larger animals, we are filled with wonder at the elegant disposition of their limbs, the inimitable order of their muscles, and the regular direction of their veins, arteries, and nerves; to what an height is our astonishment raised, when we discover all these parts arranged in the least, in the same regular manner. How is it possible but we must stand amazed when we reflect that those animalcules, whose little bodies are smaller than the finest point of our dissecting knife, have muscles, veins, arteries, and every other part common to the larger animals? Creatures so very diminutive, that our hands are not delicate enough to manage, or our eyes sufficiently acute to see, them; insomuch that we are almost excluded from anatomizing their parts, in order to come at the knowledge of their interior construction. Thus, what we know of the fabrick of those creatures reaches no farther than to a simple enumeration of the parts which we have before observed in larger creatures. We are not only thus in the dark, in attempting a discovery of the construction of the least animalcules, but we even gain very little knowledge of the wonderful texture of the viscera of the largest animals: for as the point of our dissecting knife is not minute enough to separate the tender parts of the small animals, it is not less unfit to be used in discovering the extremities of the nerves and veins in the larger.

As our knowledge of both species of animals is so far limited by our ignorance, and as we have not hitherto had such a sufficient number of experiments as are necessary to form a proper judgment of their elegant structure, and the admirable disposition of their parts, we may easily see how rash and precipitate their opinion is, who esteem the larger creatures only as perfect, and the less as scarce worthy to be classed with animals; but, as they say, produced by chance, or generated from putrefaction; rendering, by such reasoning, the constant order of nature subject to chance. But as it happens to the smallest of animals, for instance, to those produced from the egg of the Acarus which is so minute, as scarcely to be visible, so also it is with the largest animals; their origin is not more obvious or more visible, perhaps it is rather more obscure, and they derive their being from a less visible beginning. Nor let any man imagine that I say this without conviction, since I have found by diligent inquiry that the largest animal is not in its first formation bigger than the rudiment of an Ant; and therefore, unless the Great Creator had set certain bounds to the growth of every kind, which it cannot exceed, I see no reason why the Ant might not surpass in bulk the largest. Perhaps, their sizes proceed in proportion to the greater or less strength of the heart, by which the parts must be extended, against the pressure of the atmosphere. Notwithstanding the smallness of Ants, nothing hinders our preferring them to the largest animals, if we consider either their unwearied diligence, their wonderful strength, or their inimitable propensity to labour; or, to say all in one word, their amazing and incomprehensible love to their young, whom they not only carry daily to such places as may afford them food, but, if by accident they are killed, and even cut into pieces, they, with the utmost tenderness, will carry them away piecemeal in their arms. Who can shew such an example among the largest animals, which are dignified with the title of perfect? Who can

find an instance in any other creature, that may come in competition with this?
But in the entrance of this work it is not my intention to explain the form and
wonderful propagation of animalcules, which seem to be exsanguious or to
have no blood: I shall treat in general of the manner of their surprising metamorphoses; and at the same time shew, that they not only resemble other animals in the increase of their parts, but that they exceed them by infinite
degrees.   . . .

Though, amongst all the mutations of nature which deserve our attention,
none appears more surprizing to the generality of mankind, than that by which
a Caterpillar assumes the form of a winged animal, it in reality deserves no
more admiration, than any other change in the forms of Bees, or the transformation observable in plants.   This will evidently appear to any one, who,
having examined the real nature of such metamorphoses, will observe how exactly they agree, not only with the growth of animals which undergo no such
change; but also with the shooting or budding out of plants and flowers.
Whatever difficulty we find in this, is merely an effect of our own mistaken
notions; and our admiration arises from our ignorance of the nature of the
Nymph or Chrysalis.   In this the little animal lies, like the flower in its bud.
Before I proceed farther on this head, it may be proper to observe, that these
words, Nymph and Chrysalis, signify the same thing, and that there is no difference in the nature of the subjects to which they are applied.   . . .

That we may succeed the better in examining the nature of this Nymph,
or Chrysalis, upon which, as upon an immoveable basis, the doctrine of all the
changes observable in insects is so evidently founded, that the jarring opinions
of all the naturalists who have hitherto wrote upon the subject, must appear
utterly vain; it is necessary to observe, that the Nymph, or Chrysalis, is nothing more than a change of the Caterpillar or worm; or, to speak more properly, an accretion, growth, or budding of the limbs and parts of the Caterpillar or worm, containing the embryo of the winged animal that is to proceed
from it.   The Nymph, or Chrysalis, may even be considered as the winged
animal itself hid under this particular form.   From whence it follows, that in
reality the Caterpillar, or worm, is not changed into a Nymph or Chrysalis;
nor, to go a step further, the Nymph or Chrysalis into a winged animal; but
that the same worm or Caterpillar, which, on casting its skin, assumes the
form of a Nymph or Chrysalis, becomes afterwards a winged animal.   Nor,
indeed, can it be said that there happens any other change on this occasion,
than what is observed in chickens, from eggs which are not transformed into
cock or hens, but grow to be such by the expansion of parts already formed.
In the same manner the Tad-pole is not changed into a Frog, but becomes a
Frog, by an unfolding and increasing of some of its parts.

Hence it follows, that in the Aurelia, and more particularly in the Nymph,
so called by Aristotle with the greatest propriety, there are not only all the

parts and limbs of the little winged animal itself; but, what is more surprising, though 'till now unnoticed by any author I have met with, all these parts, or limbs, are to be discovered, and may be shewn in the worm itself, on stripping off its skin in a careful manner. If therefore we retain the name of Nymph, used by Aristotle, the worm at this period may be considered as marriageable, and, if we may make use of these expressions, entering into the connubial state. We may further shew this, by considering that the worms, after the manner of the brides in Holland, shut themselves up for a time, as it were to prepare, and render themselves more amiable, when they are to meet the other sex in the field of Hymen. . . .

There are four orders which comprehend the whole class of insects, so that we cannot see one, which may not be referred to one or other of them, especially if we can see its change.

The first order will comprehend those insects, which, with all their limbs and parts, proceed instantly out of the egg, and grow insensibly, until they attain a proper size; after which they are changed into the Nymph, which undergoes no other change but that of its skin.

Of the second order are those hatched with six legs, and which, when the wings are gradually perfected, are also changed into Nymphs.

The third order is, when the Worm or Caterpillar comes forth from the egg either without any legs, or with six or more, and its limbs afterwards grow under the skin, in a manner imperceptible to our sight, until at length it casts that skin and resembles the Nymph, or Chrysalis.

The fourth order is, when the Worm likewise proceeds from the egg, either without any, or with six, or more legs, and in an invisible manner grows in its limbs and parts under the skin, and does not shed this skin, but acquires the form of a Nymph under it.

REVIVAL OF EPIGENESIS

WOLFF, Caspar Friedrich (German embryologist, 1733–1794). Two fragments from *Theoria generationis*, Halle, 1759; tr. by M. F. Hamburger for this volume.

The concept of epigenesis did not begin with Wolff nor was he first to seek empirical verification for it (see Harvey). Despite the extreme difficulties of the *emboîtement* theory, however, predelineation largely dominated the thought of the seventeenth and eighteenth centuries, being promoted by Leibnitz, Swammerdam (*q.v.*), and Haller. As to the exact value of Wolff's own observations, historians disagree.* Some at least of the

---

* Contrast Nordenskjöld's statement that Wolff's observations "rival in absurdity most of what had been perpetrated in their sphere" (*Die geschichte der biologie*, p. 253, Jena, 1926) with Wheeler's description of Wolff and Darwin as "ideal investigators, patterns for all time" (*Biological lectures from the Marine Biological Laboratory*, p. 278, Boston, 1899). See also A. Kirchhoff, *Friedrich Wolff, etc.*, in *Zeitschrift für Medizin und Naturwissenschaften*, vol. 4, p. 193, Jena, 1868.

derogatory criticism stems from a misunderstanding of Wolff's strongly logical and somewhat Aristotelian interpretive methods, which occasionally give the illusion of prejudice. His findings were, at any rate, sufficiently productive so that, when they came gradually to be known and received, his theories also prevailed, and a new era of detailed observation in embryology began. In 1766, at odds with the professors at the *Collegium Medicum* and others, Wolff, personally of gentle and amiable character, accepted an invitation from Empress Catherine to come and spend the rest of his life in St. Petersburg.

[Wolff describes the youngest chick embryo observed by him (28 hours of incubation) as] a mass which is characterized only by its external shape and its position but which otherwise consists merely of spherules [cells—Ed.], little connected with each other and simply heaped one upon another, transparent, mobile, and almost liquid, and which shows neither heart nor vessels, nor traces of red blood.

In general, one cannot well say that what is not accessible to our senses is therefore not existent. Yet, applied to these observations, this principle is more sophistic than true. The particles of which all animal organs are composed, in their primordial condition are spherules always discernible with a microscope of medium power of magnification. How could one maintain that one is unable to see a body because of its smallness when the parts of which it is composed are easily recognizable? Nobody has yet discovered with the help of a stronger lens parts not also visible with the help of weaker magnification. Either one cannot see them at all or they appear sufficiently large. Therefore, that parts are hidden because of their infinite smallness and that they then only gradually come forth is a fable. The way in which nature produces organic parts can be very well recognized in the developmental history of the extremities and the kidneys. . . .

The first primordia of the protuberances which indicate the extremities, are little elevations raised above the other cell substances. The cellular substance, however, which surrounds the vertebral column, and the adjacent substance, furnish the raw material for the elevation which will be structurally organized later on. There can be no doubt about this for one who himself has observed the successive transformation of this substance.

It is therefore intelligible why the formation of a part and its organization are not accomplished in the individual by one and the same act, so that a formed part would be *eo ipso* organized; rather, a part is first formed, and then organized. . . . Furthermore, the organization of the separate part, which is a process different from its first formation is continually perfected by the rise of parts still to be organized. . . .

EARLY EXPERIMENTAL STUDY OF REGENERATION

SPALLANZANI, Lazzaro (Italian priest, philosopher, and naturalist, 1729–1799).
From *An essay upon animal reproductions;* tr. from the MS by M. Maty, London, 1769.

Other chapters in this tract describe regeneration of other organs of amphibia and of worms and snails and a remarkable "discovery of tadpoles existing in the eggs of frogs before fecundation by the male." Spallanzani studied law, took orders, taught philosophy (at Modena, later Pavia) and conducted researches on generation and regeneration.

REPRODUCTION OF THE TAIL IN THE TADPOLE

The tadpoles are those aquatic animals, which grow into frogs or toads. The reproduction of their tail could not but take up a great part of my time, as much was to be learned from thence. The extreme transparency of the membranes is equivalent to the finest and most accurate dissection; since, besides shewing the texture of the solids, it gives the clearest view of the circulation of the fluids. On viewing therefore through a lens the tail newly produced, we have the advantage of examining how the fibres of the old part unite themselves with those of the new; at what time, by what means, and how the circulation passes on from the trunk to the reproduction; and lastly, what order is observed by nature in the growth of these fibres, and the addition of the fluids. Every body must be sensible of the great importance of all these things in the present subject.

The circulation of the blood in the new-born tadpole shews itself sooner in the bronchial vessels, or organs of respiration, than in the tail. This blood is then composed of small globules, of a pale yellow hue; this is likewise the color of the liquor which soon begins to run through the arteries of the tail; but the course of circulation is different. One half of the length of the tail is an aggregate of oblique muscles, parallel to each other, but converging towards the axis. The sides are composed of a membranous skin, spotted here and there in a very elegant manner.

Small rivulets, at first but few, afterwards in greater number, issue from the muscles, make many serpentine turns in the membrane, and, by fresh windings, conceal themselves behind the muscles. A dark veil does not permit the eye to observe the origin of these rivulets. The tadpole being somewhat older, the veil disappears, and the source of these ramifications shews itself in two real vessels; the one arterial, the other venous. The first takes its origin from the root of the tail, and runs to the top; where, after some turns upwards, it forms the second. Both run in a longitudinal direction all along, and very near the middle of, the tail. The vein issues forth before the artery.

The ramifications grow more and more numerous, and in a short time fill the whole tail. The sight of these numberless rivulets of blood affords real delight to the philosopher. This blood comes from the two great vessels; and after a greater or lesser number of turns, is brought back to them.

These ramifications at first appear few, afterwards copious, and lastly crowded; but were they successively formed, or did they exist from the first, and require nothing but to be gradually unfolded?

A portion of the tail being taken off by a section perpendicular to the axis, we discover wonderful phenomena about the circulation, both in the part cut off, and in what remains of it. These will be described in my book: I shall confine myself at present to some of the effects of the reproduction.

If the whole tail, or very near the whole, be cut off, the tadpoles go to the bottom of the water, and there lie down and perish. But if a lesser part be taken off, not one of them dies; and all without exception recover what they lost.

Nature observes the following laws in the growth of these reproductions. They are more considerable, when a great part of the tail is taken off; not so large after a lesser section; and least of all, when a very small bit has been cut off. The greatest length seems however rather to take place, when the tail is divided in the middle, than when the section is higher.

If the tadpoles were very young when cut, the reproduction appears very soon. In one summer day it makes the most rapid progress; and in a short time the new part not only equals that which was cut off, but the new part of the tail and the old one joined equal in every dimension the tail of unmutilated tadpoles born at the same time. The reproduction, being arrived at this height, continues to increase in the same proportion as the tail of similar animals, to which nothing has been done.

When, therefore, this operation is performed at different periods upon tadpoles of the same species, the reproduction of the second period is equally quick with that of the first.

But if the tadpoles be greatly advanced, the beginning of the reproduction is retarded; and all other circumstances being the same, its progress will be slower. Hence follows this law, which I always found unvaried; that the quickness of the reproduction, both in its beginning and growth, is in an inverse ratio to the age of the tadpole.

This rule equally takes place in the second, third, fourth, &c. reproductions, which constantly follow upon a second, third, &c. section; in a word, these successive regenerations are never found to fail as long as the tadpole keeps its tail.

The differences observed in the manifestation and increase of the new-produced part, are analogous to what is found in the trunk. In the most advanced state, the old part does not grow in the least; in the middle state or in youth it increases but little; but in infancy the growth is very rapid.

The tadpoles, to which no kind of nourishment is given, do not grow in size, at least sensibly; the legs do not come forth, nor are the membranes of the infant state cast off. I have kept some in that state of abstinence during the greatest part of the summer; and when these were still no bigger than a small pea, the tadpoles born at the same time, and continually fed, were, at least, ten times fuller and bigger than their fasting friends; nay the greatest part of the first had already got clear of their first envelopes, and were converted into frogs. Hence the want of food retards in a frog the progress towards the state of full growth; that is, in other words, it lengthens the periods of life in these animals, in the same manner that cold operated upon the caterpillars of Réaumur, who were slower in becoming crysallids; and, when crysallids, longer in becoming butterflies. But yet I was not a little surprised to observe, that, in these abstemious tadpoles, the tail was still reproduced, and considerably increased.

Hitherto we have seen the phenomena of reproduction, as they appear with the naked eye in the tadpole; let us now take a microscopical view of these appearances.

When any piece of the tail is separated by a section perpendicular to the axis, the sides, which, as we have seen, are formed of a membranous skin, are often the first that appear. The reproduction presents itself to the eye as a prolongation of the old membrane; it is only somewhat finer and more transparent.

Not long after this, a blackish thread issues forth from the axis or center of the trunk. Upon viewing it with a glass of a very great power, it appears to be nothing else but a contexture of longitudinal fibres, parallel to one another.

The blood of the great artery does not as yet reach the reproduction; but it comes close to the section by means of several ramifications, opening into the great vein, into which it discharges itself.

The issue of longitudinal threads increases in the mean while, by the addition of many more fibres arising from the sides; and growing larger in every dimension, it soon unites to the membranous skin. It then assumes the form of a small slip or pyramid; the extremity of which is that of the tender newborn reproduction, and the basis remains engrafted on the trunk.

The arterial blood then begins to pass the limits of the section, and to advance a little way among the new fibres; but it soon takes a turn to the part whence it came, re-enters the trunk, and by other branches gets into the large venous vessel. In proportion as the reproduction increases in bulk, the large artery throws more and more blood into it, by means of the increased number of its ramifications, which after some days become very considerable. The greatest part of these branches having been carried on to the extremity of the tail, they all turn up again towards the trunk: from arterial, they become venous branches; and having been distributed by many circumvolutions throughout the whole extent of the reproduction, they discharge all their blood, as usual, into the great vein. The same process is afterwards continued by the aforesaid ramifications; except only, that as their diameters increase, they carry, in consequence,.a greater quantity of blood.

Hence arises a very considerable difference between the circulation of the blood in the reproduction, and that in the original part. For although the two real vessels in the original part, viz. the artery and vein, do send off from their sides similar and very fine ramifications, yet they both keep themselves quite distinct from the root to the extremity of the tail; and, besides their direction, have a much larger size. This happens whatever be the age of the tadpole. On the contrary, these two canals, in passing from the old to the new parts, become less, and dividing, as was before said, into a vast number of serpentine ramifications, occupy the greatest space of the new produced tail.

It is also necessary to observe that this irregularity in the circulation does not only take place in the first reproduction, but is likewise observed in all the succeeding new tails of the same tadpole, when mutilated over and over again.

On considering the new organization with regard to the solids, the following phenomena present themselves. As to the membranous skin, we have already said that the new one only seems a continuation of the old, and this, at least in appearance, is likewise the case with the longitudinal and parallel fibres. On the untouched part of the tail, the oblique muscles, which unite in an angle at the axis, form at the same time a large bundle of fibres running towards the lower part of the tail, in a direction parallel to the axis, and this bundle remains cut by the mutilation of the tadpole. Now, if we examine the reproduction, when still growing, besides the evident regeneration of the oblique muscles, we shall find that the new longitudinal *fabrillae* coincide and join so well with the old divided fibres, that the first have all the appearance of being continued from the last.

Nevertheless it sometimes happens that a small fold or deviation from the right line shews itself at the point of union between the old and new fibres; but this blemish in time is either removed, or is at least not so apparent; and it is indeed very astonishing to see the effect of time, in making the new and the original unmutilated tails similar to each other.

When the tadpole is sufficiently advanced, the increased opacity of the natural tail prevents a microscopical view of the viscera inclosed in it; and the same obstacle presents itself in the part where the reproduction has been. This part being formed upon a trunk of some thickness, is likewise pretty large in its origin, and therefore not an object of microscopical observation. Dissection however, here steps in to our assistance, and shews that nature proceeds in the reproductions formed on the trunks of tails in the more advanced tadpole, in the same unalterable way as in those that are young, and whose tails are still tender.

If, instead of cutting off the tail in this manner, the membranous skin be taken from the tadpole at any age, without touching the muscles, another membrane, exactly like the former, succeeds, and upon removing this a third. The order only and position of the venous and arterial ramifications, differ from the situation of those, which are found winding in the skins of unmutilated tails.

The tadpoles I examined, are such as are changed into frogs and toads. Notwithstanding the diversity of species of these two animals, the organization of the tail is essentially alike, and the issue of my experiments was also the same.

DOCTRINE OF PREFORMATION

BONNET, Charles (Swiss naturalist, 1720–1793). From *Contemplation de la nature*, Amsterdam, 1769; tr. by J. Wesley in vol. **II** of his *A survey of the wisdom of God in the creation, etc. containing an abridgement of that beautiful work*, The contemplation of nature, *By Mr. B., of Geneva. Also an extract from Mr. Deuten's* Inquiry into the origin of discoveries attributed to the ancients.

Bonnet's preformation doctrine found support, or so it seemed to him, in his own studies of parthenogenesis (*q.v.*) and was related by him to his general philosophy of progress in nature and human nature. While Bonnet made many directly empirical studies, the actual origins of most of his ideas on this subject were traditional and speculative. A thoroughly objective approach could scarcely at this relatively late date (see Swammerdam and Wolff) have come up with such unsophisticated concepts as Bonnet's doctrines of hybridization and incapsulation.

We may easily comprehend, that all the parts of an animal have such strict and indissoluble connexions between them, that they must necessarily have always co-existed together. The arteries imply veins; both of these imply nerves; the latter the brain; this the heart; and all of them suppose a multitude of other organs.

In the germ of a chick there is at first perceived a vital point, whose constant motion attracts the attention of the observer. The alternate and quick contractions and dilatation of the living point, sufficiently indicate that it is the heart. But this heart seems to be without any covering, and to be placed on the outside of the body. Instead of appearing in the form of a minute pyramidical mass, it bears the resemblance of a semicircle. The other viscera appear successively, and range themselves after each other, round the living speck. We cannot as yet discover any general folding; all is transparent or nearly so; and we only perceive by little and little those teguments which are appointed to cover all the parts.

In its first beginnings the animal is almost entirely fluid. It assumes by degrees the consistence of a jelly. All the parts have at that time situations, forms, and proportions that differ greatly from those they will afterwards acquire. Their minuteness, softness, and transparency, serve to strengthen the illusion. We persuade ourselves that a bowel is naked, because the transparency of its coverings prevent our seeing them.

Would you have a short and easy demonstration of this? When the lungs of the chick are first perceivable, their size is but the thousandth part of an inch. It would have been visible at the fourth part of these dimensions, were it not endued with the most perfect transparency. The liver is much greater at its first appearance; its transparency alone renders it invisible. It is the same with respect to the kidneys; whilst they do not appear even to exist, they separate the urine. The heart forces the blood into the arteries sooner than we could imagine, and it can only be perceived by the growth of the embryo, which is never more accelerated than at the very beginning.

Many other facts concur with these to establish the pre-existence of organical wholes. We are now sensible that many insects multiply, like plants, by slips. We cut them into pieces, and each piece regenerates, and becomes a perfect animal. Earth worms are ranked in the number of these insects that are reproduced from their disjoined parts; and being very large, the phenomena of their regeneration is very perceptible. The piece that is cut off never acquires any growth; it always remains as the section left it; only it falls away in a greater or lesser degree. But after some time there appears a very small whitish pimple at its extremity, which increases by degrees in bulk and length. There are soon discovered rings, which are at first very small and very close. They spread themselves insensibly every way. New lungs, a new heart, a new stomach, disclose themselves, and with them a number of other organs. This piece, which is newly produced, is extremely slender, and altogether disproportioned to the part on which it grew. We may imagine that we see a worm growing, that it is grafted at the end of this stump, endeavouring to lengthen it. This little vermiform appendage unfolds itself slowly. At length it equals in thickness the piece from which it was cut, and exceeds it in length. It can no longer be distinguished from it but by its colour, which is somewhat fainter.

Here then is a new organical whole, which grows from an ancient one, and constitutes the same body: there is an animal slip that grows, and expands itself on the stump of an animal, as a vegetable slip does on the trunk of a tree. Remark that the flesh of the piece cut off does not in the least contribute to the formation of the part regenerated; the stump only nourishes the bud; it being the soil in which the latter vegetates. The part then that is reproduced passes through all the degrees of growth, by which the entire animal itself had before passed. It is a real animal, which pre-existed in a very minute form in the great animal that served for a matrix.

Vegetable productions exhibit to us the same consequences. If a tree be topped, that does not lengthen the trunk of it; but it sends forth a multitude of buds, in each of which a little tree is comprised; for the bud or branch that springs from it is a tree that is grafted on the trunk that nourishes it.

Every seed, in like manner, comprises a plant in miniature. On a very slight inspection, we may very easily discover the stalk, leaves, and root of this little plant. But the curious rise much higher, and distinguish in a bulbous root or growing bud those flowers that do not blow till the ensuing year.

When the evolution commences in an organized whole, its form differs so prodigiously from that which it will afterwards assume, that we should be apt to mistake it, were it not to accompany it in all its progress. Observe how the parts of a plant are folded together, entwined, or concentred in the seed or bud. Is this that majestic tree which will ere long overshadow a large space of ground? This the flower that will so gracefully display itself? This the fruit that will assume such a regular figure? You can now only perceive an unformed mass of knotted filaments; yet this little chaos may already contain in it a world, where all is organized and symmetrical.

You have seen frogs in their first state. They appear at that time to consist only of a large head and a long tail. Such is the chick when it begins to expand itself. A very slender tail, stretched in a straight line, is joined to a large head; and the tail contains all the rudiments of the composition; nay, is the very composition itself; and the transparent fluid in which it floats, constitutes the whole of those soft parts with which it is afterwards covered.

The same revolutions, therefore, which occasion the heart of the chick to be transformed from its semicircular shape to that of a pyramid, bring the chick itself to a state of perfection. If we were permitted to penetrate to the foundation of the mechanism whereby these successive changes are effected, what a degree of certainty would our knowledge of animal economy acquire? We should contemplate in an egg, the mysteries of the two kingdoms. And how greatly would our imagination of that adorable wisdom be increased, which by the simplest means ever attains the most noble ends?

Thus the more we ascend to the origin of organized beings, the more we are persuaded of their having pre-existed before their first appearance; not such as they first appear to us, but disguised; and were it possible for us to trace them still higher, we should undoubtedly find them still more disguised, and should be at a loss to conceive how they could afterwards acquire that form under which they present themselves to our view.

We can then form no idea of the primitive state of organized beings; that state which I conceive to be given them by the hand of Him who has ordained all things from the beginning.

The forms of vegetables and animals, which are so elegantly varied, are, in the system of this admirable pre-ordination, only the last results of that multitude of successive revolutions, they have been liable to, and which perhaps commenced at their first creation. How great would be our astonishment, could we penetrate into these depths, and pry into the abyss! We should there discover a world very different from ours, whose strange decorations would infinitely embarrass us. The state, in which we conceive all organized bodies to have been at first, is the germ state; and the germ contains in miniature all the parts of the future animal or vegetable. It does not then acquire organs which it had not before: but those organs which did not hitherto appear, begin now to be visible. We do not know the utmost limits of the division of matter; but we see that it has been divided in a prodigious degree. From the elephant to the mite, from the globe of the sun to a globule of light, what an inconceivable multitude of intermediate degrees are there! This animalcule enjoys the light; it penetrates into its eye; it there traces the image of objects; how extremely minute must this image be! And how much more minute must that of a globule of light be, when several thousands, and perhaps millions, enter at the same time into this eye! But great and small are nothing in themselves, and have no reality but in our imagination. It is possible, that all the germs of the same kind were originally joined or linked into each other, and that they are only unfolded from generation to generation, according to that progression which geometry endeavours to assign them. . . .

The yolk has its liquors, which are conveyed to it by the arteries belonging to it. They circulate, and without veins there is no circulation. But the arteries and veins of the yolk take their origin from the mesenteric arteries and veins of the foetus: the heart of this latter therefore is the principle of that circulation which is performed in the yolk. At the time of fecundation the foetus does not weigh the hundredth part of a grain. The yolk at that time weighs a dram. It has vessels proportioned to its size. Now if the germ existed entire before fecundation, that which we stile generation is not the same thing with it; but is only the beginning of an evolution, which will by degrees bring to open day such parts as were before hid in impenetrable darkness.

But the germ cannot be unfolded in an egg which has not been fecundated, and incubation would only accelerate its eruption. What does it then want to enable it to continue to grow? It has all the organs necessary for evolution. It has even already attained to a certain degree of growth, for eggs grow in young pullets; their ovaries contain them of all sizes. The germ grows there likewise. Why cannot it enfold itself more than it does? What secret force retains it within the limits of invisibility?

Growth depends on the impulsion of the heart. A greater degree of growth depends on a greater impulsion. This degree of impulsion, consequently, is wanting in the heart of the germ that has not been fecundated.

This demonstrates a certain resistance in the parts of the germ. As it grows, this resistance augments in proportion. Some resist more than others; the bony parts, or such as will hereafter become so, more than the membranous, or those that always must remain so.

The heart of the germ then hath need of a determinate strength to surmount this resistance. Its strength is in its irritability, or in the power it has of contracting itself on the touch of some liquid. Wherefore to augment the irritability of the heart, is to augment its impulsive force.

Fecundation, without doubt, increases this force, and that can alone increase it; since it is only by the intervention of it that the germ passes over the narrow limits that it retained in its first state. . . .

But if an ass cover a mare, there will be produced from this commerce an animal that will not properly be a horse, but a mule. Nevertheless a horse was delineated in miniature in the egg of a mare: how then was it transformed into a mule? Whence did it acquire these long ears and slender tail so different from those of the horse? Dissection increases the difficulty; that informs us that this kind of transformation does not only affect the exterior part of the animal, but the interior likewise. The voice of the mule is very like that of the ass, and does not at all resemble the neighing of a horse. The organ of the ass's voice is an instrument that is very much compounded. A drum of a singular structure, lodged within the larynx, is the principal part of this instrument. This drum does not exist in the horse, but is found in the mule.

The liquor furnished by the male consequently penetrates the germ, since it there produces such great changes. But these relations of the prolific liquor to the male that furnishes it, must necessarily depend on the organs that prepare it.

There are then in these organs vessels that separate the molecules relative to different parts of the great whole. These molecules are carried to the corresponding parts of the germ, since these parts are modified by the action of the prolific liquor. Therefore it incorporates itself with the germ, and is the first aliment of it, as I said above.

The organs of generation in the ass have then a relation to his ears and larynx; for they prepare a liquor which modifies the ears and larynx of the little horse enclosed in the egg. The prolific liquor creates nothing, but it may change what already exists. It does not engender the chick, which existed before fecundation.

Growth depends on nutrition; the latter on incorporation. At the same time that a part grows, it acquires solidity. An excess of growth in a part, then, supposes a super abundance of nutricious juices, or such as are more active. The excessive growth which the ears of the horse acquire by the influence of the liquor of the ass, indicates that this liquor contains more molecules, appropriated to the unfolding of the ears, than that of the horse, or that the molecules of the first are more active than those of the second.

The extreme softness, I should rather say, fluidity of the germ, renders every part of it extremely modifiable. Those changes which you cannot conceive in an adult, depend here on the slightest causes.

But if the fecundating liquor modifies the germ, this latter in its turn, modifies the action of that liquor. By virtue of its organization, it tends to preserve its primitive state, resist more or less every new arrangement, and never gives way without always retaining something of its primitive form. . . .

EARLY OPPOSITION TO SPONTANEOUS GENERATION

SPALLANZANI, Lazzaro (Italian priest, philosopher, and naturalist, 1729–1799). From *Dissertazione* (in later editions often *Opuscoli*) *di fisica animale e vegetable*, Modena, 1780; tr. by J. G. Dalyell as *Tracts on the natural history of animals· and vegetables*, Edinburgh, 1803.

Most Greek scientists believed in generation of living beings from nonliving matter. Redi (1628–1698) disproved this for maggots but accepted it for parasitic worms and gall flies. Acceptance likewise characterized the thinking of Buffon (1707–1788), Needham (1713–1781), Lamarck (1744–1829), and Pouchet (1800–1872). Swammerdam (1637–1680) and Spallanzani (see below) were among earlier opponents, their views acquiring final confirmation in the studies of Pasteur (1822–1895).

WHETHER, ACCORDING TO A NEW THEORY OF GENERATION, ANIMALCULA ARE PRODUCED BY A VEGETATIVE POWER IN MATTER. INFUSIONS AND INFUSED SUBSTANCES EXPOSED TO HEAT.

Nothing is more common with philosophers who have invented any theory, or given a new form to one already established, and universally known, than to republish it on some other occasion, corrected, improved, or illustrated, with additional information. If we would review our discoveries, if we would examine them profoundly and with impartiality, we should in general find defects unnoticed before, which arise from the want of connection in sentiment, from the want of a necessary and laudable perspicuity, or because they are discordant with more recent discoveries.

A certain vegetative power some have conceived to reside in matter, appropriated to the formation and regulation of organised existence; that by it are the numberless combinations of the animal machine effected; the operation of nutrition and perspiration, the variety of constitution, the animal appetites and dimensions of the human frame. By the same means has it been explained why a blind or a maimed person may have children vigorous and entire; because the vegetative power will restore to them the members defective in the parent.

Not only has it been supposed to be destined for the organization of matter in animated beings, but that it might change an animal to the vegetable state, and the vegetable again to an animal; that it acts on plants while living, and when dead regenerates them in new beings; these are the animalcula of infusions, which cannot strictly be called animals, but beings simply *vital*.

One proof adduced in support of this hypothesis, is derived from the origin of animalcula. We are told they must either come from specific seeds, or be produced by the vegetative power; that the first cannot take place, because they are found in close vessels subjected to the action of heat, equally as in open vessels, whereas the included germs, if there were any, ought not to survive. Therefore, they must originate from the vegetative power alone. Nothing has been omitted to obtain favourable arguments for this opinion, and to give it that clearness, elegance, and simplicity most likely to gain converts.

Nineteen vessels, containing infused substances, were hermetically sealed, and kept an hour in boiling water. Being opened at a proper time, not a single animalcula was to be seen. To this experiment of mine, it was objected that the long continuance of heat had perhaps entirely destroyed the vegetative power of the infused substances, or materially injured the elasticity of the air remaining included in the vessels; thus, it was not surprising if animalcula did not appear.

To estimate the weight of these objections, I conceived an experiment apparently decisive; which was, to make nineteen infusions, and boil some of them a short time, others longer, and the rest very long. If it was founded, the number of animalcula would be less according to the duration of boiling, if not, the number would be alike in all cases.

Vegetable seeds, being the most fit for producing animalcula, were preferred to other substances, and those that never failed to produce them though they had experienced the influence of heat. White kidney beans, vetches, buckwheat, barley, maize, the seeds of mallows and beets were infused; and, that the experiment might be the more accurate, I endeavoured as much as possible to take each species of seed from the same plant. As the yolk of an egg in maceration abounds with animalcula, one was also infused.

Experiment has demonstrated, that the heat of boiling water is not always the same, but greater, if the atmosphere is heavier; and less, if lighter: therefore, water will acquire more heat at one time than another, which will be proportioned to the state of the atmosphere. In this, and my other experiments, the seven different kinds of seeds, and the yolk, were all boiled an equal time, that they might acquire the same degree of heat. Here the experiment was diversified, by boiling a certain quantity of each infusion half an hour; another quantity, an hour; a third, an hour and a half; and a fourth, two hours. Thus, four classes of infusion, and the egg, could be formed. The same water, in which the seeds had boiled, was taken for the infusions, and what had boiled half an hour alone taken for the seeds that had boiled half an hour. The like proportions of time were preserved in the water for the other three classes of infusions; that is, an hour, one and a half, and two hours.

Each of the four classes was marked with a different number, to avoid all hazard of confusion or error: and, because an equal temperature was most essential, all were deposited in the same place. The vessels, containing the infusions, were not hermetically sealed, but loosely stopped with corks; the only object of this examination being to discover, whether long protracted ebullition would prejudice or destroy the property of infused substances in producing animalcula; if it did, there would be no difference whether the vessels were open or close.

The examination of one, or of few drops, will often induce an observer to suppose the infusion quite deserted, or very thinly inhabited, while the observation of many drops proves it to be otherwise. I was not content with one drop only, but uniformly took a considerable number from each infusion.

The surface of infusions is generally covered with a gelatinous scum, thin at first, and easily broken, which, in process of time, acquires consistence. Here, animalcula are always most numerous, as may be seen by a method I have constantly practiced, examining with a magnifier a portion placed in a strong light.

Where the animalcula are minute, or rare, the thickness of the infusion often prevents the observer from distinguishing whether any are there or not. It is then necessary to dilute the drops with water. Elsewhere it has been remarked, that distilled water was taken to make the infusions; common water might introduce some latent animalcule. In the course of these observations and experiments, distilled water has also been employed for dilution, when required; and, for greater security, examined with a magnifier before being used. In particular cases, the accidental concealment of a single animalcule might vitiate the truth of the experiment.

I conceive it my duty to mention precautions so essential, and to put it in every individual's power to judge not only of the experiments and observations themselves, but of the mode of conducting them in matters so nice and important.

On the 15 of September, I made thirty-two infusions; and on the 23 examined them for the first time. Animalcula were in all; but the number and species different in each. In the maize infusions, they were smaller, and proportionally more rare, according to the duration of boiling.

From this it may seem, that although long continued heat had not prevented the production of animalcula, it had contributed to diminish the number, or alter the kind. But with the rest of the infusions it was otherwise: the kidney beans, vetches, barley, and mallow seeds, were in a better condition, after sustaining the violent impression of heat two hours, than those that had been exposed to it less. Let us enter on that detail which the subject merits.

In the infusion of kidney beans, boiled two hours, were three species of animalcula; very large; middle sized; and very small. The figure of the first, partly umbellated and attached to long filaments dragged along in their progress; the second were cylindrical; and the third, globular. All three were incredibly numerous.

In the infusion boiled two hours, were animalcula of the largest and smallest class, but few in number; still fewer, in that boiled an hour; and fewest of all, in that boiled half an hour.

The infusion of mallows, boiled two hours, produced middle sized circular animalcula; and some very large, with the head extremity hooked. In two infusions, boiled an hour, and an hour and a half, the number and species were the same: and though they might be surpassed by those of the infusions boiled two hours, still they were much more numerous than in those boiled half an hour.

In vetches, boiled half an hour, was an immense number of semicircular bell-shaped animalcula, all of considerable size, while in those boiled an hour and a half, they were small and rare. Some bell-shaped animalcula might be seen in an infusion boiled an hour, but it gave the eye pain to discover a few, and these most minute, when it had boiled only half an hour.

Those in a barley infusion boiled two hours were numerous beyond description, and large; part of an elliptic figure, others oblong. The infusions boiled an hour and a half had but a moderate number of animalcula very minute; and some appeared when boiled half an hour.

There was no fixed rule with the remaining infusions. In buck-wheat boiled an hour and a half were many more animalcula than in any other infusions of it. This also happened in the egg and beet seed boiled an hour; but it is to be remarked, that fewer animalcula were in these two infusions boiled half an hour than in any of the rest.

Hitherto, the figure of these legions of animalcula has been cursorily alluded to. A circumstantial account is in my Dissertations, and it will be spoken of more at large in the course of the Tract.

Thus, it is clearly evident, that long boiling of seed infusions does not prevent the production of animalcula; and, notwithstanding the maize does not seem to favour it, four infusions strongly corroborate the fact.

What is the cause that infusions boiled least have fewest animalcula? I cannot think myself mistaken in assigning the following reason. That animalcula should appear, it is necessary that the macerating substances give some indication of the dissolution of their parts; and, in proportion as dissolution advances, at least for a limited time, the number of animalcula will increase. The uniformity of this has been shewn in another place, and would be confirmed, was it requisite, by further experiments and observations, in these new inquiries. Now, as seeds have boiled a shorter time, so are they less invested and penetrated by the dissolving power of heat; therefore, when set apart to macerate, they are not so soon decomposed as those longer boiled. Thus, there is no occasion for surprise if some infusions swarm with animalcula while others have very few: And this I do believe the reason why, when two infusions are made at the same time, one of unboiled, the other of boiled seeds, animalcula are frequently observed much sooner in the latter than in the former. A little boiling will not decompose vegetable seeds, for decomposition is effected by slow and gradual maceration.

Some days after these experiments, the number of animalcula always became greater; and towards the middle of October increased so much, that each of the thirty-two infusions was equally swarming. The only difference was in size, figure, and motion: I enjoyed this pleasing microscopic scene uninterrupted until the 10 of November; and it might have amused me longer had I continued to examine the infusions.

It ought not to be omitted, that experiments exactly similar were soon afterwards made with pease, lentils, beans, and hemp seed. Except in the beans, the result so far corresponded, that a greater number of animalcula appeared in the infusions that had boiled most.

It is a fact established by the universal concurrence of philosophers, that, after water has come to the state of ebullition, it cannot acquire a greater degree of heat, however much the action of the fire may be augmented, provided it can evaporate. Therefore, when I say the seeds boiled longest have acquired greater heat, I mean it to be understood in *time* and not *intensity*, by supposing that the duration of boiling encreased the intensity of heat the seeds would be exposed to.

Recourse was had to another experiment to learn whether an encrease of heat would obstruct the production of animalcula. The eleven species of seeds were slowly heated in a coffee roaster till they became pretty well roasted, and eleven infusions formed of them with water previously boiled as usual. But this heat, so much more intense, neither prevented the origin of animalcula nor lessened the number. They were rare at first; but about the middle of October, that is, twenty days after making the infusions, the fluid was so full as absolutely to appear animated.

The constancy of their appearing even here, excited my curiosity to augment the heat still more. The seeds were burnt and ground the same as we burn and grind coffee. Of the dust, which resembled soot, I made as many infusions as different kinds of seed: likewise, an infusion was made of the yolk of an egg, which by the thermometer had suffered 279° of heat. What followed? Animalcula equally appeared in these infusions, only a little more time elapsed before they became so numerous, because the weather was colder; and they uniformly inhabit infusions sooner or later according to the temperature of the atmosphere.

Vegetable seeds were exposed to trials more severe: they were exposed to the greatest heat that can be excited by common fires, or fire augmented by art. Burning coals, and the flame of the blow pipe, were the two agents exercising their power on them. And, in the first place, I kept them on an iron plate above burning coals until entirely consumed by the violence of the flames, and converted to a dry cinder, which was reduced to powder, and as many infusions formed as there were seeds. A cinder was also made by the blow pipe, which, besides excessive aridity, had acquired considerable hardness. I must acknowledge I did not in the least expect to find animalcula in this new infusion. After viewing them once and again, hardly able to credit my eyes, I repeated the experiment twice. Some suspicion arose that the animalcula might come from the water used rather than the burnt seeds; therefore, on repeating the experiment, the same as what formed the infusions was put in other vessels. Both times, however, they re-appeared in the burnt seeds, while not one was seen in the water.

These facts fully convinced me, that vegetable seeds never fail to produce animalcula, though exposed to any degree of heat; whence arises a direct conclusion, that the *vegetative power* is nothing but the work of imagination; and if no animalcula appear in vessels hermetically sealed and kept an hour in boiling water, their absence must proceed from some other cause.

MATERIALISM, EVOLUTION, AND EPIGENESIS

DARWIN, Erasmus (English naturalist, 1731–1801). From *Zoonomia, or the laws of organic life*, London, 1794 (vol. 1), 1796 (vol. 2).

Erasmus Darwin's *Zoonomia* displays a lively concern for most of the central questions of late eighteenth-century biology but cannot justly be said to have advanced their solution materially. Darwin's analysis of organic nature was mechanico-causal in emphasis with epigenesis its preeminent theme. He was stimulated to think and publish on this subject chiefly by the recent revival of preformation in the writings of Bonnet. Erasmus was grandfather of Charles Robert Darwin and Francis Galton.

OF MOTION.

The WHOLE OF NATURE may be supposed to consist of two essences or substances; one of which may be termed spirit, and the other matter. The former of these possesses the power to commence or produce motion, and the latter to receive and communicate it. So that motion, considered as a cause, immediately precedes every effect; and considered as an effect, it immediately succeeds every cause. And the laws of motion therefore are the laws of nature.

The MOTIONS OF MATTER may be divided into two kinds, primary and secondary. The secondary motions are those, which are given to or received from other matter in motion. Their laws have been successfully investigated by philosophers in their treatises on mechanic powers. These motions are distinguished by this circumstance, that the velocity multiplied into the quantity of matter of the body acted upon is equal to the velocity multiplied into the quantity of matter of the acting body.

The primary motions of matter may be divided into three classes, those belonging to gravitation, to chemistry, and to life; and each class has its peculiar laws. Though these three classes include the motions of solid, liquid, and aerial bodies; there is nevertheless a fourth division of motions; I mean those of the supposed ethereal fluids of magnetism, electricty, heat, and light; whose properties are not so well investigated as to be classed with sufficient accuracy. . . .

The third class includes all the motions of the animal and vegetable world; as well those of the vessels, which circulate their juices, and of the muscles, which perform their locomotion, as those of the organs of sense, which constitute their ideas.

This last class of motion is the subject of the following pages; which, though conscious of their many imperfections, I hope may give some pleasure to the patient reader, and contribute something to the knowledge and to the cure of diseases. . . .

LAWS OF ANIMAL CAUSATION.

I. The fibres, which constitute the muscles and organs of sense, possess a power of contraction. The circumstances attending the exertion of this power of CONTRACTION constitute the laws of animal motion, as the circumstances attending the exertion of the power of ATTRACTION constitute the laws of motion of inanimate matter.

II. The spirit of animation is the immediate cause of the contraction of animal fibres, it resides in the brain and nerves, and is liable to general or partial diminution or accumulation.

III. The stimulus of bodies external to the moving organ is the remote cause of the original contractions of animal fibres.

IV. A certain quantity of stimulus produces irritation, which is an exertion of the spirit of animation exciting the fibres into contraction.

V. A certain quantity of contraction of animal fibres, if it be perceived at all, produces pleasure; a greater or less quantity of contraction, if it be perceived at all, produces pain; these constitute sensation.

VI. A certain quantity of sensation produces desire or aversion; these constitute volition.

VII. All animal motions which have occurred at the same time, or in immediate succession, become so connected, that when one of them is reproduced, the other has a tendency to accompany or succeed it. When fibrous contractions succeed or accompany other fibrous contractions, the connexion is termed association; when fibrous contractions succeed sensorial motions, the connexion is termed causation; when fibrous and sensorial motions reciprocally introduce each other, it is termed catenation of animal motions. All these connexions are said to be produced by habit, that is, by frequent repetition. These laws of animal causation will be evinced by numerous facts, which occur in our daily exertions; and will afterwards be employed to explain the more recondite phaenomena of the production, growth, diseases, and decay of the animal system.

Generation.

From this account of reproduction it appears, that all animals have a similar origin, viz. from a single living filament; and that the difference of their forms and qualities has arisen only from the different irritabilities and sensibilities, or voluntarities, or associabilities, of this original living filament; and perhaps in some degree from the different forms of the particles of the fluids, by which it has been at first stimulated into activity. And that from hence, as Linnaeus has conjectured in respect to the vegetable world, it is not impossible, but the great variety of species of animals, which now tenant the earth, may have had their origin from the mixture of a few natural orders. And that those animal and vegetable mules, which could continue their species, have done so, and constitute the numerous families of animals and vegetables which now exist; and that those mules, which were produced with imperfect organs of generation, perished without reproduction, according to the observation of Aristotle; and are the animals, which we now call mules. . . .

Secondly, when we think over the great changes introduced into various animals by artificial or accidental cultivation, as in horses, which we have exercised for the different purposes of strength or swiftness, in carrying burthens or in running races; or in dogs, which have been cultivated for strength and courage, as the bull-dog; or for acuteness of his sense of smell, as the hound and spaniel; or for the swiftness of his foot, as the greyhound; or for his swimming in the water, or for drawing snow-sledges, as the rough-haired dogs of the north; or lastly, as a play-dog for children, as the lap-dog; with the changes of the forms of the cattle, which have been domesticated from the greatest antiquity, as camels, and sheep; which have undergone so total a transformation, that we are now ignorant from what species of wild animals they had their origin. Add to these the great changes of shape and colour, which we daily see produced in smaller animals from our domestication of them, as rabbits, or pigeons; or from the difference of climates and even of seasons; thus the sheep of warm climates are covered with hair instead of wool; and the hares and partridges of the latitudes, which are long buried in snow, become white during the winter months; add to these the various changes produced in the forms of mankind, by their early modes of exertion; or by the diseases occasioned by their habits of life; both of which became hereditary, and that through many generations. Those who labour at the anvil, the oar, or the loom, as well as those who carry sedan-chairs, or who have been educated to dance upon the rope, are distinguishable by the shape of their limbs; and the diseases occasioned by intoxication deform the countenance with leprous eruptions, or the body with tumid viscera, or the joints with knots and distortions. . . .

When we consider all these changes of animal form, and innumerable others, which may be collected from the books of natural history; we cannot but be convinced, that the fetus or embryon is formed by apposition of new parts, and not by the distention of a primordial nest of germes, included one within another, like the cups of a conjurer.

Fourthly, when we revolve in our minds the great similarity of structure which obtains in all the warm-blooded animals, as well quadrupeds, birds, and amphibious animals, as in mankind; from the mouse and bat to the elephant and whale; one is led to conclude, that they have alike been produced from a similar living filament. In some this filament in its advance to maturity has acquired hands and fingers, with a fine sense of touch, as in mankind. In others it has acquired claws or talons, as in tygers and eagles. In others, toes with an intervening web, or membrane, as in seals and geese. In others it has acquired cloven hoofs, as in cows and swine; and whole hoofs in others, as in the horse. While in the bird kind this original living filament has put forth wings instead of arms or legs, and feathers instead of hair. In some it has protruded horns on the forehead instead of teeth in the fore part of the upper jaw; in others tushes instead of horns; and in others beaks instead of either. And all this exactly as is daily seen in the transmutations of the tadpole, which acquires legs and lungs, when he wants them; and loses his tail, when it is no longer of service to him.

Fifthly, from their first rudiment, or primordium, to the termination of their lives, all animals undergo perpetual transformations; which are in part produced by their own exertions in consequence of their desires and aversions, of their pleasures and their pains, or of irritations, or of associations; and many of these acquired forms or propensities are transmitted to their posterity. . . .

From thus meditating on the great similarity of the structure of the warm-blooded animals, and at the same time of the great changes they undergo both before and after their nativity; and by considering in how minute a portion of time many of the changes of animals above described have been produced; would it be too bold to imagine, that in the great length of time, since the earth began to exist, perhaps millions of ages before the commencement of the history of mankind, would it be too bold to imagine, that all warm-blooded animals have arisen from one living filament, which THE GREAT FIRST CAUSE endued with animality, with the power of acquiring new parts, attended with new propensities, directed by irritations, sensations, volitions, and associations; and thus possessing the faculty of continuing to improve by its own inherent activity, and of delivering down those improvements by generation to its posterity, world without end? . . .

If this gradual production of the species and genera of animals be assented to, a contrary circumstance may be supposed to have occurred, namely, that some kinds by the great changes of the elements may have been destroyed. This idea is shewn to our senses by contemplating the petrifactions of shells, and of vegetables, which may be said, like busts and medals, to record the history of remote times. Of the myriads of belemnites, cornua ammonis, and numerous other petrified shells, which are found in the masses of limestone, which have been produced by them, none now are ever found in our seas, or in the seas of other parts of the world, according to the observations of many naturalists. Some of whom have imagined, that most of the inhabitants of the sea and earth of very remote times are now extinct; as they scarcely admit, that a single fossil shell bears a strict similitude to any recent ones, and that the vegetable impressions or petrifactions found in iron-ores, clay, or sandstone, of which there are many of the fern kind, are not similar to any plants of this country, nor accurately correspond with those of other climates, which is an argument countenancing the changes in the forms, both of animals and vegetables, during the progressive structure of the globe, which we inhabit.

This idea of the gradual formation and improvement of the animal world accords with the observations of some modern philosophers, who have supposed that the continent of America has been raised out of the ocean at a later period of time than the other three quarters of the globe, which they deduce from the greater comparative heights of its mountains, and the consequent greater coldness of its respective climates, and from the less size and strength of its animals, as the tygers and allegators compared with those of Asia or Africa. And lastly, from the less progress in the improvements of the mind of its inhabitants in respect to voluntary exertions.

This idea of the gradual formation and improvement of the animal world seems not to have been unknown to the ancient philosophers. Plato having probably observed the reciprocal generation of inferior animals, as snails and worms, was of the opinion, that mankind with all other animals were originally hermaphrodites during the infancy of the world, and were in process of time separated into male and female.[1] The breasts and teats of all male quadrupeds, to which no use can be now assigned, adds perhaps some shadow of probability to this opinion. Linnaeus[2] excepts the horse from the male quadrupeds, who have teats; which might have shewn the earlier origin of his existence; but Mr. J. Hunter asserts,[3] that he has discovered the vestiges of them on his sheath, and has at the same time enriched natural history with a very curious fact concerning the male pigeon; at the time of hatching the eggs both the male and female pigeon undergo a great change in their crops; which thicken and become corrugated, and secrete a kind of milky fluid, which

---

[1] *Symposium*, 190–191.

[2] See p. 31.

[3] *Observations on certain parts of the Animal Economy.* London, 1762.

coagulates, and with which alone they for a few days feed their young, and afterwards feed them with this coagulated fluid mixed with other food. How this resembles the breasts of female quadrupeds after the production of their young! and how extraordinary, that the male should at this time give milk as well as the female!

The late Mr. David Hume, in his posthumous works,[4] places the powers of generation much above those of our boasted reason; and adds, that reason can only make a machine, as a clock or a ship, but the power of generation makes the maker of the machine; and probably from having observed, that the greatest part of the earth has been formed out of organic recrements; as the immense beds of limestone, chalk, marble, from the shells of fish; and the extensive provinces of clay, sandstone, ironstone, coals, from decomposed vegetables; all which have been first produced by generation, or by the secretions of organic life; he concludes that the world itself might have been generated, rather than created; that is, it might have been gradually produced from very small beginnings, increasing by the activity of its inherent principles, rather than by a sudden evolution of the whole by the Almighty fiat.— What a magnificent idea of the infinite power of THE GREAT ARCHITECT! THE CAUSE OF CAUSES! PARENT OF PARENTS! ENS ENTIUM!

For if we may compare infinities, it would seem to require a greater infinity of power to cause the causes of effects, than to cause the effects themselves. This idea is analogous to the improving excellence observable in every part of the creation; such as in the progressive increase of the solid or habitable parts of the earth from water; and in the progressive increase of the wisdom and happiness of its inhabitants; and is consonant to the idea of our present situation being a state of probation, which by our exertions we may improve, and are consequently responsible for our actions.

EMBRYOLOGY AND EVOLUTION

von BAER, Karl Ernst (German embryologist, 1792–1876). From *Ueber entwicke-lungs-geschichte der thiere*, Königsberg, 1828; tr. by T. H. Huxley as *On the development of animals, with observations and reflections* (= part II in) *Fragments relating to philosophical zoology. Selected from the works of K. E. von B.*, in *Scientific memoirs selected from the transactions of foreign academies of science and from foreign journals*, London, 1853.

Baer's important comparative embryological studies came at a time when two general biological theories vied for acceptance: the doctrine of types (see Goethe, Oken) and the doctrine of descent. From the vantage point of modern biology, the latter doctrine seems to cry out for recognition at every step of Baer's argument. The doctrine of types, however, was by no means without explanatory value, as the following excerpt shows. It is extremely difficult to discover from Baer's writings exactly to what extent he ad-

---

[4] *Dialogues Concerning Natural Religion.*

mits evolution as a fact; but it is clear that he felt it was often unwarrantedly invoked*
in situations where it was dispensable. Other achievements of von Baer were the discovery of mammalian egg† and the germ layer theory.‡

The further back we trace development, so much the more agreement do
we find among the most widely different animals, and thus we are led to the
question,—Are not all animals essentially similar at the commencement of
their development—have they not all a common primary form? We have
just remarked, that a distinct germinal disc probably exists in all true ova;
so far as we are acquainted with the development of germ-granules[1] (*Keim-
körner*), it seems to be wanting in them. They appear to be originally solid;
however it may be, that on their first separation from their parent, they have
an internal cavity like the central cavity of the yelk, which only escapes microscopic observation on account of the thickness of the often somewhat opake
wall. Supposing, however, they are at first solid, and eventually become hollow, as seemed to me to be the case with the germ-granules[1] of the Cercariae
and Bucephali, yet we perceive that the first act of their vital activity is to
acquire a cavity, whereby they become thick-walled, hollow vesicles. The
germ in the egg is also to be regarded as a vesicle, which in the Bird's egg
only gradually surrounds the yelk, but from the very first is completed as an
investment by the vitellary membrane; in the Frog's egg it has the vesicular form before the type of the Vertebrata appears, and in the Mammalian
from the very first it seems to surround the small mass of the yelk. Since,
however, the germ is the rudimentary animal itself, it may be said, not without
reason, that the simple vesicle is the common fundamental form from which
all animals are developed, not only ideally, but actually and historically. The
germ-granule passes into this primitive form of the independent animal immediately by its own power; the egg, however, only after its feminine nature
has been destroyed by fecundation. After this influence, the differentiation of
germ and yelk, or of body and nutritive substance, arises. The excavation of
the germ-granule is nothing else. In the egg, however, there is at first a solid
nutritive matter (the yelk), and a fluid in the central cavity; yet the solid
nutritive matter soon becomes fluid.

We remarked above, that to find a correspondence between two animal
forms, we must go back in development the further the more different these
two forms are; and we deduce thence, as the law of individual development,—

1. *That the more general characters of a large group of animals appear
earlier in their embryos than the more special characters.*

---

* Especially in J. G. Meckel, *Beytrage zur vergleichenden anatomie*, vol. 2, p. 11,
Leipzig, 1811.

† *De ovi mammalium et homini genesi*, Leipzig, 1827.

‡ In the same volume from which the present excerpt is taken.

[1] The reference is to germinal rudiments involved in asexual reproduction.

With this it agrees perfectly, that the vesicle should be the primitive form; for what can be a more general character of all animals than the contrast of an internal and an external surface?

2. *From the most general forms the less general are developed, and so on, until finally the most special arises.*

This has been rendered manifest above by examples from the Vertebrata, especially of the Birds, and also from the Articulata. We bring it forward again here only to append, as its immediate consequences, the following propositions concerning the object of investigation:—

3. *Every embryo of a given animal form, instead of passing through the other forms, rather becomes separated from them.*

4. *Fundamentally, therefore, the embryo of a higher form never resembles any other form, but only its embryo.*

It is only because the least developed forms of animals are but little removed from the embryonic condition, that they retain a certain similarity to the embryos of higher forms of animals.

This resemblance, however, if our view be correct, is nowise the determining condition of the course of development of the higher animals, but only a consequence of the organization of the lower forms.

The development of the embryo with regard to the type of organization, is as if it passed through the animal kingdom after the manner of the so-called *methode analytique* of the French systematists, continually separating itself from its allies, and at the same time passing from a lower to a higher stage of development. We represent this relation by the annexed Table:—

In detail it holds good as little as any other representation of organic relations upon a surface. Thus the single features adduced must pass for the whole characters, *e.g.*, the formation of wings and air-sacs for the whole character of Birds. The exposition, again, can only be very imperfect, since for most animals the investigation has hardly been commenced.

This scheme is only meant to bring clearly before the mind, how the first decisive distinction is whether the first rudiment is a true egg or a germ-granule; how, in the germs of ova, all animals are at first alike; how then the principal type becomes defined (which is called, origin of the embryo); whereby it remains undecided whether any radiate animal is developed from a true egg. If now the type of the vertebrate animal appears, the embryo is at first nothing but one of the Vertebrata without any particular characteristics. Chorda dorsalis, dorsal and abdominal tubes, gill-clefts, gill-vessels, and a heart with a single cavity, are formed in all. Then commences a differentiation. In a few, gill-laminae and no allantois are developed; in others, on the other hand, the gill-clefts coalesce, and an allantois buds forth. The former are aquatic animals, though not all permanently so: the others lead an aerial existence. The latter all acquire lungs. Let us follow out the former

SCHEME OF THE PROGRESS OF DEVELOPMENT

Highest grade of development

The animal rudiment is either:

a germ-granule (itself germ), or an ovum with a germ. In this arises:

? Radiate development................ ? Animals of the peripheral type.
Spiral development................ Animals of the massive type.
Symmetrical development................ Animals of the elongated type.

Doubly symmetrical development ... *Vertebrata.* They have a chorda dorsalis, dorsal plates, visceral plates, nerve tubes, gill-clefts, and acquire ...

Gills

No true lungs formed.
  The skeleton does not ossify...... *Cartilaginous fishes.*
  The skeleton ossifies...... *Osseous fishes.*

Lungs formed
  Amphibia. The gills.
    persist...... *Sirenidae.*
    do not { remain external...... *Urodela.*
    persist { become enclosed...... *Anura.*

No umbilical cord.
  No wings nor air-sacs...... *Reptilia.*
  Wings and air-sacs...... *Aves.*

A much-developed allantois.

An umbilical cord, *Mammalia,*

which falls off early
  without union with the parent?... *Monotremata.*
  after a short union with the parent *Marsupialia.*

which persists longer. The yelk-sac.
  grows for a long time
    very little...... *Rodentia.*
    moderately...... *Insectivora.*
    much...... *Carnivora.*
  The allantois grows

  grows little
    little...... *Quadrumana.*
    Umbilical cord. *Man*
    very long

  The allantois grows
    very long
    Placenta
      in scattered masses...... *Ruminantia.*
      evenly distributed... *Pachydermata Cetacea*

series first however. The embryos for a long time retain a great similarity; they push out long tails and scull about with them in the water. On the other hand, their extremities are developed very feebly and late, in relation to those of other embryos. They either never acquire true lungs, and so become fish, or else true lungs are formed. Among the latter the lungs are either feebly developed, in which case the gills are permanent and the animals become Sirenidae; or the lungs are better formed, and the gills either remain free until they cease to act (Salamanders), or they become covered over, the tail disappears, and with it all resemblance to a fish (tail-less Batrachia). In the second series of the Vertebrata, which never has external gills, the most essential distinction is perhaps this,—that in some a simple umbilicus is formed (Reptiles and Birds), in others this umbilicus is prolonged into a cord, after, as it seems, being altogether more rapidly formed.

In what manner Birds become separated from the Amphibia has already been shown. Probably a difference also arises very soon in the vascular system, whose metamorphoses in the Amphibia, however, are not yet known. While the gill-clefts in the Lizards are still open, the heart has just the same appearance as in Birds at the same period. Now just as in the Bird the special characters of the family and of the genus arise, so is it in the Mammalia. The Dog and the Pig are at first very much alike, and have short human faces. Still longer does the resemblance persist between the Pig and the Ruminant, whose lateral toes are at first almost as long as the two median ones. For the rest we are by no means sufficiently acquainted with the embryos of the Mammalia to state how and at what periods they become distinguishable from one another. We are best acquainted with the differences in the form and structure of the ova. Since these are very manifold in their form and in their relation to the parent, I have ventured, in order not to leave the Mammalia out of the Scheme, to divide them according to their ova. The embryos, in fact, may be distinguished into those which are born early and those which come into the world in a fully developed condition. Among the former the ova of the Monotremata are probably born undisturbed. In the Marsupialia the embryo has burst its membranes. The ova, which are retained longer, may be reduced to three principal divisions. In the first I place ova, in which the yelk-sac continues to grow for a long time. They yield Mammals with narrow hook-like nails (claws). In some the allantois is early arrested in its growth, and the placenta is limited to one spot, or two-lobed (Rodentia). In others the allantois is developed to a moderate extent (Insectivora): in all others it grows over the whole amnion transversely, and the placenta is annular (Carnivora). A second division of long-retained ova is formed by those in which the yelk-sac and the allantois are small; the placenta is one-sided, and is, as it would seem, in the opposite position of that of the Rodentia; the amnion and the umbilical cord are here largest. These ova produce animals with flat nails and three-lobed cerebral hemispheres. A third division has a yelk-sac which soon disappears, but an allantois which grows out immensely at its two extremities. These ova produce ungulated and finned animals; if the placenta is distributed over the whole ovum, but is collected in particular masses, we have animals with cleft hoofs; if it is distributed homogeneously, we have other Ungulata and Cetacea. Hence the principal differences of the Mammalia are marked very early in the ovum, for according as the allantois is much developed, or otherwise, does the ovum become long or short. In the former case, the embryo not only acquires a broader horny covering upon its fingers, but also a more complex stomach, and, in connexion therewith, long jaws, a flat articulation of the jaw, usually complex teeth, incapability of seizing and climbing, etc. It is the *plastic* series among the Vertebrata.

I must advert to an objection against the whole view here set forth, which may be based upon the circumstance that in some cases the embryos of nearly allied animals exhibit considerable differences at an early period. The embryos of the Ophidia, for instance, are very early rolled up, and so may be readily enough distinguished from Lizards. This plainly arises from the excessive length to which in this case the vertebrate type is drawn out.

Dissection, however, exhibits a great harmony in the internal structure; and since the posterior extremity of the Lizards also forms a spiral, the difference probably lies merely in this, that the vertebrate type in the Ophidia is more elongated, and it *seems*, in fact, to be greater than it *is*, because it presents itself so nakedly. Thus also the larvae of many families of Insects are in their external appearance very different in different families. Much probably depends in this case upon their shorter or longer sojourn in the egg. However, this objection, the only one which I have been able to discover against the view in question, can have little weight so long as no internal differences in the larvae have been demonstrated.

For the simple reason, that the embryo never passes from one principal type to another, it is impossible that it can pass successively through the whole animal kingdom. Our Scheme, however, shows at once that the embryo never passes through the form of any other animal, but only through the condition of indifference between its own form and others; and the further it proceeds, the smaller are the distinctions of the forms between which the indifference lies. In fact, the Scheme shows that the embryo of a given animal is at first only an indeterminate Vertebrate, then an indeterminate Bird, and so forth. Since at the same time it undergoes internal modification, it becomes in the whole course of its development a more and more perfect animal.

However, it may be objected here, if this be the true law of development, how comes it that so many good reasons could be adduced for that which has been previously in vogue? This may be explained readily enough. In the first place, the difference is not so great as it looks at first sight; and in the second, I believe that an assumption was made in the latter view, and it was afterwards forgotten that it had not been demonstrated; but especially, sufficient stress was not laid upon the distinction between type of organization and grade of development.

Since, in fact, the embryo becomes gradually perfected by progressive histological and morphological differentiation, it must in *this respect* have the more resemblance to less perfect animals the younger it is. Furthermore, the different forms of animals are sometimes more, sometimes less remote from the principal type. The type itself never exists pure, but only under certain modifications. But it seems absolutely necessary that those forms in which animality is most highly developed should be furthest removed from the fundamental type. In all the fundamental types, in fact, if I have discovered the true ones, there exists a symmetrical (*gleichmässige*) distribution of the organic elements. If now predominant central organs arise, especially a central part of the nervous system, according to which we must principally measure the extent of perfection, the type necessarily becomes considerably modified. The Worms, the Myriapoda, have an evenly annulated body, and are nearer the type than the Butterfly. If then the law be true, that in the course of the development of the individual the principal type appears first, and subsequently its modifications, the young Butterfly must be more similar to the perfect Scolopendra, and even to the perfect Worm, than conversely the young Scolopendra, or the young Worm, to the perfect Butterfly. Now if we leave out of sight the peculiarities of the Worm, the red blood, etc., which it attains at a later period, we may readily say that the Butterfly is at first a Worm. The same thing is obvious in the Vertebrata. Fishes are less distant from the fundamental type than Mammalia, and especially than Man with his great brain. It is therefore very natural that the Mammalian embryo should be more similar to the Fish than the embryo of the Fish to the Mammalian. Now if one sees nothing in the Fish but an imperfectly developed Vertebrate (and that is the baseless assumption to which we referred), the Mammalian must be regarded as a more highly developed Fish; and then it is quite logical to say that the embryo of a vertebrate animal is at first a Fish. Hence it was that I asserted above, that the view of the uniserial progression of animals was necessarily connected with the prevailing idea as to the law of development. But the Fish is not merely an imperfect vertebrate animal; it has besides its proper ichthyic characters, as development clearly shows.

But enough! I have attempted, in embodying the course of development, to show also, that the embryo of Man is unquestionably nearer to the Fish than conversely, since he diverges further from the fundamental type; and upon this ground alone has much been inserted that is problematical, as the umbilical attachment of the Monotremata. In detail, this representation can as little exhibit all the relations justly, as any other representation of organic relations upon a plane surface—even if the investigation were complete, instead of being just begun.

Let us sum up the contents of this section as its conclusion. The development of an individual of a certain animal form is determined by two conditions:—1st, by a progressive development of the animal by increasing histological and morphological differentiation; 2ndly, by the metamorphosis of a more general form into a more special one.

GENERAL RESULT.

. . . the most general result of these investigations and considerations may well be expressed thus:—

The history of the development of the individual is the history of its increasing individuality in all respects.

This general result is indeed so simple, that it would seem to need no demonstration, but to be cognizable *a priori*. But we believe that this simplicity is only the stamp and evidence of its truth. If the nature of the history of development had been from the first recognized as we have just expressed it, it would and must have been a deduction thence, that the individual of any particular animal form attains it by passing from the more general to the more special form. But experience everywhere teaches that deductions become much more certain if their results are previously made out by observation. Man must have received a greater spiritual endowment than he actually possesses for it to be otherwise.

If, however, the general result which has just been expressed be well based and true, then there is *one* fundamental thought which runs through all forms and grades of animal development, and regulates all their peculiar relations. It is the same thought which collected the masses scattered through space into spheres, and united them into systems of suns; it is that which called forth into living forms the dust weathered from the surface of the metallic planet. But this thought is nothing less than Life itself, and the words and syllables in which it is expressed are the multitudinous forms of the Living.

GERM LAYERS AS HOMOLOGIES

HUXLEY, Thomas Henry (British zoologist and educator, 1825–1895). From *On the anatomy and the affinities of the family of the medusae. By T. H. H. Esq., assistant surgeon of H.M.S. Rattlesnake, etc. Communicated by the Bishop of Norwich, F.R.S.,* in *Philosophical Transactions of the Royal Society of London,* p. 413, 1849.

We see Huxley here as original scientist rather than in his more familiar roles as commentator and educator. Even Huxley's excellent original anatomical investigations, however, emphasize his success as opponent or proponent rather than originator. In this paper, for example, he applies the germ-layer theory of von Baer (without alluding to it, however) to the problem of homologies among the celenterates. It was his thinking along these lines through which he became an opponent of the theory of archetypes and, later, a proponent of the doctrine of descent. Note, *e.g.*, his recurrent reference to "real affinities."

1. Perhaps no class of animals has been so much investigated with so little satisfactory and comprehensive result as the family of the *Medusae,* under which name I include here the *Medusae, Monostomatae* and *Rhizostomidae;* and this, not for the want of patience or ability on the part of the observers (the names of Ehrenberg, Milne-Edwards, and De Blainville, are sufficient guarantees for the excellence of their observations), but rather because they have contented themselves with stating matters of detail concerning particular genera and species, instead of giving broad and general views of the whole class, considered as organized upon a given type, and inquiring into its relations with other families.

2. It is my intention to endeavour to supply this want in the present paper —with what success the reader must judge. I am fully aware of the difficulty of the task, and of my own incompetency to treat it as might be wished; but, on the other hand, I may perhaps plead that in the course of a cruise of some months along the east coast of Australia and in Bass's Strait I have enjoyed peculiar opportunities for investigations of this kind, and that the study of other families hitherto but imperfectly known, has done much towards suggesting a clue in unravelling many complexities, at first sight not very intelligible.

3. From the time of Peron and Lesueur downwards, much has been said of the difficulties attending the examination of the Medusae. I confess I think that they have been greatly exaggerated; at least, with a good microscope and a good light (with the ship tolerably steady), I never failed in procuring all the information I required. The great matter is to obtain a good *successive* supply of specimens, as the more delicate oceanic species are usually unfit for examination within a few hours after they are taken.

Section I.—Of the Anatomy of the Medusae.

4. A fully-developed Medusa has the following parts:—1. A disc. 2. Tentacles and vesicular bodies at the margins of this disc. 3. A stomach and canals proceeding from it; and 4. Generative organs, either ovaria or testes. The tentacula vary in form and position in different species, and may be absent; the other organs are constantly present in the adult animal.

5. Three well-marked modifications of external structure result from variations in the relative position of these organs. There is either—1st, a simple stomach suspended from the centre of a more or less bell-shaped disc, the disc being traversed by canals, on some part of which the generative organs are situated, *e.g. Geryonia, Thaumantias;* or 2ndly, a simple stomach suspended from the centre of a disc; but the generative organs are placed in cavities

formed by the pushing in, as it were, of the stomachal wall, *e.g. Aurelia, Phacellophora;* or 3rdly, the under surface of the disc is produced into four or more pillars which divide and subdivide, the ultimate divisions supporting an immense number of small polype-like stomachs; small apertures lead from these into a system of canals which run through the pillars, and finally open into a cavity placed under the disc; the generative organs are attached to the under wall of the cavity, *e.g. Rhizostoma, Cephea.*

6. To avoid circumlocution I will make use of the following terms (employed by ESCHSCHOLTZ for another purpose) to designate these three classes, viz. CRYPTOCARPAE for the first, PHANEROCARPAE for the second, and RHIZOSTOMIDAE for the third.

7. In describing the anatomy of the Medusae it will be found most convenient to commence with the stomach, and trace the other organs from it.

Of the Stomach. This organ varies extremely both in shape and in size in the Cryptocarpae and Phanerocarpae. But whatever its appearance, it will be always found to be composed of two membranes, an inner and an outer. These differ but little in structure; both are cellular, but the inner is in general softer, less transparent and more richly ciliated, while it usually contains but few thread-cells. The outer, on the other hand, is dense, transparent, and either distinctly cellular or developed into a muscular membrane. It may be ciliated or not, but it is usually thickly beset with thread-cells, either scattered through its substance or concentrated upon more or less raised papillae developed from its surface.

8. I would wish to lay particular stress upon the composition of this and other organs of the Medusae out of *two distinct membranes,* as I believe that it is one of the essential peculiarities of their structure, and that a knowledge of the fact is of great importance in investigating their homologies. I will call these two membranes as such, and independently of any modification into particular organs, "foundation membranes."

9. When the stomach is attached to the disc, the outer membrane passes into the general substance of the disc, while the inner becomes continuous with the lining membrane of the canals. There is a larger or smaller space between the inner aperture of the stomach and the openings of the canals, with which both communicate, and which I will therefore call the "common cavity."

10. In the Rhizostomidae the structure of the stomachs is fundamentally the same, but they are very minute, and are collected upon the edges and extremities of the ramuscules of a common stem; so that the Rhizostomidae, *quoad* their digestive system, have the same relation to the Monostome Medusae as the Sertularian Polypes have to the Hydrae, or the Coralline Polypes to the Actiniae.

11. If one of the ultimate ramuscules be examined, it will be found to con-

sist of a thick transparent substance, similar in constitution to that of the mass of the disc, through which there runs, nearer one edge than the other, a canal with a distinct membranous wall ciliated internally. From this "common canal" a series of parallel diverticula are given off at regular intervals, and run to the edge of the branch, where they terminate by rounded oblique openings. It is not always easy to see these apertures, but I have repeatedly satisfied myself of their presence by passing a needle or other delicate body into them.

12. The difficulty in seeing the openings arises in great measure from the presence of a membrane which surrounds and overlaps them, and being very irritable, contracts over them on being touched. The membrane consists of two processes, one from each side of the perforated edge of the branch. In *Rhizostoma* these two processes generally remain distinct, so that their bases form a common channel into which all the apertures open; but in *Cephea* they are frequently united in front of and behind each aperture so as to form a distinct polype-like cell.

13. Each membranous process is composed of two membranes; the outer of these is continuous with and passes into the thick transparent outer substance above mentioned (11); the other is less transparent, more richly ciliated, and continuous with the lining membrane of the canals through the apertures. The two membranes are continuous at the free edge of the fold, and are here produced into numerous tentacula. The latter are beset with great numbers of thread-cells, and are in constant motion while the part retains its vitality.

14. *Of the Disc.* In the *Medusae monostomatae* the outer membrane of the stomach is, as I have said, continuous with the thick transparent mass of the disc, as the inner membrane is with the lining membrane of the canals which traverse it. The disc, therefore, is composed of two membranes inclosing a cavity variously shaped.

15. I have examined the minute structure of the disc in *Rhizostoma*. The outer surface of the transparent mass is covered with a delicate epithelium composed of polygonal nucleated cells joined edge to edge. Among these there are many thread-cells. Beneath this there is a thick gelatinous mass which is made up of an apparently homogeneous substance containing a multitude of delicate fibres interlacing in every direction, in the meshes of which lie scattered nucleiform bodies. On the lower surface of the disc, the only difference appeared to be that the epithelium was replaced by a layer of parallel muscular fibres.

16. It might be said that the gelatinous substance here described is a new structure, and not a mere thickening of the outer membrane; but a precisely similar change is undergone by the outer membrane in the Diphydae, and here it can be easily traced, *e.g.* in the formation of the bracts and in the development of muscular fibre in the outer wall of the common tube.

17. The structure of the inner membrane of the disc and its canals resem-

bles that of the corresponding tissue in the stomach, &c., but in the ultimate ramifications of the canals it becomes more delicate.

In these points there exists no difference between the Monostome and Rhizostome Medusae.

18. The three divisions, however, vary somewhat in the arrangement of the cavities and canals of the disc.

In the Cryptocarpae, the common cavity may be either small (*Thaumantias*) or large (*Oceania*); from it there proceed a number of straight unbranching canals which open into a circular canal running round the margin of the disc.

In the Phanerocarpae the general arrangement is similar, but the canals frequently branch (*Medusa aurita, Phacellophora*) and anastomose in a reticulate manner.

In many of the Monostome Medusae the centre of the under surface of the disc projects into the "common cavity" as a rounded boss, and according to its form and size will seem to divide the former more or less into secondary cavities. This appears to me to be the origin of the multiple stomachs of *Medusa aurita* as described by EHRENBERG.

19. In the Rhizostomidae, the canals of the branched processes unite and open by four (*Rhizostoma, Cephea*) or eight (*Cassiopea?*) distinct trunks into a wide curiously-shaped cavity, from whence anastomosing canals are given off to all parts of the disc. The circular vessel exists, but is not particularly obvious in consequence of anastomosing branches being given off beyond it.

20. In very many of the Cryptocarpae (*Carybdoa, Oceania, Polyxenia*) there is a circular, valvate, muscular membrane developed from the inner and under edge of the disc. In the Phanerocarpae such a membrane does not seem to be present, but in *Rhizostoma* and *Cephea* it is evidently replaced by the inflexed edge of the disc. . . .

SECTION II.—OF THE AFFINITIES OF THE MEDUSAE.

56. Certain general conclusions are deducible from the facts stated in the preceding section. It would appear,—

1st. That a Medusa consists essentially of two membranes inclosing a variously-shaped cavity, inasmuch as its various organs are so composed (7, 8, 14, . . .).

2ndly. That the generative organs are external, being variously developed processes of the two membranes; and

3rdly. That the peculiar organs called thread-cells are universally present (7, 15, &c.).

Now in these particulars the Medusae present a striking resemblance to certain other families of Zoophytes. These are the Hydroid and Sertularian Polypes, the Physophoridae and Diphydae, with all of which the same three propositions hold good.

57. But in order to demonstrate that a real affinity exists among different classes of animals, it is not sufficient merely to point out that certain similarities and analogies exist among them; it must be shown that they are constructed upon the same anatomical type, that, in fact, their organs are homologous.

Now the organs of two animals or families of animals are homologous when their structure is identical, or when the differences between them may be accounted for by the simple laws of growth. When the organs differ considerably, their homology may be determined in two ways, either—1, by tracing back the course of development of the two until we arrive by similar stages at the same point; or, 2, by interpolating between the two a series of forms derived from other animals allied to both, the difference between each term of the series being such only as can be accounted for by the laws of growth. The latter method is that which has been generally employed under the name of *Comparative Anatomy*, the former being hardly applicable to any but the lower classes of animals. Both methods may be made use of in investigating the homologies of the Medusae.

58. A complete identity of structure connects the "foundation membranes" of the Medusae with the corresponding organs in the rest of the series; and it is curious to remark, that throughout, the outer and inner membranes appear to bear the same physiological relation to one another as do the serous and mucous layers of the germ; the outer becoming developed into the muscular system and giving rise to the organs of offence and defence; the inner, on the other hand, appearing to be more closely subservient to the purposes of nutrition and generation.

59. The structure of the stomach in the Medusae is in general identical with that of the same organ in the rest of the series. The Rhizostomidae offer an apparent difficulty, but it appears to me that the marginal folds in them answer to the stomachal membrane of the Monostome Medusae; the apertures to the inner orifice of their stomach, and the common canal to their "common cavity." Just as in a polygastric Diphyes the common tube answers to the chamber into which the stomach of a monogastric Diphyes opens; and in *Cephea Wagneri* (WILL) these resemblances are still more striking. He says that each cotyledon "has at its apex a small round opening, the mouth, which leads to an ovate cavity, occupying the whole interior of the cotyledon. I consider this as the proper digestive or stomachal cavity, and believe that the cotyledons have the same relation to the vessels as the so-called suckers (*Sangröhren*) of the Diphydae to the common tube (*Saftröhre*)."

60. The disc of a Medusa is represented by the natatorial organ among the Diphydae and Physophoridae. Take for instance the disc of *Oceania* or *Cy-*

taeis. It is here a more or less bell-shaped body, traversed by radiating canals, lined by a distinct membrane, united by a circular canal at the margin. In the centre the radiating canals communicate freely with the chamber into which the stomach opens. The inner margin of the disc is provided with a delicate, circular, valvate membrane. The same description applies, word for word, to the natatorial organs of the Diphydae and Physophoridae; the only difference being, that in the latter the stomach is *outside* the cavity of the organ, instead of being, as in the Medusae, suspended from its centre *inside.* And even if the different texture of the two organs should give rise to any doubt, the genus *Rosacea,* in which the natatorial organ is perfectly soft and gelatinous, furnishes the needful intermediate form.

61. The disc of the Medusae has no representative among the Hydrae and Sertulariadae. The cell of the Sertularian Polype rather resembles the "bract" of the Diphydae than the "natatorial organ" in its structure and function, and in this manner the Diphydae form a connecting link between the Medusae and the Physophoridae.

62. Of the two kinds of tentacles of the Medusae, the first is represented, in the Physophoridae and Diphydae, by the thickenings, richly beset with thread-cells, that frequently occur in the lip of the stomach; in the Sertularian Polypes (*Plumularia, Campanularia*) by the tentacles of the margin of the mouth, which precisely resemble the tentacles of the fringe of *Rhizostoma,* or the marginal tentacles of *Thaumantias,* foundation membranes, the generative elements being developed between them.

67. In the Diphydae (and as I have good reason for believing in the Physophoridae also) the generative organ commences as a simple process of the common tube, and undergoing great changes of form in the course of its development, it becomes at last exactly similar to an ordinary natatorial organ with a sac composed of two membranes suspended from its centre. In external form it greatly resembles such a *Medusa* as *Cytaeis,* and this resemblance is much heightened when, as in some cases, it becomes detached and swims freely about. The ova or spermatozoa, as the case may be, are developed between the two membranes of the sac, the inner of which at any rate is a continuation of the inner membrane of the common tube.

68. The ovarium of the *Plumularia* above mentioned, commences as a dilatation of the apex of its pedicel, which again is a process of the common stem. It then becomes lenticular with a horny outer wall, glassy and transparent externally, but internally coloured by pigment masses. Internally it has an oval cavity communicating with that of the stem and lined by a distinct membrane. Between the two membranes is a thick layer of ova, more or less oval in shape, and about $\frac{1}{356}$th of an inch in diameter, with a germinal spot about $\frac{1}{2400}$th of an inch in diameter, seated in the middle of a clear space about twice that size, which doubtless represents the germinal vesicle.

69. The account given by LÖWEN of the generative organs of *Campanularia* differs considerably from the foregoing. After all however his "female polypes" may be nothing more than ovaria similar to those of *Diphyes* or *Coryne*, but having the production of tentacles from the margin carried to a greater extent than in the latter. If this be a correct explanation, the idea promulgated by STEENSTRUP, that there is an "alternation of generations" among the Sertularian Polypes, must be given up.

70. In *Hydra*, the ova are developed in similar processes of the lower part of the body. But among the Hydroid Polypes the ovaries of *Coryne*, *Syncorine* and *Corymorpha*, as described by SARS, LÖWEN and STEENSTRUP, are most interesting. They commence as tubercles of the stem, afterwards become bodies, precisely resembling the ovaria of the Diphydae, and finally detaching themselves develope regular tentacles from their margin. The ova are formed between the two membranes of the inner sac.

71. What has now been advanced will perhaps be deemed evidence sufficient to demonstrate,—1st, that the organs of these various families are traceable back to the same point in the way of development; or 2ndly, when this cannot be done, that they are connected by natural gradations with organs which are so traceable, in which case, according to the principles advanced in 57, the various organs are homologous, and the families have a real affinity to one another and should form one group.

72. Perhaps the view that I have taken will be more clear if I throw it into a tabular form, placing opposite one another those organs in the different families, for the homologies of which there is, I think, sufficient evidence, thus:—[1]

73. It appears then that these five families are by no means so distinct as has hitherto been supposed, but that they are members of one great group, organized upon one simple and uniform plan, and even in their most complex and aberrant forms, reducible to the same type. And I may add, finally, that on this theory it is by no means difficult to account for the remarkable forms presented by the Medusae in their young state. The Medusae are the most perfect, the most *individualized* animals of the series, and it is only in accordance with what very generally obtains in the animal kingdom if in their early condition they approximate towards the simplest forms of the group to which they belong.

74. I have purposely avoided all mention of the Beroidae in the course of the present paper, although they have many remarkable resemblances to the animals of which it treats: still such observations as I have been enabled to make upon them have led me to the belief, that they do not so much form a part of the present group as a link between it and the Anthozoic Polypes. But I hope to return to this point upon some future occasion.

---

[1] See p. 407.

STOMACH IDENTICAL IN STRUCTURE THROUGHOUT.

Medusae.	*Physophoridae.*	*Diphydae.*	*Sertularidae.*	*Hydrae.*
Disc.	Natatorial organ.	Natatorial organ.		
Canals.	Canals of natatorial organ.	Canals of natatorial organ.		
Common cavity.	Common tube.	Sacculus and common tube.	Cavity of stem.	
Canals of branches. (*Rhiz.*)		Bract.	Polype-cell	
Tentacles. 1.	Thickened edge of stomach		Oval tentacles.	
2.	Prehensile organs.		Clavate organs.	Tentacles (?)
Generative organs.	Generative sac.	Generative organ.	Generative organ.	Generative organ.
	Natatorial organ of generative sac.			Natatorial organs (Coryne).
Marginal vesicle.	?	?		?

DESCRIPTIVE EMBRYOLOGY COMES OF AGE

BALFOUR, Francis Maitland (English embryologist, 1851–1882). From *A treatise on comparative embryology*, London, 1880.

Balfour* died, when thirty-one years old, as the result of a climbing accident in the Alps. Just previously, on declining professorships at Edinburgh and Oxford, he had been named to a chair created especially for him at Cambridge. His comprehensive *Comparative Embryology* may be said to usher in the adulthood of this science. Its offspring, developmental mechanics, was born almost simultaneously (see Roux). In this sense, 1880 marks a nodal point in the history of embryology. The following passage defines the lines along which the study of comparative ontogeny afterwards developed with little change. It includes a statement on the question of biogenesis and recapitulation which displays greater maturity and detachment than the contemporary effusions of Häckel (*q.v.*) on the same subject.

The marvellous phenomenon of the evolution of a highly complicated living being from a simple undifferentiated germ in which it needs the aid of the most modern microscopical appliances to detect any visible signs of life, has not unnaturally attracted the attention of biologists from the very earliest periods. Before the establishment of the cell theory the origin of the organism from the germ was not known to be an occurrence of the same nature as the growth of the fully formed individual, and Embryological investigations were mixed up with irrelevant speculations on the origin of life.

The difficulties of understanding the formation of the individual from the structureless germ led anatomists at one time to accept the view "according to which the embryo preexisted, even though invisible, in the ovum, and the changes which took place during incubation consisted not in a formation of parts, but in a growth, *i.e.* in an expansion with concomitant changes of the already existing germ."

Great as is the interest attaching to the simple and isolated life histories of individual organisms, this interest has been increased tenfold by the generalizations of Mr Charles Darwin.

It has long been recognized that the embryos and larvae of the higher forms of each group pass, in the course of their development, through a series of stages in which they more or less completely resemble the lower forms of the group. This remarkable phenomenon receives its explanation on Mr Darwin's theory of descent. There are, according to this theory, two guiding, and in a certain sense antagonistic principles which have rendered possible the present order of the organic world. These are known as the laws of heredity and variation. The first of these laws asserts that the characters of an organism at all stages of its existence are reproduced in its descendants at corre-

---

*Arthur (first Earl) Balfour (prime minister from 1902 to 1905), brother of the embryologist, was born 3 years earlier, lived 50 years longer.

sponding stages. The second of these laws asserts that offspring never exactly resemble their parents. By the common action of these two principles continuous variation from a parent type becomes a possibility, since every acquired variation has a tendency to be inherited.

The remarkable law of development enunciated above, which has been extended, especially by the researches of Huxley and Kowalevsky, beyond the limits of the more or less artificial groups created by naturalists, to the whole animal kingdom, is a special case of the first of the above laws. This law, interpreted in accordance with the theory of descent, asserts that each organism in the course of its individual ontogeny repeats the history of its ancestral development. It may be stated in another way so as to bring out its intimate connection with the laws of inheritance and variation. Each organism reproduces the variations inherited from all its ancestors at successive stages in its individual ontogeny which correspond with those at which the variations appeared in its ancestors. This mode of stating the law shews that it is a necessary consequence of the law of inheritance. The above considerations clearly bring out the fact that Comparative Embryology has important bearings on Phylogeny, or the history of the race or group, which constitutes one of the most important branches of Zoology.

Were it indeed the case that each organism contained in its development a full record of its origin, the problems of Phylogeny would be in a fair way towards solution. As it is, however, the law above enunciated is, like all physical laws, the statement of what would occur without interfering conditions. Such a state of things is not found in nature, but development as it actually occurs is the resultant of a series of influences of which that of heredity is only one. As a consequence of this, the embryological record, as it is usually presented to us, is both imperfect and misleading. It may be compared to an ancient manuscript with many of the sheets lost, others displaced, and with spurious passages interpolated by a later hand. The embryological record is almost always abbreviated in accordance with the tendency of nature (to be explained on the principle of survival of the fittest) to attain her ends by the easiest means. The time and sequence of the development of parts is often modified, and finally, secondary structural features make their appearance to fit the embryo or larva for special conditions of existence. When the life history of a form is fully known, the most difficult part of his task is still before the scientific embryologist. Like the scholar with his manuscript, the embryologist has by a process of careful and critical examination to determine where the gaps are present, to detect the later insertions, and to place in order what has been misplaced.

The aims of Comparative Embryology are two-fold: (1) to form a basis for Phylogeny, and (2) to form a basis for Organogeny or the origin and evolution of organs. The justification for employing the results of Comparative

Embryology in the solution of the problems in these two departments of science is to be found in the law above enunciated, but the results have to be employed with the qualifications already hinted at; and in both cases a knowledge of Comparative Anatomy is a necessary prelude to their application.

In accordance with the above objects Comparative Embryology may be divided into two departments.

The scientific method employed in both of these departments is that of comparison, and is in fact fundamentally the same as the method of Comparative Anatomy. By this method it becomes possible with greater or less certainty to distinguish the secondary from the primary or ancestral embryonic characters, to determine the relative value to be attached to the results of isolated observations, and generally to construct a science out of the rough mass of collected facts. It moreover enables each observer to know to what points it is important to direct his attention, and so prevents that simple accumulation of disconnected facts which is too apt to clog and hinder the advance of the science it is intended to promote.

In the department of Phylogeny the following are the more important points aimed at.

(1) To test how far Comparative Embryology brings to light ancestral forms common to the whole of the Metazoa. Examples of such forms have been identified by various embryologists in the ovum itself, supposed to represent the unicellular ancestral form of the Metazoa: in the ovum at the close of segmentation regarded as the polycellular Protozoon parent form: in the two-layered gastrula, etc., regarded by Haeckel as the ancestral form of all the Metazoa.

(2) How far some special embryonic larval form is constantly reproduced in the ontogeny of the members of one or more groups of the animal kingdom; and how far such larval forms may be interpreted as the ancestral type of those groups.

As examples of such forms may be cited the six-limbed nauplius supposed by Fritz Müller to be the ancestral form of the crustacea; the trochosphere larva of Lankester, which he considers to be common to the Mollusca, Vermes, and Echinodermata: the planula of the Coelenterata, etc.

(3) How far such forms agree with living or fossil forms in the adult state; such an agreement being held to imply that the living or fossil form in question is closely related to the parent stock of the group in which the larval form occurs. It is not easy to cite examples of a very close agreement of this kind between the larval forms of one group and the existing or fossil forms of another. The larvae of some of the Chaetopoda with long provisional setae resemble fossil Chaetopods. The Rotifers have many points of resemblance to the trochosphere, especially to that form of trochosphere characteristic of the Mollusca. The Turbellarians have some features in common with the Coel-

enterate planula. Some of the Gephyrea in the presence of a praeoral lobe resemble certain trochosphere types. The larva of the Tunicata has the characters of a simple type of the Chordata.

Within the limits of a single group agreements of this kind are fairly numerous. In the Craniata the tadpole of the Anura has its living representative in the Pisces and perhaps especially in the Myxinoids. The larval forms of the Insecta approach Peripatus. The stalked larva of Comatula is reproduced by the living Pentacrinus and Rhizocrinus etc. Numerous examples of the same phenomenon are found amongst the Crustacea.

(4) How far organs appear in the embryo or larva which either atrophy or become functionless in the adult state, and which persist permanently in members of some other group or in lower members of the same group. Cases of this kind are of the most constant occurrence, and it is only necessary to cite such examples as the gill-slits and Wolffian body in the embryos of higher Craniata to illustrate the kind of instance alluded to. The same conclusions may be drawn from them as from the cases under the previous heading.

(5) How far organs pass in the course of their development through a condition permanent in some lower form. Phylogenetic conclusions may be drawn from instances of this character, though they have a more important bearing on Organology than on Phylogeny.

The considerations which were used to show that the ancestral history is reproduced in the ontogeny of the individual apply with equal force to the evolution of organs. The special questions in Organology, on which Comparative Embryology throws light, may be classified under the following heads.

(1) The origin and homologies of what are known as the germinal layers; or the layers into which the embryo becomes divided immediately after the segmentation.

(2) The origin of primary tissues, epithelial, nervous, muscular, connective, etc., and their relation to the germinal layers.

(3) The origin of organs. The origin of the primitive organs is intimately connected with that of the germinal layers. The first differentiation of the segmented ovum results in the cells of the embryo becoming arranged as two layers, an outer one known as the epiblast and an inner one as the hypoblast. The outer of these forms a primitive sensory organ, and the inner a primitive digestive organ.

(4) The gradual evolution of the more complicated organs and systems of organs.

This part of the subject, even more than that dealing with questions of Phylogeny, is intimately bound up with Comparative Anatomy; without which indeed it becomes quite meaningless.

EXPERIMENTAL ANALYSIS OF DEVELOPMENTAL MECHANICS

ROUX, Wilhelm (German embryolgoist, 1850–1925). From *Beiträge zur entwick-lungsmechanik des embryo, Einleitung,* in *Zeitschrift für biologie von W. Kühne und C. Voit,* p. 411, Munich and Leipzig, 1885; *ibid.,* No. 5, *Ueber die künstliche hervor-bringung halber embryonen durch zerstörung einer der beiden ersten furchungszellen sowie über die nachentwickelung (postgeneration) der fehlenden körperhälfte,* in *Vir-chow's Archiv für Pathologische Anatomie und Physiologie und Klinische Medizin,* vol. 114, p. 133, 1888; tr. by M. and V. Hamburger and T. S. Hall for this volume.

With the rise of organic chemistry and of the cell and germ plasm theories, a new formulation of the old controversy over epigenesis vs. preformation became necessary. Prompted by the studies of Pflüger and His, Roux supplied in the following paper the necessary restatement, giving the problem a form in which it was susceptible of experimental exploration. With this, developmental mechanics was established as a concept, and serious experimental embryology commenced. After a few early years of active experimentation, Roux devoted himself increasingly to the promotion, by editorial and literary enterprises, of the science for whose foundation he was so largely responsible.

Since the end of the last century, descriptive embryology has, through the indefatigable industry and ingenuity of many investigators, been advanced to a point where, for almost every organ of the vertebrates, and many invertebrates, we know with a certain degree of exactness, the *form changes* through which, beginning with the fertilized egg, it progressively shapes itself.

After one has achieved an approximate survey of these changes during development, one is justified in taking a further step aimed at a knowledge of the *processes* which produce them.

This further aim can be conceived in two different ways: first, in so far as mere *form-building processes* are to be distinguished and these presented descriptively, in a morphological sense. One would designate as the final goal of this endeavor a complete knowledge of the way taken by every particle of the fertilized ovum as it pursues its separate path until final utilization in the construction of the organism, plus a knowledge of the paths of all particles taken in from outside and in any way used in building at any time until the completion of development. Only if particles were eliminated should we relinquish our observation of them before the end result of development has been achieved. Prerequisite to this investigation would be a knowledge of the space relations of all parts of the egg at the commencement of development.

This, then, in descriptive terms, would be the definition of the further task before us, briefly, the complete description of all, even the minutest, developmental processes as material movements of the parts of the egg and parts taken in by it up to the completion of individual development. This being based upon a complete knowledge of the arrangement and external character of each minutest particle of the fertilized egg: a *kinematics of development*, if we adopt Ampère's division of mechanics, as seems desirable.

If we had this knowledge we would be in a position to present all embryonic development purely descriptively and to treat it, therefore, as a descriptive science. However, not only shall we never reach this goal but we shall not even be able to get appreciably nearer than has already been done by mere observation of normal events, for the reason that the movements of the particles which collectively produce the externally visible individual form changes, as well as those which produce so-called qualitative changes, are in the main concealed from direct observation.

Nevertheless, one cannot say in advance that we would have to renounce for good this knowledge, for there is another way of learning about these things: namely deductive and inductive reasoning on causalogical grounds.

It is evident that the developmental movements of egg particles at the beginning of development, if they are independent at all, i.e., if after development first begins they follow their own inertia alone, do so only during a very short period and for a minimal distance; at the next moment, mutual effects must occur which in the changes thereby produced actually constitute development.

Furthermore, it is evident that if we knew the juxtapositions of all particles of the egg at the moment development begins, as well as the accelerations imparted to each and the individual forces inherent in them, in other words if all the inner causes of a single moment of development were known to us, as well as all factors added from without during the entire course of development, then we could derive from this the future developmental movements of all particles and thus fill in the gap in our direct observations. Such an embryology would merit the name *kinetics of development*.

We shall not see either of these two sciences completed, but we shall have always to pursue both simultaneously in order to approach our goal along both paths: necessary combination of these two sciences we could call: *Developmental-Mechanics of the Embryo*. It is the nature of the circumstances that of the two parts comprised in this term, *kinematics*, or purely *descriptive* mechanics, will be more and more relegated to the role of an auxiliary science by *kinetics*, or *causal* mechanics.

Inasmuch as in addition to the development of the fertlized egg many other developmental processes occur and since it will be useful and instructive to include them for comparison in our considerations, we shall distinguish between a "general" and a special developmental mechanics, and the latter will be the special subject of our investigations.

By *developmental mechanics* in the general sense, with emphasis on its kinetic part we designate the *science of the nature and effect of the energy complexes which produce development*.

By development in its ordinary connotation we understand the *rise of visible manifoldness*. With regard to the visibility of the arising manifoldness, this notion contains a subjective element which obliges us, for further insight, to divide it into two separate parts: the *actual production* of manifoldness, and the mere *transformation of an invisible into a visible and perceptible* manifoldness.

The two types of development thus distinguished stand to each other in a relation which reminds one of the old antithesis of epigenesis vs. evolution: i.e., of the alternative which existed at the period when it was our one task and only possibility to establish, first of all, the formed products of formative processes: the externally visible changes of form. In this descriptive investigation of development of form, epigenesis, the successive production of new forms, achieved a complete victory over evolution, the mere becoming-visible of details of form which existed from the outset.

If we penetrate more deeply into the formative processes, however, as is necessary in a causal investigation, we are again faced with the same alternative and at the same time induced to conceive of it in a deeper sense. If we wish to retain the previous terms, then *epigenesis* means not merely the production of manifold *forms* through the potentialities of a substrate which is *structurally* simple but perhaps inherently extraordinarily complicated; rather it means a *new formation of manifoldness* in the strictest sense, an actual increase in existing manifoldness. *Evolution* on the other hand is merely the *becoming visible of preexisting latent differences*. It is clear that according to these more general definitions processes which appear in a morphological investigation as instances of epigenesis may be in reality predominantly or purely, evolutions; and we recognize that in our intention to penetrate more deeply into the process of development we are again confronted with the question: is embryonic development epigenesis or evolution?

In an organic nature, we see these two modes of development usually occurring in combination. The deeper we penetrate into an observed developmental process, however, the more we recognize, as a rule, that a large part of what appeared on first consideration as new-formed manifoldness owes its richness in perceptible manifoldness to a metamorphosis of preexisting differences. . . .

If, from the first, we have brought our particular problem under Spinoza-Kant's concept of mechanism, the assumption implied by this was founded on the prediction that nothing metaphysical would have to be considered in connection with the material course of the developmental processes of the embryo and that these processes represent a phenomenon entirely subject to the law of causality. Only on such a prediction could we establish our effort to explore these processes. I have not derived this expectation simply from present-day world philosophy; it has cost me many years of deliberation tracing possibilities as to how, from that which is relatively or seemingly simple, without corresponding formative influences from without, there can result such a complicated and typically formed structure, as a chick from its egg. Seen from the scientific viewpoint this endeavor was unnecessary, since these questions had already been dealt with most extensively by the philosophers, and most completely perhaps by H. Lotze. This is a sphere where the scientist must first apprentice himself to the philosopher, unless he wants to waste his energies in attaining to things already known, and I cannot too strongly advise my colleagues with similar interests to study the relevant writings intensively. However, since we, unlike the philosophers, are concerned not with general possibilities but with factual truths, we shall have to guard ourselves carefully against an overestimate of the empirical value of these philosophical considerations and use them merely as heuristic principles for our own painstaking and exact investigations. Although, viewed objectively, I have wasted energy in working out, anew and independently, what had been already achieved on this subject, I do not consider this labor entirely lost. . . .

It is not necessary to give a special reason why, despite the light which the theory of evolution has shed on the structural *results* of the developmental processes at each stage, these processes themselves need special investigation. No one will question the usefulness of the eventual fruits of investigations directed at these processes. They aim to acquaint us with those forces and reactions to which we owe the origin and maintenance of our very existence, with the knowledge of which our medical activities become much more scientific, hence more productive.

The method of these investigations cannot be technically definite and uniform, as is, for example, the method of staining serial cross-sections of pure cultures which currently dominate entire fields of investigation; rather almost every new problem will require the invention of new methods, mostly of an experimental character. The only universal method is causal analytical reasoning as I have explained elsewhere. This is, however, prerequisite to such a work if we are to progress steadily and not be led astray or stopped after exploiting some more or less accidental discovery. . . .

[Beginning of second paper]

The following investigation is a contribution to the solution of the problem of self-differentiation; that is, an inquiry as to whether or to what degree the fertilized egg, as a whole and in its individual parts, is capable of development out of itself; or whether, to the contrary, normal development can proceed only under "formative" influences exerted on the fertilized egg by the outer medium and under "differentiative correlations" among egg parts separated by cell division.

I have solved this question as it concerns the egg as a whole by rotating eggs in a perpendicular plane so slowly that the centrifugal force had no directive influence yet the eggs continually changed position with respect to gravity, to the magnetic meridian, and to the source of light and heat; the result was that, in this way, their normal development was neither suspended, altered, or even delayed. From this we conclude that the rise of typical morphogenesis of developing egg and embryo does not require the formative development of the fertilized egg and may be considered as "self-differentiation." . . .

After I had gained this insight, it seemed to me to be necessary next to discover whether, for the formation of normal structures in the egg, all or many parts must interact; or to the contrary, the egg's parts, separated from each other by cleavage, are capable of developing independently of each other; or, finally what might be the role in normal development of each of two principles,—that of differentiative interaction, and that of the self-differentiation of parts. . . .

Only direct experiment on the egg can yield an entirely reliable clarification of the actual share of self-differentiation of the egg's parts in normal development, and already years ago I made efforts along this line, and showed in a general way that operations on the cleaving and cleaved egg even though they cause a loss of material do not result in cessation of development nor in general malformations of the embryo, but that normally formed embryos result which have only a circumscribed defect or a circumscribed malformation. . . .

EXPERIMENTS ON THE EFFECT OF THE DESTRUCTION OF ONE HALF OF THE EGG ON THE DEVELOPMENT OF THE REMAINDER
EXPERIMENTAL METHOD

The arrangement of the experiment was as follows:

In the first experiments, the eggs of the green frog, *Rana esculenta,* were placed in single glass dishes . . . and one of the first two blastomeres was punctured once or several times with a fine needle. . . . Since when punctured several times with a single fine needle, even despite large extraovates, the cell frequently developed normally, I from the third day warmed the needle by attaching to it a brass ball to preserve heat and warming this ball a suitable amount. In this procedure only one puncture was made, but the needle was usually left in the egg until a distinct light brown discoloration of the egg substance occurred around it. This adhered to the needle somewhat on its being withdrawn and afterwards remained as a broad, slightly projecting cone: an indication that it had become firmer, hence probably half coagulated. As a result no more extraovate emerged at the point of puncture. This gave me better results, such that in about 20% of the operated eggs, only the intact cell survived the interference; the majority of the operated cells died and only a few, in which the needle was probably too cool, developed normally. In this way I reared over 100 eggs with one half killed. . . .

PROCESSES OCCURRING IN THE UNOPERATED EGG HALF

In about 20% of the operated eggs, the unoperated cell survived. From this one might expect various consequences; for instance, that abnormal processes would occur which would lead to atypical formations; or that this egg half would develop into a whole, but proportionately smaller, individual, since it is a whole cell whose nucleus is, according to some authors, perfectly similar in nature to the first cleavage nucleus. Instead of these results which would have been remarkable, something even more remarkable happened: the one cell developed in many instances into an essentially normally formed "half-embryo" in which only in the immediate neighborhood of the operated egg half did small anomalies occur. . . .

[This is followed by detailed description of different types of partial embryos: lateral and anterior half-embryos, one-quarter, and three-quarter formations. —Ed.]

Conclusions from these Findings

With respect to general considerations, we learn from the foregoing findings that each of the first two blastomeres is capable of development independently of the other and hence may also develop independently under normal conditions, and this independent development deviates from the normal formation only in a surprisingly few details explicable in crudely mechanical terms.

This mode of development was followed up to the formation of the medullary folds, brain vesicles, notochord, and mesoblast, and to the time of segregation of the latter into somitic and lateral plates and of the separation of somitic plate into somites. . . .

The development of the frog gastrula and of the immediately ensuing embryo is, from the second cleavage on, a *mosaic work* of at least four vertical, independently developing pieces. It remains to be seen to what extent this mosaic formation of at least four pieces is modified during further development by . . . *differentiative correlations* and to what extent the independence of the parts becomes restricted.

[This is followed by a detailed description of the reorganization and postgeneration of the missing half in later stages of development.—Ed.]

The blastomere which has been deprived of its developmental capacity by the operation can be gradually revived. This *reorganization* takes place in part by the migration of a large number of nuclei into the whole yolk mass except in so far as the latter was already provided with nuclei by the derivatives of its own cleavage nucleus [which was in some cases not injured by the needle—Ed.] and by the later multiplication of these two sorts of nuclei. This *Bekernung*, or nucleization, of the operated blastomere is, later on, followed by cellulation: around each nucleus occurs a cellular segregation of yolk material . . . The reorganization of the operated egg half is followed by belated development, "postgeneration," which can result in perfect completion of the missing lateral or posterior half of the embryo. This postgeneration does not proceed in the same fashion as the normal development of the primary half. . . . The postgeneration of the germ layers in the half formed later does not proceed, as in primary development, by independent formation but rather from germ layers already formed in the developed half. . . . Gastrulation proper does not take place in the completion of the lateral halfformation. . . .

Since the different yolk materials and nuclei of the operated egg half do not have typical distribution but are determined as to their position by accidental circumstances, it cannot be assumed that typical spreading and typical results thereof in postgeneration are caused by typical arrangements of specifically qualified substances capable of "self-differentiation." Therefore, we feel justified in concluding that definite differentiative effects are exerted by already differentiated materials on adjacent, as yet less differentiated cellular matter.

While through our findings primary or direct development of the first blastomeres proved to be self-differentiation, either of themselves or of the complex of their descendants, the reorganized parts of the egg are capable of only a dependent differentiation through the influence of those parts which are already differentiated.

TOTIPOTENCY OF BLASTOMERES

DRIESCH, Hans Adolf Edward (German embryologist and philosopher, 1867–1941). From *Der werth der beiden ersten furchungszellen in der echinodermentwicklung. Experimentelle erzeugung von theil- und doppelbildungen*, in *Zeitschrift für Wissenschaftliche Zoologie*, vol. 53, p. 160, Leipzig, 1892; tr. as *The potency of the first two cleavage cells in echinoderm development. Experimental production of partial and double formations* by L. Mezger, M. and V. Hamburger, and T. S. Hall for this volume.

In the following famous paper, Driesch demonstrates the power of half-blastomeres to produce whole embryos of half size. These, with characteristic terminological abstruseness, he names "part-formations." This regulability of the blastomeres seemed to him inexplicable on a physicochemical basis, and he later postulated in embryos a vitalistic factor called, after Aristotle, "entelechy"—"something which carries its purpose within itself."

"Granting that the primordium of a part originates during a certain period, one must, for greater accuracy, describe this by stating that the material for the primordium is already present in the blastoderm while the latter is still flat but the primordium is not as yet morphologically segregated and hence not recognizable as such. By tracing it back we shall be able for every primordium to determine its exact location even in the period of incomplete or deficient morphological organization; indeed, to be consistent, we should extend this determination back to the newly fertilized, even the unfertilized, egg. The principle according to which the blastoderm contains organ primordia preformed in a flat pattern and, vice versa, every point in the blastoderm can be rediscovered in a later organ, I call the principle of organ-forming germareas."

In these words His formulated the principle so designated by him. Continuing this train of thought, Roux[1] discussed in a perceptive manner the difference between evolution, or the *metamorphosis* of manifoldness, and epigenesis, or the *new formation* of manifoldness; in his well-known experiments on "half-embryos" (of which only the first part concerns us here) he decided the question under consideration, for the frog egg, in favor of evolution.

A not very generally known work by Chabry is the only further investigation of this kind known to me. His specific explanations and figures make it clear that his results are fundamentally contrary to those of Roux. I wish to mention here that I came to know of Chabry's work only after the completion of my own experiments.

As to these, I was interested in repeating Roux's experiments on material which would be resistant, easily obtainable, and readily observable; all three of these conditions are most satisfactorily fulfilled by the Echinoids, which had already served as a basis for so many investigations. My own experiments were carried out upon Echinus microtuberculatus.

The investigations were made in March and April of 1891. They have led me to many other problems closely connected with the present one, problems whose eventual solution will deepen materially our understanding of the part already solved. Nevertheless, I present my results at this time because they have decided with certainty, for my material, the cardinal point, that is, the potency of the two first blastomeres.

MATERIALS AND METHODS

The first week of my stay in Trieste was lost, inasmuch as I obtained almost exclusively useless material. Whereas the following work follows the above mentioned experiments of Roux in content, the method was taken from the excellent cellular researches of the Hertwig brothers. These investigators, by shaking unfertilized eggs, split off pieces and raised them successfully. It is well known that Boveri used the same method for the production of his "organisms produced sexually without maternal characters," although other factors prevented him from carrying out the procedure exactly.

I therefore went to Trieste with the intention of obtaining one of the first

---

[1] *Beitrage zur Entwicklungsmechanik des Embryo.* I. *Zeitschr. f. Biol.* Bd. XXI. III. *Breslauer ärztl. Zeitschr.* 1885. V. *Virchow's Arch.* Bd. CXIV.

half-blastomeres of Echinus by shaking at the two-cell stage, in order to see, provided it lived, what would become of it.

At an average temperature of about 15° C., cleavage of Echinus eggs occurred 1½ to 2 hours after artificial fertilization. Good material, and only such was used, displayed in only a very few instances immediate division into four cells, an inevitable result, according to Fol and Hertwig, of bispermy.

Shaking was done in small glass containers 4 cm long and about 0.6 cm in diameter. Fifty to one hundred eggs were placed in a small quantity of water. In order to obtain results, one must shake as vigorously as possible for five minutes or more; even then one obtains at best only about ten isolated blastomeres and about as many eggs whose membranes are still intact but whose cells are more or less separated within these membranes.

If shaking is done at the moment of completion of first cleavage, events are, so to speak, reversed; the furrow disappears and one obtains a sausage-shaped body whose two nuclei again show connections. In these recombined eggs the furrow reappears in a short time and normal development follows. On the other hand if one shakes too late, the second cleavage occurs prematurely during the shaking. It is therefore necessary to watch carefully for the right moment.

About one half of the blastomeres are, in addition to being isolated, dead; nevertheless I obtained about fifty capable of development. This appears not unfavorable considering the strength of the mechanical treatment, and considering the fact that the isolated blastomeres are in direct contact with the water on at least one side,—a completely abnormal situation. Isolation is obviously possible only where the membrane bursts.

During cleavage the preparations were observed microscopically as often as possible, and during later development usually once every morning and evening.

One more thing about the treatment of the isolated cells. The contents of the glass used for shaking must be poured into fresh sea water as soon as possible since the water has naturally warmed and evaporated.

It was to be expected that the small quantity of water would not be exactly beneficial, nor the bacteria which were especially numerous toward the end of my experiment and were encouraged by disintegrating pieces which had died.

At any rate my method guarantees that one is observing the same pieces on successive days. Unfortunately, Boveri, in his very important experiments, did not succeed in this respect.

But here I anticipate my results. I turn now to a systematic presentation of findings starting with

CLEAVAGE

First a few words about the normal course of events as revealed in Selenka's excellent investigations.

Following two meridional cleavages there is an equatorial one and the germ now consists of eight cells of equal size. Four of these now give off, toward one pole, four smaller cells, and at the same time the others divide approximately meridionally.

The germ now consists of 16 cells and shows a marked polarity with the four small cells, easily recognized, occupying one pole. Further divisions lead to stages with 28, 32, 60, and 108 cells (Selenka). The four small cells which originated at the 16-cell stage clearly indicate the animal pole for a long time. I was unable to establish certainly any differences between the cells of the blastula. At a later stage of development, but before the epithelial flattening due to close union of cells has led to the blastula proper, the Echinus germ, especially in the half containing the smaller-celled pole, consists of cellular rings.

How, then, do the blastomeres of the first division stages after isolation by shaking accomplish cleavage, assuming they survive?

I shall first describe the behavior observed in a majority of cases. Not once did I observe a completely spherical rounding up of the isolated cell. It is true that the normally flat surface tends toward sphericalness but its radius of curvature always remains greater than that of the original free surface of the hemisphere. The cell now divides into two and then, perpendicularly to this, into four parts. Normal controls fertilized at the same time now have eight similar cells the same size as our four. Simultaneously fertilized normal controls have at this time eight similar cells.

In the Echinoids no "gliding" of cells normally occurs either in the four-cell stage nor the ½ eight-cell stage (i.e., my four-cell stage). This is significant because it facilitates considerably the interpretation of the following fact.

About 5½ hours after fertilization occurs, untreated germs have divided into 16 parts, as described above, and isolated blastomeres into 8 parts.

At this point begins the really interesting part of my experiment in that the last-mentioned division brings into existence a typical single half of the 16-cell stage as described; that is, it behaves in the way expected of it according to absolute self-differentiation; it is actually a half of what Selenka's figure shows.

I will now go on to a description of the normal division of my blastomeres, later speaking about the abnormal cases (about 25%).

I carefully followed the formation of a half-germ of 16 cells, i.e., a typical ½ 32-cell stage. Each of the normal concentric cell rings is present, but each consists of half its normal number of cells. The entire structure now presents the appearance of an open hemisphere with a polarly differentiated opening.

In the majority of cases here referred to as normal, the half-germ presented, on the evening of the day of fertilization, the appearance of a typical, many-celled, open hemisphere, although the opening often seemed somewhat narrowed. As especially characteristic, I will mention here a case upon which I chanced in doing the Roux-Chabry experiment. Instead of one of the blastomeres being isolated, it was killed by the shaking. The living one, which had developed in the above manner into a typical half-formation, was in the afternoon attached to the dead one in the shape of a hemisphere; but by evening its edges were already clearly curled inward.

The cleavage of isolated blastomeres of the two-cell stage of Echinus microtuberculatus is accordingly a half-formation as described by Roux for operated frog's eggs.

As already mentioned, this is by far the most frequent behavior. One will not be surprised to find modifications of it in view of damage caused by the strong mechanical insult due to shaking. A few words about these exceptions:

In some cases, germs consisting of about 32 cells (½ 64-cell stage) presented by late afternoon a spherical appearance; development was here more compact, so to speak, though following the typical scheme. This occurs because of a closer union of the cells and is a phenomenon possibly similar to Chabry's "gliding." Normally, the blastomeres of Echinus make contact in only small areas, until shortly before blastula formation.

In other cases—nine were observed in all—there was from the outset (i.e. from the 8 or half 16-cell stage) little to be seen of the usual scheme except as to cell number; specifically, the half germ was spherical from the very beginning, and "gliding" was even more pronounced. I wish to mention especially a case in which the eight cells (half 16) were of almost equal size. Had the role of first cleavage here been different and had I here, to put it briefly, perhaps separated the animal from the vegetal pole instead of the left from the right? By analogy with the experiments of Rauber, Hallez, etc., this seems not unlikely.

The first time I was fortunate enough to make the observations described above, I awaited in excitement the picture which was to present itself in my dishes the next day. I must confess that the idea of a free-swimming hemisphere or a half gastrula with its archenteron open lengthwise seemed rather extraordinary. I thought the formations would probably die. Instead, the next morning I found in their respective dishes typical, actively swimming blastulae of half size.

I have already described how toward the evening of the day of fertilization the, as yet not epithelial, hemisphere had a rather narrowed opening and I have emphasized that tracing of individual cells and hence of the side of the opening corresponding to the animal pole proved impossible. True, I occasionally saw two smaller cells somewhere along the edge but attached no meaning to them. The question as to the actual mode of closing of the blastula must for the time being, therefore, remain unsolved. I may perhaps be briefly permitted to indicate the significance of this.

Now another general question the solution of which I intend soon to undertake: how far does the totipotency of the blastomeres go? That is, up to what stage are blastomeres still able to produce a complete, small organism? In the future I shall call these "part-formations" in contrast to Roux's "half-formations." The polar course of the cleavage, as well as the above hypothesis concerning the closure of the blastula, suggested that perhaps elements of all concentric rings must be present; that would mean, however, that the four-cell stage would be the last from which isolated cells could produce part-formations, since the equatorial cleavage (namely, the third) divides the material into north and south polar rings, so to speak. This is, as stated, for the time being still merely a question; the totipotency of the cells of the four-cell stage seems to me probable in view of the three-quarter + one-quarter blastulae which will be briefly mentioned later. If, on the other hand, the above-mentioned assumption concerning differences in the effect of the first cleavage should prove true, the latter hypothesis, that material from all three rings is necessary for part-formation, would no longer hold.

But let us leave these conjectures and return to the facts. Thirty times I have succeeded in seeing small free-swimming blastulae arise from cleavage as described above of isolated blastomeres; the rest, about 20 cases, died during cleavage or were sacrificed so I could inspect them under higher magnification. Almost all of them at this stage were still transparent and entirely normal structurally though half-sized. I was not, by a method of estimation, able to discover any difference in size between these cells and those of the normal blastula; therefore, the number of cells is probably half the normal number, which is also to be expected from their cleavage behavior.

At the end of the second day, the fate of the experimental cases seemed to be sealed; they showed the effects of strong mechanical insult and of the small amount of water. For germs still transparent at this time, one could count on raising them further; unfortunately, this was the case with 15 specimens only, that is half the total.

The Gastrula and Pluteus

In healthy specimens invagination at the vegetal pole usually begins at the end of the second day; on the morning of the third day little gastrulae swam about actively in the dishes. As stated, I succeeded in observing 15 such specimens.

Three of the formations finally became actual plutei, differing from the normal only in size.

Therefore, these experiments show that, under certain circumstances, each of the first two blastomeres of Echinus microtuberculatus is able to produce a normally developed larva, whole in form and hence a part-, not half-, formation.

This fact is in fundamental contradiction to the theory of organ-forming germ areas, as the following simple consideration specifically demonstrates.

Imagine a normal blastula split along the median plane of the future pluteus; let us now examine one of the hemispheres preserved this way, for instance the left (see Fig. 12). The material at M_0M_u would normally supply material for the median region, that at L material for the left side. But suppose that we imagine the hemisphere closing, as explained above, to form a sphere but still maintaining polarity along BC. Then M_0 will come to lie upon M_u, and hence possibly upon the right side of the future part-formation. Or, if in closure the original median areas supplied materials for the median

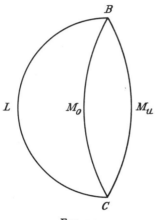

FIG. 12.

region of the part-formation, then this could be thought of only as the upper or lower median region. If it is thought of as the upper, then the lower would come from a part which would otherwise have formed the left side. However one regards it, one cannot escape the fundamental difference in the role which identical material is called upon to play depending upon whether one whole- or two part-formations arise from it,—something which can be brought about artificially. "Il n'est pas des lors permis de croire que chaque sphere de segmentation doit occuper une place et jouer un role, qui sont assignés a l'avance" (Hallez); not, at any rate, in Echinus.

That this is a particularly pleasing result one could scarcely contend; it seems almost a step backward along a path considered already established.

When compared with Roux's, my results reveal a difference in behavior in the sea urchin and frog. Yet perhaps this difference is not so fundamental after all. If the frog blastomeres were really isolated and the other half (which was probably not dead in Roux's case) really removed, would they not perhaps behave like my Echinus cells? The cohesion of the blastomeres, conforming to the law of minimal surface formation, is much greater in the frog than in my object.

I have tried in vain to isolate amphibian blastomeres; let those who are more skillful than I try their luck.

It will not have escaped the reader that the results described might throw light on at least one aspect of the theory of

Double Formation

On this subject, I am in a position to supplement what has already been said. If, from one isolated cell of the two-cell stage, a perfect embryo of half-size is formed (namely a part-formation, in contrast to mere division which yields half-formations as in the case of Roux's frog embryos), it follows unequivocally that both cells of this stage if they are isolated and kept intact will form separate embryos, or twins.

It is highly probable that the separation of blastomeres by shaking was the direct cause of double formation and that without shaking whole-formation would have resulted.

This is certain since part-formations show that an isolated blastomere, provided it lives at all, always develops into a structure which differs from the normal only in size. With other twins, the situation is different, since they are too numerous to be considered accidental formations of this kind, such having never been seen in thousands of larvae observed by the Hertwig brothers and me.

Roux's theory of double formation, must, together with the principle of organ-forming germ-areas previously discussed, be discarded, at least in its general form. I have already remarked that, for our theoretical conceptions, this might be considered a backward rather than a forward step if establishment of facts did not always constitute progress.

Whether or not mechanical isolation or separation of the first two cleavage cells is the only way to obtain twin formation will be left open at this time.

It is an old controversy whether double-formation takes place by fusion or fission; birds and fish would, as mentioned, elsewhere, provide rather unfavorable material for a solution of this problem which in its usual formulation is rather a descriptive than a fundamental one. The observations communicated by other workers as well as my own experiments establish splitting as a cause, to which I may add on the basis of my results, splitting without postgeneration.

Obviously fission- and fusion-double-formation would be two quite different things, hence twinning could be of a dual nature. It is certain that the above mentioned double-fertilization modifies cleavage in such a way that immediate four-cell formation occurs; this is in support of our position, as shown before.

What forces come into play when the blastula closes? Can perhaps part of this process be understood in physical terms? The cell mass takes the form of a sphere, the form, that is, which possesses minimal surface. Further, why is it, after all, that a strong pulling apart of the half blastomeres without destroying them results in two individuals? These and other questions present themselves, but it were futile to indulge in idle speculation without actual facts.

Summary

If one isolates one of the first two blastomeres of Echinus microtuberculatus it cleaves as if for half-formation but forms a whole individual of half-size which is a part-formation.

Therefore the principle of organ-forming germ-areas is refuted for the observed species while the possibility of artificial production of twins is demonstrated.

Addendum: In proof-reading, I will briefly add that I have just succeeded in killing one cell of the two-cell stage of Spherechinus by shaking and in raising from the other half a small pluteus after half-cleavage.

Naples, October, 1891

THE GROWTH OF ISOLATED TISSUE IN VITRO

HARRISON, Ross Granville (American embryologist, 1870–   ). *Observations on the living developing nerve fiber* in *The Anatomical Record* for June 1, 1907, vol. 1, p. 116, 1906–1908.

Few techniques in modern biology have proved more fruitful for the field to which they are adapted than has tissue culture for the study of development. Effects of both intrinsic and extrinsic factors on cellular differentiation would later be studied by Weiss,* Holtfreter,† *et al.* In other experiments, Harrison used microsurgical methods to analyze the progress of determination and differentiation in different parts of amphibian embryos.

OBSERVATIONS ON THE LIVING DEVELOPING NERVE FIBER

The immediate object of the following experiments was to obtain a method by which the end of a growing nerve could be brought under direct observation while alive, in order that a correct conception might be had regarding what takes place as the fiber extends during embryonic development from the nerve center out to the periphery.

The method employed was to isolate pieces of embryonic tissue, known to give rise to nerve fibers, as for example, the whole or fragments of the medullary tube, or ectoderm from the branchial region, and to observe their further development. The pieces were taken from frog embryos about 3 mm. long at which stage, *i.e.,* shortly after the closure of the medullary folds, there is no visible differentiation of the nerve elements. After carefully dissecting it out, the piece of tissue is removed by a fine pipette to a cover slip upon which is a drop of lymph freshly drawn from one of the lymph-sacs of an adult frog. The lymph clots very quickly, holding the tissue in a fixed position. The cover slip is then inverted over a hollow slide and the rim sealed with paraffine. When reasonable aseptic precautions are taken, tissues will live under these conditions for a week and in some cases specimens have been kept alive for

---

* *American Naturalist*, vol. 67, p. 322; *Anatomical Record*, vol. 58, p. 299; vol. 88, p. 48; *Journal of Experimental Zoology*, vol. 68, p. 293; *Physiological Reviews*, vol. 15, p. 639.

† *Roux' Archiv für Entwicklungsmechanik der Organismen*, vol. 124, p. 404; vol. 132, pp. 225, 307; vol. 138, p. 522; *Journal of Experimental Zoology*, vol. 94, p. 261; vol. 95, pp. 171, 307; vol. 98, p. 161; *Journal of Morphology*, vol. 90, p. 57.

nearly four weeks. Such specimens may be readily observed from day to day under highly magnifying powers.

While the cell aggregates, which make up the different organs and organ complexes of the embryo, do not undergo normal transformation in form, owing, no doubt, in part, to the abnormal conditions of mechanical tension to which they are subjected; nevertheless, the individual tissue elements do differentiate characteristically. Groups of epidermis cells round themselves off into little spheres or stretch out into long bands, their cilia remain active for a week or more and a typical cuticular border develops. Masses of cells taken from the myotomes differentiate into muscle fibers showing fibrillae with typical striations. When portions of myotomes are left attached to a piece of the medullary cord the muscle fibers which develop will, after two or three days, exhibit frequent contractions. In pieces of nervous tissue numerous fibers are formed, though, owing to the fact that they are developed largely within the mass of transplanted tissue itself, their mode of development cannot always be followed. However, in a large number of cases fibers were observed which left the mass of nerve tissue and extended out into the surrounding lymphclot. It is these structures which concern us at the present time.

In the majority of cases the fibers were not observed until they had almost completed their development, having been found usually two, occasionally three, and once or twice four days after isolation of the tissue. They consist of an almost hyaline protoplasm, entirely devoid of the yolk granules, with which the cell-bodies are gorged. Within this protoplasm there is no definiteness of structure; though a faint fibrillation may sometimes be observed and faintly-defined granules are discernible. The fibers are about $1.5-3\,\mu$ thick and their contours show here and there irregular varicosities. The most remarkable feature of the fiber is its enlarged end, from which extend numerous fine simple or branched filaments. The end swelling bears a resemblance to certain rhizopods and close observation reveals a continual change in form, especially as regards the origin and branching of the filaments. In fact, the changes are so rapid that it is difficult to draw the details accurately. It is clear we have before us a mass of protoplasms undergoing amoeboid movements. If we examine sections of young normal embryos shortly after the first nerves have developed, we find exactly similar structures at the end of the developing nerve fibers. This is especially so in the case of the fibers which are connected with the giant cells described by Rohon and Beard.

Still more instructive are the cases in which the fiber is brought under observation before it has completed its growth. Then it is found that the end is very active and that its movement results in the drawing out and lengthening of the fiber to which it is attached. One fiber was observed to lengthen almost $20\,\mu$ in 25 minutes, another over $25\,\mu$ in 50 minutes. The longest fibers observed were 0.2 mm. in length.

When the placodal thickenings of the branchial region are isolated, similar fibers are formed and in several of these cases they have been seen to arise from individual cells. On the other hand, other tissues of the embryo, such as myotomes, yolk endoderm, notochord, and indifferent ectoderm from the abdominal region do not give rise to structures of this kind. There can, therefore, be no doubt that we are dealing with a specific characteristic of nervous tissue.

It has not as yet been found possible to make permanent specimens which show the isolated nerve fibers completely intact. The structures are so delicate that the mere immersion in the preserving fluid is sufficient to cause violent tearing and this very frequently results in the tearing away of the tissue in its entirety from the clot. Nevertheless, sections have been cut of some of the specimens and nerves have been traced from the walls of the medullary tube but they were in all cases broken off short.

In view of this difficulty an effort, which resulted successfully, was made to obtain permanent specimens in a somewhat different way. A piece of medullary cord about four or five segments long was excised from an embryo and this was replaced by a cylindrical clot of proper length and caliber which was obtained by allowing blood or lymph of an adult frog to clot in a capillary tube. No difficulty was experienced in healing the clot into the embryo in proper position. After two, three, or four days the specimens were preserved and examined in serial sections. It was found that the funicular fibers from the brain and anterior part of the cord, consisting of naked axones without sheath cells, had grown for a considerable distance into the clot.

These observations show beyond question that the nerve fiber develops by the outflowing of protoplasm from the central cells. This protoplasm retains its amoeboid activity at its distal end, the result being that it is drawn out into a long thread which becomes the axis cylinder. No other cells or living structures take part in this process.

The development of the nerve fiber is thus brought about by means of one of the very primitive properties of living protoplasm, amoeboid movement, which, though probably common to some extent to all the cells of the embryo, is especially accentuated in the nerve cells at this period of development.

The possibility becomes apparent of applying the above method to the study of the influences which act upon a growing nerve. While at present it seems certain that the mere outgrowth of the fibers is largely independent of external stimuli, it is, of course, probable that in the body of the embryo there are many influences which guide the moving end and bring about contact with the proper end structure. The method here employed may be of value in analyzing these factors.

V. CELLULAR BIOLOGY

DESCRIPTION OF CELLS

HOOKE, Robert (English naturalist, 1635–1703). From *Micrographia*, London, 1665.

Of a personally and psychologically restless and ingenious disposition, Hooke took prophetic first steps in microscopic biology, chronometry (added balance-spring to the watch), pneumatics (built the vacuum pump Boyle used), astronomy (partially anticipated Newton on celestial mechanics and gravitation), properties of matter, and other areas of natural science. The *Micrographia*, famous especially for its description of cells in cork, contained accurate descriptions and plates of zoological subjects, including an admirable drawing of the gnat 17 inches long. Here are reprinted a part of the important and interesting Preface, and the description of cells in cork.

[From the preface.]

Toward the prosecution of this method in *Physical Inquiries*, I have here and there *gleaned* up an *handful* of observations, in the collection of most of which I made use of *Microscopes*, and some other *Glasses* and *Instruments* that improve the sense; which way I have herein taken, not that there are not multitudes of useful and pleasant Observables, yet uncollected, obvious enough without the helps of Art, but only to promote the use of Mechanical helps for the Senses, both in the surveying the already visible World, and for the discovery of many others hitherto unknown, and to make us, with the great Conqueror, to be affected that we have not yet overcome one World when there are so many others to be discovered, every considerable improvement of *Telescopes* or *Microscopes* producing new *Worlds* and *Terra-Incognita's* to our view. . . .

What the things are I observed, the following descriptions will manifest; in brief, they were either *exceeding small Bodies*, or *exceeding small Pores*, or *exceeding small Motions*, some of each of which the Reader will find in the following Notes, and such, as I presume, (many of them at least) will be *new*, and perhaps not less *strange:* Some *specimen* of each of which Heads the Reader will find the subsequent delineations, and indeed of some more then I was willing there should be; which was occasioned by my first Intentions to print a much greater number than I have since found time to compleat. Of such therefore as I had, I selected only some few of every Head, which for some particulars seem'd most observable, rejecting the rest as superfluous to the present Design.

What each of the delineated Subjects are, the following descriptions annext to each will inform, of which I shall here, only once for all, add, That in divers of them the Gravers have pretty well follow'd my directions and draughts;

and that in making of them, I indeavoured (as far as I was able) first to discover the true appearance, and next to make a plain representation of it. This I mention the rather, because of these kind of Objects there is much more difficulty to discover the true shape, then of those visible to the naked eye, the same Object seeming quite differing, in one position to the Light, from what it really is, and may be discover'd in another. And therefore I never began to make any draught before by many examinations in several lights, and in several positions to those lights, I had discovered the true form. For it is exceeding difficult in some Objects, to distinguish between a *prominency* and a *depression*, between a *shadow* and a *black stain*, or a *reflection* and a *whiteness in the colour*. Besides, the transparency of most Objects renders them yet much more difficult then if they were *opacous*. The Eyes of a Fly in one kind of light appear almost like a Lattice, drill'd through with abundance of small holes; which probably may be the Reason, why the Ingenious *Dr. Power* seems to suppose them such. In the Sunshine they look like a Surface cover'd with golden Nails; in another posture, like a Surface cover'd with Pyramids; in another with Cones; and in other postures of quite other shapes; but that which exhibits the best, is the Light collected on the Object, by those means I have already describ'd.

And this was undertaken in prosecution of the Design which the ROYAL SOCIETY has propos'd to it self. For the Members of the Assembly having before their eyes so many *fatal* Instances of the errors and falsehoods, in which the greatest part of mankind has so long wandred, because they rely'd upon the strength of humane Reason alone, have begun anew to correct all *Hypotheses* by sense, as Seamen do their *dead Reckonings* by *Coelestial Observations;* and to this purpose it has been their principal indeavour to *enlarge* and *strengthen* the *Senses* by *Medicine,* and by such *outward Instruments* as are proper for their particular works. By this means they find some reason to suspect, that those effects of Bodies, which have been commonly attributed to *Qualities,* and those confess'd to be *occult,* are perform'd by the small *Machines* of Nature, which are not to be discern'd without these helps, seeming the meer products of *Motion, Figure,* and *Magnitude;* and that made in *Looms,* which a greater perfection of Opticks may make discernable by these Glasses; so as now they are no more puzzled about them, then the vulgar are to conceive, how *Tapestry* or *flowred Stuffs* are woven. And the ends of all these Inquiries they intend to be the Pleasure of Contemplative Minds, but above all, the *ease and dispatch* of the labours of mens hands. They do indeed neglect no opportunity to bring all the *rare* things or Remote Countries within the compass of their knowledge and practice. But they still acknowledge their *most useful* Informations to arise from *common* things, and from *diversifying* their most *ordinary* operations upon them. They do not wholly reject Experiments of meer *light* and *theory;* but they principally aim at such, whose Applications will *improve* and *facilitate* the present way of *Manual Arts.*

And though some men, who are perhaps taken up about less honourable Employments, are pleas'd to censure their proceedings, yet they can shew more *fruits* of their first three years, wherein they have assembled, then any other *Society* in *Europe* can for a much larger space of time. 'Tis true, such undertaking as theirs do commonly meet with small incouragement, because men are generally rather taken with the *plausible* and *discursive,* then the real and the solid part of Philosophy; yet by the good fortune of their institution, in an Age of all others the most *inquisitive,* they have been assisted by the *contribution* and *presence* of very many of the chief *Nobility* and *Gentry,* and others, who are some of the *most considerable* in their several Professions. But that that yet farther convinces me of the *Real esteem* that the more *serious* part of men have of this *Society,* is, that several *Merchants,* men who act in earnest (whose Object is *meum* and *tuum,* that great Rudder of humane affairs) have adventur'd considerable sums of *Money,* to put in practice what some of our Members have contrived, and have continued *stedfast* in their good opinions of such Indeavors, when not one of a hundred of vulgar have believed their undertakings feasable. And it is also fit to be added, that they have one advantage peculiar to themselves, that very many of their number are *men* of *Converse and Traffick;* which is a good Omen, that their attempts will bring Philosophy from *words* to *action,* seeing the men of Business have had so great a share in their first foundation. . . .

[From the text.]

I took a good clear piece of cork, and with a Pen-knife sharpened as keen as a Razor, I cut a piece of it off, and thereby left a surface of it exceeding smooth; then examining it very diligently with a Microscope, me thought I could perceive it to appear a little porous; but I could not so plainly distinguish as to be sure that they were pores, much less what Figure they were of. But judging from the lightness and yielding quality of Cork, that certainly the texture could not be so curious but that possible, if I could use some further diligence, I might find it to be discernable with a Microscope, I with the same sharp Pen-knife, cut off from the former smooth surface an exceeding thin piece of it, and placing it on a black object Plate, because it was itself a white body, and casting light upon it with a plane-convex glass I could exceedingly plainly perceive it to be all perforated and porous, much like a Honey-comb, but that the pores of it were not regular. . . .

But to return to our Observation, I told several lines of these pores and found that there were usually about three-score of these small cells placed end ways in the eighteenth part of an inch, whence I concluded that there must be near eleven hundred of them or somewhat more than a thousand in the length of an Inch, and therefore in a square inch above a Million or 1166400 and in a Cubic Inch, above twelve hundred Million or 1259712000 a thing almost incredible did not our Microscope assure us of it by ocular demonstration; — So prodigiously curious are the works of Nature, that even these conspicuous pores of bodies, which seem to be channels or pipes through which the *succus nutritius* or natural juices of vegetables are convey'd seem to correspond to the veins, arteries, and other vessels in sensible creatures . . .

Nor is this kind of Texture peculiar to Cork onley; for upon examination into my Microscope, I have found that the pith of an Elder, or almost any other Tree, the inner pulp or pith of the Cany hollow stalks of several other Vegetables, as of Fennel, Carrots, Daucus, Bur-docks, Teasels, Fearns, and some kinds of Reeds, etc. have much such a kind of Schematism, as I have lately shewn that of Cork, save only that here the pores rang'd the long-ways, or the same way with the length of the Cane, whereas in Cork they are Transverse.

But though I could not with my Microscope, nor my breath, nor any other way I have yet try'd, discover a passage out of one of those cavities into another, yet I cannot thence conclude, that therefore there are none such, by which the *succus nutritius,* or appropriate juices of vegetables, may pass through them; for in several of these vegetables whil'st green, I have with my Microscope plainly enough discover'd these Cells or Pores [Poles in the original] fill'd with juices, and by degrees sweating them out.

Now, though I have with great diligence endeavered to find whether there be any such thing in those Microscopic pores of Wood or piths as the Valves in the heart, veins, and other passages in animals, that open and give passage to the contain'd fluid just this one way, and shut themselves, and impede the passage of such liquors back again, yet I have not hitherto been able to say anything definite in it; though, me thinks, it seems very probable that Nature has in these passages, as well as in those of Animal bodies, very many appropriated Instruments and contrivances, whereby to bring her designs and ends to pass, which 'tis not improbable but that some diligent Observer if help'd with better Microscopes may in time detect.

PROTOZOA DISCOVERED

van LEEUWENHOEK, Antony (Dutch microscopist, 1632–1723). From a communication in Dutch to the Secretary of the Royal Society of London; tr. by H. Oldenburg [?]* as *More observations from Mr. Leeuwenhoek, in a letter of Sept. 7, 1674 sent to the publishers,* in *Philosophical Transactions of the Royal Society,* vol. 9, No. 108, p. 181, London, 1674.†

Gesner had described the foraminiferan *Vaginulina* in 1565, and 100 years later Hooke published a picture of (probably) *Rotalia beccarii.* Leeuwenhoek discovered

---

* See C. Dobell, *Antony van Leeuwenhoek and his little animals,* New York, 1932.

† Reprinted here from J. F. Schill, *Antony van Leeuwenhoek's entdeckung der microorganismen,* in *Zoologischer Anzeiger,* vol. 10, p. 685, 1887.

Vorticella, 1677; *Giardia*, *Nyctotherus*, and *Opalina*, 1683; *Polystomella* and *Volvox*, 1700; *Carchesium* and (probably) *Stylonichia*, 1703.‡ We give here Leeuwenhoek's first published allusion to the protozoa, much more fully described by him in another, later letter, for part of which see pp. 435–437.

About two Leagues from this Town [Delft] there lyes an Inland-Sea, called Derkelse-Sea, whose bottom in many places is very moorish. This Winter very clear, but about beginning or in the midst of Summer it grows whitish, and there are then small green clouds permeating it, which the Country-men, dwelling near it, say is caused from the Dews then falling, and call it Hony-dew. This water is abounding in Fish, which is very good and savoury. Passing lately over this Sea at a time, when it blew a fresh gale of wind, and observing the water as above-described, I took up some of it in a Glass-vessel, which having view'd the next day, I found moving in it several Earthy particles, and some green streaks, spirally ranged, after the manner of the Copper or Tin-worms, used by Distillers to cool their distilled waters; and the whole compass of each of these streaks was about the thickness of a man-hair on his head: Other particles had but the beginning of the said streak; all consisting of small green globuls interspersed; among all which there crawled abundance of little animals, some of which were roundish; those that were somewhat bigger than others, were of an Oval figure: On these latter I saw two leggs near the head, and two little fins on the other end of their body: Others were somewhat larger than an Oval, and these were very slow in their motion, and few in number. These animalcula had divers colours, some being whitish, other pellucid; others had green and very shining little scales; others again were green in the middle, and before and behind white, others grayish. And the motion of most of them in the water was so swift, and so various, upwards, downwards, and round about, that I confess I could not but wonder at it. I judge, that some of these little creatures were above a thousand times smaller than the smallest ones, which I have hitherto seen in chees, wheaten flower, mould, and the like.

BACTERIA DESCRIBED

van LEEUWENHOEK, Antony (Dutch microscopist, 1632–1723). From a communication in Dutch dated Delfft, October 9, 1676, to the Secretary of the Royal Society of London; tr. from the MS by C. Dobell as *Letter 18* in his *Antony van Leeuwenhoek and his little animals*, New York, 1932;* by permission of Harcourt, Brace and Co., Inc.

‡ See F. J. Cole, *The history of protozoology*, London, 1926.

* An abbreviated translation of this letter was published by H. Oldenburg as *Observations, communicated to the publisher* by Mr. Antony van Leeuwenhoeck *in a Dutch letter of the 9th of Octob. 1676. here English'd: Concerning little animals by him observed in rain- well- sea- and snowwater; as also in water wherein pepper had lain infused*, in *Philosophical Transactions of the Royal Society*, vol. 12, pp. 821–831, 1677.

On Dobell's authority, this item is included here as the first unequivocal description of bacteria† based on actual observation. The rest of this, the famous eighteenth, letter describes various other protista many of which can, in- Dobell's opinion, be identified from Leeuwenhoek's descriptions. See pp. 434 and 435. We reprint here without abbreviation the first observation on pepper water.

Having made sundry efforts, from time to time, to discover, if 'twere possible, the cause of the hotness or power whereby pepper affects the tongue (more especially because we find that even though pepper hath lain a whole year in vinegar, it yet retaineth its pungency); I did now place anew about ⅓ ounce of whole pepper in water, and set it in my closet, with no other design than to soften the pepper, that I could the better study it. This pepper having lain about three weeks in the water, and on two several occasions snow-water having been added thereto, because the water had evaporated away; by chance observing this water on the 24th April, 1676, I saw therein, with great wonder, incredibly many very little animalcules, of divers sorts; and among others, some that were 3 or 4 times as long as broad, though their whole thickness was not, in my judgement, much thicker than one of the hairs wherewith the body of a louse is beset. These creatures were provided with exceeding short thin legs in front of the head (although I can make out no head, I call this the head for the reason that it always went in front during motion). This supposed head looked as if 'twas cut off aslant, in such fashion as if a line were drawn athwart through two parallel lines, so as to make two angles, the one of 110 degrees, the other of 70 degrees. Close against the hinder end of the body lay a bright pellet, and behind this I judged the hindmost part of all was slightly cleft. These animalcules are very odd in their motions, oft-times tumbling all around sideways; and when I let the water run off them, they turned themselves as round as a top, and at the beginning of this motion changed their body into an oval, and then, when the round motion ceased, back again into their former length.

The second sort of animalcules consisted of a perfect oval. They had no less nimble a motion than the animalcules first described, but they were in much greater numbers. And there was also a third sort, which exceeded both the former sorts in number. These were little animals with tails, like those that I've said were in rain-water.

The fourth sort of little animals, which drifted among the three sorts aforesaid, were incredibly small; nay, so small, in my sight, that I judged that even if 100 of these very wee animals lay stretched out one against another, they could not reach to the length of a grain of coarse sand; and if this be true, then ten hundred thousand of these living creatures could scarce equal the bulk of a coarse sand-grain.

---

† See footnote 3, p. 133, in his translation.

I discovered yet a fifth sort, which had about the thickness of the last-said animalcules, but which were near twice as long.

DESCRIPTION OF SARCODE

DUJARDIN, Félix (French microscopic zoologist, 1801–1860). From *Sur les pré-tendus estomacs des animalcules infusoires et sur une substance appelée Sarcode*, (= Part III of) *Recherches sur les organismes inférieures*, in *Annales des Sciences Naturelles: Zoologie*, 2d ser., vol. 4, p. 343, 1835, republished as part of *Histoire naturelle des zoophytes, Infusoires*, etc., Paris, 1841; tr. by T. S. Hall for this volume.

Dujardin's description of "a substance called sarcode" is of historical interest from two points of view: first, as one step in the argument with Ehrenberg (*q.v.*) concerning the internal organization of the infusoria, an argument which was to be settled by Siebold (*q.v.*); and, secondly, as a stimulus to microscopists to seek similar substances in other cells of other organisms, a quest which was to lead to the protoplasm concept of Schultze (*q.v.*) and others.

At the commencement of the study of Infusorians, an organization was attributed to them comparable to that of higher animals; for it pleases us to invoke the marvelous in this new world which the microscope has made known.

It is curious to observe Leeuwenhoek speaking of tendons, muscles, etc., contained in the tail of a sperm or in the smallest vibrios, comparing them to the tail of a rat.

O. F. Müller seemed to believe in the great simplicity of the first Infusoria and in their spontaneous generation; nevertheless in his work, which death prevented him from putting in order, he declares on page 47 that he agrees with Leeuwenhoek on the question of the organization of vibrios.

This opinion, which admits the existence of the most complex organs in the smallest of beings, has always had a number of partisans, and even today one sees muscles, nerves, and vessels whose smallness has prevented their being seen cited in physical treatises as a proof of the extreme divisibility of matter.

Lamarck, at first a botanist, had learned not to establish too big an interval between vegetable and animal life, and, commencing his study of the animal kingdom by the smallest, simplest creatures, showed himself under the influence of a more philosophical idea when he advanced the point of view that many infusoria are only accumulations of living matter, gelatinous corpuscles, without organs, without any absolutely fixed form.

M. Bory, treading the same path, saw in the microscopic beings and especially in the Gymnodes only creatures of an extreme simplicity and was not averse to the idea of spontaneous generation as a property of them; he has even written on their mode of composition, in some articles in the classical dictionary of natural history and in the encyclopedia, in such vein as to make it desirable that he undertake a more extended study.

However, several naturalists preferred to regard these little beings as repetitions of higher organisms or germs susceptible of subsequent development. M. Blainville, in his article on Zoophytes in the Dictionary of Natural Sciences, professed as practically certain that many true Infusoria were merely Planarias or young Entomostraca.

The works of M. Ehrenberg could not fail to find great favor when, seduced by appearances, this naturalist mistook for truth an ingenious hypothesis on the multiplicity of the stomachs of Infusoria (which, therefore, he called Polygastricha) endeavoring to raise them on the animal scale to the level of the Vertebrates.

These so-called "stomachs,"—Müller had seen them and had taken them for ovules or ovaries, much as the solid ovoid masses which one sees in some small animals, and yet the expressions he employs in describing them are quite suitable to giving a true idea of them and show how easy it is to make them out.

Gleichen brought to his researches on the Infusoria a truly remarkable precision and, to identify in these animals an "effective swallowing of their food" he artificially colored their interior globules with carmen; he regarded these globules as eggs; "for," said he, "when these globules are separated by interstices, they appear to be surrounded with transparent rings like those of toads' eggs." Just the same, he did not try to conceal the difficulties presented by this way of explaining things, because of the very coloration of the globules, and, although he had seen the globules outside the organisms, he said later on: "I must confess that if they are not the excrements of the animalcule, (which suffers from equal difficulties), I do not know what else can be said of them." . . .

The difficulties presented by these stomachs or globules have long seemed to me to be insoluble, and, although I made use of the most perfect possible instruments, and although I was very used to this sort of observation, I blame myself only that I was unable to discover the intestine into which these stomachs must empty, nor the anal nor buccal orifices of this intestine. The high merit which the works of M. Ehrenberg demand did not permit me to call in question a fact of such importance taken by him as a basis for his classification. In fact, I knew that other observers, clever in the use of the microscope, had been no more fortunate than I, nevertheless I would perhaps have lost courage, and abandoned this research as entirely disproportionate to my powers of insight, had I not luckily found the solution to the problem in the discovery of the properties of: *Sarcode*.

I propose to name thus what other observers have called a living jelly, this glutinous, diaphanous substance, insoluble in water, contracting into spherical masses, sticking to the dissecting needles and letting itself be drawn out like slime, and, finally, being found in all the lower animals interposed between the other structural elements.

Sarcode decomposes itself little by little in water, lessening in volume and ending by leaving but a weak irregularly granular residue. Potash does not suddenly dissolve it the way it does mucus or albumin, and seems merely to hasten its decomposition by water; nitric acid and alcohol coagulate it suddenly, turning it white and opaque. Hence its properties are very distinct from those of substances with which one might confuse it, for its insolubility in water distinguishes it from albumin which it resembles in its manner of being coagulated by nitric acid, and this coagulation along with its insolubility in potash distinguishes it from mucus, gelatin, etc.

But the strangest property of *Sarcode* is the spontaneous production in its interior of vacuoles, or little spherical cavities, occupied by the environing liquid, gradually increasing in size, and hastening the decomposition of the globules of this substance of which there no longer remains much more than a sort of open cagework and finally a weak residue.

STATUS OF THE PROTOZOA PRIOR TO ENUNCIATION OF GENERAL CELL THEORY

EHRENBERG, Christian Gottfried (German microbiologist, 1795–1876). From *Die infusionstierchen als vollkommene organismen. Ein blick in des tiefere organische leben der natur,* Leipzig, 1838; tr. by L. Mezger and M. and V. Hamburger for this volume.

Three major developments leading to the establishment of protozoology as a science were: the discovery of the protozoa (see Leeuwenhoek); the disagreement whether they should be regarded as "vollkomene" organisms (see below and Dujardin); and their assignment to a cellular status (see Siebold). Ehrenberg's famous *Infusionstierchen* bore an unlucky time relation to Schwann's cell theory which followed it shortly and necessitated a thorough revision of Ehrenberg's ideas. The following passage expresses succinctly the status of protozoology at the time, at least as it appeared to one of its most enthusiastic students. Elsewhere in the same text, we learn that the chief empirical evidence cited by Ehrenberg in proof of the "full-fledgedness" of the infusoria was a number of misinterpreted observations upon their ingestion of dyes, leading him to consider their gastric vacuoles as comparable to the alimentary apparatus of more complex animals.

[From the dedication to Frederick Wilhelm, Crown Prince of Prussia]

There is in the realm of the unseeably small a perfect organic life which immeasurably enhances the size of the large in nature. All the forms described herein, some of them amazingly important, are too small to be seen distinctly with the naked eye and far too small to be recognized by it as perfectly organized beings. They belong to an invisible but strongly active world of organism. Now that it has finally been more thoroughly explored by senses fortified artificially, and by methods which happily support the senses, its formation has made us see that even the smallest in space is therefore not simple but is so wonderfully and inconceivably endowed with manifold functional organs as to achieve equal rank, both in kind and value, with larger living forms.

The following is a summary of the particularly notable properties and relations of Infusoria established and dealt with in the text:

1) All infusoria are organized animals; many, perhaps all, highly organized. That all microscopic organisms are animals and not plants, as Buffon believed, is wrong. Many plants definitely exist as individual microscopic forms.

2) On the basis of structure, the Infusoria form two entirely natural animal classes; they can be scientifically classified according to structure; their forms do not permit of their incorporation into the same genera or families with larger animals, however similar they sometimes appear.

3) The existence of Infusoria has been demonstrated on four continents and in the ocean; identical kinds are found in the remotest parts of these regions.

4) The geographical distribution of the Infusoria over the earth follows laws already known with respect to other organisms. There are in other regions of the world a greater number of substituent and deviate forms toward the South than toward the West and East; nevertheless, they exist everywhere, and the effect of climatic differences upon form is not limited to larger beings. There are, living in the sea and salt water, many forms different from those in fresh water; yet many are the same and adapt themselves to various and different situations.

5) Most Infusoria are not visible to the naked eye, but many can be seen as little moving points; in none is the body size greater than one line. The organization of all is without exception completely invisible to the unaided eye.

6) By their immeasurable densely crowded masses, the imperceptible little Infusoria color wide expanses of water with their striking hues.

7) Themselves invisible, they effect a sort of illumination of the sea by producing their own light.

8) Individually imperceptible, they form a kind of topsoil by their closely packed living masses.

9) Since there are often more than 41,000 million single animals in one cubic inch of soil, the Infusoria constitute the numerically largest known representation of independent life; they form the main number and perhaps the main mass of animal organisms on earth.

10) The Infusoria possess the greatest reproductive potential known so far in the entire organic universe: namely, their ability to multiply from a single individual to a million in a few hours.

Since a single Vorticella or Bacillaria divides within an hour and often in another hour divides again so that in three hours one will become 4, in five hours 8, and in seven hours 16, it is possible for two animals to produce 4,096 individuals in 24 hours, 8 million in 48 hours or two days, and 140 million in four days.

In the *tripoli* of Bilin about 41,000 million Gallionellae make up 1 cubic inch of rock; therefore about 70 million make up a cubic foot. Hence, one little animal could probably produce 2 cubic feet of rock in 4 days (or by mere self division). This steadily continuing reproductivity seems to be greatly impeded by other, external conditions, but it may be stated without exaggeration that this much-reproductive force is dormant in them. Thus the trees bloom lavishly and bear only moderately or often.

11) The observed multiplication of Infusoria by self-division is responsible for their survival and their distribution in sea and air which makes useless all calculations of the possible destruction of individuals and poetically enough borders on immortality and eternal youth. One divides oneself into innumerable, always new parts to live and be young for countless years.

12) Bud-pairing which perhaps includes the still unsolved enigma of polyembryomy in all plant seeds and plant formation (the budding stems of all trees, shrubs, and plants are apparently flower stalks which resemble coral stalks) (compare also *The Origin of Mycelia*, 1820) is also encountered in the Paramecia.

13) The Infusoria by their silica shells form indestructible soils, stones and rocks which, already exceeding in antiquity the history of man, will perhaps become monuments of the formation of the earth, surpassing all limelike, more easily destroyed, organic remains.

14) From these invisible Infusoria one can make glass with lime or soda, prepare floating bricks, use them as flint, probably prepare iron from them, use them as tripoli powder to polish and fashion silver, as ochre dye, as humus and topsoil in manuring; one can also use rock meal made from them as a harmless filling against hunger.

15) The invisible Infusoria are occasionally harmful, but only, it seems, by killing fish in ponds, by making clear water turbid, by producing swamp smells and by frightening superstitious people. That they cause malaria, plague, and other diseases is improbable and has never been reliably shown. During the cholera epidemic in Berlin in 1832, I did not see any unusual phenomena in the waters nor in the atmosphere. It is true that minute itch- and pus-mites exist; but everything from Baal-Sebub and Oriental plague to Linné's Furia infernalis and the cholera animalcules consists so far only of contentions or affirmation.

16) The Infusoria as far as has been observed do not sleep.

17) The Infusoria liquefy partly during egg-laying and thus passively change their forms in many ways.

18) Infusoria comprise the *invisible*, intestinal worms of man and many animals, even if one excludes the spermatozoa.

19) The invisible Infusoria have their own 'lice' and intestinal worms, and the lice of the Infusoria have in turn their own recognizable lice.

20) The life of the Infusoria is remarkably long, even disregarding division. They may often hibernate dried out by frost and estivate, dried out by warmth but more probably they simply lie in inertia without sleep or torpor. This scarcely prolongs their life but rather shortens it.

21) Just as pine pollen annually falls from the clouds like 'sulfur-rain' so the much smaller Infusoria, lifted passively by the water vapor, and, numerous and cloudlike, appear to float, alive and unseen in the atmosphere. Less often, perhaps, they live mixed with dust. On this point, observations of them have not as yet been made in sufficient numbers nor in a strictly scientific way. The Infusoria are to be expected only at the beginning of a downpour, and before even 5 single drops have been examined, the opportunity is gone. To examine thoroughly but a thousand drops of rain takes much time, and what are 1,000 drops of one rain? This interesting field is still open for observation. Also, according to Franz Schulze's and Schwann's new experiments with artificially purified atmospheric air, the water vapor and dust-free air contains no Infusoria.

22) In general Infusoria behave toward all external influences rather similarly to large organisms. They occasionally consume strong poisons without immediate injury; nevertheless, the effect gradually is harmful. Under certain circumstances, they can withstand extreme heat and cold as can other animals and man. They can live with and without light.

23) Light as is the weight of the invisible Infusoria, it is yet computable and has been measured, and yet the softest breeze which lifts feathers will play with such little bodies as with water vapor.

24) Having become fully aware of the apparently great velocity of Infusorian movement in magnified drops, I found that *Hydatina senta* travelled 1 line in 4 seconds, *Monas punctum* 1 line in 48 seconds, and *Navicula gracilis* 1 line in 6 minutes and 24 seconds. Thus *Hydatina senta* would take 21 weeks for 1 mile, *Monas punctum* five years, and *Navicula gracilis* 40 years. A snail (*Linnaeus stagnalis*) travels ¾ lines in one second, a man at quick pace 5 feet per second, and a military horse 13 feet per second at a trot.

25) Linné stated that: all lime comes from worms (Omne calx e vermibus). Now, this invites one to ask whether all silica and all iron (that is the three main constituents of the earth) do not also come from worms or have not, at least, been consumed by them and subjected to various organic changes. Omne silex, omne ferrum e vermibus? To affirm or deny it is equally wrong at this time. Only more and more special investigations will throw light on the question.

26) All of the direct observations so far made on spontaneous generation of organic beings lack, as it seems now, a necessary exactness. The very ob-

servers who supposedly saw the sudden coming into existence of the smallest organisms from primary matter entirely overlooked their very complex structure. One cannot fail to recognize gross incongruities and a deception here is obvious.

Observations of the formation of crustacean-like animals and insects from primary matter are reminiscences of an obsolete time, when caterpillars grew from leaves. Historically, it is perfectly clear that, due to our better knowledge, spontaneous generation, first applied to aboriginal human beings, has gradually receded to the frogs, from frogs to insects, and from insects to those forms which are microscopic and barely accessible to investigation. The ground on which these supposedly rest is also falling away.

27) It has proved possible to reduce the marvelous continual form changes of some Infusoria to limits and organic laws.

28) The power of infusorian organization is strikingly displayed in a strong chewing apparatus with teeth in their mouth. They also possess perfectly distinct intellectual abilities, like other animals. Whether they have a more perfect soul, as the philosopher Crusius infers from their self-division, is open to question.

29) Observations on Infusoria have resulted in a more clearly defined concept of animals in general from which all plants and minerals are strictly and sharply separated by their lack of an animal-organic system.

30) Finally, it is evident from these investigations that experiments reveal a boundlessness of organic creation within the smallest spaces just as does the world of stars in these great spaces whose boundary is not a natural one but is set by our optical devices. Through the genera Monas, Bodo, Vibrio, and Bacterium runs a milky-way of minutest organization.

THE GENERAL CELL THEORY

SCHWANN, Theodor (German zoologist and microscopist, 1810–1882). From *Mikroskopische untersuchungen über die übereinstimmung in der struktur und dem wachstum der tiere und pflanzen*, Berlin, 1839;* tr. by H. Smith as *Microscopical researches into the structure and growth of animals and plants*, London (for the Sydenham Society), 1847.

When Schwann states that all tissues *"bilden sich aus Zellen,"*† he uses the term *aus* in its fullest sense. Schwann clearly realized that not all parts of the animal body are of cellular composition. His point was that any which are not actually cellular are nevertheless *produced by* cells or *develop out of them*. It was, in fact, largely his microscopical studies of animal *development* that brought Schwann to his conclusions concern-

---

* Republished as No. 176 of Ostwald's *Klassiker der exakten wissenschaften*, Leipzig, 1910.

† In the 1910 edition, see footnote 1, p. 163.

ing the accordance of plant and animal tissue. The following passage, part of a general retrospect at the end of the book, is followed by mechanistic speculations based on a comparison of cells with crystals.

THE CELL THEORY

When organic nature, animals and plants, is regarded as a Whole, in contradistinction to the inorganic kingdom, we do not find that all organisms and all their separate organs are compact masses, but that they are composed of innumerable small particles of a definite form. These elementary particles, however, are subject to the most extraordinary diversity of figure, especially in animals; in plants they are, for the most part or exclusively, cells. This variety in the elementary parts seemed to hold some relation to their more diversified physiological function in animals, so that it might be established as a principle, that every diversity in the physiological signification of an organ requires a difference in its elementary particles; and, on the contrary, the similarity of two elementary particles seemed to justify the conclusion that they were physiologically similar. It was natural that among the very different forms presented by the elementary particles, there should be some more or less alike, and that they might be divided, according to their similarity of figure, into fibres, which compose the great mass of the bodies of animals, into cells, tubes, globules, etc. The division was, of course, only one of natural history, not expressive of any physiological idea, and just as a primitive muscular fibre, for example, might seem to differ from one of areolar tissue, or all fibres from cells, so would there be in like manner a difference, however gradually marked between the different kinds of cells. It seemed as if the organism arranged the molecules in the definite forms exhibited by its different elementary particles, in the way required by its physiological function. It might be expected that there would be a definite mode of development for each separate kind of elementary structure, and that it would be similar in those structures which were physiologically identical, and such a mode of development was, indeed, already more or less perfectly known with regard to muscular fibres, blood-corpuscles, the ovum, and epithelium-cells. The only process common to all of them, however, seemed to be the expansion of their elementary particles after they had once assumed their proper form. The manner in which their different elementary particles were first formed appeared to vary very much. In muscular fibres they were globules, which were placed together in rows, and coalesced to form a fibre, whose growth proceeded in the direction of its length. In the blood-corpuscles it was a globule, around which a vesicle was formed, and continued to grow; in the case of the ovum, it was a globule, around which a vesicle was developed and continued to grow, and around this again a second vesicle was formed.

The formative process of the cells of plants was clearly explained by the researches of Schleiden, and appeared to be the same in all vegetable cells. So that when plants were regarded as something special, as quite distinct from the animal kingdom, one universal principle of development was observed in all the elementary particles of the vegetable organism, and physiological deductions might be drawn from it with regard to the independent vitality of the individual cells of plants, etc. But when the elementary particles of animals and plants were considered from a common point, the vegetable cells seemed to be merely a separate species, co-ordinate with the different species of animal cells, just as the entire class of cells was co-ordinate with the fibres, etc., and the uniform principle of development in vegetable cells might be explained by the slight physiological difference of their elementary particles.

The object, then, of the present investigation was to show that the mode in which the molecules composing the elementary particles of organisms are combined does not vary according to the physiological signification of those particles, but that they are everywhere arranged according to the same laws; so that whether a muscular fibre, a nerve-tube, an ovum, or a blood-corpuscle is to be formed, a corpuscle of a certain form, subject only to some modifications, a cell-nucleus, is universally generated in the first instance; around this corpuscle a cell is developed, and it is the changes which one or more of these cells undergo that determine the subsequent forms of the elementary particles; in short, that there is one common principle of development for all the elementary particles of organisms.

In order to establish this point it was necessary to trace the progress of development in two given elementary parts, physiologically dissimilar, and to compare them with one another. If these not only completely agreed in growth, but in their mode of generation also, the principle was established that elementary parts, quite distinct in a physiological sense, may be developed according to the same laws. This was the theme of the first section of this work. The course of development of the cells of cartilage and of the cells of the chorda dorsalis was compared with that of vegetable cells. Were the cells of plants developed merely as infinitely minute vesicles which progressively expand, were the circumstances of their development less characteristic than those pointed out by Schleiden, a comparison, in the sense here required, would scarcely have been possible. We endeavoured to prove in the first section that the complicated process of development in the cells of plants recurs in those of cartilage and of the chorda dorsalis. We remarked the similarity in the formation of the cell-nucleus, and of its nucleolus in all its modifications, with the nucleus of vegetable cells, and the pre-existence of the cell-nucleus and the development of the cell around it, the similar situation of the nucleus in relation to the cell, the growth of the cells, and the thickening of their wall during growth, the formation of cells within cells, and the transformation of the cell-contents just as in the cells of plants. Here, then, was a complete accordance in every known stage in the progress of development of two elementary parts which are quite distinct, in a physiological sense, and it was established that the principle of development in two such parts may be the same, and so far as could be ascertained in the cases here compared, it is really the same.

But regarding the subject from this point of view we are compelled to prove the universality of this principle of development, and such was the object of the second section. For so long as we admit that there are elementary parts which originate according to entirely different laws, and between which and the cells which have just been compared as to the principle of their development there is no connexion, we must presume that there may still be some unknown difference in the laws of the formation of the parts just compared, even though they agree in many points. But, on the contrary, the greater the number of physiologically different elementary parts, which, so far as can be known, originate in a similar manner, and the greater the difference of these parts in form and physiological signification, while they agree in the perceptible phenomena of their mode of formation, the more safely may we assume that all elementary parts have one and the same fundamental principle of development. It was, in fact, shown that the elementary parts of most tissues, when traced backwards from their state of complete development to their primary condition are only developments of cells, which so far as our observations, still incomplete, extend, seemed to be formed in a similar manner to the cells compared in the first section. As might be expected, according to this principle the cells, in their earliest stage, were almost always furnished with the characteristic nuclei, in some the pre-existence of this nucleus, and the formation of the cell around it was proved, and it was then that the cells began to undergo the various modifications, from which the diverse forms of the elementary parts of animals resulted. Thus the apparent difference in the mode of development of muscular fibres and blood-corpuscles, the former originating by the arrangement of globules in rows, the latter by the formation of a vesicle around a globule, was reconciled in the fact that muscular fibres are not elementary parts co-ordinate with blood-corpuscles, but that the globules composing muscular fibres at first correspond to the blood-corpuscles, and are like them, vesicles or cells, containing the characteristic cell-nucleus, which, like the nucleus of the blood-corpuscles, is probably formed before the cell. The elementary parts of all tissues are formed of cells in an analogous, though very diversified manner, so that it may be asserted, *that there is one universal principle of development for the elementary parts of organisms, however different, and that this principle is the formation of cells.* This is the chief result of the foregoing observations.

The same process of development and transformation of cells within a structureless substance is repeated in the formation of all the organs of an organism, as well as in the formation of new organisms; and the fundamental phenomenon attending the exertion of productive power in organic nature is accordingly as follows: *a structureless substance is present in the first instance, which lies either around or in the interior of cells already existing; and cells are formed in it in accordance with certain laws, which cells become developed in various ways into the elementary parts of organisms.*

The development of the proposition, that there exists one general principle for the formation of all organic productions, and that this principle is the formation of cells, as well as the conclusions which may be drawn from this proposition, may be comprised under the term *cell-theory,* using it in its more extended signification, whilst in a more limited sense, by theory of the cells we understand whatever may be inferred from this proposition with respect to the powers from which these phenomena result.

OMNIS CELLULA E CELLULA

VIRCHOW, Rudolph Ludwig Karl (German cytologist and pathologist, 1821–1902). From *Die cellularpathologie in ihrer begründung auf physiologische und pathologische gewebelehre. Zwanzig vorlesungen gehalten während der monate februar, märz, und april 1858 im Pathologischen Institute zur Berlin,* Berlin, 1858; tr. (from the 2d ed.) by F. Chance as *Cellular pathology as based upon physiological and pathological histology, etc.,* New York, 1860.

At least until the publication of Schwann's cell theory, cells were believed able to arise *de novo* from undifferentiated exudate of other cells—an idea accepted at that time by Schwann himself (*q.v.*). It was Virchow (see also p. 508) who established the law that cells arise only from preexisting cells. The following passage is about half of the second of his famous twenty lectures and contains an important survey of the history of ideas leading up to the pronouncement of his basic law of cellular continuity.

In my first lecture, gentlemen, I laid before you the general points to be noted with regard to the nature and origin of cells and their constituents. Allow me now to preface our further considerations with a review of the animal tissues in general, and this both in their physiological and pathological relations.

The most important obstacles which, until quite recently, existed in this quarter, were by no means chiefly of a pathological nature. I am convinced that pathological conditions would have been mastered with far less difficulty if it had not, until quite lately, been utterly impossible to give a simple and comprehensive sketch of the physiological tissues. The old views, which have in part come down to us from the last century, have exercised such a preponderating influence upon that part of histology which is, in a pathological point of view, the most important, that not even yet has unanimity been arrived at, and you will therefore be constrained after you have inspected the preparations I shall lay before you, to come to your own conclusions as to how far that which I have to communicate to you is founded upon real observation.

If you read the *Elementa Physiologiae* of Haller, you will find, where the elements of the body are treated of, the most prominent position in the whole work assigned to *fibres*, the very characteristic expression being there made use of, that the fibre (fibra) is to·the physiologist what the line is to the geometrician.

This conception was soon still further expanded, and the doctrine that fibres serve as the groundwork of nearly all the parts of the body, and that the most various tissues are reducible to fibres as their ultimate constituents, was longest maintained in the case of the very tissue in which, as it has turned out, the pathological difficulties were the greatest—in the so-called cellular tissue.

In the course of the last ten years of the last century there arose, however, a certain degree of reaction against this fiber-theory, and in the school of natural philosophers another element soon attained to honor, though it had its origin in far more speculative views than the former, namely, the *globule*. Whilst some still clung to their fibres, others, as in more recent times Milne Edwards, thought fit to go so far as to suppose the fibers, in their turn, to be made up of globules ranged in lines. This view was in part attributable to optical illusions in microscopical observation. The objectionable method which prevailed during the whole of the last and a part of the present century—of making observations (with but indifferent instruments) in the full glare of the sun—caused a certain amount of dispersion of light in nearly all microscopical objects, and the impression communicated to the observer was, that he saw nothing else than globules. On the other hand, however, this view corresponded with the ideas common amongst natural philosophers as to the primary origin of everything endowed with form.

The globules (granules, molecules) have, curiously enough, maintained their ground, even in modern histology, and there are but few histological works which do not begin with the consideration of elementary granules. In a few instances, these views as to the globular nature of elementary parts have, even not very long ago, acquired such ascendancy, that the composition, both of the primary tissues in the embryo and also of the later ones, was based upon them. A cell was considered to be produced by the globules arranging themselves in a spherical form, so as to constitute a membrane, within which other globules remained, and formed the contents. In this way did even Baumgärtner and Arnold contend against the cell theory.

This view has, in a certain manner, found support even in the history of development—in the so-called *investment theory* (Umhüllungstheorie)—a doctrine which for a time occupied a very prominent position. The upholders of this theory imagined, that originally a number of elementary globules existed scattered through a fluid, but that, under certain circumstances, they gathered together, not in the form of vesicular membranes, but so as to constitute a compact heap, a globe (mass, cluster—Klümpchen), and that this globe was the starting point of all further development, a membrane being formed outside and a nucleus inside, by the differentiation of the mass, by apposition, or intussusception.

At the present time, neither fibres, nor globules, nor elementary granules, can be looked upon as histological starting-points. As long as living elements were conceived to be produced out of parts previously destitute of shape, such as formative fluids, or matters (*blastic matter, blastema, cytoblastema*), any one of the above views could of course be entertained, but it is in this very particular that the revolution which the last few years have brought with them has been the most marked. Even in pathology we can now go so far as to establish, as a general principle, *that no development of any kind begins de novo, and consequently as to reject the theory of equivocal* [spontaneous] *generation just as much in the history of the development of individual parts as we do in that of entire organisms.* Just as little as we can now admit that a taenia can arise out of saburral mucus, or that out of the residue of the decomposition of animal or vegetable matter an infusorial animalcule, or an alga, can be formed, equally little are we disposed to concede either in physiological or pathological histology, that a new cell can build itself up out of any noncellular substance. Where a cell arises, there a cell must have previously existed (*omnis cellula e cellula*) just as an animal can spring only from an animal, a plant only from a plant. In this manner, although there are still a few spots in the body where absolute demonstration has not yet been afforded, the principle is nevertheless established, that in the whole series of living things, whether they be entire plants or animal organisms, or essential constituents of the same, an eternal law of *continuous development* prevails. There is no discontinuity of development of such kind that a new generation can of itself give rise to a new series of developmental forms. No developed tissue can be traced either to any large or small simple element, unless it be unto a cell.

ESTABLISHMENT OF THE CLASSICAL CONCEPT OF PROTOPLASM

SCHULTZE, Max (German cytologist, 1825–1874). From *Ueber muskelkörperchen und das, was man eine zelle zu nennen habe,* in *Müller's archiv für anatomie und physiologie und für wissenschaftliche medizin,* p. 1, 1861; tr. by M. and V. Hamburger and T. S. Hall for this volume.

The protoplasm concept represents but one chapter in a very old attempt of scientists to associate vital phenomena with matter in some special mode of its existence, beginning at least with Anaximines.* Purkinje† and Dujardin (*q.v.*) had suggested that the viscid contents of the cell might supply the necessary substratum for life but to Schultze (see

* See T. S. Hall, *Scientific origins of the protoplasm concept,* in *Journal of the History of Ideas,* vol. XI, p. 339, 1950.

† J. Purkinje in *Uebersicht über die arbeiten und veränderungen d. Schlesischen gesellschaft fur vanterländische kultur,* p. 1, Dresden, 1840.

below), and in botany to Cohn, belongs the credit for securing a very general and long-lived acceptance of this idea and for the notion that the substance in question was identical in its essential aspects in all cells. Actually, this generalization concerning protoplasm is not the central thought in Schultze's paper, the chief object of which was to clear up a number of uncertainties as to the definition of cell.

The uncertainty and difference of opinion which prevails in proper evaluation of structures such as muscle corpuscles is undoubtedly due in greater part to the lack of agreement among histologists concerning *what should be designated as a cell.* There was a time when the prevailing definition "vesicular structures with membrane, nucleus, and contents" was sufficient. The endless controversies in the separate fields of histology all of which center more or less around the cell concept teach us that this is no longer the case. This would be the time to make for once, an attempt to substitute something new for the old. It need not be primarily a matter of reporting new and hitherto unknown structural relations, nor a revolution in hitherto existing ideas, but is only a matter of placing foremost what is truly characteristic, of separating the essential from the unessential, of positing something as a theory whose foundations, in so far as they are needed to prove the propositions, are already generally known, of clothing in words something which, perhaps in less definite form, had long been in the minds of many.

What is the most important thing about a cell? Since there are very different types of cells we must begin the answer with another question: *which cells are the most important?* The most important cells, those in which are mirrored both the grandeur of cell life and an unlimited power as regards tissue formation, are obviously those which result from the division of the egg cell and which are so to speak not yet combined into definite tissues: namely, the embryonic cells (or, if we wish, the egg cells themselves). In these cells resides the future of the whole organism. They are capable of unlimited reproduction by ever renewed division; in them are located all the forces necessary for the construction of the tissues and various organs. It is these, and no one will deny it, which we may look upon as the true archetype of cells. From them can come, and will come, everything which exists in the normal as well as in the pathologically affected organism.

What, then, are the constituents of these cells? As is known, their center is, without exception, occupied by a *nucleus,* an almost homogeneous, spherical, moderately firm body, inside which lies a highly refractive nucleolus. The nucleus is surrounded by the cell substance proper, a viscous *"protoplasm"* which is nontransparent on account of its being densely filled with granules of a protein and fatty nature, and which can be *subdivided into a vitreous, transparent ground substance* of a viscosity peculiar to protoplasm as a whole, and *numerous imbedded granules.* This protoplasm represents a spherical lump held together *by its own inherent consistency.* Its consistency differs usually to some extent, at different depths, the quantity of the granules which fill it varying usually at different strata; especially, the outermost consists frequently only of the homogeneous, vitreous, transparent *ground substance* of the cytoplasm; but these cells do not possess a *membrane* chemically different from the protoplasm. *They are membraneless little lumps of protoplasm* with a nucleus. . . .

Our definition of what should be designated as a cell, is therefore as follows: *A cell is a lump of protoplasm inside of which lies a nucleus.* The nucleus, and the protoplasm as well, are division products of like constituents of another cell. This must be added in order to adhere to the concepts of nucleus and cell as distinct from other structures which perhaps look similar.

The cell leads, so to speak, a life of its own, the vehicle of which is again primarily the *protoplasm,* although the nucleus also definitely plays a significant role, but one which cannot yet be described in detail.

One must *emphatically* keep in mind, first of all, that the protoplasm is closed off on the outside by nothing but *its own peculiar consistency* differing from that of the surrounding watery fluid—it does not mix with water—and furthermore, if I may say so, by its centripetal life, by its property of forming a whole along with its nucleus, on which it is in a certain way dependent. Thus, the cell *without a membrane,* may remain to a certain extent independent of outside influences. Such closely packed cells like the blastomeres, which, enclosed by the common egg membrane, remain in intimate contact, *do not flow together* although the protoplasm of one is contiguous to that of the other.

Recently it has been confirmed by Kühne that true small amebae of fresh or salt water, which can scarcely be interpreted other than as unicellular organisms, as small lumps of protoplasm (with nucleus and all the properties of a cell) raised to complete individual existence, often fuse with each other. That this cannot be done with equal ease in all cases and all species cannot surprise us, if we consider the difference in consistence and resistance characterizing the contractile protoplasmic substance of the rhizopods. I say *protoplasmic substance of the rhizopods* and herewith have entered a field in which an agreement concerning terminology and viewpoints is very necessary. It would lead us too far from our subject if I should elaborate here on the organization of the rhizopods. However, since we speak of the contractile protoplasm and its properties, I consider it to be of advantage to dwell for a moment on the organizational relations of said creatures, which reveal to us the most striking vital phenomena of protoplasm. It is a matter of nothing less than a final solution of the question what the *unformed contractile substance* of the protozoa really is. The term *sarcode* has fallen into discredit, and justly so, because of the much too broad use, which many investigators, and Dujardin first, have made of it. In order to avoid misunderstandings it is desirable that it should be used no longer. Nevertheless one must give Dujardin credit for inventing a term for a substance which exists essentially in the same way he imagined it. A contractile substance which cannot be further divided into cells, and which does not contain any other contractile form elements, as fibers, etc. Such a substance is the protoplasm of the cells, contained in plant and animal cells, not the watery liquid part which fills the greatest part of the cell space in large cells, mainly in plants, but the viscous, mucous mass, densely filled with granules, which is always present at least around the nucleus and on the inner surface of the cell wall and which in this latter case usually forms manifold threadlike strands in order to connect more distant parts. In most animal cells, especially in all small young ones, but also in larger cells, if they have especially important functions to fulfill, as ganglion cells etc., the contractile protoplasm fills the entire cell cavity or, since such cells have usually no membrane, and one can therefore not speak of a cell cavity, it rather forms the entire cell. As I have already indicated at another place, there can be barely a doubt that according to observations of protoplasmic movements inside the cells, of those, for instance, of the filament-hairs of Tradescantia, we are dealing here with a substance which is contractile in the same sense as that forming the body of many rhizopods. There I mentioned that botanists (Ferd. Cohn, Unger) have already expressed a similar opinion. . . .

Thus the term *sarcode* is replaced by the term *protoplasm*, and since we attach to the latter, a very definite meaning which had never succeeded with the former, the advantage is not small. If membraneless cells which consist of protoplasm and nucleus only, execute ordinary ameboid movements throughout their life even though they are parts of another organism, or if the protoplasm inside cells with rigid walls, where outer changes of shape are no longer possible, can execute very complicated independent movements, then we must not be surprised if even more active movements are displayed in cells which singly or fused together in small groups represent a whole organism which must search for food by itself and has to undergo entirely alone all the metabolic variations requisite for life, which are otherwise delegated to definite cell groups. Thus it is with the amebae, thus it is with all, even the most complicated forms of Polythalamia and Radiolaria. This motile mass extensible in the form of fine threads and forming the cortical substance of the body of these animals, which I have described in detail for the mono- and poly-thalamic rhizopods, the Monothalamia and the Polythalamia, and which for want of a better name was called *sarcode*, must in the future be called *protoplasm*. The organisms are naked *lumps of protoplasm,* surrounded by perforated shells (or if one wishes: *cell membranes,* somewhat analogous to the thick zona pellucida of the egg cell), and originated either from *one* cell or by the fusion of *several* cells—this may vary in different species. In them a usually colored, firmer, inactive *inner* part has become distinguishable from a colorless, especially motile cortical layer, as already suggested by many small rhizopods, Amebae, *Actinophrys*, and other. The cortical layer protrudes everywhere through the holes of the shell in the form of fine strands. These are *naked protoplasmic strands*. They no more mix with water than do the protoplasmic strands of the Tradescantia cell with the adjacent watery cell content. *Where they touch, they flow together*. This fact, often doubted, even declared to be simply impossible, although exceedingly clearly observable, loses all its enigmatic character, if we use here the same standard which we apply in judging the special properties of protoplasm. Many species are distinguished from others by a softer, less viscous protoplasm. Just as there are active amebae (*A. diffluens*) and inactive, extremely slowly moving ones, and as the cortical substance of the former is finely granular and very labile whereas in the latter it is entirely hyaline and relatively resistant against acids and alkalies, so there are among the larger rhizopods also many differences in the type of motile protoplasm present. This difference in the substance of the filaments stands out most conspicuously in the two Gromian species described by me, *Gr. oviformis* and *Gr. Dujardinii*. The latter, which is also very frequent in the sea around Helgoland, becomes a torment to the waiting observer who would like to see the play of the filaments. The motion of the protoplasm which is here entirely hyaline, almost rigid, and very resistant against acids and alkalies is extraordinarily slow, and the tendency of the filaments to coalesce, which is so conspicuous in *Gromia oviformis, is here entirely lacking*. I mention this only in order to demonstrate again with indubitable cases of *naked protoplasm* differences in motion, consistence, chemical composition, and tendency to coalesce with an adjoining similar substance, differences which we also meet in naked cells of the tissues of higher animals. I have repeatedly called attention to the fact that one must not think of protoplasm as a very easily diffluent substance, which diffluence one could use to demonstrate the necessity of a membrane for all cells. The protoplasm has considerable resistance against external influences, especially where it occurs in dense masses as in the blastomeres or in the large ganglion cells of the brain and spinal chord. To be sure, such cells as the last-mentioned ganglion cells which I consider also as membraneless masses of protoplasm are difficult to dissect out of their surroundings, unless they have been hardened before, particularly since they very easily flow apart after death; during their life time, however, one need not worry about the nakedness of these entities in the brain. By their own consistence, and the remarkably delicate, reticulate and viscous connective tissue in their surroundings they are sufficiently protected. . . .

The point which I wanted to make was to demonstrate that *membraneless masses of protoplasm* play a great role in nature and that a membrane does not necessarily belong to the concept of a cell.

We have arrived at the result that important cells, in fact the most important of all, are *membraneless* and consist of a nucleus and around it only a small accumulation of protoplasm.   I would like to add here, that the proposition could even be defended that the formation of a chemically different membrane on the surface of the protoplasm is a sign of beginning *regression,* so little would the cell membrane belong to the concept of cell that it ought even to be considered as a sign of approaching senility or else at least of a stage in which the cell has already suffered considerable reduction in the vital activities with which it was originally endowed.   I would here only recall that a cell with membrane *can no longer divide* as a whole.   Only the *protoplasm* enclosed within the membrane divides, as, for instance, in the cartilage cells. Thereby, of course, a not unimportant check has been set upon the cell as regards tissue formation.   If the protoplasm knows how to free itself from the narrow prison of its enclosing membrane, as occurs, for instance, in ossification, by breaking-up of the cartilage cavity after resorption of the ground substance, then the old life starts merrily all over again.   Division follows division, *one* group of the young brood becomes marrow, *another* connective and vascular tissue, a *third* condenses in its cortical layer to bone ground substance, while, in the star-shaped cavities, the protoplasm and the nuclei return again to the composed and calm course subserving metabolism only, already pursued by the cartilage cells.   A cell with a membrane chemically different from protoplasm is like an encysted infusorian, like an imprisoned monster.   Comparable to the unfertilized egg, the protoplasm within the rigid membrane may divide a few times, but the process remains defined to minimal space and expires without any influence on its surroundings.   However, let the impetuously dividing protoplasm, goaded on ever anew by the still more impetuous nucleus, burst open its envelope, or take it away from it by resorption, or render it harmless in any way like the zona pellucida of the fertilized mammal—and the unchained protoplasm will make use of its freedom to the dismay of many.   . . .

   To the concept of cell belong two things: *nucleus* and *protoplasm,* and both must be division products of the same components of another cell.   Both constituents are of equal importance; disappearance of one or of the other destroys the cell concept.   Even though the small colored disks of mammalian and human blood possessed a nucleus in earlier stages, they cannot claim the designation *cell* in later stages since they lack the nucleus and have lost by this deficiency at least one of the most characteristic properties of cell life, the *capacity to reproduce.*   Likewise a naked nucleus, such as might possibly be left over and remain enclosed after resorption or other use of the protoplasm in the muscular primitive bundle, must never be called a cell.   I repeat: to the nucleus must be added the protoplasm.   And if, as in muscles, a capacity to reproduce can be demonstrated, then the concept of the cell is unshakably and firmly established.

   Particularly through the botanists . . . have we come to the assumption that the *vesicular nature* belongs to the concept of the cell. This holds quite true for plant tissues in so far as in them the cells acquire a membrane relatively *very early* and, from the time they combine to form tissues, always have one. Hence the relatively great *simplicity* of the plant tissues, where everywhere, even in the most complicated *internal changes* of protoplasm, the primary membranes can usually be recognized at first sight. The situation is different in the *animal body*. Here also *membraneless* cells combine to form tissues—and this is the main difference between animal and plant tissues. The whole difficulty in interpreting many of the latter rests on this fact. As long as we remain prejudiced that the *membrane* which belongs to the concept of vesicle, is also necessary for that of *cell*, we shall make little progress in the knowledge of cell life and of cellular metamorphoses in the formation of animal tissues.

If we free ourselves from this prejudice, then it follows automatically that we abandon the idea that the so-called cell content is a *fluid*. The cell content is a *watery liquid* only in large, old cells which are endowed with an indubitable membrane and which are physiologically of little importance, and in the latter, in most instances, the liquidity does also not seem to apply to the protoplasm proper; rather, the latter will keep itself *separate from* the absorbed water, as in plant cells, above all by surrounding the nucleus protectively, also by extending through the cavity of the cell in the form of threads, and spreading as a thin layer upon the inner surface of the membrane. In all younger cells, however, and in the animal body in most cells throughout life, the so-called cell content is a substance comparable to *viscous mucus, immiscible with water*, and in its consistency resembling more a soft wax than water; this alone is protoplasm. It holds its form by itself without requiring an outer membrane formed of a firmer substance. . . .

Many of those who are used to the old scheme of vesicular structures which are sharply delineated and, above all, supplied with definite membranes will find it difficult to follow our conceptions which have as their foundation the assumption that to the concept of a cell belongs exclusively a small and naked lump of protoplasm and a nucleus, and that the membrane surrounding it is something secondary, an attribute of some cells but lacking in others and definitely not necessary. And the designation *"cell"* which is derived from a *vesicle endowed with a distinct wall*, cannot at all be harmonized with *our* definition, according to which we are dealing only with a solid ball of protoplasm with a nucleus. However, I have as little intention of changing the name as changing the basic concepts concerning cell life. What I consider important is only this: that, as regards the conceptual determination of what one must designate as a cell, one should try to take the position outlined above and to study the development of tissues from the indicated points of view. Agreeable and satisfying results cannot fail to follow.

CHROMOSOME CYCLE CLARIFIED

van BENEDEN, Edouard (Belgian embryologist and cytologist, 1845–1910). From *Recherches sur la maturation de l'oeuf et la fécondation*, in *Archives de Biologie*, vol. IV, pp. 265–640, 1883; tr. by T. S. Hall for this volume.

Mitosis had been described by Strasburger for plants and by Flemming with possibly greater refinements for animals. It remained for van Beneden to demonstrate the constancy of the chromosome number for the species, its reduction in maturation, and its restoration during events preceding the first cleavage. The following is the conclusion from an extensive monograph covering researches in this area. Van Beneden was professor at Liége and, with van Bembeken, cofounder of the *Archives*.

I believe my work constitutes a forward step in the knowledge of the phenomena of fertilization in that it shows:

1. That not only the chromatic nucleus of the sperm but also the achromatic substance in which it is imbedded takes part in the formation of the male pronucleus.

2. That the germinal vesicle provides the female pronucleus not only with chromatic elements but also with an achromatic body.

3. That the two pronuclei can, without commingling, attain to the nature of ordinary nuclei by virtue of the progress of events of maturation.

4. That in the ascaris of the horse no unique fusion nucleus is produced at the expense of the two pronuclei; that there exists no "Furchungskern" in Hertwig's sense. The essence of fertilization resides not in the conjugation of the two nuclear elements but in the formation of these elements inside the female gamete. One of these nuclei derives from the egg, the other from the sperm. The nuclear elements expelled as polar globules are replaced by the male pronucleus and once the two half-nuclei, one male, the other female, have been established, fertilization has already been accomplished.

5. That in the course of a series of transformations which the nucleus undergoes,—transformations which are, moreover, identical with those occurring in a nucleus in the course of division—each pronucleus gives rise to two chromatic loops [chromosomes—Ed.].

6. That the four chromatic loops take part in the formation of the chromatic star (nuclear plaque) but they remain distinct. Each of them divides longitudinally into a pair of secondary loops.

7. That the nuclei of the two first blastomeres each receives one moiety of each primary loop; that is, they receive four secondary loops, two male and two female.

There therefore occurs no fusion of male and female chromatin at any stage in the first cleavage. If any intermingling occurs, this could only be in the nuclei of the first two blastomeres. There are reasons for believing that even in these nuclei, the male chromatin remains separate from the female.

What is certain is that the male chromatin never mingles with the female in the primary embryonic [zygotic—Ed.] nucleus. Thus one cannot say, as does Hertwig, that fertilization consists in the conjugation of two nuclei, one male and one female. In ascaris, no such conjugation occurs. If it operates in other living beings, we may feel sure that it is not fundamental to the phenomenon. Fertilization implies essentially a substitution, i.e., the replacement of a part of the germinal vesicle with nuclear elements provided by the sperm, and possibly also of part of the egg protoplasm (perivitelline substance) with sperm protoplasm.

The original male and female elements do not combine to form a cleavage nucleus and perhaps remain distinct in all the derivative nuclei.

The study of the maturation of the egg and of fertilization and cell division confirms me in the belief that cell nuclei are hermaphroditic, particularly in view of the fact that the male and female nuclei do not fuse. If the male and female pronuclei deserve to be so called, this implies their sexuality, whence the cellular nuclei are manifestly hermaphroditic. The tissue cells resemble in this respect the protozoa and protophyta.

I do not look upon fertilization as reproduction. This characteristic phenomenon of cellular life consists in an exchange and not in the genesis of a new cellular individuality. These replacements of certain elements of a cell by similar parts furnished by another cell have as their consequence the perpetuation of life; they make further division possible. One cannot conceive of reproduction except as multiplication. It seems very likely that there actually exists but one method of multiplication, namely division. In fertilization there is no increment in the number of individuals: this is quite evident when fertilization appears in the form of conjugation or even in the character which it assumes in the Vorticellas. A Vorticella A resolves itself into a macrospore M (female gamete) and a microspore m (male gamete); another, B, undergoes similar division into M' and m'. Then M fuses with m'; M' with m. From this, two individuals result, A' and B'. But no increment in the number of individuals has occurred. Fertilization as it occurs in metazoa follows the same scheme. An egg [*sic*] A divides to form a female gamete G and polar bodies g; a spermatocyte resolves itself into a cytophoral part C and a spermatozooid s. Then we behold G unite with s to give rise to a rejuvenated cell, the first cell of the embryo. Theoretically C and g should be able to engender a second cell; actually, these elements atrophy with the result that after fertilization instead of two cells capable of further division only one exists. Clearly here, as in Vorticella, there is no question of a multiplication. It appears that the faculty cells possess for self-duplication is limited: there comes a time when they are no longer capable of further division unless they undergo rejuvenation by the act of fertilization. In plants and animals the only cells capable of rejuvenation are eggs; the only cells able to rejuvenate them, sperm. All other parts of the individual are doomed. Fertilization is the necessary condition for the continuity of life. By it the reproducing unit escapes death. This is the hypothesis which I have taught since 1876, and it is one in support of which I find new arguments in the study of the ascaris of the horse.

SIGNIFICANCE OF SEXUALITY

MAUPAS, Emile François (French protozoologist, 1842–1916). *Théorie de la sexu-alité des infusoires ciliés,* in *Comptes Rendus de l'Académie des Sciences,* vol. 105, No. 7, p. 365, Paris, 1887; tr. by T. S. Hall for this volume.

Maupas' researches on the sexuality of protozoa were contemporary with the classical studies in Germany on nuclear phenomena during cell division and fusion. His postula-tion of fertilization as essential to the indefinite continuation of a cell strain proved an important stimulus to subsequent investigation not only upon protozoa (notably the stud-ies of Calkins and Woodruff) but also in related areas, e.g., parthenogenesis (Hertwig, Loeb, *et al.*) and the indefinite culture of isolated tissue fragments (see Harrison). Maupas divided his time between his scientific studies and work as a librarian in Africa.*

In my previous communication I tried to give a complete formula for the morphological phenomena accompanying the conjugatory coupling of the Cili-ates. I review them here as briefly as possible.

The micronucleus constitutes an hermaphroditic sexual mechanism. It is the only organ whose activity plays a role essential to conjugation. First, it passes through a growth phase, A, followed by two phases of division, B and C, culminating in the elimination of excess bodies. These phases, B and C, correspond, then, to the two divisions of the germinal vesicle which effect the elimination of polar globules in metazoa. Phase D, which follows immedi-ately, is another division phase producing the differentiation of a male pronu-cleus and a female pronucleus. During phase E, the conjugants carry out a reciprocal exchange of their male pronuclei which unite and fuse with the fe-male pronucleus of their new host, thus constituting a new nucleus of mixed origin. Here the essential part of fertilization ends. The two phases of divi-sion which follow have as their end the reestablishment of nuclear dualism peculiar to the Ciliates. Finally, during the last phase, H, the reconstitution phase, the exconjugants reassume their normal structure and organization, and then undergo their first fission. The old nucleus has been deorganized and eliminated by resorption.

These definite facts having been ascertained, what, now, is their physiologi-cal meaning? Since the beautiful work of Engelman and of Bütschli, we know that they are not followed by any production of new individuals distinct from the exconjugants. Hence, authors who have spoken and still speak, of this as sexual reproduction make an obvious mistake.

---

* For biography see *Bulletin de la Société d'Histoire Naturelle de l'Afrique du Nord,* vol. 7, 1915.

The observations according to which a supposed increase in fission occurred after conjugation seem to me to prove nothing. I have isolated, immediately after conjugation, individuals of several species. During successive generations of descendants, they underwent fission without showing the least acceleration.

One could even support the idea that, far from contributing to the multiplication of Ciliates, conjugation is one of the most active causes of their destruction. During conjugation and particularly during the long period of inactivity characteristic of the reconstitution phase, they are much more exposed to risks and dangers of the struggle for existence. Further, while not conjugating, they would have gone on dividing: e.g., an Onychodromus grandis would have produced from 40,000 to 50,000 descendants within the time elapsing during a simple conjugation ending with simple division into two. One thus would not speak of conjugation as necessary and inevitable; from all my experiments it comes out, quite the other way, that the Ciliates, at periods of sexual maturity, couple only when stimulated by special conditions which it would require too long to describe here.

But, whether or not the conjugation is a cause of the destruction of individuals, it is at any rate an indispensable factor for the conservation of the species; and this is, I believe, its sole use. This conclusion comes out of the following experiments.

On November 1, 1885, I isolated a Stylonichia pustulata and placed it in the usual culture. I observed and recorded the uninterrupted generations of its descendants until the end of March 1886, at which time this culture was extinguished by exhaustion of the strain, the individuals having lost the faculty of feeding and reproducing. The number of asexual generations during the total existence of the culture was 215. Individuals which I withdrew from it and allowed to intermingle with descendants from a progenitor of foreign origin supplied me with numerous instances of conjugation.

March 1, 1886, I isolated an exconjugant provided by one of the interminglings just mentioned. Its culture, followed through and watched like the preceding one, lasted till July 10, the period when it was likewise extinguished by exhaustion of the strain, after an uninterrupted series of 315 fissions. During the entire period, I effected numerous interminglings with foreign individuals. From these interminglings, I obtained numerous couplings beginning with the 130th generation. The couplings were successful and the exconjugants which emerged were normally reorganized. On the other hand, individuals nearly related and not intermingled, which had lived together without conjugating up to the 180th generation from then on conjugated frequently. But all these latter conjugations were unsuccessful, the exconjugants dying off slowly without regaining their normal organization.

I likewise followed through to the period when they became exhausted, similar cultures of Onychodromus grandis, two of Stylonichia mytilus, one of

Leucophrys patula and one of an Oxytrich of unknown species. Extinction occurred around the 330th generation with Onychodromus, around the 320th with Stylonichia, around the 330th with Oxytrich, and around 660th with the Leucophrys. In the nonintermingled preparations from these long cultures no coupling occurred; while in preparations involving withdrawal and admixture with foreign strains, I obtained many, in the case of the Onychodromus and the Leucophrys. For a reason which eludes me Stylonichia mytilus absolutely refused to conjugate. I had no foreign Oxytrichs to use in effecting admixtures.

The obvious conclusion from these long and tiring experiments is that the life of a strain is with Ciliates arranged in evolutive cycles each having as its point of departure an individual regenerated and rejuvenated by a sexual coupling. This result takes us back to an interpretation of conjugation such as has already been given by Bütschli. Sexual fecundation, which previously we visualized as so indissolubly linked with reproduction, has with the Ciliates remained quite distinct and independent. Reproduction here is always agamous, while sexual fertilization causes a simple rejuvenation, a reorganization of the individual conjugants. The reorganization manifests itself especially and probably uniquely in the nuclear mechanism. The latter, when the series of agamous generations is unduly prolonged, undergoes a degeneration and disorganization which I shall describe elsewhere. If conjugation does not intervene in time to check the destructive effect of this degenerescence, death inevitably ensues.

This is the true natural death from old age declared by certain authors not to exist among Protozoa to which they attribute a presumed immortality imposed upon eternal youth.

PROTOPLASMIC MODELS

BÜTSCHLI, Otto (German general physiologist and protozoologist, 1848–1920). A letter entitled *Professor Bütschli's experimental imitation of protoplasmic movement*, in *Quarterly Journal of Microscopical Science*, n.s., vol. 31, p. 99, 1890; tr. by E. R. Lankester; by arrangement with the Clarendon Press.

Bütschli was perfectly conscious of the limitation of the present approach in the effort to understand biological phenomena, namely, that models *portray* these phenomena without necessarily *duplicating* them. He contends, however, that through the construction of models, activities similar to those ordinarily encountered only in living systems can be caused to emerge in nonliving ones. This fact demonstrates at least the *possibility* that such emergence may account for biological activities too. The need for the supravention of a vital principle is thus obviated.

Professor Bütschli, of Heidelberg, has recently made some extremely interesting observations upon a substance which simulates in a remarkable way the

appearance and movements of the protoplasm of an Amoeba, or of the plasmodium of Mycetozoa. He has been kind enough to send to me some oil in a
suitable condition for use, with directions as to the exact details of the experiment. In my laboratory, by following his directions, the movements described by him have been observed in a satisfactory manner. In order to obtain the best results some experience and care is requisite, and probably they
cannot always be obtained by a single experiment. The subject is so interesting, and so fitted for further investigation by all who have leisure and a taste
for the study of the vital phenomena of the Protozoa and of living protoplasm
in general, that I think it will be of advantage to readers of this Journal to
have Professor Bütschli's directions, which he has permitted me to publish,
placed in their hands.

E. Ray Lankester,
March, 1890.

Heidelberg, February 1st, 1890.

You have kindly asked me how I prepare the protoplasma-like drops which
I have described. As you yourself feel greatly interested in this discovery, and
presumably a like interest exists among other English biologists and microscopists, I hasten to satisfy your desire, and to explain somewhat more fully the
methods which I have described in a previous publication.

As you well know already, I use in the preparation of these globules—showing protoplasma-like streaming—ordinary olive oil. My first experiments
were made with a small quantity of olive oil which had been standing for a
long time in my laboratory in a small bottle. By some happy chance this oil
had just the right properties which are necessary for the success of the experiment, for not every sort of olive oil is suitable. As far as my experience goes,
it tends to show that the ordinary oil cannot be directly used, because it is too
thin, or is perhaps deficient in other qualities on which the success of the experiment depends. In order, therefore, to prepare a suitable oil, I proceed in the
following manner:—A medium-sized watch-glass or flat dish is filled with a
thin layer of common olive oil, and is placed on a water-bath or in a small cup-
board, such as are used for embedding in paraffine, at a temperature of about
50° C. Under the influence of the higher temperature the oil gradually loses
its yellow colour and becomes thicker. The great point now is to select the
right moment at which the oil will have attained the proper degree of thickness and viscosity, as also the other properties which at present I am not able
to define more exactly, but on which much of the success seems to depend.
The exact moment can, however, only be found out by systematic trials.
After the oil has been thickening for three or four days a trial should be made
with a drop of it in the manner described below. Should the drop not become
finely vesiculate, and exhibit little or no streaming, continue the heating proc-

ess and experiment again on the following day. If the oil should have become too thick it will form good frothy drops, but will scarcely show any streaming. In this case mix it with a small quantity of ordinary olive oil, and thus render it more liquid. If it has become much too thick it will form a good froth, but the latter dissolves very rapidly in glycerine.

You see thus that the process to obtain the suitable oil is somewhat slow, but I do not at present know of any other method by which the result can be arrived at more quickly and surely.

To prepare the vesiculate drops I proceed in the following way:—In a small agate mortar I grind a small quantity of pure dry carbonate of potash (K_2CO_3) to a fine powder. I then breathe on to the salt till it becomes slightly moist, and with a glass rod add to it a drop of oil, mixing the two constituents to a thickish paste. The success of the experiment depends, however, more upon the nature of the oil than upon the proportions of oil and salt in this mixture. Then with a glass rod or a needle I place a few drops of the paste, about the size of a pin's head or smaller, on a cover-glass, the corners of which are supported by small pegs of soft paraffine. I then place on a slide a drop of water, and put the cover-glass over this in such a manner that the drops of the paste are immersed in the water, but are not much compressed, to which end the corners of the cover-glass have been supported by the paraffine. The preparation is then placed in a damp chamber, and remains there about twenty-four hours. The drops have now a milk-white and opaque appearance. The preparation is then well washed out with water by applying blotting-paper to one edge of the cover-glass, and supplying water at the other edge from a capillary tube.

If the drops have turned out well, they will begin almost immediately after this to move about rapidly, and change their shape continuously. The water under the cover-glass must now be displaced by glycerine, diluted with an equal bulk of water, and the drops will then exhibit a vigorous streaming and forward movement, becoming gradually quite transparent. The amoeboid movements are generally more distinct if the drops are somewhat compressed. If the drops do not show the streaming movement you may succeed in producing it by tapping the cover-glass slightly, by applying gentle pressure, or sometimes by breaking up the drops. For it seems as if at times incrustations were formed on the surface of the drops, which prevent or impede the streaming movement, and which can, in part at least, be removed by the above-mentioned manipulations.

It is especially interesting to see how fast and beautifully the drops creep to and fro in water, or in half-diluted glycerine, even when they are not compressed. The streaming movement, on the other hand, is better seen if the drops are somewhat compressed, which may be done by inserting under the cover-glass a piece of a broken cover-glass of medium thickness, and then removing the paraffine pegs. Then draw away the liquid until the necessary pressure is obtained. This streaming movement is best demonstrated twenty-four hours after the addition of the glycerine, as the drops will then be thoroughly cleared and transparent. Further, it is interesting to note that a progression of the drops takes place in the direction in which the streaming moves.

As this forward movement is rather slow in compressed drops, it is necessary to use a micrometer ocular to satisfy oneself of the advance.

Unfortunately the oils which I have prepared since my first experiments do not move and stream so well or so rapidly as those I employed then. The movement and streaming show themselves much more markedly and distinctly if they are examined on a warmed stage at a temperature of 50° C. If you should be in a position at your demonstrations to conduct the experiment at this temperature, the phenomena will certainly be much more evident.

From the preceding description you will see that it will be necessary, to obtain good results, to gradually get hold of the methods, and you must not doubt the correctness of the phenomena which I have described if the first trials do not give the desired results.

At all events, you will have at first to make some experiments so as to obtain an insight into the conditions and sort of phenomena, but I do not doubt that you will succeed in observing the appearances and in demonstrating them to others, though perhaps in not so vigorous a degree as I might desire.

I have lately made some trials to render olive oil suitable for these experiments by heating it more rapidly. Although at present I have no entirely reliable results, it seems to me that by heating ordinary olive oil to 80°–90° C. for twelve or twenty-four hours, a suitable medium may be obtained.

Finally, I would like to remark that I am the last person to defend the view that these drops, exhibiting protoplasma-like movements, are directly comparable to protoplasm. Composed as they are of oil, their substance is entirely different from protoplasm. They may be, however, compared with the latter, in my opinion, firstly with regard to their structure, and secondly with regard to their movements. But as the latter depend on the former, we may assume that the amoeboid movement of protoplasm itself depends on a corresponding physical constitution.

These drops, too, resemble organisms inasmuch as they continue for days to exhibit movements, due to internal causes, which depend on their chemical and physical structure. I do not believe that up to this time any substance has been artificially prepared which in these two points, viz. structure and movement, has so much resemblance to the most simple form of life as have these vesiculate drops. I hope, therefore, that my discovery will be a first step towards approaching the problem of life from the chemico-physical side, and towards passing from vague and general hypotheses of molecular constitution to the surer ground of concrete conceptions of a physical and chemical nature.

It is, however, a special satisfaction to me to hear that in your country, which has given rise to so many and so celebrated men in biological science, my investigations are followed with interest and sympathy.

With friendly greetings,

Yours sincerely,

O. Bütschli.

MECHANISM OF SEX–DETERMINATION

McCLUNG, Clarence Erwin (American cytologist, 1870–1946). From *The accessory chromosome—sex determinant?* in *Biological Bulletin*, vol. 3, Nos. 1 and 2, p. 43, Woods Hole, 1902; by permission of the editor.

The omitted portion of this paper, entitled *Observations and Comparisons*, summarizes pertinent contributions, chiefly of other investigators, which formed the background of the thinking represented in the following passages.

THEORETICAL CONSIDERATIONS.

In seeking an explanation for the unusual phenomena connected with the history of the accessory chromosome in the male germ cells, it is most natural to surmise the existence of a phylogenetic significance. In the spermatagonia what amounts to practically two nuclei in each cell is strongly suggestive, in mere general features, of the appearances manifested in the Protozoa where both macro- and micronuclei are present. The accessory chromosome might be homologized with the micronucleus which serves as a medium of exchange between the organisms during the act of fertilization, but it would be extremely difficult to trace any parallelism between the macronucleus and the real chromosomic vesicle of the spermatogonia. I do not, therefore, believe that we can look in this direction for an explanation of the peculiar character exhibited by the accessory chromosome. . . .

In offering a theory to account for the function of the accessory chromosome, I do so with considerable reluctance, for I realize how little real general knowledge we have of this structure. It seems to me, however, that something is necessary to concentrate the interest of spermatologists upon the fundamental character of this most suggestive chromatin element, and I know no better way of aiding in this than by publishing the working hypothesis with which I have attacked the problem.

This has led me into the field of theories concerning sex and its determination, but I have tried to avoid any more extensive discussion than is necessary to outline, in a preliminary way, the opinion I hold concerning the meaning of the accessory chromosome. Even with this reservation I have nevertheless been obliged to go further afield than I should desire with our present knowledge as a guide. I can only hope that my excursions may accomplish a measure of the purpose for which they were undertaken.

Briefly stated, then, my conception of the function exercised by the accessory chromosome is that it is the bearer of those qualities which pertain to the male organism, primary among which is the faculty of producing sex cells that have the form of spermatozoa. I have been led to this belief by the favorable response which the element makes to the theoretical requirements conceivably inherent in any structure which might function as a sex determinant.

These requirements, I should consider, are that: (*a*) The element should be chromosomic in character and subject to the laws governing the action of such structures. (*b*) Since it is to determine whether the germ cells are to grow into the passive, yolk-laden ova or into the minute motile spermatozoa, it should be present in all the forming cells until they are definitely established in the cycle of their development. (*c*) As the sexes exist normally in about equal proportions, it should be present in half the mature germ cells of the sex that bears it. (*d*) Such disposition of the element in the two forms of germ cells, paternal and maternal, should be made as to admit of the readiest response to the demands of environment regarding the proportion of the sexes. (*e*) It should show variations in structure in accordance with the variations of sex potentiality observable in different species. (*f*) In parthenogenesis its function would be assumed by the elements of a certain polar body. It is conceivable, in this regard, that another form of polar body might function as the non-determinant bearing germ cell.

(*a*) If we accept the theory that the chromatin is the bearer of hereditary qualities, there could be little doubt regarding the necessary chromosomic character of a sex determinant. Sex being an elementary characteristic of protoplasm, it would be firmly established in the hereditary basis along with metabolic activity, irritability, etc., and if any argument were needed at all it would be a general one, not concerned immediately with the question under discussion, but with the broader one suggested. It will therefore be assumed that the chromatin is this basis. This being true, it will only be necessary to point out that the work of a majority of investigators definitely proves that the accessory chromosome *is* a chromosome, and its standing in this respect is established.

(*b*) With regard to what would theoretically be required of a chromosome whose function should be the determination of sex, it is probable that almost every investigator would hold an opinion differing in some respects from those entertained by others. What I can suggest in this connection will therefore be merely tentative and an expression of my own views. One thing, however, would seem to be necessary; *i.e.*, that the determinant should exist in the cells until they are definitely established as elements of either an ovary or of a testis.

If it be that the production of male elements is a sign of catabolic conditions, or, in other words, of those that make a greater demand of energy expenditure upon the developing cell, then it would seem most natural that the determiner should be for the purpose of carrying the transformation beyond the production of ova to spermatozoa. It would therefore be a necessary content of the cells until they had passed through the stages of development beyond that at which they might pause and become laden with yolk or, in other ways, postpone the period of maturation. It is conceivable that the production of four functional cells from one spermatogonium would call for the employment of more energy than would the formation of one functional egg from an oögonium, especially since many cells contribute their substance or support in the upbuilding of the egg.

Accordingly, it would be most reasonable to expect the presence of the determinant in the latest possible stage consistent with its equal distribution to half the spermatozoa. This we find to be the case with the accessory chromosome which regularly occurs in all the cell generations up to the last and is only withheld, finally, from half of the spermatids. By its consistent course in this respect, the accessory chromosome plainly manifests its intimate influence upon the germ cells of which it is a part, and most strongly suggests a relation to sex determination. It may further be pointed out in reference to this relation, that during the multiplied spermatogonial divisions, the accessory chromosome exhibits a somewhat distant attitude toward the remainder of the chromatin, and it is only at the time of the definitive spermatocyte divisions that it comes to be an intimate member of the cell nucleus. In what manner it is borne from the fertilized egg to the testis of the embryo we do not know, and, lacking this knowledge, are placed at a considerable disadvantage for a proper appreciation of its real character.

(*c*) A most significant fact, and one upon which almost all investigators are united in opinion, is that the element is apportioned to but one half of the spermatozoa. Assuming it to be true that the chromatin is the important part of the cell in the matter of heredity, then it follows that we have two kinds of spermatozoa that differ from each other in a vital matter. We expect, therefore, to find in the offspring two sorts of individuals in approximately equal numbers, under normal conditions, that exhibit marked differences in structure. A careful consideration will suggest that nothing but sexual characters thus divides the members of a species into two well-defined groups, and we are logically forced to the conclusion that the peculiar chromosome has some bearing upon this arrangement.

I must here also point out a fact that does not seem to have the recognition it deserves; viz, that if there is a cross division of the chromosomes in the maturation mitoses there must be two kinds of spermatozoa regardless of the presence of the accessory chromosome. It is thus possible that even in the absence of any specialized element a preponderant maleness would attach to one half the spermatozoa, due to the "qualitative" division of the tetrads.[1]

(*d*) As I elsewhere suggest, it is most appropriate that the sex determinant should have its locus in the spermatozoa. These elements are most commonly freed from any close relation to the parent organism at maturity, and thus lose the opportunity to receive from it any bias toward the production of an unusual proportion of the one sex or the other as environmental conditions might require. It is otherwise with the ova. They are usually retained by the maternal organism in such intimate relation to it that surrounding conditions might easily imprint their demands upon them. Even up to the time of fertilization the female elements are so placed as to react readily to stimuli from the mother. Here they are approached by the wandering male elements from which they may choose—if we may use such a term for what is probably chemical attraction—either the spermatozoa containing the accessory chromosome or those from which it is absent. In the female element, therefore, as in the female organism, resides the power to select that which is for the best interest of the species.

(*e*) The strength with which sex is established in different species of animals is variable. Moreover it is a fact of common observation that all cell elements vary widely in different animals. We should not be surprised to find, then, that a determinant would exhibit marked varieties of form which might even be carried to the extreme of its entire suppression as a definite element. Incomplete as are the observations upon the behavior of the accessory chromosome in various species, enough evidence is forthcoming to show wide departures from anything that might be considered a typical form. And here it is that it may be possible to secure more or less definite information with regard to the meaning of the accessory chromosome. If a large number of observations show variations that parallel well-marked instances of unusual sex characters, then greatly increased probability will attach to the theory I have advanced.

(*f*) Concerning the bearing of parthenogenesis upon the problem of sex determination, we know little. In eggs, no structure comparable to the accessory chromosome has yet been observed and the presence of any such element is extremely improbable. But it is known that different sexes come from parthenogenetic eggs, and in the familiar example of the aphides, these are produced in strict response to environmental demands.

Parthenogenesis, however, is regarded as a degenerate method of sexual reproduction in which polar bodies perform the function of the spermatozoa. Sex might, therefore, be determined by the particular polar body that restored

---

[1] It is suggestive that in all those cases where there appears to be no cross division of the chromosomes in maturation, nothing like the accessory chromosome has been noted. This would seem to be some indication that there might be two types of division.

the needed amount of chromatin to the egg, for these, like the spermatozoa, would be of two kinds where a reduction division took place in the process of maturation. These facts would indicate an element of truth in Minot's view regarding the meaning of the polar bodies. In respect to this matter, however, we have only theory to guide us and must wait for more thorough study of the question.

The suggested hypothesis affords a reasonable basis for a number of theories that have been advanced and supported upon empirical data. Among these are Thury's and Düsing's on the time of fertilization; the ones relating to the nutrition of the parents and embryo; and possibly others in which age or "comparative vigor" is assigned as the influential factor.

In general, I would point out, my theory confirms these by showing that the condition of the ovum determines which sort of spermatozoon shall be allowed entrance into the egg substance. In this we see an extension, to its ultimate limit, of the well-known *rôle* of selection on the part of the female organism. The ovum is thus placed in a delicate adjustment with regard to surrounding conditions and reacts in such a way as to best subserve the interest of the species. To it come the two forms of spermatozoa from which selection is made in response to environmental necessities. Adverse conditions demand a preponderance of males, unusually favorable circumstances induce an excess of females, while normal environments apportion an approximately equal representation to each of the sexes.

Those theories regarding sex determination which contain any element of truth within them will be found dependent upon this principle. It is expressed by Geddes and Thompson in these words: "But the general conclusion is tolerably secure—that in the determination of the sex, influences inducing katabolism tend to result in production of males, as those favoring anabolism similarly increase the probability of females." The authors just cited clearly recognize that we must consider the sexual elements in the light of their elemental structure and function when the final explanation of sex is sought. They say: "That the final physiological explanation is, and must be, in terms of protoplasmic metabolism, we must again, however, remind the reader."

The *rôle* that I have suggested for the accessory chromosome in no way changes the ordinary conception of the part played in sex determination by the various observed factors, but it does offer some tangible means by which to correlate these and to fix the nature of their participation.

The conception of two forms of sexual elements which would be operative in the determination of sex is not new. It has been assumed on purely theoretical grounds that there are two kinds of ova, one of which, in the event of fertilization develops into a male organism while the other under similar conditions gives rise to a female. This theory is dismissed by Geddes and Thompson on the ground that the two forms of ova have never been observed and for the further reason that later influences might possibly change the earlier tendency.

The latter objection would prove fatal to any theory which located the determination of sex in a structural difference of the germinal elements. I do not consider this position well taken for reasons that I will give later. The more serious objection lies in the fact that, so far as observation has gone, all eggs of a species are practically alike. It is also to be depreciated because of the fact that it reverses the ordinary relations of the elements and removes the power of choice from the female.

We have in the case of the spermatozoa, however, the observed fact that there are two essentially different forms and that they are present in equal proportions. No other feature, save sex, separates the resulting offspring into two approximately equal groups. By exclusion then, it would seem that the determination of this difference is reposed in the male element. . . .

A further proof, although inferential, is that afforded by "true" twins, in which case it appears that the sex of the two individuals is always the same. If sex were established at the time of fertilization of the ovum, then sex would be shared along with the other qualities possessed by the normal individual that would have developed from the ovum under ordinary conditions. In case sex were not established at the time of impregnation, it would be natural to expect the two sexes to be occasionally represented in one birth because of the inequality of nutrition in the embryos or for other reasons.

Sex, then, *is* determined sometimes by the act of fertilization and can not be subsequently altered. But between this extreme and the other of marked instability there may be found all degrees of response to environment. It must accordingly be granted that there is no hard-and-fast rule about the determination of sex, but that specific conditions have to be taken into account in each case. The objection that Geddes and Thompson raise against the possibility of two forms of eggs, viz., that it is a useless adaptation on account of the fact that subsequent conditions may determine sex in some cases, is not a valid one in general. Such *may* be the case in some instances, but such *is* not the case in others.

Finally, with respect to the evidence to be derived from parthenogenesis, it should be remembered that we are here dealing with a practical suppression of sexuality and it is to be expected that extensive modifications of the ordinary process will follow. If the egg takes upon itself all the functions commonly exercised by it in conjunction with the spermatozoön, it must be that the determination of sex is included. This, in some instances, is a final choice on the part of the ovum and ever afterward one sex only is produced by it; again, however, it maintains a responsive attitude toward environments and gives rise to the sex most needed by the species. It is to be hoped that the very promising field opened up by the work on artificial parthenogenesis will throw much light upon these vexed problems.

VI. PATHOLOGY

Contributions to parasitology and to the biological founda-
tions of pathology

GERM THEORY OF DISEASE

KIRCHER, Athanasius (German, later Italian, natural scientist and priest, 1602–1680). From *Scrutinium physico-medicum contagiosae luis, quae pestis dicitur,* Rome, 1658; tr. (from the Widmanstad edition, 1740) by T. S. Hall for this volume.

The notion that diseases are propagated by living agents of small size appears to have occurred, with varying degrees of clarity, to Columella (first century B.C.), Agricola (1494–1555), Fracastorius (1483–1553) Mercurialis (1530–1606), Kircher, Plenciz (1705–1786), and others.* As for Kircher, who antedated Pasteur by more than two centuries, he was not primarily a biologist. For his *Scrutinium pestis* he took time off from researches on magnetism,† optics,‡ music,§ geology,|| geography and linguistics.¶ Kircher's "germ theory" has been called the first such theory to be connected with actual observation of the supposed agents; this has also been vigorously denied.**

Every natural compound exhales certain outflowings of its essential nature. These should not at this point be assumed to correspond to the qualities themselves, nor to be something propagated by the object in question as if by accident. They are really, strictly speaking, little bodies of exceedingly small size, incapable of perception by even the most powerful vision,—carriers, so to speak, of essential and nonessential properties emanating from the body in question, and identical in nature with the entirety of the thing they flow out of.

In this way, the earth's substance, and that of the stars' celestial orbs, spreads each and every one about its own globe an atmosphere as it were, which is really nothing but a sort of vaporous outflowing from said globe, composed of very minute and imperceptible little bodies, aqueous or igneous according to the nature of their globe.

All the other sorts of compound outflowings concur with the nature of their respective spheres; hence, if anything odoriferous, savory, ill-smelling, or disagreeable strikes the senses, this may be said to be an outflowing from a compound of the same sort,—going by the name of 'exhalation' in the case of dry bodies, and of 'vapor' in the case of wet ones.

---

* See H. B. Torrey, *Athanasius Kircher and the progress of medicine,* in *Osiris,* vol. 5, p. 246, 1938.
† *Magnes; sive, de arte magnetica,* Rome, 1641.
‡ *Ars magna lucis et umbrae,* Rome, 1646.
§ *Musurgia universalis,* Rome, 1650.
|| *Mundus subterraneus,* Rome?, 1664.
¶ *Polygraphia, seu artificium linguarum,* Rome, 1663.
** See first footnote, above.

Hence from fiery *bodies* is produced an outflowing constituted of fiery *little bodies*, of which it may be assumed that the more crowded they are the intenser the heat they cause; and, to the extent that they are carried a little farther out and so spread farther apart, to that extent they manifest a lower grade of heat,—until, reaching the limit of the sphere ordained them by nature, they either pass directly out of it or else are sucked back in by an attraction of the compound.

What we have said concerning fiery bodies must be said also about all other mineral and metallic bodies and precious stones; likewise about substances of vegetable or sensitive nature, and about species falling by form or property either among grasses, plants and trees, or among animals of any order whatever.

Furthermore, such little bodies are nothing but a number of breatheable particles of a compound vapor identical in properties and in kind with the object as a whole. These particles, rarefied either by the external heat bathing the object or by violent friction, are, since they then require more space, poured off into the air outside; or, in case a true chill encompasses the object, they return into that from which they flowed out.

But if, because the condition of the body makes it imperative, they do not go back, then the particles of the air nearby, replacing those which flowed out because of a certain natural attraction, are transformed into a native germ substance of the compound, endowed with the same specific properties; whence, reduction and consumption of the whole body, by a continued outflow of this sort over a period, need occasion no special concern. These things assumed, let us proceed to things pertinent to our general undertaking.

Now this sort of contagion's greatest power is manifested in cadavers. For, after the native heat has been driven off and the rule of the natural spirits overthrown, and when decay alone prevails over the lifeless body, the following situation is brought about: namely, power of jurisdiction throughout the body is now exercised by that very thing which, by spreading to all the internal and external organs, causes the cadaver as a whole to dissolve into decay; but beneath this decay lie the true seeds of the plague. And these, activated by the evil effect of decay, from within or without, or of the heat residing in the air around, are propelled in all directions by the above-mentioned outflowings of little bodies and soon dispense contagion in proportion to their endowment of vigor and effectiveness; on the matter of whether little bodies of this sort are for the most part actually alive or not,—they are soon, through the action of the heat which bathes them and is already infected with a like filth, separated off as the countless germs of imperceptible little worms. Whence as many of the little bodies as are taken in with the outflowings, an equal number of worms is sure to be born, from which it may be said not that the outflowings lack life but rather that they are animate.

To the reader these things will perhaps seem paradoxical. Yet when he has witnessed experiments checked through the most delicate microscopes over a period of many years, there is something I feel sure of: viz., not so much that he will think these things hold true of themselves, but rather that they will have been established as demonstrated by truth from experiments in addition to what we have mentioned.

Firstly then, in caves and mountain caverns we see that, from some unknown corruption in its inner parts, the earth begets out of moisture and a varied mixture of virulent refuse, not only insects of all kinds, but also a wonderful variety of poisonous animals, such as snakes, brush-toads, and lizards; from a flowing together of various parts of the earth these things, by the external heat of the air around, are separated off from standing waters, lakes, and seas. Nay but daily experience teaches, both on sea voyages and within the walls of our homes, that water, shut up in a vessel, as soon as exposed is quickened into worms.

Is anyone unaware that, within the viscera of the human body, out of the decay contracted with spoiled food, worms soon will swarm? And that, as decay creeps within the subcutaneous passages, and the body fluid is deeply corrupted, our whole substance, gradually coming alive with worms (whence phtiriasis is caused) is devoured and consumed by small serpent-like beings,— (really worms, attacking the back region and emerging in the head)?

As soon as a man's internal spirits are lost, along with his native heat, the plague-bringing poison, which he has attracted and absorbed, by its virulence disposes his native humor to decay. After this there follows a stench by which are corrupted any who approach either the afflicted man or clothes already infected by his exhalation. The exhalation, however, is nothing but an evaporation of the decayed humor; and this evaporation (actually composed of innumerable insensible little bodies), soon expands when it reaches the freer air, and infects all around by the virulent power of its contagion. These little bodies (since they are possessed of the same virulent power as is the decayed matter of which they are particles), being either breathed into the body, or creeping in from their inmost lurking places among the clothes, soon produce in the subject the same effects as in him they flowed out of.

In cadavers, where the whole body actually dissolves into decay, these outflowings of little bodies do not so much infect those near; rather they become transformed into the animate germ of very minute and imperceptible little animals. This germ remains for a time in woodwork, linens, and clothes as well as other porous matter or matter of low density; breathed in later, they defile the latent humor within by mixing it with their own substance; whence it comes about that upon the very first contact—to be specific, upon contact with the oil [*i.e.*, of the body?—Ed.]—after twisting their way in through the pores of the hands and fingers, they communicate their virulence to the one who has established contact with them; or where people have used clothes contaminated by such offspring, the latter, activated by heat, and absorbed through the skin pores elsewhere on the body as well as with the intake of air, produce those very effects to which, by reason of the plague's great injuriousness to them, such individuals are exposed.

PATHOLOGY AS OBJECTIVE BIOLOGY

SYDENHAM, Thomas (English physician, 1624–1689). From *Observationes medicae circa morborum acutorum historiam et curationem*, London, 1676; tr. by R. G. Latham as *Medical observations concerning the history and cure of acute diseases*, in *The works of Thomas Sydenham, M.D.*, vol. I, London (for the Sydenham Society), 1848.

Bacon's famous precepts for the sanity of science (one of which is quoted below) were conveyed by various successors to all branches of natural philosophy; to the medical branch, by Thomas Sydenham. As for the latter's specific scientific contributions, he is commonly credited with the introduction of laudanum, with a helpful treatment (refrigeration) of smallpox, and with the first diagnosis of scarlatina. Of immeasurably greater significance was his importation into medicine of a new standard of scientific objectivity. By his "taxonomy" of diseases he effectively rejected the view of these as exceptions to natural law. With this step, pathology took its place as a full-fledged biological science. The practical merits of his system, described below, he proceeded to prove by using them in a radical and highly influential reanalysis of the nature of diseases prevalent at the time.

. . . how great soever the efforts of others may have been, I, for my own part, have always considered that the breath of life would have been to me a vain gift, unless I, working in the same mine with them, contributed my mite to the treasury of physic. Wherefore, after long meditation, and the diligent and faithful observations of many years, I at length determined—firstly, to state my opinion as to the means by which the science of medicine was to be advanced; secondly, to publish a sample of my endeavours in that department.

I conceive that the advancement of medicine lies in the following conditions:

There must be, in the first place, a history of the disease; in other words, a description that shall be at once graphic and natural.

There must be, in the second place, a *Praxis*, or *Methodus*, respecting the same, and this must be regular and exact.

To draw a disease in gross is an easy matter. To describe it in its history, so as to escape the censure of the great Bacon, is far more difficult. Against some pretenders in this way, he launches the following censure—"*We are well aware that there existeth such a thing as a Natural History; full in bulk, pleasant from its variety, often curious from its diligence. Notwithstanding, whoever would take away from the same the citations of authors, the empty discussions, and, finally, the book-learning and ornaments, which are fitter for*

the convivial meetings of learned men than for the establishment of a Philoso-
phy, would find that it dwindled into nothing. Such a natural history is far
distant from the one we contemplate."

In like manner it is exceedingly easy to propound some common-place cure
for a complaint. It is far harder, however, to translate your words into actions, and to square your results with your promises. This is well known to
those who have learned that there occur in practical writers numerous diseases,
which neither the authors themselves, nor any persons else besides, have been
able to cure.

In respect to the histories of a disease, any one who looks at the case carefully, will see at once that an author must direct his attention to many more
points than are usually thought of. A few of these are all that need be noticed at present.

In the first place, it is necessary that all diseases be reduced to definite and
certain *species,* and that, with the same care which we see exhibited by botanists in their phytologies; since it happens, at present, that many diseases, although included in the same genus, mentioned with a common nomenclature,
and resembling one another in several symptoms, are, notwithstanding, different in their natures, and require a different medical treatment.

We all know that the term *thistle* is applied to a variety of plants; nevertheless, he would be a careless botanist, indeed, who contented himself with the
general description of a *thistle;* who only exhibited the marks by which the
class was identified; who neglected the proper and peculiar signs of the species,
and who overlooked the characters by which they were distinguished from
each other. On the same principle, it is not enough for a writer to merely
note down the common phenomena of some multiform disease; for, although
it may be true that all complaints are not liable to the same amount of variety,
there are still many which authors treat alike, under the same heads, and without the shadows of a distinction, whilst they are in their nature as dissimilar as
possible. This I hope to prove in the forthcoming pages.

More than this—it generally happens that even where we find a *specific*
distribution, it has been done in subservience to some favorite hypothesis which
lies at the bottom of the true phenomena; so that the distinction has been
adapted not to the nature of the complaint, but to the views of the author and
the character of his philosophy. Many instances prove the extent to which
medicine has been injured by a want of accuracy upon this point. We should
have known the cures of many diseases before this time if physicians, whilst
with all due good-will they communicated their experiments and observations,
had not been deceived in their disease, and had not mistaken one species for another. And this, I think, is one reason why the Material Medica has grown
so much and produced so little.

In writing the history of a disease, every philosophical hypothesis whatsoever, that has previously occupied the mind of the author, should lie in abeyance. This being done, the clear and natural phenomena of the disease should be noted—these, and these only. They should be noted accurately, and in all their minuteness; in imitation of the exquisite industry of those painters who represent in their portraits the smallest moles and the faintest spots. No man can state the errors that have been occasioned by these physiological hypotheses. Writers, whose minds have taken a false colour under their influence, have saddled diseases with phenomena which existed in their own brains only; but which would have been clear and visible to the whole world had the assumed hypothesis been true. Add to this, that if by chance some symptom really coincide accurately with their hypothesis, and occur in the disease whereof they would describe the character, they magnify it beyond all measure and moderation; they make it all and in all; the molehill becomes a mountain; whilst, if it fail to tally with the said hypothesis, they pass it over either in perfect silence or with only an incidental mention, unless, by means of some philosophical subtlety, they can enlist it in their service, or else, by fair means or foul, accommodate it in some way or other to their doctrines.

Thirdly; it is necessary, in describing any disease, to enumerate the peculiar and constant phenomena apart from the accidental and adventitious ones: these last-named being those that arise from the age or temperament of the patient, and from the different forms of medical treatment. It often happens that the character of the complaint varies with the nature of the remedies, and that symptoms may be referred less to the disease than to the doctor. Hence two patients with the same ailment, but under different treatment, may suffer from different symptoms. Without caution, therefore, our judgment concerning the symptoms of disease is, of necessity, vague and uncertain. Outlying forms of disease, and cases of exceeding rarity, I take no notice of. They do not properly belong to the histories of disease. No botanist takes the bites of a caterpillar as a characteristic of a leaf of sage.

Finally, the particular seasons of the year which favour particular complaints are carefully to be observed. I am ready to grant that many diseases are good for all seasons. On the other hand, there is an equal number that, through some mysterious instinct of Nature, follow the seasons as truly as plants and birds of passage. I have often wondered that this disposition on the part of several diseases, obvious as it is, has been so little observed; the more so, as there is no lack of curious observations upon the planets under which plants grow and beasts propagate. But whatever may be the cause of this supineness, I lay it down as a confirmed rule, that the knowledge of the seasons wherein diseases occur is of equal value to the physician in determining their species and in effecting their extirpation; and that both these results are less satisfactory when this observation is neglected.

These, although not the only, are the main points to be attended to in drawing up the history of a disease. The practical value of such a history is above all calculation. By the side thereof, the subtle discussions, and the minute refinements wherewith the books of our new school are stuffed full, even *ad nauseam*, are of no account. What short way—what way at all—is there towards either the detection of the morbific cause that we must fight against, or towards the indications of treatment which we must discover, except the sure and distinct perception of peculiar symptoms? Upon each of these points the slightest and most unimportant circumstances have their proper bearings. Something in the way of variety we may refer to the particular temperament of individuals; something also to the difference of treatment. Notwithstanding this, Nature, in the production of disease, is uniform and consistent; so much so, that for the same disease in different persons the symptoms are for the most part the same; and the selfsame phenomena that you would observe in the sickness of a Socrates you would observe in the sickness of a simpleton. Just so the universal characters of a plant are extended to every individual of the species; and whoever (I speak in the way of illustration) should accurately describe the colour, the taste, the smell, the figure, &c., of one single violet, would find that his description held good, there or thereabouts, for all the violets of that particular species upon the face of the earth.

For my own part, I think that we have lived thus long without an accurate history of diseases, for this especial reason; viz. that the generality have considered that disease is but a confused and disordered effort of Nature thrown down from her proper state, and defending herself in vain; so that they have classed the attempts at a just description with the attempts to wash blackamoors white.

To return, however, to our business. As truly as the physician may collect points of diagnosis from the minutest circumstances of the disease, so truly may he also elicit indications in the way of therapeutics. So much does this statement hold good, that I have often thought, that provided with a thorough insight into the history of any disease whatsoever, I could invariably apply an equivalent remedy; a clear path being thus marked out for me by the different phenomena of the complaint. These phenomena, if carefully collated with each other, lead us, as it were, by the hand to those palpable indications of treatment which are drawn, not from the hallucinations of our fancy, but from the innermost penetralia of Nature.

By this ladder, and by this scaffold, did Hippocrates ascend his lofty sphere —the Romulus of medicine, whose heaven was the empyrean of his art. He it is whom we can never duly praise. He it was who then laid the solid and immoveable foundation for the whole superstructure of medicine, when he taught that *our natures are the physicians of our diseases*. By this he ensured a clear record of the phenomena of each disease, pressing into his service no hypothesis, and doing no violence to his description; as may be seen in his books

'De Morbis,' 'De Affectionibus,' &c. Besides this, he has left us certain rules, founded on the observation of the processes of Nature, both in inducing and removing disease. Of this sort are the 'Coacae Praenotiones,' the 'Aphorisms,' &c. Herein consisted the theory of that divine old man. It exhibited the legitimate operations of Nature, put forth in the diseases of humanity. The vain efforts of a wild fancy, the dreams of a sick man, it did *not* exhibit.

Now, as the said theory was neither more nor less than an exquisite picture of Nature, it was natural that the practice should coincide with it. This aimed at one point only—it strove to help Nature in her struggles as it best could. With this view, it limited the province of medical art to the support of Nature when she was enfeebled, and to the coercion of her when she was outrageous; the attempt on either side being determined by the rate and method whereby she herself attempted the removal and the expulsion of disease. The great sagacity of this man had discovered that Nature by herself *determines diseases, and is of herself sufficient in all things against all of them.* This she is, being aided by the fewest and the simplest forms of medicine. At times she is independent of even these.

The other method whereby, in my opinion, the art of medicine may be advanced, turns chiefly upon what follows, viz. that there must be some fixed, definite, and consummate *methodus medendi,* of which the commonweal may have the advantage. By *fixed, definite,* and *consummate,* I mean a line of practice which has been based and built upon a sufficient number of experiments, and has in that manner been proved competent to the cure of this or that disease. I by no means am satisfied with the record of a few successful operations, either of the doctor or the drug. I require that they be shown to succeed universally, or at least under such and such circumstances. For I contend that we ought to be equally sure of overcoming such and such diseases by satisfying such and such intentions, as we are of satisfying those same intentions by the application of such and such sorts of remedies; a matter in which we generally (although not, perhaps, always) can succeed. To speak in the way of illustration, we attain our ends when we produce stools by senna, or sleep by opium.

I am far from denying that a physician ought to attend diligently to particular cases in respect to the results both of the method and of the remedies which he employs in the cure of disease. I grant, too, that he may lay up his experiences for use, both in the way of easing his memory and of seizing suggestions. By so doing he may gradually increase in medical skill, so that eventually, by a long continuance and a frequent repetition of his experiments, he may lay down and prescribe for himself a *methodus medendi,* from which, in the cure of this or that disease, he need not deviate a single straw's breadth. . . .

An objection against me will be made by the vulgar and unthinking only,

viz., that of having renounced the proper pomp of physic, and of having recommended medicines so plain and simple as not to be reducible to the Materia
Medica. Wise men know this—whatever is useful is good. They know also
that Hippocrates recommended bellows for the colic, and nothing at all for
the cancer. They know, too, that similar treatment is to be discovered in every page of his writings; and withal that his merits in medicine are as great as
if he had loaded his pages with the most pompous formulae.

FOUNDATIONS OF MODERN MEDICAL PHYSIOLOGY

BOERHAAVE, Hermann (Dutch physician and physiologist, 1668–1738). From *In-
stitutiones medicae, in usu annuae exercitationis domesticos digestae*, Leyden, 1708; tr. as
Academical lectures on the theory of physic. Being a genuine translation, etc., London,
1751.

Despite ill health, Boerhaave rose from comparative obscurity to become one of the
most famous and affluent physicians of the continent. Greater practical consequences resulted for medicine from his systematization of physiology as a discipline than from any
particular factual contribution. His physiological philosophy is mechanistic, yet retains
mind, viewed as conforming to principles not operative in the *material realm*. As such,
it constitutes a synthesis of the most highly developed thought of its time, both biological
and philosophical, and displays the influence of Borelli, Descartes, and Spinoza. In the
following exposition, Boerhaave proceeds with admirable clarity to derive the details of
his physiological system from its relation to natural philosophy as a whole.

We are to consider, (1.) That Man is composed of a *Body*[1] and *Mind*,[2]
united[3] to each other; (2.) that the *Nature*[4] of these are very different, and

[1] By the Body we understand that Part of us which is extended in three Dimensions,
has a Form, and is fitted for Motion or Rest, &c.

[2] By the Mind we understand that Being which thinks, and perceives itself thinking
and the thing thought of.

[3] The Union of the Body and Mind is such, that the Mind cannot resist forming to itself the ideas of Pleasure and Pain, when the Body is in a particular manner affected;
nor can the healthy Body refuse to obey the Action of the Mind under particular Circumstances.

[4] By the Nature of the Body or Mind, we understand every thing which we are satisfied belong to each. The essential Nature of the Mind is to be conscious, or to think;
but to think of this and that particular thing, is accidental to it. The essential Nature
of the Body is Extension and Resistance. These Attributes have nothing in common to
each other, nor ought one to conclude from Similitude, that two Beings are reducible to
one general Class. When I think of Extension, it does not infer any thing of Thought;
and when I reflect upon Thought, I can perceive no Connexion of it with Extension;
therefore the Idea of the Body has nothing in common with that of the Mind, and the reverse. In the same manner, there is no Connexion between the common Ideas of Time,
Sound, Gravity, Light, &c. *Socrates* made a proper Answer to *Crito*, when he was ask'd
in what Place he should chuse to be buried? viz. "You will not find *Socrates* when you
prepare my Tomb, nor shall I be sensible of what you then do for me." Nor are there

that therefore, (3.) each has a *Life,*[5] *Actions*[6] and Affections differing from
the other; yet (4.) that there is such a reciprocal Connection and Consent between the particular Thoughts and Affections of the Mind and the Body, that
a Change in one always produces a Change in the other, and the reverse; also,
(5.) that the Mind performs some Actions by mere Thought, without any
Effect upon the Body; and that it has other Thoughts which arise barely from
some Change in the Condition of the Body; on the other hand also, (6.) that
there are some Actions performed by the Body without the Attention, Knowledge, or Desire of the Mind, which is neither concerned therein as the Cause
or Effect of those Actions; that there are also some Ideas formed in the Mind
of a Person in Health by its past Actions; and lastly, that there are other Ideas

---

Reasons wanting to prove from the present Condition of the Mind, that it may live hereafter without any Commerce with its Body. The incomparable Mathematician *Vietus,*
who first restored Algebra to us, received the Enemies Letters from his King, to expound
their mystical Signs; while he was studying to explain their Meaning, he was taken up
with the most profound Meditation for three whole Days and Nights, insomuch that he
was not the least sensible of what had been transacted without his Knowledge, taking no
more Concern for his Body, than if it had been long deserted as an Enemy by his Mind.
In like manner, we find *Archimedes* in a Consternation when he first was ordered to answer King *Hieronus* concerning the mix'd Gold in the Crown, till at last lighting upon
the Experiment, *i.e.* going into the Bath, he cry'd out Victory. And in the same manner
a *Roman,* who was in a deep Consternation or Extasy, being not at all terrified at the
formidable Advances of the *Syracutians* in Battle, made a great Conquest without once
breaking his Lines.

[5] The Life of the Body is, 1. To generate Motion under particular Circumstances, as
the Loadstone approaches to Iron. 2. For its constituent Parts to attract each other,
from whence proceeds that Resistance to the Force of external Bodies, or *Vis inertia.*
3. To gravitate, or tend towards the Center of its Planet. And then, 4. comes the Affections proper to particular Bodies. The Life of the Mind is, 1. To perceive the Appearances of all external Objects, by the Changes they make in the Organs of Sensation.
2. To judge or compare the nature of two Ideas with each other, and then to deduce some
Consequence, as that they are of the same Kind, or different; as we conclude from our
Notions of a Circle and Triangle, that a Triangle is not a Circle. 3. To will any thing.
In a word, the Life of the Mind is, to be conscious. These are all the functions of the
Mind; for past Actions are uncertain, and they may be all referr'd to the single Act of
its Consciousness.

[6] The Action of the Body is to communicate Motion to other Bodies; the Passion of
it is to receive some Change in itself from another Body or a Mind. The Action of the
Mind is Volition, which every Body is acquainted with, but no one can explain. The
Passions of the Mind are the Changes it receives from external Objects by the Senses.
Suppose the Mind to be thinking of a Circle, and in the interim a Cannon to go off, it
will lose the Idea of a Circle, and acquire that of Sound; this is the Sufferance of the
Mind, because it can neither retain the Idea of a Circle, nor resist that of a Sound.
There are also some Affections in the Mind different from the preceding, such as violent
Passions, or involuntary Commotions, which the Mind cannot resist, and the Faculty by
which it moves and determines the several Parts of a human Body, agreeable to its
Inclination.

compounded both of the past and present. That, (7.) whatever we observe to arise from Thought in the human Body, is to be only ascribed to the Mind as the Cause. But (8.) that every Appearance which has Solidity, Figure, or Motion, is to be ascribed to the Body and its Motion for a Principle, and ought to be demonstrated and explained by their properties. That, (9.) we cannot understand or explain the Manner in which the Body and Mind reciprocally *act upon each other*[7] from any consideration of their Nature separate; we can only (10.) remark by Observation their Effects upon each other, without explaining them; and when any Difficulty or Appearance has been traced so far, that it only remains to explain the manner of their reciprocal Action, we are to suppose such account *Satisfactory*,[8] both because it may be

---

[7] We cannot understand why two Principles, which have no Agreement in Power, should thus concur in the same Functions, tho' there have been three Hypotheses framed to explain the Intercourse of the Body and Mind; the first is, by the *physical Influx*, which supposes the thing thought of, and the Thought itself, to be one and the same; which we shall hereafter demonstrate to be absurd, in as much as our Mind is ignorant of its own Nature. The second is the System of *occasional Causes;* and the third supposes a Harmony establish'd by God, taking it for an infallible Rule, that determinate Actions of the Mind must be necessarily attended with corresponding Motions in the Body, and the contrary; and this last seems to be the truest Opinion, but it leaves us equally in the Dark with the other.

[8] If any Action is to be explain'd which is compounded both of the Faculties of the Mind as well as of the Body, such as Walking, Pain, voluntary Respiration, &c. a just Account ought to be first given how far, and in what manner, the Body is concerned in the Action, and then also of the Mind; if this can be done, it is enough, without diving into the manner of Connexion between the different Actions; the Explication of the corporeal Actions appertains to the Physician, and those of the Mind to the Philosopher; but their Connexion can be explained by no Man. Heat may be conceiv'd to arise in Bodies without any relation to a thinking Mind, as Mill-stones grow hot in their grinding; but Motion is not explicable from the Affections of the Body, nor even from the Properties of the Mind, therefore Heat and Motion are not accountable from the Mind; and if you should say that the voluntary Motions of the Muscles proceed from the Act of Volition in the Mind, you explain the thing not in the least, because there is nothing in the Idea of Motion which is also to be found in any Affection of the Mind. We call an Explanation of a thing the Demonstration of Agreement or Relation between its own Properties and the same in another; but this is here not only impossible, but also quite useless to a Physician; for the great Business of a Physician is to be acquainted with the Means of restoring lost Health, and no Cure can be effected by him, but through some Change made in the human Body by the Application of others; therefore this Search after the Connexion between the Body and Mind not appertaining to a Physician, is to be rejected among those which are useless to the Art. The Physician, who cures Diseases of the Body is not sollicitous about those of the Mind; for when the first is set to rights, the latter will quickly return to its Office. Thus when the Eye is blinded with a Cataract, the Mind cannot perceive sensible Objects by it, the Aid of Physic is therefore call'd in to couch the Cataract, or depress the opake crystalline *Lens;* after which the Rays of Light finding a free Admission to the *Retina,* the Mind will be sensible of visible Objects by it; and thus the Business of Physic will be done without the assistance of Optics. When a Person is in a *Delirium,* or swoon, the Physician cannot recall the

sufficient for all the Purposes of the Physician, and as it is impossible for him to search any further.

We may also affirm, that the *primary physical Causes,*[9] in what manner, and the ultimate *metaphysical Causes,*[10] for what End, the most general Appearances are in a determinate manner affected, are neither possible, useful, or necessary to be investigated by a Physician; such as the Origin of primitive and *Seminal Forms,*[11] of *Motion,*[12] the *Elements,*[13] &c.

Mind, which has no relation to his Business; but by applying Vinegar, or other Volatiles to the Nose, he can restore the sick Machine to its former Motions, and then the Mind will also exhibit its former Actions, and this full as well as if he understood the manner of Connexion between the Actions of the Body and those of the conscious Mind.

[9] *Primary Causes* are those productive of secondary ones; but we always meet with God in our Search after these, and this puts a Stop to our further Knowledge; for God is an infinite Being, and if we compare the whole Universe with him, it will be found almost nothing.—In our Search after *physical Causes,* we should not be over sollicitous to determine every thing in which Experiment will not assist us; for we never can be certain of the Truth of such Discoveries, and if we were, it would be of little or no Use to Mankind; we are thus wholly ignorant of the Origin and Communication of Motion in Bodies; for Motion is no more essential to the Idea of Body, than a Circle is to that of the Mind. Let those Philosophers appear, who hold that an Assembly of Gods joined together to form the Universe, and explain by one simple and universal Experiment, why any Body in motion communicates part of its motion to the next which it touches; an ingenious Person would answer, God made it so. We ought therefore to rest upon Experiment, and lay aside useless Attempts to explain the most general Laws and Principles observed in Nature; taking Example by the wise Ignorance of the Chemists, who barely relating the Appearances offer'd to them, are not concerned about the first Cause. *Barthol. Schwartz* having discovered the surprising Experiment of producing Thunder and Lightning, by the Application of Fire to a Powder made of Nitre, Sulphur, and Wood-coals, mix'd in a certain Proportion, never enquired into the Cause of that Phaenomenon by which almost the whole Face of the habitable World has been chang'd. The Moderns have found, that two Grains of Gold dissolved in three times as much *Aqua regia,* and precipitated with half that Quantity of Oil of Tartar *per deliquium,* forms a Powder, which applied to a certain degree of Fire, will blow up a hundred Weight. The Chemist stops at the bare Appearance; but the Philosopher taking a Course very different from the Experiments of the Chemist, studies the Formation of a mechanical Engine, by which two Grains will raise a Weight of a hundred Pounds; and thus each of them obtain their Ends by different Means.

[10] By metaphysical Causes, are meant those general Attributes of Beings which are abstractedly essential to them as Beings; which are therefore very universal, and remote from Action.

[11] Some of the Chemists acknowledge besides Matter, Form and Vacuum, a seminal Principle; which so determines the Structure of vegetable Bodies in their Growth, that they can appear in such a particular Form, and no other. If an Aniseed be sowed in a pure Earth, moistened with Rain-water, and forwarded with a Heat equal to that of a setting Hen, it will produce the Plant Anise, whose Smell, Taste, and Structure, differs from all other Plants in the Universe; and in the Vegetation of the Plant there is also a new Production of Seeds, each of which is capable, under proper Circumstances, of producing the like Plant; if these Seeds were wanting, the whole united Power of Nature

But a Physician may, and ought to furnish himself with, and reason from, such Things as are demonstrated to be true in *Anatomy*,[14] *Chemistry*[15] and *Mechanics*,[16] with natural and experimental Philosophy, provided he confines his Reasoning within the Bounds of Truth and simple Experiment.

---

together could never produce the same Plant; therefore, according to the Opinion of the Chemists, this Seed must contain a Principle, which from Earth and Water always produces that particular Plant, which no other Seed can produce. In like manner they suppose Metals to be formed of a seminal Substance, which grows or vegetates in the Bowels of the Earth with a subterraneous Heat, by means of a particular Juice; which Opinion is confirmed by philosophical Experiments, and supported by many Reasons.

[12] The Origin of Motion is to be look'd for in God; if we substitute any other primary Cause, we do him Injustice. I may say that it becomes a true Philosopher to confess his Ignorance of first Causes, which he is never likely to attain to; but notwithstanding secondary Causes may be used to as good Purposes as if we were acquainted with their first. If I learn by Experiment the virtues of any Plant for the Cure of Diseases, I may do as much Service with it in Physic as if I had created the Plant. If everything useless to the Art was to be in this manner expunged, as we in this Section advise, Physic would lose nine Parts out of ten, and be by that means purged of its Dross, and restored to its native Simplicity.

[13] An Element is the Matter of which a Body is originally composed, and into which it may be ultimately resolved. Great has been the Controversy in all Ages about the Elements. Some contend for Water only, others for Air, and others again for Water and Fire; but the greater Number are for the four Peripatetic Elements; tho' the Chemists also build upon their Salt, Sulphur, and Mercury; but neither of these can be properly an Element, for it is essential to an Element to have its Part absolutely simple and homogenous; but then how can Matter thus homogenous form the great Variety of Bodies we meet with? If you retreat to the *Monades*, or Atoms of *Pythagoras*, and universal Matter, you do not take our Eyes with you to convince us; nor can we be certain whether there are such or no, since you tell us of things from which the Mind can never receive any real Ideas.

[14] He that desires to learn Truth, should teach himself by Facts and Experiments; by which means he will know more in a Year, than by abstract Reasoning in an Age. Proper Experiments have always Truth to defend them; also Reasoning join'd with Mathematical Evidence, and founded upon Experiment, will hold equally true; but should it be true, without those Supports it must be altogether useless. Nature distributes the Faculty of Reason to all Men equally alike, but he will excel in Reasoning who has made the best Use of Experiments, having consider'd the Structure, Situation, Figure, Size, and other Peculiarities, obvious to our Senses in the several Parts of the human Body.

[15] Chemistry acquaints us with those Changes which arise in Bodies from Mixture, and the Application of them to Fire. Suppose one Substance of a particular kind to be mix'd with another, and applied to a determinate degree of Fire, the Consequence will be a Production of new Appearances, which is the Business of the Chemist to remark; nor does ever Chemistry deceive us, if it proceeds no farther than real Experiments, and their Effects; upon the Addition of the best Oil of Cloves to rectified Oil of Vitriol, they run into a violent Commotion, and exhale clouds as thick as Pitch, which quickly turn into Flames.

[16] Mechanics teach us to apply the general Laws of Motion to all Kinds of Bodies. Every Body is extended, resists Motion, is moveable, capable of Form, &c. The Effects

It is necessary for the Physician, in furnishing himself with these Principles and Experiments, to begin first with such as are most simple, certain and easy to be understood; after which he may proceed to those which are more compounded, and so by degrees to the most complex, obscure, and difficult.

He that would learn by Experiments, ought to proceed from Particulars to Generals; but the Method of instructing academically, proceeds from General to Particulars; which is the Method we shall observe.

A Professor skill'd in the Science which he teaches, first lays down general Rules, by which the Nature of each particular Subject is to be defined; but an Inventor of Discoveries ought to learn the Properties of every particular Body by proper Experiments, that he may afterwards reduce them into Classes, according to their Affinity: The first Method is in the Schools termed Analytical, the other Synthetical. The Inventor, *Aristotle*, when he observed that Oxen, who had Horns, wanted fore Teeth in the upper jaw, and finding they were also wanting in Stags, Goats, Sheep, and other Animals with which he was acquainted, took occasion to affirm, that all Animals that had Horns wanted upper Teeth. But *Ray*, teaching the Nature of Animals, lays this down for an Axiom, from which he infers, that neither the Ox, Stag, nor Range Deer, have Teeth in their upper Jaw because they are horned.

From these Considerations appears the Order of our Doctrine; for in the first Place we are to consider Life[17]; then Health, afterwards Diseases; and lastly their several Remedies.

Hence the first general Branch of Physic in our Institutions is termed PHYSIOLOGY, or the Animal Oeconomy; demonstrating the several Parts of the human Body, with their Mechanism and Actions; together with the Doctrines of Life, Health, and their several Effects, which result from the Mechanism and Actions of the Parts. The Objects of this Branch have been usually denominated *Res naturales*, Things natural or according to Nature.

---

of all these general Qualities, and the moving Powers thence arising, are applicable to every particular Body; nor can we be deceived therein, if the Body to which they are applied be distinctly and carefully considered in all those Respects. Mechanics therefore supposes a previous Knowledge of the Structure of all the Parts in the human Body, to which we would apply mechanical Laws; and in this Sense Physic is no more than the Knowledge of such Things as are transacted in the human Body, either by the common Affections of Bodies, or by the determinate and particular Structure of the Parts in the human Body. It therefore appears that Mechanicians, ignorant of the Structure of the Parts whose Actions they would express by Numbers, must run into the Excesses of Error; which Defect has been charged upon ourselves, for what has been formerly advanced in an Oration *de usu Mechanices in Medicina*; tho' there are some, Enemies to the very Name of Mechanics, who assert, that our Bodies are not subject to the same Laws with all others.

[17] Life is the Sum or Aggregate of all the Actions resulting from the Structure of the several Parts in the human Body; when all those Actions are performed with Ease and Perfection, it is called Health.

The second Branch of Physic is called PATHOLOGY, treating of Diseases, their Differences, Causes and Effects, or Symptoms; by which the human Body is known to vary from its healthy State. This Branch is distinguished into (1.) *Diagnostic* Pathology, so far as it describes the Diseases of the Body; (2.) *Aetiologic,* when it treats of their Causes; (3.) *Diatritic,* when it considers their Differences and future Events; and lastly, (4.) the *Symptomatologic* Part of Pathology, is that which explains the various Effects or Symptoms of Diseases.—The Objects hereof are termed *res contra naturam,* Things preternatural, or contrary to Nature.

The third Part of Physic is termed SEMIOTICA, which shews the *Signs* distinguishing between Sickness and Health, Diseases, and their Causes in the human Body; it also imports the State and Degrees of Health and Diseases, and presages their future Events. The Objects of this Branch are the *Non-naturals* as well as the *Naturals* and *Preter-naturals.*

The fourth general Branch of Physic is termed HYGIENE, or *Prophylaxis;* which teaches us what Remedies are proper, and how they are to be used to preserve Life and present Health; and, as much as possible, to prevent Distempers. The chief Object hereof is the *Non-naturals,* or *Res non-naturalis.*

The fifth, and last Part of Physic, is called THERAPEUTICA; which instructs us in the Nature, Preparation, and Uses of the *Materia Medica;* and the Methods of applying the same, in order to cure Diseases and restore lost Health. This Branch is called *Methodus Medendi,* so far as it points out the Means and Cure; which are comprized under three Heads: (1.) *Pharmacy,*[18] or the Preparation and internal Use of Medicines; (2.) *Dietetics,*[19] or Regimen, respecting a Regulation of the Diet, Air, &c. And (3.) *Surgery,*[20] comprehending manual Operation with Instruments, and topical Remedies.

Having thus distributed Physic under its proper Heads, agreeable to the Nature of the Art itself, as well as the most convenient Method of teaching and learning the same, which is also approved by the established Custom of the Professors through many Ages past; we shall next proceed to treat of the several Branches separately in that Order.

---

[18] By the *Materia Medica* we here intend all Remedies, taken as well from Diet as Pharmacy; in which ample Signification *Dioscorides* has described the *Materia Medica.*

[19] Natural Remedies, as they come first to our Hands, are very often unfit for the Stomach, too strong in their Action, nauseous to a Patient, or else not sufficiently exalted in their Virtues. Physicians have therefore industriously contrived to render them more innocent, grateful, and efficacious, by subjecting them to various preparations, Compositions, and Changes; and this is the Business of Pharmacy, whether Galenical or Chemical.

[20] The *Methodus Medendi* points out to us the curative Indications, with the Time and Method of applying Remedies, being the immediate Foundation of the extemporaneous Prescription of Medicines, and of the general Rules to be given by the Physicians for the Patient's Recovery.

FROM HUMORAL TO ANATOMICAL PATHOLOGY

MORGAGNI, Giovanni Battista (Italian pathologist, 1682–1771). From *De sedibus et causis morborum per anatomen indigatis, etc.*, Padua, 1765; tr. by B. Alexander as *The seats and causes of diseases investigated by anatomy, etc.*, London, 1769.

The title of this book gives the key to its historic significance. By establishing connections between etiology and abnormal anatomy, Morgagni attacks the humoral theory of disease and lays the foundations of modern pathology. Two fragments are given here. In the first, Morgagni outlines his method. In the second, he applies this method to a specific case, an instance of apoplexy. Morgagni held the famous professorship of anatomy at Padua. His expressions of indebtedness to Bonetus should be noted.

As when a young man, I had not omitted to testify publicly, to the first academy of sciences which had admitted me, the feelings of a grateful mind on that occasion, and had seen that testimony receiv'd by them with the same degree of condescension, wherewith they had formerly conferr'd so many benefits, as are mention'd by that very celebrated man Francesco Maria Zanotti, who is one of the committee to that body, and to the Institution of Sciences at Bologna; why should I now, that I am grown old, suffer myself to die under the influence of ingratitude to five other of the most noble academies of sciences in all Europe, which had, afterwards, very condescendingly and very honourably, chosen me into the number of their fellows? Therefore, as I had nothing, nor could hope to have any-thing, whereby I might shew myself to have a grateful sense of their favours, in the best manner I was able, unless I should depute persons to wait upon each of them, to assure them of my gratitude and duty towards them, and, at the same time, present them with a copy of this work, and request that they would each of them accept it, such as it was, with their well-known condescension, and consider the intention rather than the thing; I did not think that I ought to lose such an opportunity.

And that this might be known to all of them, it very conveniently happen'd, that the number of books, into which these letters were naturally, and of themselves, divided, exactly corresponded to the number of academies; so that I could prefix to each of the books that very letter, wherein I should signify what I would wish to have said, to each of those respectable bodies, in my name. These letters I have prefix'd without observing any other order, than that of the time in which I was chosen into their celebrated societies: and that they might be the more read by every-one, I added several other things to the testimonies of a grateful and respectful mind, and of those five letters made so many prefaces, as it were, in which I might demonstrate how great an advantage there is arising from the dissections of dead bodies.

In the first, therefore, having argued against some persons, who have been presumptuous enough to call this utility into question, I have shewn in what manner the deceptions, which have been made use of as objections to the practice, may be avoided by those who dissect bodies, and who prove both the seat and the cause of the disease, which are, for the most part, easily demonstrated from the dissection. In the second I have confirmed the same utility, by the full and ample consent of almost all physicians, particularly those who have flourish'd amongst the most polite and cultivated nations, from the most ancient times, speaking of the merits of each nation in regard to this question, and mentioning the name of most of the physicians in order; and especially of those who, from their own observations, or even the observations of others, wish'd to have compil'd a Sepulchretum before the time of Bonetus.[1] In the third an answer is particularly given to those, who, because dissections are of no use in order to detect the first and most hidden causes of diseases, and such as are entirely inaccessible to the senses, think that it is, therefore, quite needless to prosecute the practice, as if they did not thereby detect any evident internal causes, or the knowledge of these causes were of no advantage, because, even where they are known, a great number of disorders are, nevertheless, still uncur'd.

In the fourth I make this enquiry, whether it is more useful to dissect the bodies of those who died of the more rare, (for some of these also I have dissected) or of the more common diseases.

In the fifth, finally, it is shewn, that, although the anatomy both of sound bodies, and of those that are carried off by disease, is useful, the latter is, nevertheless, by far the more useful.

And as all these circumstances ought, some for one reason, and some for another, not to be pass'd by; so if they had been all thrown together into this preface, they would have made that discourse, which is already long, in consequence of the many things that were necessarily to be spoken of, extremely long and prolix.

[From Book II.]

Of disorders of the head. An old man, who had been us'd for a long time past, by reason of a large ulcer in one of his legs, to sit begging at the gate of St. Anthony's church, being accustom'd to eat very plentifully, as I hear most of these people do, and using very little or no exercise for the reason above mention'd, was seiz'd with an apoplexy, whereby his internal senses, the use of his tongue and left side, were entirely taken away; so that he died within three or four days.

As the body, by reason of its putrid smell, and the bad colour of the intestines, which had a mixture of red and brown, was unfit for the anatomical demonstrations that I gave at the hospital, in the year 1741, I order'd it to be buried, preserving only the head. When the head was to be open'd, in the

---

[1] Bonet, Théophile. *Sepulchretum sive anatomia practica,* etc., Geneva, 1679.

presence of a number of learned men, and young students, to find out the cause of the apoplexy, I by chance observ'd a slight contusion on the anterior borders of the temporal muscle, on the left side. Having enquir'd into it, I found that this contusion was the consequence of his falling from his seat, when the apoplexy had seiz'd him: upon which I did not hesitate immediately to foretel, that if the cause of this apoplexy should fall under the notice of the senses, and had not its origin from serum, it would, according to a certain conjecture of mine, (hinted at in the last letter) be found in the opposite, that is, in the right side of the cranium. And in this conjecture I was much more confirm'd, when it was also added, that the man had been paralytic on his left side, as I have already said; which by chance till then I had not heard. At length the skull being cut through, and a little water having flow'd out in the operation, all the parts beneath immediately appear'd more full of blood than they generally do. Having drawn aside the dura mater, in whose upper sinus a little polypous concretion was found; not only the vessels of the pia mater were more tumid with blood on the right than on the left side, but also on the right hemisphere of the brain; and on that only, appear'd some half-concreted blood, which seem'd to have come from some of those tumid vessels, and to have flow'd downwards. For under the basis of the anterior part of the posterior lobe of the right hemisphere, a little more of the same kind of blood appear'd, in like manner, betwixt the two meninges. Nor was there any other extravasation of blood within the cranium, but that which I have mention'd; and this was about the quantity of two spoonfuls. These things being demonstrated thus to all who were present, other things also were shewn, which, though they seem'd of less consequence, yet were relative to the present enquiry. The summary of them is this. A kind of gelatinous humour was seen to shine through the substance of the pia mater. The vessels were not only distended in the medullary part of the brain, as a number of points, starting with blood here and there, testify'd, but also the vessels which creep through the surface of the lateral ventricles. These ventricles did not contain much water; yet at the posterior part of the choroid plexusses were a great number of vesicles, though not of the largest kind; but those in the right were somewhat less than in the left; and less water also was contain'd in the former than the latter. The plexusses, however, were so far from having lost their colour, that they were even more fill'd with blood than usual. Last of all, at the anterior basis of the pineal gland, was found something yellowish, but not hard.

ARTIFICIAL ACTIVE IMMUNIZATION

JENNER, Edward (British physician, 1749–1823). *An inquiry into the causes and effects of the variolae vaccinae, or cow-pox*, London, 1798.

Edward Jenner, vicar's son from Gloucestershire, was another instance of illustrious disciple of illustrious teacher (John Hunter). Many years after going back to practice in his native town of Berkeley, Jenner finally "selected a healthy boy" (James Phipps) and on him performed the following famous experiment. According to Garrison, "The mere idea of inoculation is as old as the hills. Jenner's task was to transform a local country tradition into a viable prophylactic principle." Within nine years after the appearance of this paper, at least one country, Bavaria, took the legal step (compulsory vaccination) which would finally bring this scourge under nearly complete control.

The deviation of man from the stage in which he was originally placed by nature seems to have proved to him a prolific source of diseases. From the love of splendour, from the indulgences of luxury, and from his fondness for amusement he has familiarised himself with a great number of animals, which may not originally have been intended for his associates.

The wolf, disarmed of ferocity, is now pillowed in the lady's lap. The cat, the little tiger of our island, whose natural home is the forest, is equally domesticated and caressed. The cow, the hog, the sheep, and the horse, are all, for a variety of purposes, brought under his care and dominion.

There is a disease to which the horse, from his state of domestication, is frequently subject. The farriers have called it the grease. It is an inflammation and swelling in the heel, from which issues matter possessing properties of a very peculiar kind, which seems capable of generating a disease in the human body (after it has undergone the modification which I shall presently speak of), which bears so strong a resemblance to the smallpox that I think it highly probable it may be the source of the disease.

In this dairy country a great number of cows are kept, and the office of milking is performed indiscriminately by men and maid servants. One of the former having been appointed to apply dressings to the heels of a horse affected with the grease, and not paying due attention to cleanliness, incautiously bears his part in milking the cows, with some particles of the infectious matter adhering to his fingers. When this is the case, it commonly happens that a disease is communicated to the cows, and from the cows to the dairymaids, which spreads through the farm until the most of the cattle and domestics feel its unpleasant consequences. This disease has obtained the name of the cow-pox. . . .

Thus the disease makes its progress from the horse to the nipple of the cow, and from the cow to the human subject.

Morbid matter of various kinds, when absorbed into the system, may produce effects in some degree similar; but what renders the cow-pox virus so extremely singular is that the person who has been thus affected is forever after secure from the infection of the smallpox; neither exposure to the variolous effluvia, nor the insertion of the matter into the skin, producing this distemper.

In support of so extraordinary a fact, I shall lay before my reader a great number of instances. . . .

Case II. Sarah Portlock, of this place, was infected with the cow-pox when a servant at a farmer's in the neighbourhood, twenty-seven years ago.

In the year 1792, conceiving herself, from this circumstance, secure from the infection of the smallpox, she nursed one of her own children who had accidentally caught the disease, but no indisposition ensued. During the time she remained in the infected room, variolous matter was inserted into both her arms, but without any further effect than in the preceding case. . . .

Case XVII [Entire]. The more accurately to observe the progress of the infection I selected a healthy boy, about eight years old, for the purpose of inoculation for the cow-pox. The matter was taken from a sore on the hand of a dairymaid, who was infected by her master's cows, and it was inserted, on the 14th of May, 1796, into the arm of the boy by means of two superficial incisions, barely penetrating the cutis, each about half an inch long.

On the seventh day he complained of uneasiness in the axilla, and on the ninth he became a little chilly, lost his appetite, and had a slight headache. During the whole of this day he was perceptibly indisposed, and spent the night with some degree of restlessness, but on the day following he was perfectly well.

The appearance of the incisions in their progress to a state of maturation were much the same as when produced in a similar manner by variolous matter. The only difference which I perceived was in the state of the limpid fluid arising from the action of the virus, which assumed rather a darker hue, and in that of the efflorescence spreading round the incisions, which had more of an erysipelatous look than we commonly perceive when variolous matter has been made use of in the same manner; but the whole dies away (leaving on the inoculated parts scabs and subsequent eschars) without giving me or my patient the least trouble.

In order to ascertain whether the boy, after feeling so slight an affection of the system from the cow-pox virus, was secure from the contagion of the smallpox, he was inoculated the 1st of July following with variolous matter, immediately taken from a pustule. Several slight punctures and incisions were made on both his arms, and the matter was carefully inserted, but no disease followed. The same appearances were observable on the arms as we commonly see when a patient has had variolous matter applied, after having either the cow-pox or smallpox. Several months afterwards he was again inoculated with variolous matter, but no sensible effect was produced on the constitution.

Here my researches were interrupted till the spring of the year 1798, when, from the wetness of the early part of the season, many of the farmers' horses in this neighbourhood were affected with sore heels, in consequence of which the cow-pox broke out among several of our dairies, which afforded me an opportunity of making further observations upon this curious disease.

A mare, the property of a person who keeps a dairy in a neighbouring parish, began to have sore heels the latter end of the month of February, 1798, which were occasionally washed by the servant men of the farm, Thomas Virgoe, William Wherret, and William Haynes, who in consequence became affected with sores in their hands, followed by inflamed lymphatic glands in the arms and axillae, shiverings succeeded by heat, lassitude, and general pains in the limbs. A single paroxysm terminated the disease; for within twenty-four hours they were free from general indisposition, nothing remaining but the sores on their hands. Haynes and Virgoe, who had gone through the smallpox from inoculation, described their feelings as very similar to those which affected them on sickening with that malady. Wherret never had had the smallpox. Haynes was daily employed as one of the milkers at the farm, and the disease began to shew itself among the cows about ten days after he first assisted in washing the mare's heels. Their nipples became sore in the usual way, with bluish pustules; but as remedies were early applied, they did not ulcerate to any extent. . . .

It is singular to observe that the cow-pox virus, although it renders the constitution unsusceptible of the variolous, should nevertheless, leave it unchanged with respect to its own action. . . .

It is curious also to observe that the virus, which with respect to its effects is undetermined and uncertain previously to its passing from the horse through the medium of the cow, should then not only become more active, but should invariably and completely possess those specific properties which induce in the human constitution symptoms similar to those of the variolous fever, and effect in it that peculiar change which for ever renders it unsusceptible of the variolous contagion. . . .

In some of the preceding cases I have noticed the attention that was paid to the state of the variolous matter previous to the experiment of inserting it into the arms of those who had gone through the cow-pox. This I conceived to be of great importance in conducting these experiments, and, were it always properly attended to by those who inoculate for the smallpox, it might prevent much subsequent mischief and confusion. . . .

Should it be asked whether this investigation is a matter of mere curiosity, or whether it tends to any beneficial purpose, I should answer that, notwithstanding the happy effects of inoculation, with all the improvements which the practice has received since its first introduction into this country, it not very unfrequently produces deformity of the skin, and sometimes, under the best management, proves fatal.

These circumstances must naturally create in every instance some degree of painful solicitude for its consequences. But as I have never known fatal effects arise from the cow-pox, even when impressed in the most unfavourable manner, producing extensive inflammations and suppurations on the hands; and as it clearly appears that this disease leaves the constitution in a state of perfect security from the infection of the smallpox, may we not infer that a mode of inoculation may be introduced preferable to that at present adopted, especially among those families which, from previous circumstances, we may judge to be predisposed to have the disease unfavourably? It is an excess in the number of pustules which we chiefly dread in the smallpox; but in the cow-pox no pustules appear, nor does it seem possible for the contagious matter to produce the disease from effluvia, or by any other means than contact, and that probably not simply between the virus and the cuticle; so that a single individual in a family might at any time receive it without the risk of infecting the rest or of spreading a distemper that fills a country with terror.

DISEASE AS THE RESULT OF DIETARY DEFICIENCY

BUDD, George (English physician, 1808–1882). From *Scurvy*, in A. Tweedie (Ed.), *A system of practical medicine*, vol. 5, Philadelphia, 1841.

In the sixteenth century Richard Hawkins reported, "That which I have done most fruitfull for this sicknesse [scurvy], is sower Oranges and Lemmons, and a water which amongst others (for my particular provision) I carryed to the Sea, called Doctor Stevens his Water."* Nearly two and a half centuries were required to produce a useful hypothesis concerning the biological significance of Hawkins' treatment. In the meantime it seemed "easier for the mind to believe that ill is caused by some positive EVIL AGENCY, rather than by the mere ABSENCE of any beneficial property."† Finally, however, Budd obtained a clear concept of dietary deficiency disease on the basis chiefly of historical studies. Established experimentally by Grijns (*q.v.*), explored biochemically by Hopkins (*q.v.*), and established as a general theory by Funk, the concept grew into the modern science of vitamins.

Preventives. We come now to speak more in detail of the means by which scurvy may be prevented; and shall first mention as the chief of these means, the use of oranges, lemons, or limes; and, we believe, we might add, shaddocks, and all fruits which botanists have included in the order *Aurantiaceae*.

The efficacy of oranges in preventing and curing scurvy was discovered before the disease had been described by physicians. Rousseus, one of the

---

* In his *Observations in his voyage to the South Sea*, London, 1593.

† L. J. Harris, *Vitamins in theory and practise*, p. 8, Cambridge, 1935. (Note: this work contains outstanding historical notes on vitamin research.)

earliest writers on scurvy, in a work published in 1564, observes that seamen in long voyages cure themselves of it by the use of oranges. He conjectures that Dutch sailors, afflicted with scurvy on their return from Spain with a cargo of these fruits, had by chance discovered their efficacy.

Albertus, in a treatise on Scurvy, published in 1593, recommends the juice of oranges, and of sour and austere plants. He advises that this juice should be put into soups, and that meat, while roasting, should be sprinkled with it. In the same year, the virtues of lemon juice in the cure of scurvy were experienced by Sir R. Hawkins, whose crew, while within the tropics, were affected with it in an extreme degree.

We have already given an instance of the extraordinary efficacy of lemon juice as a preventive of scurvy, in the first voyage for the establishment of the East India Company in 1600. After this it seems to have been pretty generally used in the company's ships; and, in a medical work published in this country in 1636, it is recommended as the best remedy for scurvy.

From this time it is recommended by a series of writers who have treated of this subject; and instances which show its extraordinary efficacy are to be frequently met with in our naval annals.

When Admiral Sir C. Wager commanded our fleet in the Baltic, in 1726, his sailors were dreadfully afflicted with scurvy. He had recently come from the Mediterranean, and had on board a great quantity of lemons and oranges, which he had taken in at Leghorn. Having often heard of the efficacy of these fruits, he ordered a chest of each to be brought upon deck, and opened every day. The men, besides eating what they liked, mixed the juice with their beer. It was also their constant diversion to pelt one another with the rinds, so that the deck was always strewed with them, and wet with the fragrant liquor: the happy result was, that he brought his sailors home in good health. (*Mead on Scurvy.*)

Most of these proofs of the efficacy of oranges and lemons were collected by Dr. Lind, and published in his justly celebrated work on Scurvy in 1757. His earnest recommendation for the general employment of these fruits in the navy was, however, not acted upon for some time: the disease continued to depopulate our fleets, offering a striking example of the delay which sometimes attends the practical application of most important truths. To the cause of delay in the present instance, we shall allude particularly hereafter (see chap. on *Diagnosis*); at present we only mention the fact, as one of the most singular and instructive in the history of the disease. We have already noticed the prevalence of scurvy in our fleet in the West Indies in the years 1780–1–2, and in the Channel fleet in 1795. The history of these fleets afford numerous proofs of the efficacy of the fruits in question; but in 1794 an experiment was made which established it beyond doubt. The *Suffolk*, of 74 guns, sailed from England for Madras on the 2d of April, 1794. She was provided with lemon juice; and two-thirds of a liquid ounce of this juice, together with two ounces of sugar, were mixed with each man's daily allowance of grog. The Suffolk was twenty-three weeks and one day on the passage, during which she had no communication with land. Scurvy showed itself in a few men in the course of the voyage, but soon disappeared on an additional quantity of lemon juice being given them; and the ship arrived at Madras, without the loss of a single man, and with her crew entirely exempt from scurvy.

It is to the representations of Dr. Blair and Sir G. Blane, in their capacity of commissioners for the relief of sick and wounded seamen, enforced by the result of this experiment in the *Suffolk,* that we owe the systematic introduction of lemon juice into nautical diet, in 1795, by order of the Admiralty. We have already spoken of the improvement in the health of the navy consequent on this wise measure: but we may be permitted to mention the following circumstances which show how completely it has realised the expectations of its proposers.

In 1780, 1457 cases of scurvy were admitted into Haslar Hospital: in 1810, one of the physicians of that hospital informed Sir G. Blane that he had not seen a case of it for seven years; and, in the four years preceding 1810, only two cases were received into the naval hospital at Plymouth. At present, there are many surgeons in the navy who have never seen a case of scurvy, which has, in fact, been expunged from the list of diseases incident to seamen in the navy.

The present allowance of lemon juice in the navy consists of a fluid ounce, which, after ships have been a fortnight at sea, is served daily with an ounce and half of sugar to each of the men.

Dr. Lind recommended a *rob,* formed by evaporating the juice, by a slow heat, to the consistence of thick syrup. This was found to be very inferior in efficacy to the fresh fruit (*Diseases of Seamen,* p. 56; *Med. Nautica,* vol. i. p. 425); and Sir G. Blane, in consequence, advised that the juice should be preserved by the addition of a small quantity of spirit, without the aid of heat; a plan now generally adopted. The juice with which the navy is supplied is brought from Sicily, and kept good by the addition of one part of strong brandy to ten of the juice. When preserved in this manner, its virtues seem unimpaired.

These fruits, when employed in the treatment of scurvy, combine all the good qualities we can desire in a remedy. They have a specific influence in curing the disease, but produce no other sensible effect, except a small increase in some of the secretions; and the eating of them is attended wtih great pleasure. Dr. Lind tells us that he has often observed, upon seeing scorbutic people landed at our hospitals, that the eating of these fruits was attended with a pleasure more easily imagined than described; and his testimony is confirmed by that of other naval physicians.

Oranges, lemons, and limes, seem to have nearly equal efficacy; and perhaps the same may be said of shaddocks, and all fruits of a like kind. Dr. Lind, however, from some comparative trials, was led to give oranges a preference to lemons. It is probable that the state of the fruit, as to maturity, has considerable influence on its virtues. That such is the case with the guava, appears clearly from an experiment made by Dr. Trotter. Having repeatedly observed scorbutic slaves throw away ripe guavas, while they devoured green ones with much avidity, he resolved to try if any difference could be remarked in their effects. For this purpose he selected nine blacks affected with scurvy in nearly equal degree. To three of these he gave limes, to three green guavas, and to three ripe guavas. They were kept under the half-deck, and served by himself two or three times a day. They lived in this manner for a week; at the end of which those restricted to ripe guavas, were in much the same state as before the experiment, while the others were almost well.

Most sour fruits are in all probability antiscorbutic. The good effects of unripe grapes were noticed by Fodéré in the French army of the Alps, in 1795 and, in 1824, when scurvy prevailed among our troops at Rangoon, in India, great benefit was derived from giving the men the fruit of the *Phyllanthus Emblica,* or Anola; which, when dry, as sold in bazaars, has a rich and strongly acid taste, with a flavour resembling that of tamarinds. (*Quarterly Journal of the Med. and Phys. Society of Calcutta,* vol. i. p. 306.) The efficacy of apples, as a preventive of scurvy, was alluded to by Sir J. Pringle in an address to the Royal Society, in 1776; and the following proof of their curative virtues is given by Dr. Trotter:—When Lord Bridport's fleet arrived at Spithead on the 19th of September, 1795, almost every man in the fleet was more or less affected with scurvy. Large supplies of vegetables were provided, and lemon juice being scarce in consequence of the previous great consumption, fifty baskets of unripe apples were procured at the Isle of Wight for the use of the fleet. The *Royal Sovereign,* in particular, derived great benefit from them; and the cure of the disease was everywhere so speedy, that little remained to show Earl Spencer, when he visited the fleet at the end of the month. (*Med. Naut.,* vol. i. p. 420.)

As the expense of lemon juice offers great impediment to the employment of it in the commercial marine of this country, to the extent necessary for complete extinction of scurvy, it deserves to be ascertained whether the juice of apples preserved, like that of lemons, by the addition of a certain proportion of spirit, would not be an effective substitute.

All succulent vegetables that are wholesome, are perhaps, as well as fruits, more or less antiscorbutic; but this property seems to be possessed in the highest degree, by plants comprised in the order *Cruciferae,* in which most of the vegetables in common use, as the cabbage, turnip, radish, water-cress, &c., are included.

In the earliest notices of scurvy, mention is made of the efficacy of herbs of this class in its treatment.

Rousseus, writing in 1564, informs us, that the common people cured themselves by scurvy-grass, brook-lime, and water-cresses. W. Cockburn, in a work published in 1796, entitled *Sea Diseases*, remarks the extraordinary efficacy of vegetables in the treatment of this distemper. As a proof of it he mentions the following circumstance:—When Lord Berkeley commanded the fleet in Torbay, Mr. Cockburn prevailed on his lordship to erect tents for the sick on shore. Above a hundred of the most afflicted scorbutic patients, perfect moving skeletons, hardly able to get out of their ships, were landed, and fresh provisions, including carrots, turnips, and other vegetables were given them. In a week they were able to crawl about; and before the fleet sailed, they returned healthy to their ships. The subsequent history of scurvy abounds with instances equally decisive; but the strongest proof of the efficacy of vegetables of this class, is derived from the fact that the disease, when it occurred on land, uniformly disappeared during summer and autumn; and that it gradually became less frequent as the consumption of vegetables increased.

There seems to be no country naturally destitute of remedies for scurvy. The fruits of tropical and temperate climates are replaced in countries within the polar circle by herbs of almost equal efficacy. We are told that in Greenland, where scurvy was formerly very common, the natives employed sorrel and scurvy-grass together, and that by these herbs, which were put into broths, the most advanced cases were cured in a surprisingly short time.
[There follows a passage dealing with the antiscorbutic properties, or their absence, in grasses, herbs, cress, vegetables, pickles, kraut, turnips, fir tips, pines, spruces, potatoes, malt infusions, melasses [*sic*], fermented liquirs [*sic*], flour, oatmeal.—Ed.]

We have already given examples of the occurrence of scurvy, in the highest degree, in persons well supplied with fresh animal food; and instances are not wanting, which show that food of this kind is without much efficacy as a remedy. Dr. Lind tells us that in the *Salisbury*, during a Channel cruise in 1746, the scorbutic people, by the liberality of their commander, were daily supplied with fresh provisions, such as mutton broth and fowls, and even meat from his own table; yet, at the expiration of ten weeks, they brought into Plymouth eighty men, more or less afflicted with scurvy, out of a complement of 350. (*Lind*, p. 66; see also *Lind*, p. 137, and *Dis. of Seamen*, p. 462.)

The opinion that scurvy can be prevented, or cured by fresh meat, is however still held by persons, by whom it is of the utmost importance that correct notions on this subject should be entertained. We have known the most fatal effects result from the erroneous opinions of captains of merchant vessels on this point. During the course of the present year, the captain of a vessel trading to the Mauritius furnished his men, while they stayed at the island, with a plentiful supply of fresh beef, which, being imported from Madagascar, is procured at considerable expense; but neglected to provide them with vegetables or limes, which abound in the island, and are sold at a price scarcely worth naming. The consequence was that scurvy broke out soon after they set sail; and before the ship arrived in this country, one-half the men before the mast had died of it, and the rest were totally disabled.

Portable soup was much used by Captain Cook, and has been extensively employed by Sir Edward Parry, and other modern navigators. Its antiscorbutic properties must depend chiefly on the vegetables it contains.

The facts we have adduced seem to lead to the following general conclusions:

1. That antiscorbutic properties reside exclusively in substances of vegetable origin.

2. That these properties are possessed in very different degrees by different families of plants; and that vegetables and fruits, which are farinaceous, possess them in the lowest degree; while all those, which possess them in a very high degree, are succulent.

3. That the antiscorbutic virtue resides in the juices of the plant; that it is, in general, considerably impaired by the action of strong heat, and by the process of vinous fermentation; and that it varies, in some degree, with the state of maturity of the plant from which it is derived.

4. That these properties of vegetables are not destroyed, but in some instances seem even to be developed by the process of acetous fermentation.

We are ignorant of the essential element, common to the juices of antiscorbutic plants, on which the properties in question depend; but shall, probably, not be deemed too sanguine, if we anticipate that the study of organic chemistry, and the experiments of physiologists, will at no distant period throw some light on this subject. . . .

ALTERNATION OF GENERATIONS AND HOSTS

von SIEBOLD, Carl Theodor Ernst (German zoologist and parasitologist, 1808–1884). From *Ueber die band und blasen würmer nebst einer einleitung über entstehung der eingeweide würmer,* Leipzig, 1854; tr. by T. H. Huxley as *On tape and cystic worms* (with G. F. H. Küchenmeister, *On animal and vegetable parasites, etc.*), London (for the Sydenham Society), 1857.

Alternation of generations had been observed by Chamisso* and definitively described by Steenstrup.† The doctrine was extended to many species by Siebold‡ and

---

* A. Chamisso, *De animalibus quibusdem e classe vermium, etc.,* 1819.

† J. J. S. Streenstrup, *On the alternation of generations; or, the propagation and development of animals, etc.;* tr. by G. Burk, from German version (1842) of C. H. Lorenzen, London (for the Ray Society), 1845.

‡ In *Wagner's Handwörterbuch der physiologie,* Band ii, p. 645, 1844.

used by him (see below) to overthrow one of the last strongholds of spontaneous generation, namely, the apparently otherwise inexplicable origin of intestinal parasites.
Siebold contributed importantly to protozoology and to our understanding of the life
cycle of the honey bee. Siebold's discourse is livened by a thread of innocuous anthro-
pomorphism and closes with the pious hope that the citizenry may some day be as aware
of metamorphoses in worms as they are now of those in frogs and insects!

AUTHOR'S PREFACE.

Investigations into the natural history of the entozoa, continued for many
years, have taught me that it is impossible to obtain a complete view of the different stages of existence through which these parasites pass, if one's observations are restricted to but a few of the localities in which they are found. At
an early period of my researches, it became evident that the same entozoon, in
its young state, may have a very different habitation from that in which it is
found in its adult condition; for these animals undergo the most remarkable
metamorphoses, and their habits varying with their changes in form and age,
they are necessitated repeatedly to change their residence.

These peculiarities in the natural history of the entozoa, often most difficult
of investigation, have rendered the task of the helminthologist, in seeking to
obtain a just conception of their genera and species, a very difficult one. It
has only too frequently happened that the different stages of development of
the same species of entozoa, have been described as so many distinct species
or genera; and thus the systematic arrangement of the group has been built
upon a faulty foundation. Hence, again, a difficulty has arisen in the way of
attaining correct ideas with regard to the modes of propagation of the intestinal
worms, and this obstacle could only be removed by determining, in defiance
of the authority of the older helminthologists, to give up many genera and
species established upon what they supposed to be independent forms.

The investigation of the natural history of the intestinal worms, at the
same time, opened up a channel through which their mode of origin could
be traced; and indicated a way in which the attacks of those parasites which
are dangerous or troublesome to man and animals, could be prevented; an
object, in certain cases, of the highest importance, since the morbid changes
induced by many entozoa in the organs which they infest, are not always re-
movable.

For a long period, I have been at much pains to inquire into the origin of
the entozoa found in man and the domestic animals; and in the present essay
I lay before physicians, veterinarians, and breeders, a summary of the results
of the observations and experiments which I have made upon the production
and development of these creatures. My chief attention has been directed
to the destructive cystic-worms, and I believe that the conclusions at which I

have arrived are not merely a gain for science, but promise to be of useful practical application.

Münich: *March 30th*, 1854

Upon the Origin of Intestinal Worms.

Having been occupied for many years with inquiries into the natural history of the intestinal worms, a subject involved in much obscurity, I have gradually arrived at the decided conclusion that these parasites do not originate, as has been commonly believed, by "equivocal generation," from substances of a dissimilar nature. With the usual exaggeration and misuse of language, the doctrines of equivocal generation has been applied both to the infusoria and to the intestinal worms. It was difficult, at first sight, to account for the origin and reproduction of these animals, and, even upon closer investigation, they presented many phenomena which could not be recognized in the organization and vital manifestations of other, especially of the higher animals; but instead of seeking for the cause of these exceptional peculiarities, people, accommodating themselves to the usually accepted views as to the natural history of these lower creatures, set the matter straight in their minds, by supposing that the unusual phenomena occurred somehow or other in that way; thus allowing the imagination to indulge in fancies of the wildest description, and even in opposition to the most important laws of nature. It was in this manner that physicians and naturalists thought themselves justified in assuming, that the parasitic worms in the intestines of men and animals owed their origin to ill-digested nutriment, or that they were developed in the most widely different organs from corrupt juices. They took it for granted, that certain morbid processes in any organ were competent to give rise to parasites, assuming that the elementary constituents of an organ affected by disease, mechanically separated themselves from their natural connection, and not perishing, but transforming themselves into independent organisms, became parasites. Clothed in fine phrases, this idea was everywhere received with favour, and took such deep hold of the public mind, that it is now a matter of no small trouble to eradicate what has, with many, become an article of faith, and to substitute the laws of nature, drawn from experience, for the creation of their fancy. It was certainly more convenient and enticing to give free scope to one's thoughts, and to fill up the frequent gaps left in our knowledge of the origin and multiplication of the lower animals, with pure hypotheses, than as now, renouncing this faulty method of inquiry into nature, to attain, by troublesome researches and careful experiments, a secure insight into her hidden workings. . . .

For a long time the origin of the thread-worm, known as *Filaria Insec-*

torum, that lives in the cavity of the bodies of adult and larval insects could not be accounted for. Shut up within the abdominal cavity of caterpillars, grasshoppers, beetles, and other insects, these parasites were supposed to originate by equivocal generation, under the influence of wet weather or from decayed food. Helminthologists were obliged to content themselves with this explanation, since they were unable to find a better. Those who dissected these thread-worms and submitted them to a careful inspection, could not deny the probability of the view that they arose by equivocal generation, since it was clear that they contained no trace of sexual organs. But on directing my attention to these entozoa, I became aware of the fact that they were not true *Filariae* at all, but belonged to a peculiar family of thread-worms, embracing the genera *Gordius* and *Mermis.* Furthermore, I convinced myself that these parasites wander away when full grown, boring their way from within through any soft place in the body of their host, and creeping out through the opening. How many a butterfly-collector, keeping caterpillars for the breeding of fine specimens of butterflies, must have seen one or more yellowish white thread-worms winding their way out of them! These parasites do not emigrate because they are uneasy, or because the caterpillar is sickly, but from that same internal necessity which constrains the horse-fly to leave the stomach and intestine of the horse where he has been reared, or which moves the larva of the gad-fly to work its way out of the boils on the skin of oxen. The larvae of both these insects creep forth in order to become chrysalises and thence to proceed to their higher and sexual condition. This desire to emigrate is implanted in very many parasitic insect larvae, and has long been a well-known fact in entomology. Now I have demonstrated, that the perfect, full-grown, but sexless thread-worms of insects are, in like manner, moved by this desire to wander out of their previous homes in order to enter upon a new period in their lives which ends in the development of their sexual organs. It is true that in the boxes and other receptacles, in which one is generally accustomed to keep caterpillars, these creatures perish; they roll themselves together, and from the absence of the necessary moisture, they in a short time dry up. But their fate is very different when the infested insects remain under natural conditions; the thread-worms, as they leave the bodies of their hosts, then fall to the ground, and crawl away into the deeper and moister parts of the soil. Thread-worms found in the damp earth, in digging up garden-beds and cutting ditches in the fields, have often been brought to me, which presented no external distinctions from the thread-worms of insects externally. This suggested to me that the wandering thread-worms of insects might be instinctively necessitated to bury themselves in damp ground, and I therefore instituted a series of experiments with such entozoa (which I procured in numbers from the caterpillars of a moth, *Yponomenta evonymella*), by placing the newly emigrated worms in flower-pots filled with damp earth. To my delight, I soon perceived that these worms began to bore with their heads into the earth, and by degrees drew themselves entirely in. For many months (through the whole winter) I kept the earth in the flower-pots moderately moist, and on examining the worms from time to time I found, to my great astonishment, that the sexual apparatus became gradually developed in them, and that, after a time, eggs were formed and were eventually deposited by hundreds in the earth. Towards the conclusion of winter I could succeed in detecting the commencing development of the embryo in these eggs. By the end of spring they were fully formed, and many of them, having by this time left their shells, were to be seen creeping about the earth in the flower-pots, which I still carefully kept damp. I now conjectured that these young worms would be impelled by their instincts to pursue a parasitic existence and to seek out an animal to inhabit and grow to maturity in, and it seemed not improbable that the brood I had reared would, like their parents, thrive best in the caterpillar. In order, therefore, to induce my young brood to immigrate, I procured a number of very small caterpillars of *Yponomenta,* of half a line in length, which the first spring sunshine had just called into life. For the purpose of my experiment I filled a watch-glass with damp earth, taking it from amongst the flower-pots where the thread-worms had wintered, and of course satisfying myself that it contained a number of lively young of the *Mermis albicans.* Upon this I placed several of the young caterpillars of the *Yponomenta* in order that the worms might gratify their immigrative propensities. I must explicitly remark, that before experimenting with the caterpillars, I carefully examined each with the microscope, in order to ascertain whether it was not already inhabited by young thread-worms. From their softness and transparency, I could ascertain this point with certainty, without in the least injuring them. The event proved that this inspection was necessary, for out of twenty-five individuals which I at first selected, three contained a thread-worm embryo, which was excessively like those in the flower-pots. I published the results of these experiments a year or two back, in an essay upon the thread-worms of insects, from which I quote the following:

"From amongst those caterpillars which microscopic inspection clearly demonstrated to be free from thread-worms, thirteen were placed in a watch-glass filled with damp earth containing many lively *Mermis*-embryos. After eighteen hours I was able to discover *Mermis*-embryos in five of the caterpillars. On a second occasion, three-and-thirty of the caterpillars of *Yponomenta cognatella,* likewise carefully examined and found free from parasites, were in the same way placed in a watch-glass filled with damp earth containing *Mermis*-embryos. After four-and-twenty hours, fourteen contained *Mermis*-embryos. Six of these little caterpillars each contained two small worms, whilst in two others there were as many as three worms. I also employed other caterpillars (of three lines in length) of *Pontia Crataegi, Liparis chrysorrhoea,* and *Gastropacha Neustria,* which I took out of coccoons where they had passed the winter. They were, in like manner, placed in a watch-glass upon moist earth containing *Mermis*-embryos. On the next day, among fourteen caterpillars thus treated, I found ten infested with *Mermis*-embryos; five of these contained two worms each, and into one even three worms had wandered." It was clear that these young thread-worms had bored their way through the soft skin into the interior of the young caterpillars. . . .

It was in the year 1833, whilst fulfilling my duties as district medical officer (kreisphysicus) at Heilsberg, in East Prussia, that I had occasion to examine a large number of specimens of those *Trematoda* known to Helminthologists by the name of *Monostomum mutabile,* which were very commonly found in the geese of that locality, in the cavities which lie underneath the eyeballs. I convinced myself that these parasites, belonging to the order *Trematoda,* bring forth living young, which assume the form of *Infusoria,* and swim about in the water by means of the cilia which cover the whole surface of their bodies. After some time, I observed that these embryos apparently died, their bodies seeming to break up and gradually disappearing, but always leaving behind a sharply defined, mobile, cylindrical body, provided with two short, lateral processes. In all the embryos, without exception, this body was visible through their parietes while they still lived. To my great astonishment, upon further observation of these contractile remains of the *Monostomum* embryos, I discovered that they agreed precisely in form, structure, and movement with certain young *Cercaria*-sacs. Hence I ventured to conclude that the *Cercaria*-sacs proceed from *Trematoda.* At the same time, these observations seemed to indicate how it was possible for the inert, helpless *Cercaria*-sacs to make their way into snails and mussels. The *Monostomum mutabile* is known to reside in such cavities of the body of wading and swimming birds, as possess natural external apertures; when, therefore, the embryos of a *Monostomum mutabile* are born, they will issue without much difficulty from the animal infested by their parents, each carrying its *Cercaria*-sac within its body; and the habits of the infested animals are usually such that the embryos will at once pass into water, in which they can, by means of their cilia, swim swiftly about.

In this element, the infusory *Monostomum* embryos will, instinctively and immediately, seek out those animals that are fit to serve as a nidus for the further development of the *Cercaria*-sacs enclosed within them. After the *Cercaria*-sacs have thus passively entered, by the natural apertures, into their appropriate animals, their carriers, the ciliated embryos which have hitherto enclosed them, die off. As a sort of animated covering to the *Cercaria*-sacs, they have performed their office; and it is now left to the young that have just been released, to work themselves deeper into their new habitation by their own efforts, and to seek out those places which will afford them the necessary nourishment for further growth, and for the development of their brood of *Cercariae*.

I have not yet been able absolutely to witness this process of immigration of *Monostomum* embryos containing *Cercaria*-sacs, and as I have filled up the gaps in observation with my own ideas on the subject, what really occurs may be somewhat different; still, the immigration of the *Monostomum* embryo, which is the principal point, must take place, since the singular relations of the infusorial *Monostomum* embryos and the young *Cercaria*-sacs they contain, point distinctly to this conclusion.

Every one will understand, that the knowledge of even such a small fragment of the history of the development of the *Monostomum mutabile* as this, was of the utmost value, since it afforded the key to the long inexplicable mode of origin of the *Cercaria*-sacs. There now only remained the question as to what became of the *Cercariae*, and in what relation they stood to the *Trematoda*. It was an old idea that there was great similarity between the bodies of the *Cercariae* and certain *Trematoda*, viz., *Monostomata* and *Distomata*, and the force of the comparison was strengthened by the fact that the *Cercariae* cast off their tails after leaving the sacs, and thus become still less different from these *Trematoda*. Many *Distomata* whose bodies are encircled with spines at their anterior extremity, for example, *Distomum trigonocephalum,* *echinatum, uncinatum,* and *militare,* are so like certain *Cercariae,* that when the latter have thrown off their tails, any unprejudiced person would take them for the young of these *Distomata*. In fact, in their whole organization, the *Cercariae* are really no other than young *Trematoda*. The circumstance that one never finds sexual organs in the *Cercariae* is strongly corroborative of the notion that they are young *Trematoda* not yet sexually developed. Here again we have to do with parasites destined to emigrate and immigrate, that in some other situation they may arrive at sexual maturity. The course which the *Cercariae* take in their wanderings is, however, a much longer and more complicated one, than that followed by the sexless *Gordiacei*. These need only leave the insects they have hitherto infested and withdraw into damp ground, where fully grown as they are, and provided with the necessary store of fat in their bodies, they can quietly await the development of their sexual organs. On the other hand, the emigrating *Cercariae* are destined to enter vertebrate animals, since it is only in the intestinal canal of certain mammals, birds, reptiles, or fishes, that they can grow and mature their sexual organs.

Many of my readers may be unable to conceive how it is possible for *Cercariae* living in water, to enter into the intestines of such mammals and birds as live far away from water, or, at any rate, never come into proximity with the waters in which *Cercariae* live. I can, however, offer a solution of this apparent mystery, having surprised many *Cercariae* in the act of migrating. Before I say anything more about this, I must mention a peculiarity which is to be noticed in most of the *Cercariae* after they have left their sacs. This is their habit of encysting themselves, a process which is effected in the following manner. After a *Cercaria* has been for some time in the water, first creeping and then swimming about with manifest restlessness, it gathers itself up into a ball, and emits from its whole surface a mucous secretion which soon hardens, and since inside of this mucous mass the worm, coiled up into a little ball, turns round without stopping, invests it as it were in an egg-shell. During this process of encysting the *Cercaria* invariably casts off its tail, so that the capsule eventually encloses the body merely. For a long time I vainly wondered what could be the object of this process, and, never understood what its signification in *Cercarian* life was, until, in dissecting some insects, I met with a fact which suggested how I might gain the knowledge I sought for. In the larvae of a great number of various kinds of aquatic insects, of *Libellulidae*, *Ephemeridae*, *Perlidae*, *Phryganidae*, I found encysted *Cercariae*, which I again discovered in the same animals, after they had left the water, and had been transformed into winged insects. Not one of these encysted *Cercariae* lodged in an insect, was either full-grown or possessed sexual organs. I only observed one other slight step towards their further development; the sexual apparatus, viz., the testis, the germarium, and the copulatory organs, were already faintly indicated. As, however, perfectly full-grown and sexually developed *Trematoda* are never met with in insects, I decided, after the discovery of the encysted *Cercariae* in them, that they merely sought out the insects as a temporary resting place. Most of the sexually developed *Trematoda* are parasitic upon the higher *Vertebrata*, the *Cercariae* being, in fact, nothing else than young sexless *Trematoda*, whose instinct it is, to pass out from the inferior animals where they are produced, into the higher forms in which they attain the power of sexual reproduction. Should those *Cercariae* which are generated in aquatic molluscs, be able to attain their sexual maturity in the intestines of insectivorous birds or mammals alone, they can only reach the latter locality by entering the larvae of aquatic insects, and then becoming encysted in the manner already described. In this condition they remain, until the new animal in which they have established themselves, having undergone its metamorphosis, leaves the water and is swallowed by some insectivorous vertebrate.

In the act of digestion the body of the insect is destroyed, together with the capsule of the imprisoned *Cercaria*, which in this manner finds itself transplanted into those new circumstances which are alone fitted to permit of its further change into a sexual Trematode.

That this instinctive impulse of the *Cercariae* to encyst themselves after emigration, is accompanied by a desire to pass into insect larvae, I assured myself by ocular demonstration. I had procured a large number of specimens of *Cercaria armata* which had emigrated from the common *Lymnaeus stagnalis*, and put them into a watch-glass filled with water, in company with several live Neuropterous larvae (of the families of the *Ephemeridae* and *Perlidae*). I soon observed, with the microscope, that the *Cercariae*, which at first, flapping their tails, moved freely about in the water, at last betook themselves to the insect larvae, and crept restlessly about them. It was easy to see from their movements that the little worms had some object in view. The *Cercaria armata*, as is well known, is provided with a spine-like weapon, pointing forward from the centre of the animal's head. I could readily perceive, that these *Cercariae* which I was observing, frequently paused in their inspection of the insects, and inserted this weapon into their bodies as they crept over them. This probing experiment, for it was clearly nothing else, was repeated again and again, until the larva had discovered one of the soft places between the segments of the insect's body; this being reached, it never moved from the spot, but worked incessantly with its spine, until a way was bored through the soft place it had fastened on. Scarcely was the point of the spine fairly through, ere the supple worm inserted his thin anterior extremity into the wound, widened the opening a little, and by degrees drew in his whole body, which became wonderfully slender under the operation. The tail of the *Cercaria* was not drawn inside the insect, but remained hanging outside the puncture, being doubtlessly seized and nipped off, by the sudden closing of the wound when the body of the *Cercaria* had slipped through. Having selected very young and delicate Neuropterous larvae for my inquiry, the transparency of their bodies enabled me to continue to observe the tail-less *Cercariae* after their entrance; they forthwith lay still, drew themselves up into balls, and surrounded themselves with a cyst. During the process of encysting, the frontal spine fell off from the body of the *Cercaria*, and lay apart by its side, but enclosed within the cyst. This weapon, therefore, undergoes the same fate as the tail of these animals, each apparatus being cast aside after fulfilling its intended end.

The impulse to immigrate and become encysted is so strong in all the *Cercariae*, that their efforts appear to be occasionally over hasty, and perhaps lead them altogether astray.

I have found in *Aselli* and *Gammari*, encysted *Cercariae*, which in every way resembled those which had passed into insects. Now if these *Cercariae* can only attain their sexual maturity in the warm-blooded vertebrate animals, which devour insects, and therefore seek their food in the air or on land alone, the *Cercariae*, that had established themselves in the *Aselli* and *Gammari*, would wait in vain for the time to arrive when they should be transported into the air, since the animals in which they were domiciled would never quit the water. Again, many *Cercariae*, in their haste, become encysted incautiously at so early a period, that the purpose of the process is defeated.

I have already shown that the emigrated *Cercaria ephemera* attaches itself to water-plants, or any other objects in the water, by means of the cyst which it elaborates; other *Cercariae* even become encysted before they quit the body of the aquatic snail in which they were generated; whilst some, again, have even been found encysted within the *Cercaria*-sacs. Steenstrup takes this to be a normal phenomenon; I should only consider it such, provided that the encysted *Cercariae* in the snails are intended to attain their sexual maturity, in the intestines of fishes or of water-birds feeding on snails.

Although the various facts I have communicated can only be regarded as fragments of the natural history of certain *Trematoda,* they are yet capable of being connected into a whole, if the theory of the Alternation of Generations be extended to them. . . .

In this chapter I have expressed myself somewhat at large upon the wanderings and alternation of generations of the intestinal worms, in order that I may be fully understood in the ensuing ones, when I have occasion to refer to this generation by agamozooids. The history of the propagation of certain parasites, in the foregoing pages, may seem new and astonishing to many readers, and yet the alternation of generations is not more wonderful than metamorphosis. We have been so long acquainted with the way in which metamorphosis takes place in the higher and lower members of the animal kingdom, that we no longer wonder at the various transformations of the frog, nor gaze with surprise when a caterpillar becomes a chrysalis, and after a certain time flies off in the shape of a butterfly. The many to whom the metamorphosis of frogs and insects is a common appearance, forget that there was once a time when it was unknown, and when the multiplication of grubs and larvae was ascribed to equivocal generation, their true origin being unsuspected. It is to be hoped that a time will also arrive when the complicated alternation of generations will not be known to naturalists alone.

FOUNDATION OF CELLULAR PATHOLOGY

VIRCHOW, Rudolph Ludwig Karl (German cytologist and pathologist, 1821–1902). From *Die cellularpathologie in ihrer begründung auf physiologische und pathologische gewebelehre. Zwanzig vorlesungen gehalten während der monate februar, märz, und april 1858 im Pathologischen Institute zur Berlin,* Berlin, 1858; tr. (from the 2d ed.) by F. Chance as *Cellular pathology as based upon physiological and pathological histology, etc.,* New York, 1860.

The rise of the general cell theory was marked by a corresponding reform of many branches of biology, especially histology and embryology (see Kölliker), protozoology (see Siebold), and pathology. Cellular pathology arose gradually between the time of establishment of Virchow's *Archiv* (1847)* and the appearance of his definitive *Cellularpathologie* some ten years later. Virchow was politically active, supporting liberal

---

* See especially *Ueber die reform der pathologischen und therapeutischen anschauung durch die mikroskopischen untersuchungen,* in *Virchow's Archiv für Pathologische Anatomie und Physiologie und für Klinische Medecin,* vol. 1, p. 207, 1847.

causes, and was largely responsible for the reform of the medical system, of public health, and of medical education in Germany which became in these respects the model for the rest of Europe. The following passage, which is the main substance of the first of his famous twenty lectures, is the classical statement concerning the reconstitution of pathology as a cellular science.

The history of medicine teaches us, if we will only take a somewhat comprehensive survey of it, that at all times permanent advances have been marked by anatomical innovations, and that every more important epoch has been directly ushered in by a series of important discoveries concerning the structure of the body. So it was in those old times, when the observations of the Alexandrian school, based for the first time upon the anatomy of man, prepared the way for the system of Galen; so it was, too, in the Middle Ages, when Vesalius laid the foundations of anatomy, and therewith began the real reformation of medicine; so, lastly, was it at the commencement of this century, when Bichat developed the principles of general anatomy. What Schwann, however, has done for histology, has as yet been but in a very slight degree built up and developed for pathology, and it may be said that nothing has penetrated less deeply into the minds of all than the cell-theory in its intimate connection with pathology.

If we consider the extraordinary influence which Bichat in his time exercised upon the state of medical opinion, it is indeed astonishing that such a relatively long period should have elapsed since Schwann made his great discoveries, without the real importance of the new facts having been duly appreciated. This has certainly been essentially due to the great incompleteness of our knowledge with regard to the intimate structure of our tissues which has continued to exist until quite recently, and, as we are sorry to be obliged to confess, still even now prevails with regard to many points of histology to such a degree, that we scarcely know in favour of what view to decide.

Especial difficulty has been found in answering the question, from what parts of the body action really proceeds—what parts are active, what passive; and yet it is already quite possible to come to a definitive conclusion upon this point, even in the case of parts the structure of which is still disputed. The chief point in this application of histology to pathology is to obtain a recognition of the fact, that the cell is really the ultimate morphological element in which there is any manifestation of life, and that we must not transfer the seat of real action to any point beyond the cell. Before you, I shall have no particular reason to justify myself, if in this respect I make quite a special reservation in favour of life. In the course of these lectures you will be able to convince yourselves that it is almost impossible for any one to entertain more mechanical ideas in particular instances than I am wont to do, when called upon to interpret the individual processes of life. But I think that we must look upon this as certain, that, however much of the more delicate interchange of matter, which takes place within a cell, may not concern the material structure as a whole, yet the real action does proceed from the structure as such, and that the living element only maintains its activity as long as it really presents itself to us as an independent whole.

In this question it is of primary importance (and you will excuse my dwelling a little upon this point, as it is one which is still a matter of dispute) that we should determine what is really to be understood by the term cell. Quite at the beginning of the latest phase of histological development, great difficulties sprang up in crowds with regard to this matter. Schwann, as you no doubt recollect, following immediately in the footsteps of Schleiden, interpreted his observations according to botanical standards, so that all the doctrines of vegetable physiology were invoked, in a greater or less degree, to decide questions relating to the physiology of animal bodies. Vegetable cells, however, in the light in which they were at that time universally, and as they are even now also frequently regarded, are structures, whose identity with what we call animal cells cannot be admitted without reserve.

When we speak of ordinary vegetable cellular tissue, we generally understand thereby a tissue, which, in its most simple and regular form is, in a transverse section, seen to be composed of nothing but four- or six-sided, or, if somewhat looser in texture, of roundish or polygonal bodies, in which a tolerably thick, tough wall (*membrane*) is always to be distinguished. If now a single one of these bodies be isolated, a cavity is found, enclosed by this tough, angular, or round wall, in the interior of which very different substances, varying according to circumstances, may be deposited, *e.g.* fat, starch, pigment, albumen (*cell-contents*). But also, quite independently of these local varieties in the contents, we are enabled, by means of chemical investigation, to detect the presence of several different substances in the essential constituents of the cells.

The substance which forms the external membrane, and is known under the name of cellulose, is generally found to be destitute of nitrogen, and yields, on the addition of iodine and sulphuric acid, a peculiar, very characteristic, beautiful blue tint. Iodine alone produces no colour; sulphuric acid by itself chars. The contents of simple cells, on the other hand, do not turn blue; when the cell is quite a simple one, there appears, on the contrary, after the addition of iodine and sulphuric acid, a brownish or yellowish mass, isolated in the interior of the cell-cavity as a special body (*protoplasma*), around which can be recognised a special, plicated, frequently shrivelled membrane (*primordial utricle*). Even rough chemical analysis generally detects in the simplest cells, in addition to the non-nitrogenized (external) substance, a nitrogenized internal mass; and vegetable physiology seems, therefore, to have been justified in concluding, that what really constitutes a cell is the presence within a non-nitrogenized membrane of nitrogenized contents differing from it.

It had indeed already long been known, that other things besides existed in the interior of cells, and it was one of the most fruitful of discoveries when Robert Brown detected the nucleus in the vegetable cell. But this body was considered to have a more important share in the formation than in the maintenance of cells, because in very many vegetable cells the nucleus becomes extremely indistinct, and in many altogether disappears, whilst the form of the cell is preserved.

These observations were then applied to the consideration of animal tissues, the correspondence of which with those of vegetables Schwann endeavoured to demonstrate. The interpretation, which we have just mentioned as having been put upon the ordinary forms of vegetable cells, served as the starting-point. In this, however, as after-experience proved, an error was committed. Vegetable cells cannot, viewed in their entirety, be compared with all animal cells. In animal cells, we find no such distinctions between nitrogenized and non-nitrogenized layers; in all the essential constituents of the cells nitrogenized matters are met with. But there are undoubtedly certain forms in the animal body which immediately recall these forms of vegetable cells, and among them there are none so characteristic as the cells of cartilage, which is, in all its features, extremely different from the other tissues of the animal body, and which, especially on account of its non-vascularity, occupies quite a peculiar position. Cartilage in every respect stands in the closest relation to vegetable tissue. In a well-developed cartilage-cell we can distinguish a relatively thick external layer, within which upon very close inspection, a delicate membrane, contents, and a nucleus are also to be found. Here, therefore, we have a structure which entirely corresponds with a vegetable cell.

It has, however, been customary with authors, when describing cartilage, to call the whole of the structure of which I have just given you a sketch a cartilage-corpuscle, and in consequence of this having been viewed as analogous to the cells in other parts of animals, difficulties have arisen by which the knowledge of the true state of the case has been exceedingly obscured. A cartilage-corpuscle, namely, is not, as a whole, a cell, but the external layer, the *capsule,* is the product of a later development (secretion, excretion). In young cartilage it is very thin, whilst the cell also is generally smaller. If we trace the development still farther back, we find in cartilage, also, nothing but simple cells, identical in structure with those which are seen in other animal tissues, and not yet possessing that external secreted layer.

You see from this, gentlemen, that the comparison between animal and vegetable cells, which we certainly cannot avoid making, is in general inadmissible, because in most animal tissues no formed elements are found which can be considered as the full equivalents of vegetable cells in the old signification of the word; and because in particular, the cellulose membrane of vegetable cells does not correspond to the membrane of animal ones, and between this, as containing nitrogen, and the former, as destitute of it, no typical distinction is presented. On the contrary, in both cases we meet with a body essentially of a nitrogenous nature, and, on the whole, similar in composition. The so-called membrane of the vegetable cell is only met with in a few animal tissues, as for example, in cartilage; the ordinary membrane of the animal cell corresponds, as I showed as far back as 1847, to the primordial utricle of the vegetable cell. It is only when we adhere to this view of the matter, when we separate from the cell all that has been added to it by an after-development, that we obtain a simple, homogeneous, extremely monotonous structure, recurring with extraordinary constancy in living organisms. But just this very constancy forms the best criterion of our having before us in this structure one of those really elementary bodies, to be built up of which is eminently characteristic of every living thing—without the pre-existence of which no living forms arise, and to which the continuance and the maintenance of life is intimately attached. Only since our idea of a cell has assumed this severe form —and I am somewhat proud of having always, in spite of the reproach of pedantry, firmly adhered to it—only since that time can it be said that a simple form has been obtained which we can everywhere again expect to find, and which, though different in size and external shape, is yet always identical in its essential constituents.

In such a simple cell we can distinguish dissimilar constituents, and it is important that we should accurately define their nature also.

In the first place, we expect to find a *nucleus* within the cell; and with regard to this nucleus, which has usually a round or oval form, we know that, particularly in the case of young cells, it offers greater resistance to the action of chemical agents than do the external parts of the cell, and that, in spite of the greatest variations in the external form of the cell, it generally maintains its form. The nucleus is accordingly, in cells of all shapes, that part which is the most constantly found unchanged. There are indeed isolated cases, which lie scattered throughout the whole series of facts in comparative anatomy and pathology, in which the nucleus also has a stellate or angular appearance; but these are extremely rare exceptions, and dependent upon peculiar changes which the elements has undergone. Generally, it may be said that, as long as the life of the cell has not been brought to a close, as long as cells behave as elements still endowed with vital power, the nucleus maintains a very nearly constant form.

The nucleus, in its turn, in completely developed cells, very constantly encloses another structure within itself—the so-called *nucleolus*. With regard to the question of vital form, it cannot be said of the nucleolus that it appears to be an absolute requisite; and, in a considerable number of young cells, it has as yet escaped detection. On the other hand, we regularly meet with it in fully developed, older forms; and it, therefore, seems to mark a higher degree of development in the cell. According to the view which was put forward in the first instance by Schleiden, and accepted by Schwann, the connection between the three coexistent cell-constituents was long thought to be on this wise: that the nucleolus was the first to shew itself in the development of tissues, by separating out of a formative fluid (*blastema, cytoblastema*), that it quickly attained a certain size, that then fine granules were precipitated out of the blastema and settled around it, and that about these there condensed a membrane. That in this way a nucleus was completed, about which new matter gradually gathered, and in due time produced a little membrane (the celebrated watchglass form). This description of the first development of cells out of free blastema, according to which the nucleus was regarded as preceding the formation of the cell, and playing the part of a real cell-former (*cytoblast*), is the one which is usually concisely designated by the name of the *cell-theory* (more accurately, theory of *free* cell-formation),—a theory of development which has now been almost entirely abandoned, and in support of the correctness of which not one single fact can with certainty be adduced. With respect to the nucleolus, all that we can for the present regard as certain, is, that where we have to deal with large and fully developed cells, we almost constantly see a nucleolus in them; but that, on the contrary, in the case of many young cells it is wanting.

You will hereafter be made acquainted with a series of facts in the history of pathological and physiological development, which render it in a high degree probable that the nucleus plays an extremely important part within the cell—a part, I will here at once remark, less connected with the function and specific office of the cell, than with its maintenance and multiplication as a living part. The specific (in a narrower sense, animal) function is most distinctly manifested in muscles, nerves, and gland-cells; the peculiar actions of which—contraction, sensation, and secretion—appear to be connected in no direct manner with the nuclei. But that, whilst fulfilling all its functions, the element remains an element, that it is not annihilated nor destroyed by its continual activity—this seems essentially to depend upon the action of the nucleus. All those cellular formations which lose their nucleus, have a more transitory existence; they perish, they disappear, they die away or break up. A human blood corpuscle, for example, is a cell without a nucleus; it possesses an external membrane and red contents; but herewith the tale of its constituents, so far as we can make them out, is told, and whatever had been recounted concerning a nucleus in blood-cells, has had its foundation in delusive appearances, which certainly very easily can be, and frequently are, occasioned by the production of little irregularities upon the surface. We should not be able to say, therefore, that blood-corpuscles were cells, if we did not know that there is a certain period during which human blood-corpuscles also have nuclei; the period, namely, embraced by the first months of intra-uterine life. Then circulate also in the human body nucleated blood-cells, like those which we see in frogs, birds, and fish throughout the whole of their lives. In mammalia, however, this is restricted to a certain period of their development, so that at a later stage the red blood-cells no longer exhibit all the characteristics of a cell, but have lost an important constituent in their composition. But we are also all agreed upon this point, that the blood is one of those changeable constituents of the body, whose cellular elements possess no durability, and with regard to which everybody assumes that they perish, and are replaced by new ones, which in their turn are doomed to annihilation, and everywhere (like the uppermost cells in the cuticle, in which we also can discover no nuclei, as soon as they begin to desquamate) have already reached a stage in their development, when they no longer require that durability in their more intimate composition for which we must regard the nucleus as the guarantee.

On the other hand, notwithstanding the manifold investigations to which the tissues are at present subjected, we are acquainted with no part which grows or multiplies, either in a physiological or pathological manner, in which nucleated elements cannot invariably be demonstrated as the starting-points of the change, and in which the first decisive alterations which display themselves, do not involve the nucleus itself, so that we often can determine from its condition what would possibly have become of the elements.

You see from this description that, at least, two different things are of necessity required for the composition of a cellular element; the membrane, whether round, jagged or stellate, and the nucleus, which from the outset differs in chemical constitution from the membrane. Herewith, however, we are far from having enumerated all the essential constituents of the cell, for, in addition to the nucleus, it is filled with a relatively greater or less quantity of *contents*, as is likewise commonly, it seems, the nucleus itself, the contents of which are also wont to differ from those of the cell. Within the cell, for example, we see pigment, without the nucleus containing any. Within a smooth muscular fibre-cell, the contractile substance is deposited, which appears to be the seat of the contractile force of muscle; the nucleus, however, remains a nucleus. The cell may develop itself into a nerve-fibre, but the nucleus remains, lying on the outside of the medullary substance, a constant constituent. Hence it follows, that the special peculiarities which individual cells exhibit in particular places, under particular circumstances, are in general dependent upon the varying properties of the cell-contents, and that it is not the constituents which we have hitherto considered (membrane and nucleus), but the contents (or else the masses of matter deposited without the cell, intercellular), which give rise to the functional (physiological) differences of tissues. For us it is essential to know that in the most various tissues these constituents, which in some measure, represent the cell in its abstract form, the nucleus and membrane, recur with great constancy, and that by their combination a simple element is obtained, which, throughout the whole series of living vegetable and animal forms, however different they may be externally, however much their internal composition may be subjected to change, presents us with a structure of quite a peculiar conformation, as a definite basis for all the phenomena of life.

According to my ideas, this is the only possible starting-point for all biological doctrines. If a definite correspondence in elementary form pervades the whole series of all living things, and if in this series something else which might be placed in the stead of the cell be in vain sought for, then must every more highly developed organism, whether vegetable or animal, necessarily, above all, be regarded as a progressive total, made up of larger or smaller number of similar or dissimilar cells. Just as a tree constitutes a mass arranged in a definite manner, in which, in every single part, in the leaves as in the root, in the trunk as in the blossom, cells are discovered to be the ultimate elements, so is it also with the forms of animal life. *Every animal presents itself as a sum of vital unities*, every one of which manifests all the characteristics of life. The characteristics and unity of life cannot be limited to any one particular spot in a highly developed organism (for example, to the brain of man), but are to be found only in the definite, constantly recurring structure, which every individual element displays. Hence it follows that the structural composition of a body of considerable size, a so-called individual, always represents a kind of social arrangement of parts, an arrangement of a social kind, in which a number of individual existences are mutually dependent, but in such a way, that every element has its own special action, and, even though it derive its stimulus to activity from other parts, yet alone effects the actual performance of its duties.

I have therefore considered it necessary, and I believe you will derive benefit from the conception, to portion out the body into *cell-territories* (Zellenterritorien). I say territories, because we find in the organization of animals a peculiarity which in vegetables is scarcely at all to be witnessed, namely, the development of large masses of so-called *intercellular substance*. Whilst vegetable cells are usually in immediate contact with one another by their external secreted layers, although in such a manner that the old boundaries can still always be distinguished, we find in animal tissues that this species of arrangement is the more rare one. In the often very abundant mass of matter which lies between the cells (*intermediate, intercellular substance*), we are seldom able to perceive at a glance, how far a given part of it belongs to one or another cell; it presents the aspect of a homogeneous intermediate substance.

According to Schwann, the intercellular substance was the cytoblastema, destined for the development of new cells. This I do not consider to be correct, but, on the contrary, I have, by means of a series of pathological observations, arrived at the conclusion that the intercellular substance is dependent in a certain definite manner upon the cells, and that it is necessary to draw boundaries in it also, so that certain districts belong to one cell, and certain others to another. You will see how sharply these boundaries are defined by pathological processes, and how direct evidence is afforded, that any given district of intercellular substance is ruled over by the cell, which lies in the middle of it and exercises influence upon the neighbouring parts.

It must now be evident to you, I think, what I understand by the territories of cells. But there are simple tissues which are composed entirely of cells, cell lying close to cell. In these there can be no difficulty with regard to the boundaries of the individual cells, yet it is necessary that I should call your attention to the fact that, in this case, too, every individual cell may run its own peculiar course, may undergo its own peculiar changes, without the fate of the cell lying next it being necessarily linked with its own. In other tissues, on the contrary, in which we find intermediate substance, every cell, in addition to its own contents, has the superintendence of a certain quantity of matter external to it, and this shares in its changes, nay, is frequently affected even earlier than the interior of the cell, which is rendered more secure by its situation than the external intercellular matter. Finally, there is a third series of tissues, in which the elements are more intimately connected with one another. A stellate cell, for example, may anastomose with a similar one, and in this way a reticular arrangement may be produced, similar to that which we see in capillary vessels and other analogous structures. In this case it might be supposed that the whole series was ruled by something which lay who knows how far off; but upon more accurate investigation, it turns out that even in this chainwork of cells a certain independence of the individual members prevails, and that this independence evinces itself by single cells undergoing in consequence of certain external or internal influences, certain changes confined to their own limits, and not necessarily participated in by the cells immediately adjoining.

That which I have now laid before you will be sufficient to show you in what way I consider it necessary to trace pathological facts to their origin in known histological elements; why, for example, I am not satisfied with talking about an action of the vessels, or an action of the nerves, but why I consider it necessary to bestow attention upon the great number of minute parts which really constitute the chief mass of the substance of the body, as well as upon the vessels and nerves. It is not enough that, as has for a long time been the case, the muscles should be singled out as being the only active elements; within the great remainder, which is generally regarded as an *inert mass,* there is in addition an enormous number of active parts to be met with.

Amid the development which medicine has undergone up to the present time, we find the dispute between the humoral and solidistic schools of olden times still maintained. The humoral schools have generally had the greatest success, because they have offered the most convenient explanation, and, in fact, the most plausible interpretation of morbid processes. We may say that nearly all successful, practical, and noted hospital physicians have had more or less humoro-pathological tendencies; aye, and these have become so popular, that it is extremely difficult for any physician to free himself from them. The solido-pathological views have been rather the hobby of speculative inquirers, and have had their origin not so much in the immediate requirements of pathology, as in physiological and philosophical, and even in religious speculations. They have been forced to do violence to facts, both in anatomy and physiology, and have therefore never become very widely diffused. According to my notions, the basis of both doctrines is an incomplete one; I do not say a false one, because it is really only false in its exclusiveness; it must be reduced within certain limits, and we must remember that, besides vessels and blood, besides nerves and nervous centres, other things exist, which are not a mere theatre (Substrat) for the action of the nerves and blood, upon which these play their pranks.

Now, if it be demanded of medical men that they give their earnest consideration to these things also; if, on the other hand, it be required that, even among those who maintain the humoral and neuro-pathological doctrines, attention at last be paid to the fact, that the blood is composed of many single, independent parts, and that the nervous system is made up of many active individual constituents—this is, indeed, a requirement which at the first glance certainly offers several difficulties. But if you will call to mind that for years, not only in lectures, but also at the bedside, the activity of the capillaries was talked about—an activity which no one has ever seen, and which has only been assumed to exist in compliance with certain theories—you will not find it unreasonable, that things which are really to be seen, nay are, not unfrequently, after practice, accessible even to the unaided eye, should likewise be admitted into the sphere of medical knowledge and thought. Nerves have not only been talked about where they had never been demonstrated; their existence has been simply assumed, even in parts in which, after the most careful investigations, no trace of them could be discovered, and activity has been attributed to them in parts where they absolutely do not penetrate. It is therefore certainly not unreasonable to demand, that the greater part of the body be no longer entirely ignored; and if no longer ignored, that we no longer content ourselves with merely regarding the nerves as so many wholes, as a simple, indivisible apparatus, or the blood as a merely fluid material, but that we also recognise the presence within the blood and within the nervous system of the enormous mass of minute centres of action.

DETERMINISM AND MEDICINE

BERNARD, Claude (French physician and physiologist, 1813–1878). From *Introduction à l'étude de la médecine expérimentale*, Paris, 1865; tr. by H. C. Greene as *An introduction to the study of experimental medicine*, New York, 1927; by permission of The Macmillan Company, publisher.

It has been the occasional practice of medical men and biologists to postulate in the living organism phenomena allegedly not subject to the law of cause and hence not strictly predictable. Here the great physiologist condemns this tradition and lays down a deterministic epistemology for medical physiologists.

The Necessary Conditions of Natural Phenomena Are Absolutely Determined in Living Bodies as Well as in Organic Bodies

We must acknowledge as an experimental axiom that in living beings as well as in inorganic bodies the necessary conditions of every phenomenon are absolutely determined. That is to say, in other terms, that when once the conditions of a phenomenon are known and fulfilled, the phenomenon must always and necessarily be reproduced as the will of the experimenter. Negation of this proposition would be nothing less than negation of science itself. Indeed, as science is simply the determinate and the determinable, we must perforce accept as an axiom that, in identical conditions, all phenomena are identical and that, as soon as conditions are no longer the same, the phenomena cease to be identical. This principle is absolute in the phenomena of inorganic bodies as well as in those of living beings, and the influence of life, whatever view of it we take, can nowise alter it. As we have said, what we call vital force is a first cause analogous to all other first causes, in this sense, that it is utterly unknown. It matters little whether or not we admit that this force differs essentially from the forces presiding over manifestations of the phenomena of inorganic bodies, the vital phenomena which it governs must still be determinable; for the force would otherwise be blind and lawless, and that is impossible. The conclusion is that the phenomena of life have their special law because there is rigorous determinism in the various circumstances constituting conditions necessary to their existence or to their manifestations; and that is the same thing. Now in the phenomena of living bodies as in those of inorganic bodies, it is only through experimentation, as I have already often repeated, that we can attain knowledge of the conditions which govern these phenomena and so enable us to master them.

Everything so far said may seem elementary to men cultivating the physico-chemical sciences. But among naturalists and especially among physicians, we find men who, in the name of what they call vitalism, express most erroneous ideas on the subject which concerns us. They believe that study of the phenomena of living matter can have no relation to study of the phenomena of inorganic matter. They look on life as a mysterious supernatural influence which acts arbitrarily by freeing itself wholly from determinism, and they brand as materialists all who attempt to reconcile vital phenomena with definite organic and physico-chemical conditions. These false ideas are not easy to uproot when once established in the mind; only the progress of science can dispel them. But vitalistic ideas, taken in the sense which we have just indicated, are just a kind of medical superstition,—a belief in the supernatural. Now, in medicine, belief in occult causes, whether it is called vitalism or is otherwise named, encourages ignorance and gives birth to a sort of unintentional quackery; that is to say, the belief in an inborn, indefinable science. Confidence in absolute determinism in the phenomena of life leads, on the contrary, to real science, and gives the modesty which comes from the consciousness of our little learning and the difficulty of science. This feeling incites us, in turn, to work toward knowledge; and to this feeling alone, science in the end owes all its progress.

I should agree with the vitalists if they would simply recognize that living beings exhibit phenomena peculiar to themselves and unknown in inorganic nature. I admit, indeed, that manifestations of life cannot be wholly elucidated by the physico-chemical phenomena known in inorganic nature. I shall later explain my view of the part played in biology by physico-chemical sciences; I will here simply say that if vital phenomena differ from those of inorganic bodies in complexity and appearance, this difference obtains only by virtue of determined or determinable conditions proper to themselves. So if the sciences of life must differ from all others in explanation and in special laws, they are not set apart by scientific method. Biology must borrow the experimental method of physico-chemical sciences, but keep its special phenomena and its own laws.

In living bodies, as in inorganic bodies, laws are immutable, and the phenomena governed by these laws are bound to the conditions on which they exist, by a necessary and absolute determinism. I use the word determinism here as more appropriate than the word fatalism, which sometimes serves to express the same idea. Determinism in the conditions of vital phenomena should be one of the axioms of experimenting physicians. If they are thoroughly imbued with the truth of this principle, they will exclude all supernatural intervention from their explanations; they will have unshaken faith in the idea that fixed laws govern biological science; and at the same time they will have a reliable criterion for judging the often variable and contradictory appearance of vital phenomena. Indeed, starting with the principle that immutable laws exist, experimenters will be convinced that phenomena can never be mutually contradictory, if they are observed in the same conditions; and if they show variations, they will know that this is necessarily so because of the intervention or interference of other conditions which alter or mask phenomena. There will be occasion thenceforth to try to learn the conditions of these variations, for there can be no effect without a cause. Determinism thus becomes the foundation of all scientific progress and criticism. If we find disconcerting or even contradictory results in performing an experiment, we must never acknowledge exceptions or contradictions as real. That would be unscientific. We must simply and necessarily decide that conditions in the phenomena are different, whether or not we can explain them at the time.

I assert that the word exception is unscientific; and as soon as laws are known, no exception indeed can exist, and this expression, like so many others, merely enables us to speak of things whose causation we do not know. Every day we hear physicians use the words: ordinarily, more often, generally, or else express themselves numerically by saying, for instance: nine times out of ten, things happen in this way. I have heard old practitioners say that the words "always" and "never" should be crossed out of medicine. I condemn neither these restrictions nor the use of these locutions if they are used as empirical approximations about the appearances of phenomena when we are still more or less ignorant of the exact conditions in which they exist. But certain physicians seem to reason as if exceptions were necessary; they seem to believe that a vital force exists which can arbitrarily prevent things from always happening alike; so that exceptions would result directly from the action of mysterious vital force. Now this cannot be the case; what we now call an exception is a phenomenon, one or more of whose conditions are unknown; if the conditions of the phenomena of which we speak were known and determined, there would be no further exceptions, medicine would be as free from them as is any other science. For instance, we might formerly say that sometimes the itch was cured and sometimes not; but now that we attack the cause of this disease, we cure it always. Formerly it might be said that a lesion of the nerves brought on paralysis, now of feeling, and again of motion; but now we know that cutting the anterior spinal nerve paralyzes motion only. Motor paralysis occurs consistently and always, because its condition has been accurately determined by experimenters.

The certainty with which phenomena are determined should also be, as we have said, the foundation of experimental criticism, whether applied to one's self or to others. A phenomenon, indeed, always appears in the same way if conditions are similar; the phenomenon never fails if the conditions are present, just as it does fail to appear if the conditions are absent. Thus an experimenter who has made an experiment, in conditions which he believes were determined, may happen not to get the same results in a new series of investigations as in his first observation; in repeating the experiment, with fresh precautions, it may happen again that, instead of his first result, he may encounter a wholly different one. In such a situation, what is to be done? Should we acknowledge that the facts are indeterminable? Certainly not, since that cannot be. We must simply acknowledge that experimental conditions, which we believe to be known, are not known. We must more closely study, search out and define the experimental conditions, for the facts cannot be contradictory one to another; they can only be indeterminate. Facts never exclude one another, they are simply explained by differences in the conditions in which they are born. So an experimenter can never deny a fact that he has seen and observed, merely because he cannot rediscover it.

INSECT TRANSMISSION OF DISEASE

MANSON, Patrick (English pathologist, 1844–1922). *On the development of Filaria sanguinis hominis, and on the mosquito considered as a nurse*, in *Journal of the Linnaean Society of London—Zoology*; vol. 14, p. 304, 1878; by permission of the Secretary of the Linnaean Society.

For the significance of this paper in the history of parasitology, see p. 534. Manson proved the "necessary intermediacy of a blood sucking insect"[*] in filariasis. He carried on, simultaneously, extensive practice, research (on surgery in filarial tumor, on the guinea worm, etc.), and organizational work (*e.g.*, he established and administered a Chinese medical school in Hong Kong), and suggested the mosquito as the probable vector of malaria.[†]

Development cannot progress far in the Host containing the Parent Worm. Fortunately it is an almost universal law, in the history of the more dangerous kinds of Entozoa, that the egg or embryo must escape from the host inhabited by the parent worm before much progress can be made in development. Were it possible for animals so prolific as *Filaria immitis* of the dog, or *Filaria sanguinis* of man, to be born and matured and to reproduce their kind again in an individual host, the latter would certainly be overwhelmed by the first swarm of embryos escaping into the blood, as soon as they had made any progress in growth. If, for example, the brood of embryo *Filariae*, at any one time free in the blood of a dog moderately well charged with them, were to begin growing before they had each attained a hundredth part of the size of the mature *Filaria*, their aggregate volume would occupy a bulk many times greater than the dog itself. I have calculated that in the blood of certain dogs and men there exist at any given moment more than two millions of embryos. Now the individuals of such a swarm could never attain any thing approaching the size of the mature worm without certainly involving the death of the host. The death of the host would imply the death of the parasite before a second generation of *Filariae* could be born, and this, of course,

---

[*] H. A. Bayliss, *Manson's part in helminthological discovery*, in *Journal of Tropical Medicine and Hygiene*, vol. 25, p. 164, 1922.

[†] Biography in *Proceedings of the Royal Society of London*, vol. 94, p. xliii, 1923.

entails the extermination of the species; for in such an arrangement reproduction would be equivalent to the death of both parent and offspring, an anomaly impossible in nature.

The Embryo must escape from the original Host. It follows therefore that the embryo, in order to continue its development and keep its species from extermination, must escape from the first host in some way. After accomplishing this it either lives an independent existence for a time, during which it is provided with organs for growth not possessed by it hitherto; or it is swallowed by another animal which treats it as a nursling[1] for such time as is necessary to fit it with an alimentary system. The former arrangement obtains in the *Filariae* inhabiting the intestinal canal, the *Lumbricus* and thread-worm; the latter is followed by the several species of tapeworm, and also by other kinds of Entozoa.

I find that in cases where embryo *Filariae* are not in great abundance in the blood, we may infer that there are only one or two parent worms; they often disappear completely for a time, to reappear after the lapse of a few days or weeks. From this circumstance I infer—1st, that reproduction is of an intermitting and not of a continuous character; and, 2ndly, that the embryos, after a certain time, are either disintegrated in the blood or are voided in the excretions. The latter does occur, I know from personal investigation, in the urine; and we have Dr. Lewis's testimony that he has found the animals in the tears. In this way they may have an opportunity of continuing development either free (as in the case of the *Lumbricus*) in the media into which the excretions are voided, or in the body of another animal which has intentionally or accidentally fed on these (as in the case of the tapeworms). Man, in his turn, may then swallow this hypothetic animal or other thing containing the embryo suitably perfected, and so complete the circle. This is the history of many Entozoa; but I have evidence to adduce that, if it be one way in which *F. sanguinis hominis* is nursed, it is not the only way, and therefore probably not the way at all.

The Mosquito found to be the Nurse. It occurred to me that, as the first step in the history of the haematozoon was in the blood, the next might happen in an animal who fed on that fluid. To test this idea I procured mosquitos that had fed on the patient Hinlo's blood (Case No. 46, published in 'Med. Times & Gaz.' for March), and, examining the expressed contents of their abdomens from day to day with the microscope, I found that my idea was correct, and that the haematozoon which entered the mosquito as a simple structureless animal, left it, after passing through a series of highly interesting metamorphoses, much increased in size, possessing an alimentary canal, and being otherwise suited for an independent existence.

---

[1] [Throughout this memoir Dr. Manson employs the term "nurse" in the same sense as that in which helminthologists use the term "intermediate host."—T. S. COBBOLD.]

History of the Mosquito after feeding on Human blood. I may mention that my observations have been made exclusively on the females of one species of mosquito. I have never, in many hundreds of specimens, met with a male insect charged with blood. This is explained by the arrangement of the appendages and proboscis of the male mosquito, which prevents it from penetrating the skin. As the male is provided with a complete alimentary apparatus, it is presumed that he feeds on the juices and exudations of plants and fruits. There are two species of mosquito found during the summer here: one quite a large insect about half an inch long, with a black thorax and black-and-white banded abdomen; the other about half that size and of a dingy brown colour. The former is rare comparatively; the latter is very common, and is the insect my remarks apply to. After a mosquito has filled itself with blood (which it can do, if not disturbed, in about two minutes), it is evidently much embarrassed by the weight of its distended abdomen, so that it no longer can wheel about in the air. It accordingly attaches itself to some surface, if possible near stagnant water, where it remains in a comparatively torpid condition, digesting the blood, excreting yellow gamboge-looking faeces, and maturing its ova. In the course of from three to five days these processes are completed, and the insect now betakes itself to the water, where the eggs are deposited, and on the surface of which they float in a dark-brown mass, looking like a flake of soot. The eggs do not take long to hatch (they are beautifully shaped objects, like an Etruscan vase); and the embryo emerges by forcing open a sort of lid placed at the broad end of the shell. The larvae now escape into the water, where they swim about and feed, and become the "jumpers" we are familiar with, found in every stagnant pool.

If the contents of the abdomen are examined before the mosquito has fed, or after the food has been absorbed, the following parts can easily be distinguished:—two ovisacs containing from sixty to a hundred ova, two large glandular masses (intestine and oesophagus), and a very delicate transparent fibrous bag, the stomach. If the blood contained in the dilated stomach is examined soon after ingestion, the blood-corpuscles are seen quite distinct in outline, and behaving very much as when drawn in the ordinary way; but changes rapidly occur. First, the corpuscles lose their distinctness in outline, then crystals of haematin appear; corpuscles and crystals give place to large oil-globules, and the mass is deprived of its fluidity, and before the eggs are deposited all colouring-matter disappears; the white material is absorbed or expelled, and by the time the eggs are deposited the stomach is quite empty but for the embryo *Filariae* it may contain.

How to procure Mosquitos containing embryo Filariae. It may be useful to those who wish to repeat and test my observations to know the plan I found most successful in procuring *Filaria*-bearing mosquitos, and how their bodies were afterwards treated for microscopic observation. Such details may appear frivolous and unimportant; but by following them the observer will be spared disappointment, and economize his time and patience.

I persuaded a Chinaman, in whose blood I had already ascertained that *Filariae* abounded, to sleep in what is known as a mosquito-house, in a room where mosquitos were plentiful. After he had gone to bed a light was placed beside him, and the door of the mosquito-house kept open for half an hour. In this way many mosquitos entered the "house"; the light was then put out, and the door closed. Next morning the walls of the "house" were covered with an abundant supply of insects with abdomens thoroughly distended. They were then caught below a wineglass, paralyzed by means of a whiff of to-bacco-smoke, and transferred to small phials, into some of which a little water had been poured. A cover providing for ventilation was then placed over the mouth of the phial. The effect of the tobacco-smoke, if it has not been ap-plied too long, is very evanescent, and seems to have no prejudicial influence on the posture of the mosquito. From the phials they may be removed from time to time, as required, by again paralyzing with tobacco and seizing them by the thorax with a fine pincers. The abdomen is then torn off, placed on a glass slide, and a small cylinder, such as a thin penholder, rolled over it from the anus towards the severed thoracic attachment. In this way the contents are safely and efficiently expressed, and observation is not interfered with by the almost opaque integument. If the contents are white and dry a little water should be added and mixed carefully with the mass, so as to allow of the easy separation of the two large ovisacs. These can be removed in this way by the needle, and transferred to another slide for separate examination. A thin covering-glass should be placed over the residue, which will be found to con-tain the *Filariae* either within the walls of the stomach, or, if these have been ruptured by too rough manipulation, floating in the surrounding water.

Large proportion of Filariae *ingested by the Mosquito.* The blood in the stomach of a mosquito that has fed on a *Filaria*-infested man usually contains a much larger proportion of *Filariae* than does an equal quantity of blood ob-tained from the same man in the usual way by pricking the finger. Thus six small slides, equivalent to about one drop of blood from the man on whom most of my observations were made, would contain from ten to thirty Haema-tozoa; whereas the blood drawn by a single mosquito, about as much as would fill one slide only, contained from twenty to thirty as a rule, and sometimes many more. One slide, in which I had the curiosity to count them, had up-wards of a hundred and twenty specimens. From this it would appear that the mosquito has the faculty of selecting the embryo *Filariae;* and in this strange circumstance we have an additional reason for concluding that this insect is the natural nurse of the parasite.

All Embryos do not attain maturity. By far the greater number die and are disintegrated, or are expelled in the faeces undeveloped. At the end of the third, fourth, or fifth day, when the stomach is quite empty as far as food is concerned, and an embryo could not easily be overlooked, only from two to six are found in the same or slightly different stages of the metamorphosis, which I will now attempt to describe.

The Metamorphosis of the Embryo. The embyro for a short time after entering the stomach of the mosquito retains all the appearances and habits which characterized it when in the human body; that is, it is a long snake-like animal, having a perfectly transparent structureless body enclosed in a delicate and, for the most part, closely applied tube, within which it shortens and extends itself, giving rise, from the collapse of the tube when the body is re-tracted at either end, to the appearance of a lash at the head and tail. In a very few hours changes commence. The tube first separates from the body by an appreciable interval, giving the appearance of a distinct double outline, and the body itself becomes covered with a delicate but distinct and closely set transverse striation. Oral movements are now very evident, not that they did not exist before, but because the slight increase of shading from the stria-tion renders them more apparent. The indication of a viscus seen in some specimens vanishes at this stage. Presently the tube or sheath is either digested by the gastric juices of the mosquito, or it is cast off as a snake does its skin, and the animal swims about naked, and without any trace of a head- or tail-lash. The striation becomes very marked; but gradually as the blood thick-ens, and the movements of the embryo become in consequence less vigorous, these markings completely disappear, giving place to a peculiar spotted appear-ance. Each spot is dark or luminous, according to the focusing of the micro-scope, and probably depends on some oily material now collecting in the body of the animal.

This concludes the first stage of the metamorphosis, and has taken about thirty-six hours to complete. During all this time the original proportions of the animal have been preserved and vigorous movement maintained. Now, however, it enters on a sort of chrysalis condition, during which nearly all movement is suspended, and the outline and dimensions very much altered. Hitherto the body was long and of graceful contour, but now it becomes shorter and broader, the extreme tail alone not participating in the change. The large spots in the body disappear, gradually giving place to what seems to be a fluid holding numerous minute particles in suspension. I have once or twice detected to-and-fro movements in these. The tail continues to be flexed and extended vigorously, but only at long intervals, whilst all oral movements cease. By the end of the third day the animal has become much shorter and broader, the small terminal portion of the tail still retaining its original dimen-sions, and appearing to spring abruptly from the end of the sausage-shaped body. Large cells occupy the previously homogeneous-looking body, and sometimes something like a double outline can be traced. Indications of a mouth present themselves; and if a little pressure is applied to the covering-glass, granular matter and cell-like bodies escape from an orifice placed a little in advance of the tail. The animal now begins to increase in length, and in some specimens to diminish in breadth, the growth seeming to be principally in the oral end of the body. The structure of the mouth is sometimes very evident; it is four-lipped, the lips being either open or pursed up. From the mouth a delicate line can be distinctly traced, passing through the whole length of the body to the opening already referred to as existing near the caudal extremity. Feeble movement may still sometimes be detected in the caudal appendix; but when the now growing body has attained a certain length the tail gradually disappears.

After this point, specimens of the *Filaria* in its third and last stage are difficult to procure. Most mosquitos die about the fourth or fifth day after feeding; and if their bodies, which fall into the water, are examined, they are soft and sodden and without *Filariae,* these having either decomposed or escaped. Sometimes, however, ovulation does not proceed rapidly, and the mosquito survives to the fifth or sixth day; or perhaps death may not occur, as it usually does, soon after the eggs have been laid, and the insect may survive this operation for two or three days. In such the last stage of the metamorphosis can be studied: four to six days seem necessary for its completion. Out of hundreds of mosquitos watched, I have been successful in finding *Filariae* in this last stage in four instances only. In one of these there was quite a number of embryos in regular gradation, from the passive chrysalis up to the mature and very active embryo, so that there can be no doubt of the relationship of the latter to the former, though their appearances differ so much. Owing to the small number of specimens I have examined, I am not quite certain about the details of this stage of the metamorphosis. As far as I can make out, the body gradually elongates from the hundredth to the fortieth or thirtieth of an inch, and when mature it measures fully a fifteenth of an inch in length by the five hundredth of an inch in breadth.

When at the above stage large cells occupy the interior; but as development advances these become reduced in size, and accumulate round the dark line I have already mentioned as running from the mouth to the caudal extremity. In this way an alimentary tube is fashioned, and the peculiar and characteristic valve-like termination of the oesophagus in the intestine, seen in the *Filariae,* is developed. The mouth may now be seen open and funnel-shaped, and the tail is reduced to a mere stump. Movements, first of a swaying-to-and-fro character, but afterwards brisker, now begin. The body gradually elongates and becomes perhaps slightly thinner; all cellular appearance vanishes, and, owing to the increasing transparency of the tissues, the details can no longer be made out. A vessel of some sort is seen in the centre running nearly the whole length of the body, and opening close to one extremity; this extremity is slightly tapered, and is crowned with three, or perhaps four, papillae; but whether this is the head or tail, and whether the vessel opening near it is the alimentary canal or vagina, I cannot say; the other extremity is also slightly tapered, but has no papillae. There can be no doubt which is mouth and which tail, but the intermediate steps I have failed to trace satisfactorily. There is a stage between these two in which the mouth is closed, and the oesophagus can be seen running from it. If the body is compressed, that tube can be forced through the skin and distinctly seen; but about that time the tissues become so transparent that their exact relations cannot be made out.

I cannot say if the three or four papillae round one extremity of the developed embryo constitute the perfected boring-apparatus of the worm, or if it is the boring-apparatus at all; but comparing this with what is found in other species of the same genus, I think it very probable that it either is or will become the piercing-apparatus. Some time ago I operated on an Australian horse for this worm, and had the satisfaction of finding the parasite not very much injured after removal: it was an unimpregnated female possessing all the typical structures of the *Filariae*. Its head was armed with a five- or six-toothed saw, the teeth arranged, like those in some kinds of old-fashioned trephines, in a circle round the mouth. I removed a worm from the same eye of the same horse about three or four weeks previously; the cornea had healed, and the cloudiness cleared up before the second worm appeared. I infer from this, from the very perfect boring-apparatus, and from the female being unimpregnated, that the eye is not the resting- or breeding-place of the *Filaria* found in it, but that it is sometimes accidentally entered by the worm on its travels in search of the suitable spot. From the fact that one worm succeeded the other I infer that the sexes are brought together in this way (as in the case of *Filaria sanguinolenta* of the dog): when a wandering worm comes across the tract of another, it follows it up; thus several may be found together at the end of the burrow.

Probably, then, these papillae are the boring-apparatus to be used in penetrating the tissues of man and escaping from the mosquito. At this (presumably the final stage of the *Filaria's* existence in the mosquito) it becomes endowed with marvellous power and activity. It rushes about the field, forcing obstacles aside, moving indifferently at either end, and appears quite at home, and in no way inconvenienced by the water in which it has just been immersed. This formidable-looking animal is undoubtedly the *Filaria sanguinis hominis* equipped for independent life and ready to quit its nurse the mosquito.

Future history of the Filaria. There can be little doubt as to the subsequent history of the *Filaria*, or that, escaping into the water in which the mosquito died, it is through the medium of this fluid brought in contact with the tissues of man, and that, either piercing the integuments, or, what is more probable, being swallowed, it works its way through the alimentary canal to its final resting place. Arrived there, its development is perfected, fecundation is effected, and finally the embryo *Filariae* we meet with in the blood are discharged in successive swarms and in countless numbers. In this way the genetic cycle is completed.

IMMUNITY VIA ARTIFICIALLY ATTENUATED VIRUS

PASTEUR, Louis (French biochemist, bacteriologist, and immunologist, 1822–1895). *Méthode pour prévenir la rage après morsure*, in *Comptes rendus de l'Académie des Sciences*, vol. 101, p. 765, Paris, 1885; tr. by T. S. Hall for this volume.

The central problem in establishing a science of immunology was to discover methods of lowering the *pathogenicity* of the antigens while preserving their *immunogenicity*. In the case of smallpox (see Jenner) this was done, according to the accepted interpretation, by utilizing strains accidentally attenuated through animal passage. In the present famous paper, Pasteur shows how, for a disease of wide distribution among mammals, attenuation may be accomplished artificially.

Prophylaxy in rabies, as I announced it in my name and in the name of my collaborators, in some previous notes, surely constituted real progress in the study of this disease; scientific progress, however, rather than practical. Its application lay open to accidents. Of twenty dogs treated, I could not surely claim to have rendered more than 15 or 16 refractory to rabies.

It was, furthermore, expedient to end the treatment with a final very virulent inoculation, the inoculation of a control virus, so as to fortify and reinforce the refractory state. Besides, prudence required that one keep the dogs under observation during a period longer than the period of incubation of the disease produced by the direct inoculation of this last virus. From which it follows that no less than three or four months must pass before we can feel sure that the state is one of refractoriness toward rabies.

Such requirements would have limited greatly the application of the method.

Finally, only with difficulty could one apply the method to an immediate commencement of treatment, a condition made necessary by the element of the accidental and unforeseen in rabid dog bites.

What was needed, were it possible, was to effect a method more rapid and capable of conferring upon dogs what I would presume to term perfect security.

For how other than by advancing thus far dare one make any trial at all upon man?

After what might be called innumerable trials I achieved a prophylactic method, both prompt and practical, the success of which in dogs has been so frequent and so sure that I have confidence in the applicability of it generally to all animals and to man himself.

This method rests essentially on the following facts:

The inoculation into a rabbit by trepanation, under the dura mater, of a rabic spinal cord of a dog with street rabies, always gives to these animals a case of rabies after an average incubation period of around a fortnight.

If one passes the virus from this first rabbit to a second, from the second to a third, and so on, according to the foregoing procedure there soon is manifested a tendency, increasing in distinctness, toward a diminution of the incubation period of rabies in the rabbits successively inoculated.

After twenty to twenty-five passages from rabbit to rabbit one encounters incubation periods of eight days which remain during a new series of twenty to twenty-five passages. Then one attains an incubation period of seven days which recurs with striking regularity during a new series of up to ninety passages. At any rate it is at this figure that I am at the moment. And it is scarcely if at all that there at present appears a tendency toward an incubation period of a little less than seven days.

Experiments of this type, begun in November 1882, are already three years old, without the series ever having been interrupted and without our ever having recourse to a virus other than those of these rabbits successively dead of rabies. Nothing is easier, thus, than to have constantly at hand a rabic virus of perfect purity, always identical or practically so. Upon this hinges the *practicability* of the method.

Spinal cords of these rabbits are rabic throughout their entire length with constancy of virulence.

If one detaches portions of these cords several centimeters long, utilizing the greatest precautions for purity which it is possible to realize, and if one suspends these in dry air, gradually their virulence diminishes until it finally disappears entirely. The time for extinction of virulence varies very little with the thickness of the pieces, but markedly with the external temperature. The results constitute the *scientific* point in our method.

These facts established, here is a method for rendering a dog refractory to rabies in a relatively short time.

In a series of flasks in which the air is kept dry by pieces of potash placed on the bottom of the vessel, one suspends each day, a section of fresh rabic cord from a rabbit dead of rabies, developed after seven days' incubation. Likewise daily, one inoculates under the skin of a dog one full Pravaz syringe of sterilized bouillon in which one has dispersed a small fragment of one of these desiccated cords, commencing with a cord whose order number places it sufficiently far from the day of operation so that we are sure that it is not at all virulent. Previous experiments have established what may be considered safety in this matter. On the following days one proceeds in the same way with more recent cords, separated by an interval of two days, until one arrives at a last very virulent cord which has only been in the flask for a day or two.

At this time the dog is refractory to rabies. One is able to inoculate him with rabic virus under the skin or even on the surface of the brain without rabies declaring itself.

By applying this method, I finally had fifty dogs of every age and breed all refractory without encountering a single failure when unexpectedly there presented themselves at my laboratory upon the sixth of last July three persons from Alsace:

Theodore Vone, a petty merchant grocer of Meissengott near Schlestadt, bitten in the arm July 4th by his own dog which had gone mad;

Joseph Meister, aged nine years, likewise bitten on the 4th of July at 8 o'clock in the morning by the same dog. This child, thrown to the earth by the dog, had numerous bites on his hand, legs, and buttocks, some of them very deep and making walking difficult. The chief wounds had been cauterized with carbolic acid only twelve hours after the accident, at 8 in the evening of July 4th, by Dr. Weber of Ville;

The third person who had herself received no bites was the mother of little Joseph Meister.

At the autopsy of the dog, struck down by its master, the stomach of the dog was found full of straw, hay, bits of wood. The dog had been quite mad. Joseph Meister had been dragged from under the dog covered with saliva and blood.

M. Vone had on his arm some marked contusions, but he assured me that his shirt had not been pierced by the fangs of the dog. Since he had nothing to fear I told him he could leave for Alsace the same day, and he did so. But I kept with me little Joseph Meister and his mother.

Now it happened that it was also on exactly July 6th that the Academy of Science was holding its weekly meeting; I saw there our colleague M. le Dr. Vulpian and told him what had happened. M. Vulpian, as well as Dr. Grancher, were good enough to accompany me at once to Joseph Meister and to confirm the condition and number of his wounds. He had no less than fourteen of them.

The opinions of our colleague and of Dr. Grancher were to the effect that, from the intensity and number of his bites, Joseph Meister had been practically fatally exposed to the inception of rabies. I then communicated to M. Vulpian and M. Grancher the new results I had obtained in the study of rabies since my lecture in Copenhagen a year before.

The death of this child appearing to be inevitable, I decided, not without deep and acute anxiety, as you will well imagine, to try with Joseph Meister the method which had so often proved successful with dogs.

My fifty dogs, it is true, had not been bitten before I commenced to make them refractory, but I knew that this circumstance need not preoccupy me for I had previously already obtained the refractory state in dogs after their being bitten in a large number of cases. I had already presented evidence this year to the rabies Commission of this new and important forward step.

Therefore, on July 6th, at 8 in the evening, sixty hours after the bites of July 4th, and in the presence of Drs. Vulpian and Grancher, Joseph Meister was inoculated under the skin of the hypochondrium, right side, with one half of a Pravaz syringe of cord of a rabbit dead of rabies June 21st and since then preserved in a flask of dry air, namely, for fifteen days.

The days following, new inoculations were made, always in the hypochondrial region according to the conditions which I give in the following table:

July	Time	Cord taken	Dried for
7th	9 am	June 23rd	14 days
7th	6 pm	June 25th	12 days
8th	9 am	June 27th	11 days
8th	6 pm	June 29th	9 days
9th	11 am	July 1st	8 days
10th	11 am	July 3rd	7 days
11th	11 am	July 5th	6 days
12th	11 am	July 7th	5 days
13th	11 am	July 9th	4 days
14th	11 am	July 11th	3 days
15th	11 am	July 13th	2 days
16th	11 am	July 15th	1 day

In this way I carried to 13 the number of inoculations and to 10 the number of days of treatment. I will explain later that a fewer number of inoculations would have sufficed. But you will understand that in this first attempt I felt obliged to act with the most particular circumspectness.

Each of the cords employed was also inoculated by trepanation into two new rabbits so as to follow the virulence of the materials we were injecting into the human subject.

Study of these rabbits made it possible to verify that the cords of July 6th, 7th, 8th, 9th, and 10th were not virulent, for they did not produce rabies in the rabbits. The cords of July 11th to 16th were all virulent, and the proportion of virulent substance present grew daily greater. Rabies declared itself after seven days' incubation for rabbits of July 15th and 16th; after eight days for those of the 12th and 14th; after fifteen days for those of July 11th.

During the last days, I had inoculated into Joseph Meister the very most virulent strain of rabies virus, a strain which taken from dogs and reinforced by many passages from rabbit to rabbit gives rabies to rabbits in seven days, to dogs in eight to ten. I felt justified in undertaking to do so because of what happened to the fifty dogs to which I previously referred.

When the immune state has been attained one can without fear inoculate the strongest virus in any quantity whatever. It has always seemed to me that this would have no other effect than to consolidate the refractory state.

Thus, Joseph Meister escaped not only the rabies to which his bites might have given rise but also the virus which I injected as a control upon the immunity got by treatment, and the rabies which I injected was stronger than the street strain.

The final inoculation of this very virulent material has likewise the advantage of cutting down the time during which one need be apprehensive about the aftereffects of a bite.   For if the rabies is going to be able to appear, then it will appear much sooner from the injected virus than from the weaker virus of the bite.   After the middle of August, I looked forward with confidence toward the future soundness of Joseph Meister.   Today as well, after a lapse of three months and three weeks since the accident, his health leaves nothing to be desired.

What interpretation is to be made of this new method which I have just made known for preventing rabies after the bite?   I do not intend today to treat this question in a complete fashion.   I wish to limit myself to some preliminary details having to do with the understanding of the meaning of the experiments which I am now pursuing in an effort to focus upon the best of the interpretations possible.

Returning now to the method for progressive attenuation of a fatal virus and the prophylaxy which one can derive therefrom, the influence of air upon this attenuation being given, the first idea which presents itself in explanation of the effect of the method is that the sojourn of the cords in contact with dry air progressively diminishes the intensity of their virulence to the disappearing point.

From this, one would be persuaded that the prophylactic procedure in question rests upon the use first of a virus without appreciable virulence, next weak ones, and then stronger and stronger ones.

I will show finally that the facts do not agree with this way of looking at the matter.   I will prove that the delays in the incubation period of the rabies given to rabbits day by day, such as I was speaking about above, given to check the state of the virulence of the cords desiccated in contact with air, are an effect of the impoverishment of the quantity of the virus present in these cords and not an effect of the impoverishment of the virulence.

Could one admit that the introduction of a virus of constant virulence would produce immunity—if one used very small quantities increasing them daily? This is an interpretation of my method which I have studied from the experimental point of view.

One can give the new method another interpretation, an interpretation certainly very foreign to the foregoing, but one which merits full consideration because it is in harmony with certain already achieved results in connection with vital phenomena in certain lower organisms, notably different pathogenic microbes.

Many microbes appear to give rise in their culture media to substances injurious to their own development.

In 1880, I began researches to establish that the microbe of chicken cholera produces a sort of self-poisoning substance.   I did not succeed in demonstrating such a substance; but I feel today that the study should be taken up again—nor shall I fail to do so—operating in the presence of pure carbonic acid gas. . . .

M. Raulin, my old teacher, now professor of the Faculty of Lyon, has established, in a very remarkable thesis which he defended at Paris, March 22, 1870, that the growth of *aspergillus niger* develops a substance which partially arrests the growth of this mold when the culture medium does not contain iron salts.

Could it not be that the rabies virus is composed of two distinct substances so that in addition to that which is living and able to reproduce in the nervous system, there is another having the faculty when present in suitable amounts of stopping the development of the foregoing? I will examine this third interpretation of the method of prophylaxy in rabies with all the attention it deserves in my next communication.

I scarcely need to mention in closing that perhaps the most important question to be resolved at this time is that of what interval to observe between the instant of the bite and the commencement of the treatment. That interval in the case of Joseph Meister was two and a half days. But we must be prepared for situations in which it is much longer.

Last Tuesday, October 20th, with the obliging assistance of MM. Vulpian and Grancher, I had to start treating a youth of 15 years, bitten six days before on both hands and in very serious condition.

I shall not delay in informing the Academy of the outcome from this new attempt.

Probably it will not be without emotion that the Academy will hear the account of the courageous act and presence of mind of this boy whom I started treating Tuesday last. He is a shepherd, fifteen years old, named Jean-Baptiste Jupille, from Villers-Farlay (Jura) who, on seeing a dog of suspicious actions and great size leap upon a group of six small friends of his, all younger than he was, lept forward, armed with his whip, in front of the animal. The dog seized Jupille's left hand in its mouth. At this, Jupille knocked the dog over, held it down, opened its mouth with his right hand, meanwhile receiving several new bites, and then, with the thong of his whip bound the dog's muzzle and dispatched the animal with one of his wooden shoes.

[Abstracts from comments succeeding the reading of Pasteur's paper]

VULPIAN: The Academy will not be astonished if, as a member of the Section on Medicine and Surgery, I seek the chair's recognition to express the feelings of admiration which M. Pasteur's communication inspires in me. The sentiments will be shared, I am sure, by the entire medical profession. Etc.

LARREY: . . . I therefore take the honor of moving that the Academy recommend to the French Academy that this young shepherd has, in giving so generous an example of courage, rendered himself deserving of a "prize of virtue."

BOULEY (President of the Academy): The Communication which we have just heard permits the Academy, and I might even say humanity, to conceive new hopes.

Likewise the date of October 27, 1885, will remain as one of the most memorable, if it is not the most memorable, in the history of scientific conquests and in the Annals of the Academy.

INSECTS AS VECTORS OF DISEASE

SMITH, Theobald (American parasitologist, 1859–1934). From *Investigations into the nature, causation, and prevention of southern cattle fever*, in *Eighth and Ninth Annual Reports of the Bureau of Animal Industry*, Washington, 1893; by permission of the Chief of the Bureau.

Belding[*] and Kelly[†] give this paper credit for being the first to prove an insect vector in disease transmission. Whether Garrison's[‡] contention is correct that credit should go to Manson (*q.v.*) is partly a definitional question, since Manson's filarial parasite, returning to the human host through drinking water, might better be classed as a case of alternation of hosts. Smith's paper is a monumental analysis of the disease as a whole, its etiology, transmission, epidemiology, etc. Excerpts are presented from which the reader may obtain summaries of all important experimental results plus illustrative examples of a limited number of actual procedures and observations.

It has been a more or less prevalent theory of cattle-owners in the districts occasionally invaded by Texas fever from the South that ticks are the cause of the disease. Mr. J. R. Dodge, in his historical report of this plague, mentions the fact that in 1869 an outbreak in Chester County, Pa., was believed to be caused by ticks. Gamgee in 1868 states: "The tick theory has acquired quite a renown during the past summer, but a little thought should have satisfied anyone of the absurdity of the idea." The officers of the Metropolitan Board and most subsequent observers seem to have entertained the same view of the harmlessness of the cattle tick as a carrier of the infection. In fact, few observers have given it any thought. In the entire report of F. S. Billings we find no reference whatever to these pests. Paquin states that he has "found the parasites also in *ticks bloated with blood of infectious Southern cattle.* So this must be added to the list of sources." But the ubiquity of this "germ" rather predisposes one against any belief in its existence if we did not have suf-

---

[*] D. L. Belding, *Textbook of clinical pathology*, p. 2, New York, 1942.

[†] E. C. Kelly, in *Medical Classics*, vol. 1, No. 5, p. 344 (reprints the Smith paper in extenso), 1937.

[‡] F. H. Garrison, *An introduction to the history of medicine*, p. 583, Philadelphia, 1929.

ficient positive evidence that bacteria have nothing to do with the disease. The statement thus depends simply upon the finding of a "germ" in adult ticks resembling that found in diseased cattle, and in fact everywhere else (waters, soil, manures from the South, urine, bile, liver, spleen, kidneys, etc., of infectious Northern stock). Experiments to demonstrate the relation which ticks bear to Texas fever were not made.

Nothing positive was thus contributed to the elucidation of the action of ticks in carrying the disease until the subject was taken up at the Experiment Station of the Bureau near Washington, in 1889. Here it was found by experiments to be detailed in the remainder of this report that the disease can be produced by ticks hatched artificially in the laboratory, without the presence of Southern cattle. . . .

Field Experiments to Determine the Precise Relation between the Cattle Tick and Texas Fever

These experiments were begun in the summer of 1889, and have been continued up to the present. They have been carried on in three different directions:

(1) Ticks were carefully picked from Southern animals, so that none could mature and infect the ground. The object of this group of experiments was to find out if the disease could be conveyed from Southern to Northern stock on the same inclosure without the intervention of ticks.

(2) Fields were infected by matured ticks and susceptible cattle placed on them to determine whether Texas fever could be produced without the presence of Southern cattle.

(3) Susceptible Northern cattle were infected by placing on them young ticks hatched artificially, i.e., in closed dishes in the laboratory.

These three lines were not followed simultaneously, because, for instance, the fact that the disease can be produced by placing young ticks on cattle was discovered in 1890, and hence only tried then and thereafter. In giving the details of the various experiments we shall adhere . . . to the chronological order in which the experiments were performed. This is necessary in order to describe successively the experiments of the same year, which were more or less connected with one another, and also to show the process by which the various facts concerning the cattle tick came to our knowledge.

The disease was introduced into one field each year by North Carolina cattle brought here for this purpose. In 1890 a field was infected by cattle from Texas.

The field experiments were all conducted on the experiment station of the Bureau of Animal Industry, within half a mile of the limits of the city of Washington. The arrangement of the various experimental fields is shown for each year on a plat of the station grounds. The isolated condition of the field in use in any given season may be seen by an inspection of these plats. They are either separated from one another by a piece of ground remaining permanently free from infection, or by a lane or by a strip of ground purposely fenced off between them. No two fields in use are thus separated merely by a fence. In every case, with the exception to be noted, a strip of ground intervenes which is at least 36 feet wide. A small brook passes through a portion of the grounds, as is shown in the various plats, and the space between the fields along this brook is about 20 feet wide.

EXPERIMENTS OF 1889

To carry on the experiments in the early part of the season of 1889, seven head of cattle were collected in Craven County, N. C., which is a portion of the permanently infected territory. On June 25 they were shipped by steamer from New Berne, N. C., and they arrived at the station near Washington June 27. They had thus been two days on the way. These animals were rather thin and a large number of cattle ticks (*Boöphilus bovis*) in various stages of development were attached to them. Only a few were full grown.

Experiment 1 (exposure to Southern cattle with ticks). Of these seven head four were placed in field I (see Fig. 13) on the day of arrival, June 27. The field contains about nine-sixteenths of an acre. The soil is a dry, gravelly loam. A small stream passes through it, from which the cattle obtain their drinking water.

The history of the native cattle placed in this field may be briefly summarized.

(a) North Carolina cattle with ticks:

No. 12, placed in this field June 27, removed August 17.
No. 40, placed in this field June 27, removed August 17.
No. 42, placed in this field June 27, removed August 17.
No. 45, placed in this field June 27, removed August 17.

(b) Native cattle:

June 27.—No. 7 (cow, 6 years) placed in this field. Dead[1] August 23.
June 27.—No. 8 (cow, 1½ years) placed in this field. Killed[2] August 27.
June 27.—No. 75 (calf of No. 8, 4 months) placed in this field. Recovered.
June 27.—No. 9 (bull, 1½ years) placed in this field. Died August 31.
June 27.—No. 10 (calf of No. 7, 4 months) placed in this field. Died August 31.

---

[1] Unless otherwise stated the cause of sickness and death is Texas fever.
[2] With one exception (No. 163) all native animals reported killed in this report were in a dying condition at the time.

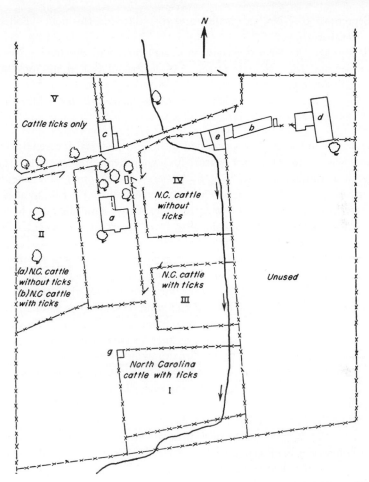

Fig. 13. Field enclosures for 1889. Scale, ¼ in. = 33 ft. *a*, dwelling house; *b*, station laboratory; *c*, horse stable; *d*, cow stable; *e*, breeding pens; *f*, tool house; *g*, shed in field.

June 27.—No. 11 (calf of No. 7, 4 months) placed in this field. Killed Sept. 10.

August 20.—No. 46 (heifer, 1⅔ years) placed in this field. Killed Sept. 10.

August 24.—No. 43 (steer, 3 years) placed in this field. Dead September 13.

August 24.—No. 44 (steer, 4 years) placed in this field. Dead September 17.

September 6.—No. 53 (heifer, 1½ years) placed in this field. Recovered.

September 6.—No. 54 (heifer, 2 years) placed in this field.   Killed September 20.

September 14.—No. 57 (cow, 9 years) placed in this field.   No result.

September 30.—No. 70 (steer, 2½ years) placed in this field.   Died October 19.

October 19.—No. 71 (heifer, 3½ years) placed in this field.   Probably no disease.

The disease in this field was designed to furnish material for general investigation as well as to serve as a control for experiment 2 below.   It illustrates admirably a number of important characters of this remarkable disease and demonstrates once again the frequently observed fact that cattle, to all appearances healthy, may become the cause of an extensive fatal disease when transferred in the warmer seasons of the year from a certain permanently infected area to territories north of this area.

The first high morning temperature appeared August 15, or thirty-nine days after the native and Southern cattle were placed on this field together. The first death occurred August 23, or forty-seven days after this same date. In other words, the cattle exposed at this time died not less than forty-seven days after the beginning of the exposure.   After a certain time, however, death follows more speedily after exposure, as may be seen when we consult those cases exposed August 20 and thereafter, for which this period was only fourteen to twenty-three days.   The field remained infected so as to cause death as late as October 19.   The later the exposure the less likely is the disease to end fatally.

Omitting the last case, No. 71, as having been exposed too late, we have ten deaths from thirteen cases exposed, or 76.8 per cent.   It should be noted that, although the Southern cattle were removed from the field August 17, the infection on the field remained unimpaired.   . . .

Experiment 6 (exposure to cattle ticks only).   This experiment was carried on in field V, an inclosure consisting of about three-eighths of an acre.   The soil is a heavy clay loam, and contains neither running nor standing water. On September 13 several thousand, mostly full-grown ticks, were scattered over the ground in this field.   These ticks had been collected from cattle near New Berne, N. C., September 9 and 10.   There were placed in this field, September 14, four natives:

No. 48 (cow, 2½ years).
No. 83 (calf of No. 48, 2 months).
No. 64 (steer, 2½ years).
No. 65 (heifer, 2½ years).

Of these, Nos. 48, 64, and 65 contracted Texas fever.   No. 83 was not examined as to its blood, but it showed no external symptoms of disease.   No.

48 was killed in a dying condition, October 21. The autopsy, as well as the examination of the blood before death, demonstrated Texas fever. Nos. 64 and 65 recovered.

Summary of the Experiments for 1889

The first series go far toward demonstrating that a field must be infected with ticks before Texas fever can appear among natives. The second series confirms the first as far as it goes. The advanced season gave rise only to what has been called the mild or autumnal type of the disease, characterized by the presence in the blood corpuscles of the peripheral coccus-like stage of the Texas-fever parasite. If we bring together the results of the four experiments we find that in the field containing the ticks only, and in which Southern cattle at no time entered, all three exposed adult natives took the disease. In the field containing Southern cattle from which the ticks had been picked no disease appeared. Finally, in the two fields which contained Southern cattle and ticks together three out of six natives became diseased. In these experiments the great importance of the method of blood examinations as described in the first part of this volume is plain. To rely solely upon external symptoms in mild attacks is out of the question. The counting of the red corpuscles, the changes going on in the latter, and the presence of the Texas-fever parasite as determined by microscopical examination are indispensable in determining whether Texas fever is present or not.

EXPERIMENTS OF 1890

The experiments of this year were chiefly occupied with the relation of ticks to Texas fever. The experiments of last year were repeated, and in addition ticks were hatched artificially and placed on cattle with the result that Texas fever appeared in every case. Southern cattle were obtained as before from North Carolina and also from Texas.

Experiment 7 (to ascertain whether the infection of 1889 survived the winter). For this purpose fields I and III of 1889 were thrown together by removing the intervening fences and the whole designated field I. The little stream was likewise fenced off in July to prevent any infection from field VI reaching it. A number of animals were pastured on this field.

> May 26, 1890.—No. 74 (heifer, 2 years). Transferred to field II September 25.
> May 26, 1890.—No. 91 (heifer, 3 years). Transferred to field VI October 1.
> July 4, 1890.—(Stream fenced off, as field VI is now used for the first time).
> July 9, 1890.—No. 130 (cow, 5 years).
> August 25, 1890.—No. 97 (bull, 1 year).

During the summer no ticks appeared in this field, so that it was evident that they had not survived the winter. No disease appeared in any of the animals exposed.   . . .

Summary of the Experiments of 1890

The discovery of 1889 that ticks also are sufficient to infect a field was confirmed this year. The experiment designed to test the theory that Southern cattle are infectious only through the ticks they carry failed this year, for the field became infected with ticks after all. Lastly, the demonstration of the important fact that the infection is conveyed by the young tick, and is probably introduced by it into the blood, was a very great stride in advance in our understanding of the external characters of the infection.

In field IX several natives were exposed to North Carolina soil without becoming diseased.

On the station grounds field V was infected with the blood and spleen pulp of cattle which had succumbed to Texas fever. The exposed natives did not become infected.

In field IV during this same year a number of sick natives were brought together and some healthy natives added. The latter had a mild attack late in the season, only detected by the microscopic examination of the blood.

These three experiments will be fully discussed farther on, and we simply refer to them here to show that the animals not exposed under certain conditions did not become infected, although pastured not far from Texas-fever cases during the summer.

EXPERIMENTS OF 1891

The arrangement of the fields for this year and the uses to which they were put are indicated on the accompanying plat. A tract of land adjoining the station grounds on the north was added to the territory in use. On this tract were situated a dwelling house and a number of unused sheds. For the purpose of carrying on the various experiments, cattle were collected near New Berne, N. C., as in previous years, and shipped by steamer from New Berne, June 30. They arrived at the station July 2, having been but two days on the way.

Experiment 14 (*exposure to North Carolina cattle with ticks*). The general control experiment of producing the disease in the natural way was conducted, as before, by exposing natives to Southern animals on the same field. For this purpose inclosure VI was again selected. In this experiment not only unexposed natives but also recovered natives were reëxposed to test any acquired immunity. Similarly Southern animals, kept for one or two years on the station, were reëxposed to determine any loss of immunity. These collateral experiments will be discussed in dealing with these subjects. In this place we simply summarize the results of the exposure of fresh natives.

The animals placed in this comprised the following:

(a) North Carolina cattle:[3]
    July 2.—No. 172 (cow, 6 years), from farm No. 6.
    July 2.—No. 174 (cow, 3 years), from farm No. 5.
    July 2.—No. 177 (cow, 5 years), from farm No. 3.
    July 2.—No. 178 (cow, 4 years), from farm No. 2.

(b) Natives.
    July 2.—No. 104 (cow, 4 years). Very sick; recovered.
    July 2.—No. 159 (heifer, 2 years). Not very sick; recovered.
    July 2.—No. 163 (cow, 6 years). Very sick; killed August 25.
    September 1.—No. 169 (cow, 8 years). Died September 14.
    September 1.—No. 181 (cow, 2½ years). Killed in dying condition
      September 19.

GENERAL SUMMARY OF THE FIELD EXPERIMENTS RELATING TO THE CATTLE TICK

We are now in a position to review the results of the field work of the past four summers and determine how far they enable us to draw definite conclusions. In addition to the general control experiments by which Texas fever was produced in the natural way in natives which pastured on the same ground with Southern (North Carolina and Texas) cattle, experiments have been carried on in the three directions outlined.

(1) Experiments with Southern cattle from which the ticks were picked off were made every year. Those made in 1889 and 1892 were successful. Those made in 1890 and 1891 failed because young ticks appeared subsequently. The conclusion from these experiments that the tick is necessary to cause infection in Northern cattle may be regarded as demonstrated.

(2) Experiments to show that fields may be infected by cattle ticks alone were made in 1889 and 1890. In both Texas fever was produced.

(3) Experiments to show that young ticks artificially hatched produce Texas fever when placed on susceptible cattle were made in 1890, 1891, and 1892. These were uniformly successful in the summer and fall months.

It was observed, however, that the disease induced by such ticks is less fatal than that produced in the fields in the natural way. We are not prepared to account for this difference, unless it be the mode of incubation. The artificial condition of heat and moisture under which the eggs are kept may lead to a

---

[3] Eight animals were brought North, two from each farm, and divided equally between this and the following experiment.

speedy destruction of the microparasites which are in some unknown way associated with them. . . .

THE RELATION OF THE CATTLE TICK TO THE MICROÖRGANISM OF TEXAS FEVER

The hypothesis which seemed most plausible after the experiments of 1889 was that the tick, while withdrawing the blood from Southern cattle, drew out in it the Texas-fever parasite, which, entering into some more resistant state, perhaps some spore state, was disseminated over the pastures when the body of the mother tick became disintegrated. These spores were then supposed to enter the alimentary tract with the food and infect the body from this direction. The later experiments, however, completely demolished this conception. Neither the feeding of adult ticks and tick eggs nor the feeding of grass from infected pastures gave any positive results. On the other hand, the unmistakable outcome of the experiments was that the young tick introduced the infection into the body. This fact implies two possibilities. Either the tick is a necessary or a merely accidental bearer of the microparasite. If a necessary bearer of the infection, we must assume that the latter undergoes certain migrations and perhaps certain changes of state in the body of the adult tick and finally becomes lodged in the ovum. Subsequently it may become localized in certain glands of the young tick and discharged thence into the blood of cattle. This hypothesis assumes a complex symbiosis between the tick and the parasite on the one hand and the cattle and the tick on the other. According to another simpler hypothesis the tick would be merely an accidental bearer of the infection. The parasite entering the body of the tick with the blood of cattle may be already in the spore state or about to enter upon such a state. The young ticks, as they are hatched near the dead body of the female, may become infected from this. This infection, clinging to their mouth parts, is introduced into the blood of the cattle to which they subsequently attach themselves. Further investigations are necessary before the probable truth of one or the other of these hypotheses can be predicted with any degree of certainty.

It should be stated that the contents of the bodies of ticks in various stages of growth have been examined microscopically with considerable care. The abundant particles resulting from the breaking up of the ingested blood corpuscles obscured the search so that nothing definite has thus far resulted from it. The very minute size of the microörganism renders its identification wellnigh impossible, and any attempts will be fraught with great difficulties.

A question of much interest, but one upon which we have no information, is the relation of the cattle tick to the enzoötic Texas fever area. Is the distribution of the tick coextensive with that of the Texas fever microparasite, or does their distribution obey different laws? This question could be solved by a thorough investigation of a small portion of the border line of the enzoötic territory. This border line probably depends on the mean annual temperature, and hence we can not expect to find it very sharply defined. Ticks may extend farther north during some seasons than others, and hence there may be a belt or strip on which cattle are partially insusceptible because of former repeated attacks, although for the time being ticks may be absent. The entire subject is at present speculative, and is simply referred to here to arouse the attention of those who are in a position to record observations concerning it.

GENERAL ZOOLOGICAL INTERPRETATION OF INFLAMMATION

METSCHNIKOFF (many spellings), Ilia (Elie,* etc.) (Russian-born† comparative embryologist and pathologist, 1845–1916). From *Leçons sur la pathologie comparée de l'inflammation,* Paris, 1892; tr. by F. A. and E. H. Starling as *Lectures on the comparative anatomy of inflammation,* London, 1893.

It might perhaps be questioned whether a paper on this subject (immunopathology) merits a place in a source book of animal biology. Actually, it is precisely its *general zoological cast* which most sharply distinguishes Metschnikoff's approach to pathology. It was nine years after first publishing‡ on phagocytic action, discovered by him in Italy,§

---

* Russian equivalent for Elijah; Metschnikoff was nicknamed "The Prophet" by affectionate younger brothers and sisters-in-law.

† Metschnikoff was attached to the Institut Pasteur from 1888 until his death.

‡ *Untersuchungen ueber die intracellulaere Verdauung bei wirbellosen Tieren,* in *Arbeiten aus dem zoologischen institut der universität Wien,* vol. 5, p. 141, 1883.

§ "I was resting from the shock of events which provoked my resignation from the university [*i.e.,* of Odessa—Ed.] and indulging enthusiastically in researches in the splendid setting of the straits of Messina.

"One day when the whole family had gone to a circus to see some extraordinary performing apes, I remained alone with my microscope, observing the life in the mobile cells of a transparent starfish larva, when a new thought suddenly flashed across my brain. It struck me that similar cells might serve in the defense of the organism against intruders. Feeling that there was something in this of surpassing interest, I felt so excited that I began striding up and down the room and even went to the seashore to collect my thoughts.

"I said to myself that, if my supposition was true, a splinter introduced into the body of a starfish larva, devoid of blood vessels or of a nervous system, should soon be surrounded by mobile cells as is to be observed in a man who runs a splinter into his finger. This was no sooner said than done.

"There was a small garden to our dwelling, in which we had a few days previously organised a 'Christmas tree' for the children on a little tangerine tree; I fetched from it a few rose thorns and introduced them at once under the skin of some beautiful starfish larvae as transparent as water.

"I was too excited to sleep that night in the expectation of the result of my experiment, and very early the next morning I ascertained that it had fully succeeded.

"That experiment formed the basis of the phagocyte theory, to the development of which I devoted the next twenty-five years of my life." O. Metschnikoff, *Life of Elie Metchnikoff, 1845–1916,* p. 116, Boston, 1921.

that he gave a famous series of lectures at the Institut Pasteur, rue Dutot. From one of these lectures we draw the following statement of Metschnikoff's general biological theory of inflammation. Other works included the interpretation of invertebrate embryology as illustrative of the germ-layer theory, studies on anthropology, geriatrics, etc. A bibliography appears in the important biography by his (second) wife.‖ See also his philosophical notion of "orthobiosis."¶ In 1908 he shared the Nobel Prize with Ehrlich.

The study of inflammation from the point of view of comparative pathology proves first of all that this phenomenon is essentially reactive in its nature. The organism, threatened by some injurious agency, protects itself by the means at its disposal. Since, as we have seen, even the lowest organisms, instead of passively submitting to the attacks of morbid agents, struggle against them, why should not the more highly developed organisms, such as man and mammals, act in the same manner? We must conclude then that the invaded organism fights against the injurious cause, but in what way? As the evolution of inflammation shows, it is this phenomenon itself which is both the most general and the most active means of defense among the members of the animal kingdom.

The essential factor in the inflammatory reaction is an endeavour on the part of the protoplasm to digest the harmful object. This digestive action, in which the whole or almost the whole organism of the Protozoa takes part, is undertaken by the entire plasmodic mass of the Myxomycetes, while, from the Sponges upwards, it is confined to the mesoderm. In those cases where the victory remains with the invaded organism, the phagocytic cells of this layer assemble, englobe and destroy the injurious agent. This phagocytic reaction, in the lower scale of animal life, is slow owing to the progression of these cells towards the injurious body being dependent solely on their amoeboid movements; but as soon as a circulatory or vascular system makes its appearance in the course of evolution, it becomes much more rapid. By means of the blood-current the organism can at any given moment send along to the threatened spot a considerable number of leucocytes to avert the evil. When the circulation is partially carried on by a lacunar system there is nothing to intercept the movement of the leucocytes towards the seat of the injury. But when these cells are enclosed within the vessels, they are obliged to adapt themselves specially to fulfill their object, which they do by passing through the vascular wall.

If we accept this conclusion that inflammation in the higher animals is a salutary reaction of the organism and that diapedesis and its accompaniments form part of this reaction, several details of inflammatory phenomena will appear clear to us. For instance the lobed and polymorphous shape of the nucleus of the pus-corpuscles has been remarked. This particular shape is pe-

---

‖ O. Metschnikoff, *Life of Elie Metchnikoff, 1845–1916*, Boston, 1921.
¶ E. Metschnikoff, *Etudes sur la nature humaine*, Paris, 1903.

culiar to the polynuclear leucocytes, which represent the vast majority (75 per cent) of the total number of white cells. As it was noticed that a quantity of pus-corpuscles died in the exsudation, this fact became associated with the curious form of the nucleus; it was said, and is still maintained, that the polynuclear leucocytes are cells predestined to perish and incapable of any considerable activity. On the contrary these leucocytes are precisely the most active cells in the organism. The shape of their nucleus may be more adequately explained as a special adaptation for passing through the vessel-wall. If the process of diapedesis be watched, the difficulty experienced by the nucleus in getting through will at once be noticed. Directly this has occurred, the rest of the protoplasm follows rapidly. It is obvious that a nucleus divided into several lobes can pass through the wall more easily than one not so separated. Hence in pus the polynuclear leucocytes are more numerous than the mononuclear leucocytes, and hence the lobed shape of the nucleus is found only in the leucocytes adapted for diapedesis and does not occur among the invertebrata (except in a few Cephalopoda).

If the irritating agent be outside the vessels, it provokes a typical inflammation, accompanied by diapedesis; if the same agent be within the vessels, no diapedesis takes place but the lecucoytes fight against the microbes in the blood itself. For instance in recurrent fever, the spirilla undoubtedly act upon the vascular wall without bringing about diapedesis. But the leucocytes increase in number; the leucocytosis is followed by a struggle which is ended by the leucocytes devouring the spirilla. We have here a case of inflammation accompanied by diapedesis; the conflict between the phagocytes and the spirilla takes place in the blood itself. Although no diapedesis occurs there is in recurrent fever great elevation of temperature as well as other symptoms which prove it to be an inflammatory disease. It is apparently a case of inflammation in the blood itself, a sort of 'hemitis' as Piorry considered it might be many years ago. We find the same conditions in animals whose vascular system and general body cavity are in communication. Thus as we have seen in the case of Daphnia, caused by the Monospora, the leucocytes often collect in large numbers around the spores of this parasite—their assemblage taking place in the body-cavity.

As another instance we may cite tuberculosis. If inoculated subcutaneously, the tubercle bacilli produce inflammation accompanied by considerable diapedesis. But if the same bacilli be injected direct into the blood no diapedesis occurs, but the phagocytes will gather round the bacilli within the vessels and form intravascular tubercles. It cannot be said that in the first case (extravascular inoculation) there is inflammation and that in the second (intravascular inoculation) there is none, especially as the same tubercles are formed in both instances. This is another example of an inflammation of the blood itself.

All these cases of intravascular inflammation without diapedesis, as well as the inflammatory phenomena in the young larvae of Axolotls and Tritons (where it is the migratory cells that collect at the seat of the injury), in fact the whole series of reactive phenomena in so many of the invertebrates, prove clearly that the essential and primary element in typical inflammation consists in a reaction of the phagocytes against a harmful agent. If the latter be in the general body-cavity, which is filled with blood, the phagocytes will collect here; if in the interior of the vessels, as in recurrent fever or in intravascular tuberculosis, the phagocytes will assemble in the blood itself; if on the contrary the injurious agent is outside the body cavity or outside the vessels, the phagocytes will emigrate towards the threatened spot—an emigration without diapedesis in the invertebrata and young larvae of Urodela, or with diapedesis in the vertebrata.

Before phagocytic reaction can take place, these cells must be excited positively. Negative sensibility may also serve as a means of defense in a mobile organism, such as the plasmodium of the Myxomycetes, which retires from the offending cause. In the cases where the latter has penetrated into the organism, negative sensibility on the part of the phagocytes will leave the field of battle to the parasite, so that, as frequently happens, the death of the organism results. Hence as we rise in the scale, we are met by a progressive evolution of positive sensibility in the leucocytes. In Daphnia the observer is struck by the number of diseases in which phagocytosis is entirely or almost entirely absent. By the time we reach the amphibia, positive chemiotaxis is already very marked and, as Gabritchewsky has shown, it is still more highly developed in rabbits. And yet among the rodents, as among the small laboratory animals generally, there occurs a certain number of rapidly fatal diseases (such as chicken cholera, hog cholera, vibrionian septicaemia of birds), in which phagocytosis is often completely absent. In man and the higher mammals, similar diseases are much less frequent.

But, in addition to the mobile phagocytes adapted by their sensibility to move towards the offending object, there are also fixed phagocytes. A good example of the latter, which are especially developed in the higher vertebrates, is furnished by the endothelial cells of the vessels. Since these cells are contractile and phagocytic, it is natural to conclude that they must also be possessed of sensibility. Thus, if we assume a chemiotactic sensibility of the endothelial cells, we may easily explain the remarkable power of reciprocal attraction possessed by the protoplasmic processes of developing capillaries, which enables them to meet and form a new vascular loop. We may apply this explanation to account for the fact that in many neoplasms, as in pannus, the vessels penetrate and branch freely in the affected tissue, whereas in the granulomata, such as tubercle, leprosy and actinomycosis, blood-vessels are absent. In the former case there is a positive chemiotactic influence attracting the vascular loops, in the latter a negative chemiotaxis or other form of negative sensibility of the endothelial cells. The cooperation of these cells in the inflammatory process would be also directed by their sensibility, at any rate so far as their active contraction is concerned.

There is one more form of sensibility which we must mention, namely that of the nervous system, which aids the phagocytic and vascular mechanisms in their reaction against deleterious agents.

To sum up: Inflammation generally must be regarded as a phagocytic reaction on the part of the organism against irritants. This reaction is carried out by the mobile phagocytes sometimes alone, sometimes with the aid of the vascular phagocytes or of the nervous system.

The theory here indicated might be termed the biological or comparative theory of inflammation, since it is founded on a comparative study of the pathological phenomena presented by living cells.

"SUBSTANCES WHICH CANNOT BE ABSENT WITHOUT INJURY"

GRIJNS, Gerrit (Dutch biochemist and physician, 1865-1944). From *Over polyneuritis gallinarum*, in *Geneeskundig tijdschrift voor Nederlandsch-Indië*, vol. 41, p. 3, 1901; tr. "under the direction of the Working Committee" in *Prof. Dr. G. Grijns' researches on vitamins 1900—1911*, Gorinchen, 1935; with the permission of J. Noorduijn en Zoon N.V., publisher.

Grijns' paper is interesting because it implements experimentally the doctrine brought forward by Budd (*q.v.*), although with a different disease, and because it illustrates the special form of experiment required to establish the absence of something as causal factor. The term *vitamine* was introduced by Funk.

In judging the suitability of a food, we have not finished when we have determined the quantity of albumen (which is generally only calculated from the determination of nitrogen), fat, carbohydrates and salts, even when we have applied the corrections for the digestibility. We can indeed calculate from this whether a balance of nitrogen will be possible with it and whether the work which must be performed both internally and externally, can be obtained from it, but not whether permanent health is possible.

The determinations of metabolism made hitherto have taught us to understand the metabolism of the whole human being or of the whole animal, but, as the metabolism of the muscles and of the large glands is so great in comparison with that of other organs, the influence of these organs comes within the observation errors and within the individual differences of the metabolism of the former. This is the reason why we, in spite of the large number of tests of all kinds of foods, know practically nothing about the metabolism of the nervous system, while we only assume the metabolism in the peripheral nerves by analogy.

Most of the experiments on the balance of nitrogen were only continued for a few days, so that only an unchanged body weight and balance of nitrogen prove that the great metabolism (as I shall call it) was satisfied, but this did not show that the requirements of the more modest organs were met.

Not only are there a large number of different albuminoid substances, fats, etc., but there are also a number of composite substances, which no doubt play an important part, although their action is not fully explained. We need only remember the peculiar fact that scurvy, which usually develops from the lack of fresh food, which sometimes occurs on long sea voyages, is usually cured when the patients can again obtain fresh meat and fresh greens.

We must therefore bear in mind that still unknown substances may be in question.

There were two ways of gaining an insight into the matter. We might try to isolate different substances from the silver skin or we might seek for other food-stuffs which had the same effect and try to get at the truth by comparative analyses.

Experiments were made in both directions.

For the first series it had to be ascertained how much silver skin or fine rice bran (dedek) was necessary in order to prevent the neuritis, if only polished rice was given besides. Fowls were fed with boiled polished rice, to which carefully weighed quantities of dedek were added. Eijkman had already stated that fowls could be kept healthy with rice and fine rice bran and even be cured if already ill. He expressly states that the quantity of silver skin must not be too small.

INTRODUCTION

When, in July 1896, the Government commissioned me to investigate the physiological and pharmacological properties of the tannin contained in red rice, and possibly other constituents of this kind of rice which might require consideration in relation particularly to beri-beri, Dr. Eijkman, who had not yet published the results of his most recent experiments on fowl-polyneuritis, had left for Europe and Dr. Vorderman was on his tour of enquiry through Java and Madura. Moreover, a few days after my return from Atjeh, Mr. Roll, deputy director of this Institute, was absent from Batavia for a considerable period owing to an epizootic at Tegal. I had, therefore, to proceed on what I learned from letters on the subject and from information obtained at the laboratory for pathological anatomy and bacteriology.

It is, therefore, not surprising that at the beginning of my investigation I could not immediately utilise all that Dr. Eijkman's researches on the connection between polyneuritis and diet had brought to light.

The above-mentioned tannin had not yet been isolated from the silver skin, and I was, therefore, faced with the question from what point of view I should approach the problem.

A personal interview with the Director of Education, Religion and Industry had convinced me that it was not the intention of the Government that I should keep strictly to the letter of the commission, but that my investigation should aim, by discovering the connection between diet and polyneuritis and by extending the results obtained by Dr. Eijkman, to get an insight into the etiology of neuritis, which would make it possible to indicate the direction in which the rational control of neuritis should be sought.

At the same time it was to be ascertained how far the pigment found in the red rice might be considered as a curative or a preventive remedy against beri-beri.

It seemed to me that I should not confine myself to the red pigment, but should keep the whole silver skin in view, for it would be a strange chance if it were just the substance most easily chemically demonstrated which was the one we were seeking.

Besides, Mr. Roll had told me that according to the last, then still unpublished experiments of Dr. Eijkman, not only red, but all semi-decorticated rice prevented polyneuritis in fowls.

There were two ways open to me: either to continue the study of polyneuritis gallinarum or to carry out feeding experiments on a large scale in prisons or army corps where there was a great deal of beri-beri.

As to the latter, it was only rational to await the results of Dr. Vorderman's enquiry and to make proposals in connection with it. Besides, it was obvious that such experiments would be difficult and expensive, so that I resolved first to try and get more light on the connection between polyneuritis of fowls and diet.

I have now continued these fowl experiments almost uninterruptedly for more than three years, and think I have made sufficient progress for it to be worth while publishing the results already obtained.

Five cocks (Nos. 86–91) were fed for 16 days with boiled polished white rice.

One had distinct symptoms of polyneuritis, of which it died on the 20th day (No. 89). Another (No. 90) showed slight paresis and cyanosis. The three others had pale combs.

After the sixteenth day, at first 8 and later 10 grammes of fine dedek (rice bran) was mixed with their rice.

Of these, No. 87 died at the end of 2 months from a croupous inflammation of the throat and eye. No degeneration was found in the nerves. No. 90 progressed at first, but died after 72 days from a general infectious disease. Only a few remains of earlier degeneration were found in the nerves.

No. 88 had no symptom of neuritis after 5 months, but was suffering from a skin disease, accompanied later by a croupous affection of the throat, nose and eye, and died 6 months after the beginning of the experiment. A few old degenerated bundles were found in the nerves.

No. 86 was healthy at first, but after 3 months showed a diminution of muscular strength. The quantity of dedek given daily was therefore raised to 12 gr. when the paralytic symptoms disappeared. After 7 months it died from croupous inflammation of the throat and eye and some degenerated bundles were found in its nerves.   . . .

RECAPITULATION

If we examine the foregoing results, we come to the following considerations.

We had in polyneuritis gallinarum, as described by Eijkman, a disease which, both in its clinical symptoms and in the changes which it produces in the peripheral nervous system, very much resembles beri-beri. The cause of both was hitherto unknown.

There was one remarkable difference that, while a very close, direct connection appeared to exist between polyneuritis of fowls and the nature of the food, beri-beri could not be so directly connected with the feeding; indeed except for the results of Vorderman's enquiry, there were not many observations which appeared to support this connection.

Various foods were held responsible for causing beri-beri, but there was much lacking in the arguments for these assertions and even Vorderman's enquiry was disputed.

It seems to me that through my results, the analogy between fowl neuritis and beri-beri is much greater than it appeared to be at first; and this strengthens my conviction that the further study of the etiology of this disease will open up new aspects of that of beri-beri.

From the fact that polyneuritis also occurred among fowls when fed with gaba and foods which do not contain any carbohydrates, it appears that here also no *direct* connection exists between the occurrence of the disease and food.

I believe that the results of tests with sterilised and boiled meat entitle us to exclude a harmful constituent present in the food. The comparison of the results with sterilised and non-sterilised katjang hidjoe and polished rice also support this and the more so, because, owing to the development of polyneuritis after feeding with potato flour and milk sugar, the reasons were removed for assuming the presence of a specific harmful substance in certain sorts of starch (one might consider this as a previously formed poison or assume that during digestion a poison has been formed from it).

In my opinion an explanation of the peculiar symptoms which occur in the disease in question can be sought in but two directions. Either we can presume a deficiency, a partial starvation, or we can imagine that there is an agent distributed in nature, which exercises a degenerative influence on the nerves and that it depends on the nature of the food, whether the peripheral nervous system has enough power of resistance to get the better of this influence. In the latter case, it is most likely that the harmful agent is a micro-organism.

That the condition of nourishment of the tissues has a great influence on their resistant power against infectious organisms is already well known; one has only to think of the development of abscesses in tissues whose condition of nourishment (owing to previous ischaemia etc.) has suffered from the infection of staphylococci into the blood. That chemical substances can also modify the susceptibility for certain infectious diseases, appears from the tests of Charin, Guillemonot and Levaditi, who found that by repeated injections of organic acids (oxalic, lactic and citric acids) the power of resistance of animals to bacterium pyocyaneum was lessened, while injections with sulphuric soda, phosphoric soda and potash and with salt had the contrary effect.

The negative result of feeding tests by Eijkman in the Zoological Gardens at Amsterdam supported such a theory. However this author informed me in a private communication that in later tests with compulsory rice feeding he saw polyneuritis develop. So the possibility of an infection is not excluded, but the principal fact in its favour is removed.

There is also much to be said for the other explanation that we have to do with a partial starvation.

I mentioned in Section 4 that we know very little of the metabolism of the peripheral nervous system. From the comparatively small number of blood vessels in the nerves, we may however assume a low metabolism. If for the maintenance of the peripheral nervous system, a certain substance or group of substances is indispensable which are immaterial for the metabolism of the muscles, then it may be assumed that very little of them is necessary. We cannot be surprised if such substances have hitherto escaped observation i.e. chemical analysis. (I only need to remind you of the fact, that we do not yet know the form in which compounds of phosphorus and calcium, must be absorbed in our body in order to benefit us, to show how great are the deficiences in our knowledge of this matter.)

When therefore in certain foods the substances indispensable for the nervous system are lacking or are present in insufficient quantity, in the first place any reserve supply, which is present either in the nerve itself or in the blood or in some other organ, will be used up. When this occurs, disturbances will develop, just as in albumen starvation the circulating albumen is used up first and then that of the organs. Therefore from the time the deficiency begins, a certain time must elapse before symptoms set in in the nervous system; there will be a sort of incubation period. This will be the longer, the smaller the deficiency of the nutritive substances, which are indispensable for the nerve, in the food taken.

If besides these "protective substances," as I have named the still unknown compounds which I have mentioned now and then without any prejudice, albumen is also withheld, which is the case with an absolute diet, then, as is well-known, the albumen indispensable for metabolism is drawn from the organs, especially the muscles.   As the muscular substance also possesses protective qualities, it is probable that with the albumen from the muscles enough of the substances are liberated and become available for the nerves, to prevent polyneuritis.

This is a simple explanation why, when there is absolute starvation no polyneuritis is observed.

I admit that there is a third possibility.   One may assume the presence of a nerve-degenerating poison, which is able to originate in the intestinal canal and of an antidote, which neutralises the poison or, at any rate, its action.   The absence of this antidote would then open the door for the development of polyneuritis and in that case, the development of the disease would depend on the occurrence or non occurrence of the poison.   This theory appears to give an easy explanation of the fact that with the same diet, some fowls fall ill and others do not.

However, my investigations have raised doubts about this theory.   As the development of polyneuritis does not appear to be connected with the presence of starch or carbohydrates, but also occurred with an exclusively albuminoid diet, it is difficult to accept the hypothesis that with some foods, special conditions are present in the intestinal canal, favourable to the development of certain micro-organisms, which produce a nerve poison, as the conditions for the growth of these organisms would always be present.   The acceptance of such a widespread antidote also seems rather rash, as we find no example of this in the case of the known antitoxins.

Besides, the above-mentioned explanation is not as simple as it looks.   Why does the poison develop in some of the fowls fed in the same place and in the same way, and not in others.   And we cannot thus escape the necessity of taking into account individual differences, of which we have no explanation and which we generally call predisposition.   If, however, we take individual differences into account, we can, if we assume partial starvation, very easily understand why, with the same food, one fowl falls ill and another does not. We know that metabolism shows quantitative differences in individuals of the same kind.   We see that one person needs a much larger quantity of food than another to maintain his physical equilibrium, while doing the same work, than another.   We see one child pining away with a certain diet, while another grows and flourishes.   If therefore the total metabolism shows important quantitative and perhaps small qualitative differences, there is no reason why, in the separate tissues which together furnish the total metabolism, we should not assume individual quantitative differences.   And if we are justified in doing this, there is nothing absurd in the supposition that a food which contains just enough of the still unknown nerve nutritive substances for one fowl contains too little for another.

Whether we envisage the influence of diet on the development of polyneuritis gallinarum as a change in the predisposition or as a purely nutritive disturbance owing to partial starvation, we always come to the following conclusion:

There occur in various natural foods, substances, which cannot be absent without serious injury to the peripheral nervous system. The distribution of these substances in the different food-stuffs is very unequal. Of those examined, phaseolus radiatus and cajanus indicus were the richest, and polished rice the poorest in these substances.

The separation of these substances meets with the difficulty that they are so easily disintegrated. This disintegration, which takes place in a damp warm place, shows that they are very complex substances. They cannot be replaced by simple chemical compounds.

VII. EVOLUTION AND HEREDITY

Papers related to the production and conservation of new types

BEGINNINGS OF MODERN PALEONTOLOGY

STENO, (Steensen, Stensen, Stenson), Nicolaus (Nils, etc.) (Danish physiologist, mathematician, mineralogist, and priest, 1638–1686). From *De solida intra solidum contento dissertatio prodromus*, Florence, 1669; tr. by J. Winter as *The prodromus of N. S.'s dissertation*, New York, 1916; by permission of the director of the University of Michigan Press.

Steno's intellectual development presents a contradiction in that the further he moved from religious doctrine in his scientific speculations, the nearer he drew to it in his religious affirmations. Thus, despite this prophetic study, in which he showed how to read the history of earth's past from its present state, Steno became converted, took orders, and, after a stormy ecclesiastical career, died as a result of self-imposed asceticism. His other scientific work was chiefly in muscular physiology.*

Concerning the matter of the strata the following can be affirmed:

1. If all the particles in a stony stratum are seen to be of the same character, and fine, it can in no wise be denied that this stratum was produced at the time of the creation from a fluid which at that time covered all things; and Descartes also accounts for the origin of the earth's strata in this way.

2. If in a certain stratum the fragments of another stratum, or the parts of animals and plants are found, it is certain that the said stratum must not be reckoned among the strata which settled down from the first fluid at the time of the creation.

3. If in a certain stratum we discover traces of salt of the sea, the remains of marine animals, the timbers of ships, and a substance similar to the bottom of the sea, it is certain that the sea was at one time in that place, whatever be the way it came there, whether by an overflow of its own or by the upheaval of mountains.

4. If in a certain stratum we find a great abundance of rush, grass, pine cones, trunks and branches of trees, and similar objects, we rightly surmise that this matter was swept thither by the flooding of a river, or the inflowing of a torrent.

5. If in a certain stratum pieces of charcoal, ashes, pumice-stone, bitumen, and calcined matter appear, it is certain that a fire occurred in the neighborhood of the fluid; the more so if the entire stratum is composed throughout of ash and charcoal, such as I have seen outside the city of Rome, where the material for burnt bricks is dug.

6. If the matter of all the strata in the same place be the same, it is certain

---

* *De musculis et glandulis, etc.,* Copenhagen, 1664; and *Elementorum myologiae specimen,* 1667.

that that fluid did not take in fluids of a different character flowing in from different places at different times.

7. If in the same place the matter of the strata be different, either fluids of a different kind streamed in thither from different places at different times (whether a change of winds or an unusually violent downpour of rains in certain localities be the cause) or the matter in the same sediment was of varying gravity, so that first the heavier particles, then the lighter, sought the bottom. And a succession of storms might have given rise to this diversity, especially in places where a like diversity of soils is seen.

8. If within certain earthy strata stony beds are found, it is certain either that a spring of petrifying waters existed in the neighborhood of that place, or that occasionally eruptions of subterranean vapors occurred, or that the fluid, leaving the sediment which had been deposited, again returned when the upper crust had become hardened by the sun's heat.   . . .

I pass to the more particular investigation of those solids dug from the earth which have given rise to many disputes; especially incrustations, deposits, angular bodies, the shells of marine animals, of mollusks, and the forms of plants. Under incrustations belong rocks of every kind consisting of layers, whose two surfaces are indeed parallel but not extended in the same plane.   The place where incrustations are formed is the entire common boundary of fluid and solid; and the result is that the form of the layers or crusts corresponds to the form of the place, and it is easy to determine which of them hardened first, which last.   For if the place was concave, the outer layers were formed first; if convex, the inner; if the place was uneven because of various larger projections, the new layers were produced in the larger spaces when the narrower spaces had been filled with the formation of the first layers.

From this fact it is easy to account for all the differences of form which are seen in sections of similar rocks, whether they show the round veins of a tree cut transversely, or resemble the sinuous folds of serpents, or run along, curved in any other way, without law.   Nor is it surprising that agates and other kinds of incrustations seem, so far as regards their outer surface, rough like ordinary stones, since the outer surface of the outer layer imitates the roughness of the place.   In torrents, however, incrustations of this kind are more frequently found outside of the place of their production, because the matter of the place has been scattered by a breaking up of the strata.

Concerning the manner in which particles of the layers which are to be added to a solid are separated from the fluid, the following at least is certain:

1. That there is in it no place for buoyancy or gravity.

2. That the particles are added to surfaces of every kind, since surfaces smooth, rough, plane, curved, and consisting of several planes at different angles of inclination, are found overspread by the layers.

3. That movement of the fluid causes them no hindrance.

Whether the substance under consideration which flows from a solid, be different from that substance which moves the parts of the fluid, or whether something else is to be sought, I leave undecided.

Different kinds of layers in the same place can be caused either by a difference of the particles which withdraw from the fluid one after the other, as this same fluid is gradually disintegrated more and more, or from different fluids carried thither at different times. From this fact it follows that the same arrangement of layers sometimes recurs in the same place, and often evident traces revealing the entrance of new matter remain. But all the matter of the layers seems to be a finer substance emanating from the stones, as will further appear in the following. . . .

OTHER PARTS OF ANIMALS

What has been said concerning shells must also be said concerning other parts of animals, and animals themselves buried in the earth. Here belong the teeth of sharks, the teeth of the eagle-fish, the vertebrae of fishes, whole fish of every kind, the crania, horns, teeth, femurs, and other bones of land animals; since all these are either wholly like true parts of animals, or differ from them only in weight and color, or have nothing in common with them except the outer shape alone.

A great difficulty is caused by the countless number of teeth which every year are carried away from the island of Malta; for hardly a single ship touches there without bringing back with it some proofs of that marvel. But I find no other answer to this difficulty than:

1. That there are six hundred and more teeth to each shark, and all the while the sharks live new teeth seem to be growing.

2. That the sea, driven by winds, is wont to thrust the bodies in its path toward some one place and to heap them up there.

3. That sharks come in shoals and so the teeth of many sharks can be left in the same place.

4. That in lumps of earth brought here from Malta, besides different teeth of different sharks, various mollusks are also found, so that even if the number of teeth favors attributing their production to the earth, yet the structure of these same teeth, the abundance in each animal, the earth resembling the bottom of the sea, and the other sea objects found in the same place, all alike support the opposite view.

Others find great difficulty in the size of the femurs, crania, and teeth, and other bones, which are dug from the earth. But the objection, that an extraordinary size makes it necessary to conclude the size to be beyond the powers of Nature, is not of so great moment, seeing that:

1. In our own time bodies of men of exceedingly tall stature have been seen.

2. It is certain that men of unnatural size existed at one time.

3. The bones of other animals are often thought to be human bones.

4. To ascribe to Nature the production of truly fibrous bones is the same as saying that Nature can produce a man's hand without the rest of the man.

There are those to whom the great length of time seems to destroy the force of the remaining arguments, since the recollection of no age affirms that floods rose to the place where many marine objects are found to-day, if you exclude the universal deluge, four thousand years, more or less, before our time. Nor does it seem in accord with reason that a part of an animal's body could withstand the ravages of so many years, since we see that the same bodies are often destroyed completely in the space of a few years. But this doubt is easily answered, since the result depends wholly upon the diversity of soil; for I have seen strata of a certain kind of clay which by the thinness of their fluid decomposed all the bodies enclosed within them. I have noticed many other sandy strata which preserved whole all that was entrusted to them. And by this test it might be possible to come to a knowledge of that fluid which disintegrates solid bodies. But that which is certain, that the formation of many mollusks which we find to-day must be referred to times coincident with the universal deluge, is sufficiently shown by the following argument.

It is certain that before the foundations of the city of Rome were laid, the city of Volterra was already powerful. But in the exceedingly large stones which are found in certain places (the remains of the oldest walls) at Volterra, shells of every kind are found, and not so very long ago there was hewn from the midst of the forum a stone packed full of striated shells; hence it is certain that the shells found to-day in the stones had already been formed at the time when the walls of Volterra were being built.

And in order that no one may say that the shells only have turned into stone, or that having been enclosed within the stone they have suffered no destruction from the tooth of time, we may remark that the whole hill upon which the most ancient of Etruscan cities is built, rises from the deposits of the sea, placed one above the other, and parallel to the horizon; and in these deposits many strata, not of stone, abound in mollusks that are real and have suffered no change at all; so it is possible to affirm that the unchanged shells which we dig from them to-day were formed three thousand and more years ago. From the founding of Rome to our own times, we reckon two thousand four hundred and twenty years and more; who will not grant that many ages elapsed from the time the first men transferred their homes to Volterra until it grew to the flourishing size it possessed at the time of the founding of Rome? And if to these centuries we add the time which intervened between the first sedimentary deposit of the hill of Volterra, and the time when that same hill was left by the sea and strangers flocked to it, we shall easily go back to the very times of the universal deluge. . . .

Six distinct aspects of Tuscany we therefore recognize, two when it was fluid, two when level and dry, two when it was broken; and as I prove this fact concerning Tuscany by inference from many places examined by me, so do I affirm it with reference to the entire earth, from the descriptions of different places contributed by different writers. But in order that no one may be alarmed by the novelty of my view, in a few words I shall set forth the agreement of Nature with Scripture by reviewing the chief difficulties which can be urged regarding the different aspects of the earth.

In regard to the first aspect of the earth Scripture and Nature agree in this, that all things were covered with water; how and when this aspect began, and how long it lasted, Nature says not, Scripture relates. That there was a watery fluid, however, at a time when animals and plants were not yet to be found, and that the fluid covered all things, is proved by the strata of the higher mountains, free from all heterogeneous material. And the form of these strata bears witness to the presence of a fluid, while the substance bears witness to the absence of heterogeneous bodies. But the similarity of matter and form in the strata of mountains which are different and distant from each other, proves that the fluid was universal. But if one say that the solids of a different kind contained in those strata were destroyed in course of time, he will by no means be able to deny that in that case a marked difference must have been noticed between the matter of the stratum and the matter which percolated through the pores of the stratum, filling up the spaces of the bodies which had been destroyed. If, however, other strata which are filled with different bodies are, in certain places, found above the strata of the first fluid, from this fact nothing would follow excepting that above the strata of the first fluid new strata were deposited by another fluid, whose matter could likewise have refilled the wastes of the strata left by the first fluid. Thus we must always come back to the fact that at the time when those strata of matter unmixed, and evident in all mountains, were being formed, the rest of the strata did not yet exist, but that all things were covered by a fluid free from plants and animals and other solids. Now since no one can deny that these strata are of a kind which could have been produced directly by the First Cause, we recognize in them the evident agreement of Scripture with Nature.

Concerning the time and manner of the second aspect of the earth, which was a plane and dry, Nature is likewise silent, Scripture speaks. As for the rest Nature, asserting that such an aspect did at one time exist, is confirmed by Scripture, which teaches us that the waters welling from a single source overflowed the whole earth.

When the third aspect of the earth, which is determined to have been rough, began, neither Scripture nor Nature makes plain. Nature proves that the unevenness was great, while Scripture makes mention of mountains at the time of the flood. But when those mountains, of which Scripture in this connection makes mention, were formed, whether they were identical with mountains of the present day, whether at the beginning of the deluge there was the same depth of valleys as there is to-day, or whether new breaks in the strata opened new chasms to lower the surface of the rising waters, neither Scripture nor Nature declares.

The fourth aspect, when all things were sea, seems to cause more difficulty, although in truth nothing difficult is here presented. The formation of hills from the deposit of the sea bears witness to the fact that the sea was higher than it is now, that too not only in Tuscany but in very many places distant enough from the sea, from which the waters flow toward the Mediterranean; nay, even in those places from which the waters flow down into the ocean. Nature does not oppose Scripture in determining how great that height of the sea was, seeing that:

1. Definite traces of the sea remain in places raised several hundreds of feet above the level of the sea.

2. It cannot be denied that as all the solids of the earth were once, in the beginning of things, covered by a watery fluid, so they could have been covered by a watery fluid a second time, since the changing of the things of Nature is indeed constant, but in Nature there is no reduction of anything to nothing. But who has searched into the formation of the innermost parts of the earth, so that he dare deny that huge caverns may exist there, filled sometimes with a watery fluid, sometimes with a fluid akin to air?

3. It is wholly uncertain what the depth of valleys at the beginning of the deluge was; reason, however, may urge that in the first ages of the world smaller cavities had been eaten out by water and fire, and that in consequence not so deep breaks of strata followed from this cause; while the highest mountains of which Scripture speaks were the highest of those mountains which were in existence at that time, not of those which we see to-day.

4. If the movement of a living being can bring it to pass that places which have been overwhelmed with waters are arbitrarily made dry, and are again overwhelmed with waters, why should we not voluntarily grant the same freedom and the same powers to the First Cause of all things?

In regard to the time of the universal deluge, secular history is not at variance with sacred history, which relates all things in detail. The ancient cities of Tuscany, of which some were built on hills formed by the sea, put back their birthdays beyond three thousand years; in Lydia, moreover, we come nearer to four thousand years: so that it is possible thence to infer that the time at which the earth was left by the sea agrees with the time of which Scripture speaks.

As regards the manner of the rising waters, we could bring forward various agreements with the laws of Nature. But if some one say that in the earth the centre of gravity does not always coincide with the centre of the figure, but recedes now on one side, and now on the other, in proportion as subterranean cavities have formed in different places, it is possible to assign a simple reason why the fluid, which in the beginning covered all things, left certain places dry, and returned again to occupy them.

The universal deluge may be explained with the same ease if a sphere of water, or at least huge reservoirs, be conceived around a fire in the middle of the earth; thence, without the movement of the centre, the pouring forth of the pent-up water could be derived. But the following method also seems to me to be very simple, whereby both a lesser depth of the valleys and a sufficient amount of water are obtained without taking into account the centre, or figure, or gravity. For if we shall have conceded (1) That by the slipping of fragments of certain strata, the passages were stopped through which the sea penetrating into hollow places of the earth sends forth the water to bubbling springs; (2) That the water undoubtedly enclosed in the bowels of the earth, was, by the force of the known subterranean fire in part driven toward springs, and in part forced up into the air through the pores of the ground which had not yet been covered with water; that, moreover, the water which not only is always present in the air but also was mixed with it in the manner previously described, fell in the form of rain; (3) That the bottom of the sea was raised through the enlarging of subterranean caverns; (4) That the cavities remaining on the surface of the earth, were filled with earthy matter washed from the higher places by the constant falling of rains; (5) That the very surface of the earth was less uneven, because nearer to its beginnings—if we shall have granted all this, we shall have admitted nothing opposed to Scripture, or reason, or daily experience.

What happened on the surface of the earth while it was covered with water, neither Scripture nor Nature makes clear; this only can we assert from Nature, that deep valleys were formed at that time. This is (1) because the cavities, made larger by the force of subterranean fires, furnished room for greater downfalls; (2) because a return passage had to be opened for the water into the deeper parts of the earth; (3) because to-day, in places far from the sea are seen deep valleys filled with many marine deposits.

As for the fifth aspect, which revealed huge plains after the earth had again become dry, Nature proves that those plains existed, and Scripture does not gainsay it. For the rest, whether the entire sea presently receded, or whether, indeed, in the course of ages new chasms opening afforded opportunity for disclosing new regions, it is possible to determine nothing with certainty, since Scripture is silent, and the history of nations regarding the first ages after the deluge is doubtful in the view of the nations themselves, and thought to be full of myths. This, indeed, is certain, that a great amount of earth was carried down every year into the sea (as is easily clear to one who considers the size of rivers, and their long courses through inland regions, and the countless number of mountain streams, in short, all the sloping places of the earth), and that the earth thus carried down by rivers, and added day by day to the shore, left new lands suited for new habitations.

This is in fact confirmed by the belief of the ancients, in accordance with which they called whole regions the gifts . . . of rivers of like name, as also by the traditions of the Greeks, since they relate that men, descending little by little from the mountains, inhabit places bordering on the sea that were sterile by reason of excessive moisture, but in course of time became fertile.

The sixth aspect of the earth is evident to the senses; herein the plains left by the waters, especially by reason of erosion, and at times through the burning of fires, passed over into various channels, valleys, and steep places. And it is not to be wondered at that in the historians there is no account as to when any given change took place. For the history of the first ages after the deluge is confused and doubtful in secular writers; as the ages passed, moreover, they felt constrained to celebrate the deeds of distinguished men, not the wonders of Nature. Nevertheless the records, which ancient writers mention, of those who wrote the history of the changes which occurred in various places, we do not possess. But since the authors whose writings have been preserved report as marvels almost every year, earthquakes, fires bursting forth from the earth, overflowings of rivers and seas, it is easily apparent that in four thousand years many and various changes have taken place.

Far astray, therefore, do they wander, who criticize the many errors in the writings of the ancients, because they find there various things inconsistent with the geography of to-day. I should be unwilling to put credence in the mythical accounts of the ancients; but there are in them also many things to which I would not gainsay belief. For in those accounts I find many things of which the falsity rather than the truth seems doubtful to me. Such are the separation of the Mediterranean Sea from the western ocean; the passage from the Mediterranean into the Red Sea; and the submersion of the island Atlantis. The description of various places in the journeys of Bacchus, Triptolemus, Ulysses, Aeneas, and of others, may be true, although it does not correspond with present day facts. Of the many changes which have taken place over the whole extent of Tuscany embraced between the Arno and the Tiber, I shall adduce evident proofs in the Dissertation itself; and although the time, in which the individual changes occurred, cannot be determined, I shall nevertheless adduce those arguments from the history of Italy, in order that no doubt may be left in the mind of anyone.

And this is the succinct, not to say disordered, account of the principal things which I had decided to set forth in the Dissertation, not only with greater clearness but also with greater fulness, adding a description of the places where I have observed each thing.

EVOLUTION AS DETERIORATION

de BUFFON, (Count) George(s) Louis Leclerc (French natural philosopher, 1707–1788). From *Histoire naturelle, générale et particulière*, Paris, 1749; tr. by W. Smellie* as *The natural history of man and quadrupeds*, in *Natural history, general and particular, by the Count de Buffon*, London, 1791.

Buffon's *Natural History*, in particular the famous chapter on *Degeneration*, poses a majority of the problems which were to dominate evolutionary thought for the succeeding century and a half. Isolation, distribution, transmutation, correlation, variation—all are recognized and discussed. In the *Natural History* as a whole, the amount of amassed detail, both intellectual and empirical, is impressive. Buffon, like other early modern evolutionists, excludes the possibility of progressive change. This appears less strange when one considers that ancient ideas revived during the Renaissance discredited such a possibility: progressive evolution ceased to receive serious consideration in Greece after the introduction of "ideal forms" by Democritus and Socrates; moreover, according to Biblical tradition, it was to be doubted whether God would have created species which were imperfectly adapted.

OF THE DEGENERATION OF ANIMALS.

Whenever man began to change his climate, and to migrate from one country to another, his nature was subjected to various alterations. In temperate countries, which we suppose to be adjacent to the place where he was originally produced, these alterations have been slight; but they augmented in proportion as he receded from this station: and, after many ages had elapsed; after he had traversed whole continents, and intermixed with races already degenerated by the influence of different climates; after he was habituated to the scorching heats of the south, and the frozen regions of the north; the changes he underwent became so great and so conspicuous, as to give room for suspecting, that the Negro, the Laplander, and the White, were really different species, if, on the one hand, we were not certain, that one man only was originally created, and, on the other, that the White, the Laplander, and the Negro, are capable of uniting, and of propagating the great and undivided family of the human kind. Hence those marks which distinguish men who inhabit different regions of the earth, are not original, but purely superficial. It is the same identical being who is varnished with black under the Torrid Zone, and tawned and contracted by extreme cold under the Polar Circle. This circumstance is alone sufficient to show, that the nature of man is endowed with greater strength, extension, and flexibility, than that of any other terrestrial being; for vegetables, and almost all the animals, are confined to particular soils and climates. This extension of our nature depends more on the qualities of the mind than on those of the body. It is by the mind that man has been enabled to find those resources which the delicacy of his body required, to brave

---

* With minor emendations by the editor.

the inclemency of the sky, and to conquer the rigidity and barrenness of the earth. He may be said to have subdued the elements: by an exertion of his intellect, he produced the element of fire, which had no existence on the surface of the earth. His sagacity taught him how to clothe his body, and to build houses for defending himself against every external attack. By the powers of genius he supplied all the qualities which are wanting in matter. Without possessing the strength, the magnitude, or the robustness of most animals, he knew how to conquer, to tame, and to confine them: he made himself master of those regions which Nature seemed to have resigned to them as an exclusive possession.

The earth is divided into two great continents. The antiquity of this division exceeds that of all human monuments; and yet man is more ancient; for he is the same in both worlds. The Asiatic, the European, and the Negro, produce equally with the American. Nothing can be a stronger proof that they belong to the same family, than the facility with which they unite to the common stock. The blood is different; but the germ is the same. The skin, the hair, the features, and the stature, have varied, without any change in the internal structure. The type is general and common: and if, by any great revolution, man were forced to abandon those climates which he had invaded, and to return to his native country, he would, in the progress of time, resume his original features, his primitive stature, and his natural colour. But the mixture of races would produce this effect much sooner. A white male with a black female, or a black male with a white female, equally produce a mulatto, whose colour is brown, or a mixture of black and white. This mulatto intermixing with a white, produces a second mulatto, less brown than the former; and, if the second mulatto unites with a white, the third will have only a slight shade of brown, which will entirely vanish in future generations. Hence, by this mixture, 150, or 200 years, are sufficient to bleach the skin of a Negro. But, to produce the same effect by the influence of climate alone, many centuries would perhaps be necessary. Since the Negroes were transported to America, which is about 200 years ago, the Negro families, who have preserved themselves from mixture, seem not to have lost any shade of their original colour. The climate of South America, it is true, being sufficiently hot to tawn its inhabitants, we ought not to be surprised that the Negroes there continue black. To put the change of colour in the human species to the test of experiment, some Negroes should be transported from Senegal to Denmark, where the inhabitants have generally white skins, golden locks, and blue eyes, and where the difference of blood and opposition of colour are greatest. These Negroes must be confined to their own females, and all crossing of the breed scrupulously prevented. This is the only method of discovering the time necessary to change a Negro into a White, or a White into a Black, by the mere operation of climate. . . .

In brute animals, these effects are greater and more suddenly accomplished; because they are more nearly allied to the earth than man; because their food, being more uniformly the same, and nowise prepared, its qualities are more decided, and its influence stronger; and because the animals, being unable to clothe themselves, or to use the element of fire, remain perpetually exposed to the action of the air, and all the inclemencies of the climate. For this reason, each of them, according to its nature, has chosen its zone and its country: for the same reason, they remain there, and, instead of dispersing themselves, like man, they generally continue in those places which are most friendly to their constitutions. But, when forced by men, or by any revolution on the globe, to abandon their native soil, their nature undergoes changes so great, that, to recognise them, recourse must be had to accurate examination, and even to experiment and analogy. If to these natural causes of alteration in free animals, we add that of the empire of man over those which he has reduced to slavery, we shall be astonished at the degree to which tyranny can degrade and disfigure Nature; we shall perceive the marks of slavery, and the prints of her chains; and we shall find, that these wounds are deeper and more incurable, in proportion to their antiquity; and that, in the present condition of domestic animals, it is perhaps impossible to restore their primitive form, and those attributes of Nature which we have taken from them.

Thus the temperature of the climate, the quality of the food, and the evils produced by slavery, are the three causes of the changes and degeneration of animals. The effects of each merit a separate examination; and their relations, when viewed in detail, will exhibit a picture of Nature in her present condition, and of what she was before her degradation.

Let us compare our pitiful sheep with the mouflon, from which they derived their origin. The mouflon is a large animal. He is fleet as a stag, armed with horns and thick hoofs, covered with coarse hair, and dreads neither the inclemency of the sky, nor the voracity of the wolf. He not only escapes from his enemies by the swiftness of his course, but he resists them by the strength of his body, and the solidity of the arms with which his head and feet are fortified. How different from our sheep, who subsist with difficulty in flocks, who are unable to defend themselves by their numbers, who cannot endure the cold of our winters without shelter, and who would all perish, if man withdrew his protection. In the warmest climates of Asia and Africa, the mouflon, who is the common parent of all the races of this species, appears to be less degenerated than in any other region. Though reduced to a domestic state, he has preserved his stature and his hair; but the size of his horns is diminished. Of all domestic sheep, those of Senegal and India are the largest, and their nature has suffered least degradation. The sheep of Barbary, Egypt, Arabia, Persia, Calmuck, &c., have undergone greater changes. In relation to man, they are improved in some articles, and vitiated in others. But, with regard to Nature, improvement and degeneration are the same thing; for they both imply an alteration of original constitution. Their coarse hair is changed into fine wool. Their tail, loaded with a mass of fat, has acquired a magnitude so incommodious, that the animals trail it with pain. While swollen with superfluous matter, and adorned with a beautiful fleece, their strength, agility, magnitude, and arms, are diminished: these long tailed sheep are half the size only of the mouflon. They can neither fly from danger, nor resist the enemy. To preserve and multiply the species, they require the constant care and support of man.

The degeneration of the original species is still greater in our climates. Of all the qualities of the mouflon, our ewes and rams have retained nothing but a small portion of vivacity, which yields to the crook of the shepherd. Timidity, weakness, resignation, and stupidity, are the only melancholy remains of their degraded nature. To restore their original size and strength, our Flanders sheep should be united with the mouflon, and prevented from propagating with inferior races; and, if we would devote the species to the more useful purposes of affording us good mutton and wool, we should imitate some neighbouring nations in propagating the Barbary race of sheep, which, after being transported into Spain, and even into Britain, have succeeded very well. Strength and magnitude are male attributes; plumpness and beauty of skin are female qualities. To obtain fine wool, therefore, our rams should have Barbary ewes; and, to augment the size, our ewes should be served with the male mouflon.

Our goats might be managed in the same manner. By intermixing them with the goat of Angora, their hair might be changed, and rendered equally useful as the finest wool. In our climate the species of the goat is not so much degenerated as that of the sheep. It appears to be still more degenerated in the warm countries of Africa and India. The smallest and weakest goats are those of Guiney, Juda, &c., and yet these countries produce the largest and strongest sheep. . . .

The wild animals, not being under the immediate dominion of man, are not subject to such great changes as the domestic kinds. Their nature seems to vary with different climates; but it is no where degraded: if they were capable of choosing their climate and their food, the changes they undergo would be still less. But, as they have at all times been hunted and banished by man, or even by the strongest and most ferocious quadrupeds, most of them have been obliged to abandon their native country, and to occupy lands less friendly to their constitution. Those whose nature had ductility enough to accommodate themselves to this new situation, have diffused over vast territories; while others have had no other resource than to confine themselves in the deserts adjacent to their own country. There is no animal which, like man, has spread over the whole surface of the earth. A great number of species are limited to the southern regions of the Old, and others to the same regions of the New Continent. Others, though fewer, are confined to the northern regions, and, instead of extending to the south, have passed from the one continent to the other, by routes which are still undiscovered. Other species inhabit only particular mountains or valleys: and the changes in their nature are, in general, less sensible, the more they are confined to a small circle.

As climate and food have little influence on wild animals, and the empire of man still less, the chief varieties amongst them proceed from another cause. They depend on the number of individuals of those which produce, as well as of those that are produced. In those species in which the male attaches himself to one female, as in that of the roebuck, the young demonstrate the fidelity of their parents by their entire resemblance to them. In those, on the contrary, the females of which often change the male, as in that of the stag, the varieties are numerous: and as, through the whole extent of Nature, there is not one individual perfectly similar to another, the varieties among animals are proportioned to the number and frequency of their produce. In species, the females of which bring forth five or six young three or four times a year, the number of varieties must be much greater than in those which produce but a single young once a year. The small animals, accordingly, which produce oftener, and in greater numbers than the larger kinds, are subject to greater varieties. Magnitude of body, which appears to be a relative quality only, possesses positive rights in the laws of Nature. Magnitude is as fixed as minuteness is changeable. Of this fact we shall soon be convinced by the following enumeration of the varieties which take place in the large and small animals.

In Guiney the boar has acquired long ears lying along the back; in China, a large, pendant belly and quite short legs; at Cape Verde and elsewhere, heavy tusks curled like a bull's horns; wherever domesticated, it has acquired semi-pendant ears, and, in cold or temperate climates, white bristles.

The roe deer in mountainous, dry, and hot countries, like Sardinia and Corsica, has lost part of its tail and acquired brown hair and brownish horns; in cold humid countries, as Bohemia and the Ardennes, its stature has increased, its coat and horns become very dark brown, its hair lengthened to the point of forming a long beard at the chin; in the north of the other continent [*i.e.*, North America—Ed.] the horns are more spread out and branched, with curved antlers. Under domestication the coat changes from brown to white, and unless the animal has its liberty and plenty of space its legs become curved and deformed.

The elephant species is the only one which servitude and domestication has not affected, since under these conditions they refuse to reproduce and hence fail to transmit to their kind the wounds and defects occasioned by their condition.

The dog, the wolf, the fox, the jackal, and the isatis, form another genus, the different species of which are so similar, particularly in their internal structure and the organs of generation, that it is difficult to conceive why they do not intermix. From the experiments I made with regard to the union of the dog with the wolf and fox, the repugnance to copulation seemed to proceed rather from the wolf than the dog, that is, from the wild, and not from the domestic animal; for the bitches which I put to the trial would have willingly permitted the fox and wolf; but the she-wolf and female fox would never suffer the approaches of the dog. The domestic state seems to render animals less faithful to their species. It likewise makes them more ardent and more fertile; the bitch generally produces twice in a year: but the she-wolf and the female fox produce only once in the same period: and it is probably, that the dogs who have become wild and have multiplied in the island of Juan Fernandes, and in the mountains of St. Domingo, produce but once a year, like the fox and wolf. Were this fact ascertained, it would fully establish the unity of genus in these three animals, who resemble each other so much in structure, that their repugnance to intermixing must be solely ascribed to some external circumstances. . . .

Beyond this brief glance which we have cast over these varieties, indicating to us the alterations peculiar to each species, a more important consideration presents itself, with a broader aspect, namely that of the alteration of the species themselves, this more ancient decline which seems to have occurred since time immemorial in each family, or, if one prefers, in each genus within which we can include related species. We have over the entire earth only a few isolated species which, as in man's case, constitute species and genus simultaneously; the elephant, the rhinoceros, the hippopotamus, the giraffe form genera or simple species propagating in a direct line only with no collateral branches. All the rest appear to form families in which we may recognize ordinarily a chief common stock from which there appear to have emerged different species in proportion to the smallness and size and the fertility of the individuals concerned.

DARWIN, Erasmus (English naturalist, 1731–1801). *Zoonomia, or the laws of organic life*, London, 1794–1896.

For the views of Erasmus Darwin on the subject of descent, see his discussion of materialism and epigenesis, pp. 387–392.

POPULATION AND NUTRITION

MALTHUS, Thomas Robert (English political economist, 1766–1834). From *An essay on the principle of the population; or, a view of its past and present effects on human happiness; with an inquiry into our prospects respecting the future removal or mitigation of the evils which it occasions*, London, 1803.*

The dangers inherent in rapidly expanding population were brought into focus partly in connection with the political liberalism of eighteenth-century France and its rapid spread in Western Europe at this period. Malthus's doctrine is said to have been developed by him during conversations with his father on the relations of human happiness to political theory and economic fact. Darwin states that he developed his own selection hypothesis immediately after reading Malthus (in 1838), and the Darwinian theory may be considered, in part, an extension of Malthusian concepts to embrace the organic world as a whole. When with the industrial revolution the use of machinery greatly increased the yield of the soil per man-hour, Malthus's ideas fell into disfavor. Recently, they have been revived.

Ratios of the Increase of Population and Food

In an inquiry concerning the improvement of society, the mode of conducting the subject which naturally presents itself, is—

1. To investigate the causes that have hitherto impeded the progress of mankind towards happiness; and,

2. To examine the probability of the total or partial removal of these causes in future.

To enter fully into this question, and to enumerate all the causes that have hitherto influenced human improvement, would be much beyond the power of an individual. The principal object of the present essay is to examine the effects of one great cause intimately united with the very nature of man; which, though it has been constantly and powerfully operating since the commencement of society, has been little noticed by the writers who have treated this subject. The facts which establish the existence of this cause have, indeed, been repeatedly stated and acknowledged; but its natural and necessary effects have been almost totally overlooked; though probably among these effects may be reckoned a very considerable portion of that vice and misery, and of that unequal distribution of the bounties of nature, which it has been the unceasing object of the enlightened philanthropist in all ages to correct.

The cause to which I allude, is the constant tendency in all animated life to increase beyond the nourishment prepared for it.

It is observed by Dr. Franklin, that there is no bound to the prolific nature of plants or animals but what is made by their crowding and interfering with

* The work first appeared anonymously in shorter form as *An essay on the principle of population as it affects the future improvement of society, with remarks on the speculations of Mr. Godwin, M. Gondorcet, and other writers*, London, 1798.

each other's means of subsistence. Were the face of the earth, he says, vacant of other plants, it might be gradually sowed and overspread with one kind only, as, for instance, with fennel: and were it empty of other inhabitants, it might in a few ages be replenished from one nation only, as, for instance, with Englishmen.

This is incontrovertibly true. Throughout the animal and vegetable kingdoms Nature has scattered the seeds of life abroad with the most profuse and liberal hand; but has been comparatively sparing in the room and the nourishment necessary to rear them. The germs of existence contained in this earth, if they could freely develop themselves, would fill millions of worlds in the course of a few thousand years. Necessity, that imperious, all-pervading law of nature, restrains them within the prescribed bounds. The race of plants and the race of animals shrink under this great restrictive law; and man cannot by any efforts of reason escape from it.

In plants and irrational animals, the view of the subject is simple. They are all impelled by a powerful instinct to the increase of their species, and this instinct is interrupted by no doubts about providing for their offspring. Wherever, therefore, there is liberty, the power of increase is exerted, and the superabundant effects are repressed afterwards by want of room and nourishment.

The effects of this check on man are more complicated. Impelled to the increase of his species by an equally powerful instinct, reason interrupts his career, and asks him whether he may not bring beings into the world for whom he cannot provide the means of support. If he attend to this natural suggestion, the restriction too frequently produces vice. If he hear it not, the human race will be constantly endeavouring to increase beyond the means of subsistence. But as, by that law of our nature which makes food necessary to the life of man, population can never actually increase beyond the lowest nourishment capable of supporting it, a strong check on population, from the difficulty of acquiring food, must be constantly in operation. This difficulty must fall somewhere, and must necessarily be severely felt in some or other of the various forms of misery, or the fear of misery, by a large portion of mankind.

That population has this constant tendency to increase beyond the means of subsistence, and that it is kept to its necessary level by these causes, will sufficiently appear from a review of the different states of society in which man has existed. But, before we proceed to this review, the subject will perhaps be seen in a clearer light, if we endeavour to ascertain what would be the natural increase of population, if left to exert itself with perfect freedom; and what might be expected to be the rate of increase in the productions of the earth, under the most favourable circumstances of human industry.

It will be allowed that no country has hitherto been known, where the manners were so pure and simple, and the means of subsistence so abundant, that no check whatever has existed to early marriages from the difficulty of providing for a family, and that no waste of the human species has been occasioned by vicious customs, by towns, by unhealthy occupations, or too severe labour. Consequently in no state that we have yet known, has the power of population been left to exert itself with perfect freedom.

Whether the law of marriage be instituted or not, the dictate of nature and virtue seems to be an early attachment to one woman and where there were no impediments of any kind in the way of a union to which such an attachment would lead, and no causes of depopulation afterwards, the increase of the human species would be evidently much greater than any increase which has been hitherto known. . . .

According to a table of Euler, calculated on a mortality of 1 in 36, if the births be to the deaths in the proportion of 3 to 1, the period of doubling will be only twelve years and four-fifths. And this proportion is not only a possible supposition, but has actually occurred for short periods in more countries than one.

Sir William Petty supposes a doubling possible in so short a time as ten years.

But, to be perfectly sure that we are far within the truth, we will take the slowest of these rates of increase, a rate in which all concurring testimonies agree, and which has been repeatedly ascertained to be from procreation only.

It may safely be pronounced, therefore, that population, when unchecked, goes on doubling itself every twenty-five years, or increases in a geometrical ratio. . . .

That we may be the better able to compare the increase of population and food, let us make a supposition, which, without pretending to accuracy, is clearly more favourable to the power of production in the earth than any experience we have had of its qualities will warrant.

Let us suppose that the yearly additions which might be made to the former average produce, instead of decreasing, which they certainly would do, were to remain the same; and that the produce of this island might be increased every twenty-five years, by a quantity equal to what it at present produces. The most enthusiastic speculator cannot suppose a greater increase than this. In a few centuries it would make every acre of land in the island like a garden.

If this supposition be applied to the whole earth, and if it be allowed that the subsistence for man which the earth affords might be increased every twenty-five years by a quantity equal to what it at present produces, this will be supposing a rate of increase much greater than we can imagine that any possible exertions of mankind could make it.

It may be fairly pronounced, therefore, that considering the present average state of the earth, the means of subsistence, under circumstances the most favourable to human industry, could not possibly be made to increase faster than in an arithmetical ratio.

The necessary effects of these two different rates of increase, when brought

together, will be very striking. Let us call the population of this island eleven
millions; and suppose the present produce equal to the easy support of such a
number. In the first twenty-five years the population would be twenty-two
millions, and the food being also doubled, the means of subsistence would be
equal to this increase. . . . In the next twenty-five years, the population
would be forty-four millions, and the means of subsistence only equal to the
support of thirty-three millions. In the next period the population would be
eighty-eight millions, and the means of subsistence just equal to the support of
half that number. And, at the conclusion of the first century, the population
would be a hundred and seventy-six millions, and the means of subsistence
only equal to the support of fifty-five millions, leaving a population of a hundred and twenty-one millions totally unprovided for.

Taking the whole earth, instead of this island, emigration would of course
be excluded; and, supposing the present population equal to a thousand millions, the human species would increase as the numbers, 1, 2, 4, 8, 16, 32, 64,
128, 256; and subsistence as, 1, 2, 3, 4, 5, 6, 7, 8, 9. In two centuries the
population would be to the means of subsistence as 256 to 9; in three centuries as 4096 to 13, and in two thousand years the difference would be almost
incalculable.

In this supposition no limits whatever are placed to the produce of the earth.
It may increase for ever, and be greater than any assignable quantity; yet still
the power of population being in every period so much superior, the increase of
the human species can only be kept down to the level of the means of subsistence by the constant operation of the strong law of necessity acting as a
check upon the greater power.

Of the General Checks to Population, and the Mode of Their Operation

The ultimate check to population appears then to be a want of food, arising
necessarily from the different ratios according to which population and food
increase. But this ultimate check is never the immediate check, except in
cases of actual famine.

The immediate check may be stated to consist in all those customs, and all
those diseases, which seem to be generated by a scarcity of the means of subsistence; and all those causes, independent of this scarcity, whether of a moral
or physical nature, which tend prematurely to weaken and destroy the human
frame.

These checks to population, which are constantly operating with more or
less force in every society, and keep down the number to the level of the means
of subsistence, may be classed under two general heads—the preventive and
the positive checks.

The preventive check, as far as it is voluntary, is peculiar to man, and arises from that distinctive superiority in his reasoning faculties which enables him to calculate distant consequences. The checks to the indefinite increase of plants and irrational animals are all either positive or, if preventive, involuntary. But man cannot look around him, and see the distress which frequently presses upon those who have large families; he cannot contemplate his present possessions or earnings, which he now nearly consumes himself, and calculate the amount of each share, when with very little addition they must be divided, perhaps, among seven or eight, without feeling a doubt whether, if he follow the bent of his inclinations, he may be able to support the offspring which he will probably bring into the world. In a state of equality, if such can exist, this would be the simple question. In the present state of society other considerations occur. Will he not lower his rank in life, and be obliged to give up in great measure his former habits? Does any mode of employment present itself by which he may reasonably hope to maintain a family? Will he not at any rate subject himself to greater difficulties, and more severe labour than in his single state? Will he not be unable to transmit to his children the same advantages of education and improvement that he had himself possessed? Does he even feel secure that, should he have a large family, his utmost exertions can save them from rags and squalid poverty, and their consequent degradation in the community? And may he not be reduced to the grating necessity of forfeiting his independence, and of being obliged to the sparing hand of charity for support?

These considerations are calculated to prevent, and certainly do prevent, a great number of persons in all civilised nations from pursuing the dictate of nature in an early attachment to one woman. . . .

The positive checks to population are extremely various, and include every cause, whether arising from vice or misery, which in any degree contribute to shorten the natural duration of human life. Under this head, therefore, may be enumerated all unwholesome occupations, severe labour and exposure to the seasons, extreme poverty, bad nursing of children, large towns, excesses of all kinds, the whole train of common diseases and epidemics, wars, plague, and famine.

On examining these obstacles to the increase of population which are classed under the heads of preventive and positive checks, it will appear that they are all resolvable into moral restraint, vice, and misery.

Of the preventive checks, the restraint from marriage which is not followed by irregular gratifications may properly be termed moral restraint.

Promiscuous intercourse, unnatural passions, violations of the marriage bed, and improper arts to conceal the consequences of irregular connections, are preventive checks that clearly come under the head of vice.

Of the positive checks, those which appear to arise unavoidably from the laws of nature, may be called exclusively misery; and those which we obviously bring upon ourselves, such as wars, excesses, and many others which it would be in our power to avoid, are of a mixed nature. They are brought upon us by vice, and their consequences are misery.

OF MORAL RESTRAINT, AND OUR OBLIGATION
TO PRACTISE THIS VIRTUE

As it appears that in the actual state of every society which has come within our review the natural progress of population has been constantly and powerfully checked, and as it seems evident that no improved form of government, no plans of emigration, no benevolent institutions, and no degree or direction of national industry can prevent the continued action of a great check to population in some form or other, it follows that we must submit to it as an inevitable law of nature; and the only inquiry that remains is how it may take place with the least possible prejudice to the virtue and happiness of human society.

All the immediate checks to population which have been observed to prevail in the same and different countries seem to be resolvable into moral restraint, vice, and misery; and if our choice be confined to these three, we cannot long hesitate in our decision respecting which it would be most eligible to encourage.

In the first edition of this essay I observed that as from the laws of nature it appeared that some check to population must exist, it was better that this check should arise from a foresight of the difficulties attending a family and the fear of dependent poverty than from the actual presence of want and sickness. This idea will admit of being pursued farther; and I am inclined to think that from the prevailing opinions respecting population, which undoubtedly originated in barbarous ages, and have been continued and circulated by that part of every community which may be supposed to be interested in their support, we have been prevented from attending to the clear dictates of reason and nature on this subject.

Natural and moral evil seem to be the instruments employed by the Deity in admonishing us to avoid any mode of conduct which is not suited to our being, and will consequently injure our happiness. If we are intemperate in eating and drinking, our health is disordered; if we indulge the transports of anger, we seldom fail to commit acts of which we afterwards repent; if we multiply too fast, we die miserably of poverty and contagious diseases. The laws of nature in all these cases are similar and uniform. They indicate to us that we have followed these impulses too far, so as to trench upon some other law, which equally demands attention. The uneasiness we feel from repletion, the injuries that we inflict on ourselves or others in anger, and the inconveniences we suffer on the approach of poverty, are all admonitions to us to regulate these impulses better; and if we heed not this admonition, we justly incur the penalty of our disobedience, and our sufferings operate as a warning to others. . . .

It is evidently therefore regulation and direction which are required with regard to the principle of population, not diminution or alteration. And if moral restraint be the only virtuous mode of avoiding the incidental evils arising from this principle, our obligation to practise it will evidently rest exactly upon the same foundation as our obligation to practise any of the other virtues.

Whatever indulgence we may be disposed to allow to occasional failures in the discharge of a duty of acknowledged difficulty, yet of the strict line of duty we cannot doubt. Our obligation not to marry till we have a fair prospect of being able to support our children will appear to deserve the attention of the moralist, if it can be proved that an attention to this obligation is of most powerful effect in the prevention of misery; and that if it were the general custom to follow the first impulse of nature and marry at the age of puberty, the universal prevalence of every known virtue in the greatest conceivable degree, would fail of rescuing society from the most wretched and desperate state of want, and all the diseases and famines which usually accompany it.

Of the Effects Which Would Result to Society from the Prevalence of Moral Restraint

. . . If, for the sake of illustration, we might be permitted to draw a picture of society in which each individual endeavoured to attain happiness by the strict fulfilment of those duties which the most enlightened of the ancient philosophers deduced from the laws of nature, and which have been directly taught and received such powerful sanctions in the moral code of Christianity, it would present a very different scene from that which we now contemplate. Every act which was prompted by the desire of immediate gratification, but which threatened an ultimate overbalance of pain, would be considered as a breach of duty, and consequently no man whose earnings were only sufficient to maintain two children would put himself in a situation in which he might have to maintain four or five, however he might be prompted to it by the passion of love. This prudential restraint, if it were generally adopted, by narrowing the supply of labour in the market, would in the natural course of things soon raise its price. The period of delayed gratification would be passed in saving the earnings which were above the wants of a single man, and in acquiring habits of sobriety, industry, and economy, which would enable him in a few years to enter into the matrimonial contract without fear of its consequences. The operation of the preventive check in this way, by constantly keeping the population within the limits of the food though constantly following its increase, would give a real value to the rise of wages and the sums saved by labourers before marriage, very different from those forced advances in the price of labour or arbitrary parochial donations which, in proportion to their magnitude and extensiveness, must of necessity be followed by a proportional advance in the price of provisions. As the wages of labour would thus be sufficient to maintain with decency a large family, and as every married couple would set out with a sum for contingencies, all abject poverty would be removed from society, or would at least be confined to a very few who had fallen into misfortunes against which no prudence or foresight could provide.

EVOLUTION THROUGH ENVIRONMENTALLY PRODUCED MODIFICATION

Chevalier de LAMARCK, Jean Baptiste Pierre Antoine de Monet (French naturalist, 1744–1829). Chapter VII in *Philosophie zoologique, ou expositions des considérations relatives à histoire naturelle des animaux*, Paris, 1809; tr. by H. Elliot as *Zoological philosophy; an exposition with regard to the natural history of animals*, London, 1914; by permission of The Macmillan Company, publisher.

Fortunately for the purposes of this volume, Lamarck summarized his views on the mechanism of evolution in the seventh chapter of the *Zoological Philosophy*. This chapter is reprinted here in its entirety. Lamarck's other works included a successful dichotomous key to the flora of France,* meteorological and geological studies, and unsuccessful attacks on the pneumatic chemistry of Lavoisier.† Introduction of the terms *invertebrata* and *biology* is attributed to him. Listed in the records of the Parisian Museum of Natural History as "Professor of Zoology, Insects, Worms, and Microscopic Animals," his interests eventually shifted almost entirely to zoology. Six years after the *Philosophie Zoologique*, his famous *Histoire Naturelle des Animaux sans Vertèbres* began to appear.

Of the Influence of the Environment on the Activities and Habits of Animals, and the Influence of the Activities and Habits of These Living Bodies in Modifying Their Organisation and Structure.

We are not here concerned with an argument, but with the examination of a positive fact—a fact which is of more general application than is supposed, and which has not received the attention that it deserves, no doubt because it is usually very difficult to recognise. This fact consists in the influence that is exerted by the environment on the various living bodies exposed to it.

It is indeed long since the influence of the various states of our organisation on our character, inclinations, activities and even ideas has been recognised;

---

* *Flore française, ou description succincte de toutes les plantes qui croissent naturellement en France, etc.*, Paris, 1778.

† *Réfutation de la théorie pneumatique ou de la nouvelle doctrine des chimistes modernes, etc.*, Paris, "an. iv" (1795–1796).

but I do not think that anyone has yet drawn attention to the influence of our activities and habits even on our organisation. Now since these activities and habits depend entirely on the environment in which we are habitually placed, I shall endeavour to show how great is the influence exerted by that environment on the general shape, state of the parts and even organisation of living bodies. It is, then, with this very positive fact that we have to do in the present chapter.

If we had not had many opportunities of clearly recognising the result of this influence on certain living bodies that we have transported into an environment altogether new and very different from that in which they were previously placed, and if we had not seen the resulting effects and alterations take place almost under our very eyes, the important fact in question would have remained for ever unknown to us.

The influence of the environment as a matter of fact is in all times and places operative on living bodies; but what makes this influence difficult to perceive is that its effects only become perceptible or recognisable (especially in animals) after a long period of time.

Before setting forth to examine the proofs of this fact, which deserves our attention and is so important for zoological philosophy, let us sum up the thread of the discussions that we have already begun.

In the preceding chapter we saw that it is now an unquestionable fact that on passing along the animal scale in the opposite direction from that of nature, we discover the existence, in the groups composing this scale, of a continuous but irregular degradation in the organisation of animals, an increasing simplification in their organisation, and, lastly, a corresponding diminution in the number of their faculties.

This well-ascertained fact may throw the strongest light over the actual order followed by nature in the production of all the animals that she has brought into existence, but it does not show us why the increasing complexity of the organisation of animals from the most imperfect to the most perfect exhibits only an *irregular gradation,* in the course of which there occur numerous anomalies or deviations with a variety in which no order is apparent.

Now on seeking the reason of this strange irregularity in the increasing complexity of animal organisation, if we consider the influence that is exerted by the infinitely varied environments of all parts of the world on the general shape, structure and even organisation of these animals, all will then be clearly explained.

It will in fact become clear that the state in which we find any animal, is, on the one hand, the result of the increasing complexity of organisation tending to form a regular gradation; and, on the other hand, of the influence of a multitude of very various conditions ever tending to destroy the regularity in the gradation of the increasing complexity of organisation.

I must now explain what I mean by this statement: *the environment affects the shape and organisation of animals,* that is to say that when the environment becomes very different, it produces in course of time corresponding modifications in the shape and organisation of animals.

It is true if this statement were to be taken literally, I should be convicted of an error; for, whatever the environment may do, it does not work any direct modification whatever in the shape and organisation of animals.

But great alterations in the environment of animals lead to great alterations in their needs, and these alterations in their needs necessarily lead to others in their activities. Now if the new needs become permanent, the animals then adopt new habits which last as long as the needs that evoked them. This is easy to demonstrate, and indeed requires no amplification.

It is then obvious that a great and permanent alteration in the environment of any race of animals induces new habits in these animals.

Now, if a new environment, which has become permanent for some race of animals, induces new habits in these animals, that is to say, leads them to new activities which become habitual, the result will be the use of some one part in preference to some other part, and in some cases the total disuse of some part no longer necessary.

Nothing of all this can be considered as hypothesis or private opinion; on the contrary, they are truths which, in order to be made clear, only require attention and the observation of facts.

We shall shortly see by the citation of known facts in evidence, in the first place, that new needs which establish a necessity for some part really bring about the existence of that part, as a result of efforts; and that subsequently its continued use gradually strengthens, develops and finally greatly enlarges it; in the second place, we shall see that in some cases, when the new environment and the new needs have altogether destroyed the utility of some part, the total disuse of that part has resulted in its gradually ceasing to share in the development of the other parts of the animal; it shrinks and wastes little by little, and ultimately, when there has been total disuse for a long period, the part in question ends by disappearing. All this is positive; I propose to furnish the most convincing proofs of it.

In plants, where there are no activities and consequently no habits, properly so-called, great changes of environment none the less lead to great differences in the development of their parts; so that these differences cause the origin and development of some, and the shrinkage and disappearance of others. But all this is here brought about by the changes sustained in the nutrition of the plant, in its absorption and transpiration, in the quantity of caloric, light, air and moisture that it habitually receives; lastly, in the dominance that some of the various vital movements acquire over others.

Among individuals of the same species, some of which are continually well fed and in an environment favourable to their development, while others are in an opposite environment, there arises a difference in the state of the individuals which gradually becomes very remarkable. How many examples I might cite both in animals and plants which bear out the truth of this principle! Now if the environment remains constant, so that the condition of the ill-fed, suffering or sickly individuals becomes permanent, their internal organisation is ultimately modified, and these acquired modifications are preserved by reproduction among the individuals in question, and finally give rise to a race quite distinct from that in which the individuals have been continuously in an environment favourable to their development.

Where in nature do we find our cabbages, lettuces, etc., in the same state as in our kitchen gardens? and is not the case the same with regard to many animals which have been altered or greatly modified by domestication?

How many different races of our domestic fowls and pigeons have we obtained by rearing them in various environments and different countries; birds which we should now vainly seek in nature?

Those which have changed the least, doubtless because their domestication is of shorter standing and because they do not live in a foreign climate, none the less display great differences in some of their parts, as a result of the habits which we have made them contract. Thus our domestic ducks and geese are of the same type as wild ducks and geese; but ours have lost the power of rising into high regions of the air and flying across large tracts of country; moreover, a real change has come about in the state of their parts, as compared with those of the animals of the race from which they come.

Who does not know that if we rear some bird of our own climate in a cage and it lives there for five or six years, and if we then return it to nature by setting it at liberty, it is no longer able to fly like its fellows, which have always been free? The slight change of environment for this individual has indeed only diminished its power of flight, and doubtless has worked no change in its structure; but if a long succession of generations of individuals of the same race had been kept in captivity for a considerable period, there is no doubt that even the structure of these individuals would gradually have undergone notable changes. Still more, if instead of a mere continuous captivity, this environmental factor had been further accompanied by a change to a very different climate; and if these individuals had by degrees been habituated to other kinds of food and other activities for seizing it, these factors when combined together and become permanent would have unquestionably given rise imperceptibly to a new race with quite special characters.

Where in natural conditions do we find that multitude of races of dogs which now actually exist, owing to the domestication to which we have reduced them? Where do we find those bull-dogs, grey-hounds, water-spaniels, spaniels, lap-dogs, etc., etc.; races which show wider differences than those which we call specific when they occur among animals of one genus living in natural freedom?

No doubt a single, original race, closely resembling the wolf, if indeed it was not actually the wolf, was at some period reduced by man to domestication. That race, of which all the individuals were then alike, was gradually scattered with man into different countries and climates; and after they had been subjected for some time to the influences of their environment and of the various habits which had been forced upon them in each country, they underwent remarkable alterations and formed various special races. Now man travels about to very great distances, either for trade or any other purpose; and thus brings into thickly populated places, such as a great capital, various races of dogs formed in very distant countries. The crossing of these races by reproduction then gave rise in turn to all those that we now know.

The following fact proves in the case of plants how the change of some important factor leads to alteration in the parts of these living bodies.

So long as *Ranunculus aquatilis* is submerged in the water, all its leaves are finely divided into minute segments; but when the stem of this plant reaches the surface of the water, the leaves which develop in the air are large, round and simply lobed. If several feet of the same plant succeed in growing in a soil that is merely damp without any immersion, their stems are then short, and none of their leaves are broken up into minute divisions, so that we get *Ranunculus hederaceus,* which botanists regard as a separate species.

There is no doubt that in the case of animals, extensive alterations in their customary environment produce corresponding alterations in their parts; but here the transformations take place much more slowly than in the case of plants; and for us therefore they are less perceptible and their cause less readily identified.

As to the conditions which have so much power in modifying the organs of living bodies, the most potent doubtless consist in the diversity of the places where they live, but there are many others as well which exercise considerable influence in producing the effects in question.

It is known that localities differ as to their character and quality, by reason of their position, construction and climate: as is readily perceived on passing through various localities distinguished by special qualities; this is one cause of variation for animals and plants living in these various places. But what is not known so well and indeed what is not generally believed, is that every locality itself changes in time as to exposure, climate, character and quality, although with such extreme slowness, according to our notions, that we ascribe to it complete stability.

Now in both cases these altered localities involve a corresponding alteration in the environment of the living bodies that dwell there, and this again brings a new influence to bear on these same bodies.

Hence it follows that if there are extremes in these alterations, there are also finer differences: that is to say, intermediate stages which fill up the interval. Consequently there are also fine distinctions between what we call species.

It is obvious then that as regards the character and situation of the substances which occupy the various parts of the earth's surface, there exists a variety of environmental factors which induces a corresponding variety in the shapes and structure of animals, independent of that special variety which necessarily results from the progress of the complexity of organisation in each animal.

In every locality where animals can live, the conditions constituting any one order of things remain the same for long periods: indeed they alter so slowly that man cannot directly observe it. It is only by an inspection of ancient monuments that he becomes convinced that in each of these localities the order of things which he now finds has not always been existent; he may thence infer that it will go on changing.

Races of animals living in any of these localities must then retain their habits equally long: hence the apparent constancy of the races that we call species,—a constancy which has raised in us the belief that these races are as old as nature.

But in the various habitable parts of the earth's surface, the character and situation of places and climates constitute both for animals and plants environmental influences of extreme variability. The animals living in these various localities must therefore differ among themselves, not only by reason of the state of complexity of organisation attained in each race, but also by reason of the habits which each race is forced to acquire; thus when the observing naturalist travels over large portions of the earth's surface and sees conspicuous changes occurring in the environment, he invariably finds that the characters of species undergo a corresponding change.

Now the true principle to be noted in all this is as follows:

1. Every fairly considerable and permanent alteration in the environment of any race of animals works a real alteration in the needs of that race.

2. Every change in the needs of animals necessitates new activities on their part for the satisfaction of those needs, and hence new habits.

3. Every new need, necessitating new activities for its satisfaction, requires the animal, either to make more frequent use of some of its parts which it previously used less, and thus greatly to develop and enlarge them; or else to make use of entirely new parts, to which the needs have imperceptibly given birth by efforts of its inner feeling; this I shall shortly prove by means of known facts.

Thus to obtain a knowledge of the true causes of that great diversity of shapes and habits found in the various known animals, we must reflect that the infinitely diversified but slowly changing environment in which the animals of each race have successively been placed, has involved each of them in new needs and corresponding alterations in their habits. This is a truth which, once recognised, cannot be disputed. Now we shall easily discern how the new needs may have been satisfied, and the new habits acquired, if we pay attention to the two following laws of nature, which are always verified by observation.

FIRST LAW

In every animal which has not passed the limit of its development, a more frequent and continuous use of any organ gradually strengthens, develops and enlarges that organ, and gives it a power proportional to the length of time it has been so used; while the permanent disuse of any organ imperceptibly weakens and deteriorates it, and progressively diminishes its functional capacity, until it finally disappears.

SECOND LAW.

All the acquisitions or losses wrought by nature on individuals, through the influence of the environment in which their race has long been placed, and hence through the influence of the predominant use or permanent disuse of any organ; all these are preserved by reproduction to the new individuals which arise, provided that the acquired modifications are common to both sexes, or at least to the individuals which produce the young.

Here we have **two permanent truths**, which can only be doubted by those who have never observed or followed the operations of nature, or by those who have allowed themselves to be drawn into the error which I shall now proceed to combat.

Naturalists have remarked that the structure of animals is always in perfect adaptation to their functions, and have inferred that the shape and condition of their parts have determined the use of them. Now this is a mistake: for it may be easily proved by observation that it is on the contrary the needs and uses of the parts which have caused the development of these same parts, which have even given birth to them when they did not exist, and which consequently have given rise to the condition that we find in each animal.

If this were not so, nature would have had to create as many different kinds of structure in animals, as there are different kinds of environment in which they have to live; and neither structure nor environment would ever have varied.

This is indeed far from the true order of things. If things were really so, we should not have race-horses shaped like those in England; we should not have big draught-horses so heavy and so different from the former, for none such are produced in nature; in the same way we should not have basset-hounds with crooked legs, nor grey-hounds so fleet of foot, nor water-spaniels, etc.; we should not have fowls without tails, fantail pigeons, etc; finally, we should be able to cultivate wild plants as long as we liked in the rich and fertile soil of our gardens, without the fear of seeing them change under long cultivation.

A feeling of the truth in this respect has long existed; since the following maxim has passed into a proverb and is known by all, *Habits form a second nature.*

Assuredly if the habits and nature of each animal could never vary, the proverb would have been false and would not have come into existence, nor been preserved in the event of any one suggesting it.

If we seriously reflect upon all that I have just set forth, it will be seen that I was entirely justified when in my work entitled *Recherches sur les corps vivans*,[1] I established the following proposition: "It is not the organs, that is to say, the nature and shape of the parts of an animal's body, that have given rise to its special habits and faculties; but it is, on the contrary, its habits, mode of life and environment that have in course of time controlled the shape of its body, the number and state of its organs and, lastly, the faculties which it possesses."

If this proposition is carefully weighed and compared with all the observations that nature and circumstances are incessantly throwing in our way, we shall see that its importance and accuracy are substantiated in the highest degrees.

Time and a favourable environment are as I have already said nature's two chief methods of bringing all her productions into existence: for her, time has no limits and can be drawn upon to any extent.

As to the various factors which she has required and still constantly uses for introducing variations in everything that she produces, they may be described as practically inexhaustible.

The principal factors consist in the influence of climate, of the varying temperatures of the atmosphere and the whole environment, of the variety of localities and their situation, of habits, the commonest movements, the most frequent activities, and, lastly, of the means of self-preservation, the mode of life and the methods of defence and multiplication.

Now as a result of these various influences, the faculties become extended and strengthened by use, and diversified by new habits that are long kept up. The conformation, consistency and, in short, the character and state of the parts, as well as of the organs, are imperceptibly affected by these influences and are preserved and propagated by reproduction.

---

[1] Paris, 1802.

These truths, which are merely effects of the two natural laws stated above, receive in every instance striking confirmation from facts; for the facts afford a clear indication of nature's procedure in the diversity of her productions.

But instead of being contented with generalities which might be considered hypothetical, let us investigate the facts directly, and consider the effects in animals of the use or disuse of their organs on these same organs, in accordance with the habits that each race has been forced to contract.

Now I am going to prove that the permanent disuse of any organ first decreases its functional capacity, and then gradually reduces the organ and causes it to disappear or even become extinct, if this disuse lasts for a very long period throughout successive generations of animals of the same race.

I shall then show that the habit of using any organ, on the contrary, in any animal which has not reached the limit of the decline of its functions, not only perfects and increases the functions of that organ, but causes it in addition to take on a size and development which imperceptibly alter it; so that in course of time it becomes very different from the same organ in some other animal which uses it far less.

The permanent disuse of an organ, arising from a change of habits, causes a gradual shrinkage and ultimately the disappearance and even extinction of that organ.

Since such a proposition could only be accepted on proof, and not on mere authority, let us endeavour to make it clear by citing the chief known facts which substantiate it.

The vertebrates, whose plan of organisation is almost the same throughout, though with much variety in their parts, have their jaws armed with teeth; some of them, however, whose environment has induced the habit of swallowing the objects they feed on without any preliminary mastication, are so affected that their teeth do not develop. The teeth then remain hidden in the bony framework of the jaws, without being able to appear outside; or indeed they actually become extinct down to their last rudiments.

In the right-whale, which was supposed to be completely destitute of teeth, M. Geoffroy has nevertheless discovered teeth concealed in the jaws of the foetus of this animal. The professor has moreover discovered in birds the groove in which the teeth should be placed, though they are no longer to be found there.

Even in the class of mammals, comprising the most perfect animals, where the vertebrate plan of organisation is carried to its highest completion, not only is the right-whale devoid of teeth, but the anteater (*Myrmecophaga*) is also found to be in the same condition, since it has acquired a habit of carrying out no mastication, and has long preserved this habit in its race.

Eyes in the head are characteristic of a great number of different animals, and essentially constitute a part of the plan of organisation of the vertebrates.

Yet the mole, whose habits require a very small use of sight, has only minute and hardly visible eyes, because it uses that organ so little.

Olivier's *Spalax*[2] (*Voyage en Égypte et en Perse*[3]), which lives underground like the mole, and is apparently exposed to daylight even less than the mole, has altogether lost the use of sight: so that it shows nothing more than vestiges of this organ. Even these vestiges are entirely hidden under the skin and other parts, which cover them up and do not leave the slightest access to light.

The *Proteus*, an aquatic reptile allied to the salamanders, and living in deep dark caves under the water, has, like the *Spalax*, only vestiges of the organ of sight, vestiges which are covered up and hidden in the same way.

The following consideration is decisive on the question which I am now discussing.

Light does not penetrate everywhere; consequently animals which habitually live in places where it does not penetrate, have no opportunity of exercising their organ of sight, if nature has endowed them with one. Now animals belonging to a plan of organisation of which eyes were a necessary part, must have originally had them. Since, however, there are found among them some which have lost the use of this organ and which show nothing more than hidden and covered up vestiges of them, it becomes clear that the shrinkage and even disappearance of the organ in question are the results of a permanent disuse of that organ.

This is proved by the fact that the organ of hearing is never in this condition, but is always found in animals whose organisation is of the kind that includes it: and for the following reason.

The substance of sound,[4] that namely which, when set in motion by the

---

[2] A genus of burrowing rodents: the mole rat, bamboo rat.—Ed.

[3] Olivier, G. A., *Voyage dans l'empire othoman, l'Égypte et la Perse, etc.*, Paris an. ix-xii (1800–1804).

[4] Physicists believe and even affirm that the atmospheric air is the actual substance of sound, that is to say, that it is the substance which, when set in motion by the shocks or vibrations of bodies, transmits to the organ of hearing the impression of the concussions received.

That this is an error is attested by many known facts, showing that it is impossible that the air should penetrate to all places to which the substance producing sound actually does penetrate.

See my memoir *On the Substance of Sound*, printed at the end of my *Hydrogéologie*, p. 225, in which I furnished the proofs of this mistake.

Since the publication of my memoir, which by the way is seldom cited, great efforts have been made to make the known velocity of the propagation of sound in air tally with the elasticity of the air, which would cause the propagation of its oscillations to be too slow for the theory. Now, since the air during oscillation necessarily undergoes alternate compressions and dilatations in its parts, recourse has been had to the effects of the caloric squeezed out during the sudden compressions of the air and of the caloric absorbed during the rarefactions of that fluid. By means of these effects, quantitatively determined by convenient hypotheses, geometricians now account for the velocity with shock or the vibration of bodies, transmits to the organ of hearing the impression received, penetrates everywhere and passes through any medium, including even the densest bodies: it follows that every animal, belonging to a plan of organisation of which hearing is an essential part, always has some opportunity for the exercise of this organ wherever it may live. Hence among the vertebrates we do not find any that are destitute of the organ of hearing; and after them, when this same organ has come to an end, it does not subsequently recur in any animal of the posterior classes.

It is not so with the organ of sight; for this organ is found to disappear, re-appear and disappear again according to the use that the animal makes of it.

In the acephalic molluscs, the great development of the mantle would make their eyes and even their head altogether useless. The permanent disuse of these organs has thus brought about their disappearance and extinction, although molluscs belong to a plan of organisation which should comprise them.

Lastly, it was part of the plan of organisation of the reptiles, as of other vertebrates, to have four legs in dependence on their skeleton. Snakes ought consequently to have four legs, especially since they are by no means the last order of the reptiles and are farther from the fishes than are the batrachians (frogs, salamanders, etc.).

Snakes, however, have adopted the habit of crawling on the ground and hiding in the grass; so that their body, as a result of continually repeated efforts at elongation for the purpose of passing through narrow spaces, has acquired a considerable length, quite out of proportion to its size. Now, legs would have been quite useless to these animals and consequently unused. Long legs would have interfered with their need of crawling and very short legs would have been incapable of moving their body, since they could only have had four. The disuse of these parts thus became permanent in the various races of these animals, and resulted in the complete disappearance of these same parts, although legs really belong to the plan of organisation of the animals of this class.

Many insects, which should have wings according to the natural characteristics of their order and even of their genus, are more or less completely devoid of them through disuse. Instances are furnished by many Coleoptera, Orthoptera, Hymenoptera and Hemiptera, etc., where the habits of these animals never involve them in the necessity of using their wings.

But it is not enough to give an explanation of the cause which has brought

---

which sound is propagated through air. But this is no answer to the fact that sound is also propagated through bodies which air can neither traverse nor set in motion.

These physicists assume forsooth a vibration in the smallest particles of solid bodies; a vibration of very dubious existence, since it can only be propagated through homogeneous bodies of equal density, and cannot spread from a dense body to a rarefied one or *vice versâ*. Such a hypothesis offers no explanation of the well-known fact that sound is propagated through heterogeneous bodies of very different densities and kinds.

about the present condition of the organs of the various animals,—a condition that is always found to be the same in animals of the same species; we have in addition to cite instances of changes wrought in the organs of a single individual during its life, as the exclusive result of a great mutation in the habits of the individuals of its species. The following very remarkable fact will complete the proof of the influence of habits on the condition of the organs, and of the way in which permanent changes in the habits of an individual lead to others in the condition of the organs, which come into action during the exercise of these habits.

M. Tenon, a member of the Institute, has notified to the class of sciences, that he had examined the intestinal canal of several men who had been great drinkers for a large part of their lives, and in every case he had found it shortened to an extraordinary degree, as compared with the same organ in all those who had not adopted the like habit.

It is known that great drinkers, or those who are addicted to drunkenness, take very little solid food, and eat hardly anything; since the drink which they consume so copiously and frequently is sufficient to feed them.

Now since fluid foods, especially spirits, do not long remain either in the stomach or intestine, the stomach and the rest of the intestinal canal lose among drinkers the habit of being distended, just as among sedentary persons, who are continually engaged on mental work and are accustomed to take very little food; for in their case also the stomach slowly shrinks and the intestine shortens.

This has nothing to do with any shrinkage or shortening due to a binding of the parts which would permit of the ordinary extension, if instead of remaining empty these viscera were again filled; we have to do with a real shrinkage and shortening of considerable extent, and such that these organs would burst rather than yield at once to any demand for the ordinary extension.

Compare two men of equal ages, one of whom has contracted the habit of eating very little, since his habitual studies and mental work have made digestion difficult, while the other habitually takes much exercise, is often out-of-doors, and eats well; the stomach of the first will have very little capacity left and will be filled up by a very small quantity of food, while that of the second will have preserved and even increased its capacity.

Here then is an organ which undergoes profound modification in size and capacity, purely on account of a change of habits during the life of the individual.

The frequent use of any organ, when confirmed by habit, increases the functions of that organ, leads to its development and endows it with a size and power that it does not possess in animals which exercise it less.

We have seen that the disuse of any organ modifies, reduces and finally extinguishes it.   I shall now prove that the constant use of any organ, accompanied by efforts to get the most out of it, strengthens and enlarges that organ, or creates new ones to carry on functions that have become necessary.

The bird which is drawn to the water by its need of finding there the prey on which it lives, separates the digits of its feet in trying to strike the water and move about on the surface.   The skin which unites these digits at their base acquires the habit of being stretched by these continually repeated separations of the digits; thus in course of time there are formed large webs which unite the digits of ducks, geese, etc., as we actually find them.   In the same way efforts to swim, that is to push against the water so as to move about in it, have stretched the membranes between the digits of frogs, sea-tortoises, the otter, beaver, etc.

On the other hand, a bird which is accustomed to perch on trees and which springs from individuals all of whom had acquired this habit, necessarily has longer digits on its feet and differently shaped from those of the aquatic animals that I have just named.   Its claws in time become lengthened, sharpened and curved into hooks, to clasp the branches on which the animal so often rests.

We find in the same way that the bird of the water-side which does not like swimming and yet is in need of going to the water's edge to secure its prey, is continually liable to sink in the mud.   Now this bird tries to act in such a way that its body should not be immersed in the liquid, and hence makes its best efforts to stretch and lengthen its legs.   The long-established habit acquired by this bird and all its race of continually stretching and lenthening its legs, results in the individuals of this race becoming raised as though on stilts, and gradually obtaining long, bare legs, denuded of feathers up to the thighs and often higher still.   (*Système des animaux sans vertèbres,* p. 14.)

We note again that this same bird wants to fish without wetting its body, and is thus obliged to make continual efforts to lengthen its neck.   Now these habitual efforts in this individual and its race must have resulted in course of time in a remarkable lengthening, as indeed we actually find in the long necks of all water-side birds.

If some swimming birds like the swan and goose have short legs and yet a very long neck, the reason is that these birds while moving about on the water acquire the habit of plunging their head as deeply as they can into it in order to get the aquatic larvae and various animals on which they feed; whereas they make no effort to lengthen their legs.

If an animal, for the satisfaction of its needs, makes repeated efforts to lengthen its tongue, it will acquire a considerable length (anteater, greenwoodpecker); if it requires to seize anything with this same organ, its tongue will then divide and become forked. Proofs of my statement are found in the humming-birds which use their tongues for grasping things, and in lizards and snakes which use theirs to palpate and identify objects in front of them.

Needs which are always brought about by the environment, and the subsequent continued efforts to satisfy them, are not limited in their results to a mere modification, that is to say, an increase or decrease of the size and capacity of organs; but they may even go so far as to extinguish organs, when any of these needs make such a course necessary.

Fishes, which habitually swim in large masses of water, have need of lateral vision; and, as a matter of fact, their eyes are placed on the sides of their head. Their body, which is more or less flattened according to the species, has its edges perpendicular to the plane of the water; and their eyes are placed so that there is one on each flattened side. But such fishes as are forced by their habits to be constantly approaching the shore, and especially slightly inclined or gently sloping beaches, have been compelled to swim on their flattened surfaces in order to make a close approach to the water's edge. In this position, they receive more light from above than below and stand in special need of paying constant attention to what is passing above them; this requirement has forced one of their eyes to undergo a sort of displacement, and to assume the very remarkable position found in the soles, turbots, dabs, etc. (*Pleuronectes* and *Achirus*). The position of these eyes is not symmetrical, because it results from an incomplete mutation. Now this mutation is entirely completed in the skates, in which the transverse flattening of the body is altogether horizontal, like the head. Accordingly the eyes of skates are both situated on the upper surface and have become symmetrical.

Snakes, which crawl on the surface of the earth, chiefly need to see objects that are raised or above them. This need must have had its effect on the position of the organ of sight in these animals, and accordingly their eyes are situated in the lateral and upper parts of their head, so as easily to perceive what is above them or at their sides; but they scarcely see at all at a very short distance in front of them. They are, however, compelled to make good the deficiency of sight as regards objects in front of them which might injure them as they move forward. For this purpose they can only use their tongue, which they are obliged to thrust out with all their might. This habit has not only contributed to making their tongue slender and very long and contractile, but it has even forced it to undergo division in the greater number of species, so as to feel several objects at the same time; it has even permitted of the formation of an aperture at the extremity of their snout, to allow the tongue to pass without having to separate the jaws.

Nothing is more remarkable than the effects of habit in herbivorous mammals.

A quadruped, whose environment and consequent needs have for long past inculcated the habit of browsing on grass, does nothing but walk about on the ground; and for the greater part of its life is obliged to stand on its four feet, generally making only few or moderate movements. The large portion of each day that this kind of animal has to pass in filling itself with the only kind of food that it cares for, has the result that it moves but little and only uses its feet for support in walking or running on the ground, and never for holding on, or climbing trees.

From this habit of continually consuming large quantities of food-material, which distended the organs receiving it, and from the habit of making only moderate movements, it has come about that the body of these animals has greatly thickened, become heavy and massive and acquired a very great size: as is seen in elephants, rhinoceroses, oxen, buffaloes, horses, etc.

The habit of standing on their four feet during the greater part of the day, for the purpose of browsing, has brought into existence a thick horn which invests the extremity of their digits; and since these digits have no exercise and are never moved and serve no other purpose than that of support like the rest of the foot, most of them have become shortened, dwindled and, finally, even disappeared.

Thus in the pachyderms, some have five digits on their feet invested in horn, and their hoof is consequently divided into five parts; others have only four, and others again not more than three; but in the ruminants, which are apparently the oldest of the mammals that are permanently confined to the ground, there are not more than two digits on the feet and indeed, in the solipeds, there is only one (horse, donkey).

Nevertheless some of these herbivorous animals, especially the ruminants, are incessantly exposed to the attacks of carnivorous animals in the desert countries that they inhabit, and they can only find safety in headlong flight. Necessity has in these cases forced them to exert themselves in swift running and from this habit their body has become more slender and their legs much finer; instances are furnished by the antelopes, gazelles, etc.

In our own climates, there are other dangers, such as those constituted by man, with his continual pursuit of red deer, roe deer and fallow deer; this has reduced them to the same necessity, has impelled them into similar habits, and had corresponding effects.

Since ruminants can only use their feet for support, and have little strength in their jaws, which only obtain exercise by cutting and browsing on the grass, they can only fight by blows with their heads, attacking one another with their crowns.

In the frequent fits of anger to which the males especially are subject, the efforts of their inner feeling cause the fluids to flow more strongly towards that part of their head; in some there is hence deposited a secretion of horny matter, and in others of bony matter mixed with horny matter, which gives rise to solid protuberances: thus we have the origin of horns and antlers, with which the head of most of these animals is armed.

It is interesting to observe the result of habit in the peculiar shape and size of the giraffe (*Camelo-pardalis*): this animal, the largest of the mammals, is known to live in the interior of Africa in places where the soil is nearly always arid and barren, so that it is obliged to browse on the leaves of trees and to make constant efforts to reach them. From this habit long maintained in all its race, it has resulted that the animal's fore-legs have become longer than its hind legs, and that its neck is lengthened to such a degree that the giraffe, without standing up on its hind legs, attains a height of six metres (nearly 20 feet).

Among birds, ostriches, which have no power of flight and are raised on very long legs, probably owe their singular shape to analogous circumstances.

The effect of habit is quite as remarkable in the carnivorous mammals as in the herbivores; but it exhibits results of a different kind.

Those carnivores, for instance, which have become accustomed to climbing, or to scratching the ground for digging holes, or to tearing their prey, have been under the necessity of using the digits of their feet: now this habit has promoted the separation of their digits, and given rise to the formation of the claws with which they are armed.

But some of the carnivores are obliged to have recourse to pursuit in order to catch their prey: now some of these animals were compelled by their needs to contract the habit of tearing with their claws, which they are constantly burying deep in the body of another animal in order to lay hold of it, and then make efforts to tear out the part seized. These repeated efforts must have resulted in its claws reaching a size and curvature which would have greatly impeded them in walking or running on stony ground: in such cases the animal has been compelled to make further efforts to draw back its claws, which are so projecting and hooked as to get in its way. From this there has gradually resulted the formation of those peculiar sheaths, into which cats, tigers, lions, etc. withdraw their claws when they are not using them.

Hence we see that efforts in a given direction, when they are long sustained or habitually made by certain parts of a living body, for the satisfaction of needs established by nature or environment, cause an enlargement of these parts and the acquisition of a size and shape that they would never have obtained, if these efforts had not become the normal activities of the animals exerting them. Instances are everywhere furnished by observations on all known animals.

Can there be any more striking instance than that which we find in the kangaroo? This animal, which carries its young in a pouch under the abdomen, has acquired the habit of standing upright, so as to rest only on its hind legs and tail; and of moving only by means of a succession of leaps, during which it maintains its erect attitude in order not to disturb its young. And the following is the result:

1. Its fore legs, which it uses very little and on which it only supports itself for a moment on abandoning its erect attitude, have never acquired a development proportional to that of the other parts, and have remained meagre, very short and with very little strength.

2. The hind legs, on the contrary, which are almost continually in action either for supporting the whole body or for making leaps, have acquired a great development and become very large and strong.

3. Lastly, the tail, which is in this case much used for supporting the animal and carrying out its chief movements, has acquired an extremely remarkable thickness and strength at its base.

These well-known facts are surely quite sufficient to establish the results of habitual use on an organ or any other part of animals. If on observing in an animal any organ particularly well-developed, strong, and powerful, it is alleged that its habitual use has nothing to do with it, that its continued disuse involves it in no loss, and finally, that this organ has always been the same since the creation of the species to which the animal belongs, then I ask, Why can our domestic ducks no longer fly like wild ducks? I can, in short, cite a multitude of instances among ourselves, which bear witness to the differences that accrue to us from the use or disuse of any of our organs, although these differences are not preserved in the new individuals which arise by reproduction: for if they were their effects would be far greater.

I shall show in Part II., that when the will guides an animal to any action, the organs which have to carry out that action are immediately stimulated to it by the influx of subtle fluids (the nervous fluid), which become the determining factor of the movements required. This fact is verified by many observations, and cannot now be called in question.

Hence it follows that numerous repetitions of these organised activities strengthen, stretch, develop and even create the organs necessary to them. We have only to watch attentively what is happening all around us, to be convinced that this is the true cause of organic development and changes.

Now every change that is wrought in an organ through a habit of frequently using it, is subsequently preserved by reproduction, if it is common to the individuals who unite together in fertilisation for the propagation of their species. Such a change is thus handed on to all succeeding individuals in the same environment, without their having to acquire it in the same way that it was actually created.

Furthermore, in reproductive unions, the crossing of individuals who have different qualities or structures is necessarily opposed to the permanent propagation of these qualities and structures. Hence it is that in man, who is exposed to so great a diversity of environment, the accidental qualities or defects which he acquires are not preserved and propagated by reproduction. If, when certain peculiarities of shape or certain defects have been acquired, two individuals who are both affected were always to unite together, they would hand on the same peculiarities; and if successive generations were limited to such unions, a special and distinct race would then be formed. But perpetual crossings between individuals, who have not the same peculiarities of shape, cause the disappearance of all peculiarities acquired by special action of the environment. Hence, we may be sure that if men were not kept apart by the distances of their habitations, the crossing in reproduction would soon bring about the disappearance of the general characteristics distinguishing different nations.

If I intended here to pass in review all the classes, orders, genera and species of existing animals, I should be able to show that the conformation and structure of individuals, their organs, faculties, etc., etc., are everywhere a pure result of the environment to which each species is exposed by its nature, and by the habits that the individuals composing it have been compelled to acquire; I should be able to show that they are not the result of a shape which existed from the beginning, and has driven animals into the habits they are known to possess.

It is known that the animal called the *ai* or sloth (*Bradypustridactylus*) is permanently in a state of such extreme weakness that it only executes very slow and limited movements, and walks on the ground with difficulty. So slow are its movements that it is alleged that it can only take fifty steps in a day. It is known, moreover, that the organisation of this animal is entirely in harmony with its state of feebleness and incapacity for walking; and that if it wished to make other movements than those which it actually does make it could not do so.

Hence on the supposition that this animal had received its organisation from nature, it has been asserted that this organisation forced it into the habits and miserable state in which it exists.

This is very far from being my opinion; for I am convinced that the habits which the ai was originally forced to contract must necessarily have brought its organisation to its present condition.

If continual dangers in former times have led the individuals of this species to take refuge in trees, to live there habitually and feed on their leaves, it is clear that they must have given up a great number of movements which animals living on the ground are in a position to perform. All the needs of the ai will then be reduced to clinging to branches and crawling and dragging themselves among them, in order to reach the leaves, and then to remaining on the tree in a state of inactivity in order to avoid falling off. This kind of inactivity, moreover, must have been continually induced by the heat of the

climate; for among warm-blooded animals, heat is more conducive to rest
than movement.

Now the individuals of the race of the ai have long maintained this habit
of remaining in the trees, and of performing only those slow and little varied
movements which suffice for their needs. Hence their organisation will
gradually have come into accordance with their new habits; and from this it
must follow:

1. That the arms of these animals, which are making continual efforts to
clasp the branches of trees, will be lengthened;

2. That the claws of their digits will have acquired a great length and a
hooked shape, through the continued efforts of the animal to hold on;

3. That their digits, which are never used in making independent movements, will have entirely lost their mobility, become united and have preserved
only the faculty of flexion or extension all together;

4. That their thighs, which are continually clasping either the trunk or
large branches of trees, will have contracted a habit of always being separated,
so as to lead to an enlargement of the pelvis and a backward direction of the
cotyloid cavities;

5. Lastly, that a great many of their bones will be welded together, and
that parts of their skeleton will consequently have assumed an arrangement
and form adapted to the habits of these animals, and different from those
which they would require for other habits.

This is a fact that can never be disputed; since nature shows us in innumerable other instances the power of environment over habit and that of
habit over the shape, arrangement and proportions of the parts of animals.

Since there is no necessity to cite any further examples, we may now turn
to the main point elaborated in this discussion.

It is a fact that all animals have special habits corresponding to their genus
and species, and always possess an organisation that is completely in harmony
with those habits.

It seems from the study of this fact that we may adopt one or other of the
two following conclusions, and that neither of them can be verified.

Conclusion adopted hitherto: Nature (or her Author) in creating animals,
foresaw all the possible kinds of environment in which they would have to
live, and endowed each species with a fixed organisation and with a definite
and invariable shape, which compel each species to live in the places and
climates where we actually find them, and there to maintain the habits which
we know in them.

My individual conclusion: Nature has produced all the species of animals
in succession, beginning with the most imperfect or simplest, and ending her
work with the most perfect, so as to create a gradually increasing complexity
in their organisation; these animals have spread at large throughout all the

habitable regions of the globe, and every species has derived from its environment the habits that we find in it and the structural modifications which observation shows us.

The former of these two conclusions is that which has been drawn hitherto, at least by nearly everyone: it attributes to every animal a fixed organisation and structure which never have varied and never do vary; it assumes, moreover, that none of the localities inhabited by animals ever vary; for if they were to vary, the same animals could no longer survive, and the possibility of finding other localities and transporting themselves thither would not be open to them.

The second conclusion is my own: it assumes that by the influence of environment on habit, and thereafter by that of habit on the state of the parts and even on organisation, the structure and organisation of any animal may undergo modifications, possibly very great, and capable of accounting for the actual condition in which all animals are found.

In order to show that this second conclusion is baseless, it must first be proved that no point on the surface of the earth ever undergoes variation as to its nature, exposure, high or low situation, climate, etc., etc.; it must then be proved that no part of animals undergoes even after long periods of time any modification due to a change of environment or to the necessity which forces them into a different kind of life and activity from what has been customary to them.

Now if a single case is sufficient to prove that an animal which has long been in domestication differs from the wild species whence it sprang, and if in any such domesticated species, great differences of conformation are found between the individuals exposed to such a habit and those which are forced into different habits, it will then be certain that the first conclusion is not consistent with the laws of nature, while the second, on the contrary, is entirely in accordance with them.

Everything then combines to prove my statement, namely: that it is not the shape either of the body or its parts which gives rise to the habits of animals and their mode of life; but that it is, on the contrary, the habits, mode of life and all the other influences of the environment which have in course of time built up the shape of the body and of the parts of animals. With new shapes, new faculties have been acquired, and little by little nature has succeeded in fashioning animals such as we actually see them.

Can there be any more important conclusion in the range of natural history, or any to which more attention should be paid than that which I have just set forth?

GEOLOGICAL HISTORY REASONED
FROM PALEONTOLOGY

Baron CUVIER, Georges Léopold Chrétien Frédéric Dagobert (French naturalist, 1769–
1832). From *Discours sur les révolutions de la surface du globe,** Paris, 1812; tr. by
R. Jameson as *Essay on the theory of the earth*, Edinburgh, 1813.

An essential aspect of geological method is the inference of the dynamic narrative of the
past from the static evidence of the present. This method, used at least as early as the
Ionian schools, was revived in Renaissance Europe (see Steno). For bringing this form
of thought to its effective maturity Cuvier was, with many mistakes, mainly responsible.
Some of his evidence was zoological (see the *Source Book in Geology*).

The lowest and most level parts of the earth, when penetrated to a very
great depth, exhibit nothing but horizontal strata composed of various substances, and containing almost all of them innumerable marine productions.
Similar strata, with the same kind of productions, compose the hills even to
a great height. Sometimes the shells are so numerous as to constitute the entire body of the stratum. They are almost everywhere in such a perfect state
of preservation, that even the smallest of them retain their most delicate
parts, their sharpest ridges, and their finest and tenderest processes. They are
found in elevations far above the level of every part of the ocean, and in
places to which the sea could not be conveyed by any existing cause. They
are not only enclosed in loose sand, but are often incrusted and penetrated on
all sides by the hardest stones. Every part of the earth, every hemisphere,
every continent, every island of any size, exhibits the same phenomenon.
We are therefore forcibly led to believe, not only that the sea has at one period
or another covered all our plains, but that it must have remained there for a
long time, and in a state of tranquillity; which circumstance was necessary for
the formation of deposits so extensive, so thick, in part so solid, and containing
exuviae so perfectly preserved.

The time is past for ignorance to assert that these remains of organized
bodies are mere *lusus naturae,*—productions generated in the womb of the
earth by its own creative powers. A nice and scrupulous comparison of their
forms, of their contexture, and frequently even of their composition, cannot
detect the slightest difference between these shells and the shells which still
inhabit the sea. They have therefore once lived in the sea, and been deposited by it; the sea consequently must have rested in the places where the
deposition has taken place. Hence it is evident the basin or reservoir containing the sea has undergone some change at least, either in extent, or in situation,
or in both. Such is the result of the very first search, and of the most superficial examination.

---

* This is a translation of a preliminary *Discours sur les révolutions etc.* which appeared with the 2d edition of *Recherches sur les ossemens fossiles*, Paris, 1812.

The traces of revolutions become still more apparent and decisive when we ascend a little higher, and approach nearer to the foot of the great chains of mountains. There are still found many beds of shells; some of these are even larger and more solid; the shells are quite as numerous and as entirely preserved; but they are not of the same species with those which were found in the less elevated regions. The strata which contain them are not so generally horizontal; they have various degrees of inclination, and are sometimes situated vertically. While in the plains and low hills it was necessary to dig deep in order to detect the succession of the strata, here we perceive them by means of the valleys which time or violence has produced, and which disclose their edges to the eye of the observer. At the bottom of these declivities, huge masses of their *debris* are collected, and form round hills, the height of which is augmented by the operation of every thaw and of every storm.

These inclined or vertical strata, which form the ridges of the secondary mountains, do not rest on the horizontal strata of the hills which are situated at their base, and serve as their first steps; but, on the contrary, are situated underneath them. The latter are placed upon the declivities of the former. When we dig through the horizontal strata in the neighbourhood of the inclined strata, the inclined strata are invariably found below. Nay sometimes, when the inclined strata are not too much elevated, their summit is surmounted by horizontal strata. The inclined strata are therefore more ancient than the horizontal strata. And as they must necessarily have been formed in a horizontal position, they have been subsequently shifted into their inclined or vertical position, and that too before the horizontal strata were placed above them.

Thus the sea, previous to the formation of the horizontal strata, had formed others, which, by some means, have been broken, lifted up, and overturned in a thousand ways. There had therefore been also at least one change in the basin of that sea which preceded ours; it had also experienced at least one revolution; and as several of these inclined strata which it had formed first, are elevated above the level of the horizontal strata which have succeeded and which surround them, this revolution, while it gave them their present inclination, had also caused them to project above the level of the sea, so as to form islands, or at least rocks and inequalities; and this must have happened whether one of their edges was lifted above the water, or the depression of the opposite edge caused the water to subside. This is the second result, not less obvious, nor less clearly demonstrated, than the first, to every one who will take the trouble of studying carefully the remains by which it is illustrated and proved.

Proofs that such Revolutions have been numerous.

If we institute a more detailed comparison between the various strata and those remains of animals which they contain, we shall soon discover still more numerous differences among them, indicating a proportional number of changes in their condition. The sea has not always deposited stony substances of the same kind. It has observed a regular succession as to the nature of its deposits; the more ancient the strata are, so much the more uniform and extensive are they; and the more recent they are, the more limited are they, and the more variation is observed in them at small distances. Thus the great catastrophes which have produced revolutions in the basin of the sea, were preceded, accompanied, and followed by changes in the nature of the fluid and of the substances which it held in solution and when the surface of the seas came to be divided by islands and projecting ridges, different changes took place in every separate basin.

Amidst these changes of the general fluid, it must have been almost impossible for the same kind of animals to continue to live:—nor did they do so in fact. Their species, and even their genera, change with the strata; and although the same species occasionally recur at small distances, it is generally the case that the shells of the ancient strata have forms peculiar to themselves; that they gradually disappear, till they are not to be seen at all in the recent strata, still less in the existing seas, in which, indeed, we never discover their corresponding species, and where several, even of their genera, are not to be found; that, on the contrary, the shells of the recent strata resemble, as it respects the genus, those which still exist in the sea; and that in the last-formed and loosest of these strata, there are some species which the eye of the most expert naturalists cannot distinguish from those which at present inhabit the ocean.

In animal nature, therefore, there has been a succession of changes corresponding to those which have taken place in the chemical nature of the fluid; and when the sea last receded from our continent, its inhabitants were not very different from those which it still continues to support. . . .

Finally, if we examine with greater care these remains of organized bodies, we shall discover, in the midst even of the most ancient secondary strata, other strata that are crowded with animal or vegetable productions, which belong to the land and to fresh water; and amongst the most recent strata, that is, the strata which are nearest the surface, there are some of them in which land animals are buried under heaps of marine productions. Thus the various catastrophes of our planet have not only caused the different parts of our continent to rise by degrees from the basin of the sea, but it has also frequently happened, that lands which had been laid dry, have been again covered by water, in consequence either of these lands sinking down below the level of the sea, or of the sea being raised above the level of the lands. The particular portions of the earth also which the sea has abandoned by its last retreat, had been laid dry once before, and had at that time produced quadrupeds, birds, plants, and all kinds of terrestrial productions; it had then been inundated by the sea, which has since retired from it, and left it to be occupied by its own proper inhabitants.

The changes which have taken place in the productions of the shelly strata have not, therefore, been entirely owing to a gradual and general retreat of the waters, but to successive irruptions and retreats, the final result of which, however, has been an universal depression of the level of the sea.

Proofs that the Revolutions have been sudden.

These repeated irruptions and retreats of the sea have neither been slow nor gradual; most of the catastrophes which have occasioned them have been sudden; and this is easily proved, especially with regard to the last of them, the traces of which are most conspicuous. In the northern regions it has left the carcasses of some large quadrupeds which the ice had arrested, and which are preserved even to the present day with their skin, their hair, and their flesh. If they had not been frozen as soon as killed, they must quickly have been decomposed by putrefaction. But this eternal frost could not have taken possession of the regions which these animals inhabited except by the same cause which destroyed them;[1] this cause, therefore, must have been as sudden as its effect. The breaking to pieces and overturnings of the strata, which happened in former catastrophes, show plainly enough that they were sudden and violent like the last; and the heaps of *debris* and rounded pebbles which are found in various places among the solid strata, demonstrate the vast force of the motions excited in the mass of waters by these overturnings. Life, therefore, has been often disturbed on this earth by terrible events—calamities which, at their commencement, have perhaps moved and overturned to a great depth the entire outer crust of the globe, but which, since these first commotions, have uniformly acted at a less depth and less generally. Numberless living beings have been the victims of these catastrophes; some have been destroyed by sudden inundations, others have been laid dry in consequence of the bottom of the seas being instantaneously elevated. Their races even have become extinct, and have left no memorial of them except some small fragment which the naturalist can scarcely recognise.

Such are the conclusions which necessarily result from the objects that we meet with at every step of our inquiry, and which we can always verify by examples drawn from almost every country. Every part of the globe bears the impress of these great and terrible events so distinctly, that they must be

---

[1] The two most remarkable phenomena of this kind, and which must for ever banish all idea of a slow and gradual revolution, are the rhinoceros, discovered in 1771 in the banks of the *Vilhoui*, and the elephant recently found by M. Adams near the mouth of the *Lena*. This last retained its flesh and skin, on which was hair of two kinds; one short, fine, and crisped, resembling wool, and the other like long bristles. The flesh was still in such high preservation, that it was eaten by dogs.

visible to all who are qualified to read their history in the remains which they
have left behind.

But what is still more astonishing and not less certain, there have not been
always living creatures on the earth, and it is easy for the observer to discover
the period at which animal productions began to be deposited.

GEOLOGY AND ORGANIC EVOLUTION

LYELL, Sir Charles (Scotch geologist, 1795–1875). From *Principles of geology:
being an inquiry how far the former changes of the earth's surface are referrable to
causes now in operation*, 5th ed., London, 1837.

The mutual interaction between Darwin and Lyell was of central significance in the
history of the doctrine of descent. The major development in Lyell's intellectual history
is his rejection of catastrophism in favor of a theory attributing earth's past modifications
to causes operative today. Applying this generalization to organic existence, Darwin
and Lyell reached different conclusions. To Darwin it offered an explanation of organic variety considered as the result of descent with change; Lyell, on the other hand,
continued for years to combat Lamarckism and reject the whole notion of transformation. Within four years after the appearance of Darwin's *Origins*, however, Lyell publicly went over to the natural-selection theory, a step he had already started to make in
his own thinking partly through his familiarity with Darwin's ideas even before the latter were published. Our selection is from the 5th ed., since it represents the state of Lyell's theory at the time when Darwin first came under its influence.

For more than two centuries the shelly strata of the Subapennine hills
afforded matter of speculation to the early geologists of Italy, and few of them
had any suspicion that similar deposits were then forming in the neighbouring
sea. They were as unconscious of the continued action of causes still producing similar effects, as the astronomers, in the case above supposed, of the
existence of certain heavenly bodies still giving and reflecting light, and performing their movements as of old. Some imagined that the strata, so rich
in organic remains, instead of being due to secondary agents, had been so
created in the beginning of things by the fiat of the Almighty; and others
ascribed the imbedded fossil bodies to some plastic power which resided in the
earth in the early ages of the world. At length Donati explored the bed of
the Adriatic, and found the closest resemblance between the new deposits
there forming, and those which constituted hills above a thousand feet high
in various parts of the Italian peninsula. He ascertained that certain genera
of living testacea were grouped together at the bottom of the sea, in precisely
the same manner as were their fossil analogues in the strata of the hills, and
that some species were common to the recent and fossil world. Beds of
shells, moreover, in the Adriatic, were becoming incrusted with calcareous
rock: and others were recently inclosed in deposits of sand and clay, precisely
as fossil shells were found in the hills. This splendid discovery of the identity

of modern and ancient submarine operations was not made without the aid of artificial instruments, which, like the telescope, brought phenomena into view not otherwise within the sphere of human observation.

In like manner, in the Vicentin, a great series of volcanic and marine sedimentary rocks was examined in the early part of the last century; but no geologists suspected, before the time of Arduino, that these were partly composed of ancient submarine lavas. If, when these inquiries were first made, geologists had been told that the mode of formation of such rocks might be fully elucidated by the study of processes then going on in certain parts of the Mediterranean, they would have been as incredulous as geometers would have been before the time of Newton, if any one had informed them that, by making experiments on the motion of bodies on the earth, they might discover the laws which regulated the movements of distant planets.

The establishment, from time to time, of numerous points of identification, drew at length from geologists a reluctant admission, that there was more correspondence between the physical constitution of the globe, and more uniformity in the laws regulating the changes of its surface, from the remote eras to the present, than they at first imagined. If, in this state of the science, they still despaired of reconciling every class of geological phenomena to the operations of ordinary causes, even by straining analogy to the utmost limits of credibility, we might have expected, at least, that the balance of probability would now have been presumed to incline towards the identity of the causes. But, after repeated experience of the failure of attempts to speculate on different classes of geological phenomena, as belonging to a distinct order of things, each new sect persevered systematically in the principles adopted by their predecessors. They invariably began, as each new problem presented itself, whether relating to the animate or inanimate world, to assume in their theories, that the economy of nature was formerly governed by rules for the most part independent of those now established. Whether they endeavoured to account for the origin of certain igneous rocks, or to explain the forces which elevated hills or excavated valleys, or the causes which led to the extinction of certain races of animals, they first presupposed an original and dissimilar order of nature; and when at length they approximated, or entirely came round to an opposite opinion, it was always with the feeling, that they conceded what they were justified a priori in deeming improbable. In a word, the same men who, as natural philosophers, would have been most incredulous respecting any extraordinary deviations from the known course of nature, if reported to have happened in their own time, were equally disposed, as geologists, to expect the proofs of such deviations at every period of the past. . . .

Progressive development of organic life. In the preceding chapters I have considered many of the most popular grounds of opposition to the doc-

trine, that all former changes of the organic and inorganic creation are referable to one uninterrupted succession of physical events, governed by the laws
of Nature now in operation.

As the principles of our science must always remain unsettled so long as no
fixed opinions are entertained on this fundamental question, I shall proceed
to examine other objections which have been urged against the assumption of
the identity of the ancient and modern causes of change. A late distinguished
writer has formally advanced some of the most popular of these objections.
"It is impossible," he affirms, "to defend the proposition, that the present
order of things is the ancient and constant order of nature, only modified by
existing laws: in those strata which are deepest, and which must, consequently,
be supposed to be the earliest deposited, forms even of vegetable life are rare;
shells and vegetable remains are found in the next order; the bones of fishes
and oviparous reptiles exist in the following classes; the remains of birds, with
those of the same genera mentioned before, in the next order; those of quadrupeds of extinct species in a still more recent class; and it is only in the loose
and slightly consolidated strata of gravel and sand, and which are usually
called diluvian formations, that the remains of animals such as now people
the globe are found, with others belonging to extinct species. But, in none
of these formations, whether called secondary, tertiary, or diluvial, have the
remains of man, or any of his works, been discovered; and whoever dwells
upon this subject must be convinced, that the present order of things, and the
comparatively recent existence of man as the master of the globe, is as certain
as the destruction of a former and a different order, and the extinction of a
number of living forms which have no types in being. In the oldest secondary strata there are no remains of such animals as now belong to the surface;
and in the rocks, which may be regarded as more recently deposited, these
remains occur but rarely, and with abundance of extinct species;—there
seems, as it were, a gradual approach to the present system of things,
and a succession of destructions and creations preparatory to the existence of
man." [1]

In the above passages, the author deduces two important conclusions from
geological data: first, that in the successive groups of strata, from the oldest
to the most recent, there is a progressive development of organic life, from
the simplest to the most complicated forms;—secondly, that man is of comparatively recent origin. It will be easy to shew that the first of these propositions, though very generally received, has but a slender foundation in fact.
The second, on the contrary, is indisputable; and it is important, therefore,
to consider how far its admission is inconsistent with the doctrine, that the
system of the natural world may have been uniform from the beginning,

---

[1] Davy, Sir H. *Consolations in travel, or, the last days of a philosopher,* London,
1830. See Dialogue IV.

or rather from the era when the oldest rocks hitherto discovered were formed. . . .

The result, then, of our inquiry into the evidence of the successive development of the animal and vegetable kingdoms, may be stated in a few words. In regard to *plants*, if we neglect the obscure and ambiguous impressions found in some of the oldest fossiliferous rocks, which can lead to no safe conclusions, we may consider those which characterize the great carboniferous group as the first deserving particular attention. They are by no means confined to the simplest forms of vegetation, as to cryptogamic plants; but, on the contrary, belong to all the leading divisions of the vegetable kingdom; some of the more fully developed forms, both of dicotyledons and monocotyledons having already been discovered, even among the first three or four hundred species brought to light: it is therefore superfluous to pursue this part of the argument farther.

If we then examine the animal remains of the oldest formations, we find bones and skeletons of fish in the old red sandstones, and even in some transition limestones below it; in other words, we have already vertebrated animals in the most ancient strata respecting the fossils of which we can be said to possess any accurate information.

In regard to birds and quadrupeds, their remains are almost entirely wanting in *marine* deposits of every era, even where interposed freshwater strata contain those fossils in abundance, as in the Paris basin. The secondary strata of Europe are for the most part marine, and there is as yet only one instance of the occurrence of mammiferous fossils in the slate of Stonesfield, a rock unquestionably of the Oolitic period, and which appears, from several other circumstances, to have been formed near the point where some river entered the sea.

When we examine the tertiary groups, we find in the Eocene or oldest strata of that class the remains of a great assemblage of the highest or mammiferous class, all of extinct species, and in the Miocene beds, or those of a newer tertiary epoch, other forms, for the most part of lost species, and almost entirely distinct from the Eocene tribes. Another change is again perceived, when we investigate the fossils of later or of the Pliocene periods. But in this succession of quadrupeds, we cannot detect any signs of progressive development of organization,—any indication that the Eocene fauna was less perfect than the Miocene, or the Miocene, than what will be designated in the fourth book the Newer Pliocene. . . .

Doctrine of successive development not confirmed by the admission that man is of modern origin. It is on other grounds that we are entitled to infer that man is, comparatively speaking, of modern origin; and if this be assumed, we may then ask whether his introduction can be considered as one step in a progressive system, by which, as some suppose, the organic world advanced slowly from a more simple to a more perfect state? In reply to this question, it should first be observed, that the superiority of man depends not on those faculties and attributes which he shares in common with the inferior animals, but on his reason, by which he is distinguished from them. When it is said that the human race is of far higher dignity than were any pre-existing beings on the earth, it is the intellectual and moral attributes only of our race, not the animal, which are considered; and it is by no means clear, that the organization of man is such as would confer a decided pre-eminence upon him, if, in place of his reasoning powers, he was merely provided with such instincts as are possessed by the lower animals.

If this be admitted, it would by no means follow, even if there had been sufficient geological evidence in favour of the theory of progressive development, that the creation of man was the last link in the same chain. For the sudden passage from an irrational to a rational animal is a phenomenon of a distinct kind from the passage from the more simple to the more perfect forms of animal organization and instinct. To pretend that such a step, or rather leap, can be part of a regular series of changes in the animal world, is to strain analogy beyond all reasonable bounds. . . .

Recapitulation. For the reasons, therefore, detailed in this and the two preceding chapters, we may draw the following inferences in regard to the reality of *species* in nature:—

1st. That there is a capacity in all species to accommodate themselves, to a certain extent, to a change of external circumstances, this extent varying greatly, according to the species.

2dly. When the change of situation which they can endure is great, it is usually attended by some modifications of the form, colour, size, structure, or other particulars; but the mutations thus superinduced are governed by constant laws, and the capability of so varying forms part of the permanent specific character.

3dly. Some acquired peculiarities of form, structure, and instinct, are transmissible to the offspring; but these consist of such qualities and attributes only as are intimately related to the natural wants and propensities of the species.

4thly. The entire variation from the·original type, which any given kind of change can produce, may usually be effected in a brief period of time, after which no farther deviation can be obtained by continuing to alter the circumstances, though ever so gradually; indefinite divergence, either in the way of improvement or deterioration, being prevented, and the least possible excess beyond the defined limits being fatal to the existence of the individual.

5thly. The intermixture of distinct species is guarded against by the aversion of the individuals composing them to sexual union, or by the sterility of the mule offspring. It does not appear that true hybrid races have ever been perpetuated for several generations, even by the assistance of man; for the cases usually cited relate to the crossing of mules with individuals of pure species, and not to the intermixture of hybrid with hybrid.

6thly. From the above considerations, it appears that species have a real existence in nature; and that each was endowed, at the time of its creation, with the attributes and organization by which it is now distinguished.

JOINT PUBLICATION OF DARWIN'S AND WALLACE'S ARGUMENTS IN FAVOR OF NATURAL SELECTION

DARWIN, Charles Robert (English naturalist, 1809–1882), and WALLACE, Alfred Russel (British naturalist, 1823–1913). *On the tendency of species to form varieties; and on the perpetuation of varieties and species by a natural means of selection*, in the *Journal of the Linnaean Society of London*, vol. 3, p. 45, 1859.

Darwin had recorded his concept of evolution in various notes and essays,* some of them twenty years prior to the present publication, and in a letter to Asa Gray in Boston in 1857. In 1858, Wallace, then at Ternate in the Moluccas, sent Darwin for criticism an essay containing independently developed ideas strikingly resembling Darwin's own. As requested by Wallace, Darwin forwarded this paper to Lyell with the comment "your words have come true with a vengeance that I should be forestalled." Lyell, already familiar with Darwin's views, requested the latter to submit an abstract of these. When this was submitted by Darwin, the results were as follows.

Read July 1st, 1858

London, June 30th, 1858

My Dear Sir,—

The accompanying papers, which we have the honor of communicating to the Linnaean Society, and which all relate to the same subject, viz., the Laws which affect the production of Varieties, Races, and Species, contain the results of the investigations of two indefatigable naturalists, Mr. Charles Darwin and Mr. Alfred Wallace.

These gentlemen having, independently and unknown to one another, conceived the same very ingenious theory to account for the appearance and perpetuation of varieties and of specific forms on our planet, may both fairly claim the merit of being original thinkers in this important line of inquiry; but neither of them having published his views, though Mr. Darwin has for many years past been repeatedly urged by us to do so, and both authors having now unreservedly placed their papers in our hands, we think it would best promote the interests of science that a selection from them should be laid before the Linnaean Society.

Taken in the order of their dates, they consist of:—

1. Extracts from a MS. work on Species, by Mr. Darwin, which was.

---

* Charles R. Darwin, *The foundations of the origin of species; two essays written in 1842 and 1844 by C. D. Ed. by his son Francis D.*, Cambridge, 1909.

sketched in 1839, and copied in 1844, when the copy was read by Dr. Hooker, and its contents afterwards communicated to Sir Charles Lyell. The first part is devoted to "The Variation of Organic Beings under Domestication and in their Natural State"; and the second chapter of that Part, from which we propose to read to the Society the extracts referred to, is headed, "On the Variation of Organic Beings in a state of Nature; on the Natural Means of Selection; on the Comparison of Domestic Races and true Species."

2. An abstract of a private letter addressed to Professor Asa Gray, of Boston, U.S., in October 1857, by Mr. Darwin, in which he repeats his views, and which shows that these remained unaltered from 1839 to 1857.

3. An Essay by Mr. Wallace, entitled "On the Tendency of Varieties to depart indefinitely from the Original Type." This was written at Ternate in February 1858, for the perusal of his friend and correspondent Mr. Darwin, and sent to him with the expressed wish that it should be fowarded to Sir Charles Lyell, if Mr. Darwin thought it sufficiently novel and interesting. So highly did Mr. Darwin appreciate the value of the views therein set forth, that he proposed in a letter to Sir Charles Lyell, to obtain Mr. Wallace's consent to allow the Essay to be published as soon as possible. Of this step we highly approved, provided Mr. Darwin did not withhold from the public, as he was strongly inclined to do (in favor of Mr. Wallace), the memoir which he had himself written on the same subject, and which, as before stated, one of us had perused in 1844, and the contents of which we had both of us been privy to for many years. On representing this to Mr. Darwin, he gave us permission to make what use we thought proper of his memoir, &c; and in adopting our present course, of presenting it to the Linnaean Society, we have explained to him that we were not solely considering the relative claims to priority of himself and his friend, but the interests of science generally; for we feel it to be desirable that views founded on a wide deduction from facts, and matured by years of reflection, should constitute at once a goal from which others may start, and that, while the scientific world is waiting for the appearance of Mr. Darwin's complete work. Some of the leading results of his labours, as well as those of his able correspondent, should together be laid before the public.

We have the honour to be yours very obediently,

CHARLES LYELL

JOS. D. HOOKER

J. J. Bennett, Esq.
   Secretary of the Linnaean Society.

I. *Extract from an unpublished Work on Species, by* C. DARWIN, *Esq., consisting of a portion of a Chapter entitled, "On the Variation of Organic Be-*

ings in a state of Nature; On the Natural Means of Selection; on the Comparison of Domestic Races and true Species."

De Candolle, in an eloquent passage, has declared that all nature is at war, one organism with another, or with external nature. Seeing the contented face of nature, this may at first well be doubted; but reflection will inevitably prove it to be true. The war, however, is not constant, but recurrent in a slight degree at short periods, and more severely at occasional more distant periods; and hence its effects are easily overlooked. It is the doctrine of Malthus applied in most cases with tenfold force. As in every climate there are seasons, for each of its inhabitants, of greater and less abundance, so all annually breed; and the moral restraint which in some small degree checks the increase of mankind is entirely lost. Even slow-breeding mankind has doubled in twenty-five years; and if he could increase his food with greater ease, he would double in less time. But for animals without artificial means, the amount of food for each species must, *on an average*, be constant, whereas the increase of all organisms tends to be geometrical, and in a vast majority of cases at an enormous ratio. Suppose in a certain spot there are eight pairs of birds, and that *only* four pairs of them annually (including double hatches) rear only four young, and that these go on rearing their young at the same rate, then at the end of seven years (a short life, excluding violent deaths, for any bird) there will be 2048 birds, instead of the original sixteen. As this increase is quite impossible, we must conclude that birds do not rear nearly half their young, or that the average life of a bird is, from accident, not nearly seven years. Both checks probably concur. The same kind of calculation applied to all plants and animals affords results more or less striking, but in very few instances more striking than in man.

Many practical illustrations of this rapid tendency to increase are on record, among which, during peculiar seasons, are the extraordinary numbers of certain animals; for instance, during the years 1826 to 1828, in La Plata, when from drought some millions of cattle perished, the whole country actually *swarmed* with mice. Now I think it cannot be doubted that during the breeding-season all the mice (with the exception of a few males or females in excess) ordinarily pair, and therefore that this astounding increase during three years must be attributed to a greater number than usual surviving the first year, and then breeding, and so on till the third year, when their numbers were brought down to their usual limits on the return of wet weather. Where man has introduced plants and animals into a new and favorable country, there are many accounts in how surprisingly few years the whole country has become stocked with them. This increase would necessarily stop as soon as the country was fully stocked; and yet we have every reason to believe, from what is known of wild animals, that *all* would pair in the spring. In the

majority of cases it is most difficult to imagine where the checks fall—though
generally, no doubt, on the seeds, eggs, and young; but when we remember
how impossible, even in mankind (so much better known than any other animal), it is to infer from repeated casual observations what the average duration of life is, or to discover the different percentage of deaths to births in different countries, we ought to feel no surprise at our being unable to discover
where the check falls in any animal or plant. It should always be remembered, that in most cases the checks are recurrent yearly in small, regular degree, and in an extreme degree during unusually cold, hot, dry, or wet years,
according to the constitution of the being in question. Lighten any check in
the least degree, and the geometrical powers of increase in every organism
will almost instantly increase the average number of the favored species. Nature may be compared to a surface on which rest ten thousand sharp wedges
touching each other and driven inwards by incessant blows. Fully to realize
these views much reflection is requisite. Malthus on man should be studied;
and all such cases as those of the mice in La Plata, of the cattle and horses
when first turned out in South America, of the birds by our calculation, &c.,
should be well considered. Reflect on the enormous multiplying power *inherent and annually in action* in all animals; reflect on the countless seeds scattered by a hundred ingenious contrivances, year after year, over the whole
face of the land; and yet we have every reason to suppose that the average
percentage of each of the inhabitants of a country usually remains constant.
Finally, let it be borne in mind that this average number of individuals (the
external conditions remaining the same) in each country is kept up by recurrent struggles against other species or against external nature (as on the borders of the Arctic regions, where the cold checks life), and that ordinarily
each individual of every species holds its place, either by its own struggle and
capacity of acquiring nourishment in some period of its life from the egg upwards; or by the struggle of its parents (in short-lived organisms, when the
main check occurs at longer intervals) with other individuals of the *same* or
different species.

  But let the external conditions of a country alter. If in a small degree,
the relative proportions of the inhabitants will in most cases simply be slightly
changed; but let the number of inhabitants be small, as on an island, and free
access to it from other countries be circumscribed, and let the change of conditions continue progressing (forming new stations), in such a case the original
inhabitants must cease to be as perfectly adapted to the changed conditions as
they were originally. It has been shown in a former part of this work, that
such changes of external conditions would, from their acting on the reproductive system, probably cause the organization of those beings which were
most affected to become, as under domestication, plastic. Now, can it be
doubted, from the struggle each individual has to obtain subsistence, that any

minute variation in structure, habits, or instincts, adapting that individual better to the new conditions, would tell upon its vigour and health? In the struggle it would have a better *chance* of surviving; and those of its offspring which inherited the variation, be it ever so slight, would also have a better *chance*. Yearly more are bred than can survive; the smallest grain in the balance, in the long run, must tell on which death shall fall, and which shall survive. Let this work of selection on the one hand, and death on the other, go on for a thousand generations, who will pretend to affirm that it would produce no effect, when we remember what, in a few years, Bakewell effected in cattle, and Western in sheep, by this identical principle of selection?

To give an imaginary example from changes in progress on an island:— let the organization of a canine animal which preyed chiefly on rabbits, but sometimes on hares, become slightly plastic; let these same changes cause the number of rabbits very slowly to decrease, and the number of hares to increase; the effect of this would be that the fox or dog would be driven to try to catch more hares: his organization, however, being slightly plastic, those individuals with the lightest forms, longest limbs, and best eyesight, let the difference be ever so small, would be slightly favored, and would tend to live longer, and to survive during that kind of a year when food was scarcest; they would also rear more young, which would tend to inherit these slight peculiarities. The less fleet ones would be rigidly destroyed. I can see no more reason to doubt that these causes in a thousand generations would produce a marked effect, and adapt the form of the fox or dog to the catching of hares instead of rabbits, than that greyhounds can be improved by selection and careful breeding. So would it be with plants under similar circumstances. If the number of individuals of a species with plumed seeds could be increased by greater powers of dissemination within its own area (that is, if the check to increase fell chiefly on the seeds), those seeds which were provided with ever so little more down, would in the long run be most disseminated; hence a greater number of seeds thus formed would germinate, and would tend to produce plants inheriting the slightly better-adapted down.

Besides this natural means of selection, by which those individuals are preserved, whether in their egg, or larval, or mature state, which are best adapted to the place they fill in nature, there is as a second agency at work in most unisexual animals, tending to produce the same effect, namely, the struggle of the males for the females. These struggles are generally decided by the law of battle, but in the case of birds, apparently, by the charms of their song, by their beauty or their power of courtship, as in the dancing rock-thrush of Guiana. The most vigorous and healthy males, implying perfect adaptation, must generally gain the victory in their contests. This kind of selection, however, is less rigorous than the other; it does not require the death of the less successful, but gives to them fewer descendants. The struggle falls, moreover, at a time of the year when food is generally abundant, and perhaps the effect chiefly produced would be the modification of the secondary sexual characters, which are not related to the power of obtaining food, or to defense from enemies, but to fighting with or rivalling other males. The result of this struggle amongst the males may be compared in some respects to that produced by those agriculturists who pay less attention to the careful selection of all their young animals, and more to the occasional use of a choice mate.

II. *Abstract of a Letter from* C. DARWIN, *Esq., to Prof.* ASA GRAY, *Boston, U.S., dated Down, September 5th,* 1857.

1. It is wonderful what the principle of selection by man, that is the picking out of individuals with any desired quality, and breeding from them, and again picking out, can do. Even breeders have been astounded at their own results. They can act on differences inappreciable to an uneducated eye. Selection has been *methodically* followed in *Europe* only for the last half century; but it was occasionally, and even in some degree methodically followed in the most ancient times. There must have been also a kind of unconscious selection from a remote period, namely in the preservation of the individual animals (without any thought of their offspring) most useful to each race of man in his particular circumstances. The "roguing," as nurserymen call the destroying of varieties which depart from their type, is a kind of selection. I am convinced that intentional and occasional selection has been the main agent in the production of our domestic races; but however this may be, its great power of modification has been indisputably shown in later times. Selection acts only by the accumulation of slight or greater variations, caused by external conditions, or by the mere fact that in generation the child is not absolutely similar to its parent. Man, by this power of accumulating variations, adapts living beings to his wants—may be said to make the wool of one sheep good for carpets, of another for cloth, &c.

2. Now suppose there were a being who did not judge by mere external appearances, but who could study the whole internal organization, who was never capricious, and should go on selecting for one object during millions of generations; who will say what he might not effect? In nature we have some *slight* variation occasionally in all parts; and I think it can be shown that changed conditions of existence is the main cause of the child not exactly resembling its parents; and in nature geology shows us what changes have taken place, and are taking place. We have almost unlimited time; no one but a practical geologist can fully appreciate this. Think of the Glacial period, during the whole of which the same species at least of shells have existed; there must have been during this period millions on millions of generations.

3. I think it can be shown that there is such an unerring power at work in *Natural Selection* (the title of my book), which selects exclusively for the good

of each organic being. The elder De Candolle, W. Herbert, and Lyell have written excellently on the struggle for life; but even they have not written strongly enough. Reflect that every being (even the elephant) breeds at such a rate, that in a few years, or at most a few centuries, the surface of the earth would not hold the progeny of one pair. I have found it hard constantly to bear in mind that the increase of every single species is checked during some part of its life, or during some shortly recurrent generation. Only a few of those annually born can live to propagate their kind. What a trifling difference must often determine which shall live, and which perish!

4. Now take the case of a country undergoing some change. This will tend to cause some of its inhabitants to vary slightly—not but that I believe most beings vary at all times enough for selection to act on them. Some of its inhabitants will be exterminated; and the remainder will be exposed to the mutual action of a different set of inhabitants, which I believe to be far more important to the life of each being than mere climate. Considering the infinitely various methods which living beings follow to obtain food by struggling with other organisms, to escape danger at various times of life, to have their eggs or seeds disseminated, &c., &c., I cannot doubt that during millions of generations individuals of a species will be occasionally born with some slight variation, profitable to some part of their economy. Such individuals will have a better chance of surviving, and of propagating their new and slightly different structure; and the modification may be slowly increased by the accumulative action of natural selection to any profitable extent. The variety thus formed will either coexist with, or, more commonly, will exterminate its parent form. An organic being, like the woodpecker or misseltoe, may thus come to be adapted to a score of contingences—natural selection accumulating those slight variations in all parts of its structure, which are in any way useful to it during any part of its life.

5. Multiform difficulties will occur to every one, with respect to this theory. Many can, I think, be satisfactorily answered. *Natura non fecit saltum* answers some of the most obvious. The slowness of the change, and only a very few individuals undergoing change at any one time, answers others. The extreme imperfection of our geological records answers others.

6. Another principle, which may be called the principle of divergence, plays, I believe, an important part in the origin of species. The same spot will support more life if occupied by very diverse forms. We see this in the many generic forms in a square yard of turf, and in the plants or insects on any little uniform islet, belonging almost invariably to as many genera and families as species. We can understand the meaning of this fact amongst the higher animals, whose habits we understand. We know that it has been experimentally shown that a plot of land will yield a greater weight if sown with several species and genera of grasses, than if sown with only two or three species. Now,

every organic being, by propagating so rapidly, may be said to be striving its utmost to increase in numbers. So it will be with the offspring of any species after it has become diversified into varieties, or subspecies, or true species. And it follows, I think, from the foregoing facts, that the varying offspring of each species will try (only few will succeed) to seize on as many and as diverse places in the economy of nature as possible. Each new variety or species, when formed, will generally take the place of, and thus exterminate its less well-fitted parent. This I believe to be the origin of the classification and affinities of organic beings at all times; for organic beings always *seem* to branch and sub-branch like the limbs of a tree from a common trunk, the flourishing and diverging twigs destroying the less vigorous—the dead and lost branches rudely representing extinct genera and families.

This sketch is *most* imperfect; but in so short a space I cannot make it better. Your imagination must fill up very wide blanks.

C. Darwin.

III. *On the Tendency of Varieties to depart indefinitely from the Original Type.* By Alfred Russel Wallace.

One of the strongest arguments which have been adduced to prove the original and permanent distinctness of species is, that *varieties* produced in a state of domesticity are more or less unstable, and often have a tendency, if left to themselves, to return to the normal form of the parent species; and this instability is considered to be a distinctive peculiarity of all varieties, even of those occurring among wild animals in a state of nature, and to constitute a provision for preserving unchanged the originally created distinct species.

In the absence or scarcity of facts and observations as to the *varieties* occurring among wild animals, this argument has had great weight with naturalists, and has led to a very general and somewhat prejudiced belief in the stability of species. Equally general, however, is the belief in what are called "permanent or true varieties,"—races of animals which continually propagate their like, but which differ so slightly (although constantly) from some other race, that the one is considered to be a *variety* of the other. Which is the *variety* and which the original *species*, there is generally no means of determining, except in those rare cases in which the one race has been known to produce an offspring unlike itself and resembling the other. This, however, would seem quite incompatible with the "permanent invariability of species," but the difficulty is overcome by assuming that such varieties have strict limits, and can never again vary further from the original type, although they may return to it, which, from the analogy of the domesticated animals, is considered to be highly probable, if not certainly proved.

It will be observed that this argument rests entirely on the assumption, that *varieties* occurring in a state of nature are in all respects analogous to or even identical with those of domestic animals, and are governed by the same laws as regards their permanence or further variation. But it is the object of the present paper to show that this assumption is altogether false, that there is a general principle in nature which will cause many *varieties* to survive the parent species, and to give rise to successive variations departing further and further from the original type, and which also produces, in domesticated animals, the tendency of varieties to return to the parent form.

The life of wild animals is a struggle for existence. The full exertion of all their faculties and all their energies is required to preserve their own existence and provide for that of their infant offspring. The possibility of procuring food during the least favourable seasons, and of escaping the attacks of their most dangerous enemies, are the primary conditions which determine the existence both of individuals and of entire species. These conditions will also determine the population of the species; and by a careful consideration of all the circumstances we may be enabled to comprehend, and in some degree to explain, what at first sight appears so inexplicable—the excessive abundance of some species, while others closely allied to them are very rare.

The general proportion that must obtain between certain groups of animals is readily seen. Large animals cannot be so abundant as small ones; the carnivora must be less numerous than the herbivora; eagles and lions can never be so plentiful as pigeons and antelopes; the wild asses of the Tartarian deserts cannot equal in numbers the horses of the more luxuriant prairies and pampas of America. The greater or less fecundity of an animal is often considered to be one of the chief causes of its abundance or scarcity; but a consideration of the facts will show us that it really has little or nothing to do with the matter. Even the least prolific of animals would increase rapidly if unchecked, whereas it is evident that the animal population of the globe must be stationary, or perhaps, through the influence of man, decreasing. Fluctuations there may be; but permanent increase except in restricted localities, is almost impossible. For example, our own observation must convince us that birds do not go on increasing every year in a geometrical ratio, as they would do, were there not some powerful check to their natural increase. Very few birds produce less than two young ones each year, while many have six, eight, or ten; four will certainly be below the average; and if we suppose that each pair produce young only four times in their life, that will also be below the average, supposing them not to die either by violence or want of food. Yet at this rate, how tremendous would be the increase in a few years from a single pair! A simple calculation will show that in fifteen years each pair of birds would have increased to nearly ten millions! whereas we have no reason to believe that the number of the birds of any country increases at all in fifteen or in one hundred and fifty years. With such powers of increase the population must have reached its limits, and have become stationary, in a very few years after the origin of each species. It is evident, therefore, that each year an immense number of birds must perish—as many in fact as are born; and as on the lowest calculation the progeny are each year twice as numerous as their parents, it follows that, whatever be the average number of individuals existing in any given country, *twice that number must perish annually,*—a striking result, but one which seems at least highly probable, and is perhaps under rather than over the truth. It would therefore appear that, as far as the continuance of the species and the keeping up the average number of individuals are concerned, large broods are superfluous. On the average all above *one* become food for hawks and kites, wild cats and weasels, or perish of cold and hunger as winter comes on. This is strikingly proved by the case of particular species; for we find that their abundance in individuals bears no relation whatever to their fertility in producing offspring. Perhaps the most remarkable instance of an immense bird population is that of the passenger pigeon of the United States, which lays only one, or at most two eggs, and is said to rear generally but one young one. Why is this bird so extraordinarily abundant, while others producing two or three times as many young are much less plentiful? The explanation is not difficult. The food most congenial to this species, and on which it thrives best, is abundantly distributed over a very extensive region, offering such differences of soil and climate, that in one part or another of the area the supply never fails. The bird is capable of a very rapid and long-continued flight, so that it can pass without fatigue over the whole of the district it inhabits, and as soon as the supply of food begins to fail in one place is able to discover a fresh feeding-ground. This example strikingly shows us that the procuring a constant supply of wholesome food is almost the sole condition requisite for insuring the rapid increase of a given species, since neither the limited fecundity, nor the unrestrained attacks of birds of prey and of man are here sufficient to check it. In no other birds are these peculiar circumstances so strikingly combined. Either their food is more liable to failure, or they have not sufficient power of wing to search for it over an extensive area, or during some season of the year it becomes very scarce, and less wholesome substitutes have to be found; and thus, though more fertile in offspring, they can never increase beyond the supply of food in the least favorable seasons. Many birds can only exist by migrating, when their food becomes scarce, to regions possessing a milder, or at least a different climate, though, as these migrating birds are seldom excessively abundant, it is evident that the countries they visit are still deficient in a constant and abundant supply of wholesome food. Those whose organization does not permit them to migrate when their food becomes periodically scarce, can never attain a large population. This is probably the reason why woodpeckers are scarce with us, while in the tropics they are among the most abundant of solitary birds. Thus the house-sparrow is more abundant than the redbreast, because its food is more constant and plentiful,—seeds of grasses being preserved during the winter, and our farm-yards and stubble-fields furnishing an almost inexhaustible supply. Why, as a general rule, are aquatic, and especially seabirds, very numerous in individuals? Not because they are more prolific than others, generally the contrary; but because their food never fails, the sea-shores and river-banks daily swarming with a fresh supply of small mollusca and crustacea. Exactly the same laws will apply to mammals. Wild cats are prolific and have few enemies; why then are they never as abundant as rabbits? The only intelligible answer is, that their supply of food is more precarious. It appears evident, therefore, that so long as a country remains physically unchanged, the numbers of its animal population cannot materially increase. If one species does so, some others requiring the same kind of food must diminish in proportion. The numbers that die annually must be immense; and as the individual existence of each animal depends upon itself, those that die must be the weakest—the very young, the aged, and the diseased,—while those that prolong their existence can only be the most perfect in health and vigour—those who are best able to obtain food regularly, and avoid their numerous enemies. It is, as we commenced by remarking, "a struggle for existence," in which the weakest and least perfectly organized must always succumb.

Now it is clear that what takes place among the individuals of a species must also occur among the several allied species of a group,—viz. that those which are best adapted to obtain a regular supply of food, and to defend themselves against the attacks of their enemies and the vicissitudes of the seasons, must necessarily obtain and preserve a superiority in population; while those species which from some defect of power or organization are the least capable of counteracting the vicissitudes of food, supply, &c., must diminish in numbers, and, in extreme cases, become altogether extinct. Between these extremes the species will present various degrees of capacity for insuring the means of preserving life; and it is thus we account for the abundance or rarity of species. Our ignorance will generally prevent us from accurately tracing the effects to their causes; but could we become perfectly acquainted with the organization and habits of the various species of animals, and could we measure the capacity of each for performing the different acts necessary to its safety and existence under all the varying circumstances by which it is surrounded, we might be able even to calculate the proportionate abundance of individuals which is the necessary result.

If now we have succeeded in establishing these two points—1st, *that the animal population of a country is generally stationary, being kept down by a periodical deficiency of food, and other checks*; and, 2nd, *that the comparative abundance or scarcity of the individuals of the several species is entirely due to their organization and resulting habits, which, rendering it more difficult to procure a regular supply of food and to provide for their personal safety in some cases than in others, can only be balanced by a difference in the population which have to exist in a given area*—we shall be in a condition to proceed to the consideration of *varieties,* to which the preceding remarks have a direct and very important application.

Most or perhaps all the variations from the typical form of the species must have some definite effect, however slight, on the habits or capacities of the individuals. Even a change of colour might, by rendering them more or less distinguishable, affect their safety; a greater or less development of hair might modify their habits. More important changes, such as an increase in the power or dimensions of the limbs or any of the external organs, would more or less affect their mode of procuring food or the range of country which they inhabit. It is also evident that most changes would affect, either favourably or adversely, the powers of prolonging existence. An antelope with shorter or weaker legs must necessarily suffer more from the attacks of the feline carnivora; the passenger pigeon with less powerful wings would sooner or later be affected in its powers of procuring a regular supply of food; and in both cases the result must necessarily be a diminution of the population of the modified species. If, on the other hand, any species should produce a variety having slightly increased powers of preserving existence, that variety must inevitably in time acquire a superiority in numbers. These results must follow as surely as old age, intemperance, or scarcity of food produce an increased mortality. In both cases there may be many individual exceptions; but on the average the rule will invariably be found to hold good. All varieties will therefore fall into two classes—those which under the same conditions would never reach the population of the parent species, and those which would in time obtain and keep a numerical superiority. Now, let some alteration of physical conditions occur in the district—a long period of drought, a destruction of vegetation by locusts, the irruption of some new carnivorous animal seeking "pastures new"—any change in fact tending to render existence more difficult to the species in question, and tasking its utmost powers to avoid complete extermination; it is evident that, of all of the individuals composing the species, those forming the least numerous and most feebly organized variety would suffer first, and, were the pressure severe, must soon become extinct. The same causes continuing in action, the parent species would next suffer, would gradually diminish in numbers, and with a recurrence of similar unfavourable conditions might also become extinct. The superior variety would then alone remain, and on a return to favorable circumstances would rapidly increase in numbers and occupy the place of the extinct species and variety.

The *variety* would now have replaced the *species,* of which it would be a more perfectly developed and more highly organized form. It would be in all respects better adapted to secure its safety, and to prolong its individual existence and that of the race. Such a variety *could not* return to the original form; for that form is an inferior one, and could never compete with it for existence. Granted, therefore, a "tendency" to reproduce the original type of the species, still the variety must ever remain preponderant in numbers, and under adverse physical conditions *again alone survive*. But this new, improved, and populous race might itself, in course of time, give rise to new varieties, exhibiting several diverging modifications of form, any of which, tending to increase the facilities for preserving existence, must, by the same general law, in their turn become predominant. Here, then, we have *progression and continued divergence* deduced from the general laws which regulate the existence of animals in a state of nature, and from the undisputed fact that varieties do frequently occur. It is not, however, contended that this result would be invariable; a change of physical conditions in the district might at times materially modify it, rendering the race which had been the most capable of supporting existence under the former conditions now the least so, and even causing the extinction of the newer and, for a time, a superior race, while the old or parent species and its first inferior varieties continued to flourish. Variations in unimportant parts might also occur, having no perceptible effect on the life-preserving powers; and the varieties so furnished might run a course parallel with the parent species, either giving rise to further variations or returning to the former type. All we argue for is, that certain varieties have a tendency to maintain their existence longer than the original species, and this tendency must make itself felt; for though the doctrine of chances or averages can never be trusted to on a limited scale, yet, if applied to high numbers, the results come nearer to what theory demands, and, as we approach to an infinity of examples, become strictly accurate. Now the scale on which nature works is so vast —the numbers of individuals and periods of time with which she deals approach so near to infinity, that any cause, however slight, and however liable to be veiled and counteracted by accidental circumstances, must in the end produce its full legitimate results.

Let us now turn to domesticated animals, and inquire how varieties produced among them are affected by the principles here enunciated. The essential difference in the condition of wild and domestic animals is this,—that among the former, their well-being and very existence depend upon the full exercise and healthy condition of all their senses and physical powers, whereas, among the latter, these are only partially exercised, and in some cases are absolutely unused. A wild animal has to search, and often to labour, for every mouthful of food—to exercise sight, hearing, and smell in seeking it, and in avoiding dangers, in procuring shelter from the inclemency of the seasons, and in providing for the subsistence and safety of its offspring. There is no muscle of its body that is not called into daily and hourly activity; there is no sense or faculty that is not strengthened by continual exercise. The domestic animal, on the other hand, has food provided for it, is sheltered, and often confined, to guard it against the vicissitudes of the seasons, is carefully secured from the attacks of its natural enemies, and seldom even rears its young without human assistance. Half of its senses and faculties are quite useless; and the other half are but occasionally called into feeble exercise, while even its muscular system is only irregularly called into action.

Now when a variety of such an animal occurs, having increased power or capacity in any organ or sense, such increase is totally useless, is never called into action, and may even exist without the animal ever becoming aware of it. In the wild animal, on the contrary, all its faculties and powers being brought into full action for the necessities of existence, any increase becomes immediately available, is strengthened by exercise, and must even slightly modify the food, the habits, and the whole economy of the race. It creates as it were a new animal, one of superior powers, and which will necessarily increase in numbers and outlive those inferior to it.

Again, in the domesticated animal all variations have an equal chance of continuance; and those which would decidedly render a wild animal unable to compete with its fellows and continue its existence are no disadvantage whatever in a state of domesticity. Our quickly fattening pigs, short-legged sheep, pouter pigeons, and poodle dogs could never have come into existence in a state of nature, because the very first step toward such inferior forms would have led to the rapid extinction of the race; still less could they now exist in competition with their wild allies. The great speed but slight endurance of the race horse, the unwieldy strength of the ploughman's team, would both be useless in a state of nature. If turned wild on the pampas, such animals would probably soon become extinct, or under favourable circumstances might each lose those extreme qualities which would never be called into action, and in a few generations would revert to a common type, which must be that in which the various powers and faculties are so proportioned to each other as to be best adapted to procure food and secure safety,—that in which by the full exercise of every part of his organization the animal can alone continue to live. Domestic varieties, when turned wild, *must* return to something near the type of the original wild stock, *or become altogether extinct.*

We see, then, that no inferences as to varieties in a state of nature can be deduced from the observation of those occurring among domestic animals. The two are so much opposed to each other in every circumstance of their existence, that what applies to the one is almost sure not to apply to the other. Domestic animals are abnormal, irregular, artificial; they are subject to varieties which never occur and never can occur in a state of nature: their very existence depends altogether on human care; so far are many of them removed from that just proportion of faculties, that true balance of organization, by means of which alone an animal left to its own resources can preserve its existence and continue its race.

The hypothesis of Lamarck—that progressive changes in species have been produced by the attempts of animals to increase the development of their own organs, and thus modify their structure and habits—has been repeatedly and easily refuted by all writers on the subject of varieties and species, and it seems to have been considered that when this was done the whole question has been finally settled; but the view here developed renders such an hypothesis quite unnecessary, by showing that similar results must be produced by the action of principles constantly at work in nature. The powerful retractile talons of the falcon- and the cat-tribes have not been produced or increased by the volition of those animals; but among the different varieties which occurred in the earlier and less highly organized forms of these groups, *those always survive longest which had the greatest facilities for seizing their prey.* Neither did the giraffe acquire its long neck by desiring to reach the foliage of the more lofty shrubs, and constantly stretching its neck for the purpose, but because any varieties which occurred among its antitypes with a longer neck than usual *at once secured a fresh range of pasture over the same ground as their shorter-necked companions, and on the first scarcity of food were thereby enabled to outlive them.* Even the peculiar colours of many animals, especially insects, so closely resembling the soil or the leaves or the trunks on which they habitually reside, are explained on the same principle; for though in the course of ages varieties of many tints may have occurred, *yet those races having colours best adapted to concealment from their enemies would inevitably survive the longest.* We have also here an acting cause to account for that balance so often observed in nature,—a deficiency in one set of organs always being compensated by an increased development of some others—powerful wings accompanying weak feet, or great velocity making up for the absence of defensive weapons; for it has been shown that all varieties in which an unbalanced deficiency occurred could not long continue their existence. The action of this principle is exactly like that of the centrifugal governor of the steam engine, which checks and corrects any irregularities almost before they become evident; and in like manner no unbalanced deficiency in the animal kingdom can ever reach any conspicuous magnitude, because it would make itself felt at the very first step, by rendering existence difficult and extinction almost sure soon to follow. An origin such as is here advocated will also agree with the peculiar character of the modifications of form and structure which obtain in organized beings—the many lines of divergence from a central type, the increasing efficiency and power of a particular organ through a succession of allied species, and the remarkable persistence of unimportant parts such as colour, texture of plumage and hair, form of horns or crests, through a series of species differing considerably in more essential characters. It also furnishes us with a reason for that "more specialized structure" which Professor Owen states to be a characteristic of recent compared with extinct forms, and which would evidently be the result of the progressive modification of any organ applied to a special purpose in the animal economy.

We believe we have now shown that there is a tendency in nature to the continued progression of certain classes of *varieties* further and further from the original type—a progression to which their appears no reason to assign any definite limits—and that the same principle which produces this result in a state of nature will also explain why domestic varieties have a tendency to revert to the original type. This progression, by minute steps, in various directions, but always checked in balance by the necessary conditions subject to which alone existence can be preserved, may, it is believed, be followed out so as to agree with all the phenomena presented by organized beings, their extinction and succession in past ages, and all the extraordinary modifications of form, instinct, and habits which they exhibit.

Ternate, February, 1858.

THE DEFENSE OF DARWIN

MÜLLER, Fritz (Johann Friedrich Theodore) (German naturalist, 1821–1897). *Chapter XI* (chapter without title) in *Für Darwin*, Leipzig, 1864; tr. by W. S. Dallas as *Chapter XI, On the progress of evolution*, in *Facts and arguments for Darwin*, London, 1869.

Few modern books have provoked as vigorous a response, favorable and antagonistic, as Darwin's *Origin of Species*. Häckel (*q.v.*) is sometimes considered the chief one to apply the theory of descent to embryology; it was done with equal excellence, and sooner, by Fritz Müller. Like Häckel's gastraea theory, Müller's nauplius theory required modification before it could be finally accepted as evidence for the hypothesis of descent. Müller's other researches were chiefly in the field of natural history.*

On the Progress of Evolution.

From this scarcely unavoidable but unsatisfactory side-glance upon the old school, which looks down with so great an air of superiority upon Darwin's "intellectual dream" and the "giddy enthusiasm" of its friends, I turn to the more congenial task of considering the developmental history of the Crustacea from the point of view of the Darwinian theory.

Darwin himself, in the thirteenth chapter of his book, has already discussed the conclusions derived from his hypotheses in the domain of developmental history. For a more detailed application of them, however, it is necessary in the first place to trace these general conclusions a little further than he has there done.

The changes by which young animals depart from their parents, and the gradual accumulation of which causes the production of new species, genera,

---

* See the translations of his work on senses of insects in G. B. Longstaff, *Butterfly hunting in many lands*, London and New York, 1912.

and families, may occur at an earlier or later period of life,—in the young state, or at the period of sexual maturity. For the latter is by no means always, as in the Insecta, a period of repose; most other animals even then continue to grow and to undergo changes. Some variations, indeed, from their very nature, can only occur when the young animal has attained the adult stage of development. Thus the Sea Caterpillars (*Polynoë*) at first possess only a few body-segments, which, during development, gradually increase to a number which is different in different species, but constant in the same species; now before a young animal could exceed the number of segments of its parents, it must of course have attained that number. We may assume a similar supplementary progress wherever the deviation of the descendants consists in an addition of new segments and limbs.

Descendants therefore reach a new goal, either by deviating sooner or later whilst still on the way towards the form of their parents, or by passing along this course without deviation, but then, instead of standing still, advance still farther.

The former mode will have had a predominant action where the posterity of common ancestors constitutes a group of forms standing upon the same level in essential features, as the whole of the Amphipoda, Crabs, or Birds. On the other hand we are led to the assumption of the second mode of progress, when we seek to deduce from a common original form, animals some of which agree with young states of others.

In the former case the developmental history of the descendants can only agree with that of their ancestors up to a certain point at which their courses separate,—as to their structure in the adult state it will teach us nothing. *In the second case the entire development of the progenitors is also passed through by the descendants, and, therefore, so far as the production of a species depends upon this second mode of progress, the historical development of the species will be mirrored in its developmental history.* In the short period of a few weeks or months, the changing forms of the embryo and larvae will pass before us, a more or less complete and more or less true picture of the transformations through which the species, in the course of untold thousands of years, has struggled up to its present state.

One of the simplest examples is furnished by the development of the Tubicolar Annelids; but from its very simplicity it appears well adapted to open the eyes of many who, perhaps, would rather not see, and it may therefore find a place here. Three years ago I found on the walls of one of my glasses some small worm-tubes, the inhabitants of which bore three pairs of barbate branchial filaments, and had no operculum. According to this we should have been obliged to refer them to the genus *Protula*. A few days afterwards one of the branchial filaments had become thickened at the extremity into a clavate operculum, when the animals reminded me, by the barbate opercular peduncle, of the genus *Filograna,* only that the latter possesses two opercula. In three days more, during which a new pair of branchial filaments had sprouted forth, the opercular peduncle had lost its lateral filaments, and the worms had become *Serpulae.* Here the supposition at once presents itself that the primitive tubicolar worm was a *Protula,*—that some of its descendants, which had already become developed into perfect *Protulae,* subsequently improved themselves by the formation of an operculum which might protect their tubes from inimical intruders,—and that subsequent descendants of these latter finally lost the lateral filaments of the opercular peduncle, which they, like their ancestors, had developed.

What say the schools to this case? Whence and for what purpose, if the *Serpulae* were produced or created as ready-formed species, these lateral filaments of the opercular peduncle? To allow them to sprout forth merely for the sake of an invariable plan of structure, even when they must be immediately retracted again as superfluous, would certainly be an evidence rather of childish trifling or dictatorial pedantry, than of infinite wisdom. But no, I am mistaken; from the beginning of all things the Creator knew, that one day the inquisitive children of men would grope about after analogies and homologies, and that Christian naturalists would busy themselves with thinking out his Creative ideas; at any rate, in order to facilitate the discernment by the former that the opercular peduncle of the *Serpulae* is homologous with a branchial filament, He allowed it to make a *détour* in its development, and pass through the form of a barbate branchial filament.

The historical record preserved in developmental history is gradually EF-FACED *as the development strikes into a constantly straighter course from the egg to the perfect animal, and it is frequently* SOPHISTICATED *by the struggle for existence which the free-living larvae have to undergo.*

Thus as the law of inheritance is by no means strict, as it gives room for individual variations with regard to the form of the parents, this is also the case with the succession in time of the developmental processes. Every father of a family who has taken notice of such matters, is well aware that even in children of the same parents, the teeth, for example, are not cut or changed, either at the same age, or in the same order. Now in general it will be useful to an animal to obtain as early as possible those advantages by which it sustains itself in the struggle for existence. A precocious appearance of peculiarities originally acquired at a later period will generally be advantageous, and their retarded appearance disadvantageous; the former, when it appears accidentally, will be preserved by natural selection. It is the same with every change which gives to the larval stages, rendered multifarious by crossed and oblique characters, a more straightforward direction, simplifies and abridges the process of development, and forces it back to an earlier period of life, and finally into the life of the egg.

As this conversion of a development passing through different young states into a more direct one, is not the consequence of a mysterious inherent impulse, but dependent upon advances accidentally presenting themselves, it may take place in the most nearly allied animals in the most various ways, and require very different periods of time for its completion. There is one thing, however, that must not be overlooked here. The historical development of a species can hardly ever have taken place in a continuously uniform flow; periods of rest will have alternated with periods of rapid progress. But forms, which in periods of rapid progress were severed from others after a short duration, must have impressed themselves less deeply upon the developmental history of their descendants, than those which repeated themselves unchanged, through a long series of successive generations in periods of rest. These more fixed forms, less inclined to variation, will present a more tenacious resistance in the transition to direct development, and will maintain themselves in a more uniform manner and to the last, however different may be the course of this process in other respects.

In general, as already stated, it will be advantageous to the young to commence the struggle for existence in the form of their parents and furnished with all their advantages—in general, but not without exceptions. It is perfectly clear that a brood capable of locomotion is almost indispensable to attached animals, and that larvae of sluggish Mollusca or of worms burrowing in the ground, &c., by swarming briskly through the sea perform essential services by dispersing the species over wider spaces. In other cases a metamorphosis is rendered indispensable by the circumstance that a division of labor has been set up between the various periods of life; for example, that the larvae have exclusively taken upon themselves the business of nourishment. A further circumstance to be taken into consideration is the size of the eggs,—a simpler structure may be produced with less material than a more compound one,—the more imperfect the larva, the smaller the egg may be, and the larger is the number of these that the mother can furnish with the same expenditure of material. As a rule, I believe indeed, this advantage of a more numerous brood will not by any means outweigh that of a more perfect brood, but it will do so in those cases in which the chief difficulty of the young animals consists in finding a suitable place for their development, and in which, therefore, it is of importance to disperse the greatest possible number of germs, as in many parasites.

As the conversion of the original development with metamorphosis into direct development is here under discussion, this may be the proper place to say a word as to the already indicated absence of metamorphosis in fresh-water and terrestrial animals the marine allies of which still undergo a transformation. This circumstance seems to be explicable in two ways. Either species without a metamorphosis migrated especially into fresh waters, or the metamorphosis was more rapidly got rid of in the emigrants than in their fellows remaining in the sea.

Animals without a metamorphosis would naturally transfer themselves more easily to a new residence, as they had only themselves and not at the same time multifarious young forms to adapt to the new conditions. But in the case of animals with a metamorphosis, the mortality among the larvae, always considerable, must have become still greater under new than under accustomed conditions, every step toward the simplification of the process of development must therefore have given them a still greater preponderance over their fellows, and the effacing of the metamorphosis must have gone on more rapidly. What has taken place in each individual case, whether the species has immigrated after it lost the metamorphosis, or lost the metamorphosis after its immigration, will not always be easy to decide. When there are marine allies without, or with only a slight, metamorphosis, like the Lobster as the cousin of the Cray-fish, we may take up the former supposition; when allies with a metamorphosis still live upon the land or in fresh water, as in the case of *Gecarcinus*, we may adopt the latter.

That besides this gradual extinction of the primitive history, a *falsification* of the record preserved in the developmental history takes place by means of the struggle for existence which the free-living young states have to undergo, requires no further exposition. For it is perfectly evident that the struggle for existence and natural selection combined with this, must act in the same way, in change and development, upon larvae which have to provide for themselves, as upon adult animals. The changes of the larvae, independent of the progress of the adult animal, will become the more considerable, the longer the duration of the life of the larva in comparison to that of the adult animal, the greater the difference in their mode of life, and the more sharply marked the division of labour between the different stages of development. These processes have to a certain extent an action opposed to the gradual extinction of the primitive history; they increase the differences between the individual stages of development, and it will be easily seen how even a straightforward course of development may be again converted by them into a development with metamorphosis. By this means many, and it seems to me valid reasons may be brought up in favour of the opinion that the most ancient Insects approached more nearly to the existing Orthoptera, and perhaps to the wingless Blattidae, than to any other order, and that the "complete metamorphosis" of the Beetles, Lepidoptera, &c., is of later origin. There were, I believe, perfect Insects before larvae and pupae; but, on the contrary, Nauplii and Zoëae far earlier than perfect Prawns. In contradistinction to the *inherited* metamorphosis of the Prawns, we may call that of the Coleoptera, Lepidoptera, &c., an *acquired* metamorphosis.

Which of the different modes of development at present occurring in a class of animals may claim to be that approaching most nearly to the original one, is easy to judge from the above statements.

The primitive history of a species will be preserved in its developmental history the more perfectly, the longer the series of young states through which it passes by uniform steps; and the more truly, the less the mode of life of the young departs from that of the adults, and the less the peculiarities of the individual young states can be conceived as transferred back from later ones in previous periods of life, or as independently acquired.

MORPHOLOGY, EMBRYOLOGY, AND EVOLUTION

HÄCKEL, Ernst Heinrich Philipp August (German zoologist, 1834–1919). From *Natürliche schöpfungsgeschichte. Gemeinverständliche Vorträge über die entwickelungslehre, etc.*, Berlin, 1868; tr. by E. R. Lankester as *The history of creation: or, the development of the earth and its inhabitants by the action of natural causes, etc.*, London, 1876.

Häckel's life work has sometimes been accused of combining a maximum of self-expression with a minimum of self-criticism. His admitted, though limited, success among scientists may be ascribed to the fact that, however faulty his arguments, many of his central ideas (mainly borrowed) were receiving independent corroboration, in a modified form, through the work of more reliable workers. The arguments themselves often gave a superficial appearance of validity arising from their bewildering verbosity. Häckel also had wide popular successes especially among political groups with whom he sympathized and who were unable to judge him on scientific grounds. The *Natürliche Schöpfungsgeschichte* is called by Nordenskjöld "perhaps the chief source of world knowledge concerning Darwinism." An examination of it soon reveals that the thesis it presumes to establish is actually assumed throughout as a premise and that by tricky interchanges of such terms as "paleontologic" and "phylogenetic" a circular argument is produced.

Most persons even now refuse to acknowledge the most important deduction of the Theory of Descent, that is the paleontological development of man from ape-like, and through them from still lower, mammals, and consider such a transformation of organic form as impossible. But, I ask, are the phenomena of the individual development of man, the fundamental features of which I have here given, in any way less wonderful? Is it not in the highest degree remarkable that all vertebrate animals of the most different classes—fishes, amphibious animals, reptiles, birds, and mammals—in the first periods of their embryonic development cannot be distinguished at all, and even much later, at a time when reptiles and birds are already distinctly different from mammals, that the dog and the man are almost identical? Verily, if we compare those two series of development with one another, and ask ourselves which of the two is the more wonderful, it must be confessed that *ontogeny,* or the short and quick history of development of the *individual,* is much more mysterious than *phylogeny,* or the long and slow history of development of the *tribe.* For one and the same grand change of form is accomplished by the latter in the course of many thousands of years, and by the former in the course of a few months.

The two series of organic development, the ontogenesis of the individual and the phylogenesis of the tribe to which it belongs, stand in the closest causal connection with each other. I have endeavoured, in the second volume of the "General Morphology," to establish this theory in detail, as I consider it exceedingly important. As I have there shown, *ontogenesis, or the development of the individual, is a short and quick repetition* (recapitulation) *of phylogenesis, or the development of the tribe to which it belongs, determined by the laws of inheritance and adaptation;* by tribe I mean the ancestors which form the chain of progenitors of the individual concerned.

In this intimate connection of ontogeny and phylogeny, I see one of the most important and irrefutable proofs of the Theory of Descent. No one can explain these phenomena unless he has recourse to the laws of Inheritance and Adaptation; by these alone are they explicable. These laws, which we have previously explained, are *the laws of abbreviated, of homochronic, and of homotopic inheritance,* and here deserve renewed consideration. As so high and complicated an organism as that of man, or the organism of every other mammal, rises upwards from a simple cellular state, and as it progresses in its differentiation and perfection, it passes through the same series of transformations which its animal progenitors have passed through, during immense spaces of time, inconceivable ages ago. I have already pointed out this extremely important parallelism of the development of individuals and tribes. Certain very early and low stages in the development of man, and the other vertebrate animals in general, correspond completely in many points of structure with conditions which last for life in the lower fishes. The next phase which follows upon this presents us with a change of the fish-like being into a kind of amphibious animal. At a later period the mammal, with its special characteristics, develops out of the amphibian, and we can clearly see, in the successive stages of its later development, a series of steps of progressive transformation which evidently correspond with the differences of different mammalian orders and families. Now, it is precisely in the same succession that we also see the ancestors of man, and of the higher mammals, appear one after the other in the earth's history; first fishes, then amphibians, later the lower, and at last the higher mammals. Here, therefore, the embryonic development of the individual is completely parallel to the palaeontological development of the whole tribe to which it belongs, and this exceedingly interesting and important phenomenon can be explained only by the interaction of the laws of Inheritance and Adaptation.

The example last mentioned, of the parallelism of the palaeontological and of the individual developmental series, now directs our attention to a third developmental series, which stands in the closest relations to these two, and which likewise runs, on the whole, parallel to them. I mean that series of development of forms which constitutes the object of investigation in *comparative anatomy,* and which I will briefly call the *systematic developmental series of species.* By this we understand the chain of the different, but related and connected forms, which exist *side by side* at any one period of the earth's history; as for example, at the present moment. While comparative anatomy compares the different forms of fully-developed organisms with one another, it endeavours to discover the common prototypes which underlie, as it were, the manifold forms of kindred genera, classes, etc., and which are more or less concealed by their particular differentiation. It endeavours to make out the series of progressive steps which are indicated in the different degrees of perfection of the divergent branches of the tribe.

The developmental series of mature forms, which comparative anatomy points out in the different diverging and ascending steps of the organic system, and which we call the systematic developmental series, is parallel to the palaeontological developmental series, because it deals with the *result* of palaeontological development, and it is parallel to the individual developmental series, because this is parallel to the palaeontological series. If two parallels are parallel to a third, they must be parallel to one another.

The laws of inheritance and adaptation known to us are completely sufficient to explain this exceedingly important and interesting phenomenon, which may be briefly designated as the *parallelism of individual; of palaeontological, and of systematic development.* No opponent of the Theory of Descent has been able to give an explanation of this extremely wonderful fact, whereas it is perfectly explained, according to the Theory of Descent, by the laws of Inheritance and Adaptation.

If we examine this parallelism of the three organic series of development more accurately, we have to add the following special qualifications. *Ontogeny,* or the history of the individual development of every organism (embryology and metamorphology), presents us with a simple *unbranching* or graduated chain of forms; and so it is with that *portion of phylogeny* which comprises the palaeontological history of development of the *direct ancestors only* of an individual organism. But *the whole of phylogeny*—which meets us in the *natural system* of every organic tribe or phylum, and which is concerned with the investigation of the palaeontological development *of all* the branches of this tribe—forms a *branching* or tree-shaped developmental series, a veritable pedigree. If we examine and compare the branches of this pedigree, and place them together according to the degree of their differentiation and perfection, we obtain the tree-shaped, branching, *systematic developmental series of comparative anatomy.* Strictly speaking, therefore, the latter is parallel to *the whole of phylogeny,* and consequently is only partially parallel to ontogeny; for ontogeny itself is parallel only to a *portion* of phylogeny.

All the phenomena of organic development above discussed, especially the threefold genealogical parallelism, and the laws of differentiation and progress, which are evident in each of these three series of organic development,

and, further, the whole history of rudimentary organs, are exceedingly important proofs of the truth of the Theory of Descent.   For by it alone can they
be explained, whereas its opponents cannot even offer a shadow of an explanation of them.   Without the Doctrine of Filiation, the fact of organic development in general cannot be understood.   We should therefore, for this reason
alone, be forced to accept Lamarck's Theory of Descent, even if we did not
possess Darwin's Theory of Selection.

THEORY OF PANGENES

DARWIN, Charles Robert (English naturalist, 1809–1882).   From *The variations in
animals and plants under domestication,* London, 1868.

Darwin was fully aware of the limitations in Lamarck's belief in the transmission of
acquired characters,* but did not seriously question the fairly general occurrence of this
phenomenon.   Had he done so, the doctrine of pangenes might never have become necessary.   For, although Darwin uses his pangenes to explain reversion and regeneration, it
was chiefly the supposed transmission of acquired characters which made pangenesis seem
indispensable.   The whole difficulty was later cleared up by Weismann (*q.v.*).   For the
scientific origins of pangenesis, see the *Source Book of Greek Science,* p. 417.

    . . . reversion is the most wonderful of all the attributes of Inheritance.   It
proves to us that the transmission of a character and its development, which
ordinarily go together and thus escape discrimination, are distinct powers; and
these powers in some cases are even antagonistic, for each acts alternately in
successive generations.   Reversion is not a rare event, depending on some unusual or favourable combination of circumstances; but occurs so regularly
with crossed animals and plants, and so frequently with uncrossed breeds, that
it is evidently an essential part of the principle of inheritance.   We know that
changed conditions have the power of evoking long-lost characters, as in the
case of some feral animals.   The act of crossing in itself possesses this power in
a high degree.   What can be more wonderful than that characters, which
have disappeared during scores, or hundreds, or even thousands of generations,
should suddenly reappear perfectly developed, as in the case of pigeons and
fowls when purely bred, and especially when crossed; or as with the zebrine
stripes on dun-coloured horses, and other such cases?   Many monstrosities
come under this same head, as when rudimentary organs are redeveloped, or
when an organ which we must believe was possessed by an early progenitor,
but of which not even a rudiment is left, suddenly reappears, as with the fifth
stamen in some Scrophulariaceae.   We have already seen that reversion acts
in bud-reproduction; and we know that it occasionally acts during the growth
of the same individual animal, especially, but not exclusively, when of crossed

---

* See the *Recapitulation* at the end of the *Origin* where Darwin cites parthenogenetic
male insects as having, *de facto,* not inherited their male behavior from male ancestors.

parentage,—as in the rare cases described of individual fowls, pigeons, cattle, and rabbits, which have reverted as they advanced in years to the colours of one of their parents or ancestors.

We are led to believe, as formerly explained, that every character which occasionally reappears is present in a latent form in each generation, in nearly the same manner as in male and female animals secondary characters of the opposite sex lie latent, ready to be evolved when the reproductive organs are injured. This comparison of the secondary sexual characters which are latent in both sexes, with other latent characters, is the more appropriate from the case recorded of the Hen, which assumed some of the masculine characters, not of her own race, but of an early progenitor; she thus exhibited at the same time the redevelopment of latent characters of both kinds and connected both classes. In every living creature we may feel assured that a host of lost characters lie ready to be evolved under proper conditions. How can we make intelligible, and connect with other facts, this wonderful and common capacity of reversion,—this power of calling back to life long-lost characters? . . .

It is almost universally admitted that cells, or the units of the body, propagate themselves by self-division or proliferation, retaining the same nature, and ultimately becoming converted into the various tissues and substances of the body. But besides this means of increase I assume that cells, before their conversion into completely passive or "form-material," throw off minute granules or atoms, which circulate freely throughout the system, and when supplied with proper nutriment multiply by self-division, subsequently becoming developed into cells like those from which they were derived. These granules for the sake of distinctness may be called cell-gemmules, or, as the cellular theory is not fully established, simply gemmules. They are supposed to be transmitted from the parents to the offspring, and are generally developed in the generation which immediately succeeds, but are often transmitted in a dormant state during many generations and are then developed. Their development is supposed to depend on their union with other partially developed cells or gemmules which precede them in the regular course of growth. Why I use the term union, will be seen when we discuss the direct action of pollen on the tissues of the mother-plant. Gemmules are supposed to be thrown off by every cell or unit, not only during the adult state, but during all the stages of development. Lastly, I assume that the gemmules in their dormant state have a mutual affinity for each other, leading to their aggregation either into buds or into the sexual elements. Hence, speaking strictly, it is not the reproductive elements, nor the buds, which generate new organisms, but the cells themselves throughout the body. These assumptions constitute the provisional hypothesis which I have called Pangenesis. Nearly similar views have been propounded, as I find, by other authors, more especially by Mr. Herbert Spencer; but they are here modified and amplified.

It may be useful to give an illustration of the hypothesis. If one of the simplest Protozoa be formed, as appears under the microscope, of a small mass of homogeneous gelatinous matter, a minute atom thrown off from any part and nourished under favourable circumstances would naturally reproduce the whole; but if the upper and lower surfaces were to differ in texture from the central portion, then all three parts would have to throw off atoms or gemmules, which when aggregated by mutual affinity would form either buds or the sexual elements. Precisely the same view may be extended to one of the higher animals; although in this case many thousand gemmules must be thrown off from the various parts of the body. Now, when the leg, for instance, of a salamander is cut off, a slight crust forms over the wound, and beneath this crust the uninjured cells or units of bone, muscle, nerves, &c., are supposed to unite with the diffused gemmules of those cells which in the perfect leg come next in order; and these as they become slightly developed unite with others, and so on until a papilla of soft cellular tissue, the "budding leg," is formed, and in time a perfect leg. Thus, that portion of the leg which had been cut off, neither more nor less, would be reproduced. If the tail or leg of a young animal had been cut off, a young tail or leg would have been reproduced, as actually occurs with the amputated tail of the tadpole; for gemmules of all the units which compose the tail are diffused throughout the body at all ages. But during the adult stage the gemmules of the larval tail would remain dormant, for they would not meet with pre-existing cells in a proper state of development with which to unite. If from changed conditions or any other cause any part of the body should become permanently modified, the gemmules, which are merely minute portions of the contents of the cells forming the part, would naturally reproduce the same modification. But gemmules previously derived from the same part before it had undergone any change, would still be diffused throughout the organisation, and would be transmitted from generation to generation, so that under favourable circumstances they might be redeveloped, and then the new modification would be for a time or for ever lost. The aggregation of gemmules derived from every part of the body, through their mutual affinity, would form buds, and their aggregation in some special manner, apparently in small quantity, together probably with the presence of gemmules of certain primordial cells, would constitute the sexual elements. By means of these illustrations the hypothesis of pangenesis has, I hope, been rendered intelligible. . . .

The hypothesis of Pangenesis, as applied to the several great classes of facts just discussed, no doubt is extremely complex; but so assuredly are the facts. The assumptions, however, on which the hypothesis rests cannot be considered as complex in any extreme degree—namely, that all organic units, besides having the power, as is generally admitted, of growing by self-division, throw off free and minute atoms of their contents, that is gemmules.[1] These

---

[1] Note the resemblance to Kircher's "*effluvia*," p. 473.

multiply and aggregate themselves into buds and the sexual elements; their development depends on their union with other nascent cells or units; and they are capable of transmission in a dormant state to successive generations.

In a highly organised and complex animal, the gemmules thrown off from each different cell or unit throughout the body must be inconceivably numerous and minute. Each unit of each part, as it changes during development, and we know that some insects undergo at least twenty metamorphoses, must throw off its gemmules. All organic beings, moreover, include many dormant gemmules derived from their grandparents and more remote progenitors, but not from all their progenitors. These almost infinitely numerous and minute gemmules must be included in each bud, ovule, spermatozoon, and pollen-grain. Such an admission will be declared impossible; but, as previously remarked, number and size are only relative difficulties, and the eggs or seeds produced by certain animals or plants are so numerous that they cannot be grasped by the intellect.

The organic particles with which the wind is tainted over miles of space by certain offensive animals must be infinitely minute and numerous; yet they strongly affect the olfactory nerves. An analogy more appropriate is afforded by the contagious particles of certain diseases, which are so minute that they float in the atmosphere and adhere to smooth paper; yet we know how largely they increase within the human body, and how powerfully they act. Independent organisms exist which are barely visible under the highest powers of our recently-improved microscopes, and which probably are fully as large as the cells or units in one of the higher animals; yet these organisms no doubt reproduce themselves by germs of extreme minuteness, relatively to their own minute size. Hence the difficulty, which at first appears insurmountable, of believing in the existence of gemmules so numerous and so small as they must be according to our hypothesis, has really little weight.

The cells or units of the body are generally admitted by physiologists to be autonomous, like the buds on a tree, but in a less degree. I go one step further and assume that they throw off reproductive gemmules. Thus an animal does not, as a whole, generate its kind through the sole agency of the reproductive system, but each separate cell generates its kind. It has often been said by naturalists that each cell of a plant has the actual or potential capacity of reproducing the whole plant; but it has this power only in virtue of containing gemmules derived from every part. If our hypothesis be provisionally accepted, we must look at all the forms of asexual reproduction, whether occurring at maturity or as in the case of alternate generation during youth, as fundamentally the same, and dependent on the mutual aggregation and multiplication of the gemmules. The regrowth of an amputated limb or the healing of a wound is the same process partially carried out. Sexual generation differs in some important respects, chiefly, as it would appear, in an insufficient number of gemmules being aggregated within the separate sexual elements, and probably in the presence of certain primordial cells. The development of each being, including all the forms of metamorphosis and metagenesis, as well as the so-called growth of the higher animals, in which structure changes though not in a striking manner, depends on the presence of gemmules thrown off at each period of life, and on their development, at a corresponding period, in union with preceding cells. Such cells may be said to be fertilised by the gemmules which come next in the order of development. Thus the ordinary act of impregnation and the development of each being are closely analogous processes. The child, strictly speaking, does not grow into the man, but includes germs which slowly and successively become developed and form the man. In the child, as well as in the adult, each part generates the same part for the next generation. Inheritance must be looked at as merely a form of growth, like the self-division of a lowly-organised unicellular plant. Reversion depends on the transmission from the forefather to his descendants of dormant gemmules, which occasionally become developed under certain known or unknown conditions. Each animal and plant may be compared to a bed of mould full of seeds, most of which soon germinate, some lie for a period dormant, whilst others perish. When we hear it said that a man carries in his constitution the seeds of an inherited disease, there is much literal truth in the expression. Finally, the power of propagation possessed by each separate cell, using the term in its largest sense, determines the reproduction, the variability, the development and renovation of each living organism. No other attempt, as far as I am aware, has been made, imperfect as this confessedly is, to connect under one point of view these several grand classes of facts. We cannot fathom the marvellous complexity of an organic being; but on the hypothesis here advanced this complexity is much increased. Each living creature must be looked at as a microcosm—a little universe, formed of a host of self-propagating organisms, inconceivably minute and as numerous as the stars in heaven.

OCEANOGRAPHIC EVIDENCE OF GEOLOGICAL HISTORY OF LAND MASSES

AGASSIZ, Alexander (American naturalist, 1835–1910). From *Preliminary report on the Echini and starfishes dredged in deep water between Cuba and the Florida Reef, by L. F. de Pourtales,* in *Bulletin of the Museum of Comparative Zoology at Harvard College in Cambridge,* vol. 1, Nos. 1–13, p. 253, 1869; by permission of the Harvard University Press.

Earth history is one of the most ancient objects of scientific speculation. For the Greek scholars, especially Xenophanes, and again in the seventeenth century (see Steno), rock formations and fossils were the chief evidence. The doctrine of evolution made available a new form of evidence: living organisms with special reference to the relations of their geographical distribution to their comparative structure. The following is a classic application of this approach using marine animals.

The specific representation on both sides of the Isthmus of Panama is becoming every day, as far as Echinoderms are concerned, more strikingly identical. Since the list given by Mr. Verrill, several species have come to light, and the following comparative list of species on both sides of the Isthmus, extending from Peru to the Gulf of California on the Pacific, and including on the Eastern side the Gulf of Mexico, Florida, the northern coast of South America, the West Indies and Bahamas, may not be out of place. (I have examined all the species here named.) This list would undoubtedly be greatly increased by additional dredging.

EASTERN FAUNA.	WESTERN FAUNA.
(*Caribbean.*)	(*Panamic.*)
Cidaris annulata GRAY	Cidaris Thouarsii VAL.
Dorocidaris abyssicola A. AG.	
Salenocidaris varispina A. AG.	
Diadema antillarum PHIL.	Diadema mexicanum A. AG.
	Astropyga venusta VER.
Caenopedina cubensis A. AG.	
Echinocidaris punctulata DESML.	Echinocidaris stellata AG.
Podocidaris sculpta A. AG.	
Echinometra Michelini DES.	Echinometra Van Brunti A. AG.
" viridis A. AG.	" rupicola A. AG.
Echinus gracilis A. AG.	
" Flemingii BALL.	
Genocidaris maculata A. AG.	
Trigonocidaris albida A. AG.	
	Toxocidaris mexicana A. AG.
Lytechinus variegatus A. AG.	Lytechinus semituberculatus A. AG.
	Psammechinus pictus VER. is the young.
	Boletia rosea A. AG.
Tripneustes ventricosus AG.	Tripneustes depressus A. AG.
Clypeaster rosaceus LAM.	
Stolonoclypus prostratus AG.	
" Ravenellii A. AG.	Stolonoclypus rotundus A. AG.
Mellita testudinata KL.	Mellita longifissa MICH.
" hexapora AG.	" pacifica VER.
Encope Michelini AG.	Encope grandis AG.
" emarginata AG.	" micropora AG.

Eastern Fauna. (*Caribbean.*)	Western Fauna. (*Panamic.*)
	Echinoglycus Stokesi GRAY.
Echinoneus semilunaris LAM.	
Echinolampas caratomoides A. AG.	
Rhyncholampas caribbaearum A. AG.	Rhyncholampas pacificus A. AG.
Neolampas rostellatus A. AG.	
Pourtalesia miranda A. AG.	
Lissonotus fragilis A. AG.	
	Lovenia sp.
Brissus columbaris AG.	Brissus obesus VER.
Meoma ventricosa LÜTK.	Meoma grandis GRAY.
Plagionotus pectoralis AG.	Plagionotus nobilis A. AG.
Agassizia excentrica A. AG.	Agassizia scrobiculata VAL.
Brissopsis lyrifera AG.	
Echinocardium ovatum GRAY.	
" laevigaster A. AG.	
" Kurtzii GIR.	
Schizaster cubensis D'ORB.	
Moera atropos MICH.	Moera clotho MICH.

With the exception of three Panama species, all the West Coast species have representatives on the Eastern Coast. The Eastern species which have not as yet been found represented on the West Coast are the deep-water species of Mr. Pourtales's collection, and, what is very peculiar, a few species, like Clypeaster rosaceus, Echinoneus semilunaris, Echinocardium Kurtzii, and Echinolampas, belonging to genera which have a most extensive range,—in fact, an almost cosmopolitan one,—are found everywhere in the great Indo-Pacific belt, and its continuation on the West Coast of Africa, extending also to the temperate zones, on both sides of this equatorial belt.

The relation of the Caribbean Fauna with the existing geographical distribution of Echini is shown by the accompanying faunal table, including only strictly representative species.

We have in Genocidaris maculata and Trigonocidaris albida representatives of the Temnopleuridae, thus far limited almost entirely to the Indian and China seas. The littoral species having the most limited bathymetrical range are those which have the widest geographical distribution. They are Tripneustes ventricosus, Diadema antillarum, Cidaris annulata, Echinometra Michelini, Lytechinus variegatus, Mellita testudinata, Encope emarginata. Some of these species extend from the southern part of Brazil to the Bermudas. They all belong to genera having representatives in the great tropical belt sur-

Caribbean.	Panamic.	Europ. Boreal.	Mediterranean.	Senegal.	Indo-Pacific.	Chinese.	Japanese.	Patagonian.
Cidaris annulata Gray	*			i	*			
Dorocidaris abyssicola A. Ag.		*	*		*			*
Salenocidaris varispina A. Ag.								
Diadema antillarum Phil.	*		*	*	*	*	*	
Caenopedina cubensis A. Ag.								
Echinocidaris punctulata Desml.	*		*	*				
Podocidaris sculpta A. Ag.								
Echinometra Michelini Des.	*			*	*		*	
" viridis A. Ag.	*				*			
Echinus gracilis A. Ag.		*	*					
" Flemingii Ball		i	i					
Genocidaris maculata A. Ag.							*	
Trigonocidaris albida A. Ag.						*		
Lytechinus variegatus A. Ag.	*				*			
Tripneustes ventricosus Ag.	*				*			
Clypeaster rosaceus Lam.				*				
Stolonoclypus prostratus Ag.		*	*		*		*	
" Ravenellii A. Ag.	*							
Mellita testudinata Kl.	*							
" hexapora Ag.	*							
Encope Michelini Ag.	*							
" emarginata Ag.	*							
Echinoneus semilunaris Lam.					*		*	
Echinolampas caratomoides A. Ag.					*i	*		
Rhyncholampas caribbaearum A. Ag.	*							
Neolampas rostellatus A. Ag.								
Pourtalesia miranda A. Ag.								
Lissonotus fragilis A. Ag.								
Brissus columbaris Ag.	*		*		*			
Meoma ventricosa Lütk.	*							
Plagionotus pectoralis Ag.	*							
Brissopsis lyrifera Ag.		i	*		*			*
Agassizia excentrica A. Ag.	*							
Echinocardium ovatum Gray		i	i					
" laevigaster A. Ag.			*					
" Kurtzii Gir.		*	*				*	
Schizaster cubensis D'Orb.		*	*		*			
Moera atropos Mich.	*							

NOTE. — *i* denotes identity of species; * denotes representative species.

rounding the globe, formed by the Indo-Pacific, Mediterranean, Senegalian, West Indian, Panamic, and Polynesian faunae,—such as Cidaris, Diadema, Echinometra, Tripneustes, Clypeaster, Stolonoclypus, Echinolampas, Echinoneus, Brissus, the species of which have a great geographical range, and are represented by the following species:—

Cidaris metularia, Tripneustes sardicus, Echinometra lucunter, Diadema Savignyi, Clypeaster Rangianus, Stolonoclypus placunarius, Echinolampas oviformis, Echinoneus cyclostomus, Brissus carinatus, all of which have an immense geographical distribution.

The effect which currents play in shaping the geographical distribution of marine animals is very great; we have an example in the Gulf Stream and the northern branch of the Amazonian current flowing into the Gulf of Mexico, which account fully for the great range of the more common littoral species. The Japanese current makes itself felt as far as San Diego, two species of Echini extending in the Northern Pacific from the northern part of Japan along Kamtchatka, the Aleutian Islands, Sitka, Vancouver's Island, the one as far as Cape Mendocino (T. drobachiensis), the other (Dendraster excentricus) to San Diego. The Indo-Pacific equatorial current has undoubtedly been the main agent of the extensive geographical range of such species as Cidaris metularia, Echinoneus cyclostomus, Heterocentrotus mammillatus, Diadema Savignyi, Tripneustes sardicus, Echinolampas oviformis, Brissus carinatus, Stolonoclypus placunarius.

The effect of currents in thus extending the distribution of marine animals would act very differently upon the several classes of the animal kingdom, and its efficiency depends to a great extent upon the nature of their earlier stages, and upon their habits during that period. The time during which the Pluteus of Echini remains helpless at the mercy of the currents is considerable: from early spring till late in the summer is the usual time required for the full growth of the Pluteus in many species of Sea-urchins, and the distance which the young could thus be transported, even by a sluggish current, during a single season, must be considerable, even under the most unfavorable circumstances.

Various writers have attempted to retrace, in former geological periods, the probable course of the currents and their effect upon the geographical distribution of marine animals; they all agree in representing up to the cretaceous period an unbroken equatorial current, passing through Central Asia, Arabia, the northern part of Africa, and connecting with the Pacific by a narrow strait through the Isthmus of Panama. The existence of this connection in the cretaceous period is placed beyond doubt by the presence of an Ananchytes, which I am unable to distinguish from Ananchytes radiata, collected on the Isthmus of Panama, and now in the Museum of Yale College, kindly loaned me for examination by Professor Verrill. From the small number of identical species, either of Mollusca, Crustacea, or Fishes, recorded on both sides of the Isthmus, this connection must have been very imperfect at a comparatively recent geological period,—since the existence of the present Faunae.

The question naturally arises, Have we not in the different Faunae of both sides of the Isthmus a standard by which to measure the changes which these species have undergone since the raising of the Isthmus of Panama and the isolation of the two Faunae? If the upheaval of the isthmus has been gradual, it must, of course, have cut off the deep-water species on both sides of the isthmus, and gradually have isolated the more shallow, till the littoral species also became separated. As a natural consequence, the deeper we go, the farther back in time we must expect to find the representation,—a result which is strikingly confirmed by the nature of the deep-water Fauna of the West Indies. Unfortunately we have not, as in the case of the littoral Faunae, a standard of comparison. At the same time, with the gradual closing of the Isthmus of Panama, the greater part of Central Asia, of the Arabian Peninsula, and of Northern Africa was emerging from the sea, reducing the range of the equatorial current, and thus confining the course of the currents much as they are at the present time. This would thus cause a limitation in the range of the species formerly having the greatest distribution, and extend that of those which were more local.

If migration on land when continents were joined together, and subsequent variations after their isolation through submergence, has been the main agent in the distribution of the existing terrestrial Faunae, we must acknowledge a similar agency to currents in the distribution of marine Faunae; and by the submergence or rise of various portions of the continents, we shall be able, if we can trace these changes, to reconstruct within certain limits the altered courses of the main oceanic currents, and get some idea of the probable geographical distribution at different geological epochs. The greater the bathymetrical range of littoral species, the longer will such species remain unaffected, while deep-sea species may early become isolated and remain as outliers as it were,—mementos of a former condition of currents, or even of a previous geological period. The careful analysis of the Fauna of a given point, its comparison with other Faunae, and accurate bathymetrical data, would go far towards reconstructing the Natural History of the sea in former ages, and showing its relation to the present and past times.

The representative species of Echini, Echinocardium, Psammechinus, Schizaster, in the Arctic and Antarctic boreal zones would be considered as the living representatives of a cosmopolitan Fauna existing at the time when the great equatorial current flowed unbroken round the globe, sending branches north and south along Eastern North and South America, along Eastern Japan and Australia, and the eastern coast of Africa; while the tropical species of the genera Diadema, Clypeaster, Echinoneus, Echinolampas, &c., existing at that time, had a more limited equatorial geographical distribution. The subsequent period of isolation of Atlantic and Pacific currents is shown by the existence of truly Atlantic and Pacific species; while as we go down in depth we go back also in time, and find at first representatives of the genera found in our Tertiaries, while at greater depth the species are representatives of genera found in the Cretaceous. A more detailed comparison than can be given here of the Caribbean Fauna, with the fossils of the tertiary and cretaceous deposits of our coasts, would be most interesting; but unfortunately the materials thus far collected are too fragmentary, and we must await a careful geological survey, accompanied by deep dredgings of a considerable extent of coast, before we shall have the data needed to follow up the important results to be gained in this way for palaeontology and geography, of which our present incomplete materials give us such an interesting glimpse.

THAT TRANSMISSIBLE VARIATIONS ARISE FROM MODIFICATIONS OF A HEREDITARY SUBSTANCE

WEISMANN, August (German naturalist, 1834–1914). From *Ueber die vererbung*, his inaugural address as Prorector of Freiburg, June 21, 1883, published Jena, 1883; tr. by A. E. Shipley as *On heredity*, Oxford, 1891; by arrangement with the Clarendon Press.

The following paper, Weismann's first and most famous on heredity, although less massive in illustrative detail than the writings of Darwin, yet shows the same tendency to sweep a multiplicity of observations into the embrace of a single, pervading point of view. The last quarter of the nineteenth century, characterized by Whitehead as "one of the dullest stages of thought since the time of the first crusade" was at least productive in the field of heredity: (Roux's allocation of hereditary functions to chromosomes, 1883; Galton's "*Inquiries*," 1883; discovery of mutations, 1895; rebirth of Mendelism, 1900.) The excerpts given below present an outline of the main structure of Weismann's argument and foreshadow much of what he was to spend the next thirty years defending.

With your permission I wish to bring before you to-day my views on a problem of general biological interest—the problem of heredity. . . .

The word heredity in its common acceptation, means that property of an organism by which its peculiar nature is transmitted to its descendants. From an eagle's egg an eagle of the same species develops; and not only are the characteristics of the species transmitted to the following generation, but even the individual peculiarities. The offspring resemble their parents among animals as well as among men.

On what does this common property of all organisms depend?

Häckel was probably the first to describe reproduction as 'an overgrowth of the individual,' and he attempted to explain heredity as a simple continuity of growth. This definition might be considered as a play upon words, but it is more than this; and such an interpretation rightly applied, points to the only path which, in my opinion, can lead to the comprehension of heredity.

Unicellular organisms, such as Rhizopoda and Infusoria, increase by means of fission. Each individual grows to a certain size, and then divides into two parts, which are exactly alike in size and structure, so that it is impossible to decide whether one of them is younger or older than the other. Hence in a certain sense these organisms possess immortality: they can, it is true, be destroyed, but, if protected from a violent death, they would live on indefinitely, and would only from time to time reduce the size of their over-grown bodies by division. Each individual of any such unicellular species living on the earth to-day is far older than mankind, and is almost as old as life itself.

From these unicellular organisms we can to a certain extent understand why the offspring, being in fact a part of its parents, must therefore resemble the latter. The question as to why the part should resemble the whole leads us to a new problem, that of assimilation, which also awaits solution. It is, at any rate, an undoubted fact that the organism possesses the power of taking up certain foreign substances, viz. food, and of converting them into the substance of its own body.

Among these unicellular organisms, heredity depends upon the continuity of the individual during the continual increase of its body by means of assimilation.

But how is it with the multicellular organisms which do not reproduce by means of simple division, and in which the whole body of the parent does not pass over into the offspring?

In such animals sexual reproduction is the chief means of multiplication. In no case has it always been completely wanting, and in the majority of cases it is the only kind of reproduction.

In these animals the power of reproduction is connected with certain cells which, as germ-cells, may be contrasted with those which form the rest of the body; for the former have a totally different rôle to play; they are without significance for the life of the individual, and yet they alone possess the power of preserving the species. Each of them can, under certain conditions, develop into a complete organism of the same species as the parent, with every individual peculiarity of the latter reproduced more or less completely. How can such hereditary transmission of the characters of the parent take place? how can a single reproductive cell reproduce the whole body in all its details?   . . .

We thus find that the reproduction of multicellular organisms is essentially similar to the corresponding process in unicellular forms; for it consists in the continual division of the reproductive cell; the only difference being that in the former case the reproductive cell does not form the whole individual, for the latter is composed of the millions of somatic cells by which the reproductive cell is surrounded. The question, 'How can a single reproductive cell contain the germ of a complete and highly complex individual?' must therefore be re-stated more precisely in the following form, 'How can the substance of the reproductive cells potentially contain the somatic substance with all its characteristic properties?'

The problem which this question suggests, becomes clearer when we employ it for the explanation of a definite instance, such as the origin of multicel-

lular from unicellular animals.   There can be no doubt that the former have
originated from the latter, and that the physiological principle upon which such
an origin depended, is the principle of division of labour.   In the course of the
phyletic development of the organized world, it must have happened that certain unicellular individuals did not separate from one another immediately
after division, but lived together, at first as equivalent elements, each of which
retained all the animal functions, including that of reproduction.   The *Ma-
gosphaera plannula* of Häckel proves that such perfectly homogeneous cell-
colonies exist, even at the present day.   Division of labour would produce a
differentiation of the single cells in such a colony: thus certain cells would be
set apart for obtaining food and for locomotion, while certain other cells
would be exclusively reproductive.   In this way colonies consisting of somatic
and of reproductive cells must have arisen, and among these for the first time
death appeared.   For in each case the somatic cells must have perished after a
certain time, while the reproductive cells alone retained the immortality inherited from the Protozoa.   We must now ask how it becomes possible that
one kind of cell in such a colony can produce the other kind by division?   Before the differentiation of the colony each cell always produced others similar
to itself.   How can the cells, after the nature of one part of the colony is
changed, have undergone such changes in *their* nature that they can now produce more than one kind of cell?

Two theories can be brought forward to solve this problem.   We may turn
to the old and long since abandoned *nisus formativus,* or adapting the name to
modern times, to a phyletic force of development which causes the organism
to change from time to time.   This *vis a tergo* or teleological force compels
the organism to undergo new transformations without any reference to the
external conditions of life.   This theory throws no light upon the numerous
adaptations which are met with in every organism; and it possesses no value
as a scientific explanation.

Another supposition is that the primary reproductive cells are influenced by
the secondary cells of the colony, which, by their adaptability to the external
conditions of life, have become somatic cells: that the latter give off minute
particles which, entering into the former, cause such changes in their nature
that at the next succeeding cell-division they are compelled to break up into
dissimilar parts.   . . .

It is well known that Darwin has attempted to explain the phenomena of
heredity by means of a hypothesis which corresponds to a considerable extent
with that just described.   If we substitute gemmules for molecules we have
the fundamental idea of Darwin's provisional hypothesis of pangenesis.   Particles of an excessively minute size are continually given off from all the cells
of the body; these particles collect in the reproductive cells, and hence any
change arising in the organism, at any time during its life, is represented in the

reproductive cell.   Darwin believed that he had by this means rendered the transmission of acquired characters intelligible, a conception which he held to be necessary in order to explain the development of species.   He himself pointed out that the hypothesis was merely provisional, and that it was only an expression of immediate, and by no means satisfactory knowledge of these phenomena.

It is always dangerous to invoke some entirely new force in order to understand phenomena which cannot be readily explained by the forces which are already known.

I believe that an explanation can in this case be reached by an appeal to known forces, if we suppose that characters acquired (in the true sense of the term) by the parent cannot appear in the course of the development of the offspring, but that all the characters exhibited by the latter are due to primary changes in the germ.

This supposition can obviously be made with regard to the above-mentioned colony with its constituent elements differentiated into somatic and reproductive cells.   It is conceivable that the differentiation of the somatic cells was not primarily caused by a change in their own structure, but that it was prepared for by changes in the molecular structure of the reproductive cell from which the colony arose.   . . .

If then the reproductive cells have undergone such changes that they can produce a heterogeneous colony as the result of continual division, it follows that succeeding generations must behave in exactly the same manner, for each of them is developed from a portion of the reproductive cell from which the previous generation arose, and consists of the same reproductive substance as the latter.   . . .

The difficulty or the impossibility of rendering the transmission of acquired characters intelligible by an appeal to any known force has been often felt, but no one has hitherto attempted to cast doubts upon the very existence of such a form of heredity.

There are two reasons for this: first, observations have been recorded which appear to prove the existence of such transmission; and secondly, it has seemed impossible to do without the supposition of the transmission of acquired characters, because it has always played such an important part in the explanation of the transformation of species.

It is perfectly right to defer an explanation, and to hesitate before we declare a supposed phenomenon to be impossible, because we are unable to refer it to any of the known forces.   No one can believe that we are acquainted with all the forces of nature.   But, on the other hand, we must use the greatest caution in dealing with unknown forces; and clear and indubitable facts must be brought forward to prove that the supposed phenomena have a real existence, and that their acceptance is unavoidable.

It has never been proved that acquired characters are transmitted, and it has never been demonstrated that, without the aid of such transmission, the evolution of the organic world becomes unintelligible.

The inheritance of acquired characters has never been proved, either by means of direct observation or by experiment. . . .

The caverns in Carniola and Carinthia, in which the blind *Proteus* and so many other blind animals live, belong geologically to the Jurassic formation; and although we do not exactly know when for example the *Proteus* first entered them, the low organization of this amphibian certainly indicates that it has been sheltered there for a very long period of time, and that thousands of generations of this species have succeeded one another in the caves.

Hence there is no reason to wonder at the extent to which the degeneration of the eye has been already carried in the *Proteus;* even if we assume that it is merely due to the cessation of the conserving influence of natural selection.

But it is unnecessary to depend upon this assumption alone, for when a useless organ degenerates, there are also other factors which demand consideration, namely, the higher development of other organs which compensate for the loss of the degenerating structure, or the increase in size of adjacent parts. If these newer developments are of advantage to the species, they finally come to take the place of the organ which natural selection has failed to preserve at its point of highest perfection.

In the first place, a certain form of correlation, which Roux calls 'the struggle of the parts in the organism,' plays a most important part. Cases of atrophy, following disuse, appear to be always attended by a corresponding increase of other organs: blind animals always possess very strongly-developed organs of touch, hearing, and smell, and the degeneration of the wing-muscles of the ostrich is accompanied by a great increase in the strength of the muscles of the leg. If the average amount of food which an animal can assimilate every day remains constant for a considerable time, it follows that a strong influx towards one organ must be accompanied by a drain upon others, and this tendency will increase, from generation to generation, in proportion to the development of the growing organ, which is favoured by natural selection in its increased bloody-supply, etc.; while the operation of natural selection has also determined the organ which can bear a corresponding loss without detriment to the organism as a whole. . . .

Now it cannot be denied that all predispositions may be improved by practice during the course of a life-time,—and, in truth, very remarkably improved. If we could explain the existence of great talent, such as, for example, a gift for music, painting, sculpture, or mathematics, as due to the presence or absence of a special organ in the brain, it follows that we could only understand its origin and increase (natural selection being excluded) by accumulation, due to the transmission of the results of practice through a series of gen-

erations. But talents are not dependent upon the possession of special organs
in the brain. They are not simple mental dispositions, but combinations of
many dispositions, and often of a most complex nature: they depend upon a
certain degree of irritability, and a power of readily transmitting impulses
along the nerve-tracts of the brain, as well as upon the especial development
of single parts of the brain. In my opinion, there is absolutely no trustworthy
proof that talents have been improved by their exercise through the course of
a long series of generations. The Bach family shows that musical talent, and
the Bernoulli family that mathematical power, can be transmitted from generation to generation, but this teaches us nothing as to the origin of such talents. In both families the high-water mark of talent lies, not at the end of the
series of generations, as it should do if the results of practice are transmitted,
but in the middle. Again, talents frequently appear in some single member of
a family which has not been previously distinguished.

Gauss was not the son of a mathematician; Handel's father was a surgeon,
of whose musical powers nothing is known; Titian was the son and also the
nephew of a lawyer, while he and his brother, Francesco Vecellio, were the
first painters in a family which produced a succession of seven other artists
with diminishing talents. These facts do not, however, prove that the condition of the nerve-tracts and centres of the brain, which determine the specific
talent, appeared for the first time in these men: the appropriate condition surely
existed previously in their parents, although it did not achieve expression.
They prove, as it seems to me, that a high degree of endowment in a special
direction, which we call talent, cannot have arisen from the experience of
previous generations, that is, by the exercise of the brain in the same specific
direction. . . .

Lessing has asked whether Raphael would have been a less distinguished
artist had he been born without hands: we might also enquire whether he
might not have been as great a musician as he was painter if, instead of living
during the historical high-water mark of painting, he had lived, under favourable personal influences, at the time of highly-developed and wide-spread
musical genius. A great artist is always a great man, and if he finds the
outlet for his talent closed on one side, he forces his way through on the
other. . . .

We have an obvious means by which the inheritance of all transmitted peculiarities takes place, in *the continuity of the substance of the germ-cells, or
germ-plasm.* If, as I believe, the substance of the germ-cells, the germ-plasm,
has remained in perpetual continuity from the first origin of life, and if the
germ-plasm and the substance of the body, the somatoplasm, have always
occupied different spheres, and if changes in the latter only arise when they
have been preceded by corresponding changes in the former, then we can,
up to a certain point, understand the principle of heredity; or, at any rate, we

can conceive that the human mind may at some time be capable of understanding it. We may at least maintain that it has been rendered intelligible, for we can thus trace heredity back to growth; we can thus look upon reproduction as an overgrowth of the individual, and can thus distinguish between a succession of species and a succession of individuals, because in the latter succession the germ-plasm remains similar, while in the succession of the former it becomes different. Thus individuals, as they arise, are always assuming new and more complex forms, until the interval between the simple unicellular protozoon and the most complex of all organisms—man himself —is bridged over.

I have not been able to throw light upon all sides of the question which we are here discussing. There are still some essential points which I must leave for the present; and, furthermore, I am not yet in a position to explain satisfactorily all the details which arise at every step of the argument. But it appeared to me to be necessary to state this weighty and fundamental question, and to formulate it concisely and definitely; for only in this way will it be possible to arrive at a true and lasting solution of the problem. We must however be clear on this point—that the understanding of the phenomena of heredity is only possible on the fundamental supposition of the continuity of the germ-plasm. The value of experiment in relation to this question is somewhat doubtful. A careful collection and arrangement of facts is far more likely to decide whether, and to what extent, the continuity of germ-plasm is reconcilable with the assumption of the transmission of acquired characters from the parent body to the germ, and from the germ to the body of the offspring. At present such transmission is neither proved as a fact, nor has its assumption been shown to be unquestionably necessary.

HUMAN INHERITANCE

GALTON, Francis (English anthropometrist, 1822–1911). From *Natural inheritance*, London, 1889.

Galton's approach is essentially a combination of painstaking anthropometry with the method of population analysis. The results of such statistical mass calculation were temporarily well received. Unfortunately, they extended themselves to traits, *e.g.*, ill temper and genius, whose significance as entities and whose susceptibility to measurement are questionable. This, along with other defects, detracted from the effectiveness of his "laws" when they came to be used for purposes of prediction. His early study of hereditary genius can only be considered an egregious miscarriage of the whole method of the sciences. First cousin of Charles Darwin, Galton engaged in remarkably diversified studies including travel comforts, qualifications of scientists, eugenics, fingerprinting, meteorography, religious tolerance, probability, eclipses, etc.

Regression.

a. Filial: However paradoxical it may appear at first sight, it is theoretically a necessary fact, and one that is clearly confirmed by observation, that the Stature of the adult offspring must on the whole, be more *mediocre* than the stature of their Parents; that is to say, more near to the M^1 of the general Population. Table 11 [page 656] enables us to compare the values of the M in different Co-Fraternal[2] groups with the Statures of their respective Mid-Parents. Fig. 14 is a graphical representation of the meaning of the Table so far as it now concerns us. The horizontal dotted lines and the graduations at their sides, correspond to the similarly placed lines of figures and graduations in Table 11. The dot on each line shows the point where its M falls. The value of its M is to be read on the graduations along the top, and is the same as that which is given in the last column of Table 11. It will be per-

---

[1] "The accepted term to express the value that occupies the Middlemost position is 'Median,' which may be used either as an adjective or as a substantive, but it will be usually replaced in this book by the abbreviated form M. I also use the word 'Mid' in a few combinations, such as 'Mid-Fraternity,' to express the same thing. The Median, M, has three properties. The first follows immediately from its construction, namely, that the chance is an equal one, of any previously unknown measure in the group exceeding or falling short of M. The second is, that the most probable value of any previously unknown measure in the group is M. Thus if N be any one of the measures, and u be the value of the unit in which the measure is recorded, such as an inch, tenth of an inch, etc., then the number of measures that fall between $(N - \frac{1}{2} u)$ and $(N \text{ plus } \frac{1}{2} u)$, is greatest when N equal M. Mediocrity is always the commonest condition, for reasons that will become apparent later on. The third property is that whenever the curve of the Scheme is symmetrically disposed on either side of M, except that one-half of it is turned upwards, and the other half downwards, then M is identical with the ordinary Arithmetic Mean or Average. This is closely the condition of all the curves I have to discuss. The reader may look on the Median and on the Mean as being practically the same things, throughout this book.

"It must be understood that M, like the Mean or the Average, is almost always an interpolated value, corresponding to no real measure. If the observations were infinitely numerous its position would not differ more than infinitesimally from that of some one of them; even in a series of one or two hundred in number, the difference is insignificant." (*Natural inheritance*, pp. 41, 42.)

[2] "I transmute all the observations of females before taking them in hand, and thenceforward am able to deal with them on equal terms with the observed male values. For example: the statures of women bear to those of men the proportion of about twelve to thirteen. Consequently by adding to each observed female stature at the rate of one inch for every foot, we are enabled to compare their statures so increased and transmuted, with the observed statures of males, on equal terms." (*Ibid.*, p. 6.)

"As all the Adult Sons and Transmuted Daughters of the *same* Mid-Parent, form what is called a Fraternity, so all the Adult Sons and Transmuted Daughters of a *group* of Mid-Parents who have the same Stature (reckoned to the nearest inch) will be termed a Co-Fraternity. Each line in Table 11 refers to a separate Co-Fraternity and expresses the distribution of Stature among them." (*Ibid.*, p. 94.)

ceived that the line drawn through the centres of the dots, admits of being interpreted by the straight line CD, with but a small amount of give and take; and the fairness of this interpretation is confirmed by a study of the MS. chart above mentioned, in which the individual observations were plotted in their right places.

Now if we draw a line AB through every point where the graduations along the top of Fig. 14, are the same as those along the sides, the line will be

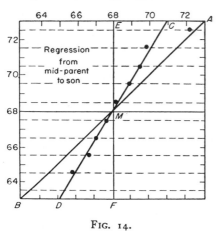

FIG. 14.

straight and will run diagonally. It represents what the Mid-Statures of the Sons would be, if they were on the average identical with those of their Mid-Parents. Most obviously AB does *not* agree with CD; therefore Sons do *not*, on the average, resemble their Mid-Parents. On examining these lines more closely, it will be observed that AB cuts CD at a Point M that fairly corresponds to the value of $68\frac{1}{4}$ inches, whether its value be read on the scale at the top or on that at the side. This is the value of P, the Mid-Stature of the population. Therefore it is only when the Parents are mediocre, that their Sons on the average resemble them.

Next draw a vertical line, EMF, through M, and let ECA be any horizontal line cutting ME at E, MC at E, and MA at A. Then it is obvious that the ratio of EA to EC is constant, whatever may be the position of ECA. This is true whether ECA be drawn above or like FDB, below M. In other words, the proportion between the Mid-Filial and the Mid-Parental deviation is constant, whatever the Mid-Parental stature may be. I reckon this ratio to be as 2 to 3: that is to say, the Filial deviation from P is on the average only two-thirds as wide as the Mid-Parental Deviation. I call this ratio of 2 to 3 the ratio of "Filial Regression." It is the proportion in which the Son is, on the average, less exceptional than his Mid-Parent.

My first estimate of the average proportion between the Mid-Filial and the Mid-Parental deviations, was made from a study of the MS. chart, and I then reckoned it as 3 to 5. The value given above was afterwards substituted, because the data seemed to admit of that interpretation also, in which case the fraction of two-thirds was preferable as being the more simple expression. I am now inclined to think the latter may be a trifle too small, but it is not worth while to make alterations until a new, larger, and more accurate series of observations can be discussed, and the whole work revised. The present doubt only ranges between nine-fifteenths in the first case and ten-fifteenths in the second.

This value of two-thirds will therefore be accepted as the amount of Regression, on the average of many cases, from the Mid-Parental to the Mid-Filial stature, whatever the Mid-Parental stature may be.

As the two Parents contribute equally, the contribution of either of them can be only one half of that of the two jointly; in other words, only one half of that of the Mid-Parent. Therefore the average Regression from the Parental to the Mid-Filial Stature must be the one half of two-thirds, or one-third. I am unable to test this conclusion in a satisfactory manner by direct observation. The data are barely numerous enough for dealing even with questions referring to Mid-Parentages; they are quite insufficient to deal with those that involve the additional large uncertainty introduced owing to an ignorance of the Stature of one of the parents. I have entered the Uni-Parental and the Filial data on a MS. chart, each in its appropriate place, but they are too scattered and irregular to make it useful to give the results in detail. They seem to show a Regression of about two-fifths, which differs from that of one-third in the ratio of 6 to 5. This direct observation is so inferior in value to the inferred result, that I disregard it, and am satisfied to adopt the value given by the latter, that is to say, of one-third, to express the average Regression from either of the Parents to the Son.

b. Mid-Parental: The converse relation to that which we have just discussed, namely the relation between the unknown stature of the Mid-Parent and the known Stature of the Son, is expressed by a fraction that is very far from being the converse of two-thirds. Though the Son deviates on the average from P only ⅔ as widely as his Mid-Parent, it does not in the least follow that the Mid-Parent should deviate on the average from P, ³⁄₂ or 1½, as widely as the Son. The Mid-Parent is not likely to be more exceptional than the son, but quite the contrary. The number of individuals who are nearly mediocre is so preponderant, that an exceptional man is more frequently found to be the exceptional son of mediocre parents than the average son of very exceptional parents. This is clearly shown by Table 11, where the very same observations which give the average value of Filial Regression when it is read in one way, gives that of the Mid-Parental Regression when it is read in another way, namely down the vertical columns, instead of along the horizontal lines. It then shows that the Mid-Parent of a man deviates on the average from P, only one-third as much as the man himself. This value of ⅓ is four and a half times smaller than the numerical converse of ³⁄₂, since 4½, or ⁹⁄₂, being multiplied into ⅓, is equal to ³⁄₂.

c. Parental: As a Mid-Parental deviation is equal to one-half of the two Parental deviations, it follows that the Mid-Parental Regression must be equal to one-half of the sum of the two Parental Regressions. As the latter are equal to one another it follows that all three must have the same value. In other words, the average Mid-Parental Regression being $\frac{1}{3}$, the average Parental Regression must be $\frac{1}{3}$ also.

As there was much appearance of paradox in the above strongly contrasted results, I looked carefully into the run of the figures in Table 11. They were deduced, as already said, from a MS. chart on which the stature of every Son and the transmuted Stature of every Daughter is entered opposite to that of the Mid-Parent, the transmuted Statures being reckoned to the nearest tenth of an inch, and the position of the other entries being in every

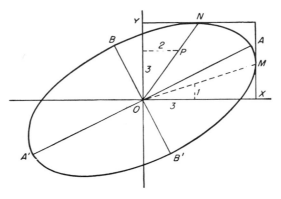

Fig. 15.

respect exactly as they were recorded. Then the number of entries in each square inch were counted, and copied in the form in which they appear in the Table. I found it hard at first to catch the full significance of the entries, though I soon discovered curious and apparently very interesting relations between them. These came out distinctly after I had "smoothed" the entries by writing at each intersection between a horizontal line and a vertical one, the sum of the entries in the four adjacent squares. I then noticed (see Fig. 15) that lines drawn through entries of the same value formed a series of concentric and similar ellipses. Their common centre lay at the intersection of those vertical and horizontal lines which correspond to the value of $68\frac{1}{4}$ inches, as read on both the top and on the side scales. Their axes were similarily inclined. The points where each successive ellipse was touched by a horizontal tangent, lay in a straight line that was inclined to the vertical in the ratio of $\frac{2}{3}$, and those where the ellipses were touched by a vertical tangent, lay in a straight line inclined to the horizontal in the ratio of $\frac{1}{3}$. It will be obvious on studying Fig. 15 that the point where each successive horizontal line touches an ellipse is the point at which the greatest value in the line will be found. The same is true in respect to the successive vertical lines. Therefore these ratios confirm the values of the Ratios of Regression, already obtained by a different method, namely those of $\frac{2}{3}$ from Mid-Parent to Son, and of $\frac{1}{3}$ from Son to Mid-Parent. These and other relations were evidently a subject for mathematical analysis and verification. It seemed clear to me that they all depended on three elementary measures, supposing the law of Frequency of Error[3] to be applicable throughout; namely (1) the value of Q[4] in the General Population, which was found to be 1·7 inch; (2) the value of Q in any Co-Fraternity, which was found to be 1·5 inch; (3) the Average Regression of the Stature of the Son from that of the Mid-Parent, which was found to be $\frac{2}{3}$. I wrote down these values, and ph asing the problem in abstract terms, disentangled from all reference to heredity, submitted it to Mr. J. D. Hamilton Dickson, Tutor of St. Peter's College, Cambridge. I asked him kindly to investigate for me the Surface of Frequency of Error that would result from these three data, and the various shapes and other particulars of its sections that were made by horizontal planes, inasmuch as they ought to form the ellipses of which I spoke.

The problem may not be difficult to an accomplished mathematician, but I certainly never felt such a glow of loyalty and respect towards the sovereignty and wide sway of mathematical analysis as when his answer arrived, con-

---

[3] "But Errors, Differences, Deviations, Divergencies, Dispersions, and individual Variations, all spring from the same kind of causes. Objects that bear the same name, or can be described by the same phrase, are thereby acknowledged to have common points of resemblance, and to rank as members of the same species, class, or whatever else we may please to call the group. On the other hand, every object has Differences peculiar to itself, by which it is distinguished from others.

"This general statement is applicable to thousands of instances. The Law of Error finds a footing wherever the individual peculiarities are wholly due to the combined influence of a multitude of 'accidents,' in the sense in which that word has already been defined. All persons conversant with statistics are aware that this supposition brings Variability within the grasp of the laws of Chance, with the result that the relative frequency of Deviations of different amounts admits of being calculated, when those amounts are measured in terms of any self-contained unit of variability, such as our Q." (*Ibid.*, p. 55.)

[4] "Suppose that we have only two Schemes, A. and B., that we wish to compare. Let $L_{.1}$, $L_{.2}$ be the lengths of the perpendiculars at two specified grades in Scheme A., and $K_{.1}$, $K_{.2}$ the lengths of those at the same grades in Scheme B.; then if every one of the data from which Scheme B. was drawn be multiplied by $\dfrac{L_{.1} - L_{.2}}{K_{.1} - K_{.2}}$, a series of transmuted data will be obtained for drawing a new Scheme B'., on such a vertical scale that its general slope between the selected grades shall be the same as in Scheme A. For practical convenience the selected Grades will be always those of 25° and 75°. They stand at the first and third quarterly divisions of the base, and are therefore easily found by a pair of compasses. They are also well placed to afford a fair criterion of the general slope of the Curve. If we call the perpendicular at 25°, $Q_{.1}$; and that at 75°, $Q_{.2}$, then the unit by which every Scheme will be defined is its value of $\frac{1}{2}$ $(Q_{.2} - Q_{.1})$, and will be called its Q. As the M measures the Average Height of the curved boundary of a Scheme, so the Q measures its general slope." (*Ibid.*, pp. 52, 53.)

firming, by purely mathematical reasoning, my various and laborious statistical conclusions with far more minuteness than I had dared to hope, because the data ran somewhat roughly, and I had to smooth them with tender caution.   His calculation corrected my observed value of Mid-Parental Regression from $\frac{1}{3}$ to $\frac{6}{17.6}$; the relation between the major and minor axis of the ellipses was changed 3 per cent; and their inclination to one another was changed less than $2°$.[5]

It is obvious from this close accord of calculation with observation, that the law of Error holds throughout with sufficient precision to be of real service, and that the various results of my statistics are not casual and disconnected determinations, but strictly interdependent.

I trust it will have become clear even to the most non-mathematical reader, that the law of Regression in Stature refers primarily to Deviations, that is, to measurements made from the *level of mediocrity* to the crown of the head, upwards or downwards as the case may be, and not from the *ground* to the crown of the head.   (In the population with which I am now dealing, the level of mediocrity is $68\frac{1}{4}$ inches (without shoes).)   The law of Regression in respect to Stature may be phrased as follows; namely, that the Deviation of the Sons from P are, on the average, equal to one-third of the deviation of the Parent from P, and in the same direction.   Or more briefly still:—If $P + (\pm D)$[6] be the Stature of the Parent, the Stature of the offspring will on the average be $P + (\pm \frac{1}{3} D)$.

---

[5]   The following is a more detailed comparison between the calculated and the observed results.   The latter are enclosed in brackets.   The letters refer to Fig. 15:—
   Given—
      The "Probable Error" of each system of Mid-Parentages = 1.22 inch.   (This was an earlier determination of its value; as already said, the second decimal is to be considered only as approximate.)
      Ratio of mean filial regression = $\frac{2}{3}$.
      "Prob. Error" of each Co-Fraternity = 1.50 inch.
Sections of surface of frequency parallel to XY are true ellipses.
      (Obs.—Apparently true ellipses.)
   MX:YO = 6:17.5, or nearly 1:3.
      (Obs.—1:3.)
   Major axes to minor axes = $\sqrt{7} : \sqrt{2} = 10:5.35$.
      (Obs.—10:5.1.)
   Inclination of major axes to OX = $26° \ 36'$.
      (Obs. $25°$.)
Section of surface parallel to XZ is a true Curve of Frequency.   [Note: No Z shown on original figure.—Ed.]
      (Obs.—Apparently so.)
   "Prob. Error," the Q of that curve, = 1.07 inch.
      (Obs.—1.00, or a little more.)
   [6] "The Median, Mid-Stature, or M of the general Population is a value of primary importance in this inquiry.   Its value will be always designated by the symbol P."
   "Every measure in a Scheme is equal to its Middlemost, or Median value, or M, *plus*

If this remarkable law of Regression had been based only on those experiments with seeds, in which I first observed it, it might well be distrusted until otherwise confirmed. If it had been corroborated by a comparatively small number of observations on human stature, some hesitation might be expected before its truth could be recognised in opposition to the current belief that the child tends to resemble its parents. But more can be urged than this. It is easily to be shown that we ought to expect Filial Regression, and that it ought to amount to some constant fractional part of the value of the Mid-Parental deviation. All of this will be made clear in a subsequent section, when we shall discuss the cause of the curious statistical constancy in successive generations of a large population. In the meantime, two different reasons may be given for the occurrence of Regression; the one is connected with our notions of stability of type, and of which no more need now be said; the other is as follows:—The child inherits partly from his parents, partly from his ancestry. In every population that intermarries freely, when the genealogy of any man is traced far backwards, his ancestry will be found to consist of such varied elements that they are indistinguishable from a sample taken at haphazard from the general Population. The Mid-Stature M of the remote ancestry of such a man will become identical with P; in other words, it will be mediocre. To put the same conclusion into another form, the most probable value of the Deviation from P, of his Mid-Ancestors in any remote generation, is zero.

For the moment let us confine our attention to some one generation in the remote ancestry on the one hand, and to the Mid-Parent on the other, and ignore all other generations. The combination of the zero Deviation of the one with the observed Deviation of the other is the combination of nothing with something. Its effect resembles that of pouring a measure of water into a vessel of wine. The wine is diluted to a constant fraction of its alcoholic strength, whatever that strength may have been.

Similarly with regard to every other generation. The Mid-Deviation in any near generation of the ancestors will have a value intermediate between that of the zero Deviation of the remote ancestry, and of the observed Deviation of the Mid-Parent. Its combination with the Mid-Parental Deviation will be as if a mixture of wine and water in some definite proportion, and not pure water, had been poured into the wine. The process throughout is one of proportionate dilutions, and the joint effect of all of them is to weaken the original alcoholic strength in a constant ratio.

The law of Regression tells heavily against the full hereditary transmission of any gift. Only a few out of many children would be likely to differ from

or *minus* a certain Deviation from M. The Deviation, or "Error" as it is technically called, is *plus* for all grades above 50°, zero for 50°, and *minus* for all grades below 50°. Thus if $(\pm D)$ be the deviation from M in any particular case, every measure in a Scheme may be expressed in the form of $M + (\pm D)$." (*Ibid.*, pp. 51, 52, & 92.)

mediocrity so widely as their Mid-Parent, and still fewer would differ as widely as the more exceptional of the two Parents. The more bountifully the Parent is gifted by nature, the more rare will be his good fortune if he begets a son who is as richly endowed as himself, and still more so if he has a son who is endowed yet more largely. But the law is even-handed; it levies an equal succession-tax on the transmission of badness as of goodness. If it discourages the extravagant hopes of a gifted parent that his children will inherit all his powers; it no less discountenances extravagant fears that they will inherit all his weakness and disease.

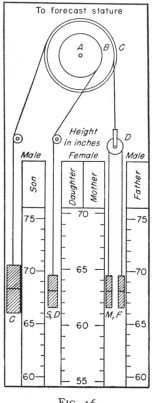

It must be clearly understood that there is nothing in these statements to invalidate the general doctrine that the children of a gifted pair are much more likely to be gifted than the children of a mediocre pair. They merely express the fact that the ablest of all the children of a few gifted pairs is not likely to be as gifted as the ablest of all the children of a very great many mediocre pairs.

The constancy of the ratio of Regression, whatever may be the amount of the Mid-Parental Deviation, is now seen to be a reasonable law which might have been foreseen. It is so simple in its relations that I have contrived more than one form of apparatus by which the probable stature of the children of known parents can be mechanically reckoned. Fig. 16 is a representation of one of them, that is worked with pulleys and weights. A, B, and C are three thin wheels

FIG. 16.

with grooves round their edges. They are screwed together so as to form a single piece that turns easily on its axis. The weights M and F are attached to either end of a thread that passes over the movable pulley D. The pulley itself hangs from a thread which is wrapped two or three times round the groove of B and is then secured to the wheel. The weight SD hangs from a thread that is wrapped two or three times round the groove of A, and is then secured to the wheel. The diameter of A is to that of B as 2 to 3. Lastly, a thread is wrapped in the opposite direction round the wheel C, which may have any convenient diameter, and is attached to a counterpoise. M refers to the male statures, F to the female ones, S to the Sons, D to the Daughters.

The scale of Female Statures differs from that of the Males, each Female height being laid down in the position which would be occupied by its male equivalent. Thus 56 is written in the position of 60·48 inches, which is equal to 56 × 1·08. Similarly, 60 is written in the position of 64·80, which is equal to 60 × 1·08.

It is obvious that raising M will cause F to fall, and *vice versâ*, without affecting the wheel AB, and therefore without affecting SD; that is to say, the Parental Differences may be varied indefinitely without affecting the Stature of the children, so long as the Mid-Parental Stature is unchanged. But if the Mid-Parental Stature is changed to any specified amount, then that of SD will be changed to ⅔ of that amount.

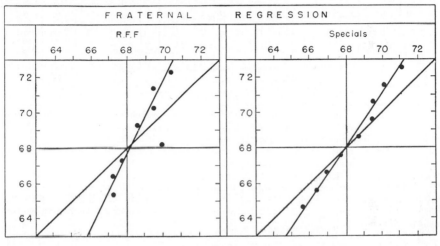

FIG. 17.

The weights M and F have to be set opposite to the heights of the mother and father on their respective scales; then the weight SD will show the most probable heights of a Son and of a Daughter on the corresponding scales. In every one of these cases, it is the fiducial mark in the middle of each weight by which the reading is to be made. But, in addition to this, the length of the weight SD is so arranged that it is an equal chance (an even bet) that the height of each Son or each Daughter will lie within the range defined by the upper and lower edge of the weight, on their respective scales. The length of SD is 3 inches, which is twice the Q of the Co-Fraternity; that is, 2 × 1·50 inch.

d. Fraternal: In seeking for the value of Fraternal Regression, it is better to confine ourselves to the Special data given in Table 13, as they are much more trustworthy than the R.F.F. data in Table 12 [not shown—Ed.].[5] By treating them in the way shown in Fig. 17, which is constructed on the same principle as Fig. 14, I obtained the value for Fraternal Regression of

---

[5] The reference is to tables in the original text showing number of brothers of various heights, relative to men of various heights; Table 12 excludes families of more than 5, Table 13 families of more than 4, brothers.

Table 11 (R.F.F. Data).*
Number of Adult Children of Various Statures Born of 205 Mid-Parents of Various Statures.

(All Female Heights have been multiplied by 1·08.)

Height of the mid-parents in inches.	Heights of the adult children.														Total number of		Medians or Values of *M*.
	Below	62·2	63·2	64·2	65·2	66·2	67·2	68·2	69·2	70·2	71·2	72·2	73·2	Above	Adult children.	Mid-parents.	
Above 72·5..	...	...	...	...	...	...	...	...	...	...	...	1	3	...	4†	5†	
72·5..	...	...	...	...	...	...	1	2	1	2	7	2		4	19	6	72·2
71·5..	...	...	...	...	1	3	4	3	5	10	4	9	2	2	43	11	69·9
70·5..	1	...	1	...	1	1	3	12	18	14	7	4	3	3	68	22	69·5
69·5..	...	...	1	16	4	17	27	20	33	25	20	11	4	5	183	41	68·9
68·5..	1	...	7	11	16	25	31	34	48	21	18	4	3	...	219	49	68·2
67·5..	...	3	5	14	15	36	38	28	38	19	11	4	...	...	211	33	67·6
66·5..	...	3	3	5	2	17	17	14	13	4	...	...	...	...	78	20	67·2
65·5..	1	...	9	5	7	11	11	7	7	5	2	1	...	...	66	12	66·7
64·5..	1	1	4	4	1	5	5	...	2	...	...	...	...	...	23	5	65·8
Below......	1	...	2	4	1	2	2	1	1	...	...	...	...	...	14	1	
Totals......	5	7	32	59	48	117	138	120	167	99	64	41	17	14	928	205	
Medians....	...	...	66·3	67·8	67·9	67·7	67·9	68·3	68·5	69·0	69·0	70·0					

Note.—In calculating the medians, the entries have been taken as referring to the middle of the squares in which they stand. The reason why the headings run 62·2, 63·2, &c., instead of 62·5, 63·5, &c., is that the observations are unequally distributed between 62 and 63, 63 and 64, &c., there being a strong bias in favour of integral inches. After careful consideration, I concluded that the headings, as adopted, best satisfied the conditions. This inequality was not apparent in the case of the mid-parents.

* "The source from which the larger part of my data is derived consists of a valuable collection of 'Records of Family Faculties' (R.F.F.), obtained through the offer of prizes. They have been much tested and cross-tested, and have borne the ordeal very fairly, so far as it has been applied." (*Ibid.*, p. 72.)

† I have reprinted this Table without alteration from that published in the *Proc. Roy. Soc.*, notwithstanding a small blunder since discovered in sorting the entries between the first and second lines. It is obvious that 4 children cannot have 5 Mid-Parents. The first line is not considered at all, on account of the paucity of the numbers it contains. The bottom line, which looks suspicious, is correct.

⅔; that is to say, the unknown brother of a known man is probably only two-thirds as exceptional in Stature as he is. This is the same value as that obtained for the Regression from Mid-Parent to Son. However paradoxical the fact may seem at first, of there being such a thing as Fraternal Regression, a little reflection will show its reasonableness, which will become much clearer later on. In the meantime, we may recollect that the unknown brother has two different tendencies, the one to resemble the known man, and the other to resemble his race. The one tendency is to deviate from P as much as his brother, and the other tendency is not to deviate at all. The result is a compromise.

As the average Regression from either Parent to the Son is twice as great as that from a man to his Brother, a man is, generally speaking, only half as nearly related to either of his Parents as he is to his Brother: In other words, the Parental kinship is only half as close as the Fraternal.

We have now seen that there is Regression from the Parent to his Son, from the Son to his Parent, and from the Brother to his Brother. As these are the only three possible lines of kinship, namely, descending, ascending, and collateral, it must be a universal rule that the unknown Kinsman, in any degree, of a known Man, is on the average more mediocre than he. Let $P \pm D$ be the stature of the known man, and $P \pm D'$ the stature of his as yet unknown kinsman, then it is safe to wager, in the absence of all other knowledge, that D' is less than D.

EVOLUTION AND VARIATION

BATESON, William (English animal geneticist, 1861–1926). From *Materials for the study of variation treated with special regard to discontinuity in the origin of species,* London, 1894; by permission of The Macmillan Co., publisher.

Early in the development of a new science, a point is frequently reached at which specific phenomena must be postulated even though direct evidence for these is to become available only later. Bateson's early studies stand at just this point in the early growth of genetics. The central phenomenon in evolution, he observed, was not the *selection* of variants, but their *production*. This led him to postulate the production and transmission of discontinuous varieties. Direct evidence would be available, after only a few years, in the discoveries of DeVries and the revived experiments of Mendel.

INTRODUCTION

All flesh is not the same flesh: but there is one kind of flesh of men, another flesh of beasts, another of fishes, and another of birds.

Section I
THE STUDY OF VARIATION

To solve the problem of the forms of living things is the aim with which the naturalist of to-day comes to his work. How have living things become what they are, and what are the laws which govern their forms? These are the questions which the naturalist has set himself to answer.

It is more than thirty years since the *Origin of Species* was written, but for many these questions are in no sense answered yet. In owning that it is so, we shall not honour Darwin's memory the less; for whatever may be the part which shall be finally assigned to Natural Selection, it will always be remembered that it was through Darwin's work that men saw for the first time that the problem is one which man may reasonably hope to solve. If Darwin did not solve the problem himself, he first gave us the hope of a solution, perhaps a greater thing. How great a feat this was, we who have heard it all from childhood can scarcely know.

In the present work an attempt is made to find a way of attacking parts of the problem afresh, and it will be profitable first to state formally the conditions of the problem and to examine the methods by which the solution has been attempted before. This consideration shall be as brief as it can be made.

The forms of living things have many characters: to solve the problem completely account must be taken of all. Perhaps no character of form is common to all living things; on the contrary their forms are almost infinitely diverse. Now in those attempts to solve the problem which have been the best, it is this diversity of form which is taken as the chief attribute, and the attempt to solve the general problem is begun by trying to trace the modes by which the diversity has been produced. In the shape in which it has been most studied, the problem is thus the problem of Species. Obscurity has been brought into the treatment of the question through want of recognition of the fact that this is really only a part of the general problem, which would still remain if there were only one species. . . .

We have to consider whether it is not possible to get beyond the present position and to penetrate further into this mystery of Specific Forms. The main obstacle being our own ignorance, the first question to be settled is what kind of knowledge would be of the most value, and which of the many unknowns may be determined with the greatest profit. To decide this we must return once more to the ground which is common to all the inductive theories of Evolution alike. Now all these different theories start from the hypothesis that the different forms of life are related to each other, and that their diversity is due to Variation. On this hypothesis, therefore, Variation, whatever may be its cause, and however it may be limited, is the essential phenomenon of Evolution. Variation, in fact, *is* Evolution. The readiest way, then, of solving the problem of Evolution is to study the facts of Variation.

Section II

ALTERNATIVE METHODS

The Study of Variation is therefore suggested as the method which is on the whole more likely than any other to give us the kind of knowledge we are wanting. It should be tried not so much in the hope that it will give any great insight into those relations of cause and effect of which Evolution is the expression, but merely as an empirical means of getting at the outward and visible phenomena which constitute Evolution. On the hypothesis of Common Descent, the forms of living things are succeeding each other, passing across the stage of the earth in a constant procession. To find the laws of the succession it will be best for us to stand as it were aside and to watch the procession as it passes by. No amount of knowledge of individual forms will tell us the laws or even the manner of the succession, nor shall we be much helped by comparison of forms of whose descent we know nothing save by speculation. To study Variation it must be seen at the moment of its beginning. For comparison we require the parent and the varying offspring together. To find out the nature of the progression we require, simultaneously, at least two consecutive terms of the progression. Evidence of this kind can be obtained in no other way than by the study of actual and contemporary cases of Variation. . . .

The same problem is nevertheless capable of being stated in the general form also. Instead of considering what has been the actual series from which a specified type has been derived, we may consider what are the characters and attributes of such series in general. It may indeed be contended that it is scarcely reasonable to expect to discover the line of descent of a given form, for the evidence is gone; but we may hope to find the general characteristics of Evolution, for Evolution, as we believe, is still in progress. It is really a strange thing that so much enterprise and research should have been given to the task of reconstructing particular pedigrees—a work in which at best the facts must be eked out largely with speculation—while no one has ever seriously tried to determine the general characters of such a series. Yet if our modern conception of Descent is a right one, it is a phenomenon now at this time occurring, which by common observations, without the use of any imagination whatever, we may now see. The chief object, then, with which we shall begin the Study of Variation will be the determination of the nature of the Series by which forms are evolved.

The first questions that we shall seek to answer refer to the manner in which differentiation is introduced in these Series. All we as yet know is the last term of the Series. By the postulate of Common Descent we take it that the first term differed widely from the last, which nevertheless is its lineal descendant: how then was the transition from the first term to the last term effected? If the whole series were before us, should we find that this transition had been brought about by very minute and insensible differences between successive terms in the Series, or should we find distinct and palpable gaps in the Series? In proportion as the transition from term to term is minimal and imperceptible we may speak of the Series as being *Continuous*, while in proportion as there appear in it lacunae, filled by no transitional form, we may describe it as *Discontinuous*. The several possibilities may be stated somewhat as follows. The Series may be wholly continuous; on the other hand it may be sometimes continuous and sometimes discontinuous; we know however by common knowledge that it is never wholly discontinuous. It may be that through long periods of the Series the differences between each member and its immediate predecessor and successor are impalpable, while at certain moments the Series is interrupted by breaches of continuity which divide it into groups, of which the composing members are alike, though the successive groups are unlike. Lastly, discontinuity may occur in the evolution of particular organs or particular instincts, while the changes in other structures and systems may be effected continuously. To decide which of these agrees most nearly with the observed phenomena of Variation is the first question which we hope, by the Study of Variation, to answer. The answer to this question is of vital consequence to progress in the Study of Life.

The preliminary question, then, of the degree of continuity with which the process of Evolution occurs, has never been decided. In the absence of such a decision there has nevertheless been a common assumption, either tacit or expressed, that the process is a continuous one. The immense consequence of a knowledge of the truth as to this will appear from a consideration of the gratuitous difficulties which have been introduced by this assumption. Chief among these is the difficulty which has been raised in connection with the building up of new organs in their initial and imperfect stages, the mode of transformation of organs, and, generally, the Selection and perpetuation of minute Variations. Assuming then that Variations are minute, we are met by this familiar difficulty. We know that certain devices and mechanisms are useful to their possessors; but from our knowledge of Natural History we are led to think that their usefulness is consequent on the degree of perfection in which they exist, and that if they were at all imperfect, they would not be useful. Now it is clear that in any continuous process of Evolution such stages of imperfection must occur, and the objection has been raised that Natural Selection cannot protect such imperfect mechanisms so as to lift them into perfection. Of the objections which have been brought against the Theory of Natural Selection this is by far the most serious.

The same objection may be expressed in a form which is more correct and comprehensive. We have seen that the differences between Species on the whole are Specific, and are differences of kind, forming a discontinuous Series, while the diversities of environment to which they are subject are on the whole differences of degree, and form a continuous Series; it is therefore hard to see how the environmental differences can thus be in any sense the directing cause of Specific differences, which by the Theory of Natural Selection they should be. This objection of course includes that of the utility of minimal Variations.

Now the strength of this objection lies wholly in the supposed continuity of the process of Variation. We see all organised nature arranged in a discontinuous series of groups differing from each other by differences which are Specific; on the other hand we see the divers environments to which these forms are subject passing insensibly into each other. We must admit, then, that if the steps by which the divers forms of life have varied from each other have been insensible—if in fact the forms ever made up a continuous series— these forms cannot have been broken into a discontinuous series of groups by a continuous environment, whether acting directly as Lamarck would have, or as selective agent as Darwin would have. This supposition has been generally made and admitted, but in the absence of evidence as to Variation it is nevertheless a gratuitous assumption, and as a matter of fact when the evidence as to Variation is studied, it will be found to be in great measure unfounded.

In what follows so much will be said of discontinuity in Variation that it will not be amiss to speak of the reasons which have led many to suppose that the continuity of Variation needs no proof. Of these reasons there are especially two. First there is in the minds of some persons an inherent conviction that *all* natural processes are continuous. That many of them do not appear so is admitted: it is admitted, for example, that among chemical processes Discontinuity is the rule; that changes in the states of matter are commonly effected discontinuously, and the like. Nevertheless it is believed that such outward and visible Discontinuity is but a semblance or mask which conceals a real process which is continuous and which by more searching may be found. With this class of objections we are not perhaps concerned, but they are felt by so many that their existence must not be forgotten. Secondly, Variation has been supposed to be always continuous and to proceed by minute steps because changes of this kind are so common in Variation. Hence it has been inferred that the mode of Variation thus commonly observed is universal. That this inference is a wrong one, the facts will show.

To sum up:

The first question which the Study of Variation may be expected to answer, relates to the origin of that Discontinuity of which Species is the objective expression. Such Discontinuity is not in the environment; may it not, then, be in the living thing itself? . . .

Section V
MERISTIC VARIATION AND SUBSTANTIVE VARIATION

It is to the origin and nature of Symmetry that the first section of the evidence of Variation will relate. That a knowledge of the modes of Variation of so universal a character is important to the general study of Biology must at once be evident, but to the particular problem of the nature of Specific Differences this importance is immense. This special importance comes from two reasons. As it is the fact first that Repetition and Symmetry are among the commonest features of organised structure, so it will be found next that it is by differences in those features that the various forms of organisms are very commonly differentiated from each other. Their forms are classified by all sorts of characters, by shape and proportions, by size, by colour, by habits and the like; but perhaps almost as frequently as by any of these, by differences in number of parts and by differences in the geometrical relations of the parts. It is by such differences that the larger divisions, genera, families, etc., are especially distinguished. In such cases of course the differences in number and Symmetry do not as a rule stand alone, but are generally, and perhaps always, accompanied by other differences of a qualitative kind; nevertheless, the differences in number and Symmetry form an integral and very definite part of the total differences, so that in any consideration of the nature of the processes by which the differences have arisen, special regard must be had to these numerical and geometrical, or, as I propose to call them, *Meristic*, changes.

In the present Introduction I do not propose to forestall the evidence more than is absolutely necessary for the purpose of making clear the principles on which the facts are grouped, but it will do the evidence no wrong if at the present stage it is stated that Meristic Variation is frequently Discontinuous, and in the case of certain classes of Repetitions is perhaps always so.

The nature of Merism and the manner in which Meristic Variations occur will be fully illustrated hereafter, but it is necessary to say this much at the present stage, since it is from this Discontinuity in the occurrence of Meristic Variations that the phenomena of Symmetry and Repetition derive their special importance in the Study of Variation.

The importance of the phenomena of Merism to the Study of Variation is thus, in the first instance, a direct one, for the Variations which have resulted in the production of Meristic Systems are a direct factor in Evolution. In addition to this direct relation to the Study of Variation, the phenomena of Merism have also an indirect relation, which is scarcely less important; for they are a factor in the estimation of the magnitude of the integral steps by which Variation proceeds. This will be more evident after the second group of Variations has been spoken of.

We have thus far spoken only of the processes by which the living body is divided into parts, and we have thus constituted a group which is to include Variations in number, Division, and geometrical position. From these phenomena of Division may be distinguished Variations in the actual constitution or substance of the parts themselves. To these Variations the name *Substantive* will be given. Under this head several phenomena may be temporarily grouped together, which with further knowledge will doubtless be found to have no real connection with each other. For the present, however, it will be convenient to constitute such a temporary group in order to bring out the relative distinctness of Variations which are Meristic. . . . Now [1] the evidence of Discontinuous Variation suggests that organisms may vary abruptly from the definite form of the type to a form of variety which has

---

[1] The train of thought continues in Section XIII from which the rest of the excerpt is taken.

also in some measure the character of definiteness. Is it not then possible that the Discontinuity of Species may be a consequence and expression of the Discontinuity of Variation? To declare at the present time that this is so would be wholly premature, but the suggestion that it is so is strong, and as a possible light on the whole subject should certainly be considered.

In view of such a possible solution of one of the chief parts of the problem of Species it will be well to point out a line of inquiry which must in that event be pursued. If it can be shown that the Discontinuity of Species depends on the Discontinuity of Variation, we shall then have to consider the causes of the Discontinuity of Variation.

Upon the received hypothesis it is supposed that Variation is continuous and that the Discontinuity of Species results from the operation of Selection. For reasons given above there is an almost fatal objection in the way of this belief, and it cannot be supposed both that all Variation is continuous and also that the Discontinuity of Species is the result of Selection. With evidence of the Discontinuity of Variation this difficulty would be removed.

It will be noted also that it is manifestly impossible to suppose that the perfection of a variety, discontinuously and suddenly occurring, is the result of Selection. No doubt it is conceivable that a race of tulips having their floral parts in multiples of four might be raised by Selection from a specimen having this character, but it is not possible that the perfection of the nascent variety can have been gradually built up by Selection, for it is, in its very beginning, perfect and symmetrical. And if it may be seen thus clearly that the perfection and Symmetry of a variety is not the work of Selection, this fact raises a serious doubt that perhaps the similar perfection and Symmetry of the type did not owe its origin to Selection either. This consideration of course touches only the part that Selection may have played in the first building up of the type and does not affect the view that the perpetuation of the type once constituted may have been achieved by Selection.

But if the perfection and definiteness of the type is not due to Selection but to the physical limitations under which Variation proceeds, we shall hope hereafter to gain some insight into the nature of these limitations, though in the present state of zoological study the prospect of such progress is small. In the observations which follow I am conscious that the bounds of profitable speculation are perhaps exceeded, and I am aware that to many this may seem matter for blame; but there is, in my judgment, a plausibility in the views put forward, sufficient at least to entitle them to examination. They are put forward in no sense as a formulated theory, but simply as a suggestion for work. It is, besides, only in foreseeing some of the extraordinary possibilities that lie ahead in the Study of Variation, that the great value of this method can be understood.   . . .

Section XIV

SOME CURRENT CONCEPTIONS OF BIOLOGY IN VIEW OF THE FACTS OF VARIATION

Enough has now been said to explain the aim of the Study of Variation, and to show the propriety of the choice of the facts of Meristic Variation as a point of departure for that study. Before leaving this preliminary consideration, reference to some cognate subjects must be made.

It has been shown that in view of the facts of Variation, some conceptions of modern Morphology must be modified, while others must be abandoned. With the recognition of the significance of the phenomena of Variation, other conceptions of biology will undergo like modifications. As to some of these a few words are now required if only to explain methods adopted in this work.

1. *Heredity.*

It has been the custom of those who have treated the subject of Evolution to speak of "Heredity" and "Variation" as two antagonistic principles; sometimes even they are spoken of as opposing "forces."

With the Study of Variation, such a description of the processes of Descent will be given up, even as a manner of speaking. In what has gone before I have as far as possible avoided any use of the terms Heredity and Inheritance. These terms, which have taken so firm a hold on science and on the popular fancy, have had a mischievous influence on the development of biological thought. They are of course metaphors from the descent of property, and were applied to organic Descent in a time when the nature of the process of reproduction was wholly misunderstood. This metaphor from the descent of property is inadequate chiefly for two reasons.

First, by emphasising the fact that the organisation of the offspring depends on material transmitted to it by its parents, the metaphor of Heredity, through an almost inevitable confusion of thought, suggests the idea that the actual body and constitution of the parent are thus in some way handed on. No one perhaps would now state the facts in this way, but something very like this material view of Descent was indeed actually developed into Darwin's Theory of Pangenesis. From this suggestion that the body of the parent is in some sort remodelled into that of the offspring, a whole series of errors are derived. Chief among these is the assumption that Variation must necessarily be a continuous process; for with the body of the parent to start from, it is hard to conceive the occurrence of discontinuous change. Of the deadlock which has resulted from the attempt to interpret Homology on this view of Heredity, I have already spoken in Section VI.[2]

Secondly, the metaphor of Heredity misrepresents the essential phenomenon

---

[2] Section VI contains a critique of current views on the phylogenetic significance of homology.

of reproduction. In the light of modern investigations, and especially those of Weismann on the continuity of the germ-cells, it is likely that the relation of parent to offspring, if it has any analogy with the succession of property, is rather that of trustee than of testator.

Hereafter, perhaps, it may be found possible to replace this false metaphor by some more correct expression, but for our present purpose this is not yet necessary. In the first examination of the facts of Variation, I believe it is best to attempt no particular consideration of the working of Heredity. The phenomena of Variation and the *origin* of a variety must necessarily be studied first, while the question of the perpetuation of the variety properly forms a distinct subject. Whenever in the cases given observations respecting inheritance are forthcoming they will be of course mentioned. But speaking of discontinuous Variation in general, the recurrence of a Variation in offspring, either in the original form or in some modification of it, has been seen in so many cases, that we shall not go far wrong in at least assuming the possibility that it *may* reappear in the offspring. At the present moment, indeed, to this statement there is little to add. So long as systematic experiments in breeding are wanting, and so long as the attention of naturalists is limited to the study of normal forms, in this part of biology which is perhaps of greater theoretical and even practical importance than any other, there can be no progress.

2. *Reversion.*

Around the term Reversion a singular set of false ideas have gathered themselves. On the hypothesis that all perfection and completeness of form or of correlation of parts is the work of Selection it is difficult to explain the discontinuous occurrence of new forms possessing such perfection and completeness. To account for these, the hypothesis of Reversion to an ancestral form is proposed, and with some has found favour. That this suggestion is inadmissible is shown at once by the frequent occurrence by discontinuous Variation, of forms which though equally perfect, cannot all be ancestral. In the case of Veronica and Linaria, for example, a host of symmetrical forms of the floral organs may be seen occurring suddenly as sports, and of these though any one may conceivably have been ancestral, the same cannot be supposed of all, for their forms are mutually exclusive. On *Veronica buxbaumii*, for instance, are many symmetrical flowers, having two posterior petals, like those of other Scrophularineae: these may reasonably be supposed to be ancestral, but if this supposition is made, it cannot be made again for the equally perfect forms with three petals, and the rest.

The hypothesis of Reversion to account for the Symmetry and perfection of modern or discontinuous Variation is made through a total misconception of the nature of Symmetry.

3. *Causes of Variation.*

Inquiry into the causes of Variation is as yet, in my judgment, premature.

4. *The Variability of "useless" Structures.*

The often-repeated statement that "useless" parts are especially variable finds little support in the facts of Variation, except in as far as it is a misrepresentation of another principle. The examples taken to support this statement are commonly organs standing at the end of a Meristic Series of parts, in which there is a progression or increase of size and degree of development, starting from a small terminal member. In such cases, as that of the last rib in Man, and several other animals, the wisdom-teeth of Man, etc., it is quite true that in the terminal member Variation is more noticeable than it is in the other members. This is, I believe, a consequence of the mechanics of Division, and has no connection with the fact that the functions of such terminal parts are often trifling. Upon this subject something will be said later on, but perhaps a rough illustration may make the meaning more clear at this stage. If a spindle-shaped loaf of bread, such as a "twist," be divided with three cuts taken at equal distances, in such a way that the two end pieces are much shorter than the middle ones, to a child who gets one of the two large middle pieces the contour-curves of the loaf will not matter so much; but to a child who gets one of the small end bits, a very slight alteration in the curves of the loaf will make the difference between a fair-sized bit and almost nothing, a difference which the child will perceive much more readily than the complementary difference in the large pieces will be seen by the others. An error in some measure comparable with this is probably at the bottom of the statement that useless parts are variable, but of course there are many examples, as the pinna of the human ear, which are of a different nature. It is unnecessary to say that for any such case in which a part, apparently useless, is variable, another can be produced in which some capital organ is also variable; and conversely, that for any case of a capital organ which is little subject to Variation can be produced a case of an organ, which, though trifling and seemingly "useless," is equally constant. With a knowledge of the facts of Variation, all these trite generalities will be forgotten.

5. *Adaptation.*

In examining cases of Variation, I have not thought it necessary to speculate on the usefulness or harmfulness of the Variations described. For reasons given in Section II, such speculation, whether applied to normal structures or to Variation, is barren and profitless. If any one is curious on these questions of Adaptation, he may easily thus exercise his imagination. In any case of Variation there are a hundred ways in which it may be beneficial or detrimental. For instance, if the "hairy" variety of the moorhen became established on an island, as many strange varieties have been, I do not doubt that ingenious persons would invite us to see how the hairiness fitted the bird in some special way for life in that island in particular. Their contention would be hard to deny, for on this class of speculation the only limitations are those of the ingenuity of the author. While the only test of utility is the success of the organism, even this does not indicate the utility of one part of the economy, but rather the net fitness of the whole.

6. *Natural Selection.*

In the view of the phenomena of Variation here outlined, there is nothing which is in any way opposed to the theory of the Origin of Species "by means of Natural Selection, or the preservation of favoured races in the struggle for life."

CHROMOSOMAL BASIS OF MENDELIAN INHERITANCE

SUTTON, Walter S. (American cytologist, 1876–1916). From *The chromosomes in heredity*, in *Biological Bulletin*, vol. 4, 1902–1903; by permission of the editor.

Sutton's major contribution was the synthesis of cytological knowledge with the recently rediscovered results of Gregor Mendel into a general cytogenetical theory of inheritance. His specific contributions included definitive demonstrations of chromosomal individuality and continuity and of the relations of chromosome size to genetic action and of tetrad splitting to independent assortment. While the same problems were the subject of simultaneous, and in some cases earlier, investigation (Boveri, Wilson, *et al.*), Sutton was probably primarily responsible for their integration into a general theoretical basis upon which the new science of cytogenetics could then be constructed.

In a recent announcement of some results of a critical study of the chromosomes in the various cell-generations of *Brachystola* the author briefly called attention to a possible relation between the phenomena there described and certain conclusions first drawn from observations on plant hybrids by Gregor Mendel in 1865, and recently confirmed by a number of able investigators. Further attention has already been called to the theoretical aspects of the subject in a brief communication by Professor E. B. Wilson. The present paper is devoted to a more detailed discussion of these aspects, the speculative character of which may be justified by the attempt to indicate certain lines of work calculated to test the validity of the conclusions drawn. The general conceptions here advanced were evolved purely from cytological data, before the author had knowledge of the Mendelian principles, and are now presented as the contribution of a cytologist who can make no pretensions to complete familiarity with the results of experimental studies on heredity. As will appear hereafter, they completely satisfy the conditions in typical Mendelian cases, and it seems that many of the known deviations from the Mendelian type may be explained by easily conceivable variations from the normal chromosomic processes.

It has long been admitted that we must look to the organization of the germ-cells for the ultimate determination of hereditary phenomena. Mendel fully appreciated this fact and even instituted special experiments to determine the nature of that organization. From them he drew the brilliant conclusion that, while, in the organism, maternal and paternal potentialities are present in the field of each character, *the germ-cells in respect to each character are pure.* Little was then known of the nature of cell-division, and Mendel attempted no comparisons in that direction; but to those who in recent years have revived and extended his results the probability of a relation between cell-organization and cell-division has repeatedly occurred. Bateson clearly states his impression in this regard in the following words: "It is impossible to be presented with the fact that in Mendelian cases the cross-bred produces on an average *equal* numbers of gametes of each kind, that is to say, a symmetrical result, without suspecting that this fact must correspond with some symmetrical figure of distribution of the gametes in the cell divisions by which they are produced."

Nearly a year ago it became apparent to the author that the high degree of organization in the chromosome-group of the germ-cells as shown in Brachystola could scarcely be without definite significance in inheritance, for, as shown in the paper already referred to, it had appeared that:

1. The chromosome group of the presynaptic germ-cells is made up of two equivalent chromosome-series, and that strong ground exists for the conclusion that one of these is paternal and the other maternal.

2. The process of synapsis (pseudo-reduction) consists in the union in pairs of the homologous members (*i.e.,* those that correspond in size) of the two series.

3. The first post-synaptic or maturation mitosis is equational and hence results in no chromosomic differentiation.

4. The second post-synaptic division is a reducing division, resulting in the separation of the chromosomes which have conjugated in synapsis, and their relegation to different germ-cells.

5. The chromosomes retain a morphological individuality throughout the various cell divisions.

It is well known that in the eggs of many forms the maternal and paternal chromosome groups remain distinctly independent of each other for a considerable number of cleavage-mitoses, and with this fact in mind the author was at first inclined to conclude that in the reducing divisions all the maternal chromosomes must pass to one pole and all the paternal ones to the other, and that the germ-cells are thus divided into two categories which might be described as maternal and paternal respectively. But this conception, which is identical with that recently brought forward by Cannon, was soon seen to be at variance with many well-known facts of breeding; thus:

1. If the germ-cells of hybrids are of pure descent, no amount of cross-breeding could accomplish more than the condition of a first-cross.

2. If any animal or plant has but two categories of germ-cells, there can be only four different combinations in the offspring of a single pair.

3. If either maternal or paternal chromosomes are entirely excluded from every ripe germ-cell, an individual cannot receive chromosomes (qualities) from more than one ancestor in each generation of each of the parental lines of descent, *e.g.*, could not inherit chromosomes (qualities) from both paternal or maternal grandparents.

Moved by these considerations a more careful study was made of the whole division-process, including the positions of the chromosomes in the nucleus before division, the origin and formation of the spindle, the relative positions of the chromosomes and the diverging centrosomes, and the point of attachment of the spindle fibers to the chromosomes. The results gave no evidence in favor of parental purity of the gametic chromatin as a whole. On the contrary, many points were discovered which strongly indicate that the position of the bivalent chromosomes in the equatorial plate of the reducing division is purely a matter of chance—that is, that any chromosome pair may lie with maternal or paternal chromatid indifferently toward either pole irrespective of the positions of the other pairs—and hence that a large number of different combinations of maternal and paternal chromosomes are possible in the mature germ-products of an individual. To illustrate this, we may consider a form having eight chromosomes in the somatic and presynaptic germ-cells and consequently four in the ripe germ-products.

The germ-cell series of the species in general may be designated by the letters A, B, C, D, and any cleavage nucleus may be considered as containing the chromosomes A, B, C, D from the father and $a, b, c, d,$ from the mother. Synapsis being the union of homologues would result in the formation of the bivalent chromosomes $Aa, Bb, Cc, Dd,$ which would again be resolved into their components by the reducing division. Each of the ripe germ-cells arising from the reduction divisions must receive one member from each of the synaptic pairs, but there are sixteen possible combinations of maternal and paternal chromosomes that will form a complete series, to wit: $a, B, C, D; A, b, C, D; A, B, c, D; A, B, C, d; a, b, C, D; a, B, c, D; a, B, C, d; a, b, c, d;$ and their conjugates $A, b, c, d; a, B, c, d; a, b, C, d; a, b, c, D; A, B, c, d; A, b, C, d; A, b, c, D; A, B, C, D.$ Hence instead of two kinds of gametes an organism with four chromosomes in its reduced series may give rise to 16 different kinds; and the offspring of two unrelated individuals may present 16×16 or 256 combinations, instead of the four to which it would be limited by a hypothesis of parental purity of gametes. Few organisms, more-over, have so few as 8 chromosomes, and since each additional pair doubles the number of possible combinations in the germ-products and quadruples that of the zygotes it is plain that in the ordinary form having from 24 to 36 chromosomes, the possibilities are immense. The table below shows the number of possible combinations in forms having from 2 to 36 chromosomes in the presynaptic cells.

Chromosomes Somatic Series	Reduced Series	Combinations in Gametes	Combinations in Zygotes
2	1	2	4
4	2	4	16
6	3	8	64
8	4	16	256
10	5	32	1,024
12	6	64	4,096
14	7	128	16,384
16	8	256	65,536
18	9	512	262,144
20	10	1,024	1,048,576
.	.	.	.
.	.	.	.
.	.	.	.
36	18	262,144	68,719,476,736

Thus if Bardeleben's estimate of sixteen chromosomes for man (the lowest estimate that has been made) be correct, each individual is capable of producing 256 different kinds of germ-products with reference to their chromosome combinations, and the numbers of combinations possible in the offspring of a single pair is 256×256 or 65,536; while *Toxopneustes*, with 36 chromosomes, has a possibility of 262,144 and 68,719, 476,736 different combinations in the gametes of a single individual and the zygotes of a pair respectively. It is this possibility of so great a number of combinations of maternal and paternal chromosomes in the gametes which serves to bring the chromosome-theory into final relation with the known facts of heredity; for Mendel himself followed out the actual combinations of two and three distinctive characters and found them to be inherited independently of one another and to present a great variety of combinations in the second generation.

The constant size-differences observed in the chromosomes of *Brachystola* early led me to the suspicion, which, however, a study of spermatogenesis could not confirm, that the individual chromosomes of the reduced series play different *rôles* in development. The confirmation of this surmise appeared later in the results obtained by Boveri in a study of larvae actually lacking in certain chromosomes of the normal series, which seem to leave no alternative to the conclusion that the chromosomes differ qualitatively and as individuals represent distinct potentialities. Accepting this conclusion we should be able to find an exact correspondence between the behavior in inheritance of any chromosome and that of the characters associated with it in the organism.

In regard to the characters, Mendel found that, if a hybrid produced by crossing two individuals differing in a particular character be self-fertilized, the offspring, in most cases, conform to a perfectly definite rule as regards the differential character. Representing the character as seen in one of the original parents by the letter A and that of the other by a, then all the offspring arising by self-fertilization of the hybrid are represented from the standpoint of the given character by the formula AA: 2Aa: aa—that is, one fourth receive only the character of one of the original pure-bred parents, one fourth only that of the other; while one half of the number receive the characters of both original parents and hence present the condition of the hybrid from which they sprang.

We have not heretofore possessed graphic formulae to express the combinations of chromosomes in similar breeding experiments, but it is clear from the data already given that such formulae may now be constructed. The reduced chromosome series in *Brachystola* is made up of eleven members, no two of which are exactly of the same size. These I distinguished in my previous paper by the letters A, B, C . . . K. In the unreduced series there are twenty-two elements which can be seen to make up two series like that of the mature germ-cells, and hence may be designated as A, B, C . . . K plus A, B, C . . . K. Synapsis results in the union of homologues and the production of a single series of double-elements thus: AA, BB, CC . . . KK, and the reducing division affects the separation of these pairs so that one member of each passes to each of the resulting germ-products.

There is reason to believe that the division-products of a given chromosome in *Brachystola* maintain in their respective series the same size relation as did the parent element; and this, taken together with the evidence that the various chromosomes of the series represent distinctive potentialities, make it probable that a given size-relation is characteristic of the physical basis of a definite set of characters. But each chromosome of any reduced series in the species has a homologue in any other series, and from the above consideration it should follow that these homologues cover the same field in development. If this be the case chromosome *A* from the father and its homologue, chromosome *a*, from the mother in the presynaptic cells of the offspring may be regarded as the physical bases of the antagonistic unit-characters *A* and *a* of father and mother respectively. In synapsis, copulation of the homologues gives rise to the bivalent chromosome, *Aa*, which as is indicated above would, in the reducing division, be separated into the components *A* and *a*. These would in all cases pass to different germ-products and hence in a monoecious form we should have four sorts of gametes,

A male a male
A female a female

which would yield four combinations,

A male plus A female equal AA
A male plus a female equal Aa
a male plus A female equal aA
a male plus a female equal aa

Since the second and third of these are alike the result would be expressed by the formula $AA: 2Aa: aa$ which is the same as that given for any character in a Mendelian case. *Thus the phenomena of germ-cell division and of heredity are seen to have the same essential features, viz., purity of units (chromosomes, characters) and the independent transmission of the same;* while as a corollary, it follows in each case that each of the two antagonistic units (chromosomes, characters) is contained by exactly half the gametes produced.

The observations which deal with characters have been made chiefly upon hybrids, while the cytological data are the result of a study of a pure-bred form; but the correlation of the two is justified by the observation of Cannon that the maturation mitoses of fertile hybrids are normal. This being the case it is necessary to conclude, as Cannon has already pointed out, that the course of variations in hybrids either is a result of normal maturation processes or is entirely independent of the nature of those divisions. If we conclude from the evidence already given that the double basis of hybrid characters is to be found in the pairs of homologous chromosomes of the presynaptic germ-cells, then we must also conclude that in pure-bred forms likewise, the paired arrangement of the chromosomes indicates a dual basis for each character. In a hypothetical species breeding absolutely true, therefore, all the chromosomes or subdivisions of chromosomes representing any given character would have to be exactly alike, since the combination of any two of them would produce a uniform result. As a matter of fact, however, specific characters are not found to be constant quantities but vary within certain limits; and many of the variations are known to be inheritable. Hence it seems highly probable that homologous chromatin-entities are not usually of strictly uniform constitution, but present minor variations corresponding to the various expressions of the character they represent. In other words, it is probable that specific differences and individual variations are alike traceable to a common source, which is a difference in the constitution of homologous chromatin-entities. Slight differences in homologues would mean corresponding, slight variations in the character concerned—a correspondence which is actually seen in cases of inbreeding, where variation is well known to be minimized and where obviously in the case of many of the chromosome pairs both members must be derived from the same chromosome of a recent common ancestor and hence be practically identical. . . .

[In the omitted portion of the paper, Sutton develops two principal themes. First he predicts genetical consequences to be expected in certain instances of anomalous chromosome behavior, especially that occurring in parthenogenesis. Second he shows how both Mendelian inheritance and apparent exceptions to it could be explained on the chromosome theory by assuming details of chromosome behavior as yet unobserved. Although not all of Sutton's detailed assumptions and predictions have been validated by subsequent observation and research, his central ideas proved sound.—Ed.]

We have seen reason, in the foregoing considerations, to believe that there is a definite relation between chromosomes and allelomorphs or unit characters but we have not before inquired whether an entire chromosome or only a part of one is to be regarded as the basis of a single allelomorph. The answer must unquestionably be in favor of the latter possibility, for otherwise the number of distinct characters possessed by an individual could not exceed the number of chromosomes in the germ-products; which is undoubtedly contrary to fact. We must, therefore, assume that some chromosomes at least are related to a number of different allelomorphs. If then, the chromosomes permanently retain their individuality, it follows that all the allelomorphs represented by any one chromosome must be inherited together. On the other hand, it is not necessary to assume that all must be apparent in the organism, for here the question of dominance enters and it is not yet known that dominance is a function of an entire chromosome. It is conceivable that the chromosome may be divisible into smaller entities (somewhat as Weismann assumes), which represent the allelomorphs and may be dominant or recessive independently. In this way, the same chromosome might at one time represent both dominant and recessive allelomorphs.

Such a conception infinitely increases the number of possible combinations of characters *as actually seen* in the individuals and unfortunately at the same time increases the difficulty of determining what characters are inherited together, since usually recessive chromatin entities (allelomorphs?) constantly associated in the same chromosome with usually dominant ones would evade detection for generations and then becoming dominant might appear as reversions in a very confusing manner.

In their experiments on *Matthiola*, Bateson and Saunders mention two cases of correlated qualities which may be explained by the association of their physical bases in the same chromosome. "In certain combinations there was close correlation between (*a*) green color of seed and hoariness, (*b*) brown color of seed and glabrousness. In other combinations such correlation was entirely wanting." Such results may be due to the association in the same chromosomes of the physical bases of the two characters. When close correlation was observed, both may be supposed to have dominated their homologues; when correlation was wanting, one may have been dominant and the other recessive. In the next paragraph to that quoted is the statement:

"The rule that plants with flowers either purple or claret arose from green seeds was universal." Here may be a case of constant dominance of two associated chromatin-entities.

CREATIVE EVOLUTION

BERGSON, Henri Louis (French philosopher, 1859–1941). From *L'evolution créatrice*, Paris, 1907; tr. by A. Mitchell as *Creative evolution*, New York, 1911; by permission of Henry Holt and Co., Inc., copyright owner.

Russell compares Bergson's universe to "a vast funicular railway in which life is the train that goes up, and matter the train that goes down." In this analogy our intellectualizations concerning reality amount essentially to observation of the descending motion by the ascending motion.* Convergence constitutes the principal empirical evidence for Bergson's special brand of vitalism. His system has been called an extension of the anti-intellectual revolt beginning in eighteenth-century France. It has definite predecessors in the Ionian schools.

Let us indicate at once the principle of our demonstration. We said of life that, from its origin, it is the continuation of one and the same impetus, divided into divergent lines of evolution. Something has grown, something has developed by a series of additions which have been so many creations. This very development has brought about a dissociation of tendencies which were unable to grow beyond a certain point without becoming mutually incompatible. Strictly speaking, there is nothing to prevent our imagining that the evolution of life might have taken place in one single individual by means of a series of transformations spread over thousands of ages. Or, instead of a single individual, any number might be supposed, succeeding each other in a unilinear series. In both cases evolution would have had, so to speak, one dimension only. But evolution has actually taken place through millions of individuals, on divergent lines, each ending at a crossing from which new paths radiate, and so on indefinitely. If our hypothesis is justified, if the essential causes working along these diverse roads are of psychological nature, they must keep something in common in spite of the divergence of their effects, as school-fellows long separated keep the same memories of boyhood. Roads may fork or by-ways be opened along which dissociated elements may evolve in an independent manner, but nevertheless it is in virtue of the primitive impetus of the whole that the movement of the parts continues. Something of the whole, therefore, must abide in the parts; and this common element will be evident to us in some way, perhaps by the presence of identical organs in very different organisms. Suppose, for an instant, that the mechanistic explanation is the true one: evolution must then have occurred

---

* B. Russell, *A history of western philosophy*, Simon and Schuster, pp. 791–810, New York, 1945.

through a series of accidents added to one another, each new accident being preserved by selection if it is advantageous to that sum of former advantageous accidents which the present form of the living being represents. What likelihood is there that, by two entirely different series of accidents being added together, two entirely different evolutions will arrive at similar results? The more two lines of evolution diverge, the less probability is there that accidental outer influences or accidental inner variations bring about the construction of the same apparatus upon them, especially if there was no trace of this apparatus at the moment of divergence. But such similarity of the two products would be natural, on the contrary, in a hypothesis like ours: even in the latest channel there would be something of the impulsion received at the source. *Pure mechanism, then, would be refutable, and finality, in the special sense in which we understand it, would be demonstrable in a certain aspect, if it could be proved that life may manufacture the like apparatus, by unlike means, on divergent lines of evolution; and the strength of the proof would be proportional both to the divergency between the lines of evolution thus chosen and to the complexity of the similar structures found in them.*

It will be said that resemblance of structure is due to sameness of the general conditions in which life has evolved, and that these permanent outer conditions may have imposed the same direction on the forces constructing this or that apparatus, in spite of the diversity of transient outer influences and accidental inner changes. We are not, of course, blind to the role which the concept of *adaptation* plays in the science of today. Biologists certainly do not all make the same use of it. Some think the outer conditions capable of causing change in organisms in a *direct* manner, in a definite direction, through physico-chemical alterations induced by them in the living substance; such is the hypothesis of Eimer, for example. Others, more faithful to the spirit of Darwinism, believe the influence of conditions works indirectly only, through favoring, in the struggle for life, those representatives of a species which the chance of birth has best adapted to the environment. In other words, some attribute a *positive* influence to outer conditions, and say that they actually *give rise* to variations, while the others say these conditions have only a *negative* influence and merely *eliminate* variations. But in both cases, the outer conditions are supposed to bring about a precise adjustment of the organism to its circumstances. Both parties, then, will attempt to explain mechanically, by adaptation to similar conditions, the similarities of structure which we think are the strongest argument against mechanism. So we must at once indicate in a general way, before passing to the detail, why explanations from "adaptation" seem to us insufficient. . . .

Let us consider the example on which the advocates of finality have always insisted: the structure of such an organ as the human eye. . . .

Let us place side by side the eye of a vertebrate and that of a mollusk such

as the common Pecten. We find the same essential parts in each, composed
of analogous elements. The eye of the Pecten presents a retina, a cornea, a
lens of cellular structure like our own. There is even that peculiar inversion
of retinal elements which is not met with, in general, in the retina of the invertebrates. Now, the origin of mollusks may be a debated question, but
whatever opinion we hold, all are agreed that mollusks and vertebrates
separated from their common parent-stem long before the appearance of an
eye so complex as that of the Pecten. Whence, then, the structural
analogy?—. . .

Let us assume, to begin with, the Darwinian theory of insensible variations,
and suppose the occurrence of small differences due to chance, and continually
accumulating. It must not be forgotten that all the parts of an organism are
necessarily co-ordinated. Whether the function be the effect of the organ or
its cause, it matters little; one point is certain—the organ will be of no use
and will not give selection a hold unless it functions. However the minute
structure of the retina may develop, and however complicated it may become,
such progress, instead of favoring vision, will probably hinder it if the visual
centers do not develop at the same time, as well as several parts of the visual
organ itself. If the variations are accidental, how can they ever agree to arise
in every part of the organ at the same time, in such way that the organ will
continue to perform its function? Darwin quite understood this; it is one of
the reasons why he regarded variation as insensible. For a difference which
arises accidentally at one point of the visual apparatus, if it be very slight, will
not hinder the functioning of the organ; and hence this first accidental variation can, in a sense, *wait for* complementary variations to accumulate and
raise vision to a higher degree of perfection. Granted; but while the insensible variation does not hinder the functioning of the eye, neither does it
help it, so long as the variations that are complementary do not occur. How,
in that case, can the variation be retained by natural selection? Unwittingly
one will reason as if the slight variation were a toothing stone set up by the
organism and reserved for a later construction. This hypothesis, so little
conformable to the Darwinian principle, is difficult enough to avoid even in
the case of an organ which has been developed along one single main line of
evolution, *e.g.*, the vertebrate eye. But it is absolutely forced upon us when
we observe the likeness of structure of the vertebrate eye and that of the
mollusks. How could the same small variations, incalculable in number, have
ever occurred in the same order on two independent lines of evolution, if
they were purely accidental? and how could they have been preserved by
selection and accumulated in both cases, the same in the same order, when
each of them, taken separately, was of no use?

Let us turn, then, to the hypothesis of sudden variations, and see whether
it will solve the problem. It certainly lessens the difficulty on one point, but

it makes it much worse on another. If the eye of the mollusk and that of the vertebrate have both been raised to their present form by a relatively small number of sudden leaps, I have less difficulty in understanding the resemblance of the two organs than if this resemblance were due to an incalculable number of infinitesimal resemblances acquired successively: in both cases it is chance that operates, but in the first case chance is not required to work the miracle it would have to perform in the second. Not only is the number of resemblances to be added somewhat reduced, but I can also understand better how each could be preserved and added to the others; for the elementary variation is now considerable enough to be an advantage to the living being, and so to lend itself to the play of selection. But here there arises another problem no less formidable, viz., how do all the parts of the visual apparatus, suddenly changed, remain so well co-ordinated that the eye continues to exercise its function? For the change of one part alone will make vision impossible, unless this change is absolutely infinitesimal. The parts must then all change at once, each consulting the others. I agree that a great number of unco-ordinated variations may indeed have arisen in less fortunate individuals, that natural selection may have eliminated these, and that only the combination fit to endure, capable of preserving and improving vision, has survived. Still, this combination had to be produced. And, supposing chance to have granted favor once, can we admit that it repeats the self-same favor in the course of the history of a species, so as to give rise, every time, all at once, to new complications marvelously regulated with reference to each other, and so related to former complications as to go further on in the same direction? How, especially, can we suppose that by a series of mere "accidents" these sudden variations occur, the same, in the same order—involving in each case a perfect harmony of elements more and more numerous and complex—along two independent lines of evolution?

The law of correlation will be invoked, of course; Darwin himself appealed to it. It will be alleged that a change is not localized in a single point of the organism, but has its necessary recoil on other points. The examples cited by Darwin remain classic; white cats with blue eyes are generally deaf; hairless dogs have imperfect dentition, etc.—Granted; but let us not play now on the word "correlation." A collective whole of *solidary* changes is one thing, a system of *complementary* changes—changes so co-ordinated as to keep up and even improve the functioning of an organ under more complicated conditions—is another. That an anomaly of the pilous system should be accompanied by an anomaly of dentition is quite conceivable without our having to call for a special principle of explanation; for hair and teeth are similar formations, and the same chemical change of the germ that hinders the formation of hair would probably obstruct that of teeth: it may be for the same sort of reason that white cats with blue eyes are deaf. In these different examples the "correlative" changes are only *solidary* changes (not to mention the fact that they are really *lesions,* namely, diminutions or suppressions, and not additions, which makes a great difference). But when we speak of "correlative" changes occurring suddenly in the different parts of the eye, we use the word in an entirely new sense: this time there is a whole set of changes not only simultaneous, not only bound together by community of origin, but so co-ordinated that the organ keeps on performing the same simple function, and even performs it better. That a change in the germ, which influences the formation of the retina, may affect at the same time also the formation of the cornea, the iris, the lens, the visual centers, etc., I admit, if necessary, although they are formations that differ much more from one another in their original nature than do probably hair and teeth. But that all these simultaneous changes should occur in such a way as to improve or even merely maintain vision, this is what, in the hypothesis of sudden variation, I cannot admit, unless a mysterious principle is to come in, whose duty it is to watch over the interest of the function. . . .

Still keeping to our comparison between the eye of vertebrates and that of mollusks, we may point out that the retina of the vertebrate is produced by an expansion in the rudimentary brain of the young embryo. It is a regular nervous center which has moved toward the periphery. In the mollusks, on the contrary, the retina is derived from the ectoderm directly, and not indirectly by means of the embryonic encephalon. Quite different, therefore, are the evolutionary processes which lead, in man and in the Pecten, to the development of a like retina. But, without going so far as to compare two organisms so distant from each other, we might reach the same conclusion simply by looking at certain very curious facts of regeneration in one and the same organism. If the crystalline lens of a Triton be removed, it is regenerated by the iris. Now, the original lens was built out of the ectoderm, while the iris is of mesodermic origin. What is more, in the *Salamandra maculata,* if the lens be removed and the iris left, the regeneration of the lens takes place at the upper part of the iris; but if this upper part of the iris itself be taken away, the regeneration takes place in the inner or retinal layer of the remaining region. Thus, parts differently situated, differently constituted, meant normally for different functions, are capable of performing the same duties and even of manufacturing, when necessary, the same pieces of the machine. Here we have, indeed, the same effect obtained by different combinations of causes.

Whether we will or no, we must appeal to some inner directing principle in order to account for this convergence of effects. Such convergence does not appear possible in the Darwinian, and especially the neo-Darwinian, theory of insensible accidental variations, nor in the hypothesis of sudden accidental variations, nor even in the theory that assigns definite directions to the evolution of the various organs by a kind of mechanical composition of the external with the internal forces. So we come to the only one of the present forms of evolution which remains for us to mention, viz., neo-Lamarckism.

It is well known that Lamarck attributed to the living being the power of varying by use or disuse of its organs, and also of passing on the variation so acquired to its descendants. A certain number of biologists hold a doctrine of this kind today. The variation that results in a new species is not, they believe, merely an accidental variation inherent in the germ itself, nor is it governed by a determinism *sui generis* which develops definite characters in a definite direction, apart from every consideration of utility. It springs from the very effort of the living being to adapt itself to the circumstances of its existence. The effort may indeed be only the mechanical exercise of certain organs, mechanically elicited by the pressure of external circumstances. But it may also imply consciousness and will, and it is in this sense that it appears to be understood by one of the most eminent representatives of the doctrine, the American naturalist Cope. Neo-Lamarckism is therefore, of all the later forms of evolutionism, the only one capable of admitting an internal and psychological principle of development, although it is not bound to do so. And it is also the only evolutionism that seems to us to account for the building up of identical complex organs on independent lines of development. For it is quite conceivable that the same effort to turn the same circumstances to good account might have the same result, especially if the problem put by the circumstances is such as to admit of only one solution. But the question remains, whether the term "effort" must not then be taken in a deeper sense, a sense even more psychological than any neo-Lamarckian supposes.

For mere variation of size is one thing, and a change of form is another. That an organ can be strengthened and grow by exercise, nobody will deny. But it is a long way from that to the progressive development of an eye like that of the mollusks and of the vertebrates. If this development be ascribed to the influence of light, long continued but passively received, we fall back on the theory we have just criticized. If, on the other hand, an internal activity is appealed to, then it must be something quite different from what we usually call an effort, for never has an effort been known to produce the slightest complication of an organ, and yet an enormous number of complications, all admirably co-ordinated, have been necessary to pass from the pigment-spot of the Infusorian to the eye of the vertebrate. But, even if we accept this notion of the evolutionary process in the case of animals, how can we apply it to plants? Here, variations of form do not seem to imply, nor always to lead to, functional changes; and even if the cause of the variation is of a psychological nature, we can hardly call it an effort, unless we give a very unusual extension to the meaning of the word. The truth, is, it is necessary to dig beneath the effort itself and look for a deeper cause. . . .

So we come back, by a somewhat roundabout way, to the idea we started from, that of an original impetus of life, passing from one generation of germs to the following generation of germs through the developed organisms which bridge the interval between the generations.   This impetus, sustained right along the lines of evolution among which it gets divided, is the fundamental cause of variations, at least of those that are regularly passed on, that accumulate and create new species.   In general, when species have begun to diverge from a common stock, they accentuate their divergence as they progress in their evolution.   Yet, in certain definite points, they may evolve identically; in fact, they must do so if the hypothesis of a common impetus be accepted.   . . .

For us, the whole of an organized machine may, strictly speaking, represent the whole of the organizing work (this is, however, only approximately true), yet the parts of the machine do not correspond to the parts of the work, because *the materiality of this machine does not represent a sum of means employed, but a sum of obstacles avoided:* it is a negation rather than a positive reality.   So, as we have shown in a former study, vision is a power which should attain *by right* an infinity of things inaccessible to our eyes.   But such a vision would not be continued into action; it might suit a phantom, but not a living being.   The vision of a living being is an *effective* vision, limited to objects on which the being can act: it is a vision that is canalized, and the visual apparatus simply symbolizes the work of canalizing.   Therefore the creation of the visual apparatus is no more explained by the assembling of its anatomic elements than the digging of a canal could be explained by the heaping-up of the earth which might have formed its banks.   A mechanistic theory would maintain that the earth had been brought cart-load by cart-load; finalism would add that it had not been dumped down at random, that the carters had followed a plan.   But both theories would be mistaken, for the canal had been made in another way.

With greater precision, we may compare the process by which nature constructs an eye to the simple act by which we raise the hand.   But we supposed at first that the hand met with no resistance.   Let us now imagine that, instead of moving in air, the hand has to pass through iron filings which are compressed and offer resistance to it in proportion as it goes forward.   At a certain moment the hand will have exhausted its effort, and, at this very moment, the filings will be massed and coordinated in a certain definite form, to wit, that of the hand that is stopped and of a part of the arm.   Now, suppose that the hand and arm are invisible.   Lookers-on will seek the reason of the arrangement in the filings themselves and in forces within the mass.   Some will account for the position of each filing by the action exerted upon it by the neighboring filings: these are the mechanists.   Others will prefer to think that a plan of the whole has presided over the detail of these elementary actions: they are the finalists. But the truth is that there has been merely one indivisible act, that of the hand passing through the filings: the inexhaustible detail of the movement of the grains, as well as the order of their final arrangement, expresses negatively, in a way, this undivided movement, being the unitary form of a resistance, and not a synthesis of positive elementary actions. For this reason, if the arrangement of the grains is termed an "effect" and the movement of the hand a "cause," it may indeed be said that the whole of the effect is explained by the whole of the cause, but to parts of the cause parts of the effect will in no wise correspond. In other words, neither mechanism nor finalism will here be in place, and we must resort to an explanation of a different kind. Now, in the hypothesis we propose, the relation of vision to the visual apparatus would be very nearly that of the hand to the iron filings that follow, canalize and limit its motion.

The greater the effort of the hand, the farther it will go into the filings. But at whatever point it stops, instantaneously and automatically the filings co-ordinate and find their equilibrium. So with vision and its organ. According as the undivided act constituting vision advances more or less, the materiality of the organ is made of a more or less considerable number of mutually coordinated elements, but the order is necessarily complete and perfect. It could not be partial, because, once again, the real process which gives rise to it has no parts. That is what neither mechanism nor finalism takes into account, and it is what we also fail to consider when we wonder at the marvelous structure of an instrument such as the eye. At the bottom of our wondering is always this idea, that it would have been possible for *a part only* of this co-ordination to have been realized, that the complete realization is a kind of special favor. This favor the finalists consider as dispensed to them all at once, by the final cause; the mechanists claim to obtain it little by little, by the effect of natural selection; but both see something positive in this co-ordination, and consequently something fractionable in its cause—something which admits of every possible degree of achievement. In reality, the cause, though more or less intense, cannot produce its effect except in one piece, and completely finished. According as it goes further and further in the direction of vision, it gives the simple pigmentary masses of a lower organism, or the rudimentary eye of a *Serpula*, or the slightly differentiated eye of the *Alciope*, or the marvelously perfected eye of the bird; but all these organs, unequal as is their complexity, necessarily present an equal co-ordination. For this reason, no matter how distant two animal species may be from each other, if the progress toward vision has gone equally far in both, there is the same visual organ in each case, for the form of the organ only expresses the degree in which the exercise of the function has been obtained.

But, in speaking of a progress toward vision, are we not coming back to the old notion of finality? It would be so, undoubtedly, if this progress required the conscious or unconscious idea of an end to be attained. But it is really effected in virtue of the original impetus of life; it is implied in this movement itself, and that is just why it is found in independent lines of evolution. If now we are asked why and how it is implied therein, we reply that life is, more than anything else, a tendency to act on inert matter. The direction of this action is not predetermined: hence the unforeseeable variety of forms which life, in evolving, sows along its path. But this action always presents, to some extent, the character of contingency; it implies at least a rudiment of choice. Now a choice involves the anticipatory idea of several possible actions. Possibilities of action must therefore be marked out for the living being before the action itself. Visual perception is nothing else: the visible outlines of bodies are the design of our eventual action on them. Vision will be found, therefore, in different degrees in the most diverse animals, and it will appear in the same complexity of structure wherever it has reached the same degree of intensity.

VIII. ZOOGEOGRAPHY

EFFECTS OF GEOGRAPHICAL ISOLATION

DARWIN, Charles Robert (British naturalist, 1809–1882). From *Journal of researches with the natural history and geology of the countries visited during the voyage around the world of H.M.S. Beagle under the command of Captain FitzRoy, R.N.*, New York, 1846.

It was directly upon termination of his formal schooling (for medicine at Edinburgh, for the church at Cambridge) that Darwin requested and received assignment as naturalist for the voyage of the *Beagle* (1832–1836). In the notebook for 1837, we read: "In July opened first notebook on Transmutation of Species. Had been greatly struck from about the month of the previous March on the character of South American fossils and species in the Galapagos Archipelago (see below). These facts (especially the latter) origin of all my views." Judging from the following account, Darwin was struck with the *fact* of selection before obtaining a satisfactory notion of its *mechanism*, a problem whose solution was facilitated by his reading (1838) of Malthus (*q.v.*). The zoological results of the voyage of the *Beagle* began to be published at government expense in 1840.*

I have not as yet noted by far the most remarkable feature in the natural history of this archipelago; it is, that the different islands, to a considerable extent, are inhabited by a different set of beings. My attention was first called to this fact by the vice-governor, Mr. Lawson, declaring that the tortoises differed from the different islands, and that he could with certainty tell from which island any one was brought. I did not for some time pay sufficient attention to this statement, and I had already partially mingled together the collections from two of the islands. I never dreamed that islands about fifty or sixty miles apart, and most of them in sight of each other, formed of precisely the same rocks, placed under a quite similar climate, rising to a nearly equal height, would have been differently tenanted; but we shall soon see that this is the case. It is the fate of most voyagers, no sooner to discover what is most interesting in any locality, than they are hurried from it; but I ought, perhaps, to be thankful that I obtained sufficient materials to establish this most remarkable fact in the distribution of organic beings.

The inhabitants, as I have said, state that they can distinguish the tortoises from the different islands, and that they differ not only in size, but in other characters. Captain Porter has described those from Charles and from the nearest island to it, namely, Hood Island, as having their shells in front thick and turned up like a Spanish saddle, whilst the tortoises from James Island are rounder, blacker, and have a better taste when cooked. M. Bibron, more-

* As *The zoology of the voyage of H.M.S. Beagle, etc.*, London.

over, informs me that he has seen what he considers two distinct species of tortoise from the Galapagos, but he does not know from which islands. The specimens that I brought from three islands were young ones, and probably owing to this cause neither Mr. Gray nor myself could find in them any specific differences. I have remarked that the marine Amblyrhynchus was larger at Albemarle Island than elsewhere; and M. Bibron informs me that he has seen two distinct aquatic species of this genus; so that the different islands probably have their representative species or races of the Amblyrhynchus, as well as of the tortoise. My attention was first thoroughly aroused by comparing together the numerous specimens, shot by myself and several other parties on board, of the mocking-thrushes, when, to my astonishment, I discovered that all those from Charles Island belonged to one species (Mimus trifasciatus); all from Albemarle Island to M. parvulus; and all from James and Chatham Islands (between which two other islands are situated as connecting links) belonged to M. melanotis. These two latter species are closely allied, and would, by some ornithologists, be considered as only well-marked races or varieties; but the Mimus trifasciatus is very distinct. Unfortunately, most of the specimens of the finch tribe were mingled together; but I have strong reasons to suspect that some of the species of the sub-group Geospiza are confined to separate islands. If the different islands have their representatives of Geospiza, it may help to explain the singularly large number of the species of this sub-group in this one small archipelago, and as a probable consequence of their numbers, the perfectly graduated series in the size of their beaks. Two species of the sub-group Cactornis, and two of Camarhynchus, were procured in the archipelago; and of the numerous specimens of these two sub-groups shot by four collectors at James Island, all were found to belong to one species of each; whereas the numerous specimens shot either on Chatham or Charles Island (for the two sets were mingled together) all belonged to the two other species: hence we may feel almost sure that these islands possess their representative species of these two sub-groups. In landshells this law of distribution does not appear to hold good. In my very small collection of insects, Mr. Waterhouse remarks that of those which were ticketed with their locality, not one was common to any two of the islands.

If we now turn to the Flora, we shall find the aboriginal plants of the different islands wonderfully different. I give all the following results on the high authority of my friend, Dr. J. Hooker. I may premise that I indiscriminately collected everything in flower on the different islands, and fortunately kept my collections separate. Too much confidence, however, must not be placed in the proportional results, as the small collections brought home by some other naturalists, though in some respects confirming the results, plainly show that much remains to be done in the botany of this group: the Leguminosae, moreover, have as yet been only approximately worked out:

Name of Island	Total No. of Species	No. of Species found in other parts of the world	No. of Species confined to the Galapagos Archipelago	No. confined to the one Island	No. of species confined to the Galapagos Archipelago but found on more than the one Island
James Island	71	33	38	30	8
Albemarle Island	46	18	26	22	4
Chatham Island	32	16	16	12	4
Charles Island	68	39 (or 29, if the probably im- ported plants be subtracted)	29	21	8

Hence we have the truly wonderful fact, that in James Island, of the thirty-eight Galapageian plants, or those found in no other part of the world, thirty are exclusively confined to this one island; and in Albemarle Island, of the twenty-six aboriginal Galapageian plants, twenty-two are confined to this one island, that is, only four are at present known to grow in the other islands of the archipelago; and so on, as shown in the above table, with the plants from Chatham and Charles Islands. This fact will, perhaps, be rendered even more striking, by giving a few illustrations: thus, Scalesia, a remarkably arborescent genus of the Compositae, is confined to the archipelago. It has six species: one from Chatham, one from Albemarle, one from Charles Island, two from James Island, and the sixth from one of the three latter islands, but it is not known from which: not one of these six species grows on any two islands. Again, Euphorbia, a mundane or widely-distributed genus, has here eight species, of which seven are confined to the archipelago, and not one found on any two islands; Acalypha and Borreria, both mundane genera, have respectively six and seven species, none of which have the same species on two islands, with the exception of Borreria, which does occur on two islands. The species of the Compositae are particularly local, and Dr. Hooker has furnished me with several other most striking illustrations of the difference of the species on the different islands. He remarks that this law of distribution holds good both with those genera confined to the archipelago and those distributed in other quarters of the world: in like manner, we have seen that the different islands have their proper species of the mundane genus of tortoise, and of the widely-distributed American genus of the mocking-thrush, as well as of two of the Galapageian subgroups of finches, and almost certainly of the Galapageian genus Amblyrhynchus.

The distribution of the tenants of this archipelago would not be nearly so wonderful, if, for instance, one island had a mocking-thrush, and a second island some other quite distinct genus; if one island had its genus of lizard, and a second island another distinct genus, or none whatever; or if the different islands were inhabited, not by representative species of the same genera of plants, but by totally different genera, as does to a certain extent hold good; for, to give one instance, a large berry-bearing tree at James Island has no representative species in Charles Island. But it is the circumstance that several of the islands possess their own species of the tortoise, mocking-thrust, finches, and numerous plants, these species having the same general habits, occupying analogous situations, and obviously filling the same place in the natural economy of this archipelago, that strikes me with wonder. It may be suspected that some of these representative species, at least in the case of the tortoise and of some of the birds, may hereafter prove to be only well-marked races; but this would be of equally great interest to the philosophical naturalist. I have said that most of the islands are in sight of each other: I may specify that Charles Island is fifty miles from the nearest part of Chatham Island, and thirty-three miles from the nearest part of Albemarle Island. Chatham Island is sixty miles from the nearest part of James Island, but there are two intermediate islands between them which were not visited by me. James Island is only ten miles from the nearest part of Albemarle Island, but the two points where the collections were made are thirty-two miles apart. I must repeat, that neither the nature of the soil, nor height of the land, nor the climate, nor the general character of the associated beings, and therefore their action one on another, can differ much in the different islands. If there be any sensible difference in their climates, it must be between the windward group (namely, Charles and Chatham Islands) and that to leeward; but there seems to be no corresponding difference in the productions of these two halves of the archipelago.

The only light which I can throw on this remarkable difference in the inhabitants of the different islands is, that very strong currents of the sea, running in a westerly and W.N.W. direction, must separate, as far as transportal by the sea is concerned, the southern islands from the northern ones; and between these northern islands a strong N.W. current was observed, which must effectually separate James and Albemarle Islands. As the archipelago is free to a most remarkable degree from gales of wind, neither the birds, insects, nor lighter seeds would be blown from island to island. And lastly, the profound depth of the ocean between the islands, and their apparently recent (in a geological sense) volcanic origin, render it highly unlikely that they were ever united; and this, probably, is a far more important consideration than any other, with respect to the geographical distribution of their inhabitants. Reviewing the facts here given, one is astonished at the amount of creative force, if such an expression may be used, displayed on these small, barren, and rocky islands, and still more so at its diverse yet analogous action on points so near each other. I have said that the Galapagos Archipelago might be called a satellite attached to America, but it should rather be called a group of satellites, physically similar, organically distinct, yet intimately related to each other, and all related in a marked, though much less, degree to the great American continent.

THE DEPTHS OF THE SEA

WYVILLE THOMSON, (Thompson) Charles (Scotch oceanographer, 1830–1882). From *The depths of the sea*, London, 1873.

Marine zoology was an integral part of ancient science, reappearing in the encyclopedic writings of the Renaissance and in the travelogues of the great explorers. Dredging for scientific purposes was systematically performed by O. F. Müller, 1779.* Although considerable evidence to the contrary was already available,† in 1859 E. Forbes postulated a bathyrhythmic limit for life around 350 fathoms.‡ This was effectively controverted in Carpenter's preliminary report on the results of the voyage of the "Lightning"§ and in Wyville Thomson's later and better-known *Depths of the Sea* from which the following excerpt has been taken. With the voyage of the "Lightning," serious deep oceanography began.

The sea covers nearly three-fourths of the surface of the earth, and, until within the last few years, very little was known with anything like certainty about its depths, whether in their physical or their biological relations. The popular notion was, that after arriving at a certain depth the conditions became so peculiar, so entirely different from those of any portion of the earth to which we have access, as to preclude any other idea than that of a waste of utter darkness, subjected to such stupendous pressure as to make life of any kind impossible, and to throw insuperable difficulties in the way of any attempt at investigation. Even men of science seemed to share this idea, for they gave little heed to the apparently well-authenticated instances of animals, comparatively high in the scale of life, having been brought up on sounding lines from great depths, and welcomed any suggestion of the animals having got entangled when swimming on the surface, or carelessness on the part of the observers. And this was strange, for every other question in Physical Geography had been investigated by scientific men with consummate patience and energy. Every gap in the noble little army of martyrs striving to extend the

---

* See his *Zoologia danica; seu, animalium Daniae et Norwegiae rariorum ac minus notorum descriptiones et historiae*, Hauniae et Lipsiae, 1779.

† Summarized in W. B. Carpenter, *Preliminary report of dredging operations in the seas to the north of the British Islands; carried on in Her Majesty's steam-vessel "Lightning,"* etc., in *Proceedings of the Royal Society of London*, vol. 17, p. 168, 1868.

‡ E. Forbes, *The natural history of the European seas*, London, 1859.

§ See second footnote above.

boundaries of knowledge in the wilds of Australia, on the Zambesi, or towards the North or South Pole, was struggled for by earnest volunteers, and still the great ocean slumbering beneath the moon covered a region apparently as inaccessible to man as the 'mare serenitatis.'

A few years ago the bottom of the sea was required for the purpose of telegraphic communication, and practical men mapped out the bed of the North Atlantic, and devised ingenious methods of ascertaining the nature of the material covering the bottom. They laid a telegraphic cable across it, and the cable got broken and they went back to the spot and fished up the end of it easily, from a depth of nearly two miles.

It had long been a question with naturalists whether it might not be possible to dredge the bottom of the sea in the ordinary way, and to send down water-bottles and registering instruments to settle finally the question of 'zero of animal life,' and to determine with precision the composition and temperature of sea-water at great depths. An investigation of this kind is beyond the ordinary limits of private enterprise. It requires more power and sea skill than naturalists can usually command. When, however, in the year 1868, at the instance of my colleague Dr. Carpenter and myself, with the effective support of the present Hydrographer to the Navy, who is deeply interested in the scientific aspects of his profession, we had placed at our disposal by the Admiralty sufficient power and skill to make the experiment, we found that we could work, not with so much ease, but with as much certainty, at a depth of 600 fathoms as at 100; and in 1869 we carried the operations down to 2,435 fathoms, 14,610 feet, nearly three statute miles, with perfect success.

Dredging in such deep water was doubtless very trying. Each haul occupied seven or eight hours; and during the whole of that time it demanded and received the most anxious care on the part of our commander, who stood with his hand on the pulse of the accumulator ready at any moment, by a turn of the paddles, to ease any undue strain. The men, stimulated and encouraged by the cordial interest taken by their officers in our operations, worked willingly and well; but the labour of taking upwards of three miles of rope coming up with a heavy strain, from the surging drum of the engine, was very severe. The rope itself, 'hawser-laid,' of the best Italian hemp, $2\frac{1}{2}$ inches in circumference, with a breaking strain of $2\frac{1}{4}$ tons, looked frayed out and worn, as if it could not have been trusted to stand this extraordinary ordeal much longer.

Still the thing is possible, and it must be done again and again, as the years pass on, by naturalists of all nations, working with improving machinery, and with ever-increasing knowledge. For the bed of the deep sea, the 140,000,-000 of square miles which we have now added to the legitimate field of Natural History research, is not a barren waste. It is inhabited by a fauna more rich and varied on account of the enormous extent of the area, and with the organisms in many cases apparently even more elaborately and delicately formed, and more exquisitely beautiful in their soft shades of colouring and in the rainbow-tints of their wonderful phosphorescence, than the fauna of the well-known belt of shallow water teeming with innumerable invertebrate forms which fringes the land. And the forms of these hitherto unknown living beings, and their mode of life, and their relations to other organisms whether living or extinct, and the phenomena and laws of their geographical distribution, must be worked out.

The late Professor Edward Forbes appears to have been the first who undertook the systematic study of Marine Zoology with special reference to the distribution of marine animals in space and in time. . . .

Although we must now greatly modify our views with regard to the extent and fauna of the zone of deep-sea corals, and give up all idea of a zero of animal life, still we must regard Forbes' investigation into the bathymetrical distribution of animals as marking a great advance on previous knowledge. His experience was much wider than that of any other naturalist of his time; the practical difficulties in the way of testing his conclusions were great, and they were accepted by naturalists generally without question.

The history of discovery bearing upon the extent and distribution of the deep-sea fauna will be discussed in a future chapter. It will suffice at present to mention in order the few data which gradually prepared the minds of naturalists to distrust the hypothesis of a zero of animal life at a limited depth, and led to the recent special investigations. In the year 1819 Sir John Ross published the official account of his voyage of discovery during the year 1818 in Baffin's Bay. At page 178 he says, "In the meantime I was employed on board in sounding and in trying the current, and the temperature of the water. It being perfectly calm and smooth, I had an excellent opportunity of detecting these important objects. Soundings were obtained correctly in 1,000 fathoms, consisting of soft mud, in which there were worms, and, entangled on the sounding line, at the depth of 800 fathoms, was found a beautiful *Caput Medusae*. These were carefully preserved, and will be found described in the appendix." This was in lat. 73° 37′ N., long. 77° 25′ W., on the 1st of September, 1818, and it is, so far as I am aware, the first recorded instance of living animals having been brought up from any depth approaching 1,000 fathoms. General Sir Edward Sabine, who was a member of Sir John Ross's expedition, has kindly furnished Dr. Carpenter with some more ample particulars of this occurrence:—" 'The ship sounded in 1,000 fathoms, mud, between one and two miles off shore (lat. 73° 37′ N., long. 77° 25′ W.); a magnificent Asterias (*Caput Medusae*) was entangled by the line, and brought up with very little damage. The mud was soft and greenish, and contained specimens of *Lumbricus tubicola*.' So far my written journal; but I can add, from a very distinct recollection, that the heavy deep-sea weight had sunk, drawing the line with it, several feet into the soft greenish mud, which still adhered to the line when brought to the surface of the water.   The star-fish had been entangled in the line so little above the mud that fragments of its arms, which had been broken off in the ascent of the line, were picked up from amongst the mud."

Sir James Clark Ross, R.N., dredging in 270 fathoms, lat. 73° 3′ S., long. 176° 6′ E., reports: "*Corallines, Flustrae,* and a variety of invertebrate animals, came up in the net, showing an abundance and great variety of animal life.   Amongst these I detected two species of *Pycnogonum; Idotea baffini,* hitherto considered peculiar to the Arctic seas; a *Chiton,* seven or eight bivalves and univalves, an unknown species of *Gammarus,* and two kinds of *Serpula* adhering to the pebbles and shells . . . It was interesting amongst these creatures to recognize several that I had been in the habit of taking in equally high northern latitudes; and although, contrary to the general belief of naturalists, I have no doubt that, from however great a depth we may be enabled to bring up the mud and stones of the bed of the ocean, we shall find them teeming with animal life; the extreme pressure at the greatest depth does not appear to affect these creatures; hitherto we have not been able to determine this point beyond a thousand fathoms, but from that depth several shell-fish have been brought up with the mud."

On the 28th of June 1845, Mr. Henry Goodsir, who was a member of Sir John Franklin's ill-fated expedition, obtained in Davis' Strait from a depth of 300 fathoms, "a capital haul,—mollusca, crustacea, asterida, spatangi, corallines, &c."   The bottom was composed of fine green mud like that mentioned by Sir Edward Sabine.

About the year 1854 Passed-midshipman Brooke, U.S.N., invented his ingenious sounding instrument for bringing up samples from the bottom.   It only brought up a small quantity in a quill.   These trophies from any depth over 1,000 fathoms were eagerly sought for by naturalists and submitted to searching microscopic examination; and the result was very surprising.   All over the Atlantic basin the sediment brought up was nearly uniform in character, and consisted almost entirely of the calcareous shells, whole or in fragments, of one species of foraminifer, *Globigerina bulloides.*   Mixed with these were the shells of some other foraminifera, and particularly a little perforated sphere, *Orbulina universa,* which in some localities entirely replaces *Globigerina;* with a few shields of diatoms, and spines and trellised skeletons of Radiolaria.   Some soundings from the Pacific were of the same character, so that it seemed probable that this gradual deposition of a fine uniform organic sediment was almost universal.

Then the question arose whether the animals which secreted these shells lived at the bottom, or whether they floated in myriads on the surface and in the upper zones of the sea, their empty shells falling after death through the water in an incessant shower.   Specimens of the soundings were sent to the eminent miscroscopists Professor Ehrenberg of Berlin and the late Professor Baily of West Point.   On the moot question these two naturalists gave opposite opinions.   Ehrenberg contended that the weight of evidence was in favour of their having lived at the bottom, while Baily thought it was not probable that the animals live at the depths where the shells are found, but that they inhabit the water near the surface, and when they die their shells settle to the bottom.   . . .

Dr. Wallich's is the only book which discusses fully and systematically the various questions bearing upon the biological relations of the sea-bed, and his conclusions are in the main correct.

In the autumn of the year 1860 Mr. Fleeming Jenkin, C.E., now Professor of Engineering in the University of Edinburgh, was employed by the Mediterranean Telegraph Company to repair their cable between Sardinia and Bona on the coast of Africa, and on January 15, 1861, he gave an interesting account of his proceedings at a meeting of the Institution of Civil Engineers.

This cable was laid in the year 1857.   In 1858 it became necessary to repair it, and a length of about 30 miles was picked up and successfully replaced. In the summer of 1860 the cable completely failed.   On taking it up in comparatively shallow water on the African shore, the cable was found covered with marine animals, greatly corroded, and injured apparently by the trawling operations in an extensive coral fishery through which it unfortunately passed.   It was broken through in 70 fathoms water a few miles from Bona. The sea-end was however recovered, and it was found that the cable which thence traversed a wide valley nearly 2,000 fathoms in maximum depth, was perfect to within about 40 miles of Sardinia.   It was then picked up from the Sardinian end, and the first 39 miles were as sound as when it was first laid down.   At this distance from the shore there was a change in the nature of the bottom, evidenced by the different colour of the mud, and the wires were much corroded.   Shortly afterwards the cable gave way in a depth of 1,200 fathoms, at a distance of one mile from the spot where the electrical tests showed that the cable had been previously broken.

With these 40 miles of cable much coral and many marine animals were brought up, but it did not appear that their presence had injured the cable, for they were attached to the sound as well as to the corroded portion.   On his return, Mr. Fleeming Jenkin sent specimens of the animals which he had himself taken from the cable, noting the respective depths, to Professor Allman, F.R.S. for determination.   Dr. Allman gives a list of fifteen animal forms, including the ova of a cephalopod, found at depths of from 70 to 1,200 fathoms.   On other portions of the cable species of *Grantia, Plumularia, Gorgonia, Caryophyllia, Alcyonium, Cellepora, Retepora, Eschara, Sali-*

cornaria, Ascidia, Lima, and *Serpula.* I observe from Professor Fleeming Jenkin's private journal, which he has kindly placed in my hands for reference, that an example of *Caryophyllia,* a true coral, was found naturally attached to the cable at the point where it gave way; that is to say, at the bottom in 1,200 fathoms water.

Some portions of this cable subsequently came into the custody of M. Mangon, Professor at the Ecole des Ponts et Chaussées in Paris, and were examined by M. Alphonse Milne-Edwards, who read a paper upon the organisms attached to them, at the Academy of Sciences, on the 15th of July, 1861. After some introductory remarks which show that he is thoroughly aware of the value of this observation as a final solution of the vexed question of the existence of animal life at depths in the sea greatly beyond the supposed 'zero' of Edward Forbes, M. Milne-Edwards gives a list of the animals which he found on the cable from the depth of 1,100 fathoms. The list includes *Murex lamellosus,* CRISTOFORI and JAN, and *Craspedotus limbatus,* PHILIPPI, two univalve shells allied to the whelk; *Ostrea cochlear,* POLI, a small oyster common below 40 fathoms throughout the Mediterranean; *Pecten testae,* BIVONA, a rare little clam; *Caryophyllia borealis,* FLEMING, or a nearly allied species, one of the true corals; and an undescribed coral referred to a new genus and species under the name of *Thalassiotrochus telegraphicus,* A. MILNE-EDWARDS.

It is right, however, to state that Prof. Fleeming Jenkin's notes refer to only one or two species, and especially to *Caryophyllia borealis,* as attached to the cable at a depth of upwards of 1,000 fathoms. From this depth he took examples of *Caryophyllia* with his own hands, but he suspects that specimens from the shallower water may have got mixed with those from the deeper in the series in the possession of M. Mangon, and that therefore M. Milne-Edwards' list is not entirely trustworthy.

Up till this time all observations with reference to the existence of living animals at extreme depths had been liable to error, or at all events to doubt, from two sources. The appliances and methods of deep-sea sounding were imperfect, and there was always a possibility, from the action of deep currents upon the sounding line or from other causes, of a greater depth being indicated than really existed; and again, although there was a strong probability, there was no absolute certainty that the animals adhering to the line or entangled on the sounding instrument had actually come up from the bottom. They might have been caught on the way.

Before laying a submarine telegraphic cable its course is carefully surveyed, and no margin of doubt is left as to the real depth. Fishing the cable up is a delicate and difficult operation, and during its progress the depth is checked again and again. The cable lies on the ground throughout its whole length. The animal forms upon which our conclusions are based are not sticking loosely to the cable, under circumstances which might be accounted for by their having been entangled upon it during its passage through the water, but they are moulded upon its outer surface or cemented to it by calcareous or horny excretions, and some of them, such as the corals and bryozoa, from what we know of their history and mode of life, must have become attached to it as minute germs, and have grown to maturity in the position in which they were found. I must therefore regard this observation of Mr. Fleeming Jenkin as having afforded the first absolute proof of the existence of highly-organized animals living at depths of upwards of 1,000 fathoms.

During the several cruises of H.M. ships 'Lightning' and 'Porcupine' in the years 1868, 1869, and 1870, fifty-seven hauls of the dredge were taken in the Atlantic at depths beyond 500 fathoms, and sixteen at depths beyond 1,000 fathoms, and in all cases life was abundant. In 1869 we took two casts in depths greater than 2,000 fathoms. In both of these life was abundant; and with the deepest cast, 2,435 fathoms, off the mouth of the Bay of Biscay, we took living, well-marked and characteristic examples of all of the five invertebrate sub-kingdoms. And thus the question of the existence of abundant animal life at the bottom of the sea has been finally settled and for all depths, for there is no reason to suppose that the depth anywhere exceeds between three and four thousand fathoms; and if there be nothing in the conditions of a depth of 2,500 fathoms to prevent the full development of a varied fauna, it is impossible to suppose that even an additional thousand fathoms would make any great difference.

The conditions which might be expected principally to affect animal life at great depths of the sea are pressure, temperature, and the absence of light which apparently involves the absence of vegetable food.

After passing a zone surrounding the land, which is everywhere narrow compared with the extent of the ocean, through which the bottom more or less abruptly shelves downwards and the water deepens; speaking very generally, the average depth of the sea is 2,000 fathoms, or about two miles; as far below the surface as the average height of the Swiss Alps. In some places the depth seems to be considerably greater, possibly here and there nearly double that amount; but these abysses are certainly very local, and their existence is even uncertain, and a vast portion of the area does not reach a depth of 1,500 fathoms.

The enormous pressure at these great depths seemed at first sight alone sufficient to put any idea of life out of the question. There was a curious popular notion, in which I well remember sharing when a boy, that, in going down, the sea-water became gradually under the pressure heavier and heavier, and that all the loose things in the sea floated at different levels, according to their specific weight: skeletons of men, anchors and shot and cannon, and last of all the broad gold pieces wrecked in the loss of many a galleon on the

Spanish Main; the whole forming a kind of 'false bottom' to the ocean, beneath which there lay all the depth of clear still water, which was heavier than molten gold.

The conditions of pressure are certainly very extraordinary. At 2,000 fathoms a man would bear upon his body a weight equal to twenty locomotive engines, each with a long goods train loaded with pig iron. We are apt to forget, however, that water is almost incompressible, and that therefore the density of sea-water at a depth of 2,000 fathoms is scarcely appreciably increased. At the depth of a mile, under a pressure of about 159 atmospheres, sea-water, according to the formula given by Jamin, is compressed by the $\frac{1}{144}$ of its volume; and at twenty miles, supposing the law of the compressibility to continue the same, by only $\frac{1}{7}$ of its volume—that is to say, the volume at that depth would be   of the volume of the same weight of water at the surface. Any free air suspended in the water, or contained in any compressible tissue of an animal at 2,000 fathoms, would be reduced to a mere fraction of its bulk, but an organism supported through all its tissues on all sides, within and without, by incompressible fluids at the same pressure, would not necessarily be incommoded by it. We sometimes find when we get up in the morning, by a rise of an inch in the barometer, that nearly half a ton has been quietly piled upon us during the night, but we experience no inconvenience, rather a feeling of exhilaration and buoyancy, since it requires a little less exertion to move our bodies in the denser medium. We are already familiar, chiefly through the researches of the late Professor Sars, with a long list of animals of all the invertebrate groups living at a depth of 300 to 400 fathoms, and consequently subject to a pressure of 1,120 lbs. on the square inch; and off the coast of Portugal there is a great fishery of sharks (*Centroscymnus coelolepis*, Boc. and Cap.), carried on beyond that depth.

If an animal so high in the scale of organization as a shark can bear without inconvenience the pressure of half a ton on the square inch, it is a sufficient proof that the pressure is applied under circumstances which prevent its affecting it to its prejudice, and there seems to be no reason why it should not tolerate equally well a pressure of one or two tons. At all events it is a fact that the animals of all the invertebrate classes which abound at a depth of 2,000 fathoms do bear that extreme pressure, and that they do not seem to be affected by it in any way. We dredged at 2,435 fathoms *Scrobicularia nitida*, MÜLLER, a species which is abundant in six fathoms and at all intermediate depths, and at 2,090 fathoms a large *Fusus*, with species of many genera which are familiar at moderate depths. Although highly organized animals may live when permanently subjected to these high pressures, it is by no means certain that they could survive the change of condition involved in the pressure being suddenly removed. Most of the mollusca and annelids brought up in the dredge from beyond 1,000 fathoms were either dead or in a very sluggish state. Some of the star-fishes moved for some time feebly, and the spines and pedicellariae moved on the shells of the urchins, but all the animals had evidently received from some cause their death-shock. Dr. Perceval Wright mentions that all the sharks brought up by the long lines from 500 fathoms in Setubal Bay are dead when they reach the surface.

DELINEATION OF ZOOGEOGRAPHICAL AREAS

WALLACE, Alfred Russel (British naturalist, 1823–1913). From *The geographical distribution of animals with a study of the relations of living and extinct faunas as elucidating the past changes of the earth's surface*, London, 1876.

The interest of Darwin and Wallace in geographical distribution exerted a powerful influence upon the development of their ideas concerning natural selection. After the joint publication of their descent hypothesis in 1858–1859, zoogeographical investigation continued to supply evidence for its validity. Wallace's *Geographical Distribution*, his definitive work on the subject, may be said to constitute the coming of age of zoogeography as a science. The work is monumental in proportions and the zones which it established (see below) remained remarkably stable in the subsequent development of research in this field.

Principles on which Zoological Regions should be formed. It will be evident in the first place that nothing like a perfect zoological division of the earth is possible. The causes that have led to the present distribution of animal life are so varied, their action and reaction have been so complex, that anomalies and irregularities are sure to exist which will mar the symmetry of any rigid system. On two main points every system yet proposed, or that probably can be proposed, is open to objection; they are,—1stly, that the several regions are not of equal rank;—2ndly, that they are not equally applicable to all classes of animals. As to the first objection, it will be found impossible to form any three or more regions, each of which differs from the rest in an equal degree or in the same manner. One will surpass all others in the possession of peculiar families; another will have many characteristic genera; while a third will be mainly distinguished by negative characters. There will also be found many intermediate districts, which possess some of the characteristics of two well-marked regions, with a few special features of their own, or perhaps with none; and it will be a difficult question to decide in all cases which region should possess this doubtful territory, or whether it should be formed into a primary region itself. Again, two regions which have now well-marked points of difference, may be shown to have been much more alike at a comparatively recent geological epoch; and this, it may be said, proves their fundamental unity and that they ought to form but one primary region. To obviate some of these difficulties a binary or dichotomous division is sometimes proposed; that portion of the earth which differs most from the rest being cut off as a region equal in rank to all that remains, which is subjected again and again to the same process.

To decide these various points it seems advisable that convenience, intelligibility, and custom, should largely guide us. The first essential is, a broadly marked and easily remembered set of regions; which correspond, as nearly as truth to nature will allow, with the distribution of the most important groups of animals. What these groups are we shall presently explain. In determining the number, extent, and boundaries of these regions, we must be guided by a variety of indications, since the application of fixed rules is impossible. They should evidently be of a moderate number, corresponding as far as practicable with the great natural divisions of the globe marked out by nature, and which have always been recognized by geographers. There should be some approximation to equality of size, since there is reason to believe that a tolerably extensive area has been an essential condition for the development of most animal forms; and it is found that, other things being equal, the numbers, variety and importance of the forms of animal and vegetable life, do bear some approximate relation to extent of area. Although the possession of peculiar families or genera is the main character of a primary zoological region, yet the negative character of the absence of certain families or genera is of equal importance, *when this absence does not manifestly depend on unsuitability to the support of the group,* and especially *when there is now no physical barrier preventing their entrance.* This will become evident when we consider that the importance of the possession of a group by one region depends on its absence from the adjoining regions; and if there is now no barrier to its entrance, we may be sure that there has once been one; and that the possession of the area by a distinct and well balanced set of organisms, which must have been slowly developed and adjusted, is the living barrier that now keeps out intruders. . . .

I now proceed to characterize briefly the six regions adopted in the present work, together with the sub-regions into which they may be most conveniently and naturally divided, as shown in our general map.

Palaearctic Region. This very extensive region comprises all temperate Europe and Asia, from Iceland to Behring's Straits and from the Azores to Japan. Its southern boundary is somewhat indefinite, but it seems advisable to comprise in it all the extra-tropical part of the Sahara and Arabia, and all Persia, Cabul, and Beloochistan to the Indus. It comes down to a little below the upper limit of forests in the Himalayas, and includes the larger northern half of China, not quite so far down the coast as Amoy. It has been said that this region differs from the Oriental by negative characters only; a host of tropical families and genera being absent, while there is little or nothing but peculiar species to characterize it absolutely. This however is not true. The Palaearctic region is well characterized by possessing 3 families of vertebrata peculiar to it, as well as 35 peculiar genera of mammalia, and 57 of birds, constituting about one-third of the total number it possesses. These are amply sufficient to characterize a region positively; but we must also consider the absence of many important groups of the Oriental, Ethiopian, and Nearctic regions; and we shall then find, that taking positive and negative characters together, and making some allowance for the necessary poverty of a temperate as compared with tropical regions, the Palaearctic is almost as strongly marked and well defined as any other.

Sub-divisions of the Palaearctic Region. These are by no means so clearly indicated as in some of the other regions, and they are adopted more for convenience than because they are very natural or strongly marked.

The first, or European sub-region, comprises Central and Northern Europe as far South as the Pyrenees, the Maritime and Dinaric Alps, the Balkan mountains, the Black Sea, and the Caucasus. On the east the Caspian sea and the Ural mountains seem the most obvious limit; but it is doubtful if they form the actual boundary, which is perhaps better marked by the valley of the Irtish, where a pre-glacial sea almost certainly connected the Aral and Caspian seas with the Arctic ocean, and formed an effective barrier which must still, to some extent, influence the distribution of animals.

The next, or Mediterranean sub-region, comprises South Europe, North Africa with the extra-tropical portion of the Sahara, and Egypt to about the first or second cataracts; and eastward through Asia Minor, Persia, and Cabul, to the deserts of the Indus.

The third, or Siberian sub-region, consists of all north and central Asia north of Herat, as far as the eastern limits of the great desert plateau of Mongolia, and southward to about the upper limit of trees on the Himalayas.

The fourth, or Manchurian sub-region, consists of Japan and North China with the lower valley of the Amoor; and it should probably be extended westward in a narrow strip along the Himalayas, embracing about 1,000 or 2,000 feet of vertical distance below the upper limit of trees, till it meets an eastern extension of the Mediterranean sub-region a little beyond Simla. These extensions are necessary to avoid passing from the Oriental region, which is essentially tropical, directly to the Siberian sub-region, which has an extreme northern character; whereas the Mediterranean and Manchurian sub-regions are more temperate in climate. It will be found that between the upper limit of most of the typical Oriental groups and the Thibetan or Siberian fauna, there is a zone in which many forms occur common to temperate China. This is especially the case among the pheasants and finches.

Ethiopian Region. The limits of this region have been indicated by the definition of the Palaearctic region. Besides Africa south of the tropic of Cancer, and its islands, it comprises the southern half of Arabia.

This region has been said to be identical in the main characters of its mam-

malian fauna with the Oriental region, and has therefore been united with it
by Mr. A. Murray. Most important differences have however been overlooked, as the following summary of the peculiarities of the Ethiopian region
will, I think, show.

It possesses 22 peculiar families of vertebrates; 90 peculiar genera of mammalia, being two-thirds of its whole number; and 179 peculiar genera of birds,
being three-fifths of all it possesses. It is further characterized by the absence
of several families and genera which range over the whole northern hemisphere, details of which will be found in the chapter treating of the region.
There are, it is true, many points of resemblance, not to be wondered at between two tropical regions in the same hemisphere, and which have evidently
been at one time more nearly connected, both by intervening lands and by a
different condition of the lands that even now connect them. But these
resemblances only render the differences more remarkable; since they show
that there has been an ancient and long-continued separation of the two
regions, developing a distinct fauna in each, and establishing marked speciali-
ties which the temporary intercommunication and immigration has not sufficed
to remove. The entire absence of such wide-spread groups as bears and deer,
from a country many parts of which are well adapted to them, and in close
proximity to regions where they abound, would alone mark out the Ethiopian
region as one of the primary divisions of the earth, even if it possessed a less
number than it actually does of peculiar family and generic groups.

Sub-divisions of the Ethiopian Region. The African continent south of the
tropic of Cancer is more homogeneous in its prominent and superficial zoological features than most of the other regions, but there are nevertheless important and deep-seated local peculiarities. Two portions can be marked off
as possessing many peculiar forms; the luxuriant forest district of equatorial
West Africa, and the southern extremity or Cape district. The remaining
portion has no well-marked divisions, and a large proportion of its animal
forms range over it from Nubia and Abyssinia, to Senegal on the one side and
to the Zambesi on the other; this forms our first or East-African sub-region.

The second, or West African sub-region extends along the coast from
Senegal to Angola, and inland to the sources of the Shary and the Congo.

The third, or South African sub-region, comprises the Cape Colony and
Natal, and is roughly limited by a line from Delagoa Bay to Walvish Bay.

The fourth, or Malagasy sub-region, consists of Madagascar and the adjacent islands, from Rodriguez to the Seychelles; and this differs so remarkably from the continent that it has been proposed to form a distinct primary
region for its reception. Its productions are indeed highly interesting; since
it possesses 3 families, and 2 sub-families of mammals peculiar to itself, while
almost all its genera are peculiar. Of these a few show Oriental or Ethiopian
affinities, but the remainder are quite isolated. Turning to other classes of

animals, we find that the birds are almost as remarkable; but, as might be expected, a larger number of genera are common to surrounding countries. More than 30 genera are altogether peculiar, and some of these are so isolated as to require to be classed in separate families or sub-families. The African affinity is however here more strongly shown by the considerable number (13) of peculiar Ethiopian genera which in Madagascar have representative species. There can be no doubt therefore about Madagascar being more nearly related to the Ethiopian than to any other region; but its peculiarities are so great, that, were it not for its small size and the limited extent of its fauna, its claim to rank as a separate region might not seem unreasonable. It is true that it is not poorer in mammals than Australia; but that country is far more isolated, and cannot be so decidedly and naturally associated with any other region as Madagascar can be with the Ethiopian. It is therefore the better and more natural course to keep it as a sub-region; the peculiarities it exhibits being of exactly the same kind as those presented by the Antilles, by New Zealand, and even by Celebes and Ceylon, but in a much greater degree.

Oriental Region. On account of the numerous objections that have been made to naming a region from the least characteristic portion of it, and not thinking "Malayan," proposed by Mr. Blanford, a good term, (as it has a very circumscribed and definite meaning, and especially because the "Malay" archipelago is half of it in the Australian region,) I propose to use the word "Oriental" instead of "Indian," as being geographically applicable to the whole of the countries included in the region and to very few beyond it; as being euphonious, and as being free from all confusion with terms already used in zoological geography. I trust therefore that it may meet with general acceptance.

This small, compact, but rich and varied region, consists of all India and China from the limits of the Palaearctic region; all the Malay peninsula and islands as far east as Java and Baly, Borneo and the Philippine Islands; and Formosa. It is positively characterized by possessing 12 peculiar families of vertebrata; by 55 genera of land mammalia, and 165 genera of land birds, altogether confined to it; these peculiar genera forming in each case about one half of the total number it possesses.

Sub-divisions of the Oriental region. First we have the Indian sub-region, consisting of Central India from the foot of the Himalayas in the west, and south of the Ganges to the east, as far as a line drawn from Goa curving south and up to the Kistna river; this is the portion which has most affinity with Africa.

The second, or Ceylonese sub-region, consists of the southern extremity of India with Ceylon; this is a mountainous forest region, and possesses several peculiar forms as well as some Malayan types not found in the first sub-region.

Next we have the Indo-Chinese sub-region, comprising South China and and Burmah, extending westward along the Himalayan range to an altitude of about 9,000 or 10,000 feet, and southward to Tavoy or Tenasserim.

The last is the Indo-Malayan sub-region, comprising the Peninsula of Malacca and the Malay Islands to Baly, Borneo, and the Philippines.

On account of the absence from the first sub-region of many of the forms most characteristic of the other three, and the number of families and genera of mammalia and birds which occur in it and also in Africa, it has been thought by some naturalists that this part of India has at least an equal claim to be classed as a part of the Ethiopian region. This question will be found fully discussed in Chapter XII. devoted to the Oriental region, where it is shown that the African affinity is far less than has been represented, and that in all its essential features Central India is wholly Oriental in its fauna.

Before leaving this region a few words may be said about Lemuria, a name proposed by Mr. Sclater for the site of a supposed submerged continent extending from Madagascar to Ceylon and Sumatra, in which the Lemuroid type of animals was developed. This is undoubtedly a legitimate and highly probable supposition, and it is an example of the way in which a study of the geographical distribution of animals may enable us to reconstruct the geography of a bygone age. But we must not, as Mr. Blyth proposed, make this hypo-thetical land one of our actual Zoological regions. It represents what was probably a primary Zoological region in some past geological epoch; but what that epoch was and what were the limits of the region in question, we are quite unable to say. If we are to suppose that it comprised the whole area now inhabited by Lemuroid animals, we must make it extend from West Africa to Burmah, South China, and Celebes; an area which it possibly did once occupy, but which cannot be formed into a modern Zoological region without violating much more important affinities. If, on the other hand, we leave out all those areas which undoubtedly belong to other regions, we reduce Lemuria to Madagascar and its adjacent islands, which, for reasons already stated, it is not advisable to treat as a primary Zoological region. The theory of this ancient continent and the light it may throw on existing anomalies of distribution, will be more fully considered in the geographical part of this work.

Australian Region. Mr. Sclater's original name seems preferable to Professor Huxley's, "Austral-Asian;" the inconvenience of which alteration is sufficiently shown by the fact that Mr. Blyth proposed to use the very same term as an appropriate substitute for the "Indian region" of Mr. Sclater. Australia is the great central mass of the region; it is by far the richest in varied and highly remarkable forms of life; and it therefore seems in every way fitted to give a name to the region of which it is the essential element. The limits of this region in the Pacific are somewhat obscure, but as so many of the Pacific Islands are extremely poor zoologically, this is not of great importance.

Sub-divisions of the Australian Region. The first sub-region is the Austro-Malayan, including the islands from Celebes and Lombock on the west to the Solomon Islands on the east. The Australian sub-region comes next, consisting of Australia and Tasmania. The third, or Polynesian sub-region, will consist of all the tropical Pacific Islands, and is characterized by several peculiar genera of birds which are all allied to Australian types. The fourth, consists of New Zealand with Auckland, Chatham, and Norfolk Islands, and must be called the New Zealand sub-region.

The extreme peculiarities of New Zealand, due no doubt to its great isolation and to its being the remains of a more extensive land, have induced several naturalists to suggest that it ought justly to form a Zoological region by itself. But the inconveniences of such a procedure have been already pointed out; and when we look at its birds as a whole (they being the only class sufficiently well represented to found any conclusion upon) we find that the majority of them belong to Australian genera, and where the genera are peculiar they are most nearly related to Australian types. The preservation in these islands of a single representative of a unique order of reptiles, is, as before remarked, of the same character as the preservation of the *Proteus* in the caverns of Carniola; and can give the locality where it happens to have survived no claim to form a primary Zoological region, unless supported by a tolerably varied and distinctly characterized fauna, such as never exists in a very restricted and insular area.

Neotropical Region. Mr. Sclater's original name for this region is preserved, because change of nomenclature is always an evil; and neither Professor Huxley's suggested alteration "Austro-Columbia," nor Mr. Sclater's new term "Dendrogaea," appear to be improvements. The region is essentially a tropical one, and the extra-tropical portion of it is not important enough to make the name inappropriate. That proposed by Professor Huxley is not free from the same kind of criticism, since it would imply that the region was exclusively South American, whereas a considerable tract of North America belongs to it. This region includes South America, the Antilles and tropical North America; and it possesses more peculiar families of vertebrates and genera of birds and mammalia than any other region.

Subdivisions of the Neotropical Region. The great central mass of South America, from the shores of Venezuela to Paraguay and Eastern Peru, constitutes the chief division, and may be termed the Brazilian sub-region. It is on the whole a forest country; its most remarkable forms are highly developed arboreal types; and it exhibits all the characteristics of this rich and varied continent in their highest development.

The second, or Chilian sub-region, consists of the open plains, pampas, and mountains of the southern extremity of the continent; and we must include in it the west side of the Andes as far as the limits of the forest near Payta, and the whole of the high Andean plateaus as far as 4° of south latitude; which makes it coincide with the range of the Camelidae and Chinchillidae.

The third, or Mexican sub-region, consists of Central America and Southern Mexico, but it has no distinguishing characteristics except the absence of some of the more highly specialized Neotropical groups. It is, however, a convenient division as comprising the portion of the North American continent which belongs zoologically to South America.

The fourth, or Antillean sub-region, consists of the West India islands (except Trinidad and Tobago, which are detached portions of the continent and must be grouped in the first sub-region); and these reproduce, in a much less marked degree, the phenomena presented by Madagascar. Terrestrial mammals are almost entirely wanting, but the larger islands possess three genera which are altogether peculiar to them. The birds are of South American forms, but comprise many peculiar genera. Terrestrial molluscs are more abundant and varied than in any part of the globe of equal extent; and if these alone were considered, the Antilles would constitute an important Zoological region.

Nearctic Region. This region comprises all temperate North America and Greenland. The arctic lands and islands beyond the limit of trees form a transitional territory to the Palaearctic region, but even here there are some characteristic species. The southern limit between this region and the Neotropical is a little uncertain; but it may be drawn at about the Rio Grande del Norte on the east coast, and a little north of Mazatlan on the west; while on the central plateau it descends much farther south, and should perhaps include all the open highlands of Mexico and Guatemala. This would coincide with the range of several characteristic Nearctic genera.

Distinction of the Nearctic from the Palaearctic Region. The Nearctic region possesses twelve peculiar families of vertebrates or one-tenth of its whole number. It has also twenty-four peculiar genera of mammalia and fifty-two of birds, in each case nearly one-third of all it possesses. This proportion is very nearly the same as in the Palaearctic region, while the number of peculiar families of vertebrata is very much greater. It has been already seen that both Mr. Blyth and Professor Huxley are disposed to unite this region with the Palaearctic, while Professor Newton, in his article on birds in the new edition of the Encyclopaedia Britannica, thinks that as regards that class it can hardly claim to be more than a sub-region of the Neotropical. These views are mutually destructive, but it will be shown in the proper place, that on independent grounds the Nearctic region can very properly be maintained.

Subdivisions of the Nearctic Region. The sub-regions here depend on the great physical features of the country, and have been in some cases accurately defined by American naturalists. First we have the Californian sub-region, consisting of California and Oregon—a narrow tract between the Sierra Nevada and the Pacific, but characterized by a number of peculiar species and by several genera found nowhere else in the region.

The second, or Rocky Mountain sub-region, consists of this great mountain range with its plateaus, and the central plains and prairies to about 100° west longitude, but including New Mexico and Texas in the South.

The third and most important sub-region, which may be termed the Alleghanian, extends eastward to the Atlantic, including the Mississippi Valley, the Alleghany Mountains, and the Eastern United States. This is an old forest district, and contains most of the characteristic animal types of the region.

The fourth, or Canadian sub-region, comprises all the northern part of the continent from the great lakes to the Arctic ocean; a land of pine-forests and barren wastes, characterized by Arctic types and the absence of many of the genera which distinguish the more southern portions of the region.

Observations on the series of Sub-regions. The twenty-four sub-regions here adopted were arrived at by a careful consideration of the distribution of the more important genera, and of the materials, both zoological and geographical, available for their determination; and it was not till they were almost finally decided on, that they were found to be equal in number throughout all the regions—four in each. As this uniformity is of great advantage in tabular and diagrammatic presentations of the distribution of the several families, I decided not to disturb it unless very strong reasons should appear for adopting a greater or less number in any particular case. Such however have not arisen; and it is hoped that these divisions will prove as satisfactory and useful to naturalists in general as they have been to the author. Of course, in a detailed study of any region much more minute sub-division may be required; but even in that case it is believed that the sub-regions here adopted, will be found, with slight modifications, permanently available for exhibiting general results.

I give here a table showing the proportionate richness and speciality of each region as determined by its *families* of vertebrates and *genera* of mammalia and birds; and also a general table of the regions and sub-regions, arranged in the order that seems best to show their mutual relations.

COMPARATIVE RICHNESS OF THE SIX REGIONS.

Regions.	Vertebrata.		Mammalia.			Birds.		
	Families.	Peculiar families.	Genera.	Peculiar genera.	Percentage.	Genera.	Peculiar genera.	Percentage.
Palaearctic ..	136	3	100	35	35	174	57	33
Ethiopian...	174	22	140	90	64	294	179	60
Oriental.....	164	12	118	55	46	340	165	48
Australian...	141	30	72	44	61	298	189	64
Neotropical..	168	44	130	103	79	683	576	86
Nearctic.....	122	12	74	24	32	169	52	31

TABLE OF REGIONS AND SUB-REGIONS.

Regions.	Sub-regions.	Remarks.
I. Palaearctic	1. North Europe.	
	2. Mediterranean (or S. Eu.)	Transition to Ethiopian.
	3. Siberia.	Transition to Nearctic.
	4. Manchuria (or Japan)	Transition to Oriental.
II. Ethiopian........	1. East Africa.	Transition to Palaearctic.
	2. West Africa.	
	3. South Africa.	
	4. Madagascar.	
III. Oriental........	1. Hindostan (or Central Ind.)	Transition to Ethiopian.
	2. Ceylon.	
	3. Indo-China (or Himalayas)	Transition to Palaearctic.
	4. Indo-Malaya.	Transition to Australian.
IV. Australian......	1. Austro-Malaya.	Transition to Oriental.
	2. Australia.	
	3. Polynesia.	
	4. New Zealand.	Transition to Neotropical.
V. Neotropical.....	1. Chili (or S. Temp. Am.)	Transition to Australian.
	2. Brazil.	
	3. Mexico (or Trop. N. Am.)	Transition to Nearctic.
	4. Antilles.	
VI. Nearctic........	1. California.	
	2. Rocky Mountains.	Transition to Neotropical.
	3. Alleghanies (or East U.S.)	
	4. Canada.	Transition to Palaearctic.

INDEX